T0227588

logic and algebra

PURE AND APPLIED MATHEMATICS

A Program of Monographs, Textbooks, and Lecture Notes

LECTURE NOTES IN PURE AND APPLIED MATHEMATICS

143. *G. Komatsu and Y. Sakane*, Complex Geometry: Proceedings of the Osaka International Conference
144. *I. J. Bakelman*, Geometric Analysis and Nonlinear Partial Differential Equations
145. *T. Mabuchi and S. Mukai*, Einstein Metrics and Yang–Mills Connections: Proceedings of the 27th Taniguchi International Symposium
146. *L. Fuchs and R. Göbel*, Abelian Groups: Proceedings of the 1991 Curaçao Conference
147. *A. D. Pollington and W. Moran*, Number Theory with an Emphasis on the Markoff Spectrum
148. *G. Dore, A. Favini, E. Obrecht, and A. Venni*, Differential Equations in Banach Spaces
149. *T. West*, Continuum Theory and Dynamical Systems
150. *K. D. Bierstedt, A. Pietsch, W. Ruess, and D. Vogt*, Functional Analysis
151. *K. G. Fischer, P. Loustaunau, J. Shapiro, E. L. Green, and D. Farkas*, Computational Algebra
152. *K. D. Elworthy, W. N. Everitt, and E. B. Lee*, Differential Equations, Dynamical Systems, and Control Science
153. *P.-J. Cahen, D. L. Costa, M. Fontana, and S.-E. Kabbaj*, Commutative Ring Theory
154. *S. C. Cooper and W. J. Thron*, Continued Fractions and Orthogonal Functions: Theory and Applications
155. *P. Clément and G. Lumer*, Evolution Equations, Control Theory, and Biomathematics
156. *M. Gyllenberg and L. Persson*, Analysis, Algebra, and Computers in Mathematical Research: Proceedings of the Twenty-First Nordic Congress of Mathematicians
157. *W. O. Bray, P. S. Milojević, and Č. V. Stanojević*, Fourier Analysis: Analytic and Geometric Aspects
158. *J. Bergen and S. Montgomery*, Advances in Hopf Algebras
159. *A. R. Magid*, Rings, Extensions, and Cohomology
160. *N. H. Pavel*, Optimal Control of Differential Equations
161. *M. Ikawa*, Spectral and Scattering Theory: Proceedings of the Taniguchi International Workshop
162. *X. Liu and D. Siegel*, Comparison Methods and Stability Theory
163. *J.-P. Zolésio*, Boundary Control and Variation
164. *M. Křížek, P. Neittaanmäki, and R. Stenberg*, Finite Element Methods: Fifty Years of the Courant Element
165. *G. Da Prato and L. Tubaro*, Control of Partial Differential Equations
166. *E. Ballico*, Projective Geometry with Applications
167. *M. Costabel, M. Dauge, and S. Nicaise*, Boundary Value Problems and Integral Equations in Nonsmooth Domains
168. *G. Ferreyra, G. R. Goldstein, and F. Neubrander*, Evolution Equations
169. *S. Huggett*, Twistor Theory
170. *H. Cook, W. T. Ingram, K. T. Kuperberg, A. Lelek, and P. Minc*, Continua: With the Houston Problem Book
171. *D. F. Anderson and D. E. Dobbs*, Zero-Dimensional Commutative Rings
172. *K. Jarosz*, Function Spaces: The Second Conference
173. *V. Ancona, E. Ballico, and A. Silva*, Complex Analysis and Geometry
174. *E. Casas*, Control of Partial Differential Equations and Applications
175. *N. Kalton, E. Saab, and S. Montgomery-Smith*, Interaction Between Functional Analysis, Harmonic Analysis, and Probability
176. *Z. Deng, Z. Liang, G. Lu, and S. Ruan*, Differential Equations and Control Theory
177. *P. Marcellini, G. Talenti, and E. Vesentini*, Partial Differential Equations and Applications: Collected Papers in Honor of Carlo Pucci
178. *A. Kartsatos*, Theory and Applications of Nonlinear Operators of Accretive and Monotone Type
179. *M. Maruyama*, Moduli of Vector Bundles
180. *A. Ursini and P. Aglianò*, Logic and Algebra

Additional Volumes in Preparation

logic and algebra

edited by

Aldo Ursini
Paolo Aglianò
University of Siena
Siena, Italy

 CRC Press
Taylor & Francis Group
Boca Raton London New York

CRC Press is an imprint of the
Taylor & Francis Group, an **informa** business

CRC Press
Taylor & Francis Group
6000 Broken Sound Parkway NW, Suite 300
Boca Raton, FL 33487-2742

First issued in hardback 2017

© 1996 by Taylor & Francis Group, LLC
CRC Press is an imprint of Taylor & Francis Group, an Informa business

No claim to original U.S. Government works

ISBN 13: 978-1-138-40194-5 (hbk)
ISBN 13: 978-0-8247-9606-8 (pbk)

Visit the Taylor & Francis Web site at
http://www.taylorandfrancis.com

and the CRC Press Web site at
http://www.crcpress.com

Library of Congress Cataloging-in-Publication Data

Logic and algebra / edited by Aldo Ursini, Paolo Aglianò.
    p. cm. — (Lecture notes in pure and applied mathematics ; 180)
  ISBN 0-8247-9606-3 (hardcover : alk. paper)
  1. Algebraic logic—Congresses. I. Ursini, Aldo. II. Aglianò, Paolo. III. Series:
Lecture notes in pure and applied mathematics : v. 180.
QA10.L63 1996
511.3'24—dc20
    96-13087
    CIP

Dedicated to the memory of Roberto Magari

Roberto Magari was born in Florence in 1934 and died in Siena in 1994. After graduating in Mathematics at the University of Florence he taught at the Universities of Florence and Ferrara. He was a Professor of Mathematics from 1967 and founded the Department of Mathematics at the University of Siena in 1972 and the School of Mathematical Logic in 1981. Since 1990 he headed the Ph.D. program in Mathematical Logic and Theoretical Computer Science. He was the author of over 60 scientific papers, giving many contributions to logic, universal algebra, algebraic logic, recursion theory and the mathematics of "metamorality."

Among his lasting scientific achievements are the theory of filtral and ideal classes of algebras, the theory of meaningful formal theories, the theory of diagonalizable algebras (now called Magari algebras). He was a master to many Italian mathematicians and logicians, and a leading force in the blossoming of research in mathematical logic in Italy. Despite all the high consideration he gained as a mathematician and an intellectual of deep insights, he kept a steady modest, gentle and tolerant mood. Only through a direct acquaintance with him did one feel the fascination and the value of this man.

PREFACE

Roberto Magari died in Siena on March 5, 1994, shortly before a conference scheduled to honor him on the occasion of his 60th birthday. The International Conference on Logic and Algebra was held in his memory in Pontignano (Siena), April 26-30, 1994. This volume is the result of that meeting.

There were 101 participants, 10 invited lectures and 36 contributed lectures. Several participants suggested that an edited collection of papers as the outcome of the conference would be a fitting tribute. Thus this book presents 13 invited papers as well as 20 contributed papers, each of which had two independent referees.

The contents clearly reflect the scientific mood of the meeting, being an attempt to put together a number of researchers in the two fields of Mathematical Logic and General Algebra. According to many participants this effort was worth undertaking. The editors, in their role of organizers of the meeting, are glad to gratefully acknowledge the financial support offered by the: CNR, CNR-GNSAGA, UNIVERSITY OF SIENA, MONTE DEI PASCHI DI SIENA, and the superb job done by the staffs of the Department of Mathematics, the Ufficio Congressi and the Certosa of Pontignano.

Finally, we express our appreciation to the publisher and to all the people involved in the preparation of this volume.

Aldo Ursini
Paolo Aglianò

ACKNOWLEDGMENTS

The following is a list of institutions and people who wanted to honor the memory of the late Professor Magari by purchasing a copy of this volume.

Università di Siena
Dipartimento di Matematica-Parma
S. Artemov
A. Berarducci
J. Berman
C. Bernardi
C. Bonotto
S. Burris
A. Cantini
A. Carboni
G. Colonna
G. Corsi
R. Cruciani
B. A. Davey
M. Emmer
M. Ernè
M. Fattorosi-Barnaba
F. Ferreira
R. Ferro
M. Forti
R. Franci
G. Gerla
G. Gnani
A. G. Grappone
B. Intrigila
A. Labella
J. Lambek
F. W. Lawvere

P. Lipparini
R. McKenzie
M. Malatesta
C. Marchini
C. Marchionna Tibiletti
S. Matthews
F. Migliorini
V. Millucci
C. Morini
F. Montagna
J. B. Nation
C. Palamidessi
A. Pixley
C. Pizzi
S. L. Read
P. Rosolini
G. Sambin
D. Schweigert
M. Servi
G. Simi
A. K. Simpson
C. Toffalori
S. Tulipani
S. Valentini
R. Willard
R. Wille
G. Zappa

CONTENTS

LIST OF PARTICIPANTS

Paolo Aglianò	Università di Siena, Italy
Gianni Aguzzi	Università di Siena, Italy
Sergei N. Artemov	Steklov Mathematical Institute, Moscow, Russia
Giulia Battilotti	Selvazzano PD, Italy
Lev D. Beklemishev	Steklov Mathematical Institute, Moscow, Russia
Dorella Bellè	Udine, Italy
Gianluigi Bellin	Universitè de Paris VII, France
Alessandro Berarducci	Università di Pisa, Italy
Claudio Bernardi	Università La Sapienza, Roma, Italy
Tom Brown	Simon Fraser University, Canada
Stanley N. Burris	University of Waterloo, Canada
Ettore Casari	Università di Firenze, Italy
Ivan Chajda	Palacký University, Czech Republic
Paolo Cintioli	Camerino MC, Italy
Saverio Cittadini	Siena, Italy
Alessandra Ciupi	Siena, Italy
Giovanna Corsi	Università di Firenze, Italy
Rosanna Cruciani	Università La Sapienza, Roma, Italy
Giovanna D'Agostino	Università di Udine, Italy
Fritjöf Dau	Universität Hannover, Germany
Brian A. Davey	La Trobe University, Australia
Dick De Jongh	University of Amsterdam, Netherlands
Mauro Di Nasso	Viareggio LI, Italy
Anotonino Drago	Università di Napoli, Italy
Jaromir Duda	Brno, Czech Republic
Josef Dudek	Wrocław University, Poland
Marcel Ernè	Universität Hannover, Germany
M. Fattorosi-Barnaba	Università La Sapienza, Roma, Italy
Fernando Ferreira	Faculdade de ciencias de Lisboa, Portugal
Isabel A. M. Ferreirim	Universidade de Lisboa, Portugal
Sandra Fontani	Siena, Italy
Marco Forti	Università di Pisa, Italy
Miriam Franchella	Università di Milano, Italy
Ervin Fried	ELTE TTK, Budapest, Hungary
Sy D. Friedman	MIT, Boston, MA
Paolo Gentilini	IMA, Genova, Italy

Silvio Ghilardi	Università di Milano, Italy
George Gratzer	University of Manitoba, Winnipeg, Canada
Petr Hajek	Czech. Acad. of Science, Prague, Czech Republic
Kalle Kaarli	University of Tartu, Estonia
Keith Kearnes	TH Darmstadt, Germany
Emil W. Kiss	ELTE TTK, Budapest, Hungary
Anna Labella	Università di Roma, Italy
Joachim Lambek	McGill University, Montreal, Canada
F. William Lawvere	SUNY at Buffalo, NY
Gianluigi Lenzi	Pisa, Italy
Ada Lettieri	Università di Napoli, Italy
Giorgio Levi	Università di Pisa, Italy
Paolo Lipparini	Università di Roma "Tor Vergata", Italy
Angus J. Macintyre	University of Oxford, England
Maria Emilia Maietti	S. Martino B.A. VR, Italy
Antonella Mancini	Siena, Italy
Piero Mangani	Università di Firenze, Italy
Carlo Marchini	Università di Parma, Italy
Annalisa Marcja	Università di Firenze, Italy
Alberto Marcone	Università di Torino, Italy
Enrico Martino	Università di Padova, Italy
Sean Matthews	Max Planck Institut, Saarbrucken, Germany
Ralph McKenzie	Vanderbilt University, Nashville, TN
Giancarlo Meloni	Università di Milano, Italy
Franco Migliorini	Università di Siena, Italy
Franco Montagna	Università di Siena, Italy
Daniela Monteverdi	Università di Parma, Italy
Ugo Moscato	Università di Milano, Italy
Daniele Mundici	Università di Milano, Italy
Sara Negri	Padova, Italy
Paolo Pagli	Università di Siena, Italy
Catuscia Palamidessi	Università di Genova, Italy
Peter P. Pálfy	Hungarian Acad. of Science, Budapest, Hungary
Giovanni Panti	Siena, Italy
Franco Parlamento	Università di Udine, Italy
M. Cristina Pedicchio	Università di Trieste, Italy
Michele Pinna	Università di Siena, Italy
Alden F. Pixley	Harvey Mudd College, Claremont, CA
Claudio Pizzi	Università di Siena, Italy
Steven Read	University of St-Andrews, Scotland
Cristina Reggiani	Università di Parma, Italy

Jurgen Reinhold	Universität Hannover, Germany
Giorgia Ricci	Siena, Italy
Giuseppe Rosolini	Università di Genova, Italy
Antonino Salibra	Università di Venezia, Italy
Giovanni Sambin	Università di Padova, Italy
E. Thomas Schmidt	Technical University of Budapest, Hungary
Dietmar Schweigert	Universität Kaiserslautern, Germany
Volodya Shavrukov	University of Amsterdam, Netherlands
Giulia Simi	Università di Siena, Italy
Alex K. Simpson	University of Edimburgh, Scotland
Dimitri Skvortsov	Moscow, Russia
Andrea Sorbi	Università di Siena, Italy
Stefano Stefani	Università di Venezia, Italy
C. Marchionna Tibiletti	Università di Milano, Italy
Elisa Tiezzi	Siena, Italy
Carlo Toffalori	Università di Camerino, Italy
Sauro Tulipani	Università di Camerino, Italy
Aldo Ursini	Università di Siena, Italy
Silvio Valentini	Università di Padova, Italy
Matthew Valeriote	McMaster University, Hamilton, Canada
Paolo Virgili	Siena, Italy
Ross D. Willard	University of Waterloo, Canada
Rudolf Wille	TH Darmstadt, Germany
Guido Zappa	Università di Firenze, Italy

logic and algebra

Logic of proofs
with complexity operators

Sergei Artëmov*   Artëm Chuprina

Abstract

The logic of proofs is the modal provability logic enriched by new operators (labeled modalities) for individual proofs. In the current paper we add to the logic of proofs also new labeled modalities which stand for the complexity of proofs. The Kripke style completeness, decidability, and arithmetical completeness theorems are obtained. The complexity logics introduced here correspond to two major classes of complexity measures: decidable and recursively enumerable ones; the completeness theorems relate either of these logics to the entire class of the relevant complexity measures.

1 Introduction.

In [1] the modal provability logic was enriched by new operators (labeled modalities) for individual proofs. The resulting *logic of proofs* was intended to meet some needs of the computer science, where not only provability is of interest but also proofs themselves. Decent algebraic models for the logic of proofs are still to be invented. Such models should probably provide a natural extension of the notion of Magari Algebras, which are also known as Diagonalizable Algebras [4].

In the current paper we add to the logic of proofs also new labeled modalities which stand for the complexity of proofs. The Kripke style completeness, decidability, and arithmetical completeness theorems are obtained.

The *complexity logics* introduced here correspond to two major classes of complexity measures: decidable and recursively enumerable ones; the completeness theorems thus relate either of these logics to the entire class of the relevant complexity measures.

---

*Supported by the grant # 93-011-16015 of the Russian Foundation for Fundamental Research and by the grant # NFQ000 from the International Science Foundation.

A logic of a specific complexity measure may depend on particular details of proof coding; the logics from the current paper may turn out to be incomplete with respect to some individual complexity measure. However, the completeness theorems from this paper give us a clear idea about what axioms one should add to these logics in order to reach the completeness in a specific case, and what sort of models (Kripke, arithmetical) are relevant to complexity measures.

In what follows we assume, for short, the Peano Arithmetic **PA** to be the basic theory for proof and provability predicates. We denote the usual Gödel proof predicate "x is a gödelnumber of a proof of the formula with the gödelnumber y" as $Proof(x, y)$ and the usual provability predicate as $Provable(y)$, i.e. $Provable(y)$ coincides with $\exists x Proof(x, y)$.

1.1 Definition.   An arithmetical formula $Prf(x, y)$ is called a **standard proof predicate** (cf.[2, 1]) iff

1. $Prf(x, y)$ is equivalent in **PA** to a recursive formula;

2. $Prf(x, y)$ numerates the theorems of **PA**:

$$\textbf{PA} \vdash \varphi \iff \text{ for some } n \in \omega \ Prf(n, \ulcorner \varphi \urcorner) \text{ is true;}[1]$$

3. the formula $Pr(y) := \exists x Prf(x, y)$ satisfies:

$$\textbf{PA} \vdash \big(Pr(x) \& Pr(x \dot\rightarrow y) \big) \rightarrow Pr(y)[2]$$

$$\textbf{PA} \vdash \sigma \rightarrow Pr(\ulcorner \sigma \urcorner)$$

for every arithmetical Σ_1^0-sentence σ;

1.2 Definition.   A standard proof predicate Prf is **functional** if for all $l, m, n \in \omega$

$$Prf(l, m) \text{ and } Prf(l, n) \implies m = n.$$

1.3 Definition.   An arithmetical formula $Cpl(x, y)$ is called a **standard complexity predicate** iff

---

[1]In this paper we do not distinguish between the natural number n and its numeral $\bar{n}$. As usual, $\ulcorner \varphi \urcorner$ denotes the gödelnumber of φ.

[2]$\dot\rightarrow$ denotes the term with two free variables, such that for any arithmetical formulas B and C, $\ulcorner B \urcorner \dot\rightarrow \ulcorner C \urcorner = \ulcorner B \rightarrow C \urcorner$.

1. $Cpl(x,y)$ is equivalent in **PA** to a recursively enumerable formula;

2. $Cpl(x,y)$ numerates the theorems of **PA**:

$$\mathbf{PA} \vdash \varphi \iff \text{for some } n \in \omega \ Cpl(n, \ulcorner \varphi \urcorner) \text{ is true,}$$

3. the formula $Prv(y) := \exists x \, Cpl(x,y)$ satisfies:

$$\mathbf{PA} \vdash \big(Prv(x) \& Prv(x \to y) \big) \to Prv(y)$$

$$\mathbf{PA} \vdash \sigma \to Prv(\ulcorner \sigma \urcorner)$$

for every Σ_1^0 arithmetical sentence σ;

4. $Cpl(x,y)$ is (provably in **PA**) monotone on the first argument, i.e.

$$\mathbf{PA} \vdash u \leq v \to \big[Cpl(u,y) \to Cpl(v,y) \big].$$

1.4 Definition. A standard complexity predicate is called **recursive** if it is equivalent in **PA** to a recursive formula.

1.5 Definition. Two predicates Prf and Cpl are called **provably compatible** if

$$\mathbf{PA} \vdash \forall y (Prv(y) \leftrightarrow Pr(y)).$$

Note that the complexity of proof and proof itself are of different character: complexity domain (the set of natural numbers) is linearly ordered, but proofs have no natural linear ordering. So for a modal description of this two notions we introduce two different sorts of variables, and assume that proof and complexity predicates are provably compatible. On the other hand, each standard decidable complexity predicate $Cpl(x,y)$ coincides with $\exists t \leq x Prf(t,y)$ for an appropriate standard proof predicate Prf. In this case we may identify proof variables and complexity variables in a natural way; while considering decidable complexity measures we will suppose that $Cpl(x,y) = \exists t \leq x Prf(t,y)$.

A labeled modal language $\mathcal{L}^{++}$ contains three sorts of variables, $p_0, p_1, \ldots$ (called *proof variables*), $\alpha_0, \alpha_1, \ldots$ (called *complexity variables*) and $S_0, S_1, \ldots$ (called *sentence variables*), symbol $\to$ for the classical implication, the truth value $\bot$ for absurdity (the usual Boolean connectives, and the truth value $\top$ for truth are defined as abbreviations), the usual modality $\Box$, for each proof variable p_i the unary modal operator $\Box_{p_i}$ and for each complexity variable α_i

the unary modal operator $\triangle_{\alpha_i}$. The set of formulas of $\mathcal{L}^{++}$ is thus generated from the **atomic** formulas $\bot, S_0, S_1, \ldots$ by $\rightarrow$ as usual, and by the modal operators as follows: if A is an $\mathcal{L}^{++}$ formula, p is a proof variable and α is a complexity variable, then $\Box A$, $\Box_p A$ and $\triangle_\alpha A$ are $\mathcal{L}^{++}$ formulas; we call formulas of the form $\Box_p A$ and $\triangle_\alpha A$ **quasiatomic**, or **q-atomic** for short. We will also use the abbreviation $\Box^+ A$ for a formula $A \wedge \Box A$, and $\Diamond^+$ will stand for $\neg\,\Box^+\,\neg$. In the sequel under a modal formula we understand a formula in the language $\mathcal{L}^{++}$. We use small letters $p, q, r, \ldots$ for proof variables, greek letters $\alpha, \beta, \ldots$ for complexity variables, capital letters $S, T, \ldots$ for sentence variables and $A, B, C, \ldots$ for modal formulas. Let $\mathcal{L}$ denote the usual modal language over $\bot, S_0, S_1, \ldots$ with the only modality $\Box$, i.e. a labeled-modalities-free fragment of $\mathcal{L}^{++}$, $\mathcal{L}^+$ denote the $\triangle_{\alpha_i}$-free fragment of $\mathcal{L}^{++}$, $\mathcal{L}^{-+}$ – the $\Box_{p_i}$-free fragment, and $\mathcal{L}^{+-}$ – the fragment of $\mathcal{L}^{++}$, where we identify proof variable p_i with the correspondent complexity variable α_i.

We assume a reader to be familiar with the general unification technique (cf. [3, 1]). In particular, by τ_{AB} we denote the most general unifier (mgu) of A and B obtained by some fixed deterministic version of the Unification Algorithm.

1.6 Definition. (cf. [1]) Let $\quad C = D \quad (\text{mod } A = B) \quad$ be an abbreviation for

$$\text{“for every substitution } \theta \; (A\theta \equiv B\theta \;\Rightarrow\; C\theta \equiv D\theta)\text{”}.$$

Apparently, if A, B are not unifiable, then $C = D \quad (\text{mod } A = B)$ holds for all C and D.

1.7 Lemma. If A and B are unifiable, then

$$C = D \quad (\text{mod } A = B) \quad \Leftrightarrow \quad C\tau_{AB} \equiv D\tau_{AB}.$$

Proof. Direction $(\Rightarrow)$ is obvious as τ_{AB} unifies A and B. Direction $(\Leftarrow)$: let $C\tau_{AB} \equiv D\tau_{AB}$ and θ be an arbitrary unifier A and B. As τ_{AB} is an mgu of A, B for some λ we have $\theta = \tau_{AB} \circ \lambda$, then

$$C\theta \equiv C\tau_{AB} \circ \lambda \equiv D\tau_{AB} \circ \lambda \equiv D\theta.$$

1.8 Corollary. The relation $\quad C = D \quad (\text{mod } A = B) \quad$ is decidable.

1.9 Definition. An **arithmetical interpretation** $*$ is a triple (Prf, Cpl, ϕ), where Prf and Cpl are provably compatible standard proof and complexity predicates and ϕ is a function which assigns:

- to each proof variable p some $n \in \omega$,

- to each complexity variable α_i some $m \in \omega$

- and to each sentence variable S a sentence of **PA**.

The arithmetical translation A^* of a modal formula A under the interpretation $*$ is the extension of ϕ to all modal formulas by:

- $\bot^* := (0 = 1)$,

- $p^* := \phi(p)$ for a proof variable p,

- $\alpha^* := \phi(\alpha)$ for a complexity variable α,

- $S^* := \phi(S)$ for a sentence variable S,

- $(\cdot)^*$ commutes with the Boolean connectives,

- $(\Box A)^* := Pr(\ulcorner A^* \urcorner)$,

- $(\Box_p A)^* := Prf(p^*, \ulcorner A^* \urcorner)$.

- $(\triangle_\alpha A)^* := Cpl(\alpha^*, \ulcorner A^* \urcorner)$,

An arithmetical interpretation (Cpl, Prf, ϕ) is called functional iff Prf is a functional proof predicate.

We'll combine the formal systems for complexity logics from the following set of axioms:

(A0) Boolean tautologies in the language $\mathcal{L}^{++}$:

(A1) $\Box(A \to B) \to (\Box A \to \Box B)$ (distributivity)

(A2) $\Box(\Box A \to A) \to \Box A$ (Löb axiom)

(A3) $\Box_p A \to A$ (q-reflexivity)

(A4) $\Box_p A \to \Box\Box_p A$ (stability)

(A5) $\neg\Box_p A \to \Box(\neg\Box_p A)$ (stability)

(A6) $\Box_p A \& \Box_p B \to (C \to D)$ if $C = D$ (mod $A = B$)   (functionality)

(A7) $\triangle_\alpha A \to A$ (q-reflexivity)

(A8) $\triangle_\alpha A \to \Box A$ (q-provability)

(A9) $\triangle_\alpha A \to \Box \triangle_\alpha A$ (stability)

(A10) $\neg \triangle_\alpha A \to \Box (\neg \triangle_\alpha A)$ (stability)

(A11) $\Box_p A \to \triangle_p A$ (in the language $\mathcal{L}^{+-}$)   (correspondence)

(A12) $\neg[(\beta_0 <_{A_0} \beta_1) \& (\beta_1 <_{A_1} \beta_2) \& \ldots \& (\beta_n <_{A_n} \beta_0)]$ (irreflexivity)
where $\beta_0, \beta_1, \ldots$ are complexity variables, $A_0, A_1, \ldots$
are formulas and $\alpha <_A \beta$ means $\Diamond^+(\triangle_\beta A \& \neg \triangle_\alpha A)$

and inference rules

(R0) $\dfrac{A \quad A \to B}{B}$

(R1) $\dfrac{A}{\Box A}$

(R2) $\dfrac{\Box A}{A}$

Axioms **A0**–**A2** and rules **R0** and **R1** axiomatize the provability logic **GL**, the system $\mathcal{B}$ is **GL** (in the language $\mathcal{L}^+$) + (**A3**–**A5**) + **R2**, the system $\mathcal{F}$ is $\mathcal{B}$ + **A6** (cf. [1]).

1.10 Definition.   (cf. [1]) A **frame** is a pair $\langle K, \prec \rangle$, where $K \neq \emptyset$ and K is finite, $\prec$ is an irreflexive tree-like ordering of K^3. In the sequel we call elements of K *nodes*, and let $\succ$ stand for $(\prec)^{-1}$. A **model** is a triple $\langle K, \prec, \Vdash \rangle$, where $\langle K, \prec \rangle$ is a frame and $\Vdash$ is a forcing relation between nodes and $\mathcal{L}^+$ formulas satisfying the following **forcing conditions**:

---

[3]In fact any **GL**-frame (i.e. transitive reverse well-founded) would fit here.

1. $\Vdash$ respects Boolean operations nodewise,

2. $x \Vdash \Box \varphi$ iff $\forall y \succ x \; y \Vdash \varphi$,

3. for every q-atomic formula $\Box_p A$, and every $x \prec y$
   $x \Vdash \Box_p A$ iff $y \Vdash \Box_p A$,
   (stability)

4. for every q-atomic formula $\Box_p A$
   $x \Vdash \Box_p A \;\Rightarrow\; x \Vdash A$,
   (q-reflexivity)

5. for every q-atomic formula $\triangle_\alpha A$
   $x \Vdash \triangle_\alpha A \;\Rightarrow\; x \Vdash \Box A$ and $\forall y \succ x \; y \Vdash \triangle_\alpha A$,

6. $\Vdash$ satisfies the axiom **A12** in each node.

1.11 Definition. By **r-model** we mean the model, restricted to the language $\mathcal{L}^{+-}$ with three additional forcing conditions for q-atomic formulas:

1. $x \Vdash \triangle_p A \;\Rightarrow\; x \Vdash A$,

2. $\forall x, y \in K \; x \prec y \;\Rightarrow\; x \Vdash \triangle_p A$ iff $y \Vdash \triangle_p A$,

3. $x \Vdash \Box_p A \;\Rightarrow\; x \Vdash \triangle_p A$.

(The two first items together are stronger, than the item 5 of the previous definition.) The r-models will serve as Kripke-like models for systems for recursive complexity measures.

1.12 Definition. We say that a modal formula A is **valid in a model** $\mathcal{K} = \langle K, \prec, \Vdash \rangle$ if A holds at every node; and $\mathcal{K}$ is a **countermodel to** A if A is not valid in $\mathcal{K}$, i.e. $\neg A$ holds at some node $x \in K$.

2   Logics for decidable complexity measures.

We formulate formal systems for decidable complexity measures in the language $\mathcal{L}^{+-}$.

2.1 Definition. System $\mathcal{BC}$ is the system $\mathcal{B}$ (in the language $\mathcal{L}^{+-}$) + **A7** + (**A9** – **A12**), and the system $\mathcal{FC}$ is the system $\mathcal{BC}$ + functionality axiom **A6**.

Note that **A8** also holds in $\mathcal{BC}$ and $\mathcal{FC}$, but it is easy to derive it from **A7** and **A9** by **R1**.

2.2 Definition.   A set X of modal formulas is **adequate** if it is closed under subformulas; $\bot \in X$; if $B \in X$ and B is not of a form $\neg C$, then $\neg B \in X$; for all p, A if there exist q, B such that $\Box_q A$ (or $\triangle_q A$) and $\Box_p B$ (or $\triangle_p B$) belong to X, then $\Box_p A \in X$ and $\triangle_p A \in X$.

Let $Ad(A)$ be the least adequate set containing A (it is easy to see that $Ad(A)$ is finite) and let $H(A) = \bigwedge \{\Box B \to B | \Box B \in Ad(A)\}$

2.3 Definition. X-model is a triple $\mathcal{K} = \langle K, \prec, \Vdash \rangle$ where $\langle K, \prec \rangle$ is a frame and the forcing relation $\Vdash$ is defined only for formulas from X and their Boolean and $\Box$-combinations and satisfies all the forcing conditions. We call an X-model A-**sound** if $H(A)$ holds at its root.

2.4 Remark.   Note that every X-model can be extended to a model by defining $x \not\Vdash \varphi$ for each node $a \in K$ and for each atomic and q-atomic formula $\varphi \notin X$. Indeed, the stability and q-reflexivity forcing conditions are clearly preserved. We also address a reader to the remark 3.3.

2.5 Theorem. $\mathcal{BC}$ is complete with respect to the class of r-models, and thus is decidable.

Proof.   The part "$\mathcal{BC} \not\vdash A \Rightarrow$ there is a countermodel" goes exactly along the lines of that for $\mathcal{BE}$ (3.5 and 3.6), but the set Y should include items corresponding to **A7** instead of **A8** and to **A10** and **A11**. The other part "$\Rightarrow$" can be easily derived from 2.6 and 2.7.
∎

2.6 Lemma. $\mathcal{BC} \vdash A \Rightarrow$ for every interpretation $*$ $\mathsf{PA} \vdash A^*$.

Proof.   By induction on a proof of A in $\mathcal{BC}$. The cases of axioms **A0** – **A2** are treated in [2], **A4**, **A5**, **A9** and **A10** hold because Prf and Cpl are recursive. Let us consider the axiom **A3**. If $Prf(p^*, \ulcorner B^* \urcorner)$ is true, then B^* is in fact provable and $\mathsf{PA} \vdash (\Box_p B \to B)^*$; If $Prf(p^*, \ulcorner B^* \urcorner)$ is false, then $(\Box_p B)^*$ is a false recursive formula, thus $\mathsf{PA} \vdash (\neg \Box_p B)^*$ and again $\mathsf{PA} \vdash (\Box_p B \to B)^*$. Now **A7** is trivial because $Cpl(x, y) = \exists t \le x \, Prf(x, y)$. Let us consider **A12**. We have

$$\mathsf{PA} \vdash u \le v \to [Cpl(u, y) \to Cpl(v, y)].$$

As $u \leq v \in \Sigma_1^0$, then $\mathbf{PA} \vdash u \leq v \rightarrow \Box(u \leq v)^4$. Hence

$$\mathbf{PA} \vdash u \leq v \rightarrow \Box^+[Cpl(u,y) \rightarrow Cpl(v,y)].$$

An interpretation of any example of $q <_A r$ is some substitution of free variables by terms in the formula $\Diamond^+[\neg Cpl(u,y) \wedge Cpl(v,y)]$; also

$$
\begin{aligned}
\Diamond^+[\neg Cpl(u,y) \wedge Cpl(v,y)] \quad &\leftrightarrow \quad \neg\Box^+[Cpl(v,y) \rightarrow Cpl(u,y)] \\
&\rightarrow \quad \neg(v \leq u) \\
&\rightarrow \quad u < v.
\end{aligned}
$$

An arithmetical interpretation of an example of irreflexivity axiom is thus equivalent in $\mathbf{PA}$ to

$$(q_0 <_{A_0} q_1)^* \& (q_1 <_{A_1} q_2)^* \& \ldots \& (q_{n-1} <_{A_{n-1}} q_n)^* \rightarrow \neg(q_n <_{A_n} q_0)^*,$$

which is provable in $\mathbf{PA}$ because

$$
\begin{aligned}
(q_0 <_{A_0} q_1)^* \& (q_1 <_{A_1} q_2)^* \& &\ldots \& (q_{n-1} <_{A_{n-1}} q_n)^* \\
&\rightarrow \quad u_0 < u_1 \& u_1 < u_2 \& \ldots \& u_{n-1} < u_n \\
&\rightarrow \quad \neg(u_n < u_0) \\
&\rightarrow \quad \neg(q_n <_{A_n} q_0)^*.
\end{aligned}
$$

Note that the proof of $\mathbf{A12}^*$ does not use the fact, that Cpl is recursive, so this proof is valid for a recursively enumerable Cpl too.

∎

2.7 Theorem. $\mathcal{BC} \not\vdash A \Rightarrow$ for some interpretation $*$ $\mathbf{PA} \not\vdash A^*$.

Proof. Let $\mathcal{BC} \not\vdash A$, and let $\mathcal{K} = (K, \prec, \Vdash)$ be an A-sound $Ad(A)$-r-model. We assume that $K = \{1, \ldots, n\}$ and 1 is the root node and define a new model $\mathcal{K}'$ by adding a node 0 to K, putting $0 \prec i$ $(1 \leq i \leq n)$, and defining $0 \Vdash B$ iff $1 \Vdash B$ for every atomic and q-atomic formula $B \in Ad(A)$. An easy induction proves that $\mathcal{K}'$ is an A-sound $Ad(A)$-r-model.

We proceed with the Solovay construction and define the Solovay function $h(t)$ and the arithmetical formulas "$l = j$" for the model $\mathcal{K}'$ and for the usual Gödel proof predicate $Proof(x,y)$, and put

$$\varphi(S_i) := \left[\bigvee_{j \Vdash S_i} \text{"}l = j\text{"} \right] \wedge i = i.$$

The following *Solovay Lemma* holds:

2.8 Lemma. [5]

---

[4]We understand here $\Box$ as $Pr(\ulcorner .^* \urcorner)$.

1. $PA \vdash$ "$0 \leq l \leq n$";

2. "$l = 0$" is true, but each of the theories $PA +$ "$l = i$" is consistent for $i = 0, 1 \ldots, n$;

3. $PA +$ "$l = i$" $\vdash Provable(\ulcorner "l \neq i" \urcorner), \quad i = 1, 2, \ldots, n$;

4. $PA +$ "$l = i$" $\vdash \neg Provable(\ulcorner "l \neq j" \urcorner), \quad i = 0, 2, \ldots, n, \ i \prec j$;

5. $PA +$ "$l = i$" $\vdash Provable(\ulcorner "l \neq j" \urcorner), \quad i = 1, 2, \ldots, n, \ i \not\prec j$.

2.9 Definition. We define now a relation **prec** on the set *Var* of all proof (= complexity) variables occurring in $Ad(A)$: p **prec** q iff there are

$$\triangle_p A_1, \triangle_{q_1} A_1, \triangle_{q_1} A_2, \ldots, \triangle_{q_{n-1}} A_n, \triangle_q A_n \in Ad(A)$$

such that

$$(p <_{A_1} q_1) \& (q_1 <_{A_2} q_2) \& \ldots \& (q_{n-1} <_{A_n} q)$$

holds in $\mathcal{K}'$.

The relation **prec** is obviously transitive and the axiom **A12** guarantees that it is also irreflexive.

2.10 Lemma. There is an injective homomorphism f of the order $(Var, \mathbf{prec})$ to the natural ordering of rationals such that

$$p \ \mathbf{prec} \ q \ \Rightarrow \ f(p) < f(q).$$

Proof. Without loss of generality we assume that $Var = \{p_0, \ldots, p_m\}$. We define f by induction on the cardinality of *Var*; $f(p_0)$ is mapped arbitrarily. Suppose $f(p_0), \ldots, f(p_{k-1})$ are defined. Let

$$H := \min\{f(p_i) \mid p_k \ \mathbf{prec} \ p_i, \ i < k\}, \quad \min \emptyset = +\infty$$

and

$$h := \max\{f(p_j) \mid p_j \ \mathbf{prec} \ p_k, \ j < k\}, \quad \max \emptyset = -\infty.$$

We claim that $h < H$. Indeed, $h = H$ is impossible because of the injectivity of f. If $H < h$, then there are $i, j < k$ such that $f(p_i) < f(p_j)$, but p_j **prec** p_k and p_k **prec** p_i. By the transitivity of **prec** we have p_j **prec** p_i which contradicts the homomorphic property of f. So $h < H$ and put $f(p_k)$ to be $h < f(p_k) < H$.

The desired inequalities are obvious: if p_i **prec** p_k, then $f(p_i) \le h$ by the definition of h, and thus $f(p_i) < f(p_k)$, etc.

∎

Again, without loss of generality, we assume that

$$f(p_0) < f(p_1) < \ldots < f(p_m),$$

put

$$Q_i = \{B \mid \triangle_{p_i} B \in Ad(A) \text{ and } \triangle_{p_i} B \text{ holds in } \mathcal{K}'\}, (0 \le i \le m),$$

$$Q = \bigcup_{0 \le i \le m} Q_i,$$

and

$$P_i = \{B \mid \square_{p_i} B \in Ad(A) \text{ and } \square_{p_i} B \text{ holds in } \mathcal{K}'\}, (0 \le i \le m),$$

and observe that $Q_i \subseteq Q_j$, whenever $i < j$, and $P_i \subseteq Q_i$ with $0 \le i \le m$. Indeed, suppose $i < j$ but there is B such that $\triangle_{p_i} B, \triangle_{p_j} B \in X$, $\triangle_{p_i} B$ holds in $\mathcal{K}'$, but $\triangle_{p_j} B$ does not. Then $\neg \triangle_{p_j} B$ is valid in $\mathcal{K}'$ and thus $p_j <_B p_i$ holds in $\mathcal{K}'$: contradiction with $i < j$. The property $P_i \subseteq Q_i$ with $0 \le i \le m$ is guaranteed by **A11**.

Put $R_i = Q_i - P_i$, define

$$p_i := 2i + 1$$

and consider the following arithmetical fixed point equation (FPE):

$$Prf(u,v) \leftrightarrow \Big[\begin{array}{llll} u = 0 & \to & \Gamma_0 & \& \\ u = 1 & \to & \Delta_0 & \& \\ u = 2 & \to & \Gamma_1 & \& \\ u = 3 & \to & \Delta_1 & \& \\ & \vdots & & \\ u = 2m & \to & \Gamma_m & \& \\ u = 2m+1 & \to & \Delta_m & \& \\ u > 2m+1 & \to & Proof(u - 2m - 2, v) \end{array} \Big]$$

where

$$\Gamma_i \equiv \begin{cases} \bigvee_{B \in R_i} (v = \ulcorner B^* \urcorner) & \text{if } R_i \ne \emptyset \\[2mm] v = \ulcorner \forall x_0 (x_0 = x_0) \urcorner & \text{otherwise,} \end{cases}$$

$$\Delta_i \equiv \begin{cases} \bigvee_{B \in P_i}(v = \ulcorner B^* \urcorner) & \text{if } P_i \neq \emptyset \\[2mm] v = \ulcorner \forall x_0(x_0 = x_0) \urcorner & \text{otherwise.} \end{cases}$$

One can easily see that Prf is recursive and $*$ is injective. Moreover,

$$\mathbf{PA} \vdash Pr(y) \leftrightarrow \left[Provable(y) \vee \bigvee_{B \in Q} (y = \ulcorner B^* \urcorner) \right].$$

We have to prove now the usual Solovay lemma:

2.11 Lemma. For every $B \in Ad(A)$ and every node j from $\mathcal{K}'$

$$j \Vdash B \quad \Rightarrow \quad \mathbf{PA} \vdash \text{``} l = j \text{''} \to B^*$$

$$j \nVdash B \quad \Rightarrow \quad \mathbf{PA} \vdash \text{``} l = j \text{''} \to \neg B^*.$$

Proof.   By induction on formulas. Basis in the case when B is a sentence variable, or a constant $\perp$ is trivial.

Let $B \equiv \Box_{p_i} D \in X$. If $j \Vdash \Box_{p_i} D$ then $D \in P_i$ and because of the FPE, $Prf(2i+1, \ulcorner D^* \urcorner)$ is true, $\mathbf{PA} \vdash Prf(2i+1, \ulcorner D^* \urcorner)$ and $\mathbf{PA} \vdash \text{``} l = j \text{''} \to B^*$. If $j \nVdash \Box_{p_i} D$, then $D \notin P_i$ and because of the FPE and the injectivity of $*$, $Prf(2i+1, \ulcorner D^* \urcorner)$ is false. In this case $\mathbf{PA} \vdash \text{``} l = j \text{''} \to \neg B^*$.

Let $B \equiv \triangle_{p_i} D \in X$. If $j \Vdash \triangle_{p_i} D$ then $D \in Q_i$ and, according to the FPE, either $Prf(2i, \ulcorner D^* \urcorner)$ is true or $Prf(2i+1, \ulcorner D^* \urcorner)$ is true. In either case

$$\mathbf{PA} \vdash \exists x \leq p_i{}^* \; Prf(x, \ulcorner D^* \urcorner)$$

and thus $\mathbf{PA} \vdash \text{``} l = j \text{''} \to B^*$.

If $j \nVdash \triangle_{p_i} D$, then $D \notin Q_i$, moreover $D \notin Q_k$ for all $k \leq i$ (thus $D \notin P_k$ and $D \notin R_k$ for all $k \leq i$). Also $D^* \not\equiv \forall x_0(x_0 = x_0)$ and, by the FPE and because of the injectivity of $*$ $Prf(k, \ulcorner D^* \urcorner)$ is false for all $k \leq 2i+1$. Thus

$$\mathbf{PA} \vdash \neg \exists x \leq p_i{}^* \; Prf(x, \ulcorner D^* \urcorner),$$

and $\mathbf{PA} \vdash \text{``} l = j \text{''} \to \neg B^*$.

Now we proceed with different inductions on formulas, first for $k > 0$, and then for $k = 0$.

Let $1 \leq k \leq n$. The induction step in case of "$\rightarrow$" is straightforward (cf.[5]). The induction step in case $B \equiv \Box H$: we proceed with the standard Solovay argument.

If $k \Vdash \Box H$, then

$$\text{for all } j \succ k \ \ j \Vdash H,$$

$$\text{for all } j \succ k \ \ \mathbf{PA} \vdash \text{``}l = j\text{''} \rightarrow H^*,$$

$$\mathbf{PA} \vdash \bigvee_{k \prec j} \text{``}l = j\text{''} \rightarrow H^*,$$

$$\mathbf{PA} \vdash \textit{Provable}(\ulcorner \bigvee_{k \prec j} \text{``}l = j\text{''} \urcorner) \rightarrow \textit{Provable}(\ulcorner H^* \urcorner), \qquad \dagger$$

$$\text{by 2.8 (5) } \mathbf{PA} \vdash \text{``}l = k\text{''} \rightarrow \bigwedge_{k \not\succ j} \textit{Provable}(\ulcorner \text{``}l \neq j\text{''} \urcorner),$$

$$\mathbf{PA} \vdash \text{``}l = k\text{''} \rightarrow \textit{Provable}(\ulcorner \bigwedge_{k \not\succ j} \text{``}l \neq j\text{''} \urcorner) \text{ (by commuting } \textit{Provable}(\cdot) \text{ and } \bigwedge),$$

$$\text{by 2.8 (1) } \mathbf{PA} \vdash \textit{Provable}(\ulcorner \bigvee_{0 \leq j \leq n} \text{``}l = j\text{''} \urcorner),$$

$$\mathbf{PA} \vdash \textit{Provable}(\ulcorner \bigwedge_{k \not\succ j} \text{``}l \neq j\text{''} \rightarrow \bigvee_{k = j, k \prec j} \text{``}l = j\text{''} \urcorner),$$

$$\text{by 2.8 (3) } \mathbf{PA} \vdash \text{``}l = k\text{''} \rightarrow \textit{Provable}(\ulcorner \text{``}l \neq k\text{''} \urcorner),$$

$$\text{finally, } \mathbf{PA} \vdash \text{``}l = k\text{''} \rightarrow \textit{Provable}(\ulcorner \bigvee_{k \prec j} \text{``}l = j\text{''} \urcorner),$$

$$\text{and by } \dagger \ \mathbf{PA} \vdash \text{``}l = k\text{''} \rightarrow \textit{Provable}(\ulcorner H^* \urcorner),$$

and since $\mathbf{PA} \vdash \forall y (\textit{Provable}(y) \rightarrow \textit{Pr}(y))$ we are done:

$$\mathbf{PA} \vdash \text{``}l = k\text{''} \rightarrow B^*.$$

If $k \not\Vdash \Box H$, then

$$\text{for some } j \succ k \ \ j \not\Vdash H,$$

$$\text{by the induction hypothesis } \mathbf{PA} \vdash \text{``}l = j\text{''} \rightarrow \neg H^*,$$

$$\mathbf{PA} \vdash H^* \rightarrow \text{``}l \neq j\text{''},$$

$$\mathbf{PA} \vdash \textit{Provable}(\ulcorner H^* \urcorner) \rightarrow \textit{Provable}(\ulcorner \text{``}l \neq j\text{''} \urcorner),$$

$$\mathbf{PA} \vdash \neg \textit{Provable}(\ulcorner \text{``}l \neq j\text{''} \urcorner) \rightarrow \neg \textit{Provable}(\ulcorner H^* \urcorner),$$

$$\text{but by 2.8 (4) } \mathbf{PA} \vdash \text{``}l = k\text{''} \rightarrow \neg \textit{Provable}(\ulcorner \text{``}l \neq j\text{''} \urcorner),$$

$$\text{thus } \mathbf{PA} \vdash \text{``}l = k\text{''} \rightarrow \neg \textit{Provable}(\ulcorner H^* \urcorner).$$

But $Pr(y)$ may differ from $Provable(y)$ only on the gödelnumbers of $*$-interpretations of the q-atomic formulas from the fixed finite set T; every formula from T is valid in each node of the model $\mathcal{K}'$, thus $H \notin T$, and

$$\mathbf{PA} \vdash Pr(\ulcorner H^* \urcorner) \to Provable(\ulcorner H^* \urcorner).$$

Finally, we have got the desired $\mathbf{PA} \vdash "l = k" \to \neg Pr(\ulcorner H^* \urcorner)$, i.e.

$$\mathbf{PA} \vdash "l = k" \to \neg B^*.$$

Let now $k = 0$. Again, the basis of the induction on formulas is done, the case of $\to$ is trivial. Let $B = \Box H$. If $0 \Vdash \Box H$, then for all $j = 1, \ldots, n$ $j \Vdash H$, by the previous induction

$$\text{for all } j = 1, \ldots, n \quad \mathbf{PA} \vdash "l = j" \to H^*,$$

$$\mathbf{PA} \vdash \bigvee_{1 \le j \le n} "l = j" \to H^*.$$

Also $1 \Vdash H \Rightarrow 0 \Vdash H$, and by the induction hypothesis $\mathbf{PA} \vdash "l = 0" \to H^*$. By 2.8 (1) $\mathbf{PA} \vdash "0 \le l \le n"$, and thus $\mathbf{PA} \vdash H^*$, $\mathbf{PA} \vdash Provable(\ulcorner H^* \urcorner)$, and $\mathbf{PA} \vdash "l = 0" \to B^*$. If $0 \not\Vdash \Box H$, then for some $j > 0$ $j \not\Vdash H$, by the previous induction $\mathbf{PA} \vdash "l = j" \to \neg H^*$, $\mathbf{PA} \vdash H^* \to "l \ne j"$,

$$\mathbf{PA} \vdash Provable(\ulcorner H^* \urcorner) \to Provable(\ulcorner "l \ne j" \urcorner),$$

$$\mathbf{PA} \vdash \neg Provable(\ulcorner "l \ne j" \urcorner) \to \neg Provable(\ulcorner H^* \urcorner),$$

$$\text{but by 2.8 (4) } \mathbf{PA} \vdash "l = 0" \to \neg Provable(\ulcorner "l \ne j" \urcorner),$$

$$\text{thus } \mathbf{PA} \vdash "l = 0" \to \neg Provable(\ulcorner H^* \urcorner).$$

The same argument as above shows that

$$\mathbf{PA} \vdash Pr(\ulcorner H^* \urcorner) \to Provable(\ulcorner H^* \urcorner),$$

and we again have

$$\mathbf{PA} \vdash "l = 0" \to \neg B^*.$$

■

2.12 Definition. A model is called **functional** if it satisfies an extra forcing condition:

$x \Vdash \neg(\Box_p B \& \Box_p C)$ if B, C are not unifiable, else
$x \Vdash \Box_p B \& \Box_p C \Rightarrow x \Vdash D \to E$ for every D, E s.t. $D = E$ $(\bmod\ B = C)$

Kripke models for $\mathcal{FC}$ are functional r-models. We skip the proof, addressing the reader to [1], Theorem 3.2 and to Lemmas 3.4 and 3.5.

2.13 Theorem.

$$\mathcal{FC} \vdash A \quad \Leftrightarrow \quad \text{for every interpretation } * \quad \mathbf{PA} \vdash A^*.$$

Proof. Correctness, i.e. the case ($\Rightarrow$). Induction on a proof of A in $\mathcal{FC}$. After Theorem 2.6 it only remains to check the correctness of the functionality axiom **A6**, which is done *de facto* in [1], theorem 3.14.

($\Leftarrow$).The proof is an easy combination of the proofs of Theorem 3.14 from [1] and 2.7. In the notations of the Theorem 2.7 let

$$Q_{i+1} - Q_i = \{A_{i,2}, \ A_{i,3}, \ldots, A_{i,n_i}\} \ (0 \le i \le m).$$

Pay attention that here $n_i = 1+$ the cardinality of R_i, so if $n_i = 1$, then $R_i = \emptyset$. Because of the functional property of the model with respect to q-atomic formulas $\Box_{p_i} A$, there is not more than one formula among $A_{i,j}$, for which $\Box_{p_i} A_{i,j}$ holds in the model; without loss of generality we may assume that it is A_{i,n_i}. Let also $M = n_0 + \ldots + n_m$. We define

$$p_i^* := n_0 + \ldots + n_i - 1$$

and consider the following fixed point equation:

$$
Prf(u, v) \ \leftrightarrow \ \Big[\ \begin{array}{llll}
u = 0 & \to & v = \ulcorner \forall x_0(x_0 = x_0) \urcorner & \& \\
u = 1 & \to & v = \ulcorner A_{0,2}{}^{\sharp} \urcorner & \& \\
u = 2 & \to & v = \ulcorner A_{0,3}{}^{\sharp} \urcorner & \& \\
& \vdots & & \\
u = n_1 - 1 & \to & v = \ulcorner A_{0,n_0}{}^{\sharp} \urcorner & \& \\
u = n_1 & \to & v = \ulcorner \forall x_0(x_0 = x_0) \urcorner & \& \\
u = n_1 + 1 & \to & v = \ulcorner A_{1,2}{}^{\sharp} \urcorner & \& \\
& \vdots & & \\
u = M - 1 & \to & v = \ulcorner A_{m,n_m}{}^{\sharp} \urcorner & \& \ \cdot \\
u \ge M & \to & Proof(u - M, v) & \ \Big]
\end{array}
$$

Now we have just to repeat the steps of the proof of Theorem 2.7 and then 3.14 from [1].

■

2.1   Comments

1. Uniform completeness of $\mathcal{BC}$ and for $\mathcal{FC}$ takes place, namely, in theorems 2.7 and 2.13 one can choose a proof predicate uniformly for all A's.

2. The case of logics of all formulas true in the standard model of arithmetic for $\mathcal{BC}$ and for $\mathcal{FC}$ can be treated exactly as that for $\mathcal{B}$ and $\mathcal{F}$ in [1]. The truth cases for $\mathcal{BC}$ and $\mathcal{FC}$ are the logics $\mathcal{BC}^\omega$ and $\mathcal{FC}^\omega$ whose axioms are all the theorems of $\mathcal{BC}$ and $\mathcal{FC}$, the axiom scheme $\square A \to A$ and the only rule is **R0**.

3. How $\mathcal{BC}$ and $\mathcal{FC}$ are related to specific complexity measures? Let us take the one determined by the usual Gödel proof predicate $Proof(x,y)$. Clearly, the logic of $Proof(x,y)$ extends $\mathcal{FC}$, but this logic itself essentially depends on some occasional details of the coding of formulas and proofs. For example, the scheme

$$\triangle_p \neg\neg A \to \triangle_p A$$

expresses the conjecture that the least proof code of any formula A is less than any proof code of $\neg\neg A$. One can easily define numerations of proofs with or without this property; the question whether it holds for a "usual" coding of proofs doesn't look relevant to real problems of the complexity of proofs. It is the main reason why unlike the paper [1] we skip the usual Gödel proof predicate case. However, for some natural complexity measures the question of their individual complexity logics might have sense; such a logic would extend $\mathcal{BC}$ or even $\mathcal{FC}$ (for a decidable complexity predicate), or $\mathcal{BE}$ from the next chapter (for a recursively enumerable complexity predicate).

■

3   Logics for recursively enumerable measures.

The system $\mathcal{E}$ is formulated in the language $\mathcal{L}^{-+}$, i.e. without modalities $\square_p$, and its axioms are **A0** – **A2** (the system **GL**) + **A8** + **A9** + **A12** and inference rules **R0** – **R2**. The system $\mathcal{BE}$ (formulated in $\mathcal{L}^{++}$) is $\mathcal{B} + \mathcal{E}$. For technical reasons we introduce an auxiliary system $\mathcal{E}^-$ that is $\mathcal{E}$ without **R2**. In the sequel "⊢" without the left argument means "**PA** ⊢". Also under modal formula we understand an $\mathcal{L}^{-+}$-formula for $\mathcal{E}$ and an $\mathcal{L}^{++}$ one for $\mathcal{BE}$. As for $\mathcal{BC}$ and $\mathcal{FC}$ we prove the following

3.1 Lemma.   For an arithmetical interpretation $*$ and a modal formula F

$$\mathcal{E} \vdash F \;\; \Rightarrow \;\; \mathbf{PA} \vdash F^*$$
$$\mathcal{BE} \vdash F \;\; \Rightarrow \;\; \mathbf{PA} \vdash F^*$$

Proof.
The proof is essentially done in Lemma 2.6. We have to prove only that $\vdash (\triangle_\alpha A \to \Box A)^*$, i.e. $\vdash Cpl(t, \ulcorner A^* \urcorner) \to \exists x\, Cpl(x, \ulcorner A^* \urcorner)$ which is obvious, and $\vdash (\triangle_\alpha A \to \Box \triangle_\alpha A)^*$, which follows from Σ_1^0-completeness of **PA**.

∎

3.2 Definition. A set X of modal formulas is **adequate** if it is closed under subformulas; $\bot \in X$; if $B \in X$ and B is not of a form $\neg C$, then $\neg B \in X$; for all α, A if there exist β, B such that $\triangle_\beta A$ and $\triangle_\alpha B$ belong to X, then $\triangle_\alpha A \in X$ and $\Box A \in X$.

The definition of an X-model $\mathcal{K}$ is the same as that for $\mathcal{BC}$ and $\mathcal{FC}$ (definition 2.3).

3.3 Remark. Every X-model can be extended to a model by defining $x \not\Vdash \varphi$ for each node $x \in K$ and for each atomic and q-atomic formula $\varphi \notin X$. The forcing conditions 1.-5. are clearly respected, we only need to verify the condition 6. **A12**. Suppose

$$x \Vdash \Diamond^+(\triangle_{\beta_1} A_0 \& \neg \triangle_{\beta_0} A_0) \& \ldots \& \Diamond^+(\triangle_{\beta_0} A_n \& \neg \triangle_{\beta_n} A_n),$$

and also suppose that there is an q-atomic formula among those mentioned explicitly here that does not belong to X; w.l.g. assume that $\triangle_{\beta_1} A_1 \notin X$. However, $x \Vdash \triangle_{\beta_1} A_0$ and $x \Vdash \triangle_{\beta_2} A_1$, thus $\triangle_{\beta_1} A_0, \triangle_{\beta_2} A_1 \in X$, hence $\triangle_{\beta_1} A_1 \in X$. It demonstrates that all q-formulas from the example of **A12** above are from X, which is impossible since we consider an X-model.

3.4 Lemma. $\mathcal{E}^- \not\vdash A \Rightarrow$ there is an $Ad(A)$-model $\mathcal{K}$ with the root $\mathbf{r}$ such that $\mathbf{r} \not\Vdash A$.

Proof. Let $\{\triangle_{\alpha_i} A_j\}_{i=0}^{n}{}_{j=0}^{m}$ be all the $\triangle_\alpha$-formulas in $Ad(A)$, and let $\{T_{ij}\}_{i=0}^{n}{}_{j=0}^{m}$ be sentence variables not occurring in $Ad(A)$. In every $B \in Ad(A)$ we replace all occurrences of $\triangle_{\alpha_i} A_j$, that are not in the scope of any labeled modality, by T_{ij}. The resulting formula B^t is in the language $\mathcal{L}$. Let Y be a set of $\mathcal{L}$-formulas, containing

1. $\Box^+(T_{ij} \to \Box A_j{}^t)$,

2. $\Box^+(T_{ij} \to \Box T_{ij})$,

3. $\Box^+(\neg(\Diamond^+(\neg T_{kr} \wedge T_{lr}) \wedge \Diamond^+(\neg T_{ls} \wedge T._s) \wedge \ldots \wedge \Diamond^+(\neg T_{k't} \wedge T_{kt})))$
   for all $(k, l \ldots k') \subset \{0 \ldots n\}^*$ and $(r, s \ldots t) \subset \{0 \ldots m\}^*$ [5].

---

[5] $\{0, \ldots, n\}^*$ denotes the set of all finite sequences of numbers from $\{0, \ldots, n\}$.

Thus the substitution of $\triangle_{\alpha_i} A_j$ for T_{ij} makes Y the set of theorems of $\mathcal{E}^-$, so if $\mathbf{GL} \vdash \bigwedge Y \to B^t$, then $\mathcal{E}^- \vdash B$, whenever $B \in Ad(A)$ (easy induction on a proof). But $\mathcal{E}^-$ does not prove A, so we can (cf. [5]) construct a $\mathbf{GL}$-countermodel $\mathcal{K}' = \langle K, \prec, \Vdash' \rangle$, which is a finite tree with the root $\mathbf{r}$ such that $\mathbf{r} \nVdash' \bigwedge Y \to A^t$, hence $\mathbf{r} \Vdash' \bigwedge Y$ and $\mathbf{r} \nVdash' A^t$. By the definition of Y and $\mathbf{r} \Vdash' \bigwedge Y$ we can define an $Ad(A)$-model $\mathcal{K} = \langle K, \prec, \Vdash \rangle$ putting

$$a \Vdash F \overset{def}{\Longleftrightarrow} a \Vdash' F^t.$$

Then $\mathbf{r} \nVdash A$.

∎

3.5 Lemma. $\mathcal{E} \nvdash A \Rightarrow$ there is an A-sound $Ad(A)$-countermodel $\mathcal{K}$ for A

Proof.   If $\mathcal{E} \nvdash A$, then for any $N \in \omega$ $\mathcal{E}^- \nvdash \square^N A$. Take $N = Card\{\square B \in Ad(A)\}$. By Lemma 3.4 there is an $Ad(A)$-model $\mathcal{K}'$ with the root $\mathbf{r}$ such that $\mathbf{r} \nVdash \square^N A$. By the forcing condition for $\square$ there is a chain $i_N \prec \ldots \prec i_0$ such that $i_k \nVdash \square^k A$, but the formula $\square B \to B$ can fail at one node of the chain at the most. So, by the pigeonhole principle, there is k such that $i_k \Vdash H(A)$. Since $i_N \preceq i_k$ the restriction of $\mathcal{K}'$ to the set $\{j \mid j \succ i_k \text{ or } j = i_k\}$ is a desired $Ad(A)$-model.

∎

3.6 Lemma. $\mathcal{BE} \nvdash A \Rightarrow$ there is an A-sound $Ad(A)$-countermodel for A.

Proof.   The proof is nearly the same as in Lemmas 3.4 and 3.5. We only need to replace $\mathbf{GL}$ by $\mathcal{B}^-$ and $\mathcal{E}^{(-)}$ by $\mathcal{BE}^{(-)}$. The Kripke model completeness for $\mathcal{B}^-$ is proved in [1].

∎

3.7 Theorem.   The systems $\mathcal{E}$ and $\mathcal{BE}$ are complete with respect to the intended classes of finite Kripke models and thus are both decidable.

Proof.   An easy combination of Lemmas 3.5 and 3.6, Theorem 3.1 and Theorem 3.8 proved below.

∎

3.8 Theorem.

$$\begin{array}{c} \mathcal{E} \vdash A \\ \mathcal{BE} \vdash A \end{array} \iff \text{for any interpretation } * \text{ } \mathbf{PA} \vdash A^*.$$

Proof. We prove the theorem for $\mathcal{BE}$ only. The proof for $\mathcal{E}$ is a straightforward restriction of that for $\mathcal{BE}$ to the language $\mathcal{L}^{-+}$.

Direction "$\Leftarrow$" by contraposition. If $\mathcal{BE} \nvdash A$, then by Lemma 3.6 there is an A-sound $Ad(A)$-countermodel $\mathcal{K}'$ for A; we can assume that $\mathcal{K}'$ is already extended to the (total) $Ad(A)$-countermodel for A. Let $\mathcal{K}'$ be $\langle K', \prec, \Vdash \rangle$. Again w.l.g. we assume, that $K' = \{1, \ldots, n\}$ and 1 is the root node, add to K' a new node 0 and define $0 \prec i$, $i \in K'$. For every atomic or q-atomic formula $F \in Ad(A)$ we define $0 \Vdash F \Leftrightarrow 1 \Vdash F$. Let K denote $\{0\} \cup K'$ and $\mathcal{K} = \langle K, \prec, \Vdash \rangle$. Again we define the Solovay function h and arithmetical formulas "$l = j$" for the model $\mathcal{K}$.

3.9 Definition.

$$\alpha < \beta \;\overset{def}{\Longleftrightarrow}\; \exists B(0 \Vdash \alpha <_B \beta)$$

$$\alpha = \beta \;\overset{def}{\Longleftrightarrow}\; \forall B \forall i \in M(i \Vdash \triangle_\alpha B \Leftrightarrow i \Vdash \triangle_\beta B)$$

$$\alpha \leq \beta \;\overset{def}{\Longleftrightarrow}\; \alpha < \beta \text{ or } \alpha = \beta$$

It is easy to see that $\leq$ is a linear quasi-ordering of the set of complexity variables. Let $\{\alpha_i\}_{i=1}^N$ be all complexity variables occurring in $Ad(A)$ and enumerated in such a way that $i \leq j \Rightarrow \alpha_i \leq \alpha_j$; let also $\{p_i\}_{i=0}^{N'-1}$ be all proof variables occurring in $Ad(A)$, and $I_i = \{l : \mathcal{M} \Vdash \Box_{p_i} C_l\}$. We define

$$\phi(S_i) = \begin{cases} (i \neq i) & \text{if } S_i \notin Ad(A) \\ \bigvee_{w \Vdash S_i} \text{``}l = w\text{''} \wedge i = i & \text{if } S_i \in Ad(A), \end{cases}$$

$$\phi(p_i) = \begin{cases} i & \text{if } i < N' \\ N' & \text{otherwise}, \end{cases}$$

$$\phi(\alpha_i) = \begin{cases} 0 & \text{if } \alpha_i \text{ does not occur in } Ad(A) \\ i & \text{otherwise}. \end{cases}$$

The predicates *Cpl* and *Prf* are defined as solutions of the following fixed point equations system

$$\vdash Cpl(x,y) \;\leftrightarrow\; \bigwedge_{k=1}^N \left\{ x = k \rightarrow \exists m \exists B \in Ad(A) \left[y = \ulcorner B^{*} \urcorner \wedge h(m) \Vdash \triangle_{\alpha_k} B \right] \right\}$$
$$\wedge \left\{ x > N \rightarrow \left[Provable(y) \wedge \forall \Box B \in Ad(A)(y \neq \ulcorner B^{*} \urcorner) \right] \right.$$
$$\left. \vee \exists m \exists B \in Ad(A) \left[y = \ulcorner B^{*} \urcorner \wedge h(m) \Vdash \Box B \right] \right\}$$

$$\vdash Prf(x,y) \;\leftrightarrow\; \bigwedge_{k=0}^{N'-1} \left\{ x = k \rightarrow \bigvee_{j \in I_k}(y = \ulcorner C_j^{*} \urcorner) \right\}$$
$$\wedge \left\{ x > N' \rightarrow Proof(x - N' - 1, y) \right\}.$$

Note that the defined interpretation $*$ is injective on $Ad(A)$. We will base the interpretation of $\square$ on Prv instead of Pr ; later we prove that both are equivalent to $Provable$.

3.10 Lemma.   For all $F \in Ad(A)$ and $w \in K$

$$\vdash \text{``}l = w\text{''} \to (\text{``}w \Vdash F\text{''} \leftrightarrow F^*)^6.$$

Proof.   We proceed with the induction on construction of F. The cases $F = S_i$, $F = \bot$ and $F = B \to C$ are treated as usual (cf. [5]).

Let F be $\square_{p_k} C_l$. If $w \Vdash F$, then $l \in I_k$, hence $Prf(p_k{}^*, \ulcorner C_l{}^{*}\urcorner)$ is true, so it is provable in **PA**. Then $\vdash \text{``}l = w\text{''} \to (\text{``}w \Vdash F\text{''} \to F^*)$. If $w \nVdash F$, then $l \notin I_k$, hence $Prf(p_k{}^*, \ulcorner C_l{}^{*}\urcorner)$ is false, then $\vdash \neg F^*$, so $\vdash \text{``}l = w\text{''} \to (\text{``}w \nVdash F\text{''} \to \neg F^*)$.

Let F be $\triangle_{\alpha_k} B$. We argue in **PA**. Suppose $\text{``}l = w\text{''}$. Then there exists $m$ such that $h(m) = w$. If $\text{``}w \Vdash F\text{''}$, then $\text{``}h(m) \Vdash \triangle_{\alpha_k} B\text{''}$ and as $\alpha_k{}^* = k$, then $Cpl(\alpha_k{}^*, \ulcorner B^*\urcorner)$. If $\text{``}w \nVdash F\text{''}$, then $\forall v \prec w(\text{``}v \nVdash F\text{''})$. But $\forall m' \leq m[h(m') \preceq h(m)]$, hence $\text{``}h(m') \nVdash \triangle_{\alpha_k} B\text{''}$. From the other side, $\text{``}l = w\text{''}$ and $\text{``}w \nVdash F\text{''}$ provide that $\forall m' \geq m[\text{``}h(m') \nVdash \triangle_{\alpha_k} B\text{''}]$. Now $\forall m \text{``}h(m) \nVdash \triangle_{\alpha_k} B\text{''}$, thus $\neg Cpl(\alpha_k{}^*, \ulcorner B^*\urcorner)$.

The case $F = \square B$ is similar: $\text{``}w \Vdash F\text{''}$ implies $Cpl(N + 1, \ulcorner B^*\urcorner)$, hence $Prv(\ulcorner B^*\urcorner)$. If $\text{``}w \nVdash F\text{''}$, then as above one can prove $\forall m \text{``}h(m) \nVdash \square B\text{''}$. But $\square B \in Ad(A)$, hence $Provable(y) \wedge \forall \square B \in Ad(A)(y \neq \ulcorner B^*\urcorner)$ is false for $y = \ulcorner B^*\urcorner$, then

$$\forall x > N \neg Cpl(x, \ulcorner B^*\urcorner) \tag{1}$$

Now, $\text{``}w \nVdash \square B\text{''} \Rightarrow \forall \alpha(\text{``}w \nVdash \triangle_\alpha B\text{''})$, hence

$$\forall x \leq N \neg Cpl(x, \ulcorner B^*\urcorner); \tag{2}$$

finally, (1) and (2) imply $\neg Prv(\ulcorner B^*\urcorner)$
∎

3.11 Lemma.   For all $\square B \in Ad(A)$ and $w \in K$

$$\vdash \text{``}l = w\text{''} \to [\text{``}w \Vdash \square B\text{''} \leftrightarrow Provable(\ulcorner B^*\urcorner)]$$

---

[6]Here and below $\text{``}w \Vdash F\text{''}$ means a natural recursive formalization of $w \Vdash F$ in **PA**.

Proof. If $w \not\Vdash \Box B$, then there is $v \succ w$ such that $v \not\Vdash B$. By the Lemma 3.10[7]

$$\vdash \text{``} l = v \text{''} \to \neg B^*$$
$$\vdash B^* \to \text{``} l \neq v \text{''}$$
$$\vdash Provable(\ulcorner B^* \urcorner) \to Provable(\ulcorner \text{``} l \neq v \text{''} \urcorner)$$
$$\vdash \text{``} l = w \text{''} \to \neg Provable(\ulcorner \text{``} l \neq v \text{''} \urcorner) \text{ by the Solovay lemma, } 2.8(4)$$
$$\vdash \text{``} l = w \text{''} \to \neg Provable(\ulcorner B^* \urcorner)$$
$$\vdash \text{``} l = w \text{''} \to [\text{``} w \not\Vdash \Box B \text{''} \to \neg Provable(\ulcorner B^* \urcorner)]$$

If $w \Vdash \Box B$, then for every $v \succ w$ we have $v \Vdash B$. Thus by the Lemma 3.10 for all $v \succ w$ we have $\vdash \text{``} l = v \text{''} \to B^*$,

$$\vdash \bigvee_{v \succ w} \text{``} l = v \text{''} \to B^*$$

If $w \succ 0$, then

$$\vdash Provable(\ulcorner \bigvee_{v \succ w} \text{``} l = v \text{''} \urcorner) \to Provable(\ulcorner B^* \urcorner)$$
$$\vdash Provable(\ulcorner \bigvee_{v \in K} \text{``} l = v \text{''} \urcorner)$$
$$\vdash \text{``} l = w \text{''} \to Provable(\ulcorner \bigwedge_{w \not\prec v} \text{``} l \neq v \text{''} \urcorner)$$
$$\vdash \text{``} l = w \text{''} \to Provable(\ulcorner \bigvee_{w \prec v} \text{``} l = v \text{''} \urcorner)$$
$$\vdash \text{``} l = w \text{''} \to Provable(\ulcorner B^* \urcorner).$$

If $w = 0$, then $1 \Vdash B$, hence $0 \Vdash B$, hence $\forall v \in K \; v \Vdash B$.

$$\vdash \bigvee_{v \in K} \text{``} l = v \text{''} \to B^*$$
$$\vdash \bigvee_{v \in K} \text{``} l = v \text{''}$$
$$\vdash B^*$$
$$\vdash Provable(\ulcorner B^* \urcorner)$$
$$\vdash \text{``} l = 0 \text{''} \to Provable(\ulcorner B^* \urcorner).$$

In any case $\vdash \text{``} l = w \text{''} \to Provable(\ulcorner B^* \urcorner)$, thus

$$\vdash \text{``} l = w \text{''} \to [\text{``} w \Vdash \Box B \text{''} \to Provable(\ulcorner B^* \urcorner)],$$

and we are done.

∎

---

[7] We use the fact that the model is described in **PA** by natural recursive formulas.

3.12 Lemma. $\vdash Prv(y) \leftrightarrow Provable(y)$

Proof.   Since $w \Vdash \triangle_\alpha B \Rightarrow w \Vdash \square B$, one can demonstrate in **PA** that *Prv* and *Provable* can differ only on gödelnumbers of some B^*'s such that $\square B \in Ad(A)$. For such B^* by Lemma 3.10

$$\vdash \text{``}l = w\text{''} \to \left[\text{``}w \Vdash \square B\text{''} \leftrightarrow Pr(\ulcorner B^* \urcorner)\right].$$

By Lemma 3.11 we get

$$\vdash \text{``}l = w\text{''} \to \left[Provable(\ulcorner B^* \urcorner) \leftrightarrow Pr(\ulcorner B^* \urcorner)\right]$$
$$\vdash \bigvee_{w \in K} \text{``}l = w\text{''} \to \left[Provable(\ulcorner B^* \urcorner) \leftrightarrow Pr(\ulcorner B^* \urcorner)\right]$$
and as
$$\vdash \bigvee_{w \in K} \text{``}l = w\text{''},$$

and we are done.

■

3.13 Lemma. $Cpl(x, y)$ is a standard complexity predicate.

Proof.   By FPE $Cpl \in \Sigma_1^0$. Let us prove

$$\vdash u \leq v \to \left[Cpl(u, y) \to Cpl(v, y)\right].$$

All the following reasoning is formalizable in **PA**.

1. $u = 0 \Rightarrow \neg Cpl(u, y)$;

2. If $0 < u \leq v \leq N$, then two cases are possible:

   - y is not equal to any B^* such that $\triangle_{\alpha_u} B \in Ad(A)$, hence $\neg Cpl(u, y)$;
   - $y = \ulcorner B^* \urcorner$ for some such B^*. Then

   $$u \leq v \Rightarrow \forall w(w \Vdash \triangle_{\alpha_u} B \Rightarrow w \Vdash \triangle_{\alpha_v} B),$$

   hence $Cpl(u, y) \to Cpl(v, y)$;

3. $u \leq N < v$. This case is similar to 2. but here we have $\forall w(w \Vdash \triangle_{\alpha_u} B \Rightarrow w \Vdash \square B)$;

4. If $u > N$, then $v > N$ and by the definition of Cpl we get $Cpl(u, y) \leftrightarrow Cpl(v, y)$.

The other conditions on a standard complexity predicate follow from Lemma 3.12 and from the properties of *Provable* .

∎

3.14 Lemma. *Prf* is a standard proof predicate and *Prf* and *Cpl* are provably compatible.

Proof. The recursiveness follows from the definition. We'll prove $\vdash Pr(y) \leftrightarrow Provable(y)$, what will imply all other conditions required. From the definition

$$\vdash Pr(y) \leftrightarrow Provable(y) \vee \bigvee_{i=0}^{N'} \bigvee_{j \in I_i} (y = \ulcorner C_j^* \urcorner).$$

But for all i, $j \in I_i$ and $w \in K$ $w \Vdash \Box_{p_i} C_j$, hence $w \Vdash C_j$, then by Lemma 3.10

$$\vdash \bigvee_{w \in K} \text{``} l = w \text{''} \to C_j^*$$
$$\vdash C_j^*$$
$$\vdash Provable(\ulcorner C_j^* \urcorner)$$

∎

Let us now complete the proof of the Theorem 3.8. Lemmas 3.13 and 3.14 provide that the interpretation $*$ is defined correctly. If $\mathcal{K}$ is a countermodel for A, then there is w such that $w \nVdash A$. Thus $\vdash \text{``} l = w \text{''} \to \neg A^*$. Now if $\mathbf{PA} \vdash A^*$, then $\mathbf{PA} \vdash \text{``} l \neq w \text{''}$, that contradicts the Solovay lemma.

∎

Comment

As usual, the truth cases for $\mathcal{E}$ and $\mathcal{BE}$ called $\mathcal{E}^\omega$ and $\mathcal{BE}^\omega$ whose axioms are all the theorems of $\mathcal{E}$ and $\mathcal{BE}$, the axiom scheme $\Box A \to A$ and the only rule is **R0**, are complete with respect to all arithmetical interpretations:

3.15 Theorem.

$$\begin{array}{c} \mathcal{E}^\omega \vdash A \\ \mathcal{BE}^\omega \vdash A \end{array} \iff \text{ for every interpretation } * \quad A^* \text{ is true.}$$

The proof goes like that in [2].

It looks now a routine exercise to incorporate the functionality property into the logic of recursively enumerable complexity predicates by combining the techniques from chapters 2 and 3.

References

[1] S. Artëmov, *Logic of Proofs*, Ann. Pure Appl. Logic **67** (1994), 29–59.

[2] D. Guaspari and R. M. Solovay, *Rosser sentences*, Ann. of Math. (2) **16** (1979), 81–99.

[3] J. Lassez, M. Maher, and K. Marriott, *Unification revisited*, Foundations of Deductive Databases and Logic Programming (J. Minker, ed.), Morgan Kaufmann Publishers, Inc., 1987, pp. 587–625.

[4] R. Magari, *The diagonalizable algebras (the algebraization of the theories which express Theor.:II)*, Boll. Un. Mat. Ital. **12** (1975), 117–125.

[5] R. M. Solovay, *Provability interpretations of modal logic*, Israel J. Math. **25** (1976), 287–304.

STEKLOV MATHEMATICAL INSTITUTE, VAVILOVA 42, MOSCOW 117966, RUSSIA
e-mail address: SERGEI@ARTEMOV.MIAN.SU
MATHEMATICAL LOGIC SECTION, DEPARTMENT OF MATHEMATICS, MOSCOW STATE UNIVERSITY, MOSCOW 119899, RUSSIA

Beyond the s-Semantics:
a Theory of Observables

Marco Comini   Giorgio Levi

Abstract

We give an algebraic formalization of SLD-trees and their abstractions (observables). We can state and prove in the framework several useful theorems (lifting, AND-compositionality, correctness and full abstraction of the denotation, equivalent top-down and bottom-up constructions) about semantic properties of various observables. Observables are represented by Galois co-insertions and can be used to model abstract interpretation. The constructions and the theorems are inherited by all the observables which can be formalized in the framework. The power of the framework is shown by reconstructing some known examples (answer constraints, finite failures, call patterns, correct call patterns and ground dependencies call patterns).

Keywords: semantics, abstract interpretation, SLD-trees, observational equivalences.

1   Introduction

SLD-trees are structures used to describe the operational semantics of logic programs. From an SLD-tree we can derive several operational properties which are useful for reasoning about programs. Examples are SLD-derivations, resultants, call patterns, partial answers, computed answers.

All these properties, that we call *observables*, can be obtained as abstractions of the SLD-tree. Let for example T be an SLD-tree with initial goal g_0. Then,

- an SLD-derivation is any path of T;

- a resultant is the formula $g_n \rightsquigarrow g_0 \vartheta_n$, if g_n and ϑ_n are the goal and the substitution associated to any node n in T;

- a call pattern is any atom selected in T;

- a partial answer is the substitution associated to any node n in T (restricted to g_0);

- a computed answer is the substitution associated to any node n, such that g_n is the empty clause.

The behavior of a program p (via selection rule r) with respect to a given observable can be understood by observing the corresponding properties for all possible goals. We know from recent results on the semantics [22, 26, 3] that we can characterize this behavior just by observing the property for some specific atomic goals, namely the "most general" atomic goals. This result was first obtained for computed answers in the s-semantics framework [18, 19] and then proved for other less abstract observables, such as resultants, call patterns and partial answers in [22, 26].

The behaviors for most general atomic goals can then be considered a program denotation. In [26] this approach is used to define a semantic framework, where one can define denotations modeling various observables, by inheriting from the framework the basic constructions and theorems. Some of the denotations enjoy additional properties, such as full abstraction. The technical tool used to define the abstractions is the definition of suitable equivalence relations.

We approach here the same problem with a different emphasis and a different technical tool. Our main objective is in fact the formalization of the observable, while our abstractions will be based on abstract interpretation techniques [14]. This will allow us to model, within the same framework, the approximation which is typically involved in the abstractions used for program analysis. A similar approach can be found in [30].

Our main results are the definition of an algebraic framework for reasoning about SLD-trees and their abstractions (observables) in the case of Constraint Logic Programs. The framework is provided with several general theorems (lifting, AND-compositionality, correctness and full abstraction, equivalent top-down and bottom-up constructions), which are valid for any abstraction, possibly in a weaker form in the case of abstract interpretation (i-observables). We give the reconstruction of several existing constructions to show the expressive power of the framework.

The paper is organized as follows. Section 2 gives some preliminaries on logic programs, abstract interpretation and constraint systems. Section 3 summarizes the basic definitions and results of the s-semantics, which was the first semantics developed from an observable, i.e., computed answers [18, 19, 20]. Section 4 characterizes the general theory in the case of SLD-trees. Section 5 formalizes the main class of observables, i.e., the s-observables for which all the general results are valid. Finally section 6 considers the *i*-observables, where the results are weaker yet meaningful from the abstract interpretation theory viewpoint. All the proofs can be found in [12].

2 Preliminaries and basic definitions

2.1 Semantics and lattices

Let $\approx$ be an equivalence relation on programs. A semantics $\mathcal{S}(p)$ is *correct* w.r.t. $\approx$, if $\mathcal{S}(p_1) = \mathcal{S}(p_2)$ implies $p_1 \approx p_2$. Furthermore $\mathcal{S}(p)$ is *fully abstract* w.r.t. $\approx$ if the vice versa holds. Semantic domains are lattices where the partial order formalizes an underlying approximation structure. In the following, we will assume familiarity with *lattice theory*.

A *poset* $P_\leq$ is a set P equipped with a partial order $\leq$. Given a poset $P_\leq$ and $X \subseteq P$, $y \in P$ is an *upper bound* for X iff $x \leq y$ for each $x \in X$. An upper bound y for X is the *least upper bound*, often denoted by $lub_\leq(X)$, iff for every upper bound $y' : y \leq y'$. A *lattice* is a poset $P_\leq$ such that for any pair of elements there exist the greatest lower bound (*meet*), and the least upper bound (*join*). A lattice L with partial ordering $\leq$, least upper bound $\sqcup$, greatest lower bound $\sqcap$, bottom element $\bot$ ($\bot = \sqcup \emptyset = \sqcap L$) and top element $\top$ ($\top = \sqcap \emptyset = \sqcup L$) is denoted by $L_\leq(\bot, \top, \sqcup, \sqcap)$. When clear from the context, we use the poset notation to denote lattices. A *sublattice* of a lattice L is a subset X of L such that $a, b \in X$ imply $a \sqcap b \in X$ and $a \sqcup b \in X$. A *complete lattice* is a poset $L_\leq$ for which all subsets have a least upper bound and a greatest lower bound. Any finite lattice is complete.

2.2 Notation and preliminaries on logic programming

The reader is assumed to be familiar with the terminology of and the basic results in the semantics of logic programs. For a comprehensive description of the semantics of (positive) logic programs see [36, 1]. Let the signature consist of a set Σ of *data constructors* (a set of function symbols), a finite set Π of *predicate symbols*, a denumerable set V of *variable symbols*. All the definitions in the following will assume a given signature (Σ, Π). The term algebra over Σ and V is denoted by *Term*. Variable-free terms are called *ground*.

A *substitution* is a (finite) mapping from $\vartheta : V \longrightarrow Term$ such that the set $D(\vartheta) = \{x \mid \vartheta(x) \neq x\}$ (*domain* of ϑ) is finite. If $W \subset V$, we denote by $\vartheta_{|W}$ the *restriction* of ϑ to the variables in W, i.e., $\vartheta_{|W}(y) = y$ for $y \notin W$. Moreover if E is any expression, we use the abbreviation $\vartheta_{|E}$ to denote $\vartheta_{|Var(E)}$. λ denotes the empty substitution. A *grounding substitution* for a variable x is a substitution such that $\vartheta(x)$ is ground. The *composition* $\vartheta\sigma$ of the substitutions ϑ and σ is defined as the functional composition, i.e., $\vartheta\sigma(x) = \sigma(\vartheta(x))$.

A *renaming* is a substitution ϱ for which there exists the inverse ϱ^{-1}, such that $\varrho\varrho^{-1} = \varrho^{-1}\varrho = \lambda$. We denote by Θ the set of all the renamings with variables in V. Note that $(\Theta, \circ, id)$ is a group, also called *group of renamings*.

The pre-ordering $\leq$ (more general than) on substitutions is such that $\vartheta \leq \sigma$ iff there exists ϑ' such that $\vartheta\vartheta' = \sigma$. The result of the application of the substitution ϑ to a term t is an *instance* of t denoted by $t\vartheta$. In general the application of a substitution ϑ to a syntactic object s is denoted by $s\vartheta$ and is defined as the syntactic object obtained by replacing each variable x in s by $\vartheta(x)$. We define $s \leq s'$ (s is more general than s') iff there exists ϑ such that $s\vartheta = s'$. A substitution ϑ is *grounding* for s if $s\vartheta$ is ground. The relation $\leq$ is a preorder. $\equiv$ denotes the associated equivalence relation (*variance*). A substitution ϑ is a *unifier* of terms t and t' if $t\vartheta = t'\vartheta$. The *most general unifier* of t_1 and t_2 is denoted by $mgu(t_1, t_2)$.

An *atom* a is an object of the form $q(t_1, \ldots, t_n)$ where $q \in \Pi_n$ and $t_1, \ldots, t_n \in$ *Term*. A *definite clause* is a formula of the form $h :\!- a_1, \ldots, a_n$ with $n \geq 0$, where h (the *head*) and $a_1, \ldots, a_n$ (the *body*) are atoms. ":−" and "," denote logic implication and conjunction respectively, and all the variables are universally quantified. If the body is empty the clause is a *unit clause*. A *positive program* is a finite set of clauses $p = \{k_1, \ldots, k_n\}$. A *positive goal* is a formula $a_1, \ldots, a_m$, where each a_i is an atom.

A *Herbrand interpretation* $\mathcal{I}$ for a program p is a subset of the *Herbrand base* $\mathcal{B}$ (the set of all ground atoms). The intersection $M(p)$ of all the Herbrand models of a positive program p is a model (least Herbrand model). $M(p)$ is also the least fixpoint $T_p \uparrow \omega$ of a continuous transformation T_p (*immediate consequences operator*) on Herbrand interpretations. The *ordinal powers* of a generic monotonic operator T_p on a complete lattice $(D, \leq)$ with bottom $\perp$ are defined as usual, namely $T_p \uparrow 0 = \perp$. $T_p \uparrow (\alpha + 1) = T_p(T_p \uparrow \alpha)$ for α successor ordinal and $T_p \uparrow \alpha = lub(\{T_p \uparrow \beta \mid \beta < \alpha\})$ if α is a limit ordinal. If g is a positive goal, $g \overset{\vartheta}{\leadsto}_{p,r} b_1, \ldots, b_n$ denotes an *SLD-derivation* of $b_1, \ldots, b_n$ from the goal g in the program p which uses the selection rule r and such that ϑ is the composition of the mgu's used in the derivation. $g \overset{\vartheta}{\longmapsto}_p \square$ denotes the *SLD-refutation* of g in the program p with computed answer substitution ϑ. A computed answer substitution is always restricted to the variables occurring in g.

The set of variables which occur in a syntactic object (e.g. terms, atoms, clauses, etc.) t is denoted by $Var(t)$. Tuples of variables and terms are sometimes denoted by $\tilde{x}, \tilde{y}, \ldots$ and $\tilde{t}, \tilde{s}, \ldots$, etc. Similarly $\tilde{b}$ denotes a (possibly empty) conjunction of atoms $b_1, \ldots, b_n$ and $(\tilde{b}, \tilde{b}')$ denotes atom tuples conjunction $b_1, \ldots, b_m, b'_1, \ldots, b'_n$. We denote by $\tilde{t}$ both the tuple and the set of corresponding syntactic objects. If $\sim$ is an equivalence relation on X, we denote by $[x]_\sim$ the equivalence class of $x \in X$. Sequences of syntactic objects are denoted by $[s_1, \ldots, s_n]$ and $::$ denotes concatenation of sequences (ε is the empty sequence).

2.3 Abstract interpretation

Galois connections are used to formalize the relation between abstract and concrete meanings of a computation. Although they can be defined on preordered sets, we define them on posets and lattices.

Let $C_\leq$ and $A_\sqsubseteq$ denote posets. A *Galois connection* is a pair of maps $\alpha : C \longrightarrow A$ and $\gamma : A \longrightarrow C$ such that for each $x \in C$ and $y \in A$: $\alpha(x) \sqsubseteq y \Leftrightarrow x \leq \gamma(y)$. The map $\alpha\,(\gamma)$ is called the *lower (upper) adjoint* or *abstraction (concretization)* in the context of abstract interpretation. The above definition of Galois connection is equivalent to the following set of conditions: α and γ are monotonic; $\forall y \in A : \alpha(\gamma(y)) \sqsubseteq y$; $\forall x \in C : x \leq \gamma(\alpha(x))$.

A Galois co-insertion of $C_\leq$ onto $A_\sqsubseteq$ is a Galois connection (α, γ) where α is surjective (iff γ is injective), or (equivalently) where $\alpha \circ \gamma = id$. When γ is not injective several distinct elements of the abstract domain A have the same meaning (by γ). A Galois connection can always be forced to a co-insertion by considering a more concise abstract domain A which is isomorphic to the quotient set $A_{/\approx_\gamma}$, such that for each $x, y \in A : x \approx_\gamma y$ iff $\gamma(x) = \gamma(y)$.

When a Galois connection has been established between the concrete and the abstract domains, any concrete fixpoint can be approximated by an abstract fixpoint by means of an approximation of the program semantic function. Thus, in addition to an abstract domain having a Galois co-insertion with the concrete semantic domain, an abstract interpretation framework is characterized by an *approximate semantic function*. One of the main results obtained by Cousot and Cousot in [14] was showing that there always exists a *best* choice for an approximate semantic function, once a Galois connection is established. Let $F : C \longrightarrow C$ be a monotonic function. Then $\alpha\,(lfp_C\,(F)) \sqsubseteq lfp_A\,(\alpha \circ F \circ \gamma)$. This result (*fixpoint abstraction theorem* [14]) has a fundamental outcome, i.e., the choice of the concrete semantics and the Galois connection between the concrete domain and the abstract one fully determines the (best) abstract semantics as $lfp_A\,(\alpha \circ F \circ \gamma)$.

2.4 Constraint systems

We consider here semi-distributive constraint systems, as defined by the generalized semantics of CLP [15, 29, 16], whose aim was to provide a framework for the abstract interpretation of CLP programs.

A *term system* is an algebra of terms provided with a binary operator which realizes substitutions.

Definition 2.1 [term system] A term system $\mathcal{T}$ is an algebraic structure $(\mathcal{T}, S, V)$, where $\mathcal{T}$ is a set of $\mathcal{T}$-terms (terms), V is a countable set of $\mathcal{T}$-variables (variables) in $\mathcal{T}$, S is a countable set of binary operations on $\mathcal{T}$, indexed by V, and the following conditions hold, for all $x, y \in V$ and $t, t', t'' \in \mathcal{T}$.

T_1 $s_x(t, x) = t$ (**identity**)

T_2 $s_x(t, y) = y$ if $x \neq y$ (**annihilation**)

T_3 $s_x(t, s_x(y, t')) = s_x(y, t')$ if $x \neq y$ (**renaming**)

T_4 $s_x(t', s_y(t'', t)) = s_y(s_x(t', t''), s_x(t', t))$
    if $x \neq y$, y *ind* t' (**independent composition**)

where a term t is *independent* (or free) on the variable x (x *ind* t), if $s_x(t', t) = t$ for any $t' \in \mathcal{T}$.

Definition 2.2 [term systems morphisms] Let $(\mathcal{T}, S, V)$ and $(\mathcal{T}', S', V')$ be term systems, with $s \in S$ and $s' \in S'$. A morphism $\kappa : \mathcal{T} \longrightarrow \mathcal{T}'$, is a function mapping $\mathcal{T}$-terms to $\mathcal{T}'$-terms such that for any $t_1, t_2 \in \mathcal{T}$ and $x \in V$: $\kappa(s_x(t_1, t_2)) = s'_{\kappa(x)}(\kappa(t_1), \kappa(t_2))$.

The process of building constraints in any evaluation of a *CLP* program is mainly based on set union and conjunction. This is modeled in [15, 29, 16] by giving an algebraic characterization of constraints over term system.

Definition 2.3 [semirings] A semiring $(\mathcal{C}, \otimes, \oplus, 1, 0)$ is an algebraic structure such that

1. $(\mathcal{C}, \oplus, 0)$ is a idempotent and commutative monoid,

2. $(\mathcal{C}, \otimes, 1)$ is a monoid and

3. 0 is an *annihilator* for $\otimes$.

We can give a natural partial ordering over a semiring, namely $c \leq c'$ iff $c + c' = c'$.

Definition 2.4 [(semi)closed semiring] A *semiclosed semiring* $(\mathcal{C}, \oplus, 0)$ is a semiring s.t.

1. If $a_1, \ldots a_n, \ldots$ is a countable sequence of elements in $\mathcal{C}$, $a_1 \oplus a_2 \oplus \ldots \oplus a_n \oplus \ldots$ exists and is unique.

2. Moreover associativity, commutativity and idempotence of $\oplus$ apply to countably infinite as well as to finite joins.

3. Finally, $\otimes$ is *left/right semi-distributive*, i.e., $(c \otimes c_1) + (c \otimes c_2) \leq c \otimes (c_1 + c_2)$ and the symmetric.

A *closed semiring* is a semiclosed semiring where $\otimes$ distributes over $\oplus$.

Semi-distributivity allows us to approximate constraints by possibly infinite joins of finite meets (also called *simple constraints*). *Cylindric algebras* provide variable projection by means of a family of "hiding" operators on the underlying algebra (a cylindric algebra is formed by enhancing a Boolean algebra by means of a family of unary operations called *cylindrifications*). They are needed here because when we solve a goal, we are interested only in constraints on the variables that appear in that goal. *Diagonal elements* provide parameter passing. The equality symbol "=" in any constraint system provides term unification. Diagonal elements represent equations on the term system $\mathcal{T}$. The result is the introduction of "term unification" (i.e., equations on terms) as distinguished elements in the algebra. Finally, a unary operator ∂_x^t, where x and t are a variable and a term respectively, extends the substitution operation to idempotent substitutions on constraints.

Definition 2.5 [semi-distributive constraint system] [29, 16] A *constraint system* $\mathcal{A}$ is an algebraic structure $(\mathcal{C}, \otimes, \oplus, 1, 0, \exists_\Delta, \partial_x^t, t=t')$, with $\{x\}, \Delta \subseteq V$. $t, t' \in \mathcal{T}$. where $\mathcal{C}$ is a set of $\mathcal{A}$-constraints generated by a set of *atomic constraints*, V is a denumerable set of variables, $\mathcal{T}$ is a term system, 0. 1, $t=t'$ are (atomic) elements of $\mathcal{C}$, for each $t, t' \in \mathcal{T}$, $\{\exists_\Delta\}_{\Delta \subseteq V}$ and $\{\partial_x^t\}_{x \in V, t \in \mathcal{T}}$ are unary operations on $\mathcal{C}$. the latter being defined for x *ind* t, $\otimes$, $\oplus$ are binary operations on $\mathcal{C}$. such that the following properties are satisfied for any $c, c' \in \mathcal{C}$; $\Delta, \Psi \subseteq V$ and $t, t', t'' \in \mathcal{T}$.

R_1. $(\mathcal{C}, \otimes, \oplus, 1, 0)$ is a semi-closed semiring,

C_1. $c \oplus \exists_\Delta c = \exists_\Delta c$,

C_2. $\exists_\Delta (c \otimes \exists_\Delta c') = \exists_\Delta (\exists_\Delta c \otimes c') = \exists_\Delta c \otimes \exists_\Delta c'$,

C_3. $\exists_\Delta \exists_\Psi c = \exists_{\Delta \cup \Psi} c$,

C_4. $\exists_\Delta$ distributes over infinite joins,

D_1. $t=t = 1$,

D_2. $\exists_{\{x\}} x=t = 1$,

D_3. $t_=t' = t'_=t$,

S_1. $\partial_x^t(c) = \exists_{\{x\}}(x_=t \otimes c)$,

S_2. $\partial_x^t(t'_=t'') = [t/x]\,t'_=[t/x]\,t''$.

S_3. $\partial_x^t(c \otimes c') = \partial_x^t c \otimes \partial_x^t c'$.

If $X \subseteq V$, we denote $\exists_{var(c)/X}(c)$ (i.e., the hiding of all the variables in c apart from those of X) by $\exists(c)_X$. We often omit brackets in cylindrifications on sets of variables. Moreover given a syntactic object s we denote $\exists(c)_{var(s)}$ by $\exists(c)_s$.

Differently from [22] we do not need distributivity explicitly for the semantic construction because we keep the full set of solutions instead of its sum over the semiring.

Definition 2.6 [constraint systems semi-morphism] Let $\mathcal{A}$ and $\mathcal{A}'$ be constraint systems where

$$\mathcal{A} = (\mathcal{C}, \otimes, \oplus, 1, 0, \exists_X, \partial_x^t, t_1=t_2),$$

$$\mathcal{A}' = (\mathcal{C}', \otimes', \oplus', 1', 0', \exists'_X, \partial_x'^t, t_1='t_2).$$

A semi-morphism $\mu_\kappa : \mathcal{A} \longrightarrow \mathcal{A}'$ is an extension of a term systems morphism $\kappa : \mathcal{T} \longrightarrow \mathcal{T}'$ such that for each $c, c_1, c_2 \in \mathcal{C}$, $C \subseteq \mathcal{C}$, $\{x\}, X \subseteq V$ and $t, t_1, t_2 \in \mathcal{T}$ with x *ind* t, the following conditions hold.

1. $\mu(\oplus C) \leq' \oplus'\mu(C), \quad \mu(0) = 0'$

2. $\mu(c_1 \otimes c_2) \leq' \mu(c_1) \otimes' \mu(c_2), \quad \mu(1) \leq' 1'$

3. $\mu(\exists_X c) \leq' \exists'_{\kappa(X)}\mu(c)$

4. $\mu(t_1=t_2) \leq' \kappa(t_1)='\kappa(t_2)$

When we have a semi-morphism $\mu : \mathcal{A} \longrightarrow \mathcal{A}'$ we say that $\mathcal{A}'$ is *correct* w.r.t. $\mathcal{A}$.

In the discussion that follows, we will denote with the same symbol both a semi-morphism and the morphism on term system.

3 The *s*-semantics

3.1 Top-down semantics and π-interpretations

The observable we consider is the *computed answer substitutions* which induces the following program equivalence $\simeq$.

Definition 3.1 Let p_1, p_2 be positive programs. $p_1 \simeq p_2$ if for every positive goal g, $g \xmapsto{\vartheta}_{p_1} \square$ iff $g \xmapsto{\vartheta'}_{p_2} \square$ and $\vartheta_{|g} = (\vartheta'\rho)_{|g}$, where ρ is a renaming.

The above observable is captured by the following operational semantics. Recall that $\tilde{x}$ denotes a tuple of distinct variables.

Definition 3.2 [s-semantics] [18, 19] Let p be a positive program.

$$\mathcal{O}(p) = \{a \mid \exists \tilde{x} \in V, \ \exists \vartheta, \ q(\tilde{x}) \xmapsto{\vartheta}_p \square, \ a = q(\tilde{x})\vartheta\}$$

In order to model $\mathcal{O}(p)$ the usual Herbrand base has to be extended to the set of all the (possibly non-ground) atoms modulo variance.

Definition 3.3 Let $\mathcal{B}$ be the quotient set of all the atoms w.r.t. variance. A π-interpretation is any subset of $\mathcal{B}$.

In the following $\mathcal{O}(p)$ will then be formally considered as a subset of $\mathcal{B}$. Moreover, we will denote the equivalence class of an atom a by a itself. Note that π-interpretations of definition 3.3 are not Herbrand interpretations, yet are interpretations defined on the Herbrand universe. These interpretations were called *canonical realizations* in [38, 32].

Theorem 3.4 shows that $\mathcal{O}$ actually models computed answer substitutions and that it is *fully abstract*, since $p_1 \simeq p_2$ implies $\mathcal{O}(p_1) = \mathcal{O}(p_2)$.

Theorem 3.4 *[18, 19] Let p_1, p_2 be positive programs. $p_1 \simeq p_2$ iff $\mathcal{O}(p_1) = \mathcal{O}(p_2)$.*

The following theorem asserts that the observable behavior of any (possibly conjunctive) goal can be derived from $\mathcal{O}(p)$, i.e. from the observable behaviors of atomic goals of the form $q(\tilde{x})$. This property is a kind of *AND*-compositionality. Similar theorems will be shown to hold for all the semantics defined according to the s-semantics style. This is also the key property which allows us to use abstractions of the semantics for goal independent abstract interpretation.

Theorem 3.5 *[18, 19] Let p be a positive program and $g = g_1, \ldots, g_n$ be a positive goal. Then $g \xmapsto{\vartheta}_p \square$ iff there exist atoms $a_1, \ldots, a_n \in \mathcal{O}(p)$ such that $\vartheta = (\gamma\rho)$ where ρ is a renaming and $\gamma = mgu((a_1, \ldots, a_n), (g_1, \ldots, g_n))_{|g}$.*

Theorem 3.5 shows that $\mathcal{O}(p)$ provides a denotation which can actually be used to simulate the program execution for any goal $g = g_1, \ldots, g_n$. Namely the answer substitutions for g can be determined by "executing g in $\mathcal{O}(p)$", i.e. by computing a most general unifier of $g_1, \ldots, g_n$ and $a_1, \ldots, a_n$, where the a_i's are renamed apart variants of atoms in $\mathcal{O}(p)$.

Let us consider now the *success set* and the *atomic logical consequences semantics* formally defined as follows.

Definition 3.6 Let p be a positive program. Then

    1. $\mathcal{O}_1(p) \ = \{a \mid a \text{ is ground and } a \overset{\varepsilon}{\longmapsto}_p \square\}$ **(success set)**

    2. $\mathcal{O}_2(p) = \{a \mid a \overset{\varepsilon}{\longmapsto}_p \square\}$ **(atomic logical consequences semantics)**

Note that the semantic domain of $\mathcal{O}_1$ is the usual Herbrand base, i.e. the set of all the ground atoms. Note also that $\mathcal{O}_2$ is the semantics considered in [7, 28] and called *c*-semantics in [19]. We will now compare the three semantics on an example.

Example 3.7 Consider the programs p_1 and p_2 on the signature S, defined by $C = \{a\backslash 0, f\backslash 1\}$.

$$
\begin{array}{ll}
p_1 = \{ & q(a). \\
& q(x). \\
& r(f(a)).\}
\end{array}
\qquad
\begin{array}{ll}
p_2 = \{ & q(x). \\
& r(f(a)).\}
\end{array}
$$

$$
\begin{array}{llllll}
\mathcal{O}(p_1) & = & \{r(f(a)), & q(x), & q(a) & \} \\
\mathcal{O}(p_2) & = & \{r(f(a)), & q(x) & & \} \\
\mathcal{O}_1(p_1) \ = \ \mathcal{O}_1(p_2) & = & \{r(f(a)), & & q(a), & q(f(a)), \ \ldots\} \\
\mathcal{O}_2(p_1) \ = \ \mathcal{O}_2(p_2) & = & \{r(f(a)), & q(x), & q(a), & q(f(x)), \ q(f(a)), \ \ldots\}
\end{array}
$$

Note that $p_1 \simeq p_2$ does not hold, since the goal $q(x)$ computes different answer substitutions in p_1 and in p_2. Note also that the denotations defined by $\mathcal{O}$ are finite, while those computed by both $\mathcal{O}_1$ and $\mathcal{O}_2$ are infinite.

Example 3.7 shows that the three semantics are different. Indeed, if we denote by $\approx_i$ the program equivalence induced by $\mathcal{O}_i$, $i = 1, 2$, the following (strict) inclusion holds [19, 20]. $\simeq \, \subset \, \approx_2 \, \subset \, \approx_1$, i.e. $\simeq$ is finer than $\approx_2$, and $\approx_2$ is finer than $\approx_1$. This shows that the success set semantics is not correct with respect to computed answers. Moreover the correctness cannot be achieved by just using interpretations consisting of sets of non-ground atoms. In fact also the *c*-semantics does not correctly model the computed answers.

    Let I be a π-interpretation. If $[I]$ denotes the set of ground instances of the atoms in I, $[I]$ is clearly a Herbrand interpretation. The following theorem relates the *s*-semantics to the success set (and therefore to the least Herbrand model).

Theorem 3.8 *[18, 19] If p is a positive program, then $\mathcal{O}_1(p) = [\mathcal{O}(p)]$.*

We have shown that the success set semantics does not correctly model the computed answers. One could still think that this is not the case in most reasonable logic programs. Which is the class of positive programs for which the success set is correct with respect to computed answers? This is the class of programs for which the *s*-semantics and the least Herbrand model semantics do coincide. Theorem 3.10 shows that this is exactly the class of *language independent* programs as defined in [39].

Definition 3.9 [39] A program p with underlying language L_p is *language independent* iff, for any extension L'_p of L_p, its least L'_p-Herbrand model is equal to its least L_p-Herbrand model.

Theorem 3.10 *[35] Let p be a program. Then p is language independent iff* $\mathcal{O}(p) = \mathcal{O}_1(p)$.

A program p belongs to this class only if any goal in p returns ground answers. It is therefore essentially the class of *allowed* positive programs [36] and does not contain any program able to compute partial data structures.

Another related useful property of the *s*-semantics is its independence from the language. This means that the denotation defined by $\mathcal{O}$ is not affected by the choice of the language signature. The language signature affects the domain of π-interpretations $\mathcal{B}$. Since $\mathcal{O}(p)$ is a subset of $\mathcal{B}$ it might also be affected. Therefore, let us denote by $\mathcal{O}^L(p)$ the denotation for a given language L. If L_p is the language underlying program p, the following theorem shows the language independence property. Note that the same property does not hold for other variable-based semantics, such as those in [7, 21].

Theorem 3.11 *[35] If p is a positive program, then $\mathcal{O}^{L_p}(p) = \mathcal{O}^{L'_p}(p)$ for any extension L'_p of L_p.*

This is the key property which makes the *s*-semantics adequate to formalize meta-programming with the non-ground meta-level representation of object level variables.

3.2 Fixpoint semantics

We will now introduce an immediate consequences operator T_p^π on π-interpretations whose least fixpoint will be shown to be equivalent to the computed answer substitutions semantics $\mathcal{O}(p)$.

Lemma 3.12 *The set of all π-interpretations $(\Im, \subseteq)$ is a complete lattice.*

Definition 3.13 [18, 19] Let p be a positive program and I be a π-interpretation.

$$T_p^\pi(I) = \{ \; a \in \mathcal{B} \mid \exists k = a' :- b'_1, \ldots, b'_n \in p,$$
$$\exists \; b_1, \ldots, b_n \text{ variants of atoms in } I \text{ and renamed apart,}$$
$$\exists \vartheta = mgu((b_1, \ldots, b_n), (b'_1, \ldots, b'_n)) \text{ and } a = a'\vartheta \; \}$$

Note that T_p^π is different from the standard T_p operator [41] in that it derives instances of the clause heads by *unifying* the clause bodies with atoms in the current π-interpretation, rather than by taking all the possible ground instances. In other words T_p^π defines a bottom-up inference based on the same rule (unification) which is used by top-down *SLD*-resolution. The following theorem allows us to define a fixpoint semantics for positive logic programs.

Theorem 3.14 *[18, 19] The T_p^π operator is continuous on $(\Im, \subseteq)$. Then there exists the least fixpoint $T_p^\pi \uparrow \omega$ of T_p^π.*

Definition 3.15 [18, 19] The fixpoint semantics of a positive program p is defined as $\mathcal{F}(p) = T_p^\pi \uparrow \omega$.

It is worth noting that, since any program p is a finite set of clauses, all the finite fixpoint approximations $T_p^\pi \uparrow n$, $n \leq \omega$ are finite. The T_p^π operator can then effectively be used for the construction of bottom-up proofs.

The equivalence between $\mathcal{F}(p)$ and $\mathcal{O}(p)$ is proved by introducing the unfolding semantics.

Definition 3.16 [33, 34] Let p and q be positive programs. Then the unfolding of p w.r.t. q is defined as

$$unf_p(q) = \{(a :- \tilde{d}_1, \ldots, \tilde{d}_n)\vartheta \mid \quad \exists a :- b_1, \ldots, b_n \in p,$$
$$\exists b_i' :- \tilde{d}_i \in q, i = 1, \ldots, n,$$
$$\text{renamed apart, such that}$$
$$\vartheta = mgu((b_1, \ldots, b_n), (b_1', \ldots, b_n'))\}$$

The unfolding rule can be applied to any atom in a clause and preserves the operational semantics, i.e. the language is closed under unfolding. Therefore it is possible to define the immediate consequences operator in terms of the unfolding rule. Theorem 3.20 was proved in [34]. An alternative proof is given in [17] by using lemma 3.19. A direct proof of $\mathcal{F}(p) = \mathcal{O}(p)$ was first given in [19].

Definition 3.17 [34, 17] Let p be a positive program. Then we define the collection of programs

$$p_0 = p$$
$$p_i = unf_{p_{i-1}}(p), \ i = 1, 2, \ldots$$

and the collection of π-interpretations $I_i(p) = \{a \mid a \in \mathcal{B} \text{ and } a \in p_i\}$. The unfolding semantics $\mathcal{U}(p)$ of the program p is defined as

$$\mathcal{U}(p) = \bigcup_{i=0,1,\ldots} I_i(p).$$

Theorem 3.18 (unfolding vs operational semantics) *[34, 17] Let p be a positive program. Then $\mathcal{U}(p) = \mathcal{O}(p)$.*

Lemma 3.19 *[17] Let p, q be positive programs. Then T_p^π is compatible with $unf_p(q)$, i.e.*

$$T_{unf_p(q)}^\pi(\emptyset) = T_p^\pi(T_q^\pi(\emptyset)).$$

Since T_p^π is compatible with the unfolding rule and by definition of the unfolding rule, $T_p^\pi(I) = unf_p(I)$ and then $T_p^\pi \uparrow (i+1) = T_{p_i}^\pi(\emptyset) = unf_{p_i}(\emptyset)$. Therefore,

Theorem 3.20 shows that $\mathcal{F}(p)$ is the fully abstract semantics w.r.t. computed answer substitutions.

Theorem 3.20 (fixpoint vs operational semantics) *[34, 17] Let p be a positive program. Then $\mathcal{F}(p) = \mathcal{U}(p) = \mathcal{O}(p)$.*

3.3 Model-theoretic semantics

The operational and fixpoint semantics of a program p define a π-interpretation I_p, which can be viewed as a syntactic notation for a set of Herbrand interpretations. $\mathcal{H}(I_p)$ denotes the set of all the Herbrand interpretations represented by I_p. In general, our aim is finding a notion of π-model such that $\mathcal{O}(p)$ (and $\mathcal{F}(p)$) are π-models and every Herbrand model is a π-model. This can be obtained by the following definition.

Definition 3.21 Given a program p and a π-interpretation I, I is a π-model of p iff p is true in all the Herbrand interpretations in $\mathcal{H}(I)$.

In order to define π-models according to definition 3.21 we have to specify the function $\mathcal{H}$ from π-interpretations to sets of Herbrand interpretations.

Definition 3.22 [20] Let I be a π-interpretation. Then $\mathcal{H}(I) = [I]$ where $[I]$ is the set of ground instances of atoms in I or, equivalently, the least Herbrand model of I.

Proposition 3.23 *[20] Let p be a program. Then every Herbrand model of p is a π-model of p. Moreover $\mathcal{O}(p), \mathcal{O}_1(p), \mathcal{O}_2(p)$ are π-models of p.*

The program p_2 of example 3.7 shows that the model intersection property does not hold any longer. In fact, $\mathcal{O}(p_2) \cap \mathcal{O}_1(p_2) \cap \mathcal{O}_2(p_2) = \{q(f(a))\}$ which is not a π-model of p_2. This is not surprising, since set theoretic operations do not adequately model the operations on non-ground atoms, which stand for all their ground instances. A more adequate partial order relation $\preceq$ on the set $\Im$ of π-interpretations was defined in [20] with the following properties

- $(\Im, \preceq)$ is a complete lattice. $\mathcal{B}$ is the top element and $\emptyset$ is the bottom element.

- If M is a set of π-models of p, then $glb(M)$ is a π-model of p.

- The least π-model $\mathcal{M}(p) = glb(\{I \in \Im \mid I$ is a π-model of $p\})$ is the least Herbrand model.

It is worth noting that, according to proposition 3.23, the s-semantics $\mathcal{O}(p)$ is simply a non-ground representation of the least Herbrand model $\mathcal{O}_1(p)$. From the Herbrand models viewpoint the two semantics are therefore equivalent. However $\mathcal{O}(p)$ contains more useful information. On one side, it correctly models computed answers. On the other side, it has nice properties also from the model-theoretic viewpoint. This can be shown by considering the properties of the (atomic logical consequences) semantics $\mathcal{O}_2$ and the relation between $\mathcal{O}$ and $\mathcal{O}_2$.

Theorem 3.24 *[19, 28] Let p be a positive program and a be a (possibly non-ground) atom. Then $p \models a$ iff $a \in \mathcal{O}_2(p)$.*

Theorem 3.25 *[19] Let p be a positive program. Then $\mathcal{O}_2(p) = \{a \mid \exists b \in \mathcal{O}(p)$ such that $a = b\vartheta\}$*

This allows us to determine from $\mathcal{O}(p)$ the *correct answer substitutions*, as shown by the following corollary, which can easily be derived from theorems 3.24 and 3.25.

Corollary 3.26 *Let p be a program and $g = g_1, \ldots, g_n$ be a goal. Then ϑ is a correct answer substitution for g in p (i.e. $p \models \forall (a_1 \wedge \ldots \wedge a_n)\vartheta$) iff all the atoms $a_i\vartheta$ are instances of atoms in $\mathcal{O}(p)$.*

Note that, as shown in [36], correct answer substitutions cannot be determined from the least Herbrand model.

4   The basic framework

When we want to formalize program execution we must take into account, in addition to the inference rules which specify how derivations are made, the properties we observe in a computation (*observables*). In the following we give a rigorous definition of observable in terms of *SLD*-trees.

We consider a version of *CLP*, where any goal ($\mathcal{A}$-goal) contains all the information on the derivation. Namely, in addition to the constraint and to the conjunction of atoms, it contains a sequence of clauses (the sequence of clauses used to derive it).

Let $\mathcal{A}$ be a semi-distributive constraint system (based on the term system $\mathcal{T}$). An $\mathcal{A}$-*goal* is a formula $c \,\Box\, b_1, \ldots, b_n \diamond ks$ with $n \geq 0$, where c is an $\mathcal{A}$-constraint, $b_1, \ldots, b_n$ is a sequence of $(\mathcal{T}, \Pi)$-atoms, and ks is a sequence of clauses of $\mathcal{K}_{\mathcal{A}}$ (the set of $CLP(\mathcal{A})$ clauses). We denote (g, g') $\mathcal{A}$-goals conjunction, i.e., $((c \,\Box\, \tilde{b} \diamond ks), (c' \,\Box\, \tilde{b}' \diamond ks')) \stackrel{def}{=} c \otimes c' \,\Box\, (\tilde{b}, \tilde{b}') \diamond ks :: ks'$. $G_{\mathcal{A}}$ is the set of $\mathcal{A}$-goals and $P_{\mathcal{A}}$ is the set of programs, i.e., $\wp(\mathcal{K}_{\mathcal{A}})$. A *selection rule* r is a function $r : G_{\mathcal{A}} \to \mathbb{N}$ such that $r(c \,\Box\, b_1, \ldots, b_n \diamond ks) \in \{1, \ldots, n\}$. Given a standard goal $c \,\Box\, \tilde{b}$, its $\mathcal{A}$-goal version is $g \stackrel{def}{=} c \,\Box\, \tilde{b} \diamond \varepsilon$. When clear from the context, we drop the $\mathcal{A}$ from the notations.

We must redefine the concepts of resolvent and derivation to take into account the additional information contained in goals.

Definition 4.1 [resolvent] Let p be a logic program, $g = c \,\Box\, a_1, \ldots, a_n \diamond ks$ be a goal and $k = h :- c_k \,\Box\, b_1, \ldots, b_m$ be a clause of p. A *resolvent* of g and k (with selected atom a_i) is the goal

$$g' = c \otimes a_i{=}h \otimes c_k \,\Box\, a_1, \ldots, a_{i-1}, b_1, \ldots, b_m, a_{i+1}, \ldots, a_n \diamond ks :: [k]$$

if the constraint $c \otimes a_i{=}h \otimes c_k$ is satisfiable.

The function $= : Atom \times Atom \to \mathcal{A}$ is defined as $q(t_1, \ldots, t_n){=}r(s_1, \ldots, s_m) \stackrel{def}{=} t_1{=}s_1 \otimes \ldots \otimes t_n{=}s_n$ if $q = r$ and $n = m$, 0 otherwise.

Definition 4.2 [SLD-derivations] An *SLD-derivation* of g in p via r is a maximal sequence of goals $g_0, \ldots, g_n, \ldots$ such that $g = g_0$ and, for each i,

1. $g_i = c_i \,\Box\, \tilde{b}_i \diamond ks_i$;

2. $c \,\Box\, \tilde{b} \diamond ks_i :: [k]$ is a resolvent of g_i and a variant k of a clause in p not sharing any variable with g_i and ks_i. Moreover the atom in g_i is selected according to r;

3. $g_{i+1} = \exists(c)_{g,\tilde{b}} \,\Box\, \tilde{b} \diamond ks_i :: [k]$.

A derivation is *successful* if the last goal has an empty body and is called *SLD*-refutation. A finite derivation which is not successful is *failed*.

If g is a goal, $g \underset{p,r}{\leadsto} c \,\Box\, \tilde{b} \diamond ks$ denotes an *SLD*-derivation of the resolvent $c \,\Box\, \tilde{b} \diamond ks$ from g in p via r. $g \underset{p,r}{\leadsto} c \,\Box\, \diamond ks$ denotes the refutation of g and c is the computed answer constraint.

Definition 4.3 [SLD-tree] The *SLD-tree* of g in p via r is the prefix tree built from all the *SLD*-derivations of g in p via r.

An *SLD*-tree of a goal g and a program p via a selection rule r is a compact notation for representing a set of derivations of g in p via r. The set of *SLD*-trees (denoted by $\mathcal{ST}_p^{g,r}$) is ordered by tree inclusion. This partial order formalizes the evolution of the computation process.

4.1 The domain of *SLD*-trees

SLD-trees are technically hard to handle. Therefore we will introduce another type of notation for representing sets of derivations, which is suitable for our needs and is more compact. We can consider the *resultants*, first introduced by [37] in the theory of partial evaluation, and later used in [1] to discuss the correctness of *SLD*-resolution.

Definition 4.4 [resultant] Let $g_0, g_1, \ldots$ be a derivation of g_0 in p via r. A *resultant* (of level i), is the expression $g_0 \rightsquigarrow g_i$. $Res_p^{r,g}$ is the set of all the resultants of g in p via r.

We can define on the set of resultants the following partial order: $g \rightsquigarrow c \,\square\, \tilde{b} \diamond ks \;\leq\; g \rightsquigarrow c' \,\square\, \tilde{b}' \diamond ks'$ iff ks is a prefix of ks'.

Definition 4.5 [well-formed resultant sets] A set A of resultants is *well-formed* if

$$\rho \in A \text{ implies } \forall \rho' \leq \rho, \; \rho' \in A.$$

It is easy to see that a well-formed set of resultants contains all the information of an *SLD*-tree, because we can identify each resultant with a node of the tree and vice versa. Thus the set of all the well-formed resultants of g and p via r, $R_p^{r,g} \overset{def}{=} \{\, A \in Res_p^{r,g} \mid A \text{ is well-formed} \,\}$ has the following property.

Proposition 4.6 (domain of *SLD*-trees) $R_p^{r,g}$ *is isomorphic to* $\mathcal{ST}_p^{g,r}$.

We can define the domain of all the well-formed sets of resultants $R \overset{def}{=} \cup_{g,r,p} R_p^{r,g}$. We have now an alternative representation of *SLD*-trees. Since we are interested in all the *SLD*-trees of a program p, we define $W^r(p) : P \longrightarrow R$ as $W^r(p) \overset{def}{=} \cup_{g \in G} R_p^{r,g}$.

The set R inherits the set inclusion partial ordering from $\mathcal{ST}_p^{g,r}$. It is easy to prove that

Proposition 4.7 (resultants lattice) $(R, \subseteq)$ *is a complete lattice.*

4.2 The observables

The aim of this paper is to define a precise notion of observable. Roughly speaking we want to express an observable as an abstraction function defined from R into a suitable domain of observable properties.

An observable property domain is a set of properties of the derivation with an ordering relation which can be viewed as an approximation structure. An observation consists in looking at an *SLD*-tree, and then extracting some property (abstraction). Therefore the observable is a function from $\mathcal{ST}_p^{g,r}$ to a suitable property domain D, which preserves the approximation structure. Such a function must be a Galois connection between $\mathcal{ST}_p^{g,r}$ and D. Because of proposition 4.6 we can give the following definition.

Definition 4.8 [observable] Let R be the domain of *SLD*-trees and D be an observable domain. $\alpha : R \longrightarrow D$ is an *observable* when there exists γ s.t. $(\alpha, \gamma) : R \longrightarrow D$ is a Galois co-insertion.

We denote by the same symbol an observable and the Galois connection it can be extended to.

A choice of the observable α induces an *observational equivalence* $=_\alpha$ on programs. Namely $p_1 =_\alpha p_2$ iff p_1 and p_2 are observationally indistinguishable according to α, i.e., iff the observable properties of p_1 are exactly those of p_2. Namely

$$p_1 =_\alpha p_2 \iff \alpha\left(W^r(p_1)\right) = \alpha\left(W^r(p_2)\right).$$

Example 4.9 If ξ denotes *computed answer constraints* we can take $CA \stackrel{def}{=} (\wp(G \times \mathcal{A}), \subseteq)$ as properties domain and extend to a connection the function $\xi : R \longrightarrow CA$

$$\xi(A) = \{ (g, c) \mid g \rightsquigarrow c \,\square\, \diamond\, ks \in A \}.$$

It is easy to see that $p_1 =_\xi p_2$ iff for any goal g, g has the same answer constraints in p_1 and in p_2.

Let us now define what it means for "an observable α to be stronger than another observable α'". The intuition is that $p_1 =_\alpha p_2$ should imply $p_1 =_{\alpha'} p_2$ (i.e., $=_\alpha$ is finer than $=_{\alpha'}$). It is easy to see that this means that all the objects of the weak observable can be expressed by the stronger, by preserving all the properties of the former. This in turn means that there exists a Galois connection between the domains of the observables.

Lemma 4.10 *Let* $(f_\alpha, f_\gamma) : A \longrightarrow B, (g_\alpha, g_\gamma) : A \longrightarrow C$ *be co-insertions. Then the following statements are equivalent.*

$$\forall x, y \in A \; f_\alpha(x) = f_\alpha(y) \Rightarrow g_\alpha(x) = g_\alpha(y)$$

$$\exists h : B \longrightarrow C \; \text{co-insertion s.t.} \; g = h \circ f.$$

Lemma 4.10 allows us to understand the following

Definition 4.11 [observables approximation] An observable $\alpha : R \longrightarrow D$ *approximates* $\alpha' : R \longrightarrow D'$ if there exists a co-insertion $\beta : D \longrightarrow D'$ s.t.

$$\alpha' = \beta \circ \alpha.$$

We can define an ordering on observables, using closure of Galois co-insertion composition, simply by defining $\alpha \leq \alpha'$ iff α' approximates α.

Lemma 4.12 (the observables lattice) *Let* $(\mathbb{O}, \leq)$ *be the set of observables ordered by approximation.* $(\mathbb{O}, \leq)$ *is a complete lattice.*

4.3   Semantic properties of *SLD*-trees

The goal we want to achieve is to develop a denotation modeling *SLD*-trees. We follow the approach in [25, 3], by first defining a "syntactic" semantic domain (π-interpretation). Our modeling of *SLD*-trees is essentially the basic denotation defined in terms of clauses in [22, 26], extended to handle constraint systems in the style of [29, 15, 16]. In the following for the sake of simplicity we consider the PROLOG leftmost selection rule (denoted by lm). All our results can be generalized to local selection rules [42].

Let us consider the equivalence relation of variance extended to resultants $\equiv$.

Definition 4.13 [π-interpretation] A π-interpretation $\mathcal{I}$ is an element of $R_{/\equiv}$. We denote by $\mathcal{R}$ the set of π-interpretations (that is $\mathcal{R} \stackrel{def}{=} R_{/\equiv}$).

Lemma 4.14 ($\mathcal{R}$ lattice) $(\mathcal{R}, \subseteq)$ *is a complete lattice.*

A denotation of the program characterizing its *SLD*-trees computed by using the rule lm might be the set of the *SLD*-trees for all the possible goals modulo variance, i.e., $W^{lm}(p)_{/\equiv}$. Because of the *AND*-compositionality theorem 4.18 below, this set can be obtained from the top-down *SLD*-trees denotation, which is the set of *SLD*-trees for the most general atomic goals.

Definition 4.15 [top-down *SLD*-trees denotation $\mathcal{O}$] Let p be a program. The *top-down SLD-trees denotation* of p according to lm is the π-interpretation

$$\mathcal{O}(p) \stackrel{def}{=} \left\{ q(\tilde{x}) \rightsquigarrow c \,\square\, \tilde{b} \diamond ks \in Res_p^{lm,q(\tilde{x})} \mid q \in \Pi_n, \; \tilde{x} \in V^n \right\}_{/\equiv}.$$

It is easy to see that $\mathcal{O}(p)$ is well-formed.

Example 4.16 For the program

$$p = \{\, q(x) :- \ x{=}ss(y) \,\square\, q(y),\ q(x) :- \ x{=}0\,\square\ \}$$

the denotation $\mathcal{O}(p)$ is the set

$$\{\ \ 1 \,\square\, q(x) \diamond \varepsilon \rightsquigarrow x{=}0 \,\square\ \diamond \big[q(x_1){:}{-}\,x_1{=}0\,\square\big],$$
$$1 \,\square\, q(x) \diamond \varepsilon \rightsquigarrow x{=}ss(0)\,\square\ \diamond \big[q(x_1){:}{-}\,x_1{=}ss(y_1)\,\square\, q(y_1),\ q(x_2){:}{-}\,x_2{=}0\,\square\big],$$
$$1 \,\square\, q(x) \diamond \varepsilon \rightsquigarrow x{=}ss(y_1)\,\square\, q(y_1) \diamond \big[q(x_1){:}{-}\,x_1{=}ss(y_1)\,\square\, q(y_1)\big],$$
$$\ldots\ \}$$

Now we prove that this denotation fully characterizes all the *SLD*-trees of p. This is obtained by first proving a lifting lemma which relates the *SLD*-trees of an atomic goal to the *SLD*-trees of the corresponding most general atomic goal. The second step, i.e., the *AND*-compositionality theorem, relates the *SLD*-trees of a conjunctive goal to the *SLD*-trees of the atomic goals.

Lemma 4.17 (lifting) *Let p be a program, $g \overset{def}{=} c_g \,\square\, q_1(\tilde{t}_1), \ldots, q_n(\tilde{t}_n) \diamond ks_g$ be a goal and $g_{\tilde{x}} \overset{def}{=} 1 \,\square\, q_1(\tilde{x}_1), \ldots, q_n(\tilde{x}_n) \diamond ks_g$ be its lifted version with $\tilde{x}_1, \ldots, \tilde{x}_n$ containing fresh distinct variables. Then*

$$g \underset{p,lm}{\rightsquigarrow} c \,\square\, \tilde{b} \diamond ks \iff g_{\tilde{x}} \underset{p,lm}{\rightsquigarrow} c_{\tilde{x}} \,\square\, \tilde{b}_{\tilde{x}} \diamond ks,$$
$$c = \exists (c_g \otimes \tilde{x}_1{=}\tilde{t}_1 \otimes \ldots \otimes \tilde{x}_n{=}\tilde{t}_n \otimes c_{\tilde{x}})_{g,\tilde{b}},$$
$$\tilde{b} \equiv \tilde{b}_{\tilde{x}} \vartheta_{\tilde{x}}^{\tilde{t}},$$

where $\vartheta_{\tilde{x}}^{\tilde{t}} \overset{def}{=} \big\{ \tilde{x}_1/\tilde{t}_1, \ldots, \tilde{x}_n, /\tilde{t}_n \big\}$.

Theorem 4.18 (*AND*-compositionality) *Let $g \overset{def}{=} c_g \,\square\, q_1(\tilde{t}_1), \ldots, q_n(\tilde{t}_n) \diamond ks_g$ be a goal and p be a program. Then*

$$g \underset{p,lm}{\rightsquigarrow} c \,\square\, \tilde{b} \diamond ks$$

if and only if there exist

1. $r_j = q_j(\tilde{x}_j) \rightsquigarrow c_j \,\square\ \diamond ks_j \in \mathcal{O}(p),\ 1 \le j < m,$

2. $r_m = q_m(\tilde{x}_m) \rightsquigarrow c_m \,\square\, \tilde{b}_m \diamond ks_m \in \mathcal{O}(p)$

such that

1. $c = \exists \Big(c_g \otimes \tilde{x}_1{=}\tilde{t}_1 \otimes \ldots \otimes \tilde{x}_m{=}\tilde{t}_m \otimes c_1 \otimes \ldots \otimes c_m \Big)_{g,\tilde{b}},$

2. $\tilde{b} = \tilde{b}_m, q_{m+1}(\tilde{t}_{m+1}), \ldots, q_n(\tilde{t}_n)$ and

3. $ks = ks_g :: ks_1 :: \ldots :: ks_m$.

The above closure property allows us to show that the semantics $\mathcal{O}(p)$ is correct and fully abstract for the identical observable.

Corollary 4.19 (correctness and full abstraction) *Let p_1, p_2 be two programs. Then*

$$p_1 =_{id} p_2 \quad \Longleftrightarrow \quad \mathcal{O}(p_1) = \mathcal{O}(p_2).$$

The restriction to local rules plays a fundamental role in the definition of the bottom-up denotation. By using local rules we are able of reconstruct "from the bottom" a derivation, because the local rule chooses only among the atoms introduced in the last derivation step and then "forgets" about the previous steps, which, in a bottom-up construction, are not available yet. The following definition specifies an immediate consequences operator for the *lm* case.

Definition 4.20 [immediate consequences operator T_p] Let $\mathcal{I}$ be a π-interpretation, p be a program. The *immediate consequences operator* of p via *lm* $T_p : \mathcal{R} \longrightarrow \mathcal{R}$ is:

$$
\begin{aligned}
T_p(\mathcal{I}) = \{ \; & q(\tilde{x}) \rightsquigarrow c \,\square\, \tilde{b} \diamond ks \mid \\
& k = q(\tilde{t}) :- c_k \,\square\, q_1(\tilde{t}_1), \ldots, q_m(\tilde{t}_m), a_{m+1}, \ldots, a_n \in p, \\
& q_j(\tilde{x}_j) \rightsquigarrow c_j \,\square\, \diamond ks_j \in \mathcal{I}, \; 1 \le j < m, \\
& q_m(\tilde{x}_m) \rightsquigarrow c_m \,\square\, \tilde{b}_m \diamond ks_m \in \mathcal{I}, \\
& c = \exists \left(\tilde{x}=\tilde{t} \otimes c_k \otimes \tilde{x}_1=\tilde{t}_1 \otimes c_1 \otimes \ldots \otimes \tilde{x}_m=\tilde{t}_m \otimes c_m \right)_{\tilde{x},\tilde{b}} \\
& \tilde{b} = \tilde{b}_m, a_{m+1}, \; \ldots, a_n, ks = [k] :: ks_1 :: \ldots :: ks_m \}.
\end{aligned}
$$

It is important to note that each resultant of level 0 belongs to each $T_p(\mathcal{I})$.

Lemma 4.21 $T_p : \mathcal{R} \longrightarrow \mathcal{R}$ *is continuous on* $\mathcal{R}$.

Definition 4.22 [fixpoint denotation $\mathcal{F}$] Let p be a program. The *fixpoint denotation* of p according to *lm* is the π-interpretation $\mathcal{F}(p) \stackrel{def}{=} T_p \uparrow \omega$.

The following theorem states the equivalence between the top-down and the bottom-up constructions, and shows that $\mathcal{F}(p)$ is also correct and fully abstract w.r.t. the identity observable.

Theorem 4.23 *Let p be a program. Then $\mathcal{F}(p) = \mathcal{O}(p)$.*

4.4   An algebraic formalization of *SLD*-trees semantic properties

The properties we found for $\mathcal{O}$ and $\mathcal{F}$ allow us to claim that we have a good denotation modeling *SLD*-trees. Our goal however is to find the same results for the denotations modeling more abstract observables. We want then to develop a theory according to which the semantic properties of *SLD*-trees shown in subsection 4.3 are inherited by the denotations which model abstractions of the *SLD*-trees.

In order to define the denotation as a function of the observable, we need a mathematical formalization where one can model the abstraction process and specify properties which have to be shared by the constructions associated to the various abstractions.

The first interesting property is the lifting one, which can be modeled only if we can instantiate variables in the derivation by means of constraints. Then we can define an operation $\cdot$ which adds a constraint to a denotation:

$$c \cdot A \stackrel{def}{=} \{ \ c \otimes c' \,\Box\, \tilde{b}' \diamond ks' \rightsquigarrow c \otimes c'' \,\Box\, \tilde{b}'' \diamond ks'' \mid$$
$$c' \,\Box\, \tilde{b}' \diamond ks' \rightsquigarrow c'' \,\Box\, \tilde{b}'' \diamond ks'' \in A \}$$

This operation is related to $\otimes$ by the following property: $\forall d \in R$ and $c_1, c_2 \in \mathcal{A}$, $(c_1 \otimes c_2) \cdot d = c_1 \cdot (c_2 \cdot d)$.

The next relevant property is *AND*-compositionality. We assume that there exists an operation $\times$, defined over the set of denotations, which computes the *AND*-composition of two denotations. In the case of *SLD*-trees denotations, the properties of $\times$ are shown by the following equation.

$$A \times B \stackrel{def}{=} \{ (g_1, g_2) \rightsquigarrow (g_1', g_2') \mid g_1 \rightsquigarrow g_1' \in A, g_2 \rightsquigarrow g_2' \in B \}.$$

$\times$ is (clearly) related to the operation $\otimes$ defined over the constraint system $\mathcal{A}$, to the conjunction of atom sequences and to the *AND*-compositionality property of theorem 4.18.

Another essential feature that we want to preserve is the *SLD*-trees branching structure. The operation which puts together two denotations nondeterministically is the union of well-formed sets of resultants. By using an algebraic notation, for each pair A, B of well-formed sets of resultants we write $A + B \stackrel{def}{=} A \cup B$. It is straightforward to realize that $+$ is well defined over R. The operation $+$ is related to the operation $\oplus$ defined over the constraint system $\mathcal{A}$ by the properties: $(c_1 \oplus c_2) \cdot v = c_1 \cdot v + c_2 \cdot v$ and $c \cdot (d_1 + d_2) = c \cdot d_1 + c \cdot d_2$. In analogy to what happens in $\mathcal{A}$ for $\oplus$ and $\otimes$, the product $\times$ is (left and right) distributive w.r.t. $+$, i.e., $d_1 \times (d_2 + d_3) = d_1 \times d_2 + d_1 \times d_3$. This property shows that the answers of conjunctive goals are all the compositions of the answers of the conjuncts.

The last issue we must be concerned with is that all the properties must hold "modulo variance". This property is usually modeled by renamings. Therefore we define a "renaming operation" ∇ on the objects of the domain. ∇ commutes with $+$ and $\times$ and is defined as a family of renaming functions ∇_ϑ depending on the renaming ϑ. In order to satisfy the usual properties of renamings, $\nabla_\vartheta \circ \nabla_{\vartheta'} = \nabla_{\vartheta \circ \vartheta'}$ must hold. This means that ∇ is a group homomorphism w.r.t. the functional composition operations. Furthermore ∇_ϑ must be an extension of the renaming operation ∂_ϑ of the constraint system $\mathcal{A}$, on which the domain is defined. For each well-formed set A we define

$$\nabla_\vartheta(A) \overset{def}{=} \{ \quad \partial_\vartheta(c) \,\Box\, \tilde{b}\vartheta \diamond ks\vartheta \rightsquigarrow \partial_\vartheta(c') \,\Box\, \tilde{b}'\vartheta \diamond ks'\vartheta \mid$$
$$c \,\Box\, \tilde{b} \diamond ks \rightsquigarrow c' \,\Box\, \tilde{b}' \diamond ks' \in A \}.$$

We have that for each $\vartheta_1, \vartheta_2 \in \Theta$ exists ϑ s.t. $\nabla_{\vartheta_1}(A) \times \nabla_{\vartheta_2}(B) = \nabla_\vartheta(A \times B)$.

We will appreciate the power of the above algebraic construction in the following section where the operations $+$, $\times$ and ∇ will play a relevant role in the definition of abstract denotations. Right now the formalization allows us to give an alternative nice formulation of theorem 4.18. First of all note that $g \underset{p,lm}{\rightsquigarrow} c \,\Box\, \tilde{b} \diamond ks$ iff $g \rightsquigarrow c \,\Box\, \tilde{b} \diamond ks \in \mathcal{W}(p)$, where $\mathcal{W}(p) \overset{def}{=} W^{lm}(p)_{/\equiv}$. Theorem 4.18 is equivalent to

Corollary 4.24 (*AND*-compositionality) *Let p be a program. Then*

$$v \in \mathcal{W}(p) \quad \Longleftrightarrow \quad \exists c \in \mathcal{A},\ e_1, \ldots, e_n \in \mathcal{O}(p)\ s.t.$$

$$v = c \cdot (\{ e_1 \} \times \ldots \times \{ e_n \}).$$

Moreover, by defining an *AND* closure operator on $\mathcal{D}$ as

$$\chi_{\mathcal{D}}(x) \overset{def}{=} \min\{ z \in \mathcal{D} \mid \quad x \leq z \text{ and } y, y' \leq z \Longrightarrow$$
$$\forall c \in \mathcal{A} : c \cdot (y \times y') \leq z \},$$

we can restate theorem 4.18 as

Corollary 4.25 ($\mathcal{O}(p)$ closure) *Let $\chi_{\mathcal{R}}$ be the AND-closure operator on $\mathcal{R}$. Then*

$$\mathcal{W} = \chi_{\mathcal{R}} \circ \mathcal{O}.$$

5 The abstraction framework

We consider two classes of observables, namely s-observables and i-observables. The s-observables (*S*emantic observables) are observables for which all the structural properties of *SLD*-trees are preserved. The corresponding denotations provide a correct and complete characterization of the (abstract) program behavior. We show how we can reconstruct within the framework some existing semantics, such as the answer constraint semantics [24] and the call patterns semantics [26, 27], thus obtaining all the relevant theorems simply by specializing the theorems which are valid in the framework.

The *i*-observables (abstract *interpretation* observables) are meant to capture the abstractions involved in abstract interpretation, where approximation is the rule of the game. Theorems valid for *i*-observables are therefore weaker and denotations provide characterizations of semantic properties which are correct in the sense of abstract interpretation theory. We show how we can reconstruct the abstract semantics defined in [23], which allows us to derive groundness relations among the arguments of procedure calls.

5.1 An algebraic formalization of observables: the *s*-observables

We consider observable domains which are meant to have properties similar to those of R. We want first to be able to derive the observable behavior of a conjunctive goal from the observable behaviors of its conjuncts. Moreover we do not want to loose the non-deterministic structure and — furthermore — the independence of the results upon renaming. We enforce all these properties by defining an *s*-domain.

Definition 5.1 [constrained domain] A nonempty set D is a *constrained domain* over a constraint system $\mathcal{A}$, and is denoted by $(D, \cdot)$ if there exist an operation $\cdot : \mathcal{A} \times D \longrightarrow D$ and an element $0 \in D$ s.t. for all $c_1, c_2 \in \mathcal{A}, d \in D$

1. $c_1 \cdot (c_2 \cdot d) = (c_1 \otimes c_2) \cdot d$,

2. $0 \cdot d = 0$,

3. $1 \cdot d = d$.

Let $(\Theta, \circ, id)$ be the group of renamings (on the variables in V).

Definition 5.2 [renaming operator] A *renaming operator* ∇ on a constrained domain $(D, \cdot)$ is an injective groups homomorphism $\nabla : (\Theta, \circ, id) \longrightarrow (D \to D, \circ, id)$ s.t. for each $c \in \mathcal{A}, d \in D$ and $\vartheta \in \Theta$

$$\nabla_\vartheta (c \cdot d) = \partial_\vartheta(c) \cdot \nabla_\vartheta(d).$$

A renaming operator induces a "variance" relation $=_\nabla$. Namely $x =_\nabla y \Longleftrightarrow \exists \vartheta : \nabla_\vartheta x = y$. Note that the renaming operator on R of subsection 4.4 induces exactly the variance relation $\equiv$, i.e., $x =_\nabla y \Leftrightarrow x \equiv y$.

Definition 5.3 [*s-domain*] A nonempty set D is an *s-domain* on a constraint system $\mathcal{A}$, and is denoted by $D(\cdot, +, \times, \nabla)$, if there exist four operations $\cdot$, $+$, $\times$ and ∇ on D, and two elements 0, 1 in D s.t.

1. $(D, \times, +, 1, 0)$ is a closed semiring,

2. $(D, \cdot)$ is a constrained domain,

3. ∇ is a renaming operator on $(D, \cdot)$.

Moreover for each $c, c_1, c_2 \in \mathcal{A}$ and $d, d_1, d_2 \in D$

1. $c \cdot (d_1 + d_2) = c \cdot d_1 + c \cdot d_2$,

2. $(c_1 \oplus c_2) \cdot d = c_1 \cdot d + c_2 \cdot d$.

Furthermore for each $\vartheta_1, \vartheta_2 \in \Theta$, there exists $\vartheta \in \Theta$ s.t.

1. $\nabla_{\vartheta_1}(d_1) \times \nabla_{\vartheta_2}(d_2) = \nabla_\vartheta(d_1 \times d_2)$,

2. $\nabla_{\vartheta_1}(d_1) + \nabla_{\vartheta_2}(d_2) = \nabla_\vartheta(d_1 + d_2)$.

We can define, for each *s-domain* D, a canonical ordering as follows: $d_1 \leq d_2$ iff $d_1 + d_2 = d_2$. It is easy to see, by using proposition 4.7, that $(D, \leq)$ is a complete lattice.

Example 5.4 The set R is an *s-domain*. The operations are defined in subsection 4.4. Therefore we are able to model the interesting properties of *SLD*-trees by means of *s-domains*.

Example 5.5 The set CA of example 4.9 is an *s-domain*. In fact the lifting and the AND-composition are

$$c \cdot A = \{ ((c \otimes c' \Box \tilde{b}), c \otimes s) \mid ((c' \Box \tilde{b}), s) \in A, c \otimes s > 0 \}$$

$$A \times B = \{ \quad ((c \otimes c' \Box \tilde{b}, \tilde{b}'), s \otimes s') \mid ((c \Box \tilde{b}), s) \in A,$$
$$((c' \Box \tilde{b}'), s') \in B, s \otimes s' > 0 \}.$$

The sum operation is set union, while the renaming operation is

$$\nabla_\vartheta(A) = \{ ((\partial_\vartheta(c) \Box \tilde{b}\vartheta), \partial_\vartheta(s)) \mid ((c \Box \tilde{b}), s) \in A \}.$$

that is a slight extension of *CLP* renamings.

Example 5.6 The call patterns of a program p for a goal g and a selection rule r are the atoms selected in any *SLD*-derivation of g in p via r. We define the s-domain $CP_A \stackrel{def}{=} \wp(G \times Atom)$. The interpretation of each $(g, c \,\Box\, a) \in CP_A$ is: "the execution of the goal g generates a procedure call a with state (constraint) c". The operations can be defined as follows.

$$c \cdot A = \{\, ((c \otimes c' \,\Box\, \tilde{b}), (c \otimes c'' \,\Box\, a)) \mid ((c' \,\Box\, \tilde{b}), c'' \,\Box\, a) \in A, c \otimes c'' > 0 \,\},$$

$$A + B = A \cup B,$$

$$A \times B = \{\ \ ((g, g'), c \,\Box\, a) \mid (g, c \,\Box\, a) \in A \,\} \cup \{\, ((g, g'), c \otimes c' \,\Box\, a') \mid$$
$$(g, c \,\Box\,) \in A, (g', c' \,\Box\, a') \in B \,\}.$$

$$\nabla_\vartheta(A) = \{\, (\partial_\vartheta(c) \,\Box\, \tilde{b}\vartheta, \partial_\vartheta(c') \,\Box\, a\vartheta) \mid (c \,\Box\, \tilde{b}, c' \,\Box\, a) \in A \,\}.$$

We want now to define a notion of (forgetful) morphism between s-domains.

Definition 5.7 [s-morphism] A *s-morphism* (a morphism between s-domains), $\alpha : D(\cdot, +, \times, \nabla) \longrightarrow \bar{D}(\bar{\cdot}, \bar{+}, \bar{\times}, \bar{\nabla})$ is a surjective application $\alpha : |D| \longrightarrow |\bar{D}|$ s.t. $\forall x, y \in D, c \in \mathcal{A}, \vartheta \in \Theta$

1. $\alpha(x + y) = \alpha(x) \,\bar{+}\, \alpha(y), \qquad \alpha(0) = \bar{0}, \qquad\qquad$ **(nondeterminism)**

2. $\alpha(x \times y) = \alpha(x) \,\bar{\times}\, \alpha(y), \qquad \alpha(1) = \bar{1}, \qquad$ **(AND-compositionality)**

3. $\alpha(c \cdot x) = c^{\bar{\cdot}}\alpha(x), \qquad\qquad\qquad\qquad\qquad\qquad\quad$ **(lifting)**

4. $\alpha(\nabla_\vartheta(x)) = \bar{\nabla}_\vartheta(\alpha(x)). \qquad\qquad\qquad\qquad\qquad$ **(renaming)**

Example 5.8 The observable ξ of example 4.9 is an s-morphism.

Example 5.9 The abstraction which allows us to obtain the call patterns is

$$\eta(A) = \{\, (g, c \,\Box\, a) \mid g \rightsquigarrow c \,\Box\, a, \tilde{b} \diamond ks \in A \,\}$$

It is easy to see that η is an s-morphism.

Additivity and surjectivity allow the s-morphism to associate the right observation in $\bar{D}$ to any concrete object in D. This is because s-morphisms are Galois co-insertions with respect to the canonical orderings.

Proposition 5.10 (s-morphisms are co-insertions) *Let $D, \bar{D}$ be s-domains. If there exists an s-morphism $\alpha : D(\cdot, +, \times, \nabla) \longrightarrow \bar{D}(\bar{\cdot}, \bar{+}, \bar{\times}, \bar{\nabla})$, then there exists a mapping $\gamma : |\bar{D}| \longrightarrow |D|$ such that (α, γ) is a Galois co-insertion of $(D, \leq)$ onto $(\bar{D}, \bar{\leq})$.*

Definition 5.11 [*s*-observable] Let $\bar{D}(\bar{\cdot}, \bar{+}, \bar{\times}, \bar{\nabla})$ be an *s*-domain. An *s*-morphism $\alpha : R(\cdot, +, \times, \nabla) \longrightarrow \bar{D}(\bar{\cdot}, \bar{+}, \bar{\times}, \bar{\nabla})$ is an observable and we call it *s*-observable. $\mathbb{O}_s$ denotes the set of *s*-observables.

Lemma 5.12 *If the s-morphism* $\alpha : R \longrightarrow D$ *approximates* $\alpha' : R \longrightarrow D'$ *then there exists an s-morphism* $\beta : D \longrightarrow D'$ *s.t.* $\alpha = \beta \circ \alpha'$.

As was the case for observables, we can order $\mathbb{O}_s$ by approximation.

Proposition 5.13 (*s*-observables lattice) $(\mathbb{O}_s, \leq)$ *is a complete sublattice of* $(\mathbb{O}, \leq)$.

5.2 Semantic properties of *s*-observables

s-domains are strongly related to semantic domains. We can map an *s*-domain $D(\cdot, +, \times, \nabla)$ in a semantic domain $\mathcal{D}(\bar{\cdot}, \bar{+}, \bar{\times}) \overset{def}{=} D(\cdot, +, \times, \nabla)_{/=\nabla}$ by using the canonical equivalence induced by ∇: $[\cdot]_\nabla : D \longrightarrow \mathcal{D}$, $[x]_\nabla \overset{def}{=} [x]_{/=\nabla}$. For each $[x]_\nabla, [y]_\nabla \in \mathcal{D}$, we define $c^{\bar{\cdot}} [x]_\nabla \overset{def}{=} [c \cdot x]_\nabla$, $[x]_\nabla \bar{+} [y]_\nabla \overset{def}{=} [x + y]_\nabla$ and $[x]_\nabla \bar{\times} [y]_\nabla \overset{def}{=} [x \times y]_\nabla$. The set $\mathcal{D}$ inherits from D all the properties we have discussed in section 4.4 for R, i.e., (the modulo variance versions of) lifting, *AND*-compositionality and nondeterminism. From now on, we will call the structure $\mathcal{D}(\bar{\cdot}, \bar{+}, \bar{\times})$ *semantic domain*, and denote by $\mathbb{D}$ the set of all the semantic domains. Moreover we denote the operations of D and $\mathcal{D}$ by the same symbol.

Each *s*-morphism $\alpha : R(\cdot, +, \times, \nabla) \longrightarrow D(\bar{\cdot}, \bar{+}, \bar{\times}, \bar{\nabla})$ can be transformed into a "semantic domains homomorphism" (that we still denote by α) from $\mathcal{R}(\cdot, +, \times)$ to the semantic domain $\mathcal{D}(\bar{\cdot}, \bar{+}, \bar{\times})$. We only need to set $\alpha ([x]_\nabla) \overset{def}{=} [\alpha(x)]_\nabla$. This homomorphism is well defined thanks to axiom 4 in definition 5.7. Thus we have a syntactic abstraction which *preserves the interesting semantic properties of* $\mathcal{R}$.

Now we show how the algebraic construction can be used to easily derive interesting properties.

Lemma 5.14 (closure operators and *s*-morphisms) *Let* $\alpha : \mathcal{D} \longrightarrow \mathcal{D}'$ *be an s-morphism and* $\chi_\mathcal{D}, \chi'_\mathcal{D}$ *be the closure operators of* $\mathcal{D}, \mathcal{D}'$ *respectively. Then*

$$\alpha \circ \chi_\mathcal{D} = \chi'_\mathcal{D} \circ \alpha.$$

We can then define the α-top-down denotation $\mathcal{O}_\alpha(p)$ of a program p.

Definition 5.15 [α-top-down denotation] Let $\alpha : \mathcal{R} \longrightarrow \mathcal{D}$ be an *s*-observable and p be a program. The α-top-down denotation of p is

$$\mathcal{O}_\alpha(p) \overset{def}{=} \alpha \left(\mathcal{O}(p) \right)$$

From lemma 5.14 we immediately derive the (abstract) *AND*-compositionality property of $\mathcal{O}_\alpha(p)$.

Corollary 5.16 ($\mathcal{O}_\alpha(p)$ closure) *Let $\alpha : \mathcal{R} \longrightarrow \mathcal{D}$ and $\chi_{\mathcal{D}}$ be the AND-closure operator on $\mathcal{D}$. Then*

$$\mathcal{W}_\alpha = \chi_{\mathcal{D}} \circ \mathcal{O}_\alpha.$$

Theorem 5.17 (abstract *AND*-compositionality) *Let p be a program, $\alpha : \mathcal{R}(\cdot, +, \times) \longrightarrow \mathcal{D}(\bar{\cdot}, \bar{+}, \bar{\times})$ be an s-observable. Then*

$$\bar{v} \in \mathcal{W}_\alpha(p) \quad \Longleftrightarrow \quad \exists c \in \mathcal{A}, \ \bar{e}_1, \dots, \bar{e}_n \in \mathcal{O}_\alpha(p) \ s.t.$$

$$\bar{v} = c^{\bar{\cdot}}(\{\bar{e}_1\} \bar{\times} \dots \bar{\times} \{\bar{e}_n\}).$$

Corollary 5.18 (correctness and full abstraction) *Let p_1, p_2 be programs and $\alpha : \mathcal{R} \longrightarrow \mathcal{D}$ be an s-observable. Then*

$$p_1 =_\alpha p_2 \quad \Longleftrightarrow \quad \mathcal{O}_\alpha(p_1) = \mathcal{O}_\alpha(p_2).$$

In the bottom-up case the best approximation for the immediate consequences operator is $\alpha \circ T_p \circ \gamma$.

Definition 5.19 [α-immediate consequences operator] Let $\alpha : \mathcal{R} \longrightarrow \mathcal{D}$ be an s-observable and p be a program. The α-immediate consequences operator of p $T_{p,\alpha} : \mathcal{D} \longrightarrow \mathcal{D}$ is

$$T_{p,\alpha} \stackrel{def}{=} \alpha \circ T_p \circ \gamma$$

If $\leq$ is the canonical ordering on $\mathcal{D}$, $T_{p,\alpha}$ is continuous on the lattice $(D, \leq)$. Indeed $(\alpha \circ T_p \circ \gamma)(\sum x_i) = (\alpha \circ T_p)(\sum \gamma(x_i)) = \alpha(\sum (T_p \circ \gamma)(x_i)) = \sum (\alpha \circ T_p \circ \gamma)(x_i)$. We can then define the α-bottom-up denotation as $\mathcal{F}_\alpha(p) \stackrel{def}{=} T_{p,\alpha} \uparrow \omega$.

Definition 5.20 [α and T_p compatibility] α is *compatible with T_p* if $T_{p,\alpha} \circ \alpha = \alpha \circ T_p$.

Theorem 5.21 (bottom-up vs top-down) *Let p be a program and α be an s-observable. Then $\mathcal{O}_\alpha(p) \leq \mathcal{F}_\alpha(p)$. Moreover if α is compatible with T_p then $\mathcal{O}_\alpha(p) = \mathcal{F}_\alpha(p)$.*

Some remarks about the above results are necessary. The top-down abstract denotation, which is defined simply as the abstraction of the top-down denotation, has exactly the same properties of the top-down *SLD*-trees denotation, namely *AND*-compositionality, correctness and full abstraction. The bottom-up abstract denotation is in general less precise. The loss of precision is due to the fact that the abstract immediate consequences operator $T_{p,\alpha}$ is obtained by specializing the general immediate consequences operator T_p ($T_{p,\alpha} \stackrel{def}{=} \alpha \circ T_p \circ \gamma$). It is exactly this specialization which may sometimes result in a loss of precision. However, if the observable α is compatible with T_p, the two constructions are equivalent. It is worth noting that most reasonable observables are indeed compatible with T_p (see the examples below). A similar relation between the top-down and bottom-up constructions was already noted for abstractions of the answer constraint in [29, 15, 16]. In that framework the equivalence was guaranteed in the case of distributive constraint systems. In such a case, our compatibility condition is always satisfied.

Let us finally note that, when the two constructions are equivalent, the bottom-up one is indeed more efficient, since abstraction is used at every step of the fixpoint construction, thus deriving only the minimal amount of information about the *SLD*-trees which is needed to characterize the observable property. The top-down construction, on the contrary, is always forced to build complete *SLD*-trees which have later to be abstracted to get the observation.

Example 5.22 We show now how to reconstruct the *CLP* version of the *s*-semantics [18, 19]. We have already shown in example 5.8 that ξ is an *s*-observable. We can than apply theorem 5.17 and the definition of resultant, to obtain the following denotations:

$$W_\xi^{lm}(p) = \left\{ (g,c) \,\middle|\, g \underset{p,lm}{\rightsquigarrow} c\,\square \diamond ks \right\},$$

$$\mathcal{O}_\xi(p) = \left\{ (q(\tilde{x}),c)_{/\equiv} \,\middle|\, 1\,\square\,q(\tilde{x}) \diamond \varepsilon \underset{p,lm}{\rightsquigarrow} c\,\square \diamond ks \right\}.$$

Note that $\mathcal{W}_\xi(p)$ contains all the answer constraints of p, while $\mathcal{O}_\xi(p)$ is *exactly* the *CLP* version of the top-down definition of the *s*-semantics [24]. Corollary 5.18 tells us that $\mathcal{O}_\xi$ is correct and fully abstract w.r.t. answer constraints. Moreover theorem 5.17 tells us that answer constraints for any goal can be derived from the answer constraints of the most general atomic goals. For the bottom-up case, by applying the construction, we obtain

$$
\begin{aligned}
T_{p,\xi}(X) = \{ \ & (q(\tilde{x}),c) \mid \exists q(\tilde{t}) :- \ c_k \,\square\, q_1(\tilde{t}_1),\ldots,q_n(\tilde{t}_n) \in p, \\
& (q_i(\tilde{x}_i),c_i) \in X, \\
& c = \tilde{x}{=}\tilde{t} \otimes c_k \otimes \tilde{x}_1{=}\tilde{t}_1 \otimes c_1 \otimes \ldots \otimes \tilde{x}_n{=}\tilde{t}_n \otimes c_n \ \}.
\end{aligned}
$$

Moreover $\xi \circ T_p = T_{p,\xi} \circ \xi$. Hence T_p is compatible with ξ, and, because of theorem 5.21, $\mathcal{O}_\xi = \mathcal{F}_\xi$. This result shows the power of our theory. In fact the proof of the same result, using classical methods [24], needs much more effort.

Example 5.23 In some artificial cases T_p and α are not compatible. Take for example an s-observable derived from ξ of example 5.22, by forgetting the information about a predicate q.

$$\zeta(A) = \{\, (g,c) \mid g \rightsquigarrow c\,\square\,\diamond\, ks \in A, q \notin pred(g) \,\}.$$

ζ is clearly an s-observable because the set of predicates Π is arbitrary. Consider the program $p = \{\, r(x) :- c_r\,\square\,q(x)\}$. In this case $T_{p,\zeta}(X) = \{\, (r(\tilde{x}), c_r)\,\}$. Hence$(T_{p,\zeta} \circ \zeta)(\emptyset) \neq (\zeta \circ T_p)(\emptyset) = \emptyset$. This shows the incompatibility between T_p and ζ. In fact $\mathcal{O}_\zeta(p) = \emptyset \neq \mathcal{F}_\zeta(p) = \{\, (r(\tilde{x}), c_r)\,\}$.

Example 5.24 We consider now the call pattern semantics defined in [27, 22, 26], based on the s-observable of example 5.9. By using the definition of resultant, we have that

$$\mathcal{W}_\eta(p) = \{\, (g, c\,\square\,a) \mid g \underset{p,lm}{\rightsquigarrow} c\,\square\,a, \tilde{b} \diamond ks \,\},$$

which is exactly the set of all the call patterns of p. The top-down denotation is

$$\mathcal{O}_\eta(p) = \{\, (q(\tilde{x}), c\,\square\,a) \mid 1\,\square\,q(\tilde{x}) \diamond \varepsilon \underset{p,lm}{\rightsquigarrow} c\,\square\,a, \tilde{b} \diamond ks \,\},$$

while the immediate consequences operator turns out to be

$$
\begin{aligned}
T_{p,\eta}(X) = \{\, (q(\tilde{x}), c\,\square\,a) \mid\ &\exists q(\tilde{t}) :- c_k\,\square\,q_1(\tilde{t}_1), \ldots, q_n(\tilde{t}_n) \in p,\ \text{and} \\
&c = \tilde{x}{=}\tilde{t} \otimes c_k,\ a = q_1(\tilde{t}_1)\ or \\
&c = \tilde{x}{=}\tilde{t} \otimes c_k \otimes \tilde{t}_1{=}\tilde{y} \otimes c', \\
&(q_1(\tilde{y}), c'\,\square\,a) \in X \,\}.
\end{aligned}
$$

Since T_p is compatible with η, $\mathcal{F}_\eta(p) = \mathcal{O}_\eta(p)$. $\mathcal{F}_\eta$ is then correct and fully abstract w.r.t. $=_\eta$.

Example 5.25 We show now how to reconstruct within the framework the *CLP* version of the ground success set [31, 24]. The observable is the set all the solutions of the answer constraints.

Let $[\![c]\!]$ denote the set of solutions of the constraint c and let *Sol* denote the set of all the solutions. We define the observable $\Xi : R \longrightarrow AS$, with $AS = \wp(G \times Sol)$, as

$$\Xi(A) = \{\, (g, \vartheta) \mid g \rightsquigarrow c\,\square\,\diamond\, ks \in A, \vartheta \in [\![c]\!] \,\}$$

Note that Ξ can be obtained as an abstraction of the s-observable ξ of example 5.8. Indeed let A be an element of CA and define $\alpha(A) = \{\, (g, \vartheta) \mid (g, c) \in A, \vartheta \in [\![c]\!] \,\}$, then $\Xi = \alpha \circ \xi$.

The set AS is an s-domain. In fact

$$c \cdot A = \{\, (c \otimes c' \,\Box\, \tilde{b}, \vartheta) \mid (c \otimes c' \,\Box\, \tilde{b}, \vartheta) \in A, \vartheta \in \llbracket c \rrbracket \,\},$$

$$A + B = A \cup B,$$

$$A \times B = \{\, ((g, g'), \vartheta) \mid (g, \vartheta) \in A, (g', \vartheta) \in B, \,\}.$$

$$\nabla_{\vartheta}(A) = \{\, (\partial_{\vartheta}(c) \,\Box\, \tilde{b}\vartheta, \sigma\vartheta) \mid (c \,\Box\, \tilde{b}, \sigma) \in A \,\}$$

It is easy to see that the abstraction α is an s-morphism. Hence Ξ is an s-observable. The denotations derived from the definitions in the framework are exactly those given in [24]. Namely,

$$\mathcal{O}_{\Xi}(p) = \alpha(\mathcal{O}_{\xi}(p)) = \left\{\, (q(\tilde{x}), \vartheta) \,\middle|\, q(\tilde{x}) \underset{p,lm}{\leadsto} c \,\Box\, \diamond ks, \vartheta \in \llbracket c \rrbracket \,\right\},$$

$$T_{p,\Xi}(X) = (\alpha \circ T_{p,\xi} \circ \gamma)(X) =$$

$$\{\, (q(\tilde{x}), \vartheta) \mid \quad \exists q(\tilde{t}) :- c_k \,\Box\, q_1(\tilde{t}_1), \dots, q_n(\tilde{t}_n) \in p$$
$$\vartheta \in \llbracket \tilde{x}{=}\tilde{t} \otimes c_k \otimes \tilde{x}_1{=}\tilde{t}_1 \otimes \dots \otimes \tilde{x}_n{=}\tilde{t}_n \rrbracket$$
$$(q_i(\tilde{x}_i), \vartheta) \in X \,\}$$

Since T_p is compatible with Ξ, $\mathcal{O}_{\Xi}(p) = \mathcal{F}_{\Xi}(p)$. All the other properties hold, i.e., $\mathcal{O}_{\Xi}(p)$ is correct and fully abstract w.r.t. the solutions of the answer constraints and is also AND-compositional. Note that the denotation $\mathcal{O}_{\xi}(p)$ is correct (but not fully abstract) w.r.t. the $=_{\Xi}$ equivalence.

We can specialize the above definitions to the case of logic programs, where $\mathcal{A}$ is the Herbrand constraint system $(\mathcal{A}_{\mathcal{H}})$ and e is a set of equations in solved form.

$$\mathcal{O}_{\Xi}(p) = \{\, (q(\tilde{x}), \vartheta) \mid q(\tilde{x}) \underset{p,lm}{\leadsto} e \,\Box\, \diamond ks, \vartheta \text{ is a solution of } e \,\},$$

$$T_{p\Xi} = \{\, (q(\tilde{x}), \vartheta) \mid \quad \exists q(\tilde{t}) :- e_k \,\Box\, q_1(\tilde{t}_1), \dots, q_n(\tilde{t}_n) \in p$$
$$\vartheta \text{ is a solution of } \{\, \tilde{x}{=}\tilde{t}, \tilde{x}_1{=}\tilde{t}_1, \dots, \tilde{x}_n{=}\tilde{t}_n \,\} \cup e_k$$
$$(q_i(\tilde{x}_i), \vartheta) \in X \,\}$$

$\mathcal{O}_{\Xi}(p)$ can be proved in to be equivalent to the success set, which is — however — defined by considering ground atomic goals only. Note that the standard definition cannot be accommodated within the framework.

The success set was also proved in [18] to be correct (and fully abstract) w.r.t. an observational equivalence based on successful derivations, according to which two programs are equivalent iff they have the same set of successful goals. Therefore, the answer constraints solutions observable defines the same observable equivalence given by the successful derivations observable. However, one can easily show that the successful derivations is not an s-observable. Hence we cannot define a good semantics based on that observable.

It is worth noting that we can derive also the so called c-semantics [7, 21, 18], by the s-observable

$$v(A) = \{ (g, c') \mid g \leadsto c \,\square\, \diamond ks \in A, c' \leq c \}.$$

5.3   Other observables

For a generic observable $\alpha : R \longrightarrow D$, the previous theorems state that if α is not an s-observable then we cannot have a correct and fully abstract denotation which is AND-compositional, because in such a case D should have the structure of an s-domain, thus contradicting the hypothesis. Instead we can have a correct AND-compositional denotation which is minimal w.r.t. the information content. In analogy to what we did in the case of observables, we can consider the set $\mathbb{S}$ of AND-compositional denotations $\mathcal{S} : \mathcal{P} \longrightarrow \mathcal{D}$ with $\mathcal{D} \in \mathbb{D}$. We can partially order $\mathbb{S}$ by approximation.

Definition 5.26 [denotation approximation] A denotation $\mathcal{S} : \mathcal{P} \longrightarrow \mathcal{D} \in \mathbb{S}$ approximates $\mathcal{S}' : \mathcal{P} \longrightarrow \mathcal{D}'$ if there exists a Galois co-insertion $\alpha : \mathcal{D} \longrightarrow \mathcal{D}'$ s.t.

$$\mathcal{S}' = \alpha \circ \mathcal{S}.$$

Lemma 5.27 (denotation lattice) *Let* $\mathcal{S} \leq \mathcal{S}'$ *iff* $\mathcal{S}'$ *approximates* $\mathcal{S}$. $(\mathbb{S}, \leq)$ *is a complete lattice.*

Note that every denotation which approximates another denotation which is correct w.r.t. an observable is still correct w.r.t. the same observable.

Since $\mathbb{O}$ is a complete lattice, we have that $\alpha \uparrow s \stackrel{def}{=} \{ \beta \in \mathbb{O}_s \mid \beta \leq \alpha \}$ has a minimum α'. Hence we can find a denotation (correct and fully abstract for α') which models correctly α and is *minimal*.

Theorem 5.28 (the denotation) *For each observable α there exists a minimal AND-compositional denotation correct w.r.t. it, i.e., which is approximated by all the other AND-compositional correct denotations.*

Example 5.29 We consider now the correct call patterns semantics defined in [22, 27]. The *correct* call patterns of a program p for a goal g and a selection rule r are the call patterns for which there exists a refutation, i.e., the atoms (procedure calls) selected in any *SLD*-refutation of g in p via r. The s-domain is that of example 5.6. The abstraction which allows us to obtain the correct call patterns is

$$\eta^c(A) = \{ (g, c \,\square\, a) \mid g \leadsto c \,\square\, a, \tilde{b} \diamond ks < g \leadsto c_* \,\square\, \diamond ks' \in A \}.$$

In this case it is not possible to define the $\times$ operation, because given two correct call patterns $(g, c \,\square\, a)$ and $(g', c' \,\square\, a')$ we cannot say if g, g' will have a refutation. The only way to use the algebraic framework is to consider a stronger observable, for example one where we add the computed answer constraint to the correct call patterns. Namely

$$\varrho(A) = \{\, (g, c \,\square\, a, c_*) \mid g \rightsquigarrow c \,\square\, a, \tilde{b} \diamond ks < g \rightsquigarrow c_* \,\square\, \diamond ks' \in A \,\}.$$

The $\cdot$ and $+$ operations are defined in the natural way while the $\times$ operation is defined as follows.

$$
\begin{aligned}
A \times B = \ & \{\, ((g, g'), c \,\square\, a, c_* \otimes c'_*) \mid (g, c \,\square\, a, c_*) \in A, \\
& (g', c' \,\square\, a', c'_*) \in B, c_* \otimes c'_* > 0 \,\} \cup \\
& \{\, ((g, g'), c_* \otimes c' \,\square\, a', c_* \otimes c'_*) \mid (g, c_* \,\square\, , c_*) \in A, \\
& (g', c' \,\square\, a', c'_*) \in B, c_* \otimes c'_* > 0 \,\}.
\end{aligned}
$$

The observable ϱ clearly approximates η^c. By using the definition of resultant, we have that the top-down goal independent denotations are

$$\mathcal{O}_{\eta^c}(p) = \{\, (q(\tilde{x}), c \,\square\, a) \mid 1 \,\square\, q(\tilde{x}) \diamond \varepsilon \underset{p,lm}{\rightsquigarrow} c \,\square\, a, \tilde{b} \diamond ks \underset{p,lm}{\rightsquigarrow} c_* \,\square\, \diamond ks' \,\};$$

$$\mathcal{O}_{\varrho}(p) = \{\, (q(\tilde{x}), c \,\square\, a, c_*) \mid 1 \,\square\, q(\tilde{x}) \diamond \varepsilon \underset{p,lm}{\rightsquigarrow} c \,\square\, a, \tilde{b} \diamond ks \underset{p,lm}{\rightsquigarrow} c_* \,\square\, \diamond ks' \,\};$$

while the ρ immediate consequences operator turns out to be

$$
\begin{aligned}
T_{p,\varrho}(X) = \ \{\ & (q(\tilde{x}), c \,\square\, a, c_*) \mid \\
& \exists q(\tilde{t}) :- c_k \,\square\, q_1(\tilde{t}_1), \ldots, q_n(\tilde{t}_n) \in p, \text{ and} \\
& c = \tilde{x} = \tilde{t} \otimes c_k \otimes \tilde{t}_1 = \tilde{y} \otimes c', \ c_* = \tilde{x} = \tilde{t} \otimes c_k \otimes \tilde{t}_1 = \tilde{y} \otimes c'_*, \\
& (q(\tilde{y}), c' \,\square\, a, c'_*) \in X \text{ or} \\
& c = \tilde{x} = \tilde{t} \otimes c_k, \ c_* = \tilde{x} = \tilde{t} \otimes c_k \otimes \tilde{t}_1 = \tilde{t} \otimes c_1, \\
& q_1(\tilde{t}') :- c_1 \,\square\, \in p \,\}.
\end{aligned}
$$

The denotation $\mathcal{O}_{\varrho}$ is *AND*-compositional while $\mathcal{O}_{\eta^c}$ is not.

Example 5.30 We consider now finite failures. A goal g is finitely failed if its *SLD*-tree is finite and has no successful derivations. The set of all finitely failed goals of p is denoted by $FF(p)$. The s-domain is $\wp(G)$ with the following operations

$$c \cdot A = \{\, c \otimes c' \,\square\, \tilde{b} \diamond ks \mid c' \,\square\, \tilde{b} \diamond ks \in A \,\},$$

$$A \times B = \{\, (g, g') \mid g \in A, g' \in B \,\},$$

$$\nabla_{\vartheta}(A) = \{\, (\partial_{\vartheta}(c) \,\square\, \tilde{b}\vartheta) \mid (c \,\square\, \tilde{b}) \in A \,\}.$$

The abstraction which allows us to obtain the finite failures is

$$\phi(A) = \{\, g \mid \text{ the set } B \text{ of the resultants of the form}$$
$$g \rightsquigarrow g' \text{ in } A \text{ is finite, non empty and}$$
$$\text{there exist no resultants of the form } g \rightsquigarrow c \,\square\, \diamond\, ks \in B \,\}.$$

ϕ is a ×-morphism, but is not a ··-morphism. Hence it is not an s-observable. The only way to use the algebraic framework is to consider a stronger observable, for example one where we add to the goal the negation of all computed answer constraints. Namely $\varphi : R \longrightarrow CA$

$$\varphi(A) = \{\, (g,c) \mid \text{ the set } B \text{ of the resultants of the form}$$
$$g \rightsquigarrow g' \text{ in } A \text{ is finite, non empty and}$$
$$c = \neg(\textstyle\bigoplus c') \,\forall c' : g \rightsquigarrow c \,\square\, \diamond\, ks \in A \,\},$$

where we define $\bigoplus(\emptyset) = 0$. The observable φ clearly approximates ϕ. The resulting top-down denotations are

$$\mathcal{O}_\phi(p) = \{\, q(\tilde{x}) \mid q(\tilde{x}) \in FF(p) \,\},$$

$$\mathcal{O}_\varphi(p) = \{\, (q(\tilde{x}), c) \mid c = \neg(\textstyle\bigoplus c'), \forall c' : 1 \,\square\, q(\tilde{x}) \diamond \varepsilon \underset{p,lm}{\rightsquigarrow} c' \,\square\, \diamond\, ks \,\};$$

while the immediate consequences operator $T_{p,\varphi}$ turns out to be

$$T_{p,\varphi}(X) = \{\, (q(\tilde{x}), c) \mid \; c = \bigotimes c', \forall c' :$$
$$\exists q(\tilde{t}) :- c_k \,\square\, q_1(\tilde{t}_1), \dots, q_n(\tilde{t}_n) \in p,$$
$$c = \neg\left(\exists(\tilde{x}_{=}\tilde{t} \otimes c_k)_{\tilde{x}}\right) \oplus$$
$$\bigoplus_{1 \le i \le m} \exists(\tilde{x}_{=}\tilde{t} \otimes c_k \otimes \tilde{x}_i{=}\tilde{t}_i \otimes c_i)_{\tilde{x}},$$
$$(q_i(\tilde{x}_i), c_i) \in X \,\}.$$

The denotation $\mathcal{O}_\varphi$ is *AND*-compositional while $\mathcal{O}_\phi$ is not. Moreover φ is compatible with T_p. Hence $\mathcal{F}_\varphi = \mathcal{O}_\varphi$.

It is worth noting that the s-observable φ models finite failures in a style which looks like constructive negation [5, 6, 40].

6 Abstract interpretation

One more motivation for our algebraic construction can be found in abstract interpretation. The essence of abstract interpretation is to give a non-standard interpretation to the language. In the *CLP* case, as shown in [11, 29], we only need to give a non-standard interpretation to constraints (and terms).

In general a constraint system is an interpretation (in a semi-closed semiring) for constraint formulas. According to the approach of [15, 29, 16] constraint systems are related by means of constraint system semi-morphisms.

The program interpretation process is expressed in terms of a set of algebraic operators which model how data objects are collected during the computation. An abstract interpretation for a given $CLP(\mathcal{A})$ program is then the semantics of an abstract program in $CLP(\mathcal{A}')$ where $\mathcal{A}'$ is a suitable constraint system correct w.r.t. $\mathcal{A}$.

The overall abstract interpretation methodology can be described as follows.

- Select an observable α, such that the property to be considered by the analysis is an abstraction of α.

- If α is an s-observable, we have a semantics $\mathcal{O}_\alpha(p)$ which is correct and fully abstract w.r.t. α. If α is not an s-observable, we have to choose an s-observable α' which approximates α, from which we have a semantics correct (but not fully abstract) w.r.t. α. If α (α') is compatible with T_p, $\mathcal{O}_\alpha(p)$ $(\mathcal{O}_{\alpha'}(p))$ can equivalently be determined by

    - A top-down construction, obtained by collecting the observables for the atomic goals of the form $q(\tilde{x})$.

    - A bottom-up construction, obtained by computing the least fixpoint of a specialized immediate consequences operator.

- Define a suitable abstraction μ (semi-morphism) of the constraint system $\mathcal{A}$, according to the approach in [15, 29, 16]. μ is extended to a homomorphism on s-domains $\mu^\mathcal{D}$. The overall abstraction is then an i-observable, i.e., the s-semi-morphism $\alpha_\mu = \alpha \circ \mu^\mathcal{D}$ resulting from the composition of $\mu^\mathcal{D}$ and α. We obtain two abstract top-down and bottom-up goal independent denotations $\mathcal{O}_{\alpha_\mu}(p)$ and $\mathcal{F}_{\alpha_\mu}(p)$, which — if α_μ is compatible with T_p — are equivalent. If we want to derive from $\mathcal{O}_{\alpha_\mu}(p)$ the result of the analysis for a specific goal g, we can only exploit an AND-semi-compositionality theorem, which allows us to derive approximations of the abstract observable.

- As first suggested in [15, 29, 16] abstract interpretation can be viewed as computation in a suitable instance of the CLP framework, by first transforming the program p into an *abstract program* $\mu(p)$, obtained by abstracting the concrete constraints. According to this approach we obtain two additional abstract denotations, $\mathcal{O}_\alpha(\mu(p))$ and $\mathcal{F}_\alpha(\mu(p))$, which are equivalent — if α is compatible with $T_{\mu(p)}$ — and which are safe approximations of $\mathcal{O}_{\alpha_\mu}(p)$. The result of the analysis for a specific goal g can be obtained by exploiting an AND-compositionality theorem. Namely, the abstract behavior of a goal g can be determined by "executing" g in $\mathcal{O}_\alpha(\mu(p))$.

Let us discuss some specific analysis problems in the framework of the above methodology.

- If we are interested in properties of the answer substitutions (such as aliasing and sharing) we have to choose a concrete semantics correct w.r.t. answer substitutions. Therefore the least Herbrand model semantics is not adequate and a semantics at least as detailed as the *s*-semantics has to be chosen. This is the approach taken in [2, 8].

- If we want to perform analysis of program components in a modular way, we need an *OR*-compositional semantics, i.e., a semantics compositional w.r.t. program union. As a matter of fact the framework in [2] has been extended to handle modularity [9], by replacing the *s*-semantics with its compositional version [4], which has clauses as semantic objects.

- If we want to determine abstract properties of the call patterns, we should use a concrete semantics which gives more information on the computation than just the computed answers. Namely, we have to model an observable consisting of all the procedure calls. This problem has been considered in [23], where the concrete semantics is the *set of resultants*. The resulting abstract semantics is goal independent and parametric w.r.t. the (local) selection rule. A similar result was obtained in [10] for the leftmost selection rule, by using a magic-set like transformational approach and the *OR*-compositional semantics in [4].

As already discussed, in our framework abstract interpretation can be viewed as the composition of a constraint system semi-morphism (the abstraction of the domain interpretation) and of an *s*-observable, which chooses an adequate abstraction of *SLD*-trees. The construction is based on semi-morphisms on constraint systems as defined by definition 2.6. A constraint system semi-morphism $\mu : \mathcal{A} \longrightarrow \mathcal{A}'$ can be extended to an *s*-domain homomorphism $\mu^D : D_{\mathcal{A}} \longrightarrow D_{\mathcal{A}'}$ in a natural way.

Definition 6.1 [*s*-semi-morphism] Let $D_{\mathcal{A}}$, $D'_{\mathcal{A}'}$ be *s*-domains, with $\mathcal{A}'$ correct w.r.t. $\mathcal{A}$. An *s-semi-morphism* $\alpha_\mu : D_{\mathcal{A}} \longrightarrow D'_{\mathcal{A}'}$, based on the constraint system semi-morphism $\mu : \mathcal{A} \longrightarrow \mathcal{A}'$, is the composition of μ^D and an *s*-morphism $\alpha : D_{\mathcal{A}'} \longrightarrow D'_{\mathcal{A}'}$, i.e., $\alpha_\mu = \alpha \circ \mu^D$.

Proposition 6.2 (*s*-semi-morphisms are co-insertions) *Let $D_{\mathcal{A}}$, $D'_{\mathcal{A}'}$ be s-domains, with $\mathcal{A}'$ correct w.r.t. $\mathcal{A}$. If there exists an s-semi-morphism $\alpha_\mu : D_{\mathcal{A}} \longrightarrow \bar{D}_{\mathcal{A}'}$, then there exists a mapping $\gamma_\mu : |\bar{D}_{\mathcal{A}'}| \longrightarrow |D_{\mathcal{A}}|$ such that (α_μ, γ_μ) is a Galois co-insertion of $(D_{\mathcal{A}}, \leq)$ onto $(\bar{D}_{\mathcal{A}'}, \bar{\leq})$.*

Since s-semi-morphisms from R are strongly related to abstract interpretation, we will call them *i-observables*. Let $\mathbb{O}_i$ denote the set of i-observables. $\mathbb{O}_i$ can be ordered by approximation, as it was the case for $\mathbb{O}_s$.

Proposition 6.3 (*i-observables lattice*) $(\mathbb{O}_i, \leq)$ *is a complete sublattice of* $(\mathbb{O}, \leq)$. *Moreover* $(\mathbb{O}_s, \leq)$ *is a complete sublattice of* $(\mathbb{O}_i, \leq)$.

Example 6.4 Let $\mathcal{A}_\mathcal{H}$ be the Herbrand constraint system. We show how to reconstruct the ground dependency analysis for call patterns described in [23]. The abstract domain of computation is *Prop* [13], consisting of propositional formulas which provide a concise representation of abstract substitutions which describe ground dependency relations among arguments of a procedure call. A propositional formula with connectives $\{\leftrightarrow, \vee, \wedge\}$ is associated with each constraint. For example the constraint $x{=}[z|y]$ is represented by the formula $x \leftrightarrow y \wedge z$. The formula $x \wedge y$ states the groundness of x and y, while the formula $(x \wedge y) \leftrightarrow z$ represents that z is ground iff both x and y are ground.

Prop can be formalized as a constraint system. Take a term system morphism μ which maps each term onto its set of free variables (the empty set for ground terms). $\mathcal{A}_{Prop}$ is the algebra of possibly existentially quantified disjunctions of formulas

$$\left(Prop, \wedge, \vee, true, false, \exists_x, \partial_x^t, \Lambda(t) \leftrightarrow \Lambda(t') \right),$$

where each t belongs to V^n (for some n) and $\Lambda(t) = x_1 \wedge \ldots \wedge x_n$ for $t = \{x_1, \ldots, x_n\}$ ($\Lambda(\emptyset) = true$).

We can define a constraint system semi-morphism $\mu : \mathcal{A}_\mathcal{H} \longrightarrow \mathcal{A}_{Prop}$ as

$$\mu(c) = \begin{cases} false & \text{if } c = 0 \\ \Lambda(var(c)) & \text{if } c \text{ is a simple constraint} \\ \Lambda(\mu(t)) \leftrightarrow \Lambda(\mu(t')) & \text{if } c = t{=}t' \\ \mu(c_1) \wedge \mu(c_2) & \text{if } c = c_1 \otimes c_2 \\ \mu(c_1) \vee \mu(c_2) & \text{if } c = c_1 \oplus c_2 \end{cases}$$

and extend it to well-formed sets of resultants as

$$\begin{aligned} \mu^\mathcal{R}(A) = \{ \quad & \mu(c) \,\square\, q_1(\mu(\tilde{t}_1)), \ldots, q_n(\mu(\tilde{t}_n)) \diamond ks \rightsquigarrow \\ & \mu(c') \,\square\, q'_1(\mu(\tilde{t}'_1)), \ldots, q'_{n'}(\mu(\tilde{t}'_{n'})) \diamond ks' \mid \\ & c \,\square\, q_1(\tilde{t}_1), \ldots, q_n(\tilde{t}_n) \diamond ks \rightsquigarrow \\ & c' \,\square\, q'_1(\tilde{t}'_1), \ldots, q'_{n'}(\tilde{t}'_{n'}) \diamond ks' \in A \}. \end{aligned}$$

Since the call pattern observable η of example 5.9 was defined for any constraint system $\mathcal{A}$, we can compose it with $\mu^\mathcal{R}$, $\eta_\mu : R_{\mathcal{A}_\mathcal{H}} \longrightarrow CP_{\mathcal{A}_{Prop}}$:

$$\eta_\mu(A) = (\eta \circ \mu^\mathcal{R})(A) = \{ (g, c \,\square\, a) \mid g \rightsquigarrow c \,\square\, a, \tilde{b} \diamond ks \in \mu^\mathcal{R}(A) \}$$

and obtain an abstract interpretation in *Prop*.

The following discussion on the abstract denotations relies on the notion of "abstracting the program". The idea is to replace the constraints (defined on the constraint system $\mathcal{A}$) in the original program with their abstract version (defined on a constraint system $\mathcal{A}'$ correct w.r.t. $\mathcal{A}$), thus obtaining a *CLP* program on a different constraint system. If $\mu : \mathcal{A} \longrightarrow \mathcal{A}'$ is the semi-morphism which formalizes the (correct) constraint system abstraction we denote by $\mu^{\mathcal{P}} : \mathcal{P}_\mathcal{A} \longrightarrow \mathcal{P}_{\mathcal{A}'}$ the homomorphism obtained by extending μ to programs in the natural way, i.e., by applying μ to the constraints and terms occurring in the program. The abstract denotation of p is simply the denotation $\mathcal{O}(\mu^{\mathcal{P}}(p))$ of the abstract program $\mu^{\mathcal{P}}(p)$. When clear from the context we denote $\mu^{\mathcal{P}}$ by μ. The following theorem generalizes a result in [29] and states that the semantics of the "abstract program" is a safe approximation of the abstraction of the semantics of the program.

Theorem 6.5 (abstract program) *Let $\mu : \mathcal{A} \longrightarrow \mathcal{A}'$ be a constraint system semi-morphism. Then $\mu^{\mathcal{R}}(\mathcal{O}(p)) \leq \mathcal{O}(\mu(p))$.*

Since an i-observable on the constraint system $\mathcal{A}$ is obtained by an s-observable on the constraint system $\mathcal{A}'$ we expect the semantics construction to be almost the same for the abstract interpretation case. Obviously the theorems are weaker, because of the lack of precision of the denotations.

Definition 6.6 [abstract denotations] Let p be a program, $\alpha_\mu : \mathcal{R}_\mathcal{A} \longrightarrow \mathcal{D}_{\mathcal{A}'}$ be an i-observable.

1. The ideal top-down denotation of an i-observable is $\mathcal{O}_{\alpha_\mu}(p) \stackrel{def}{=} \alpha_\mu(\mathcal{O}(p))$, while

2. the abstract top-down denotation is $\mathcal{O}_\alpha(\mu(p)) \stackrel{def}{=} \alpha(\mathcal{O}(\mu(p)))$.

3. The ideal immediate consequences operator is $T_{p,\alpha_\mu} \stackrel{def}{=} \alpha_\mu \circ T_p \circ \gamma_\mu$, while

4. the abstract immediate consequences operator is $T_{\mu(p),\alpha} \stackrel{def}{=} \alpha \circ T_{\mu(p)} \circ \gamma$.

5. The ideal bottom-up denotation is $\mathcal{F}_{\alpha_\mu}(p) \stackrel{def}{=} T_{p,\alpha_\mu} \uparrow \omega$, while

6. the abstract bottom-up denotation is $\mathcal{F}_\alpha(\mu(p)) \stackrel{def}{=} T_{\mu(p),\alpha} \uparrow \omega$.

The following theorem relates different abstract interpretation mechanisms, i.e., the bottom-up execution of the abstract program $\mathcal{F}_\alpha(\mu(p))$, the top-down abstract execution of the abstract program $\mathcal{O}_\alpha(\mu(p))$, the abstraction of the top-down concrete execution $\mathcal{O}_{\alpha_\mu}(p)$ and the (specialized) bottom-up execution of the concrete program $\mathcal{F}_{\alpha_\mu}(p)$. As one could expect, the ideal top-down denotation is (safely) approximated by all the other denotations. $\mathcal{F}_\alpha(\mu(p))$ is the least precise and, in the case of compatibility, we have the usual equivalence between top-down and bottom-up executions. This is shown by the following theorem.

Theorem 6.7 *Let p be a program, $\alpha_\mu : \mathcal{R}_A \longrightarrow \mathcal{D}_{A'}$ be an i-observable. Then*

1. $\mathcal{F}_\alpha(\mu(p)) \geq \mathcal{O}_\alpha(\mu(p))$,

2. $\mathcal{F}_\alpha(\mu(p)) \geq \mathcal{F}_{\alpha_\mu}(p)$,

3. $\mathcal{O}_\alpha(\mu(p)) \geq \mathcal{O}_{\alpha_\mu}(p)$,

4. $\mathcal{F}_{\alpha_\mu}(p) \geq \mathcal{O}_{\alpha_\mu}(p)$.

Moreover if α is compatible with $T_{\mu(p)}$ we have

$$\mathcal{F}_\alpha(\mu(p)) = \mathcal{O}_\alpha(\mu(p)) \geq \mathcal{F}_{\alpha_\mu}(p) = \mathcal{O}_{\alpha_\mu}(p).$$

The next theorem shows that *AND*-compositionality holds in the denotation based on the abstract program.

Theorem 6.8 (AND-semi-compositionality) *Let p be a program and $\alpha_\mu :$ $\mathcal{R}_A(\cdot, +, \times) \longrightarrow \mathcal{D}_{A'}(\bar{\cdot}, \bar{+}, \bar{\times})$ an i-observable. Then*

$$v' \in \mathcal{W}_\alpha(\mu(p)) \quad \Longleftrightarrow \quad \exists c' \in \mathcal{A}', e'_1, \ldots, e'_n \in \mathcal{O}_\alpha(\mu(p)) \ s.t.$$

$$v' = c' \cdot (\{ e'_1 \} \bar{\times} \ldots \bar{\times} \{ e'_n \}),$$

$$v' \in \mathcal{W}_{\alpha_\mu}(p) \quad \Longleftrightarrow \quad \exists c' \in \mathcal{A}', e'_1, \ldots, e'_n \in \mathcal{O}_{\alpha_\mu}(p) \ s.t.$$

$$v' \leq c' \cdot (\{ e'_1 \} \bar{\times} \ldots \bar{\times} \{ e'_n \}).$$

Example 6.9 We show now the ground dependencies call patterns denotation, based on the i-observable η_μ of example 6.4.

$$\mathcal{O}_{\eta_\mu}(p) = \{ \, (q(\tilde{x}), c_\mu \,\square\, q_1(\tilde{t}_\mu)) \mid 1 \,\square\, q(\tilde{x}) \diamond \varepsilon \underset{p, lm}{\rightsquigarrow} c \,\square\, q_1(\tilde{t}), \tilde{b} \diamond ks,$$
$$c_\mu = \mu(c), \ \tilde{t}_\mu = \mu(\tilde{t}) \, \};$$

$$\mathcal{O}_\eta(\mu(p)) = \{ \, (q(\tilde{x}), c \,\square\, a) \mid 1 \,\square\, q(\tilde{x}) \diamond \varepsilon \underset{\mu(p), lm}{\rightsquigarrow} c \,\square\, a, \tilde{b} \diamond ks \, \},$$

$$T_{p, \eta_\mu}(X) = \{ \ (q(\tilde{x}), c_\mu \,\square\, q_1(\tilde{t}_\mu)) \mid$$
$$\exists q(\tilde{t}) :- c_k \,\square\, q_1(\tilde{t}_1), \ldots, q_n(\tilde{t}_n) \in p, \text{ and}$$
$$c_\mu = \Lambda(\tilde{x}) \leftrightarrow \Lambda(\tilde{t}) \wedge \mu(c_k), \ \tilde{t}_\mu = \mu(\tilde{t}_1) \text{ or}$$
$$c_\mu = \Lambda(\tilde{x}) \leftrightarrow \Lambda(\tilde{t}) \wedge \mu(c_k) \wedge \Lambda(\tilde{t}_1) \leftrightarrow \Lambda(\tilde{y}) \wedge c'_\mu,$$
$$(q_1(\tilde{y}), c'_\mu \,\square\, q_1(\tilde{t}_\mu)) \in X \, \}.$$

$$
\begin{aligned}
T_{\mu(p),\eta}(X) \;=\; \{ \;\; &(q(\tilde{x}), c \,\square\, a) \mid \\
&\exists q(\tilde{t}) :- c_k \,\square\, q_1(\tilde{t}_1), \ldots, q_n(\tilde{t}_n) \in \mu(p), \text{ and} \\
&c = \Lambda(\tilde{x}) \leftrightarrow \Lambda(\tilde{t}) \wedge c_k, \; a = q_1(\tilde{t}_1) \text{ or} \\
&c = \Lambda(\tilde{x}) \leftrightarrow \Lambda(\tilde{t}) \wedge c_k \wedge \Lambda(\tilde{t}_1) \leftrightarrow \Lambda(\tilde{y}) \wedge c', \\
&(q_1(\tilde{y}), c' \,\square\, a) \in X \; \}.
\end{aligned}
$$

Since $T_{\mu(p)}$ is compatible with η, $\mathcal{F}_\alpha(\mu(p)) = \mathcal{O}_\alpha(\mu(p))$. $\mathcal{O}_\alpha(\mu(p))$, which is a goal independent denotation computable either top-down or bottom-up, is the semantics obtained in [23] and, through a magic set like transformation, in [10].

7   Conclusions

We have defined an algebraic framework which allows us to prove several properties of concrete and abstract *SLD*-trees. The framework provides:

- a denotation consisting of all the *SLD*-trees obtained from most general atomic goals, with an equivalent alternative fixpoint construction;

- a set of important theorems (lifting, *AND*-compositionality) which show that the denotation characterizes the *SLD*-trees for any goal;

- a mechanism (*s*-observable) for abstracting the semantics, which guarantees that the general theorems do hold for any abstraction and always leads to the best (correct and fully abstract) denotation;

- a mechanism (*i*-observable) to model abstraction by approximation, which guarantees that a weaker form of the general theorems is still valid and provides the semantic basis for abstract interpretation.

We have not considered yet two issues that could be discussed within the framework. The first one is related to *OR-compositionality* [4], which is a relevant property if we want to be able to reason about programs in a modular way. *SLD*-trees are indeed *OR*-compositional. However, the abstraction process can destroy this property. For example, two of the abstractions that we have considered, namely computed answer constraints and call patterns, are not *OR*-compositional. The theory should be extended with a characterization of *OR*-compositional observables. For non-*OR*-compositional observables there should be a result similar to the one of theorem 5.28. The second issue is related to the computation rule. The current results apply to the case of local selection rules, even if we have considered the leftmost selection rule only. The property that could be analyzed within the framework is the *independence from the selection rule*. *SLD*-trees do depend on the selection rule. However, more abstract observables, such as answer constraints, are independent from the selection rule. A second relevant extension of the framework might be the definition of conditions which guarantee that the abstractions are independent from the selection rule.

References

[1] K. R. Apt, *Introduction to Logic Programming*, Handbook of Theoretical Computer Science (J. van Leeuwen, ed.), vol. B: Formal Models and Semantics, Elsevier and The MIT Press, Amsterdam, Cambridge MA, 1990, pp. 495–574.

[2] R. Barbuti, R. Giacobazzi, and G. Levi, *A General Framework for Semantics-based Bottom-up Abstract Interpretation of Logic Programs*, ACM Transactions on Programming Languages and Systems **15** (1993), no. 1, 133–181.

[3] A. Bossi, M. Gabbrielli, G. Levi, and M. Martelli, *The s-semantics approach: Theory and applications*, J. Logic Programming **19-20** (1994), 149–197.

[4] A. Bossi, M. Gabbrielli, G. Levi, and M. C. Meo, *A Compositional Semantics for Logic Programs*, Theoret. Comput. Sci. **122** (1994), no. 1-2, 3–47.

[5] D. Chan, *Constructive Negation Based on the Completed Database*, Proc. Fifth Int'l Conf. on Logic Programming (R. A. Kowalski and K. A. Bowen, eds.), MIT Press, Cambridge MA, 1988, pp. 111–125.

[6] ――――, *An Extension of Constructive Negation and its Application in Coroutining*, Proc. North American Conf. on Logic Programming'89 (E. Lusk and R. Overbeek, eds.), MIT Press, Cambridge MA, 1989, pp. 477–493.

[7] K. L. Clark, *Predicate logic as a computational formalism*, Res. report doc 79/59, Imperial College, Dept. of Computing, London, 1979.

[8] M. Codish, D. Dams, and E. Yardeni, *Bottom-up Abstract Interpretation of Logic Programs*, Theoret. Comput. Sci. **124** (1994), no. 1, 93–125.

[9] M. Codish, S. K. Debray, and R. Giacobazzi, *Compositional Analysis of Modular Logic Programs*, Proc. Twentieth Annual ACM Symp. on Principles of Programming Languages, ACM Press, 1993, pp. 451–464.

[10] M. Codish and B. Demoen, *Analysing Logic Programs using "**prop**"-ositional Logic Programs and a Magic Wand*, Proc. 1993 Int'l Symposium on Logic Programming (D. Miller, ed.), MIT Press, Cambridge MA, 1993, pp. 114–129.

[11] P. Codognet and G. Filè, *Computations, Abstractions and Constraints*, Proc. Fourth IEEE Int'l Conference on Computer Languages, IEEE Press, 1992.

[12] M. Comini and G. Levi, *An algebraic theory of observables*, Tech. report, Dipartimento di Informatica, Università di Pisa, 1994.

[13] A. Cortesi, G. Filè, and W. Winsboroug, *Prop revisited: Propositional Formula as Abstract Domain for Groundness Analysis*, Proc. Sixth IEEE Symp. on Logic in Computer Science, IEEE Computer Society Press, 1991, pp. 322–327.

[14] P. Cousot and R. Cousot, *Systematic Design of Program Analysis Frameworks*, Proc. Sixth ACM Symp. Principles of Programming Languages, 1979, pp. 269–282.

[15] S. K. Debray, R. Giacobazzi, and G. Levi, *A Generalized Semantics for Constraint Logic Programs*, Proceedings of the International Conference on Fifth Generation Computer Systems 1992, 1992, pp. 581–591.

[16] S. K. Debray, R. Giacobazzi, and G. Levi, *Joining Abstract and Concrete Computations in Constraint Logic Programming*, Algebraic Methodology and Software Technology (AMASt'93), Proceedings of the Third International Conference on Algebraic Methodology and Software Technology (M. Nivat, C. Rattray, T. Rus, and G. Scollo, eds.), Workshops in Computing, Springer-Verlag, Berlin, 1993, pp. 111–127.

[17] F. Denis and J. P. Delahaye, *Unfolding, Procedural and Fixpoint Semantics of Logic Programs*, STACS 91 (C. Choffrut and M. Jantzen, eds.), Lecture notes in Computer Science **480**, Springer-Verlag, Berlin, 1991, pp. 511–522.

[18] M. Falaschi, G. Levi, M. Martelli, and C. Palamidessi, *A new Declarative Semantics for Logic Languages*, Proc. Fifth Int'l Conf. on Logic Programming (R. A. Kowalski and K. A. Bowen, eds.), MIT Press, Cambridge MA, 1988, pp. 993–1005.

[19] _____, *Declarative Modeling of the Operational Behavior of Logic Languages*, Theoret. Comput. Sci. **69** (1989), no. 3, 289–318.

[20] _____, *A Model-Theoretic Reconstruction of the Operational Semantics of Logic Programs*, Information and Computation **102** (1993), 86–113.

[21] G. Ferrand, *Error Diagnosis in Logic Programming, an Adaptation of E. Y. Shapiro's Method*, J. Logic Programming **4** (1977), 177–198.

[22] M. Gabbrielli, *The Semantics of Logic Programming as a Programming Language*, Ph.D. thesis, Dipartimento di Informatica, Università di Pisa, 1992.

[23] M. Gabbrielli and R. Giacobazzi, *Goal independency and call patterns in the analysis of logic programs*, Proc. SAC '94, 1994.

[24] M. Gabbrielli and G. Levi, *Modeling Answer Constraints in Constraint Logic Programs*, Proc. Eighth Int'l Conf. on Logic Programming (K. Furukawa, ed.), MIT Press, Cambridge MA, 1991, pp. 238–252.

[25] _____, *On the Semantics of Logic Programs*, Automata, Languages and Programming, 18th International Colloquium (J. Leach Albert, B. Monien, and M. Rodriguez-Artalejo, eds.), Lecture notes in Computer Science **510**, Springer-Verlag, Berlin, 1991, pp. 1–19.

[26] M. Gabbrielli, G. Levi, and M. C. Meo, *Observational Equivalences for Logic Programs*, Proc. Joint Int'l Conf. and Symposium on Logic Programming (K. Apt, ed.), MIT Press, Cambridge MA, 1992, Extended version to appear in Information and Computation, pp. 131–145.

[27] M. Gabbrielli and M. C. Meo, *Fixpoint Semantics for Partial Computed Answer Substitutions and Call Patterns*, Algebraic and Logic Programming, Proceedings of the Third International Conference (H. Kirchner and G. Levi, eds.), Lecture Notes in Computer Science **632**, Springer-Verlag, Berlin, 1992, pp. 84–99.

[28] H. Gaifman and E. Shapiro, *Fully abstract compositional semantics for logic programs*, Proc. Sixteenth Annual ACM Symp. on Principles of Programming Languages, ACM, 1989, pp. 134–142.

[29] R. Giacobazzi, *Semantic Aspects of Logic Program Analysis*, Ph.D. thesis, Dipartimento di Informatica, Università di Pisa, 1992.

[30] _____, *On the Collecting Semantics of Logic Programs*, Verification and Analysis of Logic Languages (F. S. de Boer and M. Gabbrielli, eds.), Proc. of the Post-Conference ICLP Workshop, 1994, pp. 159–174.

[31] J. Jaffar and J. L. Lassez, *Constraint Logic Programming*, Proc. Fourteenth Annual ACM Symp. on Principles of Programming Languages, ACM, 1987, pp. 111–119.

[32] G. Kreisel and J. L. Krivine, *Elements of Mathematical Logic (Model Theory)*, North-Holland, Amsterdam, 1967.

[33] G. Levi, *Models, Unfolding Rules and Fixpoint Semantics*, Proc. Fifth Int'l Conf. on Logic Programming (R. A. Kowalski and K. A. Bowen, eds.), MIT Press, Cambridge MA, 1988, pp. 1649–1665.

[34] G. Levi and P. Mancarella, *The Unfolding Semantics of Logic Programs*, Technical report tr-13/88, Dipartimento di Informatica, Università di Pisa, 1988.

[35] G. Levi and D. Ramundo, *A formalization of metaprogramming for real*, Proc. Tenth Int'l Conf. on Logic Programming (D. S. Warren, ed.), MIT Press, Cambridge MA, 1993, pp. 354–373.

[36] J. W. Lloyd, *Foundations of Logic Programming*, 2nd ed., Springer-Verlag, Berlin, 1987.

[37] J. W. Lloyd and J. C. Shepherdson, *Partial Evaluation in Logic Programming*, J. Logic Programming **11** (1991), 217–242.

[38] H. Rasiowa and R. Sikorski, *The mathematics of metamathematics*, North-Holland, Amsterdam, 1963.

[39] D. De Schreye and B. Martens, *A Sensible Least Herbrand Semantics for Untyped Vanilla Meta-Programming and its Extension to a Limited Form of Amalgamation*, Meta-Programming in Logic. Third International Workshop, META '92 (A. Pettorossi, ed.), Lecture notes in Computer Science **649**, Springer-Verlag, Berlin, 1993, pp. 192–204.

[40] P. J. Stuckey, *Constructive Negation for Constraint Logic Programming*, Proc. Sixth IEEESymp. on Logic in Computer Science, IEEE Computer Society Press, 1991, pp. 328–339.

[41] M. H. van Emden and R. A. Kowalski, *The semantics of predicate logic as a programming language*, J. Assoc. Comput. Mach. **23** (1976), no. 4, 733–742.

[42] L. Vieille, *Recursive query processing: the power of logic*, Theoret. Comput. Sci. **69** (1989), 1–53.

DIPARTIMENTO DI INFORMATICA, UNIVERSITÀ DI PISA, CORSO ITALIA 40, 56125 PISA, ITALY

e-mail address: COMINI@DI.UNIPI.IT

e-mail address: LEVI@DI.UNIPI.IT

The Logic of Commuting Equivalence Relations

David Finberg   Matteo Mainetti   Gian-Carlo Rota

Dedicated to the memory of R. Magari

1   Introduction

The lattice $L(V)$ of subspaces of a vector space V is in many ways similar to the lattice, or Boolean algebra, of subsets of a set S. This similarity has been the guiding principle of a great deal of work in both logic and combinatorics in this century. For a long time, however, the need has been felt for an intuitively graspable way of visualizing expressions in joins and meets of subspaces that might allow to learn how a polynomial in meets and joins of subspaces of a vector space may be simplified, in much the same way as the drawing of Venn diagrams simplifies the visualization of relations among sets.

The purpose of the present work is to describe one way of visualizing polynomials in joins and meets of subspaces of a vector space, and to develop the logical theory that goes with such visualization. In so doing, we are led to introduce a new system of natural deduction (in the sense of Gentzen) that applies not only to subspaces of a vector space, but to a class of lattices of frequent occurrence, that we have proposed to name **linear lattices.**

A linear lattice is a sublattice of the lattice of partitions of a set, with the property that the equivalence relations associated with any two partitions in the lattice commute, in the sense of composition of relations. Such lattices are of frequent occurrence: lattices of subspaces of a vector space, lattices of normal subgroups of a group, lattices of ideals of a ring are examples of linear lattices. Every linear lattice is also a modular lattice. This coincidence has misled logicians for at least the last hundred years. Modular lattices, which could be defined by one equation, seemed to be the next natural concept after Boolean

69

algebras. The fact is, however, that most, if not all, examples of modular lattices occurring in algebra and combinatorics enjoy the stronger property of being linear lattices. Unlike modular lattices, linear lattices cannot be defined by identities alone; as a matter of fact, there is to this day no simple way of axiomatizing linear lattices. This lack of an abstract definition is perhaps the reason why in the past the theory of linear lattices was subsumed into the theory of modular lattices.

In the present work, we show that although linear lattices do not have a simple axiomatic definition, they can nevertheless be characterized by a simple, elegant proof theory.

Such a proof theory is in several ways analogous to the classical Gentzen system of natural deduction [11] for the predicate calculus. If one views the lattice of subspaces of a vector space as a "quantum" analog of classical logic, then the present theory shows how reasoning with such a logic might actually work.

The system of natural deduction for linear lattices that is presented in the present work follows the lines of other natural deduction systems that were developed for the predicate calculus. In its simplest form, it provides a series of mechanical steps whereby, given an inequality $P \leq Q$ between two lattice polynomials P and Q, either the inequality is proved true for every linear lattice, or else a linear lattice in which the inequality does not hold is automatically constructed.

The presentation in the present paper is self contained, and requires no more knowledge than the definition of a lattice. No previous knowledge of commuting equivalence relations is required.

2   Synopsis

An equivalence relation R on a set B may be viewed either as a relation or as a partition. These two points of view, although strictly speaking identical, serve to visualize different features. Thus, relations may be composed in much the same way as functions. Equivalence relations that commute under composition can be explicitly characterized in terms of the blocks of the corresponding partitions, by a theorem due to Mme. Dubreil [7].

Commuting equivalence relations occur frequently. The typical example is found in geometry. If V is a vector space, and x is a subspace of V, then, by defining two vectors α and β of V to be equivalent when $\alpha - \beta \in x$, one associates to x an equivalence relation $R(x)$. If y is another subspace of V, then the equivalence relations $R(x)$ and $R(y)$ commute. Thus, the lattice $L(V)$ of subspaces of V is isomorphic to a lattice of commuting equivalence relations.

This representation of the lattice $L(V)$ is closely related to the geometric representation of vectors as points in projective space. Thus, expressions involving joins and meets of subspaces of a vector spaces may be translated into expressions in terms of joins and meets of partitions, whose associated equivalence relations commute.

The idea is to associate to every set of equivalence relations $\{R(a) : a \in A\}$ on a set B of vertices a graph: one simply joins by an edge labeled a two elements of B that are a-equivalent. If two vertices α and β are equivalent under two equivalence relations $R(a)$ and $R(b)$, then one draws two edges joining α and β, labeled a and b.

Now suppose we are given a graph whose edges are labeled by a or b. When can we say that this graph corresponds to two commuting equivalence relations? The answer (given below) is extremely simple. Besides an obvious property corresponding to the transitive property of equivalence, it turns out that the crucial property the graph must have is the following. If an edge labeled by the letter a has vertices α, β and if an edge labeled by b has vertices β, γ , then the graph must have a vertex δ, distinct from α, β and γ, and edges labeled b and a whose vertices are α, δ and δ, β, respectively.

The meet and join of the partitions associated with equivalence relations $R(a)$ and $R(b)$ are described graphically as follows. If α, β are vertices joined by two edges, labeled a, b, then add another edge joining the same vertices, labeled $a \wedge b$. If α, β and β, γ are vertices which are the endpoints of edges labeled a and b respectively, then add an edge labeled $a \vee b$, whose endpoints are the vertices α, γ. It is clear that, given any graph on the vertex set B with any set of edges labeled by the letters of an alphabet A, then by a process of saturation the graph can be embedded in a larger graph, in general infinite, which will be the graph of a linear lattice of equivalence relations on the set B. This process is a lattice-theoretic analog of the construction of a group defined by generators and relations.

Now suppose that we are given an inequality $P \leq Q$, where P and Q are lattice polynomials in the alphabet A, and in the symbols $\vee$ and $\wedge$. The preceding construction allows us to accomplish two objectives. First, it constructs a linear lattice in which the inequality is not true, if the inequality does not hold in all linear lattices. Second, it provides a proof of the inequality if the inequality is true. If an edge is given, labeled by a polynomial P, the correspondence between lattice operations and adding edges leads to add several edges labeled by each of the variables occurring in P. Now apply the process of "saturation" which adds all possible edges to this original graph, until a linear lattice is obtained. The inequality $P \leq Q$ will be true in this linear lattice if and only if to every edge joining, say, α and β, and bearing the polynomial P as a label, there is a corresponding edge joining the same α and β and labeled by the polynomial Q. This much is immediate from the definition of refinement of partition. What is more, the edge bearing the label Q has been obtained by "adding edges" in the process of saturation, in a finite number of steps. Each of these steps can be easily shown to correspond to a logically correct inference on commuting equivalence relations. Thus, the saturation process itself yields the proof of the inequality $P \leq Q$ if it is true; if the inequality is not true, then the saturation will have a pair of vertices joined by an edge labeled P but not by an edge labeled Q, thus the linear lattice obtained by the process of saturation will provide a counterexample to the inequality. Remarkably, a number of theorems of projective geometry can be proved by this method, as we show in the last section, for example, Desargues's theorem and a generalization to higher dimensions.

3   Commuting Equivalence Relations

Given a set S, a **relation** on S is a subset R of $S \times S$. A relation is sometimes identified with a bipartite graph, namely, with a graph on the set S of vertices whose edges are the pairs (α, β) which belong to R.

On the set $\mathcal{R}$ of relations on S, all Boolean operations among sets are defined. For example, $\cup$ and $\cap$ are the usual union and intersection; similarly, one defines the complement of a relation. The identity relation is the relation $I = \{(\alpha, \alpha) : \alpha \in S\}$. In addition, composition of relations is defined, which is the analog for relations of functional composition. If R and T are relations, define:

$$R \circ T = \{(\alpha, \beta) \in S \times S \mid \exists \gamma \in S \; s.t. \; (\alpha, \gamma) \in R, \; (\gamma, \beta) \in T\}.$$

Given a relation R, one defines the **inverse relation** as follows:

$$R^{-1} = \{(\alpha, \beta) | (\beta, \alpha) \in R\}.$$

It has been shown that this wealth of operations defined on the set of all relations on S does not suffice to give an algebraic characterization of the set of relations, and such an algebraic description remains to this day an open problem.

There is, however, a class of relations for which an extensive algebraic theory can be developed, and that is the class of commuting equivalence relations on a set S. We say that two relations R and T **commute** when $R \circ T = T \circ R$.

Recall that a relation R is an **equivalence relation** if it is reflexive, symmetric and transitive. In the notation introduced above, these three properties are expressed by $R \supseteq I$, $R^{-1} = R$, and $R \circ R \subseteq R$, where I is the identity relation. The equivalence classes of an equivalence relation form a **partition** of S.

Recall that a partition a of a set S is a family of non empty subsets of S, called blocks, which are pairwise disjoint and whose union is the set S. Clearly, every partition a of S defines a unique equivalence relation whose equivalence classes are the blocks of a. In other words, the notions of equivalence relation and partition are mathematically identical, though psychologically different.

We denote by R_a or $R(a)$ the equivalence relation associated to the partition a. We will often write $\alpha R_a \beta$ in place of $(\alpha, \beta) \in R_a$. The set of partitions of a set S, denoted by $\Pi(S)$, is endowed with the partial order of **refinement**: we say that $a \leq b$ when every block of a is contained in a block of b. The refinement partial order has a unique maximal element $\hat{1}$, namely, the partition having only one block, and a unique minimal element $\hat{0}$, namely, the partition for which every block has exactly one element.

One verifies that the partially ordered set $\Pi(S)$ is a lattice. Lattice meets and joins, denoted by $a \vee b$ and $a \wedge b$, can be described by using the equivalence relations $R(a)$ and $R(b)$ as follows.

Proposition 1
$$R(a \wedge b) = R(a) \cap R(b) \tag{1}$$

$$
\begin{aligned}
R(a \vee b) &= R_a \cup R_a \circ R_b \cup R_a \circ R_b \circ R_a \cup \cdots \cup \\
&\cup R_b \cup R_b \circ R_a \cup R_b \circ R_a \circ R_b \cup \cdots
\end{aligned}
\tag{2}
$$

Proof: (1) is immediate, while to prove (2) recall that $a \vee b$ is the smallest partition containing a and b. From the transitivity property of $R(a \vee b)$ one obtains the long expression (2).

$\square$

It can be shown that the number of iterations of compositions of relations as well as of unions in (2) cannot be bounded a priori. As a matter of fact, pairs of relations can be classified according to the minimum number of terms in it required to express the join of the corresponding partitions.

We are thereby led into the topic of the present work, namely, the study of pairs of equivalence relations for which the expression for the join has the minimum allowable number of terms. It turns out that pairs of equivalence relations in which the right side of (2) reduces to two terms are precisely commuting equivalence relations, as the following two theorems show.

Theorem 2 $R(a \vee b) = R_a \circ R_b$ *iff* $R_a \circ R_b = R_b \circ R_a$.

Proof: If $R_a \circ R_b = R_b \circ R_a$, then, by the transitive law, the right side in (2) simplifies to $R(a \vee b) = R_a \circ R_b$.

Conversely, assume $R(a \vee b) = R_a \circ R_b$. By (2) we infer at once $R_a \circ R_b \supseteq R_b \circ R_a$. Taking inverses, we infer

$$(R_a \circ R_b)^{-1} \supseteq (R_b \circ R_a)^{-1}$$

This in turn implies

$$R_b \circ R_a = R_b^{-1} \circ R_a^{-1} \supseteq R_a^{-1} \circ R_b^{-1} = R_a \circ R_b.$$

$\square$

Since $R(a \vee b)$ is the smallest equivalence relation containing $R_a \circ R_b$, the preceding theorem can be restated as follows:

Theorem 3 *Two equivalence relations R and T commute if and only if $R \circ T$ is an equivalence relation.*

$\square$

The most important example of commuting equivalence relations comes from information theory. Two equivalence relations R_a, R_b (or, equivalently, two partitions a and b) are said to be **independent** when, for any two blocks $A \in a$, $B \in b$, we have $A \cap B \neq \emptyset$. Independent relations commute, since $R(a \vee b) = R(\hat{1}) = R_b \circ R_a$.

We briefly digress on the information theoretic meaning of independence of two partitions. From the point of view of information theory, a partition of a set S can be viewed as a step in the search process for an unknown member σ of the set S. A partition determines the block in which the element σ is located. For example, one may imagine that the unknown element has a color, and the blocks of a partition would contain elements of the same color. A set of partitions of S constitutes a perfect search process whenever the meet of the partitions in the set is the partition $\hat{0}$. From this point of view, two partitions a and b are independent if knowledge of the block of a in which the unknown element is located yields no information whatsoever as to which block of b contains such an element.

It is tempting to infer that any two commuting equivalence relations are independent. Such a presumption is however false, as the following example indicates. If a is a partition of a set S, and if T is a subset of S, we write $a_{|T}$ to indicate the restriction of the partition a to the set T, that is, the partition whose blocks are the intersections of the blocks of a with the set T, whenever such an intersection is not empty.

Given partitions a_1 and a_2 on disjoint sets S_1, S_2 (respectively), we define the **disjoint sum** $a = a_1 + a_2$ to be the unique partition on $S_1 \cup S_2$ such that $a_{|S_1} = a_1$ and $a_{|S_2} = a_2$. The disjoint sum of any set of partitions defined on disjoint sets is similarly defined.

Clearly, if b_1 and b_2 are another pair of partitions on the same disjoint sets S_1, S_2, and if each of the pairs a_1 and b_1 as well as a_2 and b_2 are independent, then the disjoint sums $a = a_1 + a_2$ and $b = b_1 + b_2$ are not independent, but they commute.

Our objective is to show that this construction gives all pairs of commuting equivalence relations. To this end we require the following elementary lemmas, the first of which is just a restatement of theorem 2.

Lemma 4 *Two equivalence relations R_a and R_b on a set S commute if and only if, for any elements α, β in S such that $\alpha R(a \vee b)\beta$, there exist elements γ, δ in S such that*

$$\alpha R_a \gamma, \quad \gamma R_b \beta$$

$$\alpha R_b \delta, \quad \delta R_a \beta$$

$\square$

Lemma 5 *If R_a and R_b commute, and $a \vee b = \hat{1}$, then the equivalence relations R_a and R_b are independent.*

Proof: Let $A \in a$ and $B \in b$ be blocks. Choose elements $\alpha \in A$ and $\beta \in B$. Since $\hat{1} = a \vee b$, we have $\alpha R(a \vee b)\beta$, and since a and b commute, lemma 4 implies that there exists an element γ such that $\alpha R_a \gamma$ and $\gamma R_b \beta$; in other words $\gamma \in A$ and $\gamma \in B$. Thus, the blocks A and B meet.

$\square$

We may now proceed to the proof of the main structure theorem for commuting equivalence relations:

Theorem 6 (Dubreil-Jacotin) *Two equivalence relations R_a and R_b associated with partitions a and b commute if and only if for every block C of the partition $a \vee b$, the restrictions $a_{|C}$, $b_{|C}$ are independent partitions.*

Proof: Suppose the equivalence relations R_a and R_b commute. Then $R_{a|C}$ and $R_{b|C}$ commute too. Moreover, in the lattice $\Pi(C)$ of partitions of the block C, we have

$$a_{|C} \vee b_{|C} = (a \vee b)_{|C} = \hat{1}_C$$

by definition of the join of partitions, where $\hat{1}_C$ is the maximum element of the partition lattice $\Pi(C)$. By lemma 5, the equivalence relations R_a and R_b are independent.

The converse is obvious, as already remarked.

$\square$

It would be interesting to find an information theoretic interpretation of two commuting equivalence relations, which generalizes the interpretation of

independent equivalence relations given above. A problem which does not
seem to have been considered is the generalization of the notion of commuting
equivalence relations to more than two relations. One gleans from probability
theory that three partitions a, b, c are **3-independent** when for any blocks
$A \in a, B \in b, C \in c$, one has $A \cap B \cap C \neq \emptyset$. Let us then agree that a set of
three partitions are 3-commuting when they are disjoint sums of 3-independent
partitions. Can 3-commuting equivalence relations be characterized within
lattice theory ?

4   Natural Deduction for Linear Lattices

We shall describe a system of natural deduction whose intended models are all
linear lattices.

- **Variables.**

    Variables will be of two sorts. Variables of the first sort will range over
    a countable Roman alphabet (not capitalized) $A = (a, b, c, \ldots)$. Vari-
    ables of the second sort will range over a countable Greek alphabet (not
    capitalized) $B = (\alpha, \beta, \gamma, \ldots)$. We denote by Free($A$) the free lattice
    generated by the set A. An element of Free(L) will be called a **lattice
    polynomial** in the variables $a, b, \ldots$, and denoted by $P(a, b, c, \cdots)$.

- **Connectives.**

    There are three connectives: lattice joins $\vee$ and meets $\wedge$, which are
    binary connectives used in lattice polynomials, and a unary connective
    R.

- **Formation rules.**

    The formation rules for lattice polynomials are understood. We define
    an **equation** to be an expression of the form

    $$\alpha R(P)\beta,$$

    where α and β are any Greek letters, where P is any lattice polynomial.
    We define an **atomic equation** to be an expression of the form

    $$\alpha R(a)\beta,$$

    where a is any Roman letter.

- **Well formed formulas.**

    Any equation is a well formed formula.

    We denote by Γ, Δ, etc. sets of equations.

- **Models.**

  A **model** $\{L, f, g\}$ is a linear lattice L consisting of partitions of a set S, together with functions $f : B \to S$ and $g : A \to L$. It follows that a unique lattice homomorphism from Free(A) to L is defined. This homomorphism will also be denoted by g. An equation $\alpha R(P)\beta$ is said to **hold** in a given model, whenever $f(\alpha)R(g(P))f(\beta)$, that is, whenever the ordered pair $(f(\alpha), f(\beta))$ is an element of the relation $R(g(P))$ on the set S.

- **Validity.**

  A Pair (Γ, Δ) of sets of equations is said to be **valid** when every equation in Δ holds in every model in which every equation in Γ holds.

- **Deduction rules.**

  A **proof** is a sequence of sets of equations $\Gamma_1, \Gamma_2, \ldots, \Gamma_n$ such that

  $$\frac{\Gamma_i}{\Gamma_{i+1}}$$

  is an instance of a linear deduction rule (v. below). In such circumstance, we write

  $$\frac{\Gamma_1}{\Gamma_n},$$

  to signify that the set of sentences Γ_n can be proved from the set of sentences Γ_1.

  The set Γ_1 is the **premiss** of the deduction rule, and the set Γ_n is the **conclusion**.

- **Provability.**

  If Γ_1 and Γ_n are sets of lattice polynomials, we say that Γ_n is **provable** from Γ_1 if there exists a proof $\Gamma_1, \Gamma_2, \ldots, \Gamma_n$.

- **Linear deduction rules for the theory of linear lattices.**

  1. **Reflexivity.**

     $$\frac{\Gamma}{\Gamma, \, \alpha R(P)\alpha}$$

     where P is any lattice polynomial, and where α is an arbitrary Greek variable.

2. **Transitivity.**

$$\frac{\Gamma,\ \alpha R(P)\beta,\ \beta R(P)\gamma}{\Gamma,\ \alpha R(P)\beta,\ \beta R(P)\gamma,\ \alpha R(P)\gamma}$$

where P is any lattice polynomial, and where α , β and γ are arbitrary Greek variables.

3. **Splitting Meets.**

$$\frac{\Gamma,\ \alpha R(P \wedge Q)\beta}{\Gamma,\ \alpha R(P \wedge Q)\beta,\ \alpha R(P)\beta,\ \alpha R(Q)\beta}$$

where α and β are Greek variables, and where P and Q are arbitrary lattice polynomials.

4. **Combining Meets.**

$$\frac{\Gamma,\ \alpha R(P)\beta,\ \alpha R(Q)\beta}{\Gamma,\ \alpha R(P)\beta,\ \alpha R(Q)\beta,\ \alpha R(P \wedge Q)\beta}$$

where α and β are Greek variables, and where P and Q are arbitrary lattice polynomials.

5. **Splitting Joins.**

$$\frac{\Gamma,\ \alpha R(P \vee Q)\beta}{\Gamma,\ \alpha R(P \vee Q)\beta,\ \alpha R(P)\gamma,\ \gamma R(Q)\beta}$$

with the same provisos as in the preceding rule for linear deduction, and with the additional proviso that γ is a new symbol, that is, the Greek letter γ does not appear in Γ and is unequal to α, β.

6. **Combining Joins.**

$$\frac{\Gamma,\ \alpha R(P)\gamma,\ \gamma R(Q)\beta}{\Gamma,\ \alpha R(P)\gamma,\ \gamma R(Q)\beta,\ \alpha R(P \vee Q)\beta}$$

where α and β are Greek variables, and P and Q are arbitrary lattice polynomials.

7. **Commutativity.**

$$\frac{\Gamma,\ \alpha R(P)\gamma,\ \gamma R(Q)\beta}{\Gamma,\ \alpha R(P)\gamma,\ \gamma R(Q)\beta,\ \alpha R(Q)\delta,\ \delta R(P)\beta}\text{'}$$

where δ is again a new symbol, it does not appear in Γ, and is unequal to α, β, γ.

8. **Symmetry.**

$$\frac{\Gamma,\ \alpha R(P)\beta}{\Gamma,\ \alpha R(P)\beta,\ \beta R(P)\alpha}$$

where P is any lattice polynomial, and where α and β are arbitrary Greek variables.

9. **Weakening.**

$$\frac{\Gamma, \Delta}{\Gamma}$$

that means that any subset of a provable set is provable.

Theorem 7 (Soundness) *If*

$$\frac{\Gamma}{\Delta},$$

that is, if (Γ, Δ) *is provable, then the set of equations* Δ *holds in every linear lattice in which the set* Γ *holds.*

Proof of theorem 7 is clear and will not be provided. $\square$

By way of example of the use of linear deduction, we give a proof of the fact that every linear lattice is a modular lattice. Recall that a lattice is said to be **modular** when it satisfies the inequality

$$a \wedge (b \vee (a \wedge c)) \leq (a \wedge b) \vee (a \wedge c) \text{ for any } a, b, c \in L. \tag{3}$$

Proposition 8 *Every linear lattice is modular.*

Proof: In what follows, bear in mind the linear deduction rules **1 – 8** of previous pages.

Using **3**,

$$\frac{\alpha R(a \wedge (b \vee (a \wedge c)))\beta}{\alpha R(a \wedge (b \vee (a \wedge c)))\beta,\ \alpha R_a \beta,\ \alpha R(b \vee (a \wedge c))\beta}. \tag{4}$$

Using **5**,

$$\frac{\alpha R(b \vee (a \wedge c))\beta}{\alpha R(b \vee (a \wedge c))\beta,\ \alpha R_b \gamma,\ \gamma R(a \wedge c)\beta}. \tag{5}$$

Using **3**,

$$\frac{\gamma R(a \wedge c)\beta}{\gamma R(a \wedge c)\beta,\ \gamma R_a \beta,\ \gamma R_c \beta}. \tag{6}$$

Using **8**,

$$\frac{\gamma R_a \beta}{\gamma R_a \beta, \ \beta R_a \gamma}. \tag{7}$$

From (4) and (7), using **2**

$$\frac{\alpha R_a \beta, \ \beta R_a \gamma}{\alpha R_a \beta, \ \beta R_a \gamma, \ \alpha R_a \gamma}. \tag{8}$$

From (5) and (8), using **4**,

$$\frac{\alpha R_b \gamma, \ \alpha R_a \gamma}{\alpha R_b \gamma, \ \alpha R_a \gamma, \ \alpha R(a \wedge b)\gamma}. \tag{9}$$

From (9) and (5), using **6**,

$$\frac{\alpha R(a \wedge b)\gamma, \ \gamma R(b \wedge c)\beta}{\alpha R(a \wedge b)\gamma, \ \gamma R(b \wedge c)\beta, \ \alpha R((a \wedge b) \vee (b \wedge c))\beta}. \tag{10}$$

Therefore we have inferred

$$\frac{\alpha R(a \wedge (b \vee (a \wedge c)))\beta}{\alpha R((a \wedge b) \vee (b \wedge c))\beta}.$$

$\square$

As a matter of fact, the inequality

$$a \wedge (b \vee (a \wedge c)) \geq (a \wedge b) \vee (a \wedge c)$$

holds as well, since it holds in any lattice.

5 Linear Lattices Generated by Equations

In this section, we introduce the notion of a linear lattice generated by a set of equations. Thus, let Γ be a set of equations. The **graph** $Graph(\Gamma)$ of the set Γ is defined as follows. If the equation $\alpha R(P)\beta$ belongs to the set Γ, then the graph $Graph(\Gamma)$ has an edge labeled by the lattice polynomial P, whose adjacent vertices are the elements α and β.

We define the **saturation** of the graph $Graph(\Gamma)$ in the following steps. We define an infinite sequence of graphs $G_0, G_1, \ldots$, as follows. Set $G_0 = Graph(\Gamma)$. Having defined $G_n(P)$, we construct $G_{n+1}(P)$ by applying to $G_n(P)$ the following operations in the given order.

1. **Reflexive :** To any vertex α in G_n add loops around α , one with each of the labels in any edge of G_n.

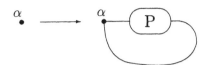

2. **Transitive :** If $\alpha R(P)\gamma$ and $\gamma R(P)\beta$ are edges of G_n, connect α and β by an edge labeled P.

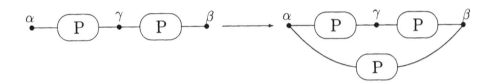

3. **Splitting meets :** For every edge having vertices α, β labeled by $P \wedge Q$ add two new edges with vertices α, β, labeled P and Q.

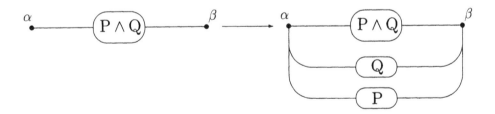

4. **Combining meets :** If E is any multiset of cardinality at most n, of edges in G_n whose endpoints are vertices α and β and whose labels are lattice polynomials $P, Q, \ldots, R$, add a new edge labeled $P \wedge Q \wedge \ldots \wedge R$.

5. **Splitting joins :** For every edge with vertices α, β labeled by $P \vee Q$, add two new edges with endpoints α, γ and γ, β, labeled P and Q, respectively, where γ is a Greek letter not appearing among the vertices already present.

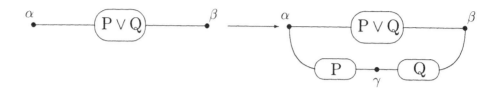

6. **Combining joins :** Given an ordered sequence of edges of cardinality at most n, whose vertices are $\alpha, \gamma, \gamma, \delta, \ldots, \rho, \sigma, \sigma, \beta$ whose labels are the lattice polynomials $P, Q, \ldots, S, T$, add an edge whose endpoints are α, β, labeled by the polynomial $P \vee Q \vee \ldots \vee S \vee T$.

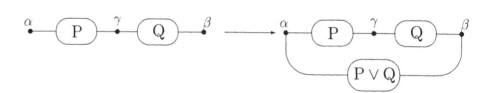

7. **Commutativity :** Given two edges whose vertices are α, γ and γ, β and whose labels are polynomials P and Q, add a new vertex, say δ , not appearing among the vertices already present, together with edges having endpoints α, δ and δ, β, and labeled Q and P, respectively.

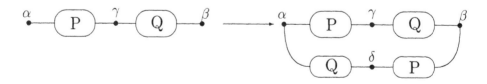

The sequence of graphs thus constructed has the property that $G_0 \subseteq G_1 \subseteq G_2 \subseteq \ldots$. We denote by $Sat(\Gamma)$ the union of these graphs. The graph $Sat(\Gamma)$ is called the **saturation** of the set of atomic equations Γ.

We note

Proposition 9 *If G_0 is a finite graph, then every G_n is also a finite graph.*

The preceding construction yields at once the following important result.

Theorem 10 (Saturation) *On the set of vertices of the saturation $Sat(\Gamma)$, define an equivalence relation $R(P)$ for every polynomial P by setting $\alpha R(P)\beta$ whenever the vertices α and β are connected by an edge labeled P. The image of $Free(A)$ under the lattice homomorphism $P \to R(P)$ is a linear lattice.*

We call such a linear lattice the **linear lattice generated by the set Γ of equalities**, and denote it by $Lin(\Gamma)$

We are now ready to prove a completeness theorem for the natural deduction system described in the preceding section.

Theorem 11 (Completeness) *If a pair (Γ, Δ) of finite sets of equations such that all variables (Roman and Greek) occurring in Δ also occur in Γ is valid, then it is provable.*

Proof: Suppose that the pair (Γ, Δ) is valid. We want to show that

$$\frac{\Gamma}{\Delta},$$

that is, that there exists a sequence of deductions $\Gamma_1, \Gamma_2, \ldots, \Gamma_n$ where $\Gamma_1 = \Gamma$ and $\Gamma_n \supseteq \Delta$.

Since (Γ, Δ) holds in every linear lattice, it holds in particular in the linear lattice $Lin(\Gamma)$. Thus, each of the equations in Δ hold in $Lin(\Gamma)$. It follows that in the sequence of graphs defining $Sat(\Gamma)$ there exists one graph, say G_n, which contains all equations in Δ. Let Γ_i be the set of equations corresponding to the edges of G_i. The construction of $Sat(\Gamma)$ shows that

$$\frac{\Gamma_i}{\Gamma_{i+1}},$$

since each of the operations by which Γ_{i+1} is constructed corresponds to the linear deduction bearing the same name. Thus, the sequence $\Gamma_1, \Gamma_2, \ldots, \Gamma_n$ provides the proof of a set of equations Γ_n of which Δ is a subset.

$\square$

We wish to apply the preceding completeness theorem to the proof of lattice inequalities holding in all linear lattices. To this end, we have

Theorem 12 *Let P and Q be lattice polynomials. The lattice inequality $P \leq Q$ holds in all linear lattices if and only if*

$$\frac{\{P\}}{\{Q\}},$$

in other words, the inequality $P \leq Q$ holds in all linear lattices if and only if the equation $\alpha R(Q)\beta$ is provable from the equation $\alpha R(P)\beta$.

□

6   Inequalities

For most applications a stronger version of the completeness theorem will be
required. In this section we extend the completeness theorem to include Horn
sentences .

Horn Sentences

A Horn Sentence is an expression of the form

$$P_1 \leq Q_1, \ldots, P_n \leq Q_n \vdash P \leq Q.$$

We say that a Horn sentence is **valid** in the theory of linear lattices when
every linear lattice satisfying inequalities $P_1 \leq Q_1, \ldots, P_n \leq Q_n$ also satisfies
the inequality $P \leq Q$. A single inequality is a special case of a Horn sentence,
one with no assumptions.

In order to define a notion of proof for Horn sentence, an additional linear
deduction has to be added to the list already given, to wit:

8. **Conditional Implication**

$$\frac{\alpha R(P_i)\beta}{\alpha R(Q_i)\beta}$$

  for $i \in I$.

Similarly, we extend the notion of saturation to include conditional implication, by adding an additional operation to the operations used in constructing
the saturation of a set of sentences, to wit:

9. **Conditional Implication:** For every edge labeled P_i with vertices α, β,
   add a new edge labeled Q_i, with the same vertices α and β.

Thus, just as before, G_n will be a finite graph for all n, and there is a one to one correspondence between deduction rules and operations in G_n. The union of this infinite increasing sequence of graphs gives a graph which may be called the **saturation** of the set of equations Γ **relative to the (finite) set of implications** $\{P_1 \leq Q_1, \ldots, P_n \leq Q_n\}$. Again, we denote this graph by $Sat(\Gamma; \{P_1 \leq Q_1, \ldots, P_n \leq Q_n\})$. As in the previous completeness theorem, the graph $Sat(\Gamma; \{P_1 \leq Q_1, \ldots, P_n \leq Q_n\})$ defines a linear lattice $Lin(\Gamma; \{P_1 \leq Q_1, \ldots, P_n \leq Q_n\})$.

The linear lattice $Lin(\Gamma; \{P_1 \leq Q_1, \ldots, P_n \leq Q_n\})$ is the linear lattice generated by the set of equations Γ and by the set of implications $\{P_1 \leq Q_1, \ldots, P_n \leq Q_n\}$.

Theorem 13 (Completeness) *The Horn sentence*

$$P_1 \leq Q_1, \ldots, P_n \leq Q_n \vdash P \leq Q$$

is provable in the theory of linear lattices by linear deductions 1-9 if and only if it is valid in the theory of linear lattices.

The proof is identical to that of the preceding completeness theorem

While the condition of finiteness of Horn sentences is not necessary, it simplifies the proof, for then the G_n's are finite. If the assumptions are infinite more care needs to be taken in the construction, but the proof can be extended to this case.

7 Examples

We are now ready to show that some theorems of projective geometry can be proved by the linear system of natural deduction developed above.

Proposition 14 *Every linear lattice is modular.*

Proof: (compare with proof of proposition (8))
What follows is the proof of the inequality

$$(a \wedge b) \vee (a \wedge c) \leq a \wedge (b \vee (a \wedge c)) :$$

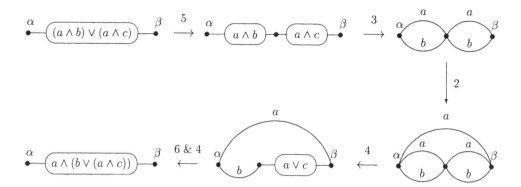

Here $5, 3, 2, 4, 6$ are the operations on the graphs corresponding to splitting joins, splitting meets, transitivity (on a), combining meets and combining joins, respectively. The possibility of taking subgraphs is always understood.

The opposite inequality,

$$a \wedge (b \vee (a \wedge c)) \leq (a \wedge b) \vee (a \wedge c)$$

is proved by the following diagram:

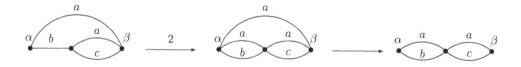

where the splitting joins and meets at the beginning (and their combining at the end) have been omitted.

□

What follows is a classical result in which Horn sentences are used.

Theorem 15 (Desargues) *Given a, b, c, a', b', c' in any linear lattice, then*

$$(a \vee a') \wedge (b \vee b') \leq c \vee c' \;\vdash\; (a \vee b) \wedge (a' \vee b') \leq ((a \vee c) \wedge (a' \vee c')) \vee ((c \vee b) \wedge (c' \vee b'))$$

Proof: See figure 1.

□

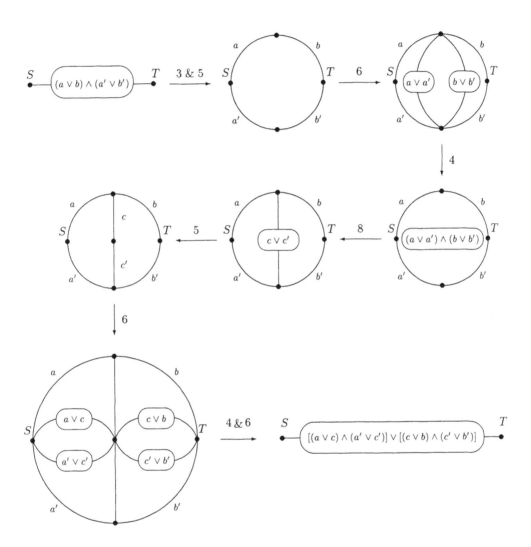

Figure 1: Proof of Desargues' theorem.

Probably most readers are better acquainted with geometric Desargues' theorem:

Theorem 16 *Given points* a, b, c, a', b', c' *in a projective plane* P, *if the three lines* aa', bb', cc' *meet in one point then the three points* $ac \cap a'c'$, $ab \cap a'b'$, $bc \cap b'c'$ *lie on a line.*

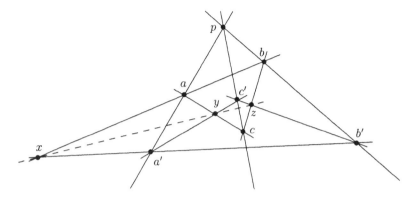

Figure 2: Geometric version of Desargues' theorem.

In the case of Desargues' theorem, statement 15 can be shown to be equivalent to the following, where we have a single inequality (often called Arguesian inequality), without implications.

Theorem 17 (Haiman) *In every linear lattice the following inequality holds:*

$$c \wedge ((c_1' \wedge c_2') \vee \{[b_2' \wedge b_1') \vee (b_2 \wedge b_1)] \wedge [a_2' \wedge a_1') \vee (a_2 \wedge a_1)]\}) \leq$$

$$a_1 \vee ((a_1' \vee c_2') \wedge \{[(b_1' \vee a_2') \wedge (b_1 \vee a_2)] \vee [(c_1' \vee b_2') \wedge (c \vee b_2)]\}).$$

The Arguesian inequality can be generalized to higher dimensions. In case of dimension 3 we have the following.

Theorem 18 *In every linear lattice the following inequality holds:*

$$d \wedge ((d_1' \wedge d_2') \vee \{[c_2' \wedge c_1') \vee (c_2 \wedge c_1)] \wedge [(b_2' \wedge b_1') \vee (b_2 \wedge b_1)] \wedge [a_2' \wedge a_1') \vee (a_2 \wedge a_1)]\}) \leq$$

$$a_1 \vee ((a_1' \vee d_2') \wedge \{[(b_1' \vee a_2') \wedge (b_1 \vee a_2)] \vee [(c_1' \vee b_2') \wedge (c_1 \vee b_2)] \vee [(d_1' \vee c_2') \wedge (d \vee c_2)]\}).$$

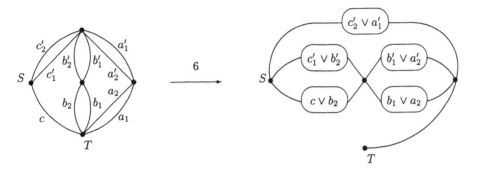

Figure 3: Proof of the Arguesian inequality.

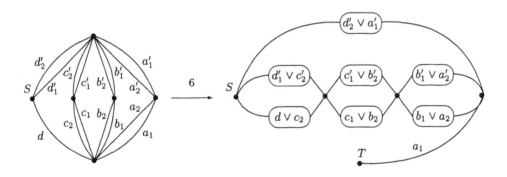

Figure 4: Proof of Arguesian inequality of dimension 3.

A proof is shown in figure 4, where the first splitting and the final combining have been omitted.

Figure 4 should shed light on theorem (18); moreover, it should be clear how to generalize the Arguesian inequality to higher dimensions, by adding four more variables f_1, f_2, f_1', f_2' and so forth.

The geometric version of the inequality in theorem (18) can be obtained as follows. In (18), set $a_1 = a_2 = a$, $a_1' = a_2' = a'$, and do the same for all the other variables b, c, d. Interpret them as points in a 3-dimensional projective space. The left side of (18) becomes

$$d \wedge (d' \vee \{[c' \vee c] \wedge [b' \vee b] \wedge [a' \vee a]\}).$$

Suppose the three lines cc', bb', aa' meet at one point, p. If we suppose also that dd' passes through p, joining p with d' and meeting with d gives d (see figure 5).

The right side of 18) reads

$$a \vee ((a' \vee d') \wedge \{[(b' \vee a') \wedge (b \vee a)] \vee [(c' \vee b') \wedge (c \vee b)] \vee [(d' \vee c') \wedge (d \vee c)]\}).$$

Call x, y, z the three points given by clauses between square brackets, namely

$$x = (b' \vee a') \wedge (b \vee a) \quad y = (c' \vee b') \wedge (c \vee b) \quad z = (d' \vee c') \wedge (d \vee c).$$

The line $a'd'$ meets the plane xyz in a point called w. Joining now with a gives a line that goes through d (recall that "$\leq$" in geometric linear lattices means "contained"). This means that

$$w = (d' \vee a') \wedge (d \vee a),$$

namely:

Theorem 19 *Given points $a, a', b, b', c, c', d, d'$ in a three dimensional projective space, if the four lines aa', bb', cc', dd' meet at one point then the four points $ab \cap a'b', bc \cap b'c', cd \cap c'd', da \cap d'a'$ lie on the same plane.*

$\square$

Previous examples were shown by Haiman in [14], while next theorem, due to Michael Hawrylycz [17], has been proved by linear lattice methods by one of the authors:

Theorem 20 (Hawrylycz) *Given points a, a', b, b', c, c' in a projective plane, than the points*

$$p_1 = b \quad p_2 = a'b' \wedge cc' \quad p_3 = aa' \wedge b'c' \tag{11}$$

lie on the same line if and only if the lines

$$l_1 = (ab \wedge a'b') \vee c \quad l_2 = (bc \wedge b'c') \vee a \quad l_3 = a'c' \tag{12}$$

meet at one point.

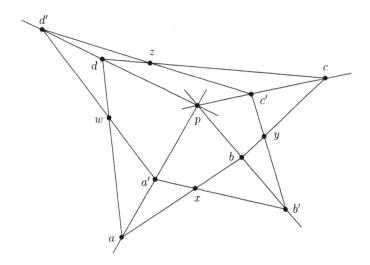

Figure 5: Geometric version of Arguesian identity of dimension 3.

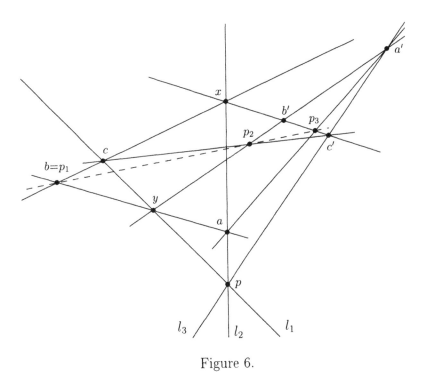

Figure 6.

Proof: This theorem is autodual under the point-line automorphism of the linear lattice of projective plane, therefore we just need to prove one implication, that will be

$$\text{"points lie on a line} \implies \text{lines meet at one point."} \tag{13}$$

To that goal we begin with looking for a lattice inequality that has (13) as a possible interpretation. Using notations in (11) and (12), the lattice inequality

$$(p_3 \vee p_2) \wedge p_1 \leq \{[(l_3 \wedge l_1) \vee a] \wedge b'c'\} \vee c \tag{14}$$

is a suitable one. In fact the left hand side, computed on a projective plane, gives the point $b = p_1$ if p_1, p_2, p_3 lie on a line, and the empty set otherwise. The right hand side gives a line through b (in fact the line bc) if and only if the lines l_1, l_2, l_3 meet at one point p. See figure 6 for a geometric insight. For a proof of inequality (14) refer to figure 7 , where $e = (ab \wedge a'b') \vee c$, $f = a'c'$, $g = e \wedge f = [(ab \wedge a'b') \vee c] \wedge a'c'$, $h = g \vee a$, $i = g \wedge b'c'$ and finally $j = i \vee c = [(\{a'c' \wedge [(ab \wedge a'b') \vee c]\} \vee a) \wedge b'c'] \vee c$, as wanted.

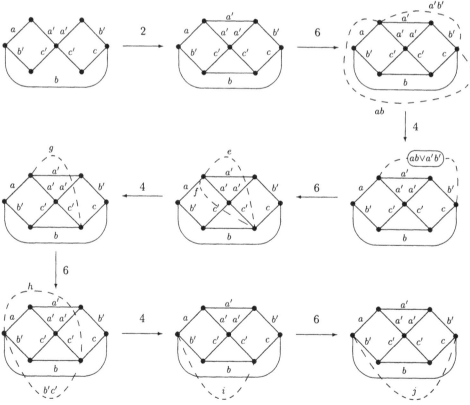

Figure 7: Proof of inequality (14).

8 Historical

The first systematic study of lattices of commutative equivalence relations is to be found in M.-L. Dubreil-Jacotin's thesis [7]. The structure theorem for pairs of commuting equivalence relations (6) is to be found in this thesis. It has been reworked and rediscovered several times since. In her thesis, as well as in her joint work with P. Dubreil, commuting equivalence relations are called "associable." In this thesis for the first time, various results that had been proved for groups are shown to be valid more generally for what we call linear lattices. Mme. Dubreil's work was not read, and is seldom quoted in the literature, perhaps because at the time modular lattices, which can be defined by a simple identity, were the center of attention in the then recently developed lattice theory. This thesis also contains the glimmerings of construction of free linear lattices, a problem that is open to this day. Birkhoff, in the third edition of his "Lattice Theory" [2, p. 69], does remark that "most" modular lattices are lattices of commuting equivalence relations. The same fact was realized by the universal algebraists, who make extensive studies of algebras with lattices of commuting congruences (for a recent survey, see the book [4]).

After Mme. Dubreil's work, two incisive papers of Bjarni Jonsson made decisive contributions to the theory of equivalence relations [24, 25]. In the first, he defined and investigated pairs of equivalence relations which are now known as being of "Jonsson type n." In particular, pairs of equivalence relations of Jonsson type 1 are commuting equivalence relations. In the second, Jonsson characterized linear lattices (also known at lattices of type one, following Jonsson's work) by an infinite sequence of implications. The conjecture remained open whether linear lattices could be characterized by the identity that is a latticial rendering of Desargues's theorem, or the generalizations of Desargues's theorem discovered by Haiman. This conjecture was disproved by Haiman in a striking paper [16]. At present, the only characterization of linear lattices is still Jonsson's, which to be sure has been simplified by several authors. The present work is independent of Jonsson's results. First example of proof theory for lattice of commuting equivalence relations is due to G. Hutchinson, in [22]. In fact, Hutchinson develops a proof they for lattices of equivalence relations of Jonsson type n, for any n.

The relationship between inequalities in linear lattices and properties of series- parallel graphs was first proposed by one of the present authors, who conjectured that an inequality would hold in a linear lattice whenever a suitable homomorphism between the series-parallel graphs associated with such inequalities could be found. Rota did not prove this conjecture, and proposed it in 1983 to Elvira Ramos. Ramos proved Proposition 2.1 in [14] (which Haiman attributes to Rota), but was not able to settle the conjecture either.

Rota proposed the same problem to Haiman, who solved it in his doctoral dissertation [13]. The present treatment is a simplified presentation of the work of Haiman. We leave it to the reader who may compare the two treatments to decide in which way the present work is simpler. We stress, however, that we have not dealt with Haiman's normal form theorem, which is the main step forward so far in establishing whether the word problem for linear lattices is solvable. In the present work we have adopted the standard notation of proof theory. We believe that this notation renders the main result, theorem 11 more transparent. There is a strong connection between the study of inequalities in linear lattices and identities holding in the double algebra of Barnabei, Brini and Rota [1]. This connection has recently been made explicit in the thesis of Michael Hawrylycz [17].

References

[1] M. Barnabei, A. Bini, and G.-C. Rota, *On the exterior calculus of invariant theory*, J. Algebra **96** (1985), no. 1, 120–160.

[2] G. Birkhoff, *Lattice theory*, third ed., AMS Colloqium Publications, vol. 25, Providence R.I., 1967.

[3] T. Brylawski, *A combinatorial model for series-parallel networks*, Trans. Amer. Math. Soc. **154** (1971), 1–22.

[4] S. Burris and H. P. Sankappanavar, *A course in universal algebra*, Springer-Verlag, New York, 1981.

[5] A. Church, *Introduction to mathematical logic*, vol. I, Princeton University Press, Princeton, N.J., 1956.

[6] P. Crawley and R. P. Dilworth, *Algebraic theory of lattices*, Prentice Hall, Eglewood Cliffs, N.J., 1973.

[7] P. Dubreil and M.-L. Dubreil-Jacotin, *Théorie algébrique des relations d'équivalence*, J. de Mathématique **18** (1939), 63–95.

[8] R. J. Duffin, *Topology of series-parallel networks*, J. Math. Anal. Appl. **10** (1965), 303–318.

[9] I. Gel'fand and V. Ponomarev, *Free modular lattices and their representations*, Russian Math. Surveys **29** (1974), 1–56.

[10] ———, *Lattices, representations, and algebras connected with them*, Russian Math. Surveys **31 and 32** (1977), 67–85,91–114.

[11] G. Gentzen, *Untersuchungen uber das logische schliessen*, Math. Z. **39** (1934), 176–210.

[12] G. Grätzer, *General lattice theory*, Berkhauser-Verlag, Basel, 1978.

[13] M. Haiman, *The theory of linear lattices*, Ph.D. thesis, Massachusetts Institute of Technology, 1984.

[14] _____, *Proof theory for linear lattices*, Adv. in Math. **58** (1985), no. 3, 209–242.

[15] _____, *Two notes on the Arguesian identity*, Algebra Universalis **21** (1985), 167–171.

[16] _____, *Arguesian lattices which are not type-1*, Algebra Universalis **28** (1991), no. 1, 128–137.

[17] M. J. Hawrylycz, *Geometric identities in invariant theory*, Ph.D. thesis, Massachusetts Institute of Technology, 1994.

[18] Ch. Herrmann, *On the word problem for the modular lattice with four free generators*, Math. Ann. **265** (1983), 513–527.

[19] A. Horn, *On sentences which are true if diret unions of algebras*, J. Symbolic Logic **16** (1951), 14–21.

[20] G. Hutchinson, *Recursively unsolvable word problems of modular lattice and diagram chasing*, J. Algebra **26** (1973), 385–399.

[21] _____, *Embedding and unsolvability theorems for modular lattices*, Algebra Universalis **7** (1977), 47–84.

[22] _____, *A complete logic for n-permutable congruence lattices*, Algebra Universalis **13** (1981), 206–224.

[23] B. Jónsson, *On the representation of lattices*, Math. Scand. **1** (1953), 193–206.

[24] _____, *Modular lattices and Desargues' thoerem*, Math. Scand. **2** (1954), 295–314.

[25] _____, *Representation of modular lattices and of relation algebras*, Trans. Amer. Math. Soc. **92** (1959), 449–464.

[26] _____, *The class of Arguesian lattices is self-dual*, Algebra Universalis **2** (1972), 396.

[27] L. Lipshitz, *The undecidability of the word problems of modular lattices and projective geometries*, Trans. Amer. Math. Soc. **193** (1974), 171–180.

[28] G. McNulty, *Fragments of first order logic. I: Universal Horn logic*, J. Symbolic Logic **42** (1976), 221–237.

[29] J. S. Mönting, *Cut elimination and word problems for varieties of lattices*, Algebra Universalis **12** (1981), 290–321.

[30] O. Ore, *Theory of equivalence relations*, Duke Math. J. **9** (1942), 573–627.

[31] D. Pedoe, *Geometry, a comprehensive course*, Dover, Mineola N.Y., 1988.

[32] D. Prawitz, *Natural deduction: A proof-theoretical study*, Stockholm Studies in Philosophy, vol. 3, Almqvist and Wiksells, Uppsala, 1965.

[33] R. Smullyan, *First-order logic*, Springer-Verlag, New York, 1968.

DEPARTMENT OF MATHEMATICS, M.I.T., 77 MASSACHUSETTS AVENUE, CAM-
BRIDGE, MA 02139-4307

DEPARTMENT OF MATHEMATICS, M.I.T., 77 MASSACHUSETTS AVENUE, CAM-
BRIDGE, MA 02139-4307
    matteo@mit.edu
DEPARTMENT OF MATHEMATICS, M.I.T., 77 MASSACHUSETTS AVENUE, CAM-
BRIDGE, MA 02139-4307
    rota@mit.edu

Proof-Nets : the parallel syntax for proof-theory

Jean-Yves Girard

Abstract

The paper is mainly concerned with the extension of proof-nets to additives, for which the best known solution is presented. It proposes two cut-elimination procedures, the *lazy* one being in linear time. The solution is shown to be compatible with quantifiers, and the structural rules of exponentials are also accommodated.

Traditional proof-theory deals with cut-elimination ; these results are usually obtained by means of sequent calculi, with the consequence that 75% of a cut-elimination proof is devoted to endless commutations of rules. It is hard to be happy with this, mainly because :

▶ the structure of the proof is blurred by all these cases ;

▶ whole forests have been destroyed in order to print the same routine lemmas ;

▶ this is not extremely elegant.

However old-fashioned proof-theory, which is concerned with the ritual question : "is-that-theory-consistent?" never really cared. The situation changed when subtle algorithmic aspects of cut-elimination became prominent : typically the determinism of cut-elimination, its actual complexity, its implementation cannot be handled in terms of sequent calculus without paying a heavy price. Natural deduction could easily fix the main drawbacks of cut-elimination, but this improvement was limited to the negative fragment of intuitionistic logic.

The situation changed in 1986 with the invention of linear logic : *proof-nets* were introduced in [3] as a new kind of syntax for linear logic, in order to cope with the problems arising from the intrinsic parallelism of linear sequent calculus. 9 years later the technology is perfectly efficient and proof-nets are now available for full linear logic [1]. Using implicit translations, proof-nets are also

---

1. Except the additive neutrals.

available for classical and intuitionistic logics, i.e. for extant logical systems. The essential result of [3] was a *sequentialization* theorem showing the equivalence of the new syntax with the traditional one. However this original result was restricted to the multiplicative fragment **MLL**, non-multiplicative features being accommodated by *boxes*, i.e. sequents under disguise. Later on, the original sequentialization proof was extended to quantifiers in [5, 7] ; the second version, the only satisfactory one, made a significant use of simplifications of the method of [3] discovered by Danos & Regnier, [1]. The problem of the extension to full linear logic, and especially to the additives (let us say to the fragment **MALL** of multiplicative-additive linear logic) has remained open for years. In fact two distinct problems must be solved :

▸ *To find the right notion of proof-net* : the notion was found quickly -say in the beginning of 1987- and since that time never really changed. The problem is to cope with the &-rule of sequent calculus, for which a superimposition of two proof-nets must be made ; by introducing for each &-link a boolean variable which

distinguishes between the two *slices* of the superimposition, one eventually gets a notion of net in which formulas and links are *weighted* by boolean polynomials. Although this notion remains the only serious candidate, it is to be noted that it is far from being absolutely satisfactory : this is because the question of determining the identity between two formulas (or two links) in two different proof-nets cannot receive a satisfactory answer, especially out of **MALL**.

▸ *To prove sequentialization* : the case of quantifiers was considered as a preliminary case, but the solution found in [5] was too acrobatic to be extended. This solution was improved in [7] by means of a new kind of switchings, determined by certain *formal* dependencies. The solution immediately extends to additives, provided certain *weight* dependencies are forbidden (this is the *dependency condition*, which requires all weights to be monomials). Since 1990 we have been trying to get rid of this technical limitation... and we must admit that this was a stupid attitude : is spite of their limitations our proof-nets obtained were extremely efficient (elimination of boxes is always a big simplification), whereas getting rid of the dependency restriction would of course make the proof-nets more intrinsic in some cases, but would still not make them absolutely satisfactory. This is the reason why we eventually decided to publish our partial results.

We shall first develop the present version of additive proof-nets, which is enough for most applications, together with a restricted (*lazy*) cut-elimination.

In an appendix we shall consider possible improvements and the general cut-elimination procedure.

Of course, proof-nets are not intended for multiplicative and additives only. In fact we shall devote some sections to the (unproblematic) extension to quantifiers, and to the structural rules in the exponential case (here with limited problems with weakening). Therefore we get rid of all boxes, but exponential boxes. This is of course a tremendous improvement over traditional sequent calculus.

1 Proof-Nets for MALL

1.1 Proof-structures

Definition 1

A link L is an expression

$$\frac{P_1, \ldots, P_n}{Q_1, \ldots, Q_m} L$$

involving n formulas (the premises of L) $P_1, \ldots, P_n$ and m formulas (the conclusions of L) $Q_1, \ldots, Q_m$

ID $-links$	:	0 premise		2 conclusions :	$A, A^\perp$
CUT $-links$	:	2 premises :	$A, A^\perp$	0 conclusion	
$\otimes$ $-links$	:	2 premises :	A, B	1 conclusion :	$A \otimes B$
$\otimes$ $-links$	:	2 premises :	A, B	1 conclusion :	$A \otimes B$
$\oplus_1$ $-links$	:	1 premise :	A	1 conclusion :	$A \oplus B$
$\oplus_2$ $-links$	:	1 premise :	B	1 conclusion :	$A \oplus B$
$\&$ $-links$	:	2 premises :	A, B	1 conclusion :	$A \& B$

The premises of $\otimes, \otimes, \&$-links are ordered : this means that we can distinguish a left premise (here A) and a right premise (here B). On the other hand the premises of a CUT-link and the conclusions of an ID-link are unordered.

Remark. — It is convenient to consider generalized axioms $\vdash A_1, \ldots, A_n$ $(n > 0)$, which are interpreted by generalized axiom links (no premise, but ordered conclusions $A_1, \ldots, A_n$). Such generalized axioms will occur in the proof of our main theorem 2 ; they also occur when one wants to accommodate other styles of syntax, which are foreign to the proof-net technology, in which case they are called *boxes*. The idea of a box is that from the outside it looks like a generalized axiom, whereas it has an inside which can be in turn another proof-net. A box freezes n formulas, and can therefore be seen as a sequent, the conclusion of a rule, whose premises are proven in the box. Traditional sequent calculus is therefore a system of proof-nets in which the only links are boxes, and all the improvement made in 9 years consist in progressively restricting the use of boxes : in this paper boxes are limited to exponential connectives (and to the neutral ⊤).

Remark. — One should never speak of formulas, but of *occurrences*, which is extremely awkward. We adopt once for all the convention that all our formulas are distinct (for instance by adding extra indices). In particular $ID, \otimes, \otimes, \oplus, \&$-links are determined by their conclusion(s), and a CUT-link is determined by its premises.

Definition 2

▸ *If L is a &-link, with A&B as its conclusion, we introduce the eigenweight p_L, which is a boolean variable. The intuitive meaning of p_L is the choice l/r between the two premises A and B of the link, p_L for "left", i.e. A, $\neg p_L$ for "right", i.e. B ; we use ϵp_L to speak of p_L or $\neg p_L$.*

▸ *If Θ is a structure involving the &-links $L_1, \ldots, L_k$ (with associated eigenweights $p_1, \ldots, p_k$), then a weight (relative to Θ) is any element of the boolean algebra generated by $p_1, \ldots, p_k$.*

Definition 3

A proof-structure Θ *consists of :*

▸ *A set of formulas (see the previous remark) ;*

▸ *A set of links ; each of these links takes its premise(s) and conclusion(s) among the formulas of Θ ;*

▸ *For each formulas A of Θ, a weight $w(A)$, i.e. a non-zero element of the boolean algebra generated by the eigenweights $p_1, \ldots, p_n$ of the &-rules of Θ) ;*

▸ *For each link L of Θ, a weight $w(L)$.*

satisfying the following conditions :

▸ *Each formula is the premise of* at most one link *and the conclusion of* at least one link *; the formulas which are not premises of some link are called the* conclusions *of Θ ;*

▸ $w(A) = \sum w(L)$, *the sum being taken over the set of links with conclusion A ;*

▸ *if A is a conclusion of Θ, then $w(A) = 1$;*

▶ *if w is any element of the boolean algebra generated by the weights occurring in Θ, and L is a $\&$-link, then $w.\neg w(L)$ does not depend on p_L, i.e. belongs to the boolean algebra generated by the eigenweights distinct from p_L ;*

▶ *if w is any weight occurring in Θ, then w is a monomial $\epsilon_1 p_{L_1} \ldots \epsilon_k p_{L_k}$ of eigenweights and negations of eigenweights[2] ;*

▶ *$w(L) \neq 0$; moreover if L is any non-identity link, with premises A and (or) B then*

    – *if L is any of $\otimes, \otimes, CUT$, then $w(L) = w(A) = w(B)$;*

    – *if L is $\oplus_1$, then $w(L) = w(A)$;*

    – *if L is $\oplus_2$, then $w(L) = w(B)$;*

    – *if L is a $\&$-link, then $w(A) = w(L).p_L$ and $w(B) = w(L).\neg p_L$, (hence $w(L) = w(A) + w(B)$).*

Remark. —

▶ Weights are in a boolean algebra, and therefore both algebraic and logical graphism can be used ; here we decide to use the product notation (instead of the intersection), but we keep $\neg w$ (instead of $1 - w$) ; when we use the sum, we of course mean the disjoint union, i.e. when I write $w(L) = w(A) + w(B)$, I implicitly mean that $w(A).w(B) = 0$.

▶ The technical condition "$w.\neg w(L)$ does not depend on p_L" says that the boolean variable p_L has no real meaning "outside $w(L)$" ; applying the condition to $\neg w(L)$, we see that $w(L)$ does not depend on p_L, in particular $w(L).\epsilon p_L \neq 0$.

▶ There are two ways to think of the dependency condition : either as a technical restriction needed for the sequentialization theorem (all our efforts to get rid of it failed) or as a nice companion to the previous condition, since both are very natural when a proof-structure is seen as a coherent space, see A.1.1.

1.2 Sequent calculus and proof-nets

Definition 4

Let Θ be a proof-structure and let L be either a CUT-link, or a link with only one conclusion, which is in turn a conclusion of Θ and such that $w(L) = 1$; we say that L is a terminal link of Θ. Given such a link, we define the removal of L in Θ which consists (provided it makes sense) in one or two proof-structures.

---

2. This is the *dependency* condition.

▸ If L is a $\otimes$-link (resp. a *CUT*-link) with premises A, B, and $\Gamma, A \otimes B$ (resp. Γ) is the set of conclusions of Θ : the removal of L consists in partitioning (if possible) the formulas of Θ distinct from $A \otimes B$ (resp. the formulas of Θ) in two subsets X and Y, one containing A, the other containing B, in such a way that, whenever a link L' distinct from L has a premise or a conclusion in X (resp. in Y), then all other premises and conclusions of L' belong to X (resp. to Y). The restrictions Θ/X and Θ/Y are defined in an obvious way, and are proof-structures with respective conclusions Γ', A and Γ'', B. Observe that $\Gamma = \Gamma', \Gamma''$.

▸ If L is a $\wp$-link with premises A, B, and $\Gamma, A \wp B$ is the set of conclusions of Θ : the removal of L consists in removing the conclusion $A \wp B$ and the link L ; this induces a proof-structure with conclusions Γ, A, B.

▸ If L is a $\oplus_1$-link with premise A, and $\Gamma, A \oplus B$ is the set of conclusions of Θ : the removal of L consists in removing the conclusion $A \oplus B$ and the link L ; this induces a proof-structure with conclusions Γ, A.

▸ If L is a $\oplus_2$-link with premise B, and $\Gamma, A \oplus B$ is the set of conclusions of Θ : the removal of L consists in removing the conclusion $A \oplus B$ and the link L ; this induces a proof-structure with conclusions Γ, B.

▸ If L is a &-link with premises A, B, and $\Gamma, A \& B$ is the set of conclusions of Θ : the removal of L consists in first removing the conclusion $A \& B$ and the link L (to get Θ') and then forming two proof-structures Θ_A and Θ_B :

  – In Θ' make the replacement $p_L = 1$, and keep only those links L' whose weight is still non-zero, together with the premises and conclusions of such links : the result is by definition Θ_A, a proof-structure with conclusions Γ, A.

  – In Θ' make the replacement $p_L = 0$, and keep only those links L' whose weight is still non-zero, together with the premises and conclusions of such links : the result is by definition Θ_B, a proof-structure with conclusions Γ, B.

Definition 5

A proof-structure Θ is sequentialisable when it can be reduced, by iterated removal of terminal rules, to identity links. In more pedantic terms :

▸ An identity link is sequentialisable ;

▸ If the result of removing the terminal link L in Θ yields sequentialisable proof-structures, then Θ is sequentialisable.

Remark. —

▸ The removal of a given terminal link is not always possible, and its result is not necessarily unique (however, for *proof-nets*, it would be easy to show, by a connectivity argument, that the removal of a $\otimes$- or CUT-link is unique).

▸ Each removal step consists in the writing down of a rule of MALL ; therefore a sequentialisable proof-structure has a *sequentialization*, which consists in a proof in MALL.

▸ Conversely, given a proof Π of $\vdash \Gamma$ in sequent calculus, one can build a proof-structure Π° with conclusions Γ, such that Π is a sequentialization of Π°. But contrarily to the situation of our previous papers [3, 5, 7] on proof-nets, Π° is no longer unique. The problem is in the interpretation of a &-rule

$$\frac{\vdash \Gamma, A \quad \vdash \Gamma, A}{\vdash \Gamma, A\&B} \&$$

applied to proofs Π_1 and Π_2 of $\vdash \Gamma, A$ and $\vdash \Gamma, B$. We must find a proof-structure Π° such that the removal of a terminal &-link yields Π_1° and Π_2°. The basic idea is to merge the two proof-structures by means of the eigenweight p_L : anything with respective weights w_1 and w_2 in Π_1° and Π_2° will now get the weight $p_L.w_1 + \neg p_L.w_2$ (we give the weight $w_1 = 0$ to something which is absent in Π_1°). This simple idea yields the following list of cases :

– If X is a formula or a link occurring in both of Π_1° and Π_2°, with the weights $w_1(X)$ and $w_2(X)$, then X will occur in Π° with the weight
$p_L.w_1(X) + \neg p_L.w_2(X)$; in particular the formulas of Γ occur in Π° with the weight 1.

– If X is a formula or a link occurring in Π_1° but not in Π_2°, with the weight $w_1(X)$, then X will occur in Π° with the weight $p_L.w_1(X)$; in particular, A will occur with the weight p_L.

– If X is a formula or a link occurring in Π_2° but not in Π_1°, with the weight $w_2(X)$, then X will occur in Π° with the weight $\neg p_L.w_2(X)$; in particular, B will occur with the weight $\neg p_L$.

and to add to this merge the formula $A\&B$ together with the &-link L whose premises are A, B and whose conclusion is $A\&B$; both L and $A\&B$ receive the weight 1. But this is not as simple as it might seem : how do we know that a formula or a link X of Π_1 is the same as another formula or link Y of Π_2 ? There is no simple answer (except in some very specific cases, see discussion A.1.4), and moreover, in cases where we feel entitled to make such identifications, the resulting weight $p_L.w_1(X) + \neg p_L.w_2(X)$ is not a monomial, which contradicts the dependency condition. However there is at least the possibility to decide that no identification between Π_1 and Π_2 is made, but for the conclusions, i.e. the formulas of Γ ; by the way the sequent calculus formulation of the &-rule stipulates that the contexts of the two premises must be equal, hence this is a clear case where there is no doubt as to the identification between a formula in Π_1 and a formula in Π_2.

Anyway, the main problem is to find a *sequentialization* theorem ; this means to give an intrinsic characterization of *sequentialisable* proof-structures. The answer (a notion of *proof-net*) to this problem is important because we want to carry our program of geometrization of proofs, introduced in [6] under the name of *geometry of interaction*. Remember that *geometry of interaction* is issued from the analysis of multiplicative proof-nets in terms of permutations, [4]. In fact the redaction of this paper was postponed until a geometry of interaction for additives was found, which is now the case, see [2].

1.3   A wrong answer : slicing

Definition 6

Let φ be a valuation for Θ, i.e. a function from the set of eigenweights of Θ into the boolean algebra $\{0,1\}$, which induces a function (still denoted φ) from the weights of Θ to $\{0,1\}$. The slice $\varphi(\Theta)$ is obtained by restricting to those formulas A of Θ such that $\varphi(w(A)) = 1$, with an obvious modification for the remaining &-links : only one premise is present.

The definition suggests a simple-minded criterion (which is anyway an approximation to the real solution) : observe that (if we neglect unary links), $\varphi(\Theta)$ is a multiplicative proof-structure. Therefore we can require $\varphi(\Theta)$ to be multiplicatively correct, i.e. to be a multiplicative proof-net. This condition is obviously necessary (in fact it is what we get by restricting definition 10 to *normal* jumps).
But the condition cannot be sufficient : it corresponds to a separate treatment of additives and multiplicatives, without real interaction between the conditions. This would mean that all multiplicatives distribute over all additives : for instance the obvious proof-structure with conclusions $(A^{\perp} \otimes B^{\perp}) \oplus (A^{\perp} \otimes C^{\perp})$ and $A \otimes (B \& C)$ has two multiplicatively correct slices, although it states a wrong instance of distributivity, and is not sequentialisable, whereas the obvious proof-structure with conclusions $(A^{\perp} \otimes B^{\perp}) \& (A^{\perp} \otimes C^{\perp})$ and $A \otimes (B \oplus C)$ (which has essentially the same slices) is sequentialisable. Our ultimate criterion must therefore force the additive and multiplicative layers to interact.

1.4   Proof-nets

Our basic idea will be to mimic our criterion of [7] ; in this paper, certain switchings for $\forall$-links were induced by the dependency of some formula upon the *eigenvariable* of the link.

Definition 7

Let φ be a valuation of Θ, let p_L be an eigenweight ; we say that the weight w (in Θ) depends on p_L (in $\varphi(\Theta)$) iff $\varphi(w) \neq \varphi_L(w)$, where the valuation φ_L is defined by :

- $\varphi_L(p_L) = \neg(\varphi(p_L))$
- $\varphi_L(p_{L'}) = \varphi(p_{L'})$ if $L' \neq L$.

A formula A of Θ is said to depend on p_L (in $\varphi(\Theta)$), if A is conclusion of a link L' such that $\varphi(w(L')) = 1$ and $\varphi_L(w(L')) = 0$. This basically means that A and L' are present in $\varphi(\Theta)$, but that changing the value of the valuation for p_L would make A (or at least L') disappear from the slice.

Definition 8

A switching $\mathcal{S}$ of a proof-structure Θ consists in :

- The choice of a valuation $\varphi_{\mathcal{S}}$ for Θ ;
- The selection of a choice $\mathcal{S}(L) \in l, r$ for all $\invamp$-links of $\varphi_{\mathcal{S}}(\Theta)$;
- The selection for each &-link L of $\varphi_{\mathcal{S}}(\Theta)$ a formula $\mathcal{S}(L)$, the jump of L, depending on p_L in $\varphi_{\mathcal{S}}(\Theta)$. There is always a normal choice of jump for L, namely the premise A of L such that $\varphi_{\mathcal{S}}(w(A)) = 1$.

Definition 9

Let $\mathcal{S}$ be a switching of a proof-structure Θ ; we define the graph $\Theta_{\mathcal{S}}$ as follows :

- The vertices of $\Theta_{\mathcal{S}}$ are the formulas of $\varphi_{\mathcal{S}}(\Theta)$;
- For all ID-links of $\varphi_{\mathcal{S}}(\Theta)$, we draw an edge between the conclusions ;
- For all generalized axiom links with conclusions $A_1, \ldots, A_n$, we draw an edge between A_1 and A_2, etc., A_{n-1} and A_n ;
- For all CUT-links of $\varphi_{\mathcal{S}}(\Theta)$, we draw an edge between the premises ;
- For all $\oplus$-links of $\varphi_{\mathcal{S}}(\Theta)$, we draw an edge between the conclusion and the premise ;
- For all $\otimes$-links of $\varphi_{\mathcal{S}}(\Theta)$, we draw an edge between the left premise and the conclusion, and between the right premise and the conclusion ;

▸ For all $\mathop{\gamma}\limits$-links L of $\varphi_S(\Theta)$, we draw an edge between the premise (left or right) selected by $\mathcal{S}(L)$ and the conclusion ;

▸ For all &-links L of $\varphi_S(\Theta)$, we draw an edge between the jump $\mathcal{S}(L)$ of L and the conclusion.

Definition 10

A proof-structure Θ is said to be a proof-net when for all switchings $\mathcal{S}$, the graph Θ_S is connected and acyclic.

We immediately get the

Theorem 1   *If Θ is sequentialisable, then Θ is a proof-net.*
PROOF. — *The proof is straightforward and uninteresting.*   □

1.5   The sequentialization theorem

Theorem 2   *Proof-nets are sequentialisable.*

We shall devote the remainder of the subsection to the proof of the theorem. We shall make some simplifying hypotheses :

1.5.1   Empires

In this subsection and in the next one, we fix one for all a valuation φ and we concentrate on the slice $\varphi(\Theta)$, where Θ is a given proof-net. All switchings $\mathcal{S}$ considered are such that $\varphi_S = \varphi$.

Definition 11

If $\mathcal{S}$ is a switching and A is a formula of $\varphi(\Theta)$, then we modify the graph Θ_S by deleting (in case A is premise of a link L) any edge induced by L and connecting A to another formula B. (There is at most such an edge, connecting A with the conclusion of L or with the other premise of L if L is a CUT-link.) This determines a partition of Θ_S into at most two connected components, and the one containing A is denoted Θ_S^A. We define the empire of A as $eA := \bigcap_S \Theta_S^A$, the intersection being taken over all switchings $\mathcal{S}$.

Lemma 2.1   *Empires are imperialistic ; this means that, as soon B_1 and B_2 are linked by L and $B_1 \in eA$, then $B_2 \in eA$, with only three exceptions :*

- B_1 *is* A *and is a premise of* L : *if* B_2 *is the conclusion of* L, *it is not in* eA ;

- L *is a* $\otimes$-*link*, B_2 *is the conclusion of the link*, B_1 *is one of the two premises of* L *and the other premise of* L *is not in* eA : *in such a case the conclusion is never in* eA ;

- L *is a* &-*link*, B_2 *is the conclusion of the link*, B_1 *is the premise of* L *(remaining in* $\varphi(\Theta)$*) and there is a formula* C *in* $\varphi(\Theta)$ *depending on* p_L *which is not in* eA : *in such a case the conclusion is never in* eA.

PROOF. — *straightforward ; see e.g.* [7], *lemma 1. We sketch the proof for the sake of self-containment :*

- *one first remarks that if* B_1 *is a conclusion of* L, *then* $B_2 \in eA$; *the typical example is that of a* $\otimes$-*link. If* $\mathcal{S}$ *is a switching inducing two components, and such that* $B_2 \notin \Theta_{\mathcal{S}}^A$, *then we see (perhaps by changing our mind about* $\mathcal{S}(L)$*) that* $B_1 \notin \Theta_{\mathcal{S}}^A$. *The other cases are handled in the same way. Observe that the same argument shows that if the conclusion* B_1 *of a* &-*link* L *is in* eA, *so is any formula whose weight depends on* p_L.

- *it only remains to show that if* L *is a* $\otimes$- *or* &-*link and all possible choices for* $\mathcal{S}(L)$ *are in* eA, *so is the conclusion* B *of* L, *provided* $B \neq A$. *But any switching* $\mathcal{S}$ *must connect* B *with one of these formulas which are in* $\Theta_{\mathcal{S}}^A$, *and this forces* $B \in \Theta_{\mathcal{S}}^A$.

$\square$

Lemma 2.2 *There is an* $\mathcal{S}$ *such that* $eA = \Theta_{\mathcal{S}}^A$; $\mathcal{S}$ *is called a* principal switching *for* eA.

PROOF. — $\mathcal{S}$ *is obtained as follows : in case* A *is the premise of a* $\otimes$- *or* &-*link, set* $\mathcal{S}$ *so as to draw an edge between* A *and the conclusion of the link ; the other* $\otimes$ *and* &-*links are switched as follows :*

- *if* L *is a* $\otimes$-*link with its conclusion* C *not in* eA, *then one of its premises, let us say* B *is not in* eA *by lemma 2.1 and we set* $\mathcal{S}$ *so as to draw an edge between* B *and* C ;

- *if* L *is a* &-*link* L *with its conclusion* C *not in* eA, *then a formula of* $\varphi(\Theta)$, *let us say* B, *depending on* p_L, *is not in* eA *by lemma 2.1 and we set* $\mathcal{S}$ *so as to draw an edge between* B *and* C.

It is immediate that $eA = \Theta_{\mathcal{S}}^A$. $\square$

Lemma 2.3   *Let A and B be distinct formulas in $\varphi(\Theta)$ and assume that $B \notin eA$; then*

- *if $A \in eB$, then $eA \subset eB$;*

- *if $A \notin eB$, then $eA \cap eB = \emptyset$.*

PROOF. — *see [7], lemma 3. We also sketch the proof : we define a principal switching of eB as in lemma 2.2 by adding another constraint : if L is a $\invamp$ or $\&$ whose conclusion is in eB, then in case A is a premise of L set $\mathcal{S}$ so as to connect the conclusion with A, and (otherwise) if some possible choice for $\mathcal{S}(L)$ is not in eA, make this very choice. Now if a formula C belongs to $eA \cap eB$, then C is connected with B (which is not in eA) inside $\Theta_{\mathcal{S}}^B = eB$, which means that some edge induced by $\mathcal{S}$ inside eB links a formula in eA with a formula outside eA. This is only possible (because of lemma 2.1) when the formula of eA is A. This proves the second part of the claim. If $A \in eB$, A must be the premise of a link L whose conclusion is not in eA, and $\mathcal{S}(L)$ has been set so as to get two connected components : $\Theta_{\mathcal{S}}^A$ and its complement. It is immediate that $\Theta_{\mathcal{S}}^A \subset \Theta_{\mathcal{S}}^B$ hence $eA \subset \Theta_{\mathcal{S}}^A \subset \Theta_{\mathcal{S}}^B = eB$.*   □

Definition 12

A formula B of eA is said to be a door of eA iff

- *Either it is the premise of a link L whose conclusion does not belong to eA ;*

- *Or it is a conclusion of Θ. Obviously A is a door of eA, the main door ; the other doors are called auxiliary doors. The set of doors of eA is the border of eA.*

Lemma 2.4   *Let C be an auxiliary door of eA which is not a conclusion of Θ ; then C is the premise of a $\invamp$- or a $\&$-rule.*

PROOF. — *immediate from lemma 2.1.*   □

1.5.2   Maximal empires

The hypotheses are the same as in the previous section ; but we now assume that A is the conclusion of a $\&$-link L and that eA is maximal w.r.t. inclusion among similar empires (i.e. that if B is the conclusion of a $\&$-link and $eA \subset eB$, then $A = B$).

Lemma 2.5   *Let L' be any &-link and let $B, B' \in \varphi(\Theta)$ be such that B, B' depends on $p_{L'}$, and assume that $B \in eA$; then $B' \in eA$.*

PROOF. — *let w be the weight of the link whose conclusion is B and such that $\varphi(w) = 1$; then the technical condition "$w.\neg w(L')$ does not depend on $p_{L'}$" ensures that $\varphi(\neg w(L')) = 0$, hence if C is the conclusion of L', that $C \in \varphi(\Theta)$. By lemma 2.1 $B \in eC$, hence $eA \cap eC \neq \emptyset$; by maximality $eA \subset eC$ is impossible, hence by lemma 2.3 we get $C \in eA$. Now by lemma 2.1 this implies $B' \in eA$.* □

Lemma 2.6   *If C is a border formula of eA and L' is any &-link, then $\varphi_{L'}(w(C)) = 1$, i.e. the border formulas are still present in $\varphi'_L(\Theta)$.*

PROOF. — *C is either a conclusion, in which case $w(C) = 1$ or C is a premise of some rule L'' : let B be the conclusion of L'' (in case L'' is a cut, let B be the other premise of L''). Then $B \notin eA$ but since B depends on L', lemma 2.5 yields $B \in eA$, a contradiction.* □

1.5.3   Stability of maximal empires

The purpose of this section is to show that if A is the conclusion of a &-link and is maximal among similar empires w.r.t. a given φ, then it remains maximal w.r.t. any other choice of φ. It will be enough to start with a given φ and to show that A is still maximal w.r.t. φ_L. We can only prove this essential fact under the dependency condition, which has the following consequence :

Lemma 2.7   *Assume that $\varphi(w(A)) = 1$ and that (w.r.t. φ) $w(A)$ depends neither on p_L and p_M ; then this remains true w.r.t. φ_L.*

PROOF. — *the monomial $w(A)$ cannot make use of ϵp_L etc.* □

This is wrong for non monomials, typically $p \cup q$ depends neither on p nor on q if $\varphi(p) = \varphi(q) = 1$, but depends on q if $\varphi(p) = 0$. This phenomenon is exactly the familiar failure of stability in the case of the *parallel or*.

Lemma 2.8   *Assume that $B \in eA$ (w.r.t. φ) and that $w(B)$ does not depend on p_L ; then w.r.t. φ_L we still have $B \in eA$.*

PROOF. — *it takes nothing to assume that A is not a conclusion (in which case eA is $\varphi(\Theta)$ and everything is trivial) ; so one can fix a formula C with the same weight as A such that C cannot belong to eA (take C to be the conclusion of the link of which A is a premise, or the other premise if this link is L a CUT). Let D be the conclusion of the link L, then :*

▸ *either $D \in eA$, hence by lemma 2.5, the formulas outside eA do not depend on p_L, hence by lemma 2.7, changing φ to φ_L does not alter any dependency outside. Assume that a switching S has been chosen, (w.r.t. φ_L) with $B \in \Theta_S^A$, $C \notin \Theta_S^A$. Then if we define a switching S' (w.r.t. φ) essentially by making the same choices outside eA (which is possible, since the dependencies are unaffected), then B and C are still connected, hence we still have $B \in \Theta_{S'}^A$, contradicting the hypothesis.*

▸ *or $D \notin eA$, hence by lemma 2.5, the formulas inside eA do not depend on p_L, hence by lemma 2.7, changing φ to φ_L does not alter any dependency inside. Assume that a switching S has been chosen, (w.r.t. φ_L) with $B \notin \Theta_S^A$. Then if we define a switching S' (w.r.t. φ_L) essentially by making the same choices inside eA (which is possible, since the dependencies are unaffected), then B and A are still not connected, hence we still have $B \notin \Theta_{S'}^A$, contradicting the hypothesis.*

$\square$

Lemma 2.9   *eA is maximal w.r.t. φ_L.*
PROOF. — *assume that (w.r.t. φ_L) $eA \subset eB$, with eB maximal. Then (still w.r.t. φ_L) $A \in eB$ but $B \notin eA$. Using the maximality of eA w.r.t. φ, of eB w.r.t. φ_L and lemma 2.8, we get (w.r.t. φ) $A \in eB$ and $B \notin eA$, which contradicts the maximality of eA w.r.t. φ.* $\square$

1.5.4   Proof of the theorem

PROOF. — by induction on the number n of $\&$-links in Θ :

▸ if $n = 0$, then we are basically in the multiplicative case and we are done. To be precise, we argue by induction on the number m of links of the proof-net :

   – if $m = 1$, then the proof-net consists in a single axiom link and we are done ;

   – if $m > 1$, then by connectedness Θ must contain some link which is not an axiom, hence some terminal link (see definition 4) exists, and

        ∗ if there is a terminal $\wp$ or $\oplus_i$-link, then we can remove it, thus getting another proof-net (immediate) with a smaller value of m, to which the induction hypothesis applies ; then Θ is sequentialisable ;

        ∗ if all terminal links are $\otimes$- or CUT-links, with premises A_i, B_i, choose one of these premises (say A_1) such that eA_1 is maximal inside the eA_i

and eB_i. If C is a border formula of eA_1 and C is not a conclusion of Θ, then C stands (hereditarily) above some A_i or B_j, let us say above B_2. Then by lemma 2.1, $C \in eB_2$, hence by lemma 2.3 $eA_1 \subset eB_2$, contradicting the maximality of eA_1. This shows that the border of eA_1 only consists of conclusions. But then it is easy to see that $\Theta_S^{A_1}$ is always equal to eA_1 (apply lemma 2.1), hence $\Theta_S^{B_1}$ is always equal to eB_1. This means that the removal of our terminal link induces a splitting of the proof-net into two connected components, which are obviously proof-nets and to which the induction hypothesis applies ; Θ is therefore sequentialisable.

▸ if $n > 0$, then choose φ (e.g. $\varphi(p_L) = 1$ for all L), and let A be a conclusion of some &-link such that eA is maximal w.r.t. φ. Now by lemma 2.9 eA remains maximal w.r.t. to any φ', and its border remains the same by lemma 2.6. Moreover, if L is any & link whose conclusion B belongs to eA w.r.t. φ, B still belongs to eA w.r.t. any φ' such that $\varphi'(w(L)) = 1$ (this is another use of the dependency condition : since $w(L)$ is a monomial, is possible to write φ' as $\varphi_{L_1 \ldots L_k}$ for a suitable sequence $L_1 \ldots L_k$ with the property that all intermediate valuations $\varphi_{L_1 \ldots L_i}$ yield the same value $\varphi_{L_1 \ldots L_i}(w(L)) = 1$. Then lemma 2.8 is enough to conclude.) This means that a given eigenweight p_L cannot occur both in eA for some valuation and in the complement of eA for some other valuation. In other terms, we can split the eigenweights into two disjoint groups, those who may occur in eA (group I) and those who may occur in its complement (group II). Now let us introduce two new proof-nets :

– a proof-net Θ' whose conclusions are those of the (constant) border of eA ; a formula B is in Θ' when $B \in eA$ for some valuation. This immediately induces a proof-structure (take those links whose conclusion is in Θ', with the weight they had in Θ). Θ' contains all occurrences of eigenvariables from group I, and no occurrence from group II. It is immediate that Θ' is a proof-net. Moreover Θ' has a terminal &-link and removing this link as in definition 4 induces two proof-nets (immediate) to which the induction hypothesis applies, and which are therefore sequentialisable. Then Θ' is sequentialisable. In case A is a conclusion of Θ, then $\Theta = \Theta'$ and we are done ; otherwise we introduce

– a proof-net Θ'' consisting of those formulas B such that $B \notin eA$ for some valuation φ' such that $\varphi'(w(B)) = 1$ and of the border formulas of eA which are not conclusions (there are some). The links of Θ'' are all the links of Θ whose conclusion is outside eA and a new axiom link whose conclusions consist in the border of eA. The links and formulas coming from Θ receive the same weight, whereas the new axiom link is weighted

1. Θ" contains all occurrences of eigenvariables from group II, and no occurrence from group I. Now it is an easy exercise in graph theory to show that Θ" is a proof-net. The induction hypothesis applies and Θ" is therefore sequentialisable. By the way observe that if we replace our new axiom link in Θ" by the proof-net Θ′, the result is the original Θ.

The sequentialization of Θ is obtained by taking the sequentialization of Θ" and replacing the sequent calculus axiom corresponding to the new axiom link with the sequentialization of Θ′.

□

1.6   Cut-elimination in MALL (lazy procedure)

There is a Church-Rosser cut-elimination procedure for additive proof-nets, which enjoys the subformula property. For practical uses this procedure is challenged by a coarser one, *lazy* cut-elimination, which can be proved to achieve the same result in some important cases. The more general procedure is described in appendix A.1.3.

1.6.1   Lazy cut-elimination

Lazy cut-elimination is concerned with only a restricted kind of cut :

Definition 13

A cut-link L is said to be ready *iff*

▸ *the weight of the cut-link is 1*

▸ *both premises of the cut are the conclusion of exactly one link (these links are therefore also of weight 1)*

Theorem 3   *Let Θ be a proof-net whose conclusions do not contain the connective & and without ready cut ; then Θ is cut-free.*
PROOF. — *if Θ contains no &-link, we are done, since all weights are 1. Otherwise, choose a &-link L in such a way that the empire eA of its conclusion is maximal among similar empires (w.r.t. any valuation). Then A is weighted 1 as well his (hereditary) conclusions. Below A sits a terminal link L' with no conclusion (otherwise A would be a subformula of this conclusion, and & would occur in a conclusion), hence L' is a cut of weight 1. In fact A sits hereditarily above the premise B of L which is in turn the conclusion of a link of weight 1 (either L or a link "below" L). Now the premise $B^\perp$ of L' might*

be the conclusion of several links, in which case there is a valuation φ and a &-link L'' (with conclusion C) such that $B^\perp$ depends on $p_{L''}$. In the slice $\varphi(\Theta)$ observe that $B^\perp \notin eA$, but $A \in eC$ (this is because $B^\perp \in eC$ and so $B \in eC$ and then $A \in eC$ by lemma 2.1). But then $eA \subset eC$ by lemma 2.3, a contradiction.   □

Remark. —

▸ The theorem therefore establishes that the lazy procedure which only removes ready cuts is enough in the absence of additive connectives. In fact since the general procedure extends the lazy one and is Church-Rosser, the lazy procedure yields the same result in this important case.

▸ The theorem still holds for full linear logic : if a proof-net is without ready cuts and its conclusions mention neither & nor existential higher order quantifiers, then it is cut-free. The restriction on existentials is just to forbid the hiding of a & by a ∃-rule, which can be the case with higher order.

Definition 14

Let L be a ready cut in a proof-net Θ, whose premises A and $A^\perp$ are the respective conclusions of links L, L'. Then we define the result Θ' of reducing our cut in Θ :

▸ *If L is an ID-link then Θ' is obtained by removing in Θ the formulas A and $A^\perp$ (as well as the cut-link and L) and giving a new conclusion to L' : the other conclusion of L, which is another occurrence of $A^\perp$ (ID-reduction).*

▸ *If L is a $\otimes$-link (with premises B and C) and L' is a $\mathfrak{R}$-link (with premises $B^\perp$ and $C^\perp$), then Θ' is obtained by removing in Θ the formulas A and $A^\perp$ as well as our cut links and L,L' and adding two new cut links with respective premises $B,B^\perp$ and $C,C^\perp$ ($\otimes$-reduction).*

▸ *If L is a &-link (with premises B and C) and L' is a $\oplus_1$-link (with premise $B^\perp$), then Θ' is obtained in three steps : first we remove in Θ the formulas A and $A^\perp$ as well as our cut link and L,L' ; then we replace the eigenweight p_L by 1 and keep only those formulas and links that still have a nonzero weight : therefore B and $B^\perp$ remain with weight 1 whereas C disappears ; finally we add a cut between B and $B^\perp$ ($\oplus_1$-reduction).*

▸ *If L is a &-link (with premises B and C) and L' is a $\oplus_2$-link (with premise $C^\perp$), then Θ' is obtained in three steps : first we remove in Θ the formulas A and $A^\perp$ as well as our cut link and L,L' ; then we* *replace the eigenweight p_L by 0 and keep only those formulas and links that still have a nonzero weight : therefore C and $C^\perp$ remain with weight 1 whereas B disappears ; finally we add a cut between B and $B^\perp$ ($\oplus_2$-reduction).*

Proposition 1

If Θ' is obtained from a proof-net Θ by lazy cut-elimination, then Θ' is still a proof-net.

PROOF. — we consider all possible cut-elimination steps :

- ► ID-reduction : the only important thing to remark is that the other conclusion of L (which is an occurrence of $A^\perp$) cannot be the same occurrence (otherwise the structure would bear a cycle).

- ► $\otimes$-reduction : let us fix a valuation φ, and let X be the set $\varphi(\Theta)$, that we can write as the union $Y \cup Z$ of $\varphi(\Theta')$ and $\{B, C, A, A^\perp, B^\perp, C^\perp\}$, so that $Y \cap Z = \{B, C, B^\perp, C^\perp\}$. Given a switching $\mathcal{S}$ of Θ', it induces a graph $\Theta'_\mathcal{S}$ on Y and we can consider its subgraph $\mathcal{G}$ (on the same Y) in which the edges between $B, B^\perp$ and $C, C^\perp$ have been removed, as well as the subgraph $\mathcal{H}$ (on $Y \cap Z$) consisting of these two edges : we have $\Theta'_\mathcal{S} = \mathcal{G} \cup \mathcal{H}$. We want to show that $\Theta'_\mathcal{S}$ is connected and acyclic. In order to do this, we "extend" our switching $\mathcal{S}$ to a switching $\mathcal{S}'$ of Θ : it is immediate that $\Theta_{\mathcal{S}'}$ can be written as $\mathcal{G} \cup \mathcal{H}'$, where $\mathcal{H}'$ is a graph on Z with one edge between B, A, one edge between C, A, one edge between $A, A^\perp$, one edge between $A^\perp, B^\perp$ (or between $A^\perp, C^\perp$, depending on the switching of L'). Since $\mathcal{G} \cup \mathcal{H}'$ is connected and acyclic, then $\mathcal{G}$ has 3 connected components, and each of them meets $Y \cap Z$. Now B, C are not in the same component (otherwise there would be a cycle in $\mathcal{G} \cup \mathcal{H}'$; $B, B^\perp$ are neither in the same component, since, if we switch L' to "left", we get a cycle in $\mathcal{G} \cup \mathcal{H}'$; the same is true, for symmetrical reasons, of $B, C^\perp$ and $C, C^\perp$. Then $B^\perp, C^\perp$ must lie in the same component. From this it is immediate that $\mathcal{G} \cup \mathcal{H}$, i.e. $\Theta'_\mathcal{S}$, is connected and acyclic.

- ► ($\oplus - 1$-reduction) : a valuation φ for Θ' can be extended to a valuation φ' for Θ by setting $\varphi'(p_L) = 1$ (and giving any value to the other eigenweights which might have disappeared when making the cut-elimination step). One can also extend a switching $\mathcal{S}$ of Θ' into a switching $\mathcal{S}'$ of Θ by setting $\mathcal{S}(L) = B$. Then it is almost immediate that $\Theta_{\mathcal{S}'}$ and $\Theta'_\mathcal{S}$ are of the same type.

▸ $(\oplus - 2\text{-reduction})$: symmetrical to the previous case.

□

Definition 15

The size *of a proof-net is defined to be the number of its links.*

Theorem 4   *Lazy cut-elimination converges to a unique lazy normal form (i.e. a proof-net without ready cuts) in a time which is linear in the size of the proof-net.*

PROOF. — *unicity of the normal form is an easy consequence of the fact that our procedure is Church-Rosser. To prove termination, observe that some of the eigenweights receive definite values $p_L = 0$ or $p_L = 1$ during the cut-elimination of Θ. Let φ be a valuation of Θ which extends these choices. Now, if $\Theta \mapsto \Theta_1 \mapsto \ldots \mapsto \Theta_n$ is a reduction sequence, we can observe the slices $\varphi(\Theta)$, $\varphi(\Theta_1), \ldots, \varphi(\Theta_n)$, and observe that the number of links in each of these slices is strictly decreasing. Then n cannot be bigger than the number of links in $\varphi(\Theta)$, hence, n is smaller than the number of links in Θ. Lazy normalization therefore takes a linear number of steps, which is not quite the same as linear time. However, it is quite easy to see how to transform this into a linear time algorithm : we keep track of the values taken by the eigenweights and we delay the substitutions occurring in the additive reductions, i.e. we perform them only in case they yield values 0 (in which case we may erase) or 1 ... When the process is completed, then we perform the remaining substitutions.*   □

Remark. — Only a limited part of the correctness criterion is actually used to prove proposition 1 ; the most important part is devoted to the absence of deadlock, i.e. theorem 3.

2   The case of quantifiers

Quantifiers have been treated in previous papers [5, 7], hence we shall be brief. Roughly speaking the role of eigenweights is taken by eigenvariables, and everything is adapted *mutatis mutandis.*

2.1   Proof-structures

$$\forall -links \; : \; 1\; premise \; : \; A[e/x] \quad 1\; conclusion \; : \; \forall x A$$
$$\exists_t -links \; : \; 1\; premise \; : \; A[t/x] \quad 1\; conclusion \; : \; \exists x A$$

∀-links make use of *eigenvariables* ; for each ∀-link L there is a specific variable e_L which is associated with this link. Each existential link comes with the name of a term, namely the term t of the premise of the ∃-link. This remark would be pure pedantism if we had not to take care of the case of fake dependencies (i.e. when x does not occur in A) where we cannot recover t from the premise. We shall say that a formula A *depends* on an eigenvariable e_L when

► either A is the premise of L

► or e_L occurs in A

► or A is the premise of a link $∃_t$ and e_L occurs in t

Proof-structures are defined as expected i.e. as in definition 3 :

► if L is a quantifier link with premise A, then $w(L) = w(A)$;

► we require that $w(A).\neg w(L) = 0$ for any formula A depending on e_L ;

► we require that no eigenvariable occurs in a conclusion.

One easily defines what it means for a sequent calculus proof to be a sequentialization of a given proof-structure, by extending definition 4 :

► If L is a ∀-link with premise $A[e_L/x]$, and $Γ, ∀xA$ is the set of conclusions of $Θ$: the removal of L consists in removing the conclusion ∀ and the link L and replace everywhere e_L with a fresh variable e ; this induces a proof-structure with conclusions $Γ, A[e/x]$.

► If L is a $∃_t$-link with premise $A[t/x]$, and $Γ, ∃xA$ is the set of conclusions of $Θ$: the removal of L consists in removing the conclusion $∃xA$ and the link L ; this induces a proof-structure with conclusions $Γ, A[t/x]$. Observe that this removal might be impossible, typically if some eigenvariable occurs in t.

2.2 Proof-nets

Definitions 8 and 9 are adapted as follows :

► S selects for each ∀-link L of $φ_S(Θ)$ a formula $S(L)$, the *jump* of L, depending on p_L in $φ_S(Θ)$. There is always a *normal* choice of jump for L, namely the premise A of L.

► In $Θ_S$ we draw an edge between the conclusion and the premise of any ∃-link, and for all ∀-links L of $φ_S(Θ)$, we draw an edge between the jump $S(L)$ of L and the conclusion of L.

The condition for being a proof-net is exactly defined as in definition 10. We have only to check the extension of theorems 1 and 2. We only give some indications as to the proof of the latter :

▸ We shall mainly be concerned with a maximal empire eA where A is the conclusion of either a & or a ∀-link.

▸ If eA is maximal, then its border formulas depend on no eigenvariable

▸ If eA is maximal and if $B, B' \in \varphi(\Theta)$ are such that e'_L occurs in B, B' ; then $B \in eA$ implies $B' \in eA$.

2.3 Cut-elimination in MALLq

Lazy cut-elimination is defined by adding a case to definition 14 :

▸ If L is a ∀-link (with premises $B[e_L/x]$) and L' is a $\exists_t$-link (with premise $B^\perp[t/x]$), then Θ' is obtained in three steps : first we remove in Θ the formulas A and $A^\perp$ as well as our cut link and L, L' ; then we replace the eigenvariable p_L by t ; finally we add a cut between $B[t/x]$ and $B^\perp[t/x]$ ($\exists_t$-reduction).

The procedure might not work for general proof-structures : if e_L actually occurs in t, then substituting t for e_L changes $B^\perp[t/x]$ into $B^\perp[t'/x]$ which does not match $B[t/x]$, and also e_L is still present. But the proof-net condition forbids this (just set $\mathcal{S}$ so that $\mathcal{S}(L) = B^\perp[t/x]$).
It is of course essential to prove that the proof-net condition is preserved through cut-elimination, which essentially amounts looking at our new step : given a switching $\mathcal{S}$ of Θ', we want to back it up into a switching $\mathcal{S}'$ of Θ, which offers no difficulty as to L (set $\mathcal{S}'(L) = B[e_L/x]$) ; but this might be problematic for some ∀-link L', since a formula C may not depend on e'_L, whereas $C[t/e_L]$ does, and if $\mathcal{S}(L') = C[t/e_L]$, we cannot set $\mathcal{S}'(L') = C$. However observe that :

▸ in such a case, e'_L must occur in t, and this forces (w.r.t. Θ) $B[t/x] \in eA'$, hence $\forall x A \in eA'$, where A' is the conclusion of L'. But since $B[t/x] \notin eA$, we also get $A' \notin eA$, hence $eA \subset eA'$

▸ C must also depend on e_L, hence $C \in eA$, which implies $C \in eA'$

▸ but it is immediate that the illegal jump $\mathcal{S}'(L') = C$ to an element of eA' will not alter the correctness criterion.

We can therefore prove that the proof-net condition is preserved by induction on the number of illegal jumps (with an induction loading : we consider proof-nets in which fake dependencies have been declared). The basis step being trivial, the induction step ($n + 1$ illegal jumps) consists in choosing L', A', C as above, and to modify our proof-net by introducing a fake dependency i.e. to pretend that e'_L occurs in C. This induces another proof-net (as observed above) in which our jump is now legal. There are only n illegal jumps in this new proof-net, and the induction hypothesis applies, yielding a connected acyclic graph.

2.4 Complexity of cut-elimination

As before, cut-elimination involves a linear number of steps. This is clearly not enough for linear time normalization, since one of these steps is a substitution, whose iteration is exponential. However, the algorithm remains linear if we do not perform the substitutions, i.e. we keep the eigenvariables in the proof-net and we use an auxiliary stack to remember which term should be substituted for them. This process is particularly efficient in the following case : we normalize a proof of A without & or $\exists$; then the lazy cut-elimination will eventually yield a cut-free proof (in which some substitutions have not been performed) in linear time. Now, imagine that we perform the delayed substitutions in a given formula B of our net : we get a subformula B' of A which is completely determined by the choice of the links below B, hence we can forget the substitutions and directly write this subformula.

3 Exponentials

We shall deal with only two rules, which are central in the study of exponentials, namely weakening and contraction. The point of the introduction of exponentials is precisely to allow weakening and contraction for certain formulas, those which are prefixed by ?. This is not enough, and in the standard version of linear logic, two additional rules, promotion and dereliction, are added. We now know that the choice of additional rules is much more open, and we shall therefore ignore them. As a consequence, it will be impossible to speak of cut-elimination, and we shall concentrate on correctness.

Our basic ingredient will be the notion of a *discharged* formula (terminology strictly inspired from natural deduction). Besides usual formulas we shall allow in a proof-net *discharged* ones, denoted $[A]$. These formulas are handled in a very specific way :

▸ If $[A]$ is the premise of L, then L is a ?-link (see below).

▶ If $[A]$ is a conclusion of the proof-net, we no longer require its weight to be 1. We therefore modify definition 10 as follows : a non-discharged conclusion has weight 1.

▶ We shall assume that $[A]$ is the conclusion of (unspecified) generalized axioms (i.e. boxes). This will handle all the additional rules we know, but usual dereliction [3].

▶ Since discharged formulas are bound to be merged by ?-links (see below), there is no need to superimpose such formulas, hence we require that discharged formulas are the conclusions of a unique link.

The only link we consider is the n-ary link $?_n$, with n unordered premises, which are all occurrences of the same discharged formula $[A]$, and one conclusion, namely $?A$. The case $n = 0$ is allowed, and accounts for weakening. The condition for being a proof-structure is the following : the weight increases, i.e. if $[A_i]$ is any premise of L, then $w([A_i]) \leqslant w(L)$. The condition for being a proof-net depends on an additional datum : for any ?-link L, a default jump jL is chosen : jL can be any formula B (discharged or not) such that $w(L) \leqslant w(B)$. Given a valuation φ, a switching $\mathcal{S}$ will select a jump for each ?-link L : this jump may be the default one jL or any premise $[A_i]$ of L such that $\varphi(w([A_i])) = 1$. We then state the usual connected-acyclic condition.

One must introduce a sequent calculus with two kinds of formulas, usual ones and discharged ones. This calculus is identical to the usual but for :

▶ the &-rule : the contexts may differ on some discharged formulas. From $\vdash \Gamma, [\Delta], A$ and $\vdash \Gamma, [\Delta'], B$, deduce $\vdash \Gamma, [\Delta], [\Delta'], A\&B$.

▶ the new rule of weakening/contraction : from $\vdash \Gamma, [A], \ldots, [A]$, deduce $\vdash \Gamma, ?A$.

Sequentialization is an easy exercise.

Appendix A

A.1 More about additives

A.1.1 An alternative to weights

Let Θ be a proof-net ; then we can define a structure of coherent space, on the set consisting of all formulas and all links of Θ : $X \frown Y$ iff $w(X).w(Y) \neq 0$.

---

3. Usual dereliction easily fits into our pattern : just allow any formula which is the conclusion of a link to be discharged.

Now, due to the fact that weights are monomials, as soon as $X_1, \ldots, X_n$ are pairwise coherent, the intersection $w(X_1) \cap \ldots \cap w(X_n)$ is nonzero. This means that maximal cliques in the coherent space correspond to slices. This also means that we can replace the weighting of Θ by the coherent space structure [4]. This alternative presentation has many advantages, in particular that we have not to name the eigenweights. This becomes crucial with the technique of spreading that we now introduce, since a spreading may introduce new eigenweights.

A.1.2   Spreading of a proof-net

Let Θ be a proof-net, let A be a formula in Θ and let p be an eigenweight. The spreading of Θ above A w.r.t. p consists in replacing every formula or link hereditarily above A (including A) and whose weight does not depend on p by two copies, X_1 and X_2. This induces a new coherent space structure :

- ▸ $X_i \mathbin{\bigcirc} Y_j$ iff $i = j$ and $X \mathbin{\bigcirc} Y$ in Θ

- ▸ $X_1 \mathbin{\bigcirc} Y$ iff $w(X).w(Y).p \neq 0$ in Θ

- ▸ $X_2 \mathbin{\bigcirc} Y$ iff $w(X).w(Y).\neg p \neq 0$ in Θ

- ▸ $X \mathbin{\bigcirc} Y$ iff $X \mathbin{\bigcirc} Y$ in Θ

The resulting coherent space needs not be a proof-structure. However, in the case we shall use this construction, this will be the case, and the resulting coherent space will indeed be a proof-net.

A.1.3   Cut-elimination

Let us say that a cut is *almost ready*, when both premises are the conclusions of a unique link. For almost ready cuts there is an obvious cut-elimination, the same as in the ready case. We shall now explain how to reduce a general cut to almost ready ones.

Assume that the cut is between A and $A^\perp$, with the same monomial weight w ; let $w_1, \ldots, w_m$ be the weights of the links with conclusion A, and $w'_1, \ldots, w'_n$ the weights of the links with conclusion $A^\perp$, so that

$w = w_1 + \ldots + w_m = w'_1 + \ldots + w'_n$; we assume that the cut is not at most ready, so $m + n > 2$. Let us say that an eigenweight p *splits* A when we can partition $w_1, \ldots, w_m$ into two non-empty subsets such that the partial sums are respectively equal to $w.p$ and $w.\neg p$; as soon as $m > 1$ there is at least one splitting for A.

---

4.   In this way the dependency condition looks very natural.

- if the eigenweight p is a common splitting for A and $A^\perp$, then we can duplicate A into two copies of respective weights $w.p$ and $w.\neg p$, the same for $A^\perp$, and replace our cut with two cuts

- otherwise, there is a splitting, of say, $A^\perp$ w.r.t. an eigenweight p ; we spread Θ above A w.r.t. p and we are back to the previous case. (The spreading makes sense mainly because $A^\perp \in eA$ w.r.t. any valuation, hence that all formulas above $A^\perp$ are in eA).

This procedure is Church-Rosser and terminating. It should be considered when theoretical questions (subformula property etc.) are at stake. Its computational value is limited, since iterated spreadings may induce exponentially many duplications.

A.1.4 Discussion

We would like to discuss several issues connected with additive proof-nets. Contrarily to the multiplicative case, the extant solution is not perfect (although it has the virtue to exist). Let us discuss the weaknesses of our solution and potential improvements :
the main question which occurs when translating sequent calculus to proof-nets is the problem of superimposition. There are two difficulties :

A.1.5 Weights

Weights must be monomials. However, weights of the form $p \cup q$ will naturally occur if we want to allow more superimpositions. The present state of affairs is as follows :

- in spite of years of efforts, I never succeeded in finding the right correctness criterion for these more liberal proof-nets

- general boolean coefficients might be delicate to represent (on the other hand, the case we consider has a natural presentation in terms of coherent spaces)

- normalization in the full case might be messy

A.1.6 Identity of formulas

Imagine that we have no problem with weights, and that we try to maximize the identifications. One idea is to adopt a binary $\oplus$-rule, which is quite natural. Besides that, we immediately stumble on a lot of problems :

▸ if I have a cut on A in both proofs. should I superimpose them... worse, if both proofs have two cuts on A, which ones should be superimposed ? The same question occurs with the contraction rule : how do we superimpose two contraction rules (remember that the premises are indiscernible).

▸ what about existential witnesses ? Should one superimpose two portions of proofs with the same structure, but different witnesses ?

▸ what about normalization ? During this process, distinct portions of proofs might become equal, hence it would be necessary to superimpose them...

These limitations do not apply if we restrict to cut-free proofs in **MALL**, and if we make the extra assumption that all identity links are atomic : there is a well defined notion of net (provided one can fix the problem of weights) which enjoys the maximum number of identifications.

If identification is difficult, its converse is easy, i.e. there is not the slightest problem to forget that two formulas are equal. This could indicate a possible theoretical way out, namely considering a proof-net as the set of its slices. The computational value of this idea is limited (exponential growth of the net), but this might be valuable for theoretical considerations. However, we have no idea how to define a correctness condition for such sets of slices.

There are other problems connected to normalization :

A.1.7   Normalization

Multiplicative proof-nets have a local cut-elimination, and this is still true for quantifiers, provided we do not compute the existential witnesses. The additive case involves a real global move, which consists in setting an eigenweight to 0 or 1, and erase everything with weight 0. This is rather brutal, and completely foreign to the parallel asynchronous spirit of proof-nets. In [2] (section 3.4.) a variant of usual sequent calculus is introduced, with a local cut-elimination, i.e. the erasing is performed in a lazy way, which means that some useless "beards", which are bound to be erased, are still hanging. The calculus with these beards is connected with the additive neutral $\top$ (which has still no satisfactory treatment), and we can expect that there is a notion of additive bearded proof-net with a local elimination. This is perhaps the most promising open question in this paper.

A.2   Multiplicative neutrals

There are two multiplicative neutrals, 1 and $\perp$, and two rules, the axiom $\vdash 1$ and the weakening rule : from $\vdash \Gamma$, deduce $\vdash \Gamma, \perp$. Both rules are handled by means of links with one conclusion and no premise ; however ⊥-links are treated like 0-ary ?-links, i.e. they must be given a default jump. Sequentialization is immediate.

At first sight, cut-elimination is unproblematic : replace a cut between the conclusions 1 and ⊥ of zero-ary links with... nothing. But we notice a new problem, namely that a cut formula A can be the default jump of a ⊥-link L, and we must therefore propose another jump for L. Usually one of the premises of the link with conclusion A works (or the jump of L' if A is the conclusion of a ⊥-link) works. Worse, this new jump is by no ways natural (if A is $B \otimes C$, the new jump can either be B or C), which is quite unpleasant. As far as we know, the only solution consists in declaring that the jumps are not part of the proof-net, but rather of some control structure. It is then enough to show that at least one choice of default jump is possible. This is not a very elegant solution : we are indeed working with equivalence classes of proof-nets and if we want to be rigorous we shall have to endlessly check that such and such operation does not depend on the choice of default jumps. In practice one can be rather sloppy...

Of course everything would be nicer without any default jump. But then a proof-net for a multiplicative combination A of occurrences of 1 and ⊥ would basically be nothing more than A itself : the correctness criterion for proof-nets without jumps encompasses the decision problem for such combinations, and this problem is known to be NP-complete by [8]... The existence of a correctness criterion of the same style as the familiar ones is therefore very unlikely [5].

This discussion is fully relevant to the exponential case : as soon as we start to normalize exponential cuts, then the same problems (with the same solution) arise.

A.3 Additive neutrals

There is still no satisfactory approach to additive neutrals, which are fortunately extremely uninteresting in practice. The only way of handling ⊤ is by means of a box or, if one prefers, by means of a second order translation : on this Kamtchatka of linear logic, the old problems of sequent calculus are not fixed. The absence of a satisfactory treatment of ⊤ calls for another notion of proof-net... presumably a solution to the wider question of bearded proof-nets.

---

5. There is still a possibility, namely to work with the necessary condition *acyclic graph with $n + 1$ connected components*, where n is the number of ⊥-links ; but we would have to check that this condition is preserved through cut-elimination (which seems likely), and we would have to restrict to proof-nets in which the formula ⊥ cannot occur in the absence of cuts, so that our necessary condition eventually becomes sufficient...

References

[1] V. Danos and L. Regnier, *The structure of multiplicatives*, Arch. Math. Logik Grundlag. **28** (1989), no. 3, 181–203.

[2] J.-Y. Girard, *Geometry of interaction III: the general case*, Proceedings of the Workshop on Linear Logic, MIT Press, submitted.

[3] _____, *Linear Logic*, Theoret. Comput. Sci. **50** (1987), no. 1, 1–102.

[4] _____, *Multiplicatives*, Rend. Sem. Mat. Univ. Politec. Torino (1987), 11–33, special issue on Logic and Computer Science.

[5] _____, *Quantifiers in linear logic*, Temi e prospettive della logica e della filosofia della scienza contemporanee, CLUEB, Bologna, 1989.

[6] _____, *Towards a Geometry of Interaction*, Categories in Computer Science and Logic, Contemporary Mathematics **92**, AMS, 1989, pp. 69–108.

[7] _____, *Quantifiers in linear logic II*, Nuovi problemi della logica e della filosofia della scienza, CLUEB, Bologna, 1991.

[8] P. Lincoln and T. Winkler, *Constant-only Multiplicative Linear Logic is NP-complete*, Theoret. Comput. Sci. **35** (1994), 155–159.

Laboratoire de Mathématiques Discrètes, UPR 9016 – **CNRS**, 163, Avenue de Luminy, Case 930, F-13288 Marseille Cedex 09
    e-mail address: GIRARD@LMD.UNIV-MRS.FR

Magari and others
on Gödel's ontological proof

Petr Hájek

Abstract

Gödel's proof of the necessary existence of God is analyzed from the point of view of modal logic together with related papers by Magari and Anderson. In particular, Magari's claim on redundance in Gödel's axioms is analyzed and shown to be only true for some extension of Gödel's system (but true for Anderson's modification, as it was shown elsewhere). Completeness of the underlying modal logic is proved; and it is shown that the "ontological" proof may use only the logic KD45 (logic of belief) instead of S5 (logic of knowledge).

1 Introduction

This paper is a continuation of my paper [4] and concentrates almost exclusively to mathematical properties of logical systems underlying Gödel's ontological proof and its variant by Anderson [1], with special care paid to Magari's criticism [6]. Since [4] is written in German, we shall try to summarize its content in such a way that knowledge of [4] will be not obligatory for reading the present paper (even it remains advantageous). Here we describe the development related to Gödel's proof in a rough way, in Section 1 (Preliminaries) we tell more formal details. Gödel's proof uses variables for individuals, variables for properties, one predicate P (positive) applicable to properties, modalities $\Box$ and $\Diamond$ and equality among individuals. There are 5 axioms and some definitions; in particular, x is *godlike*, notation $G(x)$, if x has all positive properties; a property X is an *essence* of x, notation $X Ess.x$, if $X(x)$ and each Y such that $Y(x)$ necessarily includes X; x *necessarily exists* if each essence X of x is necessarily instantiated, i.e. $\Box(\exists y)X(y)$. There are some few lemmas and Theorem 3 saying $\Box(\exists x)G(x)$ - necessarily there is a godlike individual. (See Sect.1,4 for more details). Gödel's text resembles sacral texts; it has no introduction, motivations, statement of underlying logic, and therefore needs some "hermeneutic" interpretation. In particular, it has become clear that following questions are relevant: (1) what *comprehension axiom(s)* does one have to assume, i.e. which formulas define properties? (2) what *equality axioms* are to be assumed? And (3), what *modal axioms* are to be assumed? Sobel [7] and Polívka (cf.[4]) observed that assuming full comprehension leads to *collapse of modalities*, i.e. $(\forall X)(\forall x)(X(x) \equiv \Box X(x))$ becomes provable, which means that the system trivializes. Magari [6] claims that the first three Gödel's axioms $(A1) - (A3)$ already imply the main theorem; we shall analyze his claim in Section 2 and show that his proof implicitly uses a too strong equality axiom (without which the claim is not true, as shown in [4]). Anderson [1] presented an emended system admitting full comprehension and not suffering by collapse of modalities; I proved in [4] that for Anderson's system Magari's claim becomes true; the analogs of Gödel's $A1 - A3$ prove necessary existence of a god-like individual (in the modified sense), but this does not mean that the system trivializes because Anderson's system (of his $A1 - A3$ plus full comprehension) can be presented as a conservative extension of Gödel's system $A1 - A5$ with a certain "cautious" comprehension axiom. Briefly, Anderson's system admits some "non-convertible" properties; Gödel's system results when we drop all non-convertible properties. We give more details in Section 2 (Preliminaries.) Section 2 analyzes Magari's claim; Section 4 presents a completeness theorem for the underlying logic (which shows that no hidden axioms remain to be formulated). Up to that point, the modal logic is $S5$; in Section 5 we show that a weaker logic $KD45$, known as *logic of belief* (in contradistinction to $S5$, called *logic of knowledge* in AI) is sufficient to the proof of the main theorem both in Gödel's and in Anderson's system but is not sufficient to prove that Anderson's system includes Gödel's system via restriction to convertible properties. Section 6, the only place discussing extra-mathematical things, contemplates on Magari's philosophical standpoint, as apparent from [6].

2   Preliminaries

The propositional skeleton of the ontological proof, as presented by [2], is simple: one has to establish (q standing for $(\exists x)G(x)$)

(1) $\Diamond q$ (as intuitively accepted),

(2) $q \to \Box q$ (Anselm's principle).

Then $\Diamond q \to \Diamond \Box q \to \Box q$ (e.g. by S5).

   Gödel's axioms are

(A1) $P(X) \equiv \neg P(\neg X)$

(A2) $P(X) \ \& \ \Box(\forall x)(X(x) \to Y(x)) \to P(Y)$

(A3) $P(G)$
(A4) $P(X) \to \Box P(X)$
(A5) $P(NE)$

(here $G(x) \equiv (\forall Y)(P(Y) \to Y(x))$, $XEss.x \equiv X(x)$ & $(\forall Y)(Y(x) \to \Box(\forall z)(X(z) \to Y(z)))$, $NE(x) \equiv (\forall Y)(YEss.x \to \Box(\exists z)Y(z)))$. We quickly go through Gödel's proof.

Lemma: (i) $P(X) \to \Diamond(\exists x)X(x)$; (ii) $G(x) \to GEss.x$

Proof. (i) If $\Box(\forall x)\neg X(x)$ then $\Box(\forall x)(X(x) \to \neg X(x))$, thus $P(X) \to P(\neg X)$, contradiction. (ii) One easily shows $G(x) \to (\forall Y)(P(Y) \equiv Y(x))$; now if $G(x)$ and $Y(x)$ then $P(Y)$, hence $\Box P(Y)$, hence $\Box(\forall z)(G(z) \to Y(z))$. This is $GEss.x$.

Theorem. $\Box(\exists x)G(x)$

Proof. Lemma (i) gives $\Diamond(\exists x)G(x)$; we prove $G(x) \to \Box(\exists x)G(x)$. Indeed, assume $G(x)$; since $P(NE)$ and $GEss.x$ we get $\Box(\exists z)G(z)$ by the definition of NE. The theorem follows by Hartshorne's observation above.

Remark. (i) $G(x) \to \Box G(x)$, thus $(\exists x)\Box G(x)$.
(ii) $G(x)$ & $Y(x). \to \Box Y(x)$.

Proof. (i) Let $Z(y)$ be $y = x$; then $G(x) \to \Box(\exists y)(G(y) \to Z(y))$, thus $G(a)$ & $\Box(\exists y)G(y) \to \Box(\exists y)(G(y)$ & $y = a) \to \Box G(a)$.
(ii) $G(a)$ & $Y(a) \to \Box G(a)$ & $\Box(\forall z)(G(z) \to Y(z)) \to \Box Y(a)$.

The *comprehension axiom* for a formula $\varphi(x)$ (and possibly other free variables, but not Y) is the axiom

$$(\exists Y)\Box(\forall x)(Y(x) \equiv \varphi(x)).$$

The *full comprehension scheme* (C_{full}) is the scheme of *all* comprehension axioms. We implicitly used three comprehension axioms above, e.g. to know that $G, NE, \Phi_x(y) \equiv y = x$, exist as properties. As mentioned above, Gödel's system with (C_{full}) suffers by the collapse of modalities; the following is the *cautious comprehension scheme* (C_{caut}):
$(\forall x)(G(x) \to (\Box\varphi(x) \vee \Box\neg\varphi(x))) \to$
$\to (\exists Y)\Box(\forall z)(Y(z) \equiv \varphi(z))$.
$(GO)_{caut}$ will denote the axiom system (A1)-(A5), comprehension axioms for G, NE, I_x (see below) and the schema (C_{caut}).

$$(\exists G)\Box(\forall x)(G(x) \equiv (\forall Y)(P(Y) \to Y(x)))$$

$$(\exists NE)\Box(\forall x)(NE(x) \equiv (\forall Y)(YEss.x \to \Box(\exists z)Y(z)))$$

$$(\forall x)(\exists I)\Box(\forall y)(I(y) \equiv y = x).$$

$(GO)_{caut}$ is consistent and does not suffer by collapse of modalities (see next section).

Anderson's variant: (A1) is weakened to
(AA1) $\qquad\qquad P(X) \rightarrow \neg P(\neg X),$
(AA2) is identical with (A2). The definition of a godlike object is changed to

$$H(x) \equiv (\forall Y)(P(Y) \equiv \Box Y(x))$$

(thus x is godlike in new sense iff positive properties are exactly properties necessarily applying to x).
(AA3) is $P(H)$ $\qquad\qquad$ (godlikeness is positive).
Anderson's axioms (AA4), (AA5) are irrelevant:

Theorem: $(AA1) - (AA3) \vdash \Box(\exists x)H(x)$

Proof. $\Diamond(\exists x)H(x)$ is provable as above; we prove $H(x) \rightarrow \Box H(x)$. Indeed, if $H(x)$ and $P(H)$ then H necessarily applies to x, by the definition of H.

Here we have used just the comprehension axiom for H. $(AO)_{full}$ will be (AA1)-(AA3) plus (C_{full}). $(AO)_{full}$ is consistent and does have collapse of modalities, by [1] (and by obvious models, see below). More than that: define in $(AO)_{full}$ a property X to be *convertible* if

$$(\forall x)(H(x) \rightarrow (\Box X(x) \vee \Box \neg X(x))).$$

Note that e.g. H is convertible. ("Convertible" since going from X to $\neg X$ we convert from positive to negative; nonconvertible classes are neither positive non negative. Also "convertible" should resemble "constructible"; the reason for this will be immediately clear to a reader knowing properties of the Gödel's constructible inner model of set theory).

Theorem: ([4]) Restricting properties to convertible properties, interpreting G as H and letting everything else absolute we get an interpretation # of $(GO)_{caut}$ in $(A5)_{full}$. This interpretation is faithful. i.e. for each closed formula φ of $(GO)_{caut}$,

$$(GO)_{caut} \vdash \varphi \text{ iff } (AO)_{full} \vdash \varphi^{\#}$$

This is proved by constructing an interpretation $\flat$ of $(AO)_{full}$ in $(GO)_{caut}$ such that the composition $\# * \flat$ is identical.

3 The status of Magari's claim

In [6], Magari presents a passionate criticism of Gödel's proof (and proofs of God in general - see Sect.5). Here we shall analyze his claim, mentioned above, that Gödel's axioms (A1)-(A3) are enough to prove $\Box(\exists x)G(x)$. His proof is semantical: he describes a class of models of the underlying language and then shows that each model of this class which satisfies (A1)-(A3) satisfies also $\Box(\exists x)G(x)$. The difficulty is that the system is not complete with respect to this class; we shall show which additional axiom, true in all Magari models, suffices to make his claim valid, then we refer to [4] for another semantics which is enough to show that without any additional axiom Magari's claim is false. (But note again the remarkable fact explained in the preceding section that for Anderson's variant Magari's claim does hold.)

Define for a moment: a *Magari model* (see [6]) is a tuple $\langle W, (M_w)_{w \in W},$ $(B_w)_{w \in W}, (P_w)_{w \in W}\rangle$ where W is a set of possible worlds and for each $w \in W$: $M_w \neq \emptyset$ is a set of individuals, B_w is a Boolean algebra of subsets of M_w, $P_w \subseteq B_w$ is a system of positive sets. Each object variable is interpreted as a mapping f_x such that $f_x(w) \in M_w$ for all w, similarly $f_X(w) \in B_w$ for call w; $w \Vdash X(x)$ iff $f_x(w) \in F_X(w)$, $w \Vdash P(X)$ iff $f_X(w) \in P_w$. One defines $w \Vdash X = Y$ iff $w \Vdash (\forall x)(X(x) \equiv Y(x))$, i.e. $f_X(w) = f_Y(w)$. But observe that under this definition the following axiom becomes true in each Magari model (pointwise equality axiom for P):

(PEP) $\qquad\qquad P(X) \ \& \ X = Y. \rightarrow P(Y).$

Theorem (Magari revised) (A1)-(A3), in presence of (PEP), (and some comprehension) prove $\Box(\exists x)G(x)$.

Proof. We first show $P(X) \rightarrow (\exists x)X(x)$. Assume $P(X)$ and $(\forall x)\neg X(x)$. Then for all Y, $Y = X \cup Y$ (the existence of a property being (necessarily) the union of two gives properties is the comprehension axiom needed and valid in all Magari models); we have $\Box(X \subseteq X \cup Y)$ and therefore $P(X \cup Y)$ by (A2). Using PEP (!) we get $P(Y)$, thus $(\forall Y)P(Y)$ which contradicts (A1).

In particular, by (A3) we have $P(G)$, hence $(\exists x)G(x)$, which gives $\Box(\exists x)G(x)$ by necessitation.

Note that similar semantics is considered by Czermak [3]. Now we recall the semantics of [4]; we just speak on models.

A *model* is a tuple $K = \langle W, M, Prop, \mathcal{P}\rangle$ where $W \neq \emptyset$ is a set of possible worlds, $M \neq \emptyset$ is a set of individuals (common for all possible worlds), $Prop$ is a non-empty set of mappings $F : W \times M \rightarrow \{0,1\}$ and $\mathcal{P} : W \times Prop \rightarrow \{0,1\}$. Individual variables x are interpreted by elements m_x of M, property variables X by elements F_X of $Prop$; $w \Vdash X(x)$ iff $F_X(w, m_x) = 1$. $w \Vdash P(X)$ iff $\mathcal{P}(w, F_x) = 1$; $w \Vdash x = y$ iff $m_x = m_y$. The rest is obvious (in particular, $w \Vdash \Box\varphi$ iff for all $w' \in W$, $w' \Vdash \varphi$). Note that here we also have a "hidden" axiom, namely $x = y \rightarrow \Box(x = y)$ (objects are fixed); this could be removed but we prove completeness (in the next section) and it is some simplification to have just one universe of individuals.

Fact. Let $K = \langle W, M, Prop, \mathcal{P} \rangle$ be a model and let $g \in M$. (1) Assume that $Prop$ consists of all $F : M \times W \rightarrow \{0, 1\}$ such that F is constant as $W \times \{g\}$ and that $\mathcal{P}(w, F) = 1$ iff (independently of w) for all $v \in W$, $F(v, g) = 1$. Then K is a model of $(GO)_{cau}$.

(2) Assume that $Prop$ consists of *all* mappings $F : M \times W \rightarrow \{0, 1\}$; let $\mathcal{P}$ be as above. Then K is a model of $(AO)_{full}$.

(3) Let $w_0 \in W$ let $Prop$ be as in (2) and let $\mathcal{P}(w, F) = 1$ iff $F(w_0) = 1$ (independently of w). Then $K \Vdash (A1)-(A3)$ and $\Diamond(\exists x)G(x)$ and $\Diamond\neg(\exists x)G(x)$. This refutes Magari's claim in its original form. (All these examples are from [4].)

4 A completeness theorem

We prove, in a standard manner, a completeness theorem giving completeness of systems like $(AO)_{full}$, $(GO)_{caut}$ as a particular case.

The system.
Variables: object variables $x, y, \ldots$ and property variables $X, Y, \ldots$.
Atomic formulas: $X(x)$ (these can be written as $x \in X$), $P(X)$, $x = y$ and $X = Y$.

Axioms: Two-sorted *predicate calculus* (examples of axioms: $(\forall x)\varphi(x) \rightarrow \varphi(y)$, $(\forall X)\varphi(X) \rightarrow \varphi(Y)$ - under the usual substitutability conditions); *equality axioms* for objects (reflexivity, symmetry, transitivity); $\varphi(x) \;\&\; x = y. \rightarrow \varphi(y)$; definition of *equality for properties:* $X = Y \equiv (\forall x)(X(x) \equiv Y(x))$ (extensionality); *equality axiom* for P: $P(X) \;\&\; \Box(X = Y). \rightarrow P(Y)$. *Modal axioms* (S5):
$\Box(\varphi \rightarrow \psi) \rightarrow (\Box\varphi \rightarrow \Box\psi)$, $\Box\varphi \rightarrow \varphi$,
$\psi \equiv \Box\psi$ for ψ boxed (i.e. resulting from formulas of the form $\Box\chi$ using connectives and quantifiers).
Necessary equality for objects: $x = y \rightarrow \Box(x = y)$.

Global and local theories. A *global theory* Q is given by some special axioms; φ is *provable* in Q if there is a proof of φ from logical axioms above and special axioms of Q using modus ponens, generalization and necessitation as deduction rules. Similarly for *local theories* T, but the only deduction rules are modus ponens and generalization. $Q \vdash \varphi$ means global provability, $T \vdash^o \varphi$

local provability. $(AO)_{full}$ and $(GO)_{caut}$ are examples of global theories, i.e. we are interested in their global consequences. $Cn(Q)$ is the set of all global consequences of Q.

Models are structures as above, i.e. $K = \langle W, M, Prop, \mathcal{P} \rangle$ with $Prop$ being a set of mappings $X : (W \times M) \to \{0, 1\}, \mathcal{P} : (W \times Prop) \to \{0, 1\}$.

Completeness theorem. If Q is a global theory α a closed formula and $Q \not\vdash \neg\alpha$ then there is a model $K = \langle W, M, Prop, \mathcal{P} \rangle$ of Q (i.e. each axiom of Q is true in each world of K) and a possible world $w \in W$ such that $w \Vdash \alpha$. The rest of the section contains a proof of the completeness, together with some auxiliary lemmas.

Lemma. $(1) \vdash \Box(\forall x)\varphi \equiv (\forall x)\Box\varphi$
$(2) \vdash (\exists x)\Box\varphi \to \Box(\exists x)\varphi$
Proof. $\vdash \Box(\forall x)\varphi \to \varphi$, $\vdash \Box(\Box(\forall x)\varphi \to \varphi)$, $\vdash \Box\Box(\forall x)\varphi \to \Box\varphi$, $\vdash \Box(\forall x)\varphi \to \Box\varphi$, $\vdash (\forall x)(\Box(\forall x)\varphi \to \Box\varphi)$, $\vdash \Box(\forall x)\varphi \to (\forall x)\Box\varphi$.
The rest is similar.

Notation. Q and α are fixed; T_{-1} is $Cn(Q) \cup \{\alpha\}$ each $n = 0, 1, \ldots$, $Const_n$ is a set of object and property constants (infinitely many of either kind), for $m \neq n$ $Const_m$ and $Const_n$ are disjoint. $Const_\infty = \bigcup_n Const_n$; $Const_{<n} = \bigcup_{i<n} Const_i$. $\{\varphi_n\}$ is an enumeration of all formulas of $L(Const_\infty)$ (the language enriched by all the constants) such that for each n, $\varphi_n \in L(Const_{<n})$, $\varphi_0 = \bot$. We assume that each closed formula φ occurs infinitely many times in the enumeration; n is a *starter* if φ_n occurs first time, otherwise n is a *repeater* and $t(n)$ is the biggest $m < n$ such that φ_n is the same formula as φ_m.

Construction. T_0 is a complete Henkin extension of T_{-1} using $Const_0$ as Henkin constants; i.e. for each closed formula $(\exists x)\varphi(x)$ of $L(Const_0)$ there is a constant $c \in Const_0$ such that $T_0 \vdash^0 (\exists x)\varphi(x) \to \varphi(c)$, and similarly for $(\exists X)\varphi(X)$.

For each T, $B(T) = \{\Box\psi \mid \psi \text{ closed}, T \vdash \Box\psi\}$. Note that if $T \supseteq Cn(Q)$ is complete and ψ is closed then either $(\Box\psi) \in B(T)$ or $(\Box\Diamond\neg\psi) \in B(T)$ and $B(T) \vdash^0 \Diamond\neg\psi$.

We construct complete Henkin local theories T_n; for each n we shall have $T_n \supseteq Cn(Q) \cup \{\alpha\} \cup B(T_{n-1})$; if n is a repeater we shall have $T_n \supseteq T_{t(n)}$. The definition is as follows:
Let T_{n-1} be defined $(n > 0)$. We define an auxiliary theory T':
Case 1. n is a starter and $(\Box\neg\varphi_n) \in B(T_{n-1})$; put $T' = T_{n-1}$
Case 2. n is a starter and $(\Box\Diamond\varphi_n) \in B(T_{n-1})$; put $T' = Cn(Q) \cup B(T_{n-1}) \cup \{\varphi_n\}$.
Case 3. n is a repeater: put $T' = Cn(Q) \cup B(T_{n-1}) \cup T_{t(n)}$.

Claim 1. T' is consistent (as a local theory). (Proof below). Now let T_n be a complete Henkin local theory using $Const_n$ as witnessing constants. This completes the construction.

For each closed φ, let $w_\varphi = \bigcup \{T_n \mid \varphi_n = \varphi\}$.

Claim 2. w_φ is a complete Henkin local theory in $L(Const_\infty)$; $B(w_\varphi)$ is the some set B for all φ's . (Obvious.)

Definition of the model. For each object constant c, let $[c] = \{d \mid (\Box(c = d)) \in B\}$; let M be the set of all $[c]$. W is the set of all w_φ. For each property constant C, let F_C be the mapping on $W \times M$ such that $F_C(w_\varphi, [c]) = 1$ iff $w_\varphi \vdash^0 C(c)$ (clearly, the definition does not depend on the choice of a representative c of $[c]$). *Prop* is the set of all mappings F_C; $\mathcal{P}(w_\varphi, F_C) = 1$ iff $w_\varphi \vdash^0 P(C)$. This completes the definition of $\langle W, M, Prop, \mathcal{P} \rangle$.

Claim 3. $(F_C = F_D \ \& \ \mathcal{P}(w_\varphi, C) = 1) \to \mathcal{P}(w_\varphi, D) = 1$.

Claim 4. For each closed φ and each $w \in W$, $w \Vdash \varphi$ iff $w \vdash^0 \varphi$. In particular, $w_\perp \Vdash \alpha$, which completes the proof of completeness. Since soundness of our semantics is evident we have the usual of Q.

Corollary. $Q \vdash \alpha$ iff α is true in each world of each model.

Remark. $(GO)_{caut}$ and $(AO)_{full}$ are examples of global theories, thus we have completeness for them. Note that both theories prove $(\forall Y)(P(Y) \to \Box P(Y))$, thus they prove $(\forall Y)(P(Y) \equiv \Box P(Y))$ and we may confine us to models $\langle W, M, Prop, \mathcal{P} \rangle$ where $\mathcal{P}$ does not depend on possible worlds, i.e. $\mathcal{P} : Prop \to \{0, 1\}$.

Appendix to Section 4: Proofs of claims.

Proof of claim 1. Case 2: Let $Cn(Q) \cup B(T_{n-1}) \cup \{\varphi_n\}$ be inconsistent (as a local theory); then $Cn(Q) \cup B(T_{n-1}) \vdash^0 \neg\varphi_n$, thus for some $(\Box\psi) \in B(T_{n-1})$, $Cn(Q) \vdash^0 \Box\psi \to \neg\varphi_n$, thus $Q \vdash \Box\psi \to \varphi_n$, $Q \vdash \Box\psi \to \Box\neg\varphi_n$, thus $T_{n-1} \vdash \Box\neg\varphi_n$, a contradiction (with the assumption of Case 2).

Case 3. Assume $Cn(Q) \cup B(T_{n-1}) \cup T_{t(n)}$ inconsistent, thus $Cn(Q) \cup B(T_{n-1}) \cup T_{t(n)} \vdash^0 \perp$. Assume axioms of $T_{t(n)}$ to be closed. Then there is a $\chi \in T_{t(n)}$ and a $(\Box\psi) \in B(T_{n-1})$ such that $Cn(Q) \cup \{\Box\psi\} \vdash^0 \neg\chi$, thus $Q \vdash \Box\psi \to \neg\chi$, $Q \vdash \Box\psi \to \Box\neg\chi$, $T_{n-1} \vdash^0 \Box\neg\chi$, but $T_{t(n)} \vdash^0 \psi$, $T_{t(n)} \vdash^0 \Diamond\psi$, $(\Diamond\psi) \in B_{t(n)} \subseteq T_{n-1}$, thus T_{n-1} is contradictory.

Proof of claim 3. To prove claim 3 it is enough to show that if $F_C = F_D$ then $\Box(\forall x)(C(x) \equiv D(x)) \in B$. Indeed, if $F_C = F_D$ then for all object constants c and all w_φ, $w_\varphi \vdash^0 C(c) \equiv D(c)$; since w_φ is Henkin, we get $w_\varphi \vdash (\forall x)(C(x) \equiv D(x))$. And if the formula $\Box(\forall x)(C(x) \equiv D(x))$ were not in B then for the n which is the starter for this formula we would get $(\exists x)(C(x) \not\equiv D(x))$, a contradiction.

Proof of claim 4. We prove $w \Vdash \varphi$ iff $w \vdash^0 \varphi$ by induction on the complexity of the closed formula φ. (For non-closed formulas one works with satisfaction by a given evaluation e of object and property variables.) The claim is obvious for atomic formulas $c = d$, $C(c)$, $P(C)$; the induction step for connectives and for quantifiers is also obvious (using completeness and existence of Henkin constants). Consider $\Box\varphi$, φ being closed:

$(\forall w)(w \Vdash \Box\varphi)$ iff $(\forall w)(w \Vdash \varphi)$ iff $(\forall w)(w \vdash^0 \varphi)$ iff $(\forall w)(w \vdash^0 \Box\varphi)$;

in the last equivalence, $\Leftarrow$ is obvious and $\Rightarrow$ uses the fact that if $(\exists w)(w \vdash^0 \Diamond\neg\varphi)$ then $(\exists w')(w' \vdash^0 \neg\varphi)$: take n to be the starter for $\neg\varphi$, then $w_n \vdash^0 \neg\varphi$ since $(\Box\varphi) \notin B(T_{n-1})$ (because $(\Box\Diamond\neg\varphi) \in B$). This completes the proof.

5   Weakening the modal logic

Let us now delete the axiom $\Box\varphi \to \varphi$, i.e. let us weaken the modal logic from S5 (logic of knowledge) to KD45 (logic of belief) and let us see what happens with our theories. (Cf. e.g. [8] for information on KD45.) $(AO)^-_{full}$ and $(GO)^-_{caut}$ denote the theories with the same special axioms as $(A5)_{full}$ and $(GO)_{caut}$ respectively but with the weakened logic.

Observation. $(GO)^-_{caut} \vdash \Box(\exists x)G(x)$ and $(AO)^-_{full} \vdash \Box(\exists x)H(x)$, by the same proofs as in Section 1. (Just check that the deleted modal axiom is not used.)

This means that with this modal logic both system prove that it is *believed* that a godlike individual exists. We consider some models of both systems and exhibit some unprovabilities.

Models for $(GO)^-_{caut}$: Consider models of the form

$$K_1 = \langle W, M, Prop, \mathcal{P}, W_0, g \rangle$$

where W, M are as above, $\emptyset \neq W_0 \subseteq W$, $g \in M$, $Prop$ consists of all mappings $X : (W \times M) \to \{0,1\}$ constant on $W_0 \times \{g\}$; $\mathcal{P} : Prop \to \{0,1\}$, $\mathcal{P}(X) = 1$ iff X equals 1 on $W_0 \times \{g\}$; $w \Vdash \Box\varphi$ iff for all $w' \in W_0$, $w'' \Vdash \varphi$. Then K_1 is a model of $(GO)^-_{caut}$, $K \Vdash \Box G(g)$, but for $w \in W - W_0$, $w \Vdash (\forall x)\neg G(x)$.

Corollary 1. $(GO)^-_{caut} \nvdash (\exists x)G(x)$.

Models for $(AO)^-_{full}$: If we modify the model above just letting $Prop$ to be the set of all $X : (W \times M) \to \{0,1\}$ but let the definition of $\mathcal{P}$ unchanged then we get a model K_2 of $(AO)^-_{full}$ in which for each w, $w \Vdash H(g)$ (since the truth-value of $H(g)$ is world-independent in this model). But we may get other models of $(AO)^-_{full}$, not satisfying $(\forall Y)(\mathcal{P}(Y) \equiv \Box P(Y))$, as follows: $K = \langle W, M, Prop, \mathcal{P}, W_0, g \rangle$, $Prop$ consists of all $X : (W \times M) \to \{0,1\}$, $\mathcal{P} : (W \times Prop) \to \{0,1\}$ and the following holds:

$W \in W_0$ implies $\mathcal{P}(w, X) = 1$ iff x has constantly value 1 on $W_0 \times \{g\}$, $w \in W - W_0$ implies $\mathcal{P}(w, X) = 0$.

Note that necessity is defined using $W_0 : w \Vdash \Box\varphi$ iff $(\forall w' \in W_0)(w' \Vdash \varphi)$.

In this model we have $K \Vdash \Box H(g)$, (i.e. for each $w \in W_0$, $w \Vdash H(g)$), but for $w \in W - W_0$ we have $w \Vdash (\forall x)\neg H(x)$.

Corollary 2. $(AO)^-_{full} \nvdash (\exists x)H(x)$.

Corollary 3. $(AO)^-_{full} \nvdash G^\#(x) \equiv H(x)$.

Indeed, in K_2 $w \Vdash H(g)$ for each w, but if $w \in W - W_0$ then $w \Vdash \neg G^\#(g)$ (note that $\#$ defines K_1 in K_2). Thus $\#$ is not an interpretation of $(GO)^-_{caut}$ in $(AO)^-_{full}$. On the positive side, we have the following:

Observation. (1) $(AO)^-_{full} \vdash \Box(G^\#(x) \equiv H(x))$

(2) $(AO)^-_{full} + (\forall Y)(P(Y) \equiv \Box P(Y)) \vdash (\exists x)H(x)$

Proof. (1) follows from the fact that $KD45 \vdash \Box(\Box\varphi \to \varphi)$.

(2) Let g be such that $\Box H(g)$, thus

$\Box((\forall Y)(P(Y) \equiv \Box Y(g)))$,

$(\forall Y)(\Box P(Y) \equiv \Box Y(g))$,

$(\forall Y)(P(Y) \equiv Y(g))$.

which gives $H(g)$.

Summarizing, the removal of the axiom $\Box\varphi \to \varphi$, i.e. replacement of S5 by KD45 does not affect proofs of the main theorem in either system, but the existence of a godlike object becomes unprovable and the nice relation between both systems is lost. A deeper investigation of the weakened systems (including some completeness theorem) could shed more light to them.

6   Logica, teofilia e morte

Professor Magari starts his paper [6] by saying "Teofili hanno spesso fornito ingegnosi argomenti... esistono anche teofobi (io lo sono di tutto cuore)", thus "theophils have collected ingenious arguments [for the existence of God]; there are also theophobs (as me, by my whole hearth)". I cannot share Magari's "theophoby" (being a Christian); but, as I remarked in [4], I agree with him saying "non è più facile ammettere gli assiomi che ammettere direttamente il teorema" (thus "it is not easier to accept the axioms [of Gödel] than to accept directly the theorem"). Quotations from [5] will not be repeated here; but let me state explicitly that I regret very much that I have postponed serious discussion with Magari - until it was too late.

Quoting from the end of [6], I think that both "teofili" and "teofobi" may agree with him saying "occorre in ogni caso stare molto in guardia contro tutto ciò che può essere suggerito dal desiderio di credere" (in any case one has to be very sensitive against everything possibly suggested with the desire to believe)

- just because one's faith is not and should not be a result of somebody else's demand. The immediate continuation of the last quotation documents, in my opinion, Magari's tolerance and preference of objectivity against his own feelings: "Debbo ammettere, con una certa riluttanza, che analogamente va trattato il desiderio di non credere" (I have to admit, with a certain reluctance, that the desire not to believe has to be treated analogously). The end of the quotation reads "...che però mio sembra assai più raro" (but the latter seems to me rather more rare); I comment that during the communist system in my country the official atheist propaganda and "desiderio di non credere" was very strong and frequent. But this does not concern the main topic of discussion.

Professor Magari is dead and we remain grateful for his contribution to logic and various other domains of mathematics.

References

[1] C.A. Anderson, *Some Emendations to Gödel's Ontological Proof*, Faith and Philosophy **7** (1990).

[2] R. Brecher, *Hartshorne's Modal Argument for the Existence of God*, Ratio **XVII** (1975).

[3] J. Czermak, *Abriss des ontologischen Argumentes*, Wahrheit und Beweisbarkeit – Leben und Werk K. Gödels (Buld, Köhler, and Schimanowich, eds.), to appear in 1995.

[4] P. Hájek, *Der Mathematiker und die Frage der Existenz Gottes (betreffend Gödels ontologischen Beweis)*, Wahrheit und Beweisbarkeit – Leben und Werk K. Gödels (Buld, Köhler, and Schimanowich, eds.), to appear in 1995.

[5] H. Küng, *Existiert Gott?*, Piper Verlag, Stuttgart, 1978.

[6] R. Magari, *Logica e teofilia*, Notizie di logica VII **4** (1988).

[7] J.H. Sobel, *Gödel's Ontological Argument*, On Being and Saying (Hersg. von Thompson), MIT Press, Cambridge MA, 1987.

[8] F.P.M.J. Vorbraak, *As for as I know (Epistemic logic and uncertainty)*, Ph.D. thesis, Univ. Utrecht, 1993.

INSTITUTE OF COMPUTER SCIENCE, ACADEMY OF SCIENCES, 182 07 PRAGUE, CZECH REPUBLIC
e-mail address: hajek@uivt.cas.cz

Finitely Generated Magari Algebras and Arithmetic

Lex Hendriks*   Dick de Jongh

Abstract

Some consequences are studied of Shavrukov's theorem regarding the Magari algebras (diagonalizable algebras) that are embeddable in the Magari algebra of formal arithmetical theories. Semantic characterizations of faithfully interpretable modal propositional theories in a finite number of propositional letters are given, in particular for finitely axiomatizable ones. Supported by this theory computer aided calculations on the theories of lowest complexity in one propositional letter were executed leading to a complete list of 62 formulas that axiomatize such theories under which the 8 maximal ones of particular interest.

1 Introduction

This paper discusses Magari algebras (often called diagonalizable algebras) over a finite number of generators. Magari algebras are the algebras corresponding to the provability logic L (GL in [1], PRL in [10]). According to Solovay's theorem [11] on provability interpretations the theorems of the provability logic L are precisely those modal formulas that are provable in PA under arbitrary arithmetical interpretations (interpreting $\Box$ as the formalized provability predicate in PA). Here, we are concerned with finitely generated Magari algebras that are embeddable in the Magari algebra of Peano Arithmetic. Shavrukov [9] characterized these subalgebras, which are recursively enumerable, as having the so-called strong disjunction property.

In this paper the terminology of propositional theories (i.e. sets of propositional modal formulas closed under Modus ponens and Necessitation) is more convenient. Rephrased in that terminology, we study those propositional theories T over L in a finite number of propositional variables that are (faithfully)

*Research supported by the Netherlands Organization for Scientific Research (NWO) interpretable in PA. Theories correspond to τ-filters in the free Magari algebras and interpretability to embeddability as a subalgebra. Interpretable theories T in $p_1, \cdots, p_n$ are those propositional theories in $p_1, \cdots, p_n$ for which there is a sequence of arithmetical sentences $A_1, \cdots, A_n$ such that an L formula ψ is an L consequence of T iff ψ^* is a theorem of PA in the arithmetical interpretation $*$ in which the atomic formula p_i is interpreted as A_i (see e.g. [11], [1] or [10]). Written out: T axiomatizes an arithmetically interpreted theory:

$$\{\psi \mid T \vdash_L \psi\} = \{\psi \mid \vdash_{PA} \psi^*(A_1, \cdots, A_n)\}.$$

The faithfully interpretable propositional theories T in L^n (i.e., L restricted to the language of $p_1, \cdots, p_n$) are according to Shavrukov the consistent recursively enumerable (r.e.) theories that satisfy *the strong disjunction property*: $T \vdash_L \Box\psi \vee \Box\chi$ implies $T \vdash_L \psi$ or $T \vdash_L \chi$. (Parenthetically: interpretable theories in infinitely many propositional variables need not be r.e.) The strong disjunction property may be thought of as being composed out of the simple disjunction property: $T \vdash_L \Box\psi \vee \Box\chi$ implies $T \vdash_L \Box\psi$ or $T \vdash_L \Box\chi$, and ω-*consistency*: $T \vdash_L \Box\psi$ implies $T \vdash_L \psi$.

An older concept to which this can be related is the concept of *exact provability* introduced in [2] (see also [3]): in the terminology used here a formula can be defined to be exactly provable if it axiomatizes an interpretable theory. That means that an exactly provable (or *exact*) formula of L is a formula ϕ which axiomatizes an arithmetically interpreted propositional theory:

$$\{\psi \mid \phi \vdash_L \psi\} = \{\psi \mid \vdash_{PA} \psi^*(A_1, \cdots, A_n)\}.$$

One of the objects of our research is to get an overview of exact formulas of low complexity aided by computerized calculations. For that purpose different semantic characterizations of interpretable theories and exact formulas in terms of Kripke-models have been developed which are of interest in their own right. It turns out that an important role is played by *maximal exact formulas*, i.e. exact formulas that are not implied by any other exact formula, and, more in general, by *maximal theories with the strong disjunction property*. The characterizations of these concepts discussed in this paper make heavy use of the relationship between exactly provable formulas in provability logic and sets of finite *types* of modal formulas (see [9], first introduced as *characters* in [5]). The paper is built up as follows. After a preliminary section 2, characterizations of interpretable theories and exact formulas are given in section 3. Then maximal theories with the strong disjunction property and their relationship to maximal interpretable theories, in general, are discussed in section 4, and maximal exact formulas, in particular, in section 5. In the last section 6, it is shown how the theory was applied to calculate the 62 exact formulas in one propositional variable of modal complexity 1, and the 8 maximal ones among them. These calculations can be seen as an extension of the research in intuitionistic logic executed with G. Renardel de Lavalette (see [4]) to provability logic. We thank him for his encouragement.

We owe considerable thanks to V. Shavrukov for advice and for heeding us from some mistakes.

2 Preliminaries

The *provability logic L* is the modal propositional logic with as its axioms the ones of classical propositional logic as well as all formulas of the forms $\Box(\phi \to \psi) \to (\Box\phi \to \Box\psi)$ and $\Box(\Box\phi \to \phi) \to \Box\phi$, and the inference rules Modus ponens and Necessitation. As usual $\Diamond\psi$ is defined as $\neg\Box\neg\psi$, and we will use the abbreviation $\boxdot\phi$ for the formula $\phi \wedge \Box\phi$. Note that $\phi \vdash_L \psi$ will mean that ψ is derivable from ϕ using the axioms and rules of L including Necessitation (but not substitution). This means that $\phi \vdash_L \psi$ is equivalent to $\vdash_L \boxdot\phi \to \psi$. We say "$\phi$ *is interderivable with* ψ" and write $\phi \equiv \psi$ for the conjunction of $\phi \vdash_L \psi$ and $\psi \vdash_L \phi$. Note that this implies that always $\phi \equiv \boxdot\phi$. We reserve the terminology "ϕ *is equivalent to* ψ" for $\vdash_L \phi \leftrightarrow \psi$. From this point on we will usually abbreviate $\phi \vdash_L \psi$ to $\phi \vdash \psi$. *Propositional theories* in L^n will be here sets of propositional formulas closed under $\vdash$. Such a propositional theory T is called *consistent* if $T \nvdash \bot$.

By its completeness theorem, L is the logic of all finite, transitive and irreflexive Kripke-models (as proved in [1] and [10]). In the sequel we will frequently use facts about the semantics of modal logic that can be found, for example, in [12]. The Kripke-models are supposed to be finite, transitive and irreflexive, unless stated otherwise. We consider our Kripke-models to be triples $\langle W, <, \Vdash \rangle$ with W its set of *worlds* or *nodes*, $<$ its *accessibility relation* and $\Vdash$ its *forcing relation*. We will sometimes use the notations: $\uparrow k = \{l \mid l > k\}$ and $\Uparrow k = \{l \mid l \geq k\}$.

Definition 1 *The level of box nesting of an L formula is denoted by the inductively defined function* $\beta(\phi)$:

p *atom:* $\beta(p) = 0$;

$\phi = \psi \circ \chi$: $\beta(\phi) = max\{\beta(\psi), \beta(\chi)\}$ *if* $\circ \in \{\wedge, \vee, \to\}$;

$\phi = \neg\psi$: $\beta(\phi) = \beta(\psi)$;

$\phi = \Box\psi$: $\beta(\phi) = \beta(\psi) + 1$.

The fragment L_m^n will be the fragment with $P^n = \{p_1, \cdots, p_n\}$ as its set of propositional variables and the nesting of the box operator restricted by the condition $\beta(\phi) \leq m$.

Fact 2 *The Lindenbaum algebra of L_m^n is a finite Boolean algebra and the Lindenbaum algebra of L^n an infinite Boolean algebra.*

We will call the atoms of these Boolean algebras *irreducible* elements. Note that ϕ is irreducible in L_m^n iff ϕ is not a contradiction and, for all $\psi, \chi \in L_m^n$, $\vdash \phi \to \psi \vee \chi$ implies $\vdash \phi \to \psi$ or $\vdash \phi \to \chi$. The Lindenbaum algebra of L^n is, of course, always a *Magari algebra* which is defined to be a Boolean algebra with an additional operator $\Box$ that satisfies the laws $\Box(\overline{\Box\alpha} \cup \alpha) = \Box\alpha$, $\Box(\alpha \cap \beta) = \Box\alpha \cap \Box\beta$ and $\Box 1 = 1$.

As noted in [9], the irreducible elements in L_{m+1}^n are precisely the satisfiable formulas of the form:

$$a = b \wedge \bigwedge \Diamond\gamma \wedge \Box \bigvee \gamma$$

where b is an irreducible element of the classical fragment CpL^n and the γ range over sets of irreducible elements of L_m^n.

Facts 3 *Writing A_m^n for the set of irreducible (equivalence classes of) formulas in L_m^n we have in every L^n Kripke-model K for every node k:*

1. *k forces exactly one of the elements of A_m^n.*

2. *If $\alpha \in A_m^n$ is forced by k,*
   then, for every L_m^n formula ϕ: $k \Vdash \phi \iff \vdash \alpha \to \phi$.

Definition 4 *The $\alpha \in A_m^n$ such that $k \Vdash \alpha$ will be denoted by $\phi_m^n(k)$.*

Fact 5 *In any L^n Kripke-model K, $\phi_{m+1}^n(k)$ depends for any $k \in K$ only on the set of atoms forced at k and the $\phi_m^n(k')$ of the successors k' of k in K.*

Definition 6 *Let k be a node in an L^n Kripke-model K. Then the n, m-type of k, $t_m^n(k)$, is defined by:*

- $t_0^n(k) = \langle \{ p \in P^n \mid k \Vdash p \}, \emptyset \rangle$;

- $t_{m+1}^n(k) = \langle \{ p \in P^n \mid k \Vdash p \}, \{ t_m^n(l) \mid l \in K \ \& \ k < l \} \rangle$.

The set of all such n, m-types is written T_m^n.

Although we prefer this semantic definition of n, m-type in the context of this paper, we like to note that the distinction between the semantic n, m-types and the $\phi_m^n(k)$ (essentially what is called the n, m-*type of k* in [9]) is just a matter of point of view.

If $t = \langle Q, X \rangle$, we write $j_0(t)$ for Q and $j_1(t)$ for X. The following facts are easily verified and reflect the close relationship between the $t_m^n(k)$ and $\phi_m^n(k)$.

Facts 7 *Let K be an L^n Kripke-model and k, l be elements of K.*

1. *k is of exactly one of the types in T_m^n.*

2. *If $t \in T_m^n$ is the type of both k and l,*
   then for every L_m^n formula ϕ: $k \Vdash \phi \iff l \Vdash \phi$.

3. *The $n, m+1$-type of k depends only on the set of atoms forced at k and the n, m-types of the successors of k.*

Facts 3, 5 and 7 combine into:

Fact 8 *The set of n, m-types corresponds exactly with the set A_m^n of irreducible elements of L_m^n in the sense that:*

$$\forall l \in K(\, l \Vdash \phi_m^n(k) \iff t_m^n(l) = t_m^n(k))$$

Another useful, easily verifiable fact is the following:

Fact 9 *The $n, m+k$-type t of $k \in K$ uniquely determines the n, m-type of $k \in K$. We write $t{\upharpoonright}m$ for this n, m-type.*

Lemma 8 immediately leads to:

Corollary 10 *If K is a Kripke model such that each n, m-type occurs exactly once in K, then the subsets of K correspond exactly with the equivalence classes of L_m^n. That is, the following function Ω is an isomorphism:*

$$\Omega(\phi) = \{k \in K \mid k \Vdash \phi\}.$$

It can be proved that such an *exact model*, in which each subset corresponds to a formula and vice versa, exists for each L_m^n.

Theorem 11 *For each n and m there exists an exact Kripke-model K for L_m^n, i.e., for each $U \subseteq K$, there is a formula ϕ in L_m^n such that $\{k \in K \mid k \Vdash \phi\} = U$, and K is n, m-complete, in the sense that for all $\phi, \psi \in L_m^n$, $\{k \in K \mid k \Vdash \phi\} = \{k \in K \mid k \Vdash \psi\}$ iff $\vdash \phi \leftrightarrow \psi$.*

Proof. We apply the so-called Henkin method to the (up to equivalence) finite set of formulas in L_m^n, which is closed under taking subformulas. This gives one a Kripke-model with the maximal consistent sets as its worlds, with $\Gamma < \Delta$ defined by: for each $\Box\gamma \in \Gamma$, both $\Box\gamma$ and γ are elements of Δ and, for some $\Box\gamma \in \Delta$, $\Box\gamma \notin \Gamma$, and $\Gamma \Vdash p_i$ by: $p_i \in \Gamma$. The maximal consistent sets can be replaced by their conjunctions which are exactly the irreducible elements of L_m^n. A subset of the model corresponds then to a disjunction of irreducibles, i.e. an arbitrary formula of L_m^n. Obviously, non-equivalent formulas are forced on different subsets of the model. $\dashv$

In general, unlike in the case of intuitionistic propositional logic (see [4], [7]), not all the exact models of L_m^n are isomorphic. For the infinite fragment L^n there is no such exact model in which all subsets determine a formula, but there is a canonical (infinite) model which is n-complete. Its construction (which was first given in [6] and [8]) is given presently (somewhat sloppily but, we hope, clearly) after the introduction of the notion of *depth* of a node. We baptize this canonical n-complete model ExL^n. It gives considerable insight into the free Magari algebra over n generators.

Definition 12 *Let K be a Kripke-model and $k \in K$. The* depth *of k, $\delta(k)$, is defined by:*

- $\delta(k) = 0$ *if k is a terminal node;*

- $\delta(k) = max\{\delta(l) \mid k < l\} + 1$ *otherwise.*

Definition 13 ExL^n *with its $<$ and $\Vdash$ is defined as the union of inductively defined ExL_m^n for $m \in \omega$.*

- $ExL_0^n = \wp(\{p_1, \cdots, p_n\})$, *the elements of ExL_0^n are all $<$-incomparable, and $Q \Vdash p \Leftrightarrow p \in Q$;*

- $ExL_{m+1}^n = \{\langle Q, X\rangle \mid Q \subseteq \{p_1, \cdots, p_n\},$
  $X \subseteq \bigcup_{i \leq m} ExL_i^n, X \cap ExL_m^n \neq \emptyset, X$ *closed upwards*$\}$,
  $\langle Q, X\rangle < Y \Leftrightarrow Y \in X$, *and* $\langle Q, X\rangle \Vdash p \Leftrightarrow p \in Q$;

- $ExL^n = \bigcup_{i \in \omega} ExL_i^n$.

The above definition is such that ExL_i^n will contain the elements of ExL^n of depth exactly i. This immediately brings along with it that ExL^n is conversely well-founded. It is obvious from the construction that each n, m-type will be realized by some $k \in ExL^n$. The latter ensures the n-completeness of ExL^n.

It is convenient to us to execute most of our constructions inside this model. Many of these constructions are applicable more generally, however.

Example 14 *The construction of ExL_0^1 and ExL_1^1 yields:*

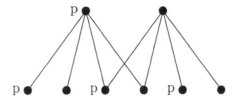

Let us write $\Box^0\phi = \phi$ and $\Box^{n+1}\phi = \Box\Box^n\phi$. The following facts about the nodes of ExL^n will be useful in the sequel.

Facts 15

1. $\delta(k) = m \quad \Leftrightarrow \quad k \Vdash \neg\Box^m\bot \wedge \Box^{m+1}\bot$;

2. *If* $\delta(k), \delta(l) \leq m$ *and* $t_m^n(k) = t_m^n(l)$, *then* $k = l$;

3. *If* $\delta(k) = m$ *and* $l \Vdash \phi_m^n(k) \wedge \neg\Box^m\bot \wedge \Box^{m+1}\bot$, *then* $k = l$.

These facts suggest a kind of normal form for the irreducible formulas corresponding to the elements of ExL^n.

Definition 16 *Let* $k \in ExL^n$ *and assume* $\delta(k) = m$. *Then* $\phi^n(k) = \phi_m^n(k) \wedge \neg\Box^m\bot \wedge \Box^{m+1}\bot$.

From the n-completeness of ExL^n we conclude that for $k \in ExL^n$ the $\phi^n(k)$ are the irreducible elements in L^n.

3   Interpretable theories and exact formulas in L^n

As stated in the introduction, Shavrukov's theorem in [9] gives a characterization of the exactly provable formulas in L (or *exact formulas* for short).

Fact 17 *A formula* $\phi \in L$ *is exact iff* ϕ *is not a contradiction and has the strong disjunction property (is s.d.):*

$$\forall\psi, \chi \in L \, (\phi \vdash \Box\psi \vee \Box\chi \; \Rightarrow \; \phi \vdash \psi \; \text{ or } \; \phi \vdash \chi)$$

The property in this fact is called *steady* by Shavrukov [9]. Whether a formula $\phi \in L_m^n$ is steady or not, Shavrukov [9] also proved, depends only on its behavior with regard to other formulas in L_m^n:

Fact 18 *A formula $\phi \in L_m^n$ is exact iff ϕ is not a contradiction and is s.d. for formulas in L_m^n:*

$$\forall \psi, \chi \in L_m^n (\phi \vdash \Box \psi \vee \Box \chi \;\Rightarrow\; \phi \vdash \psi \;\; or \;\; \phi \vdash \chi)$$

For a simple proof of this last fact see [14]. We will transform this characterization of exactness into a semantic one. That will turn out not to work for interpretable theories in general. The characterization will apply however to the maximal interpretable theories of the next section.

Definition 19 $\omega^n(\phi) = \{k \in ExL^n \mid k \Vdash \Box \phi\}$.
If T is a propositional theory in L^n, then $\omega^n(T) = \{k \in ExL^n \mid k \Vdash T\}$.

Obviously $\omega^n(\phi)$ and $\omega^n(T)$ will always be *closed upwards* in the sense that, if e.g. $k \in \omega^n(\phi)$ and $k < l$, then $l \in \omega^n(\phi)$.

Theorem 20 *A formula $\phi \in L^n$ is exact iff $\omega^n(\phi)$ is non-empty and downwards directed, i.e. $\forall k, l \in \omega^n(\phi) \, \exists h \in \omega^n(\phi) \, (h < k \,\&\, h < l)$.*

Proof. $\Rightarrow$: Let ϕ be an exact formula in L^n. As ϕ unequals the contradiction by definition, we have $\omega^n(\phi) \neq \emptyset$ by the completeness of ExL^n. To prove the second condition, let $k, l \in \omega^n(\phi)$, and $\phi^n(k), \phi^n(l)$ be (representatives of) the irreducible classes in L^n corresponding to k and l. Assume that, if $h \in \omega^n(\phi)$ and $h < k$, then $h \not< l$. Then, again by the completeness of ExL^n, we would have $\phi \vdash \Diamond \phi^n(k) \rightarrow \Box \neg \phi^n(l)$, or equivalently $\phi \vdash \Box \neg \phi^n(k) \vee \Box \neg \phi^n(l)$. As ϕ is supposed to be exact, ϕ would either prove $\neg \phi^n(k)$ or $\neg \phi^n(l)$, in contradiction with the assumption that $k, l \in \omega^n(\phi)$. Hence, there should be an $h \in \omega^n(\phi)$ such that $h < k$ and $h < l$.
$\Leftarrow$: Let $\psi, \chi \in L^n$ and $\phi \vdash \Box \psi \vee \Box \chi$, and assume there are $k, l \in \omega^n(\phi)$ such that $k \not\Vdash \psi$ and $l \not\Vdash \chi$. By the last condition of the theorem, there is an $h \in \omega^n(\phi)$ such that $h < k$ and $h < l$. As we would then have $h \not\Vdash \Box \phi \vee \Box \psi$, we obtain a contradiction. Hence, we proved that $\phi \vdash \psi$ or $\phi \vdash \chi$. $\dashv$

By the completeness of L non-interderivable ϕ and ψ give rise to distinct $\omega^n(\phi)$ and $\omega^n(\psi)$. This is in general not so for theories. An example is the theory axiomatized by p on the one hand, and the theory T_1 axiomatized by $\Box^m \bot \rightarrow p$ for each m, on the other. The sets $\omega^1(p)$ and $\omega^1(T_1)$ are the same, consisting of all nodes that together with all their successors force p, but clearly the theories are not: p is not a consequence of T_1. Similarly, the theory $T_2 = \Box^m \bot \to \Box p \vee \Box \neg p$ for each m can be shown to have the strong disjunction property. But not all pairs of nodes in $\omega^1(T_2)$ have a common predecessor in ExL^1, because $\omega^1(T_2)$ consists of those nodes that together with all their successors force p and those nodes that together with all their successors don't force p. This shows that the semantic characterization of exactness does not generalize to interpretability of non-finitely axiomatizable theories, at least if one doesn't freely use infinite models. There is a restricted class of theories that does respect the characterization.

Definition 21 *T has the* finite model property (f.m.p.) *iff $T \nvdash \phi$ implies that there is a finite Kripke-model for T on which ϕ is falsified.*

Fact 22 *T has the* finite model property (f.m.p.) *iff*
$\{\phi \mid$ *for each $k \in \omega^n(T)$, $k \Vdash \phi\} = T$.*

If T has the f.m.p., then T is determined uniquely by $\omega^n(T)$. Also the semantic characterization of exactness immediately generalizes by the same proof.

Theorem 23 *A theory T in L^n with the f.m.p. is interpretable iff $\omega^n(T)$ is non-empty and downwards directed.*

Later we will encounter the finite model property in the syntactic form exhibited in the following lemma.

Lemma 24 *An L theory T has the f.m.p. iff, for each ϕ,*
$T \vdash \phi \quad \Leftrightarrow \quad$ *for each $n \in \omega$, $T \vdash \Box^n \bot \to \phi$.*

Proof. $\Rightarrow$: Assume T has the f.m.p. First let $T \nvdash \phi$. By its f.m.p. T has a finite model with a node k falsifying ϕ. Supposing k has depth m ensures that $T \nvdash \Box^{m+1} \bot \to \phi$. The other direction is trivial.

$\Leftarrow$: If $T \nvdash \phi$, then the right hand side implies that $T \nvdash \Box^n \bot \to \phi$ for some n, i.e., there is a model for T of depth $n-1$ falsifying ϕ. In the case we are dealing with, T being a theory in a finite number of propositional variables, this reduces to a finite model. $\dashv$

One sees from the proof that actually the negative formulation of the right hand side of the above lemma is the natural one.

Note that the $\omega^n(\phi)$ of an exact $\phi \in L^n$ is infinite by the conditions of the characterization. On the other hand there is a simple correspondence between such an infinite set and a finite set of n, m-types in ExL^n:

Definition 25 *Let ϕ be an L^n formula.*
Then $T_m^n(\phi) = \{t_m^n(k) \mid k \in ExL^n, k \Vdash \square \, \phi\}$.

Lemma 26 *Let ϕ and ψ be L_m^n formulas.*
    Then $T_m^n(\phi) = T_m^n(\psi)$ iff $\phi \equiv \psi$.

Proof. For the non-trivial direction, from left to right, let $T_m^n(\phi) = T_m^n(\psi)$, and assume $k \Vdash \square \, \phi$. Then $t_m^n(k) \in T_m^n(\phi) = T_m^n(\psi)$. Hence $t_m^n(k) = t_m^n(k')$ for some k' that forces $\square \, \psi$. So, $k' \Vdash \psi$ and, since $\psi \in L_m^n$, $k \Vdash \psi$. The rest is evident. $\dashv$

Lemma 27 *Let ϕ be an L_{m+k}^n formula.*
Then $T_m^n(\phi) = \{t \restriction m \mid t \in T_{m+k}^n(\phi)\}$.

Proof. Obvious, considering fact 9. $\dashv$

Lemma 28 *For each L^n formula ϕ and each m there is a finite upwards closed subset K of $\omega^n(\phi)$ such that the elements of K exactly realize $T_m^n(\phi)$, i.e., $T_m^n(\phi) = \{T_m^n(k) \mid k \in K\}$.*

Proof. Just take any finite subset of $\omega^n(\phi)$ such that its elements exactly realize $T_m^n(\phi)$. The upward closure of this set will do, because its elements also force $\square \, \phi$. $\dashv$

To find the sets of n, m-types suitable for exact formulas ϕ, we have to translate the conditions on the $\omega^n(\phi)$ of exact ϕ into conditions on the underlying set of n, m-types. For example, for a finite $T_m^n(\phi)$ to correspond to an infinite $\omega^n(\phi)$, it is necessary that some type in $T_m^n(\phi)$ can have itself as a successor. To describe this kind of reflexivity we introduce the notion of a *reflexive type*.

Definition 29 *A type $t \in T_{m+1}^n$ is called* reflexive *if $t \restriction m \in j_1(t)$.*

The following theorem is related to lemma 5.13 of [9].

Theorem 30 *A formula $\phi \in L_m^n$ with $m > 0$ is exact iff there is a type $t \in T_m^n(\phi)$ such that $j_1(t) = T_{m-1}^n(\phi)$, which, of course, makes t a reflexive type.*

Proof. $\Rightarrow$: Let $\phi \in L_m^n$ be an exact formula. Note that $T_{m-1}^n(\phi)$ is a finite set of types. Let $K \subset \omega^n(\phi)$ be finite and closed upwards such that $\{t_m^n(k) \mid k \in K\} = T_m^n(\phi)$, as guaranteed to exist by lemma 28. According to theorem 20 we can find an $h \in \omega^n(\phi)$ below all of the elements of K. By lemma 27 this h must have a type as required.

$\Leftarrow$: Assume ϕ and t to fulfill the conditions given. As $T_m^n(\phi) \neq \emptyset$, also $\omega^n(\phi) \neq \emptyset$. Suppose $k, l \in \omega^n(\phi)$. Let K be a finite upwards closed subset of $\omega^n(\phi)$ such that $k, l \in K$ and $\{t_{m-1}^n(k') \mid k' \in K\} = T_{m-1}^n(\phi)$ (compare lemma 28). Consider a world h just below this K such that $\{p \in P^n \mid h \Vdash p\} = j_0(t)$. It will be clear that $t_m^n(h) = t$ and (since ϕ is assumed to be an L_m^n formula) this proves $h \in \omega^n(\phi)$. Of course, $h < k$ and $h < l$, so the conditions of theorem 20 apply to $\omega^n(\phi)$. $\dashv$

The theory developed in this article has enabled us to calculate the exact formulas in L_1^1. This will be explained in more detail in the last section. It will be shown that already in this very first small fragment there are 62 non-interderivable members. It turned out that it was worthwhile to single out the 8 maximal elements of these 62. More general than maximal exactness the concept of maximal s.d. theory turns out to be in itself an interesting one. The next section will be devoted to this last concept and the one after it to maximal exact formulas.

4 Maximal theories with the strong disjunction property and maximal interpretable theories

In this section it turns out to be fruitful to not restrict one's attention to r.e. s.d. theories, but also include non-r.e. ones, i.e. non-interpretable theories.

Definition 31

- *A theory T is* maximal s.d. *if T is consistent and s.d. and, for no consistent U with $T \subset U$, U is s.d.;*

- *An L^n formula ϕ is* maximal exact *if ϕ is exact and, for each exact L^n formula ψ such that $\psi \vdash \phi$, also $\phi \vdash \psi$;*

- *A theory T is* ω-consistent *if $T \vdash \Box\phi$ implies $T \vdash \phi$;*
  A theory T is maximal ω-consistent *if T is ω-consistent and, for no U with $T \subset U$, U is ω-consistent;*

- *A theory T is* of infinite height (i.h.) *if, for no $n \in \omega$, $T \vdash \Box^n \bot$;*
  A theory T is maximal i.h. *if T is i.h. and, for no U with $T \subset U$, U is i.h.*

- *A theory T is* maximal interpretable *if T is interpretable and, for no U with $T \subset U$, U is interpretable.*

Note that the concept of a maximal interpretable theory is a very natural one. Such a theory, and in particular also a maximal exact formula (as will be shown in the next section), describes a maximal condition that arithmetical sentences can be required to satisfy.

Lemma 32 *Each i.h. theory is contained in a maximal i.h. theory.*

Proof. By Zorn's lemma.   ⊣

Lemma 33 *T is maximal i.h. iff, for each ϕ, either $T \vdash \phi$, or, for some $n \in \omega$, $T \cup \{\phi\} \vdash \Box^n \bot$.*

Proof. ⇒: Assume ϕ is such that not $T \vdash \phi$, and, for no $n \in \omega$, $T \cup \{\phi\} \vdash \Box^n \bot$. Then T can be properly extended to an i.h. theory containing ϕ which means that T is not maximal i.h.
⇐: Assume T is not maximal i.h., i.e., for some i.h. U, $T \subset U$. Let ϕ be an element of U, but not of T. Then not $T \vdash \phi$, but not $T \cup \{\phi\} \vdash \Box^n \bot$ for any n either, because for no n, $U \vdash \Box^n \bot$.   ⊣

Lemma 34 *If T is s.d., then T is ω-consistent. If T is ω-consistent, then T is i.h.*

Proof. Trivial.   ⊣

Lemma 35 *If T is a maximal i.h. theory, then T is s.d. and hence ω-consistent.*

Proof. Let T be maximal i.h. To show that T is s.d., assume $T \vdash \Box\phi \vee \Box\psi$. Now apply lemma 33 to both ϕ and ψ to obtain that, either $T \vdash \phi$ or, for some $n \in \omega$, $T \cup \{\phi\} \vdash \Box^n \bot$, and the same for ψ. It is sufficient to exclude the possibility that $T \cup \{\phi\} \vdash \Box^n \bot$ for some n as well as $T \cup \{\psi\} \vdash \Box^m \bot$ for some m. In that case it would follow that $T \vdash \Box\phi \rightarrow \Box^{n+1} \bot$ and $T \vdash \Box\psi \rightarrow \Box^{m+1} \bot$ which implies $T \vdash \Box^{max(n+1,m+1)} \bot$, contradicting the infinite height of T.   ⊣

Lemma 36 *T is maximal i.h. iff T is maximal ω-consistent iff T is maximal s.d.*

Proof. From the previous lemmas.   ⊣

Corollary 37

1. *Each s.d. theory in L^n is contained in a maximal s.d. theory.*

2. *If T is r.e. and maximal s.d., then T is recursive.*

3. *Not each interpretable theory in L^n is contained in a maximal interpretable theory.*

Proof. Part 1 is trivial. Part 2 follows from lemmas 36 and 33. Part 3 follows from part 2 by considering two recursively inseparable r.e. sets of natural numbers A and B and then constructing the theory

$$U = \{\Box^{n+1}\bot \wedge \neg\Box^n\bot \to p \mid n \in A\} \cup \{\Box^{n+1}\bot \wedge \neg\Box^n\bot \to \neg p \mid n \in B\}.$$

The theory U is s.d., since any two models can be joined by a new root taking care only that the new root may need a specific valuation for p, and r.e., and hence interpretable. A maximal interpretable extension of U, which by 2 would have to be recursive, would separate A and B and, therefore, cannot exist.   ⊣

One would expect it to be easy to prove that each maximal interpretable theory has to be maximal s.d., but we have not been able to show this, so we leave it as an open question.

Open Question 38 *Is each maximal interpretable theory maximal s.d.?*

Lemma 39 *If T is maximal s.d., then T has the f.m.p.*

Proof. Assume $T \nvdash \phi$ for a maximal s.d. theory T. By lemma 33, for some $n \in \omega$, $T \vdash \phi \to \Box^n\bot$. Löb's theorem implies then that $T \nvdash \Box^{n+1}\bot \to \phi$. Now apply lemma 24.   ⊣

The next theorem uses the above to provide a semantic characterization of maximal s.d. theories.

Theorem 40 *T in L^n is maximal s.d. iff T has the f.m.p. and, for each $k \in \omega^n(T)$, there is an $m \in \omega$ such that for all $l \in \omega^n(T)$ of depth $\geq m$, $l < k$.*

Proof. $\Rightarrow$: Let T in L^n be maximal s.d. That T has the f.m.p. is lemma 39, so it suffices to check the second point. For that purpose, consider $k \in \omega^n(T)$. Obviously $\neg\phi^n(k)$ is not a theorem of T. So, for some $m \in \omega$, $T \vdash \Box\neg\phi^n(k) \to \Box^m\bot$. Equivalently, $T \vdash \neg\Box^m\bot \to \phi^n(k) \vee \Diamond\phi^n(k)$. This implies that, for all $l \in \omega^n(T)$ of depth $\geq m+1$, $l < k$, since $\phi^n(k)$ uniquely describes k.

$\Leftarrow$: It is sufficient to check the condition of lemma 33 under the assumption that the right hand side of the present theorem is satisfied by T. So, assume $T \nvdash \phi$. As T has the f.m.p., there is a finite model for T that falsifies ϕ, given as $\uparrow k$ for some $k \in \Omega^n(T)$. Since, for some $m \in \omega$, all nodes with depth $\geq m$ in $\omega^n(T)$ are below k, $T \vdash \neg\Box^m\bot \to \Diamond\neg\phi$. By contraposition, $T \vdash \Box\phi \to \Box^m\bot$ which was to be proved. $\dashv$

A natural question concerning L^1 is whether there are any maximal interpretable theories that are symmetric with respect to p and $\neg p$. Rosser-sentences are interpretations of maximal exact formulas (see the last section), but they are not examples of such symmetry, since $p \to \Box p$ is in the theory of the Σ form of the Rosser sentence, and $\neg p \to \Box\neg p$ in the dual theory. It is easy to see that no s.d. theory can contain both $p \to \Box p$ and $\neg p \to \Box\neg p$ without containing p or $\neg p$ itself (use the excluded middle). In the next section it will become clear that an exact formula in p cannot be symmetric in p and $\neg p$, and maximally exact. So, an example will necessarily have an infinite axiomatization.

Example 41 *A symmetric maximal interpretable theory in p.*
Consider the subset K of ExL^1 defined as follows:

$K_0 = \{\{p\}, \emptyset\}$, $\quad K_{m+1} = \{\langle\{p\}, \bigcup_{i\leq m} K_i\rangle, \langle\emptyset, \bigcup_{i\leq m} K_i\rangle\}$, $\quad K = \bigcup_{m\in\omega} K_m$.
The theory $T = \{\phi \mid k \Vdash \phi$ for all $k \in K\}$ is maximal s.d. (it is easily checked that it satisfies the conditions of theorem 40) and symmetric. An axiomatization of T is given by:

$\Diamond p \leftrightarrow \Diamond\neg p$
$\Diamond(p \wedge \theta(p)) \leftrightarrow \Diamond(\neg p \wedge \theta(p))$ *for each modalized $\theta \in L^1$,*

where modalized means that each occurrence of a propositional variable is in the scope of a modal operator. A sentence interpreting such a theory in PA is a kind of symmetric Rosser-sentence which can, of course, neither be Σ^0_1, nor Π^0_1. Note, in fact, that $p \to \Box p$ as well as $\neg p \to \Box\neg p$ are strongly non-provable for such a sentence, in the sense that e.g., $\Box(p \to \Box p) \to \Box\Box\bot$ is provable for it. (Note also that $\Box(p \to \Box p) \to \Box\bot$ can never be provable for an arithmetic sentence.)

Proof that the axiomatization is complete for T. Assume $T \nvdash \phi$. There exists a (possibly infinite) Kripke-model $\bar{K}$ satisfying T on which ϕ is falsified (see e.g. [12]). It suffices to show that each $1, m$-type realized in $\bar{K}$ can be realized in $K'_m = \bigcup_{i \leq m} K_i$. For $1, 0$-types this is obvious. Assume some $1, m+1$-type t is realized in $\bar{K}$. All types in $j_1(t)$ are, by induction hypothesis, realized in K'_m. Assume that of the elements in K'_m realizing types in $j_1(t)$, k is in the K_i with highest index i. Since the existence of nodes above a node k realizing all the types in $j_1(t)$ can be expressed by a modalized formula, it will be obvious that the axioms of T imply that there has to be a type present in $j_1(t)$ that is realized by the (only) other element of K_i. This implies that t is realized by an element of K_{i+1}, which will do. $\dashv$

5 Maximal exact formulas

This section will be devoted to maximal exact formulas. First we will have to sharpen our semantic characterization of exact formulas. Let us exploit the relationship between irreducible formulas and semantic types to write $\phi_m^n(t)$ for the $\phi_m^n(k)$ with $t_m^n(k) = t$.

Definition 42 *Let C be a set of n, m-types.*
Then $\phi_m^n(C) = \bigvee \{\phi_m^n(t) \mid t \in C\}$.

Recall that $T_m^n(\phi) = \{t_m^n(k) \mid k \Vdash \Box \phi\}$.

Lemma 43 *If $\phi \in L_m^n$, then $\phi \equiv \phi_m^n(T_m^n(\phi))$.*

Proof. Immediate from lemma 26 as soon as one realizes that

$$T_m^n(\phi_m^n(T_m^n(\phi))) = T_m^n(\phi).$$

$\dashv$

Lemma 44 *If $C \subseteq T_m^n$ $(m > 0)$, then $C = T_m^n(\phi)$ for an exact formula $\phi \in L_m^n$ iff*

1. *There is a finite upwards closed $K \subseteq ExL^n$ such that $C = \{t_m^n(k) \mid k \in K\}$ (we will call C upwards closed realizable)*

2. *There is $t \in C$ such that $\forall t' \in C$ $(t' \restriction (m-1) \in j_1(t))$. Such a type t will be called an enveloping type for C.*

Moreover, in that case $\phi \equiv \phi_m^n(C)$.

Proof. ⇒: If $\phi \in L_m^n$ is exact, then $T_m^n(\phi)$ will have the required property 1 by the definition of $T_m^n(\phi)$, property 2 by theorem 30, and satisfies the final requirement by lemma 43.

⇐: We prove that $\phi_m^n(C)$ is an exact formula. To apply theorem 30 to $\phi_m^n(C)$, we have to find an appropriate reflexive n, m-type. By the assumption on C, there is an n, m-type $t \in C$ such that $\forall t' \in C \, (t' \restriction (m-1) \in j_1(t))$. Let K be the upwards closed realization of the types in C as assumed in the first condition of this lemma. Note that K realizes precisely the $n, m-1$-types in $\{t' \restriction (m-1) \mid t' \in C\}$ (compare lemma 27). Let k be a (new) root immediately below K such that k forces exactly the elements of $j_0(t)$. Then $t_m^n(k) = t$. So, $k \Vdash \Box \, \phi_m^n(C)$ and, hence, t is a member of $T_m^n(\phi_m^n(C)) \subseteq C$ and a type appropriate for the application of theorem 30. ⊣

We will prove that the maximal exact formulas in L^n correspond to what we will call *tail models* in ExL^n. Clearly this is a result that is, to a large extent, bound to the particular model ExL^n.

Definition 45 $K \subset ExL^n$ *is called a tail model iff:*

1. K *is closed upwards;*

2. *there is an* $m \in \omega$ *such that* $\{k \in K \mid \delta(k) \geq m\}$ *is linearly ordered by* $<$ *and all nodes of this set force the same atoms.*

If $k \in ExL^n$, *then we write* $\uparrow k \downarrow$ *for the tail model consisting of* $\uparrow k$ *and a tail descending from* k *with the forcing of the atoms as in* k.

Our definition of tail model slightly differs from the one in [13] in that Visser's tail models are equipped with a minimal (infinite-depth) element.

Lemma 46 *If* $\phi \in L_m^n$, $k \in \omega^n(\phi)$ *and* k *has a reflexive* n, m-type, *then* $\uparrow k \downarrow \subseteq \omega^n(\phi)$.

Proof. First note that all elements of the tail have the same n, m-type as k. Hence, all these nodes force ϕ, and consequently $\Box \, \phi$. ⊣

Lemma 47 *If* $K \subset ExL^n$ *is a tail model, then* $K = \omega^n(\phi)$ *for some* ϕ *in* L^n.

Proof. Let K be $\uparrow k \downarrow$, k having depth m, and let θ be the conjunction of the propositional variables and negations of propositional variables as they are forced on k. Then $K = \omega^n(\phi)$ for ϕ defined as the conjunction of $\Box^{m+1} \bot \rightarrow \bigvee \{\phi^n(k') \mid k \leq k'\}$ and $\neg \Box^{m+1} \bot \rightarrow \theta \wedge \Diamond \, \phi^n(k)$. ⊣

Lemma 48 *If $\phi \in L^n$ and $\omega^n(\phi)$ is a tail model, then ϕ is maximal exact.*

Proof. Assume $\omega^n(\phi)$ is a tail model and $\phi \in L^n_m$. Since $\omega^n(\phi)$ is infinite and $T^n_{m-1}(\phi)$ finite it is obvious that the tail has to contain elements appropriate for an application of theorem 30. This shows that ϕ has to be exact. Assume ψ to be an exact formula such that $\psi \vdash \phi$, i.e., such that $\omega^n(\psi) \subseteq \omega^n(\phi)$. Then, because $\omega^n(\psi)$ is non-empty and downwards directed it has to contain the tail elements from a certain node downwards, and, because it is closed upwards it has to contain all other elements of $\omega^n(\phi)$, which means that ϕ and ψ are interderivable. Hence, ϕ is maximal exact. $\dashv$

Lemma 49 *If $\phi \in L^n$, then there exists a formula $\psi \in L^n$ such that $\omega^n(\psi)$ is a tail model and $\omega^n(\psi) \subseteq \omega^n(\phi)$.*

Proof. Let $\phi \in L^n$ and assume $t \in T^n_m(\phi)$ is a reflexive type with the properties guaranteed to exist by theorem 30. Now, take as in the proof of lemma 44($\Leftarrow$) $k \in \omega^n(\phi)$ with n, m-type t such that $t^n_m(\uparrow k) = T^n_m(\phi)$. By lemma 46, $\uparrow k \downarrow \subseteq \omega^n(\phi)$. By lemma 47, there exists a ψ with $\omega^n(\psi) = \uparrow k \downarrow \subseteq \omega^n(\phi)$. $\dashv$

From lemma 49 it follows immediately that any L^n formula ϕ is determined uniquely (up to interderivability) by the maximal exact L^n formulas that imply it. Certainly this does not generalize to interpretable theories. The s.d. theory T_1 axiomatized by $\Box^n \bot \to p$ for each n that was introduced after theorem 20 provides a counter-example. Its only maximal s.d. extension is the one axiomatized by p. Also, lemma 49 does not, in general imply that ϕ is equivalent to a finite disjunction of maximal exact formulas (each preceded by $\Box$), although that may very well be the case. A counter-example is provided by the formula $\top$.

Theorem 50 *If $\phi \in L^n$, then ϕ is maximal exact in L^n iff $\omega^n(\phi)$ is a tail model in ExL^n.*

Proof. The direction from right to left follows from lemma 48. The other direction from 49 using the simple fact that, if one tail model is part of another, they have to be equal. $\dashv$

From lemma 47 and theorem 50 it is clear that there is a one-one correspondence between maximal exact formulas and tail models. Moreover, the fact announced before that maximal exactness implies maximal interpretability becomes an easy corollary.

Corollary 51 *If ϕ is a maximal exact L^n formula, then $\{\psi \mid \phi \vdash \psi\}$ is a maximal interpretable L^n theory.*

Proof. If ϕ is maximal exact, then $\omega^n(\phi)$ is a tail model by theorem 50. Applying theorem 40 then immediately shows that $\{\psi \mid \phi \vdash \psi\}$ is a maximal s.d., and hence, maximal interpretable L^n theory. $\dashv$

Also from theorem 50, the remark we made above that maximal exact formulas in p cannot be symmetric with regard to p and $\neg p$ becomes immediately obvious. The tail is always asymmetric. We follow with some additional properties and problems concerning maximal exact formulas.

Theorem 52 *If a formula $\phi \in L_m^n$ is maximal exact, then there is precisely one reflexive type t in $T_m^n(\phi)$. Moreover, $T_{m-1}^n(\phi) = j_1(t)$.*

Proof. The last part follows immediately from theorem 30. Assume $\phi \in L_m^n$ with $m > 0$ is maximal exact. Assume s and s' to be two distinct n, m-types in $T_m^n(\phi)$. If k and k' in $\omega^n(\phi)$ realize s and s' respectively, then, by lemma 46, $\uparrow k\downarrow$ and $\uparrow k'\downarrow$ are two distinct tail models within $\omega^n(\phi)$. This contradicts the fact that $\omega^n(\phi)$ is a tail model. $\dashv$

Examples of non-maximal exact L_1^1 formulas with exactly one reflexive $1, 1$-type will be given in the table in the last section.

Definition 53 *An exact L_m^n formula ϕ is called n, m-maximal exact iff, for all exact $\psi \in L_m^n$ such that $\psi \vdash \phi$, $\psi \equiv \phi$.*

It will turn out in the last section that the $1, 1$-maximal exact formulas in L^1 are maximal exact. In general, however, not all the n, m-maximal exact formulas in L_m^n are maximal exact. To construct counter-examples the following insight derived from lemma 44 and the fact that, by lemma 26, L_m^n formulas are, up to $\equiv$, determined by their n, m-types was used.

Fact 54 *The m-maximal exact L_m^n-formulas are the ones with a set of types C that contains exactly one reflexive n, m-type t and for which C is minimal upwards closed realizable, in the sense that, C is upwards closed realizable, but this is not the case for any proper subset of C containing t.*

The simplest counter-example we found uses a set of $3, 2$-types with exactly one minimal enveloping type in the sense of the previous fact. Such a set of $3, 2$-types will correspond to a $3, 2$-maximal exact formula. The following two models, both built using only this set of types, show there is a real choice in ordering it.

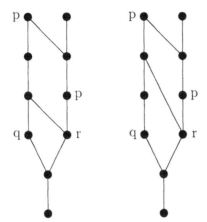

The models above can be extended to tail models corresponding to different maximal exact L_3^3 formulas. From both of these formulas the $3, 2$-maximal exact formula corresponding with the set of $3, 2$-types is derivable. Hence this $3, 2$-maximal exact formula is clearly not maximal exact.

A further conjecture is that the set of n, m-types of an arbitrary exact L_m^n formula ϕ is the union of the sets of types of the n, m-maximal exact L_m^n formulas from which ϕ is derivable. That such a union always is the set of types of an exact formula if a common enveloping type is present follows immediately from the next lemma.

Lemma 55 *If C is the union of sets $C_1, \cdots, C_k$ of n, m types corresponding to exact L_m^n formulas $\phi_1, \cdots, \phi_k$ with an enveloping type t for all of C, then there exists a $\phi \in L_m^n$ such that $T_m^n(\phi) = C$.*

Proof. It suffices to note that, if $K_1, \cdots, K_k$ are upwards closed realizations of $C_1, \cdots, C_k$, then $K_1 \cup \cdots \cup K_k$ is an upwards closed realization of C, and then to apply lemma 44. ⊣

It is certainly not true that any union of types of n, m-maximal exact formulas is the set of n, m-types of some exact L_m^n formula. A counter-example is provided by the sets of types belonging to p and to $\neg p$, both $0, 1$-maximal exact formulas, which cannot be combined to an exact formula, even for $m = 1$. A common enveloping type is needed, and is obviously not available for p and $\neg p$ (see section 6).

6   Calculating exact formulas

In this section the calculation of the exact formulas in L_1^1 will be discussed. It will be shown that already in this very first small fragment there are 62 non-interderivable members with 8 maximal elements. Of the next fragment L_2^1 even the cardinality of the set of maximal exact elements has eluded us so far. The fragment L_3^1 is definitely too large to attack in this manner.

   To calculate the exact formulas in L_1^1 we use sets of $1,1$-types. The $1,1$-types can be ordered into an exact Kripke model (this notion was introduced in the preliminary section):

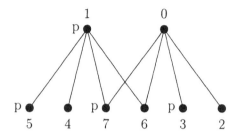

This exact model corresponds to the first two layers (ExL_0^1 and ExL_1^1) in the construction of ExL^1. In ordering the types into an exact model other choices could have been made, resulting in different models. In fact, in the calculation of exact formulas the choice of the exact model is arbitrary. In the sequel we will denote the $1,1$-types by their number in the exact model above. In the previous section we proved that $\phi \in L_1^1$ is exact iff

1. $T_1^1(\phi)$ is upwards closed realizable;

2. there is a $t \in T_1^1(\phi)$ such that $\forall t' \in T_1^1(\phi)(t' \restriction 0 \in j_1(t))$.

These criteria are easily translated into a test on a set of $1,1$-types C. The first condition of this test requires C to be upwards closed realizable (in ExL^1 not necessarily in the model above) and the second condition demands an enveloping type in C. Let $\phi \in L_1^1$ and $C = T_1^1(\phi)$. Then ϕ is exact iff

1. if $2 \in C$ or $3 \in C$, then $0 \in C$
   if $4 \in C$ or $5 \in C$, then $1 \in C$
   if $6 \in C$ or $7 \in C$, then $C \cap \{0,2,4\} \neq \emptyset$ and $C \cap \{1,3,5\} \neq \emptyset$

2. $6 \in C$ or $7 \in C$ or $C = \{0,2\}$ or $C = \{1,5\}$

The sets of $1,1$-types corresponding to exact formulas in L^1 can be found in applying the above test to the 255 non-empty subsets of T_1^1. We prefer however to calculate the exact formulas together with their corresponding sets of 1,1-types. To do so, the exact model above will be used to calculate all L_1^1 formulas in the following manner.

Our computer program generates a list of formulas and sets. It starts with the formulas $\bot$ and p and the sets $\Omega(\bot) = \emptyset$ and $\Omega(p) = \{1, 3, 5, 7\}$. (Note: Ω was defined in the preliminary section.)

The list of formulas ϕ and sets $\Omega(\phi)$ is extended by systematically applying the connectives $(\neg, \wedge, \vee, \rightarrow, \Box)$ and the corresponding set operations, adding a pair consisting of a formula and its set only if the set does not yet occur in the list. In this way we ensure that no two distinct interderivable formulas will occur in the list. A similar computation of the Lindenbaum algebra of an exact Kripke model was described before (for intuitionistic propositional logic) in [4].

In generating the list of formulas and sets the test defined above is applied to distinguish the exact formulas in L_1^1.

The exact formulas in L_1^1 have been enumerated in the list below. For each formula the corresponding set of 1,1-types is given on the right. Note that for the bracketing the priority of $\wedge$ is higher than $\vee$ and $\rightarrow$. Likewise $\vee$ has a higher priority than $\rightarrow$.

1:	$p \rightarrow p$	$\{\,0\ 1\ 2\ 3\ 4\ 5\ 6\ 7\,\}$
2:	$p \rightarrow \Box\bot$	$\{\,0\ 1\ 2\ 4\ 6\,\}$
3:	$p \vee \Box\bot$	$\{\,0\ 1\ 3\ 5\ 7\,\}$
4:	$p \rightarrow \Box p$	$\{\,0\ 1\ 2\ 4\ 5\ 6\,\}$
5:	$p \wedge \Box p$	$\{\,1\ 5\,\}$
6:	$p \vee \Box p$	$\{\,0\ 1\ 3\ 4\ 5\ 7\,\}$
7:	$p \rightarrow \Box\neg p$	$\{\,0\ 1\ 2\ 3\ 4\ 6\,\}$
8:	$p \vee \Box\neg p$	$\{\,0\ 1\ 2\ 3\ 5\ 7\,\}$
9:	$\neg p \wedge \Box\neg p$	$\{\,0\ 2\,\}$
10:	$\Box\bot \vee \neg\Box p$	$\{\,0\ 1\ 2\ 3\ 6\ 7\,\}$
11:	$\Box\neg p \rightarrow \Box p$	$\{\,0\ 1\ 4\ 5\ 6\ 7\,\}$
12:	$p \wedge \Box p \rightarrow \Box\bot$	$\{\,0\ 1\ 2\ 3\ 4\ 6\ 7\,\}$
13:	$p \wedge \Box\neg p \rightarrow \Box\bot$	$\{\,0\ 1\ 2\ 4\ 5\ 6\ 7\,\}$
14:	$\Box\neg p \rightarrow p \wedge \Box\bot$	$\{\,1\ 4\ 5\ 6\ 7\,\}$
15:	$\Box p \rightarrow (p \vee \Box\bot)$	$\{\,0\ 1\ 2\ 3\ 5\ 6\ 7\,\}$
16:	$\Box\neg p \rightarrow (p \vee \Box\bot)$	$\{\,0\ 1\ 3\ 4\ 5\ 6\ 7\,\}$
17:	$\Box p \rightarrow \neg p \wedge \Box\bot$	$\{\,0\ 2\ 3\ 6\ 7\,\}$
18:	$\Box\neg p \vee (p \rightarrow \Box p)$	$\{\,0\ 1\ 2\ 3\ 4\ 5\ 6\,\}$
19:	$\Box\neg p \vee p \wedge \neg\Box p$	$\{\,0\ 1\ 2\ 3\ 7\,\}$
20:	$\Box\neg p \vee (p \vee \Box p)$	$\{\,0\ 1\ 2\ 3\ 4\ 5\ 7\,\}$
21:	$(p \vee \Box p) \rightarrow \Box\neg p$	$\{\,0\ 1\ 2\ 3\ 6\,\}$
22:	$(p \vee \Box\bot) \wedge \neg(p \wedge \Box p)$	$\{\,0\ 3\ 7\,\}$
23:	$\Box p \vee p \wedge \neg\Box\neg p$	$\{\,0\ 1\ 4\ 5\ 7\,\}$
24:	$(p \vee \Box\neg p) \rightarrow \Box p$	$\{\,0\ 1\ 4\ 5\ 6\,\}$
25:	$\neg(p \wedge \Box p) \wedge (p \vee \Box\neg p)$	$\{\,0\ 2\ 3\ 7\,\}$
26:	$p \wedge \Box\bot \vee \neg(p \vee \Box\neg p)$	$\{\,1\ 4\ 6\,\}$
27:	$p \wedge \Box p \vee \neg(p \vee \Box\neg p)$	$\{\,1\ 4\ 5\ 6\,\}$
28:	$\Box\bot \vee \neg(p \vee \Box p)$	$\{\,0\ 1\ 2\ 6\,\}$
29:	$\Box\bot \vee p \wedge \neg\Box p$	$\{\,0\ 1\ 3\ 7\,\}$
30:	$p \wedge (\Box p \vee \Box\neg p) \rightarrow \Box\bot$	$\{\,0\ 1\ 2\ 4\ 6\ 7\,\}$
31:	$(\Box p \vee \Box\neg p) \rightarrow (p \vee \Box\bot)$	$\{\,0\ 1\ 3\ 5\ 6\ 7\,\}$
32:	$\Box\bot \vee \neg(p \vee \Box\neg p)$	$\{\,0\ 1\ 4\ 6\,\}$
33:	$\Box\bot \vee p \wedge \neg\Box\neg p$	$\{\,0\ 1\ 5\ 7\,\}$

34: $\Box\bot \vee \neg(\Box p \vee \Box\neg p)$ $\{\,0\;1\;6\;7\,\}$
35: $(p \vee \Box p) \wedge (p \wedge \Box p \to \Box\bot)$ $\{\,0\;1\;3\;4\;7\,\}$
36: $\Box\bot \vee \neg(\Box\neg p \vee p \wedge \Box p)$ $\{\,0\;1\;4\;6\;7\,\}$
37: $(p \vee \Box\neg p) \wedge (p \wedge \Box\neg p \to \Box\bot)$ $\{\,0\;1\;2\;5\;7\,\}$
38: $\Box\bot \vee \neg(\Box p \vee p \wedge \Box\neg p)$ $\{\,0\;1\;2\;6\;7\,\}$
39: $(p \vee \Box p) \wedge (\Box\neg p \to p \wedge \Box\bot)$ $\{\,1\;4\;5\;7\,\}$
40: $p \wedge \Box\bot \vee \neg(\Box\neg p \vee p \wedge \Box p)$ $\{\,1\;4\;6\;7\,\}$
41: $(p \vee \Box p) \to \Box p \wedge (p \vee \Box\bot)$ $\{\,0\;1\;2\;5\;6\,\}$
42: $(\Box p \vee \Box\neg p) \to \Box p \wedge (p \vee \Box\bot)$ $\{\,0\;1\;5\;6\;7\,\}$
43: $\Box\bot \vee (\Box p \to p) \wedge \neg(p \wedge \Box\neg p)$ $\{\,0\;1\;2\;5\;6\;7\,\}$
44: $(p \vee \Box\neg p) \to \Box\neg p \wedge (p \vee \Box\bot)$ $\{\,0\;1\;3\;4\;6\,\}$
45: $\Box\bot \vee \neg\Box p \wedge (\Box\neg p \to p)$ $\{\,0\;1\;3\;6\;7\,\}$
46: $\Box\bot \vee \neg(p \wedge \Box p) \wedge (\Box\neg p \to p)$ $\{\,0\;1\;3\;4\;6\;7\,\}$
47: $(p \vee \Box p) \to \Box\neg p \wedge \neg(p \wedge \Box\bot)$ $\{\,0\;2\;3\;6\,\}$
48: $(\Box p \vee \Box\neg p) \to (p \vee \Box\bot) \wedge \neg(p \wedge \Box p)$ $\{\,0\;3\;6\;7\,\}$
49: $(p \vee \Box p) \to (\Box\neg p \vee p \wedge \Box p)$ $\{\,0\;1\;2\;3\;5\;6\,\}$
50: $(p \vee \Box\neg p) \to (\Box p \vee p \wedge \Box\neg p)$ $\{\,0\;1\;3\;4\;5\;6\,\}$
51: $\Box\neg p \vee (p \vee \Box p) \wedge \neg(p \wedge \Box p)$ $\{\,0\;1\;2\;3\;4\;7\,\}$
52: $\Box\bot \vee p \wedge \neg(\Box p \vee \Box\neg p)$ $\{\,0\;1\;7\,\}$
53: $(p \vee \Box\neg p) \wedge (p \wedge (\Box p \vee \Box\neg p) \to \Box\bot)$ $\{\,0\;1\;2\;7\,\}$
54: $\Box p \vee (p \vee \Box\neg p) \wedge \neg(p \wedge \Box\neg p)$ $\{\,0\;1\;2\;4\;5\;7\,\}$
55: $\Box\bot \vee \neg(\Box\neg p \vee (p \vee \Box p))$ $\{\,0\;1\;6\,\}$
56: $\Box\neg p \wedge (p \vee \Box\bot) \vee \neg(\Box\neg p \vee (p \vee \Box p))$ $\{\,0\;1\;3\;6\,\}$
57: $(p \vee \Box p) \wedge (p \wedge (\Box p \vee \Box\neg p) \to \Box\bot)$ $\{\,0\;1\;4\;7\,\}$
58: $\Box p \wedge (p \vee \Box\bot) \vee \neg(\Box\neg p \vee (p \vee \Box p))$ $\{\,0\;1\;5\;6\,\}$
59: $(\Box\neg p \vee (p \vee \Box p)) \wedge (p \wedge (\Box p \vee \Box\neg p) \to \Box\bot)$ $\{\,0\;1\;2\;4\;7\,\}$
60: $(p \vee \Box p) \wedge (p \wedge \Box\bot \vee \neg(\Box\neg p \vee p \wedge \Box p))$ $\{\,1\;4\;7\,\}$
61: $(\Box\neg p \vee (p \vee \Box p)) \to (p \vee \Box\bot) \wedge (\Box p \vee \Box\neg p)$ $\{\,0\;1\;3\;5\;6\,\}$
62: $(p \to \Box\neg p \wedge \neg\Box\bot) \wedge ((\Box p \vee \Box\neg p) \to (p \vee \Box\bot))$ $\{\,0\;3\;6\,\}$

The above list of formulas is a slightly edited version of the output of a computer program. The formulas constructed by this computer program depend on its settings, like the priority in application of the connectives and a selection on the 'best' (for example shortest) formula made.

To find the $1, 1$-maximal exact formulas ϕ in the above list, one has to look for the minimal sets $T_1^1(\phi)$ (i.e. those that do not occur as a proper subset of some $T_1^1(\psi)$ in the list).

The sets of types of this kind are:

1. $\{1, 5\}$ 3. $\{0, 1, 7\}$ 5. $\{1, 4, 7\}$ 7. $\{0, 3, 7\}$
2. $\{0, 2\}$ 4. $\{0, 1, 6\}$ 6. $\{1, 4, 6\}$ 8. $\{0, 3, 6\}$

It turns out that each of these $1, 1$-maximal exact formulas is maximal exact. The corresponding tail models can be found, using the model below, by extending the submodels $\uparrow k$ downward with a tail of copies of k for each of the numbered elements.

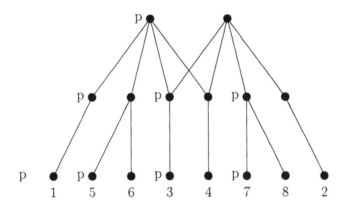

We will give these maximal exact formulas in L_1^1 a more informative form:

1. p
2. $\neg p$
3. $(\Box p \rightarrow \Box \perp) \wedge (\Box \neg p \rightarrow \Box \perp) \wedge (\neg p \rightarrow \Box \neg p)$
4. $(\Box p \rightarrow \Box \perp) \wedge (\Box \neg p \rightarrow \Box \perp) \wedge (p \rightarrow \Box p)$
5. $p \leftrightarrow \Diamond \neg p \vee \Box \perp$
6. $p \leftrightarrow \Box \neg p$
7. $p \leftrightarrow \neg \Box p$
8. $p \leftrightarrow \Box \neg p \vee \neg \Box \perp$

Formulas 1 and 2 correspond to *provable* and *refutable* sentences in PA. Formulas 6 and 7 can be (faithfully) interpreted by *Gödel-sentences* and their duals in PA. Similarly, formulas 3 and 4 correspond to *Rosser-sentences* and their duals in PA. The only small surprise is formed by formula 8 and its dual 5. It is easy to see that 8 is interderivable with $p \leftrightarrow \Box \Box \perp \wedge \neg \Box \perp$ and thus, of course, 5 with $p \leftrightarrow (\Box \Box \perp \rightarrow \Box \perp)$. These two formulas are not L_1^1, but can apparently interderivably be given as such. Note also that, by the fixed point theorem of L (see e.g., [10], [1]), there is no surprise in the fact that in the equivalences of 5 and 8 the p in the right hand side can be eliminated in favor of the $\perp$, but only in the fact that by using p instead of $\perp$ one can push down the complexity.

References

[1] G. Boolos, *The Logic of Provability*, Cambridge University Press, 1993.

[2] D.H.J. de Jongh, *Formulas of one Propositional Variable in Intuitionistic Arithmetic*, The L.E.J. Brouwer Centenary Symposium (A.S. Troelstra and D. van Dalen, eds.), 1982, pp. 51–64.

[3] D.H.J. de Jongh and L.A. Chagrova, *The Decidability of Dependency in Intuitionistic Propositional Logic*, to appear in J. Symbolic Logic.

[4] D.H.J. de Jongh, L. Hendriks, and G.R. Renardel de Lavalette, *Computations in Fragments of Intuitionistic Propositional Logic*, Journal of Automatic Reasoning **7** (1991), 537–561.

[5] Z. Gleit and W. Goldfarb, *Characters and Fixed Points in Provability Logic*, Notre Dame J. Formal Logic **31** (1990), 26–36.

[6] R.Sh. Grigolia, *Finitely generated Free Magari algebras*, Logiko Metodologicheskie Issledovaniya, Metsniereba, Tbilisi, 1983, pp. 135–149 (Russian).

[7] L. Hendriks, *Inventory of Fragments and exact Models in Intuitionistic Propositional Logic*, ILLC Prepublication ML–93–11, University of Amsterdam, 1993.

[8] V.V. Rybakov, *On Admissibility of Inference Rules in the Modal System G*, Matematicheskaya Logika i Algoritmicheskie Problemy (Novosibirsk) (Yu.L. Ershov, ed.), Trudy Instituta Matematiki, vol. 12, Nauka, 1989, pp. 120–138 (Russian).

[9] V.Yu. Shavrukov, *Subalgebras of Diagonalizable Algebras of Theories containing Arithmetic*, Dissertationes Math. (Rozprawy Mat.), vol. CCCXXIII, Warszawa, 1993.

[10] C. Smoryński, *Self-Reference and Modal Logic*, Universitext, Springer Verlag, 1985.

[11] R. Solovay, *Provability Interpretations of Modal Logic*, Israel J. Math. **25** (1976), 287–304.

[12] J.F.A.K. van Benthem, *Modal Logic and Classical Logic*, Bibliopolis, Napoli, 1983.

[13] A. Visser, *The Provability Logics of Recursively Enumerable Theories extending Peano Arithmetic at Arbitrary Theories extending Peano Arithmetic*, J. Philos. Logic **13** (1984), 97–113.

[14] D. Zambella, *Shavrukov's Theorem on the Subalgebras of Diagonalizable Algebras for Theories containing $I\Delta_0 + EXP$*, Notre Dame J. Formal Logic **35** (1994), 147–157.

DEPARTMENT OF MATHEMATICS AND COMPUTER SCIENCE, UNIVERSITY OF AMSTERDAM, PLANTAGE MUIDERGRACHT 24, 1018 TV AMSTERDAM, NETHERLANDS
e-mail address: DICKDJ@FWI.UVA.NL

The Butterfly and the Serpent

J. Lambek*

Abstract

The Butterfly Lemma was discovered by Zassenhaus to facilitate the proof of the Jordan-Hölder-Schreier Theorem for normal series in group theory. It may be deduced from an old theorem due to Goursat: if ρ is a homomorphic relation between groups G and H, then $\rho^\vee \rho$ and $\rho\rho^\vee$ are congruence relations of subgroups of G and H respectively and these subgroups become isomorphic modulo the respective congruence relations. It has become apparent that all this can be generalized from groups to a larger class of algebras, which can be characterized, for example, by the existence of a quaternary operation $hxyzt$ satisfying the identities $hxxxy = y$ and $hxxyy = hyxyx$. The Snake Lemma, which guarantees the existence of the "connecting homomorphism" in homological algebra, usually stated for modules, can also be derived from a homological version of Goursat's Theorem, the so-called Two-Square Lemma. It is now clear that this also holds in any variety of algebras with such a quaternary operation.

1 Introduction

To explain the title of this paper, let me begin by mentioning two fundamental lemmas, one in group theory and the other in homological algebra. The first, often called the Butterfly Lemma, is due to Zassenhaus, who used it to give a neat proof of the Jordan-Hölder-Schreier Theorem for normal series in group theory. The second, often called the Snake Lemma, is used to construct the so-called connecting homomorphism in homological algebra.

It so happens that both these lemmas are consequences of a still more basic result, which must be credited to Goursat [7]. Its group theoretic version asserts that any homomorphic relation between two groups (by which we mean

---

*The author acknowledges support from the Natural Sciences and Engineering Research Council of Canada and the hospitality of the University of Siena.

a binary relation whose graph is a subgroup of their direct product) gives rise to and is completely determined by an isomorphism between factor groups of subgroups of the two given groups. Its homological version [11] asserts that in the following situation:

$$
\begin{array}{ccccc}
A & \longrightarrow & B & \longrightarrow & C \\
\downarrow & & \downarrow & & \downarrow \\
D & \longrightarrow & E & \longrightarrow & F
\end{array}
$$

where the top and bottom row are exact sequences of groups or modules and where both squares commute, a certain invariant of the first square (invariant under reflection about the main diagonal) is isomorphic to another invariant of the second square.

All this can now be stated and proved in a more general context, either algebraic or categorical. As these basic and elementary results are scattered over the literature and not all of them are referred to in standard texts, it was felt that an expository presentation to a wider audience may be called for. At the same time, some unnecessary restrictions on two of these results will be removed. Considering the prejudice many mathematicians have against category theory (one that is apparently not shared by theoretical computer scientists), we shall largely confine attention here to varieties of algebras and downplay categorical arguments as well as generalizations to other categories, which will be discussed in Section 5.

By a *variety of algebras* is meant the collection of all models of an algebraic theory, presented by finitary operations and equations (identities) between them. Every mathematician is familiar with homomorphisms and congruence relations, but perhaps not with homomorphic relations in general, as these are not treated even in recent books on Universal Algebra.

A *homomorphic relation* ρ between two algebras A and B in a variety is a triple $\rho = (B, R, A)$ of algebras such that R is a subalgebra of the direct product $B \times A$. One writes

$$
b\rho a \qquad for \qquad (b, a) \in R .
$$

We call $R = |\rho|$ the *graph* of ρ and say that ρ is a relation *from A to B*. The *converse* relation $\rho^\vee$ from B to A is such that

$$
a\rho^\vee b \qquad if\ and\ only\ if \qquad b\rho a .
$$

The *composition* (relative product) of ρ with another relation σ from B to C is the relation $\sigma\rho$ from A to C such that

$$
c\sigma\rho a \qquad if\ and\ only\ if \qquad \exists_{b \in B}(c\sigma b \wedge b\rho a) .
$$

The *identity relation* 1_A from A to A is of course defined by:

$$a1_A a' \quad if \ and \ only \ if \quad a = a' .$$

Clearly, its graph is the diagonal of $A \times A$. If ρ and ρ' are two homomorphic relations from A to B, we write

$$\rho \subseteq \rho' \quad for \quad |\rho| \subseteq |\sigma| .$$

A homomorphic relation ρ from A to B is said to be

single valued	if	$\rho\rho^\vee \subseteq 1_B,$
universally defined	if	$1_A \subseteq \rho^\vee\rho,$
one-to-one	if	$\rho^\vee\rho \subseteq 1_A,$
onto	if	$1_B \subseteq \rho\rho^\vee.$

A homomorphic relation satisfying the first two of these conditions is a *homomorphism*; if it satisfies all four conditions it is an *isomorphism*. When ρ is a homomorphism, we also write

$$b = \rho a \quad for \quad b\rho a .$$

(Our notation was chosen to ensure that composition of homomorphisms agrees with composition of relations.) It should be pointed out that any homomorphic relation ρ from A to B has the form $\rho = \varphi\psi^\vee$, where φ and ψ are homomorphisms as follows:

$$\psi : |\rho| \to B \times A \to A ; \quad \varphi : |\rho| \to B \times A \to B .$$

A homomorphic relation α from A to A is said to be

reflexive	if	$1_A \subseteq \alpha,$
symmetric	if	$\alpha^\vee \subseteq \alpha,$
transitive	if	$\alpha\alpha \subseteq \alpha.$

A homomorphic relation satisfying the last two conditions only is called a *subcongruence* of A; if it satisfies all three conditions it is a *congruence relation* on A.

With any homomorphism $\varphi : A \to B$ we may associate its *image* and its *kernel*:

$$Im\varphi = \{b \in B | \exists_{a \in A} b = \varphi a\} , \quad Ker\varphi = \varphi^\vee\varphi .$$

The former is a subalgebra of B, the latter a congruence relation on A. We call φ a *surjection* if $Im\varphi = B$, an *injection* if $Im\varphi = A$. Evidently, a surjection is the same thing as homomorphisms onto. On the other hand, an injection is one-to-one, meaning $\mathrm{Ker}\varphi = 1_A$, but not conversely, except up to isomorphism. A one-to-one homomorphism μ is the same thing as a *monomorphism*, meaning that $\mu\varphi = \mu\psi$ implies $\varphi = \psi$. On the other hand, a surjection ε is an *epimorphism*, meaning that $\varphi\varepsilon = \psi\varepsilon$ implies $\varphi = \psi$, but not conversely, not even up to isomorphism. For example, in the variety (category) of commutative rings, $\mathbb{Z} \to \mathbb{Q}$ is an epimorphism.

If θ is a congruence relation on A, one can form the *quotient algebra A/θ*, whose elements are equivalence classes of elements of A modulo θ, and there is a surjection $\pi : A \to A/\theta$ sending the element a of A onto the equivalence class of a modulo θ and $\theta = Ker\pi$.

The *first isomorphism theorem* asserts that, for any homomorphism $\varphi : A \to B$,

$$Im\varphi \cong A/Ker\varphi \ .$$

There are two other isomorphism theorems, which we will state for completeness and to help the reader to get used to our notation, even though they will not be used here explicitly. As with the ten commandments, different authors do not agree on the numbering or even on the precise enunciation.

The *second isomorphism theorem* asserts that, if $\pi : A \to B$ is a surjection, then any congruence relation θ on A containing $\mathrm{Ker}\pi$ gives rise to a congruence relation $\pi\theta\pi^\vee$ on B and that every congruence relation θ' on B has this form, with $\theta = \pi^\vee\theta'\pi$; moreover $A/\theta \cong B/\pi\theta\pi^\vee$.

The *third isomorphism theorem* asserts that, if $\mu : A \to B$ is an injection and θ a congruence relation on B, then $\mu^\vee\theta\mu$ is a congruence relation on A, namely the restriction of θ to A, and $A/\mu^\vee\theta\mu \cong \theta A/\nu^\vee\theta\nu$, where $\theta A = \{b \in B | \exists_{a \in A} b\theta a\}$ is a subalgebra of B and $\nu : \theta A \to B$ is the injection of this subalgebra.

2 Maltsev and Goursat Varieties

For which varieties of algebras (in place of groups) can one carry out the program outlined at the beginning of the Introduction? It has been known for some time that this can be done for *Maltsev varieties*, which may be characterized by any number of equivalent conditions, of which we will mention only a few relevant ones here.

Proposition 2.1 *For any variety of algebras, the following conditions are equivalent:*

1. *Among the operations there is a ternary operation $fxyz$ satisfying the identities*

$$fxyy = x, \qquad fyyz = z \ .$$

2. *Any two congruence relations θ and θ' on any algebra permute, that is, $\theta\theta' = \theta'\theta$. (Equivalently: $\theta\theta'$ is the join of θ and θ' in the lattice of all congruence relations of the algebra.)*

3. *Any homomorphic relation ρ between two algebras is* difunctional, *meaning that $\rho\rho^\vee\rho = \rho$.*

4. *Any reflexive homomorphic relation on any algebra is a congruence relation.*

A variety of algebras satisfying the equivalent conditions of Proposition 2.1 is called a *Maltsev variety*. Examples of Maltsev varieties abound: groups, rings, modules, loops, quasigroups, complemented lattices, Heyting algebras etc.

This is where matters stood when Cristina Pedicchio and Aurelio Carboni realized that, in the proof of the algebraic version of Goursat's theorem, the difunctionality of ρ could be replaced by the assumption that $\rho\rho^\vee$ and $\rho^\vee\rho$ are subcongruence relations. After the dust had settled, and with due respect to the historical comments in Section 6, here is the outcome.

Proposition 2.2 *For any variety of algebras, the following statements are equivalent:*

1. *There are ternary operations $fxyz$ and $gxyz$ satisfying the identities*

$$fxyy = x, \quad fyyz = gyzz, \quad gyyz = z .$$

2. *Any two congruence relations θ and θ' on any algebra 3-permute, meaning that $\theta\theta'\theta = \theta'\theta\theta'$. (Equivalently: $\theta\theta'\theta$ is the join of θ and θ'.)*

3. *Any homomorphic relation ρ between any two algebras satisfies $\rho\rho^\vee\rho\rho^\vee = \rho\rho^\vee$.*

4. *There is a quaternary operation $hxyzt$ satisfying the identities*

$$hxxxt = t, \quad hxxtt = htxtx .$$

Proof.(1) $\Rightarrow$ (2). Suppose $a\theta b$, $b\theta'c$ and $c\theta d$; we seek u and v such that $a\theta'u$, $u\theta v$ and $v\theta'd$. Take $u = fabc$, $v = gbcd$ and calculate:

$$a = fabb \; \theta' fabc \; \theta fbbc = gbcc \; \theta gbcd \; \theta' gccd = d .$$

(2) $\Rightarrow$ (1). Consider the free algebra $F(x,y,z,t)$ generated by the set $\{x,y,z,t\}$ as well as the homomorphisms $\alpha : F(x,y,z,t) \to F(x,z)$ and

β : $F(x, y, z, t) \rightarrow F(x, y, t)$ such that $\alpha x = \alpha y = x$, $\alpha z = \alpha t = z$ and $\beta x = x$, $\beta y = \beta z = y$, $\beta t = t$. Then $x\alpha^\vee \alpha y$, $y\beta^\vee \beta z$ and $z\alpha^\vee \alpha t$. But, by (2), $\alpha^\vee \alpha \beta^\vee \beta \alpha^\vee \alpha = \beta^\vee \beta \alpha^\vee \alpha \beta^\vee \beta$, so there exist $u = pxyzt$ and $v = qxyzt$ in $F(x, y, z, t)$ such that $x\beta^\vee \beta u$, $u\alpha^\vee \alpha v$ and $v\beta^\vee \beta t$, that is,

$$x = pxyyt, \quad pxxzz = qxxzz, \quad qxyyt = t .$$

Take $fxyz = pxyzz$ and $gyzt = qyyzt$, then (1) follows.

(2) $\Rightarrow$ (3). As was pointed out in Section 1, we may write $\rho = \varphi \psi^\vee$, where φ and ψ are homomorphisms. Hence

$$\begin{aligned}
\rho\rho^\vee \rho\rho^\vee &= \varphi(\psi^\vee \psi)(\varphi^\vee \varphi)(\psi^\vee \psi)\varphi^\vee \\
&= \varphi(\varphi^\vee \varphi)(\psi^\vee \psi)(\varphi^\vee \varphi)\varphi^\vee
\end{aligned}$$

if the congruence relations $\psi^\vee \psi$ and $\varphi^\vee \varphi$ 3-permute. But $\varphi\varphi^\vee \varphi = \varphi$ and so

$$\rho\rho^\vee \rho\rho^\vee = \varphi\psi^\vee \psi\varphi^\vee = \rho\,\rho^\vee .$$

(3) $\Rightarrow$ (2). Take two congruence relations θ and θ' on an algebra A and apply (3) to $\rho = \theta\theta'$, then $\rho\rho^\vee = \theta\theta'\theta$ will be a subcongruence. But, as it is also reflexive, it is a congruence relation, which contains both θ and θ'. If θ'' is another congruence relation containing both θ and θ', then it contains also $\theta\theta'\theta$. Therefore $\theta\theta'\theta = \theta \cup \theta'$ is the join of θ and θ' in the lattice of congruence relations. But similarly $\theta'\theta\theta' = \theta \cup \theta' = \theta\theta'\theta$.

(1) $\Rightarrow$ (4). Take $hxyzt = fgxytzx$ and verify (4).

(4) $\Rightarrow$ (1). Take $fxyz = hzzyx$ and $gxyz = hxyxz$ and verify (1).

A variety of algebras satisfying the equivalent statements of Proposition 2.2 is called a *Goursat variety*.

Obviously, any Maltsev variety is a Goursat variety, since permutability of congruence relations implies 3-permutability. It may be instructive to see this also by looking at the operations. Suppose $fxyz$ is a Maltsev operation satisfying $fxyy = x$ and $fyyz = z$. Take $gxyz = z$ to be the projection onto the third coordinate, then automatically (1) holds.

To get as far away from a Maltsev variety as possible, we might take $gxyz = fzyx$. Then (1) becomes:

$$(1^*) \qquad\qquad fxyy = x, \quad fxxy = fyyx .$$

If we are looking for Goursat varieties that are not Maltsev, a good start may be to find examples satisfying (1*). For example, take $fxyz = (z \Rightarrow y) \Rightarrow x$, where $\Rightarrow$ is a binary operation. Then (1*) becomes:

$$(y \Rightarrow y) \Rightarrow x = x, \quad (y \Rightarrow x) \Rightarrow x = (x \Rightarrow y) \Rightarrow y .$$

These identities hold in any so-called *implication algebra*. Indeed, Mitschke [18] pointed out that, in the three-element implication algebra $\{1, a, b\}$ obtained from a four-element Boolean algebra by deleting the bottom element, congruence relations do not permute. Thus, if $a\theta 1$ but not $b\theta 1$ and $b\theta'1$ but not $a\theta'1$, then $a\theta\theta'b$ but not $a\theta'\theta b$.

For any subcongruence κ of A we write

$$Dom\kappa = \kappa A = \{a \in A | a\kappa a\} .$$

This is clearly a subalgebra of A on which κ acts as a congruence relation. We now obtain Goursat's Theorem:

Proposition 2.3 *If ρ is any homomorphic relation from A to B such that $\rho^\vee \rho$ and $\rho\rho^\vee$ are subcongruences of A and B respectively, in particular if ρ is any homomorphic relation whatsoever between two algebras in a Goursat variety, then*

$$Dom\rho^\vee \rho / \rho^\vee \rho \cong Dom\rho\rho^\vee / \rho\rho^\vee .$$

Proof. We may restrict ρ to a homomorphic relation ρ' from $Dom\rho^\vee \rho$ to $Dom\rho\rho^\vee$, which is universally defined and onto. Now ρ' induces a homomorphic relation φ from $Dom\rho^\vee \rho / \rho^\vee \rho$ to $Dom\rho\rho^\vee / \rho\rho^\vee$, which relates a modulo $\rho^\vee \rho$ to b modulo $\rho\rho^\vee$. This is clearly universally defined and onto, since ρ' is, but it is also single valued and one-to-one. For example, to show the former, assume $a \, mod\rho^\vee \rho = a' mod\rho^\vee \rho$, that is, $a\rho^\vee \rho a'$. If $b\rho a$ and $b'\rho a'$ then $b\rho\rho^\vee \rho \rho^\vee b'$, hence $b\rho\rho^\vee b'$ that is, $b \, mod\rho\rho^\vee = b' mod\rho\rho^\vee$. This shows that φ is single-valued. The same argument applied to $\varphi^\vee$ shows that φ is one-to-one.

3   The Butterfly Lemma

Zassenhaus originally stated this lemma as follows: if U and V are subgroups of a group G and U' and V' are normal subgroups of U and V respectively, then

$$(U' \cdot (U \cap V))/(U' \cdot (U \cap V')) \cong ((U \cap V) \cdot V')/((U' \cap V) \cdot V') .$$

Its entomological appellation presumably derives from the following picture:

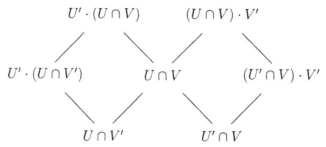

which played a role in the original proof.

One may reformulate the Butterfly Lemma using the calculus of relations. U' defines a congruence relation κ on U such that, for all $a, b \in U$,

$$a\kappa b \quad \text{if and only if} \quad ab^{-1} \in U' .$$

We may view κ as a *subcongruence* on G with domain U, satisfying $\kappa^\vee \subseteq \kappa$ and $\kappa\kappa \subseteq \kappa$, but not $1_G \subseteq \kappa$, unless $U = G$. Similarly V' defines a subcongruence relation λ of G with domain V. Since groups form a Maltsev variety, with $fxyz = xy^{-1}z$, $\kappa\lambda\kappa = (\kappa\lambda)(\kappa\lambda)^\vee$ is a subcongruence. The Butterfly Lemma then takes the form:

$$Dom\kappa\lambda\kappa/\kappa\lambda\kappa \cong Dom\lambda\kappa\lambda/\lambda\kappa\lambda ,$$

making it a consequence of Goursat's Theorem, Proposition 2.3, in view of the following observation.

Proposition 3.1 *If U and V are subgroups of a group with normal subgroups U' and V' respectively, then $\kappa\lambda\kappa$ defines the congruence relation on $U' \cdot (U \cap V)$ corresponding to the normal subgroup $U' \cdot (U \cap V')$.*

Proof. (a) Suppose $a\kappa\lambda\kappa b$, then there exist c and d in G such that $a\kappa c, c\lambda d$ and $d\kappa b$, that is,

$(*) \qquad a, c \in U, \ ac^{-1} \in U', \ c, d \in V, \ cd^{-1} \in V', \ d, b \in U, \ db^{-1} \in U' .$

Hence $ac^{-1} \in U'$, $cd^{-1} \in U \cap V'$ and $db^{-1} \in U'$. Therefore

$$ab^{-1} = (ac^{-1})(cd^{-1})(db^{-1}) \in U'(U \cap V')U' = U'(U \cap V') .$$

Moreover, a, b, c and d belong to $U'(U \cap V)$; for example, $a \in U'c \subseteq U'(U \cap V)$.

(b) Assume conversely that $a, b \in U'(U \cap V)$ and $ab^{-1} \in U'(U \cap V')$. We seek c and d such that $(*)$, namely $c \in U'a \cap V$ and $d \in U'b \cap V$ such that $cd^{-1} \in V'$.

Since $a \in U'(U \cap V)$, we may take $a \in U'\alpha$ with $\alpha \in U \cap V$, and similarly $b \in U'\beta$ with $\beta \in U \cap V$. So we want to show that

$$(U'\alpha \cap V)(\beta^{-1}U' \cap V) \quad meets \quad V' \, .$$

Thus we seek x and y in U' such that $x\alpha \in V$, $y\beta \in V$ and $x\alpha\beta^{-1}y^{-1} \in V'$. That is, we want $\alpha\beta^{-1} \in x^{-1}V'y \subseteq (U' \cap V)V'(U' \cap V) = (U' \cap V)V$. Since we know that $\alpha\beta^{-1} \in U'ab^{-1}U' \subseteq U'V'U'$ and $\alpha\beta \in U \cap V$, it suffices to show that

$$U'V' \cap V \subseteq (U' \cap V)V' \, ;$$

but this is the familiar modular law for groups.

In view of Proposition 3.1, the original Butterfly Lemma of Zassenhaus is a special case of the following.

Proposition 3.2 *If κ and λ are subcongruences of any algebra in a Goursat variety.,*

$$Dom\kappa\lambda\kappa/\kappa\lambda\kappa \cong Dom\lambda\kappa\lambda/\lambda\kappa\lambda \, .$$

Proof. Take $\rho = \kappa\lambda$, hence $\rho^{\vee} = \lambda\kappa$, and apply Proposition 2.3.

One can derive the Jordan-Hölder-Schreier Theorem for normal series in any Goursat variety from Proposition 3.2, just as Zassenhaus did for groups. But first we have to explain the terminology. If C is a subalgebra of A and κ a subcongruence of A such that $C \subseteq \kappa C = \{a \in A | \exists_{c \in C} a \kappa c\}$, we shall call κ a *subcongruence of A over C*. According to Goldie [6], a *normal series from A to C* is an m-tuple of subcongruences $\kappa_1, \cdots, \kappa_m$ of A over C such that

$$A = \kappa_1 A, \ \kappa_1 C = \kappa_2 A, \cdots, \kappa_{m-1}C = \kappa_m A, \ \kappa_m C = C \, .$$

Proposition 3.3 *If $\kappa_1, \cdots, \kappa_m$ and $\lambda_1 \cdots, \lambda_n$ are normal series from A to C in a Goursat variety, then the rectangular array $\{\kappa_i\lambda_j\kappa_i\}_{i,j}$ may be ordered by rows to give a "refinement" of the former and the array $\{\lambda_j\kappa_i\lambda_j\}_{i,j}$ may be ordered by columns to give a refinement of the latter, so that corresponding entries of the two arrays determine isomorphic quotient algebras of subalgebras of A.*

Proof. First note that, if κ and λ are subcongruences of A over C, then

$(*)$ $$\lambda\kappa\lambda A = \lambda\kappa A, \quad \lambda\kappa\lambda C = \lambda\kappa C \, ;$$

for from $\lambda A \subseteq A$ and $\kappa A \subseteq A$ we infer that

$$\lambda\kappa\lambda A \subseteq \lambda\kappa A = \lambda\kappa\lambda\kappa A \subseteq \lambda\kappa\lambda A \, ,$$

and from $C \subseteq \lambda C$ and $C \subseteq \kappa C$ we infer that

$$\lambda \kappa \lambda C \subseteq \lambda \kappa \lambda \kappa C = \lambda \kappa C \subseteq \lambda \kappa \lambda C \ .$$

Now let j be fixed and consider the subcongruences $\lambda_j \kappa_i \lambda_j$, where i ranges from 1 to m. These are not only subcongruences of A over C but also of $\lambda_j A$ over $\lambda_j C$, as is easily checked; in fact they form a normal series from $\lambda_j A$ to $\lambda_j C$, since, by $(*)$,

$$
\begin{aligned}
\lambda_j A = \lambda_j \kappa_1 A \ &= \lambda_j \kappa_1 \lambda_j A = \lambda_j \kappa_1 \lambda_j \lambda_j A \ , \\
\lambda_j \kappa_i \lambda_j \lambda_j C \ &= \lambda_j \kappa_i \lambda_j C = \lambda_j \kappa_i C = \lambda_j \kappa_{i+1} A = \lambda_j \kappa_{i+1} \lambda_j A \\
&= \lambda_j \kappa_{i+1} \lambda_j \lambda_j ; A \qquad for \quad i = 1, \cdots, m-1 \ , \\
\lambda_j \kappa_m \lambda_j \lambda_j C \ &= \lambda_j \kappa_m \lambda_j C = \lambda_j \kappa_m C = \lambda_j C.
\end{aligned}
$$

Now let j vary from 1 to n and recall that

$$A = \lambda_1 A, \quad \lambda_j C = \lambda_{j+1} C \quad for \quad j < n, \quad \lambda_n C = C \ .$$

Then, putting all the above normal series together, we get a single normal series from A to C.

Finally, we use Proposition 3.2 to obtain the asserted isomorphism.

4   The Homological Two-Square Lemma

Now let us look for a homological version of Goursat's Lemma.

Proposition 4.1 *In a Goursat variety, consider the following diagram:*

$$
\begin{array}{ccc}
B & \xrightarrow{\ \mu\ } & C \\
\big\downarrow{\scriptstyle \beta} & & \\
D \xrightarrow[\lambda]{} & E &
\end{array}
$$

and assume that

$(*)$ $$\mu^{\vee} \mu \beta^{\vee} Im\lambda \subseteq \beta^{\vee} Im\lambda \ .$$

Then

$$\frac{Im\beta \cap Im\lambda}{\beta(Ker\mu)\beta^\vee} \cong \frac{\beta^\vee Im\lambda}{Ker\mu \cup Ker\beta} .$$

Here $Ker\mu \cup Ker\beta$ denotes the join of two congruence relations on B; but the congruence relations appearing in both denominators are assumed to be restricted to the algebras in the numerators.

Proof. Take $\rho = \lambda\lambda^\vee\beta\mu^\vee\mu$, that is, $e\rho b$ if and only if $e \in Im\lambda$ and $e\beta\mu^\vee\mu b$. Then $\rho\rho^\vee = \lambda\lambda^\vee\beta\mu^\vee\mu\beta^\vee\lambda\lambda^\vee$ is the restriction of $\beta\mu^\vee\mu\beta$ to $Im\lambda$, hence to $Dom\rho\rho^\vee = Im\beta \cap Im\lambda$. Now $e\rho b$ if and only if, for some $b' \in B$,

$$e \in Im\lambda \quad and \quad e = \beta b' \quad and \quad \mu b' = \mu b .$$

But, under the assumption $(*)$, $\beta b' = e \in Im\lambda$ can be replaced by $b \in \beta^\vee Im\lambda$, so $b\rho\rho^\vee b'$ if and only if

$$b, b' \in \beta^\vee Im\lambda \quad and \quad b\mu^\vee\mu\beta^\vee\beta\mu^\vee\mu b' ,$$

that is, $\rho\rho^\vee$ is the restriction to $\beta^\vee Im\lambda$ of the join of the congruence relations $\mu^\vee\mu$ and $\beta^\vee\beta$ on B.

Corollary 4.2 *In a Goursat variety, consider the following diagram:*

$$A \;\underset{\lambda_2}{\overset{\lambda_1}{\rightrightarrows}}\; B \;\xrightarrow{\mu}\; C$$

$$\downarrow \beta$$

$$D \;\xrightarrow{\lambda}\; E \;\underset{\mu_2}{\overset{\mu_1}{\rightrightarrows}}\; F$$

and assume $(*)$ and that the top and bottom rows are exact:

$$Im(\lambda_1, \lambda_2) = Ker\mu , \quad Im\lambda = Ker(\mu_1, \mu_2) .$$

Then

$$\frac{Im\beta \cap Im\lambda}{\beta Im(\lambda_1, \lambda_2)\beta^\vee} \cong \frac{\beta^\vee Ker(\mu_1, \mu_2)}{Ker\mu \cup Ker\beta} .$$

Here

$$Ker(\mu_1, \mu_2) = \{e \in E | \mu_1 e = \mu_2 e\}$$

is the usual equalizer of the arrows μ_1 and μ_2 and

$$Im(\lambda_1, \lambda_2) = \text{ the smallest congruence relation on } B \text{ containing } \lambda_1\lambda_2^\vee.$$

Corollary 4.2 is just a rewritten form of Proposition 4.1, its advantage being that the left side of the isomorphism depends only on the left part of the diagram and the right side of the isomorphism depends only on the right part of the diagram.

For purposes of application, we introduce two additional arrows $A \rightrightarrows D$ and $C \rightrightarrows F$.

Proposition 4.3 *(Two-square lemma). In a Goursat diagram consider the following diagram:*

$$
\begin{array}{ccccc}
A & \overset{\lambda_1}{\underset{\lambda_2}{\rightrightarrows}} & B & \overset{\mu}{\longrightarrow} & C \\[2mm]
\alpha_1 \Big\| \Big\| \alpha_2 \;\; (1) & & \Big| \beta \;\; (2) & & \gamma_1 \Big\| \Big\| \gamma_2 \\[2mm]
D & \underset{\lambda}{\longrightarrow} & E & \overset{\mu_1}{\underset{\mu_2}{\rightrightarrows}} & F
\end{array}
$$

and assume that the top and bottom rows are exact and that the squares (1) and (2) quasi-commute in the sense that

$$
\beta Im(\lambda_1, \lambda_2)\beta^{\vee} = \lambda Im(\alpha_1, \alpha_2)\lambda^{\vee} = Im(A \rightrightarrows E')
$$

say, when both sides are restricted to $Im\beta \cap Im\lambda = E'$, and that

$$
\beta^{\vee} Ker(\mu_1, \mu_2) = \mu^{\vee} Ker(\gamma_1, \gamma_2) = Ker(B \rightrightarrows F) \ .
$$

Then

$$
\frac{Im(B \to E) \cap Im(D \to E)}{Im(A \rightrightarrows E')} \cong \frac{Ker(B \rightrightarrows F)}{Ker(B \to C) \cup Ker(B \to E)} \ .
$$

Proof. In view of Corollary 4.2, it suffices to verify the assumption $(*)$ of Proposition 4.1. This can be done by exploiting the quasi-commutativity of square (2). Suppose $b \in \mu^{\vee}\mu\beta^{\vee} Im\lambda$, that is, $\mu b = \mu b'$ and $b' \in \beta^{\vee} Im\lambda = \beta^{\vee} Ker(\mu_1, \mu_2) = \mu^{\vee} Ker(\gamma_1, \gamma_2)$ for some $b' \in B$. Since $\mu b = \mu b'$, also $b \in \beta^{\vee} Im\lambda$, and this shows $(*)$.

The statement of the Two-Square Lemma may be abbreviated thus:

$$
Im(1) \cong Ker(2) \ .
$$

The Two-Square-Lemma may be used to prove various results in homological algebra. Its usefulness arises from the fact that it asserts an isomorphism relating certain expressions, associated with the double squares, which are *invariant under reflection about their main diagonals*. This may be exploited by stacking double squares in staircase fashion, as is illustrated in the so-called Snake Lemma.

Proposition 4.4 *In the following diagram in a Goursat variety it is assumed that all double squares quasi-commute and that all rows and columns are exact.*

$$
\begin{array}{ccc}
A & \rightrightarrows & B \\
\big\| \Big\downarrow\ (1) & & \big\downarrow \\
C \rightrightarrows & D & \rightarrow\ E \\
\big\|\ (3) & \big\downarrow\ (2) & \big\| \\
F & \rightarrow\ G & \rightrightarrows\ H \\
\big\downarrow\ (4) & \big\| & \\
I & \rightrightarrows & J
\end{array}
$$

It is furthermore assumed that $B \longrightarrow E$ *and* $F \longrightarrow G$ *are injections and that* $D \longrightarrow E$ *and* $F \longrightarrow I$ *are surjections. Then there exists a so-called* connecting *homomorphism* $B \longrightarrow I$ *such that*

$$A \rightrightarrows B \longrightarrow I \rightrightarrows J$$

is exact, that is,

$$Im(A \rightrightarrows B) = Ker(B \longrightarrow I), \quad Im(B \longrightarrow I) = Ker(I \rightrightarrows J) .$$

(It is when we attempt to draw the arrow $B \longrightarrow I$ on the diagram that the picture of a snake appears.)

Proof. We begin by constructing two new double squares as follows:

$$
\begin{array}{ccc}
A & \rightrightarrows\ B & \rightarrow\ L \\
\big\|\ (1) & \big\downarrow\ (0) & \big\| \\
D & \rightarrow\ E & \rightrightarrows\ \top
\end{array}
\qquad , \qquad
\begin{array}{ccc}
\bot & \rightrightarrows\ F & \rightarrow\ G \\
\big\|\ (5) & \big\downarrow\ (4) & \big\| \\
K & \rightarrow\ I & \rightrightarrows\ J
\end{array}
$$

Here

$$L = B/Im(A\rightrightarrows B) , \quad K = Ker(I\rightrightarrows J) ;$$

moreover $\top$ and $\bot$ denote the terminal and initial objects in the variety respectively. Evidently, the four rows thus introduced are exact and the new squares (0) and (5) quasi-commute. One easily calculates $Ker(0) = L$ and $Im(5) = K$, hence

$$K = Im(5) \cong Ker(4) \cong Im(3) \cong Ker(2) \cong Im(1) \cong Ker(0) = L .$$

Thus we have

$$A\rightrightarrows B \longrightarrow L \cong K \longrightarrow I\rightrightarrows J ,$$

from which it follows that

$$A\rightrightarrows B \longrightarrow I\rightrightarrows J$$

is exact, as was to be proved.

5 Relations in Categories

Homomorphic relations may be studied not only in varieties of algebras, but in arbitrary categories. One way to do this is to approach a binary relation between two objects via its graph, a subobject of their direct product. To this purpose one may have to assume that the category in question is regular, as in Meisen [17], or even exact in the sense of Barr et al. [1], as in Carboni et al. [2]. There is a small problem with this approach: composition of relations is associative only up to isomorphism, and one really ought to show that this isomorphism is "coherent". For this reason, I would like to suggest a different approach here, as in [14].

Given any category $\mathcal{C}$, by a *potential homomorphic relation* $\rho : A \nrightarrow B$ between objects A and B of $\mathcal{C}$ we shall understand a family of ordinary binary relations

$$\rho_C : Hom(C, A) \nrightarrow Hom(C, B)$$

between Hom-sets, where C ranges over all objects of $\mathcal{C}$, subject to the condition that, for all (generalized elements) $a : C \to A$, $b : C \to B$ and $c : D \to C$,

$$b\rho_C a \Rightarrow (bc)\rho_D(ac) .$$

In particular, every arrow $f : A \to B$ in $\mathcal{C}$ may be viewed as a potential homomorphic relation if we write $bf_C a$ for $b = fa$. For instance, we have the *identity relation* $1_A : A \nrightarrow A$, where $a'1_A a$ means $a' = a$. Relations may be *composed* in the usual way: if $\rho : A \nrightarrow B$ and $\sigma : B \nrightarrow C$, one defines $\sigma\rho : A \nrightarrow C$ by

$$(\sigma\rho)_D = \sigma_D \rho_D \ .$$

Also one has the *converse relation* $\rho^\vee : B \nrightarrow A$, defined by

$$(\rho^\vee)_C = (\rho_C)^\vee \ .$$

By an actual *homomorphic relation* between objects of $\mathcal{C}$ we shall understand a potential homomorphic relation which is generated by arrows of $\mathcal{C}$ and their converses. Imposing a mild assumption on $\mathcal{C}$ we can ensure that every homomorphic relation has the form $fg^\vee$. We need merely assume that $\mathcal{C}$ has *weak pullbacks*, that is, given arrows $A \xrightarrow{h} C \xleftarrow{k} B$, there exist arrows $A \xleftarrow{f} R \xrightarrow{g} B$ such that $k^\vee h = fg^\vee$.

By a *subcongruence* α of A we shall understand a homomorphic relation $\alpha : A \nrightarrow A$ satisfying

$$\alpha\alpha \subseteq \alpha \ , \quad \alpha^\vee \subseteq \alpha \ .$$

Associated with α_C is the set

$$F_\alpha(C) = Dom\,\alpha_C / \alpha_C \ .$$

For each arrow $c : D \to C$ one may define the mapping $F_\alpha(c) : F_\alpha(C) \to F_\alpha(D)$ by

$$F_\alpha(c)(a \bmod \alpha_C) = ac \bmod \alpha_D \ ,$$

thus making F_α into a functor $\mathcal{C}^{op} \to$ Set, to be precise, a quotient functor of a subfunctor of the representable functor Hom $(-, A)$.

The subcongruences of objects of $\mathcal{C}$ may be viewed as objects a category $\mathcal{C}^\#$, whose arrows $(\beta, \rho, \alpha) : \alpha \to \beta$ are defined as follows. Assuming α and β to be subcongruences of A and B respectively, we require $\rho : A \nrightarrow B$ and stipulate

$$\alpha \subseteq \rho^\vee \beta \rho \ , \quad \rho\alpha\rho^\vee \subseteq \beta$$

to make sure that the induced mapping $t_\rho(C) : F_\alpha(C) \to F_\beta(C)$ is single-valued and universally defined, where

$$t_\rho(C)(a \bmod \alpha_C) = b \bmod \beta_C$$

whenever $b\rho_C a$. It is easily seen that $t_\rho : F_\alpha \to F_\beta$ is a natural transformation.

Equality of arrows, $(\beta, \rho, \alpha) = (\beta, \sigma, \alpha)$, is defined to mean

$$\alpha \subseteq \rho^\vee \beta\sigma \ , \quad \rho\alpha\sigma^\vee \subseteq \beta \ ;$$

but it is easily seen that each of these conditions implies the other. One may also verify that

$$(\beta, \rho, \alpha) = (\beta, \sigma, \alpha) \Rightarrow t_\rho = t_\sigma \; ;$$

thus $\mathcal{C}^\#$ may be identified with the closure of $\mathcal{C}$ in $\mathrm{Set}^{\mathcal{C}^{op}}$ (via the Yoneda embedding) under certain subfunctors and quotient functors.

To make the above arguments go through, one may still have to assume that our category is exact, but one must also postulate that any intersection of congruence relations is a congruence relation.

By a *Maltsev category* or *Goursat category* one may now understand such a category the homomorphic relations of which satisfy $\rho\rho^\vee\rho = \rho$ or $\rho\rho^\vee\rho\rho^\vee = \rho\rho^\vee$ respectively. Many of our results for Maltsev or Goursat varieties may be extended to such categories, provided we shift attention from algebras to subcongruences. Assuming that $\mathcal{C}$ is exact will often allow one to pass from subcongruences back to actual objects of $\mathcal{C}$. We must defer elaboration of these ideas to another occasion.

6 Historical Comments

Proposition 2.1. The term "difunctional" was introduced by Riguet [19],[20], who first studied such relations explicitly. Other authors keep on replacing this by the rather overworked term "regular"or, as I learned recently, by "rectangular". The equivalence of (1) and (2) was proved by Maltsev (= Mal'cev) [16]. The equivalence with (3) was observed in [10], the equivalence with (4) by Findlay [4]. There are a number of equivalent statements not mentioned here, e.g. by Linton, Faro and Fleischer.

Proposition2.2. The notion of n-permutability, as a generalization of the case $n = 2$ considered by Maltsev, has been part of the literature of universal algebra for some time, as I learned from Klaus Denecke. Originally, more complicated conditions appeared in place of (1) ([8], [21], [22]). Hagemann and Mitschke [9] discovered the simpler condition (1), as well as its generalization to n-permutability. Their proof of the equivalence between (1) and (2) makes use of the earlier results by Grätzer et al. The more direct proof here imitates the original proof by Maltsev. The equivalence of (2) and (3) is due to Kelly and Pedicchio. The equivalence with (4) was asserted in a postscript to [13] (though with an unfortunate misprint).

Proposition 2.3. This goes back to Goursat [7], as I learned from H.S.M. Coxeter.

Proposition 3.1. This had been asserted, even for left loops in place of groups, by quoting Laplace's Lemma: "it is easily seen". Trying to reconstruct the proof 25 years later, I had some difficulty with (b), so it was decided to include the proof here, if only for groups.

Proposition 3.2. The Butterfly Lemma was reformulated in terms of homomorphic relations in [10], where it was deduced from Goursat's Theorem (see also Appendix D in Zassenhaus [23]).

Proposition 3.3. There is also a Jordan-Hölder-Schreier Theorem for principal series, which deals with congruences rather than subcongruences. This had already been generalized to Maltsev varieties by Birkhoff. The general definition of normal series is due to Goldie ([5],[6]), who gave an even more general version of the Jordan-Hölder-Schreier Theorem, which involved the transitive closures of $\kappa\lambda$ and $\lambda\kappa$. The proof given here follows [10].

It is tempting to write $\kappa A = Im\kappa$ and $\kappa C = Ker\kappa$, then a normal series appears in the guise of an exact sequence. If C is a one-element subalgebra of A, as it is in the original case for groups, it is easy to see that a normal series satisfies $\kappa_1 \subseteq \kappa_2 \subseteq \cdots \kappa_m$. This would suggest a generalization of the Jordan-Hölder-Schreier Theorem in a different direction.

Proposition 4.1 and Corollary 4.2. I am not convinced that condition $(*)$ is necessary for these assertions though it seems to be required by the proof given here.

Proposition 4.3. The homological Two-Square Lemma was established for groups and modules in [11] and exploited extensively in [12] to derive a number of basic results in homological algebra. Leicht [15] worked out a generalization to "quasi-exact" and even more general categories in which it makes sense to speak of "normal monomorphisms", in analogy to normal subgroups. The present version involving double-squares was proposed for Maltsev varieties by the author at the Maltsev conference in Novosibirsk in 1988 (see [13]) and was subsequently generalized to Maltsev categories by Carboni et al. [3]. Unfortunately, both these articles made use of the bad definition $Im(\lambda_1, \lambda_2) = \lambda_1\lambda_2^{\vee}$ and unnecessarily insisted on strict commutativity of the double squares. As pointed out by Max Kelly (see also Carboni, Kelly and Pedicchio [2]), the resulting statement of the two-square lemma did not generalize the classical result for modules.

However, our present two-square lemma does generalize the classical one for modules (if not that for groups). Compare the diagram of Proposition 4.3 with the following diagram for modules:

$$
\begin{array}{ccccc}
A & \xrightarrow{\lambda_1 - \lambda_2} & B & \xrightarrow{\mu} & C \\
{\scriptstyle \alpha_1 - \alpha_2} \big\downarrow & & \big\downarrow {\scriptstyle \beta} & & \big\downarrow {\scriptstyle \gamma_1 - \gamma_2} \\
D & \xrightarrow{\lambda} & E & \xrightarrow{\mu_1 - \mu_2} & F
\end{array}
$$

The exactness of the top and bottom rows of this diagram immediately translates into the corresponding exactness of the diagram of Proposition 4.3. Moreover, the commutativity of the above two squares *implies* the quasi-commutativity of the two double squares in Proposition 4.3. For (2) this is obvious, but for (1) a small argument is required to show that $e\beta Im(\lambda_1, \lambda_2)\beta^\vee e'$ if and only if $e - e' \in Im(\beta\lambda_1 - \beta\lambda_2)$ and $e, e' \in Im\beta$. The 'only-if' part is clear, so suppose $e - e' = \beta(\lambda_1 - \lambda_2)a$ for some $a \in A$ and $e' = \beta b'$ for some $b' \in B$. Put $b = (\lambda_1 - \lambda_2)a + b'$, then $bIm(\lambda_1, \lambda_2)b'$ and $\beta b = \beta(\lambda_1 - \lambda_2)a + \beta b' = e - e' + e' = e$. Finally, look at the conclusion as stated in Corollary 4.2. The congruence relation $\beta Im(\lambda_1, \lambda_2)\beta^\vee$ may be replaced by the subgroup $Im(\beta\lambda_1 - \beta\lambda_2)$ and the join of the congruence relations on B by the sum of the subgroup $\mu^\vee 0$ and $\beta^\vee 0$ (the usual kernels of μ and β) yielding the usual two-square lemma.

Proposition 4.4. The Snake Lemma is a classical construction in homological algebra. It was extended to groups in [11] and to Maltsev categories in Carboni et al. [3], though the statement given there suffers from the same defect as Proposition 4.3.

The present exposition elaborates the algebraic component of a talk given at the International Conference on Category Theory at Como in 1990. The manuscript was put aside for several years, until I realized that the above mentioned flaws could be overcome quite trivially by a mere reformulation of two crucial definitions.

References

[1] M. Barr, P.A. Grillet, and D.H. van Osdal, *Exact categories and categories of sheaves*, Lecture Notes in Mathematics **236**, Springer Verlag, 1971.

[2] A. Carboni, G.M. Kelly, and M.C. Pedicchio, *Some remarks on Maltsev and Goursat categories*, Appl. Categ. Struc. **1** (1993), 385–421.

[3] A. Carboni, J. Lambek, and M.C. Pedicchio, *Diagram chasing in Mal'cev categories*, J. Pure Appl. Algebra **69** (1990), 271–284.

[4] G.D. Findlay, *Reflexive homomorphic relations*, Canad. Math. Bull. **3** (1960), 131–132.

[5] A.W. Goldie, *The Jordan-Hölder-Schreier theorem for general abstract algebras*, Proc. London Math. Soc. **52** (1950), 107–131.

[6] _____, *The scope of the Jordan-Hölder-Schreier theorem in abstract algebra*, Proc. London Math. Soc. (3) **2** (1952), 349–368.

[7] É. Goursat, *Sur les substitutions orthogonales etc..*, Ann. Sci. École Norm. Sup. (3) **6** (1889), 9–102.

[8] G. Grätzer, *Two Mal'cev type theorems in universal algebra*, J. Combin. Theory Ser. A **8** (1970), 334–342.

[9] J. Hagemann and A. Mitschke, *On n-permutable congruences*, Algebra Universalis **3** (1973), 8–12.

[10] J. Lambek, *Goursat's theorem and the Zassenhaus lemma*, Canad. J. Math. **10** (1957), 45–56.

[11] _____, *Goursat's theorem and homological algebra*, Canad. Math. Bull. **7** (1964), 597–608.

[12] _____, *Lectures on rings and modules*, Blaisdell Publ. Co., Waltham, 1966, Chelsea Publ. Co., New York, 1976, 1986.

[13] _____, *On the ubiquity of Mal'cev operations*, Contemporary Mathematics **131** (1992), 135–146.

[14] _____, *Some aspects of categorical logic*, Logic, Methodology and Philosophy of Science, Uppsala 1991 (D. Prawitz et al., ed.), Elsevier Science, 1994, pp. 69–89.

[15] J.B. Leicht, *Axiomatic proof of J. Lambek's homological theorem*, Canad. Math. Bull. **7** (1964), 609–613.

[16] A.I. Mal'cev, *On the general theory of algebraic systems*, Mat.Sbornik N.S. **35** (1954), 3–20.

[17] J. Meisen, *Relations in regular categories*, Lecture Notes in Mathematics **418**, Springer Verlag, 1974, pp. 90–92.

[18] A. Mitschke, *Implication algebras are 3-permutable and 3-distributive*, Algebra Universalis **1** (1971), 182–186.

[19] J. Riguet, *Relations binaires etc.*, Bull. Soc. Math. France **76** (1948), 114–155.

[20] _____, *Quelques propriétés des relations difonctionelles*, C. R. Acad. Sci. Paris Sér. I Math. **230** (1950), 1999–2000.

[21] E.T. Schmidt, *Kongruenzrelationen algebraischer strukturen*, Math. Forschungsberichte, vol. 25, Berlin, 1969.

[22] R. Wille, *Kongruenzklassengeometrien*, Lecture Notes in Mathematics **113**, Spriger Verlag, 1970.

[23] H. Zassenhaus, *The theory of groups*, 2nd ed., Chelsea Publ. Co., New York, 1958.

DEPARTMENT OF MATHEMATICS AND STATISTICS MCGILL UNIVERSITY, 805 SHERBROOKE STREET WEST, MONTREAL QC, H3A 2K6, CANADA

e-mail address: LAMBEK@TRIPLES.MCGILL.CA

Adjoints in and among Bicategories

F. William Lawvere

Abstract

Professor Roberto Magari is remembered for the depth and clarity
of his lectures which impressed students and colleagues alike. Indeed
he championed the deepening of logic through algebraic models and the
clarification of algebra through explicit recognition of logic. In that
spirit, I here model the basic construction of proof theory as the left
bi-adjoint to the inclusion of Posets into Categories. That simple description has a rich history, some of which I here recall. Its specific
content stems from the analysis, within such bicategories, of quantifiers
and implications as cases of Kan's notion of adjoint. The construction leads to some open algebraic problems concerning known classes
of locally-cartesian-closed categories and their role as models for proof
theory.

1 Some Basic Concepts Concerning Bicategories

A bicategory is (apart from technical considerations concerning the precise
sense in which composition is associative) a category enriched in Categories;
that is, between any two objects there is (not just a class, but) a *category*
of morphisms (whose objects are called 1-morphisms and whose morphisms
are called 2-morphisms of the bicategory) and composition is a *functor* on the
cartesian product category (for each triple of objects, or "0-morphisms", of
the bicategory). Combinatorially, if we consider an ordinary category to be
a multiplicative graph, where a graph is a conglomerate of 1-balls (arrows)
with boundary 0-spheres given as the union of two points, then analogously, a
bicategory is a bimultiplicative bigraph, where a bigraph is a conglomerate of
2-balls ("lozenges") with boundary 1-sphere given as the union of two 1-balls
intersecting in *their* boundaries; the "vertical" multiplication of 2-morphisms
is that within each hom-category, whereas their "horizontal" multiplication
is the functorial composition of the bicategory. Just as category theory can
explicitly encapsulate much more mathematics than pure set theory, while yet

remaining universal, so bicategories contain qualitatively more information than pure categories. On the other hand, the notion of tricategory (and even ∞-category), which has proven useful in homotopy theory, has the striking feature that even there the concept of bicategory is central, since it is the structure relating *any two* levels. Explicitly, the functorality of composition implies the law

$$(ab) \circ (xy) = (a \circ x)(b \circ y)$$

involving three 0-morphisms, six 1-morphisms, and four 2-morphisms a, b, x, y which are composable, where juxtaposition denotes vertical composition and circle denotes the horizontal composition. The "bifunctors", which are the "morphisms" between bicategories, are of course required to preserve the bigraph structure as well as the two compositions (but again there is considerable content in the precise sense to which the composition of 1-morphisms is preserved, since it may be true only "up to an invertible 2-morphism").

2   Cat and Adjointness

A basic concrete example is Cat, where categories, functors, and natural transformations are the 0-morphisms, 1-morphisms, and 2-morphisms respectively. Indeed, as Yoneda's lemma permits analysis of ordinary categories in terms of Set-valued functors, so Yoneda's bilemma permits analysis of general bicategories in terms of Cat-valued bifunctors. That applies in particular to the notion of adjointness. It seems that a bicategory is precisely the most general environment in which the notion of adjointness can be defined, by any of the following three equivalent conditions on a pair of 0-morphisms, a pair of 1-morphisms, and a pair of 2-morphisms in the configuration

$$U \circ F \overset{\eta}{\longleftarrow} \mathcal{A} \underset{U}{\overset{F}{\rightleftarrows}} \mathcal{B} \overset{\epsilon}{\longleftarrow} F \circ U$$

in which the two "lozenges" η and ϵ are squashed because half the boundary of η is reduced to the point $1_{\mathcal{A}}$ and the other half of the boundary of ϵ is reduced to the point $1_{\mathcal{B}}$. One condition is just that the two equations

$$(\epsilon \circ F)(F \circ \eta) = 1_F$$
$$(\eta \circ U)(U \circ \epsilon) = 1_U$$

hold in the bicategory itself; these imply that, among the 2-morphisms (constructible from the data) between the iterated loops at $\mathcal{A}$ or $\mathcal{B}$, all the "simplicial" identities (familiar from composition of order-preserving maps between finite totally-ordered sets) hold. This can be Yoneda-tested as follows: for any $\mathcal{T}$ in the bicategory, the induced functors

$$Hom(\mathcal{T}, \mathcal{A}) \quad \underset{U}{\overset{F}{\rightleftarrows}} \quad Hom(\mathcal{T}, \mathcal{B})$$

between actual categories should be adjoint, with adjunction unit and counit induced by η and ϵ. That in turn has meant for over 35 years that for each pair A, B of 1-morphisms $\mathcal{T} \longrightarrow \mathcal{A}$, $\mathcal{T} \longrightarrow \mathcal{B}$, η and ϵ induce natural bijections

$$\frac{F \circ A \longrightarrow B}{A \longrightarrow U \circ B}$$

between the indicated sets of 2-morphisms. The third equivalent condition is that for each $\mathcal{V}$ in the bicategory, there is a similarly induced pair

$$Hom(\mathcal{A}, \mathcal{V}) \quad \underset{F^*}{\overset{U^*}{\rightleftarrows}} \quad Hom(\mathcal{B}, \mathcal{V})$$

of adjoint ordinary functors.

It is the first, equational, condition which makes it obvious that all bifunctors preserve adjointness. That is a far-reaching generalization of the fact that ordinary functors preserve isomorphism, just as the theorem that adjoint 1-morphisms uniquely determine each other is a far-reaching generalization of the fact that in an ordinary category inverse isomorphisms uniquely determine each other. Here the sense of uniqueness for 1-morphisms appropriate in a bicategory is "up to an invertible 2-morphism". The explicit sense in which the two mentioned results are generalizations stems from the construction, for each ordinary category, of the bicategory which extends it by taking 2-morphisms to be just equalities between the given 1-morphisms.

3　Posets and Quantifiers

A key bicategory for predicate logic is the one in which 0-morphisms are posets, 1-morphisms are order-preserving maps, and wherein there is a 2-morphism from f to g if and only if $fx \leq gx$ in the (common) codomain for all x in the (common) domain. This bicategory is seemingly somewhat "trivial" in that any two 2-morphisms with the same domain f and the same codomain g

are themselves equal. Nonetheless, there are many non-trivial adjoints *in* it, such as the upper and lower integrals of the Darboux-McShane theory, as well as the semantic correspondents of the rules of inference usually considered in predicate logic. Indeed, these examples are of a special nature made possible by the function-space adjoint which acts *on* this bicategory of posets to give it its cartesian closed character. Explicitly, if T is any complete poset and $f : A \longrightarrow B$ is any order-preserving map, then there is an induced order-preserving map T^f which substitutes f into the various "predicates" in T^B yielding new predicates in T^A. This T^f has both a left adjoint and a right adjoint, which is the meaning of existential and universal quantification.

Often the term "quantifier" is restricted to the case of the adjoints induced by some special f's, as in first-order single-sorted predicate logic where $A = X^W$, $B = X^V$, $f = X^\sigma$, for maps $\sigma : V \longrightarrow W$ between finite sets of "variables"; even here the surjective (but not injective) σ's (so the injective f's) incorporate in their induced adjoints the rules of inference for equality predicates, whereas it is the injective (but not surjective) σ's (so the surjective f's) which induce the quantifiers which make a genuine existential or universal leap. However, most of the work on algebraic logic in the past twenty-five years has followed the more liberal interpretation according to which the operation of forming the image of definable parts of the domain of *any* term f is considered as an existential quantification, because it satisfies the appropriate rule of inference.

These two key features of the desired semantics of first-order predicate logic, namely the mutual determination of substitution and quantifications via adjointness and the role of the category of finite sets as variables, should be retained in any abstract algebraic structure which makes the universe X generic, and even in a syntactical scheme for presenting examples of such algebraic structures by generators ("atomic predicates") and relations ("axioms"). The early attempts in the 1950's to describe such algebraic structures in the form of cylindric or polyadic algebras dealt instead with the infinitude of variables by using a single infinite set, rather than the category of finite sets. This distortion of the intended content obscured the adjointness between substitution and quantification, which was left implicit and accounted for instead by special axioms which seemed peculiar to the subject, since they were at best vaguely analogous to conditions appearing in analysis, topology, or other parts of algebra. This double distortion of the content, as I pointed out at the 1963 Berkeley Model Theory symposium, meant that these two particular algebraic versions of logic were destined for relatively little application to mathematics.

Of course, the predicates $X^V \longrightarrow T$ (especially in case $T = 2$) classify certain subobjects of X^V or "relations" and the ordering given by the 2-morphisms reflects the objective inclusions between these subobjects; it is these objective inclusions and their transformations which are the content of "logic in the narrow sense", reflected subjectively as entailment relations between the formulas which name the subobjects. But inclusions between subobjects of B are established by commutative triangles

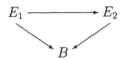

where the vertical arrows specify the insertions of the subobjects. Of course, since insertions of subobjects are monomorphisms, there is at most one $E_1 \longrightarrow E_2$ establishing the inclusion of any two given subobjects. On the other hand, in topology, geometry, combinatorics, etc. there often arises the need to compare $E_i \longrightarrow B$ which are not necessarily monomorphisms; this suggests the need for "logic in a broader sense" in which the attributes of B form a category which does not reduce to a poset. The text of my talk "Category-valued Higher-order Logic", at the 1967 Los Angeles Set Theory Symposium, was distributed to the participants of that Symposium and discussed also in the 1968 Versailles meeting on Automatic Demonstration (SLNM 125). Revised and expanded, that material was published in three papers ([3],[4],[5]). These three papers helped to popularize the new field of "proof theory", which had emerged in the work of H. Läuchli, circulated in Spring 1967 as a new complete semantics for intuitionistic logic. (The emergence had been predicted by an old remark of Curry concerning the striking analogy between modus ponens (for propositions and inferences) and the laws of functionality (for types and terms); that analogy had been objectified by my observation that both involve cartesian-closed categories ([1])). Läuchli used actual mathematical objects (namely permutation representations), rather than the ritual equivalence classes of strings of symbols, so his work was described at Versailles as "non constructive" and not seriously looked at again until very recently. The construction by D. Scott in 1970 of models for the untyped λ-calculus, using the remarkable properties of adjoint retractions in the cartesian-closed bicategory of posets with filtered sups, continued to popularize the virtues of examples constructed instead by means of inverse and direct limits from spaces; these virtues already manifested themselves in a remarkable way in the 1966 Scott-Solovay Boolean-valued models for set theory, which were seen to constitute an important corner of another bicategory, that of toposes.

4   The Curry-Läuchli Adjoint

The central problem of proof theory is the unbounded nature of "there exists a proof". Like any existential quantification, this concerns a left adjoint, but one embedded in a definite context. The essential content of this Curry-Läuchli adjoint is, I believe, independent of syntactic presentations and embodied in the following

Proposition. *The inclusion of the bicategory of posets into the bicategory of categories has a left adjoint which is a bifunctor (hence preserves adjointness) and which preserves finite cartesian products.*

Explicitly, the poset associated to a given category has as elements the objects of the category, and $A \leq B$ iff there exists a morphism $A \longrightarrow B$. Although this Curry-Läuchli adjoint has many applications, we pass immediately to the most typical. Recall that a category $\mathbf{C}$ is called locally cartesian-closed if it has pullbacks and if for each $A \to B$ in $\mathbf{C}$, the pullback functor f^* has a right adjoint $f\Pi$.

$$\mathbf{C}/A \xleftarrow{\underset{f\Pi}{\overset{f\Sigma}{\quad\quad f^* \quad\quad}}} \mathbf{C}/B$$

Here the left adjoint $f\Sigma$ is the functor which exists trivially for any category and which comes about by merely composing f with each $E \to A$ in $\mathbf{C}/A$; in case $B = 1$, the effect of $f\Sigma$ is to extract the total of an object distributed over A. The composite functor, $f\Pi$ following f^*, is the function-space construction which shows that each category $\mathbf{C}/B$ is cartesian closed.

Corollary. *If $\mathbf{C}$ is a locally-cartesian-closed category, denote by $\mathcal{P}_\mathbf{C}(A)$ the poset reflection of $\mathbf{C}/A$, for each "type" A in $\mathbf{C}$. Then each $\mathcal{P}_\mathbf{C}(A)$ is a Heyting algebra, in the sense that it has conjunction and implication related by modus ponens (as well as disjunction in case $\mathbf{C}$ has coproducts). For each "term" $A \to B$ in $\mathbf{C}$, f^* induces a substitution which is a Heyting homomorphism and has quantifiers*

$$\mathcal{P}_\mathbf{C}(A) \xleftarrow{\underset{f\forall}{\overset{f\exists}{\quad\quad f^* \quad\quad}}} \mathcal{P}_\mathbf{C}(B)$$

satisfying the correct rules of inference. Moreover, to give a "proof" in $\mathbf{C}/B$ of $(f\exists\varphi) \cdot (y)$ is to give an x for which $fx = y$ together with a proof of $\varphi \cdot (x)$, while to give a proof of $(f\forall\varphi) \cdot (y)$ is to give, for each x with $fx = y$, a proof of

$\varphi \cdot (x)$, *but the latter in a uniform manner in the sense that it is implemented by a single map in* $\boldsymbol{C}$.

[Notation: $f^*\varphi = \varphi \cdot f$, for example $\varphi \cdot (x) = x^*\varphi$, so that $(\varphi \cdot f) \cdot x = \varphi \cdot (fx)$. Here the x and y denote arbitrary $\boldsymbol{C}$-morphisms from an arbitrary test object to A and B respectively. Note that $(gf)\exists\varphi = g\exists(f\exists\varphi)$ and that if for the case $f = X^\sigma$ we write instead $f\exists\varphi = \varphi\exists\sigma$, then we have $\varphi\exists(\sigma\tau) = (\varphi\exists\sigma)\exists\tau$ for the iterated quantifiers.]

In case $\boldsymbol{C}$ is the category of (small) abstract sets, the above construction recovers the usual power sets, but the situation is strikingly different for most Grothendieck toposes, where the axiom of choice is false. For a precise comparison, let $\Omega_{\boldsymbol{C}}(A)$ denote the poset of all $\boldsymbol{C}$-subobjects of A. Then the epi-mono image factorization yields a retraction functor $\boldsymbol{C}/A \longrightarrow \Omega_{\boldsymbol{C}}(A)$, which by the universal property of the poset reflection yields a retraction of posets

$$\mathcal{P}_{\boldsymbol{C}}(A) \longrightarrow \Omega_{\boldsymbol{C}}(A).$$

The latter is an equivalence for each A iff all epimorphisms in $\boldsymbol{C}$ split; otherwise $\mathcal{P}_{\boldsymbol{C}}(A)$ is bigger. In fact, it can be a proper class, as for example if $\boldsymbol{C}$ is the topos of graphs, though Läuchli's work showed that it is small in the case where $\boldsymbol{C}$ is the topos of permutation representations of a given group.

5   Open Problems

It seems to be still an open question to determine for which Grothendieck toposes $\boldsymbol{C}$ the "proof-theoretic power set" $\mathcal{P}_{\boldsymbol{C}}(A)$ is always a small Heyting algebra.

There is a natural weakening of the stringent requirement

for the provability of "E implies F". Namely, we could first analyze the hypothesis by passing to a judiciously-chosen epimorphic cover $E' \longrightarrow E$ (which will have the same subobject of A as image) and then try for a direct proof of the conclusion from the refined hypothesis:

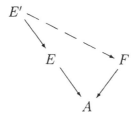

This turns out to be equivalent to the existence of an actual inclusion between the image subobjects in $\Omega_{\mathbf{C}}(A)$!

In a topos the subobjects of A are internally representable by maps $A \longrightarrow \Omega_{\mathbf{C}}$ to a fixed representing object. To achieve something like that for $\mathcal{P}_{\mathbf{C}}(A)$, we would have to replace $\mathcal{P}_{\mathbf{C}}$ by its sheafification and to verify that at least that sheaf is small. This sheafification involves a special form of the "refinement of hypotheses" idea of the previous paragraph, namely passing first to a covering of the universe A and then pulling back the hypothesis:

For which Grothendieck toposes is the sheafification of the proof-theoretic power set small, and hence internally representable?

Since directed graphs are actually diagrams of a finite shape, the combinatorial aspect of these open questions is underlined by the special case: for which finite categories does the topos of presheaves have the property that the proof-theoretic power set functor has small values?

6   The Special Role of Negation

The proof theory of negation seems at first problematic since in any cartesian-closed category with an initial object "false", any object of the form $(E \Rightarrow \text{false})$ is already in the poset of subobjects of 1, seemingly obliterating any possibility of making distinctions between proofs of a negative formula. That Läuchli nonetheless obtained a completeness result for intuitionistic predicate calculus (in spite of using only Boolean toposes in his semantics) stems from his understanding the negation of an object E in $\mathbf{C}/A$ not merely as another object, but as the functor of F in $\mathbf{C}/1$

$$(\neg E)(F) = (E \Rightarrow A^*F)$$

where A is the map from A to 1. The same idea would seem relevant in contexts other than proof theory. For example, in any topos, negation so defined becomes a monomorphism from truth-values to unary propositional operators. Similarly, in the "generalized logic" over any metric space A, a Lipschitz function E is not recoverable from its zero-set alone, but is recoverable from its family of superlevel sets, one for each constant A^*F for F real.

References

[1] S. Eilenberg and G.M. Kelly, *Closed Categories*, Proceedings of the Conference on Categorical Algebra, La Jolla 1965, Springer Verlag, 1966.

[2] H. Läuchli, Manuscript circulated in Spring 1967.

[3] F.W. Lawvere, *Adjointness in Foundations*, Dialectica **23** (1969), 281–296.

[4] _____, *Diagonal Arguments and Cartesian Closed Categories*, Lecture Notes in Mathematics **92**, 1969, pp. 134–145.

[5] _____, *Equality in Hyperdoctrines and the Comprehension Scheme as an Adjoint Functor*, Proceedings of the AMS Symposium on Pure Mathematics, XVII, 1970, pp. 1–14.

[6] _____, *Metric Spaces, Generalized Logic, and Closed Categories*, Rend. Sem. Mat. Fis. Milano **43** (1973), 135–168.

[7] M. Makkai, *The fibrational formulation of intuitionistic predicate logic: Completeness according to Gödel, Kripke and Läuchli*, Notre Dame J. Formal Logic **34** (1993), 334–337 and 471–498.

DEPARTMENT OF MATHEMATICS, SUNY AT BUFFALO, 106 DIEFENDORF HALL, BUFFALO, NY 14214, USA

e-mail address: MTHFWL@UBVMS.CC.BUFFALO.EDU

Exponential Algebra

Angus J. Macintyre

Abstract

In this paper various aspects of the algebra of the real exponential functions are discussed. These are:

- identities and Schaunel's conjecture;
- the correct notion of exponential algebraic;
- orders of infinity;
- noetherian phenomena.

Introduction

Over the last twenty five years, beginning with the work of Richardson [18], much effort went into understanding the elementary theory of the real exponential field. The breakthrough was achieved by Wilkie [22], who proved model-completeness, and since then significant further simplification has been achieved [7, 17].

In the early stages, one developed the rudiments of the algebra of exponentiation on (ordered) rings, following an approach (to model-completeness) which had been successful in the logic of real and p-adic fields [13]. This, however, proved inadequate (and even irrelevant) for the deeper works of the late 1980s and early 1990s. Despite this, the above-mentioned exponential algebra is interesting, and has been enriched by ideas from analytic geometry and differential topology, to the extent that I think it appropriate, in a paper dedicated to the memory of a distinguished algebraist, to set out my view of the achievements and prospects of the subject.

1 From Rings to E-Rings

1.1. Though rings are ubiquitous in mathematics, there seem to be very few rings with exponentiation, with most of those involving analysis or formal power series. On the other hand, there are many intriguing situations in category theory where $+$, $\cdot$ and an exponential interact. See [10, 20].

1.2. In this paper, all rings will be commutative with 1. These form an equational class, and the free object on a set X is easily identified as $\mathbf{Z}[X]$, the polynomial ring, with solvable word problem.

It is convenient to work also with the equational class of R-rings, for a fixed ring R. Here the objects are rings S equipped with a structure of unital R-algebra (with respect to their ring sum and product). Now the free object on X is $R[X]$, and this has solvable word problem if R has.

1.3.

Definition 1

    a) An E-ring S is a ring S together with a map $E : S \to S$ satisfying

        i) $E(0) = 1$;
       ii) $E(x + y) = E(x)E(y)$

    b) If R is an E-ring (wrt E_R), and S is an E-ring (wrt E_S) such that S is an R-ring, then S is an E-ring over R if

$$E_S(r.1) = E_R(r).1 \quad \text{for} \quad r \in R.$$

The class of E-rings, and the class of E-rings over R are both equational classes.

1.4. *Examples of E-rings*

1.4.1. $\mathbf{R}$, with e^x.

1.4.2. $\mathbf{C}$, with e^x.

1.4.3. $\mathbf{Z}_p$, the ring of p-adic integers with $(1 + p)^x$ if $p > 2$, and 5^x if $p = 2$.

1.4.4. Let S be an E-field of characteristic 0, $R = S[t]$, the ring of formal power series in t over S. If $f \in R$, f can be written uniquely as $r + f_1$, where r is the constant term of f. Let

$$E(f) = E(r) . \sum_{n=0}^{\infty} (f_1)^n / n!$$

1.4.5. R any ring, $\quad E(x) = 1$.

1.5. *Basic observations* (R an E-ring)

Lemma 1 E *is a homomorphism from the additive group of R to the multiplicative group of invertible elements of R.*

Corollary. If $\lambda_1, \ldots, \lambda_n$ are linearly dependent over $\mathbf{Z}$, $E(\lambda_1), \ldots, E(\lambda_n)$ are multiplicatively dependent.

1.6. *Notation.* I adopt three suggestive pieces of notation.

a) Let $e = E(1)$;
b) $\mathbf{Z}[X]^E$ is the free E-ring on X;
c) $R[X]^E$ is the free E-ring over R on X.

Notes. a) 1.4.5 shows need for caution.
b) and (c). We need an independent construction of these entities, and results relating them to $\mathbf{Z}[X], R[X]$.

1.7. *Constructing Free Objects. Power Series.*

1.7.1. It turns out to be convenient to establish now a stock of constructions involving a ring R and an additive semigroup G with 0.

The first is $R[t^G]$. Additively this is the free left R-module on elements $t^g, g \in G$. Then make it an R-algebra via

$$t^g.t^h = t^{g+h}$$

In particular, t^0 is the ring unit of $R[t^G]$, and one can embed R in $R[t^G]$ by

$$r \mapsto r.t^o.$$

If G is a group, then t^G, defined as $\{t^g : g \in G\}$ is a multiplicative subgroup of $R[t^G]$, isomorphic to G via

$$g \mapsto t^g.$$

It is worth observing

Lemma 2 $R[t^{G_1}][t^{G_2}] \cong R[t^{G_1 \oplus G_2}]$

1.7.2. Note that the above generalizes the construction of polynomial rings, even in many variables. To get $R[x]$, one simply takes $R[t^{\mathbf{N}}]$, isomorphic via $X^n \mapsto t^n$.

To get $R[x_1, \ldots, x_k]$, one takes $R[t^{\mathbf{N_1} \times \cdots \times \mathbf{N_k}}]$, each $\mathbf{N}_i = \mathbf{N}$, isomorphic via

$$x_1^{n_1} \cdots x_k^{n_k} \mapsto t^{<n_1 \cdots n_k>} >$$

1.7.3. Next, I consider infinitary power series constructions, now not aiming for maximum generality. G will be an ordered (additive) semigroup with cancellation, under $<$.

Now I describe $R((t^G))$. The underlying set consists of formal sums

$$f = \sum_{r \in G} r_\gamma . t^\gamma,$$

where each $r_\gamma \in R$, and the *support of* f, defined as $\{\gamma : r_\gamma \neq 0\}$, is *well-ordered by* $<$.

Addition is given by

$$\left(\sum_\gamma r_\gamma . t^\gamma \right) + \left(\sum_\gamma s_\gamma . t^\gamma \right)$$
$$= \sum_\gamma (r_\gamma + s_\gamma) . t^\gamma.$$

The left R-module structure is given by

$$r . \left(\sum_\gamma r_\gamma . t^\gamma \right) = \sum_\gamma (r r_\gamma) . t^\gamma.$$

The ring (and R-algebra) structure is given by:

$$\left(\sum_\gamma r_\gamma . t^\gamma \right) . \left(\sum_\gamma s_\gamma . t^\gamma \right)$$
$$= \sum_\gamma w_\gamma . t^\gamma,$$

where

$$w_\gamma = \sum_{\gamma_1 + \gamma_2 = \gamma} r_{\gamma_1} . s_{\gamma_2},$$

a finite sum, because of the constraint on supports.

As before t^0 is the ring unit, and

$$r \mapsto r . t^o$$

injects R into $R((t^G))$.

In fact, there is a natural R-module injection

$$R[t^G] \mapsto R((t^G)).$$

Let $R[[t^G]]$ be the R-submodule of $R((t^G))$ consisting of the f of nonnegative support. Then $R[t^G]$ injects into $R[[t^G]]$.

1.6.4. Again one may specialize to more familiar constructions. If $G = \mathbf{N}$, one gets the usual ring of formal power series, and if $G = \mathbf{Z}$ the Laurent series.

If one takes $G = \mathbf{N}_1 \times \cdots \times \mathbf{N}_k$ as in 1.6.2 one gets $R[[x_1, \cdots x_k]]$, the ring of formal power series in $x_1, \cdots x_k$ over R.

1.6.5. Let G_+ be the subsemigroup of positive elements of G. Then (allowing semigroups without unit in preceding definitions) $R[[t^{G_+}]]$ is an ideal in $R[[t^G]]$, with quotient ring naturally isomorphic to R.

It is of basic importance for the deeper results of the subject that each $R[[x_1, \cdots, x_k]]$ acts as a ring of functions from $R[[t^{G_+}]]$ to $R[[t^G]]$, as I now explain.

Let $H = \mathbf{N}_1 \times \ldots \times \mathbf{N}_k$ as in 1.6.2. For $\gamma \in H$, $\gamma = \langle n_1, \ldots n_k \rangle$, let

$$(y_1, \ldots, y_k)^\gamma = y_1^{n_1} \cdots y_k^{n_k}$$

for $y_1, \ldots, y_k$ in any ring.

Suppose $F \in R[[x_1, \ldots, x_k]]$,

$$F = \sum_{\gamma \geq 0} r_\gamma.(x_1, \ldots, x_k)^\gamma \qquad , r_\gamma \in R.$$

Let $\varepsilon_1, \ldots, \varepsilon_k \in R[[t^{G_+}]]$. Define

$$F(\varepsilon_1, \ldots, \varepsilon_k) = \sum_\gamma r_\gamma.(\varepsilon_1, \ldots, \varepsilon_k)^\gamma,$$

and just calculate that this is a well-defined element of $R[[t^G]]$ (or see analogous computations in [4, 7]).

One easily checks:

Lemma 3 *For each $\vec{\varepsilon} = (\varepsilon_1, \ldots, \varepsilon_k)$ the map*

$$F \mapsto F(\vec{\varepsilon})$$

is an R-algebra homomorphism from

$$R[[x_1, \ldots, x_k]] \qquad to \qquad R[[t^G]].$$

A classical special case is:

Lemma 4 *If ε is in $R[[t^{G_+}]]$, $1 + \varepsilon$ is invertible in $R[[t^G]]$.*

Proof. Take $k = 1$, and consider $F_1 = 1 + x_1,$, $F_2 = \sum_{d \geq 0} (-1)^d x_1^d$, so that $F_1.F_2 = 1$. □

Corollary. If R is a field, and G a group, $R((t^G))$ is a field.

1.6.6. In analogy with Lemma 2 one evidently has

Lemma 5 $R((t^{G_1}))((t^{G_2}))$ *is naturally isomorphic to*

$$R((t^{G_1 \oplus G_2})),$$

where $G_1 \oplus G_2$ has the anti-lexicographic ordering given by

$$\langle g_1, g_2 \rangle \geq 0 \qquad if \qquad g_2 \geq 0 \, or \, g_2 = 0 \, and \, g_1 \geq 0$$

1.7. *Free objects via power series.* I give a construction uniform for $\mathbf{Z}[X]^E$ and $R[X]^E$. It is essentially that of [5] (supplemented by [12]) but with a different notation.

$R_0 = \mathbf{Z}[X]$ (resp. $R[X]$),

$H_0 = \{0\}$ (resp. R),

$G_0 = \mathbf{Z}[X]$ (resp. the ideal generated by X,

so $R_0 = H_0 \oplus G_0$, additively

$E_0 : H_0 \to R_0$ is given by

$E_0(0) = 1$ if $R_0 = \mathbf{Z}[X]$,

and $E_0 = $ the E of R if $R_0 = R[X]$.

Now, recursively:

$R_{n+1} = R_n[t^{G_n}]$;

$H_{n+1} = R_n$ (naturally embedded in R_{n+1});

$G_{n+1} = $ the R_n-submodule in R_{n+1} generated by the t^g, $g \in G_n$, $g \neq 0$. Then clearly $R_{n+1} = H_{n+1} \oplus G_{n+1}$, and now define $E_{n+1} : H_{n+1} \to R_{n+1}$ by

$$E_{n+1}(h \oplus g) = E_n(h) \cdot t^g$$

if $h \in H_n$, and $g \in G_n$.

Let R_ω be the natural direct limit of the R_n, and E_w the natural direct limit of the E_n. Clearly R_ω is an E-ring under E_ω, and when $R_0 = R[X]$ is an E-ring over R.

It is simple to prove:

Theorem 6

a) When $R_0 = Z[X]$, R_ω is $Z[X]^E$.
b) When $R_0 = R[X]$, R_ω is $R[X]^E$.

And then:

Theorem 7

a) $Z[X]^E$ has solvable word problem;
b) $R[X]^E$ has solvable word problem if R has.

1.8. *Heights, Lengths and Degrees.* I maintain the notation of the last subsection.

Definition 2 If $y \in R_w$, let

$$
\begin{aligned}
h(y) &= \text{the height of } y \\
&= \text{the least } n \text{ such that } y \in R_n.
\end{aligned}
$$

Definition 3 Suppose $n = h(y)$. Let

$$
\begin{aligned}
\ell(y) &= \text{the length of } y \\
&= 1 \qquad if \qquad n = 0 \\
&= \text{the cardinality of the support of } y \text{ as a power series in } t \text{ over} \\
&\quad R_{n-1} \text{ with exponents in } G_{n-1}, \text{if } n \geq 1.
\end{aligned}
$$

Now R_o comes equipped with various degree maps to w. For example, if $Y \subseteq X$, d_y is the total Y-degree. As in [12], and in some sense following [5], one may extend any such degree map $d : R_o \to w$ to $d_w : R_w \to w^w$ by the following recursive procedure.

Let $d_0 = d$. If $d_n : R_n \to w^{n+1}$ is defined, define $d_{n+1} : R_{n+1} \to w^{n+2}$ as follows:
If $y \in R_{n+1}$, let y_0 be the constant term of y (i.e. coefficient of t^0), so $y_0 \in R_n$. Then let

$$
d_{n+1}(y) = w^{n+1} \cdot \ell(y) + d_n(y_0)
$$

Let d_ω = union of the d_n. It turns out that induction on d_ω is a useful generalization of the usual induction on degree in polynomial algebra, partially in situations where differentiation is involved. Before coming to that (in 1.10) I give in 1.9 some more direct uses of d_ω.

1.9 *The Schanuel Property.* First one observes

Theorem 8

a) $\mathbf{Z}[X]^E$ *is a domain.*
b) $R[X]^E$ *is a domain, if R is.*

More significantly:

Theorem 9 $\mathbf{Z}[X]^E$ *satisfies the following "Schanuel Property" (SP).*
If $\lambda_1, \cdots, \lambda_n$ are linearly independent over $\mathbf{Z}$, then the transcendence degree over $\mathbf{Q}$ of the set $\{\lambda_1, \cdots, \lambda_n, E(\lambda_1, \cdots, E(\lambda_n)\}$ is at least n.

The famous Schanuel Conjecture (SC) in transcendental number theory says that SP is true for arbitrary complex numbers $\lambda_1, \cdots, \lambda_n$. It is an amusing exercise to show that SC (assumed only for real $\lambda_1, \cdots, \lambda_n$) implies that

$$e, e^e, e^{e^e} \ldots$$

are algebraically independent over $\mathbf{Q}$, and (using $e^{\pi i} = -1$) that the complex SC implies that e and π are algebraically independent over $\mathbf{Q}$.

If one restricts the $\lambda_1, \cdots, \lambda_n$ to lie in the smallest E-subring of the reals, I showed in [14] that this restricted version of SC is *equivalent* to a neat universal algebraic statement. Precisely:

Theorem 10 *SC restricted to the smallest E-subring of the reals is equivalent to this E-subring being free on no generators.*

Corollary. If SC then the above E-subring of the reals has solvable word problem.

The corollary (due to [1]) is sometimes expressed as saying that if SC then the elementary constants problem is decidable, i.e. the problem of deciding if two closed exponential terms are equal, is decidable. In Section I will put this in a much wider setting.

Arguments similar to those for Theorem 10 give:

Theorem 11 *(Assume SC). The E-subring of $\mathbf{R}$ generated by π is free on π.*

Theorem 12 *(Assume SC). The E-subring of $\mathbf{C}$ generated by π and i is finitely presented, with defining relations $E(\pi i) = -1, i^2 = -1$.*

1.10. *Differential Structure.* The polynomial rings $\mathbf{Z}[X]$ and $R[X]$ carry the derivations $\partial/\partial x, x \in X$. This generalizes nicely to $\mathbf{Z}[E]^E$ and $R[X]^E$.

Recall that a derivation D on a ring R is a linear map $D : R \to R$ satisfying $D(uv) = uD(v) + vD(u)$. A D-*constant* is an element u with $D(u) = 0$.

In the case of E-rings, the following is natural:

Definition 4 A derivation D on an E-ring is an E-derivation if for all u, $D(E(u)) = D(u).E(u)$.

The following is easy using the power series construction of $R[X]^E$.

Theorem 13 *There is a unique set of E-derivations*

$$\frac{\partial}{\partial x} : R[X]^E \to R[X]^E$$

for $x \in X$, subject to:

a) $\frac{\partial}{\partial x}$ vanishes on R;

b) $\frac{\partial}{\partial x}(x) = 1$;

c) $\frac{\partial}{\partial x}(y) = 0$ if $y \in X, y \neq x$.

The following is harder, if not surprising [5];

Theorem 14 *If R is a domain of characteristic 0, the $\partial/\partial x$-constants are exactly the elements of $R[X \setminus \{x\}]^E$.*

Notes

   a) It is clear enough that $R[Y]^E$ is naturally embedded in $R[X]^E$, if $Y \subseteq X$.
   b) If R is a characteristic p domain, any derivation vanishes on p^{th} powers.

1.11. *E-ideals.* If R is an E-ring, and I is an ideal in R, I say I is an E-ideal if I is closed under

$$\alpha \mapsto E(\alpha) - 1.$$

It is quite trivial to see that this is equivalent to I being the kernel of an E-homomorphism, and that in fact R/I is an E-ring under

$$E(\alpha + I) = E(\alpha) + I.$$

(There is already a tacit appeal to E-ideals in Theorem 12).

The next result generalizes a result so trivial for polynomial rings that it has perhaps never been isolated. It is not trivial in the exponential setting.

Theorem 15 *Suppose R is a characteristic zero E-domain. Suppose I is an E-ideal of $R[X]^E$ closed under $\partial/\partial x$. Then either $I = 0$ or I contains a nonzero element of $R[X\setminus\{x\}]^E$.*

Corollary. If R is a characteristic zero E-ring which is a field, and I is an E-ideal closed under all $\partial/\partial x$ then $I = 0$ or $R[X]^E$.

The idea is to combine the degrees of 1.8 and the derivations, looking at a nonzero element of I of minimal degree. The idea succeeds because of an observation of G.H. Hardy [8].

Theorem 16 *Let R be an E-ring, and d_ω the degree on $R[x]^E$ coming from the x-degree on $R[x]$. Suppose $f \in R[x]^E$, $f \in R$. Then $\exists g \in R[x]^E$ with $d_\omega(g) < d_\omega(f)$ and*

$$d_\omega\left(\frac{\partial}{\partial x}(E(g).f)\right) < d_\omega(f).$$

The significance is that for any ideal I, $f \in I$ iff $E(g).f \in I$, since $E(g)$ is invertible.

1.12. *Application to Freeness.* As 1.4.5 shows, some E-rings satisfy nontrivial laws. One does not know any such laws for the classical analytical E-rings 1.4.1, 1.4.2, 1.4.3, or for the power series example 1.4.4. Note however that if e^e is rational, there would be hidden laws in 1.4.1 and 1.4.2. But, using the ideas of the preceding subsections, one can show in the above cases that all laws come from constants. Precisely:

Theorem 17 *Let R be one of the E-rings $\mathbf{R}, \mathbf{C}$ or $\mathbf{Z}_p$. Then for each $n \geq 1$ the natural E-morphism*

$$R[x_1, \cdots, x_n]^E \to R^{R^n}$$

is injective.

Proof. For $\mathbf{R}$ or $\mathbf{C}$, the morphism has image included in the C^∞ functions, and it sends the E-algebraic $\partial/\partial x_i$ to the analytic $\partial/\partial x_i$. Now consider the kernel I, and apply the corollary to Theorem 15.

For $\mathbf{Z}_p$, the argument is a little trickier, and involves, instead of E-derivations, derivations D with $D(E(u)) = \lambda.D(u).E(u)$, where λ is $\log(1+p), p \neq 2$, and $\log 5$ if $p = 2$, where $\log$ is the usual p-adic logarithm. But the essential idea is the same.

In a similar vein, using the natural $\partial/\partial x_i$ on power series rings, one gets (cf. 1.4.4)

Theorem 18 *If R is a characteristic 0 field, then for each $n \geq 1$ the natural*

$$R[x_1, \ldots, x_n]^E \to R[[x_1, \ldots, x_n]]$$

is injective.

2 Zeros and the Notion of E-algebraic

2.1. Section 1 has covered the universal algebra of exponentiation, and the link to transcendental number theory. The next natural step (following the classical line on model-completeness) is to study systems of exponential equations and their solutions.

Obviously if $R \to S$ is a morphism (resp. an embedding) of E-rings it induces a morphism (resp. an embedding) $R[X]^E \to S[X]^E$. Accordingly, the notion of evaluating an element $f(x_1, \cdots, x_n) \in R[x_1, \cdots, x_n]^E$ at a point $(\alpha_1, \cdots, \alpha_n)$ of S^n is clear, and the value is written as $f(\alpha_1, \cdots, \alpha_n)$. When $f(\bar{\alpha}) = 0$, $\bar{\alpha}$ is called a zero of f, etc.

2.2. The first difficulty one meets is for $R = \mathbf{C}$ and $f(x) = E(x) - 1$, which has as zeros exactly the points $2n\pi i$, $n \in \mathbf{Z}$. Obviously then:

Theorem 19 *The E-field $\mathbf{C}$ is undecidable.*

Using Matijasevic's Theorem one improves this to:

Theorem 20 *The existential theory of the E-field $\mathbf{C}$, with π as a distinguished constant, is undecidable.*

Problem Remove the use of π.

Note. The existential theory of the field $\mathbf{Q}$ is clearly reducible to the existential theory of the E-field $\mathbf{C}$.

2.3. Faced with the negative result above, one can take comfort in G.H. Hardy's result [8] that if $f \in \mathbf{R}[x]^E$ and $f \neq 0$ then f has only finitely many zeros in $\mathbf{R}$. For this Hardy used Rolle's Theorem and the method of Theorem 16.

It turned out that one variable exponential polynomial algebra is not enough to analyze the theory of the E-field $\mathbf{R}$. A key new idea came from Hovanskii [9].

What should E-algebraic mean? (Probably) not the following:

If R is an E-ring, and B a subset of R, α is E-algebraic over B if

$\exists f \in \mathbf{Z}[x, x_1, \cdots, x_n]^E$ and $b_1, \cdots, b_n \in B$ such that

i) $f(x, \bar{b}) \neq 0$) qua element of $R[x]^E$;

ii) $f(\alpha, \bar{b}) = 0$.

The main defect is not that $f(x, \bar{b})$ may have infinitely many zeros, but rather that one does not know even if the sum of elements E-algebraic over B is E-algebraic over B. The problem is traceable to lack of a good elimination theory.

A suggestive view of the problem is the following. Suppose $f(x, y), g(x, y) \in R[x, y]^E$ and we have a transversal intersection (α_1, α_2) with $f(\alpha_1, \alpha_2) = g(\alpha_1, \alpha_2) = 0$ and

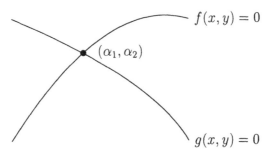

with the Jacobian

$$\left| \begin{array}{cc} \frac{\partial f}{\partial x} & \frac{\partial f}{\partial y} \\ \frac{\partial g}{\partial x} & \frac{\partial g}{\partial y} \end{array} \right| (\alpha_1, \alpha_2) \neq 0.$$

Then in commutative algebra α_1 and α_2 should be algebraic over R, but no way is known of deducing this with the preceding definition of E-algebra.

So we scrap the preceding, and build in transversality.

2.4   *The Right Definition*

Definition 5 α is E-algebraic over B if there exist n, m $b_1, \ldots, b_m$ in B, and $\alpha_1, \ldots, \alpha_n$ and $f_1, \ldots, f_n \in \mathbf{Z}[x_1, \ldots, x_n, y_1, \ldots, y_m]^E$ such that

   i) $\alpha = \alpha_1$;
   ii) $f_i(\alpha_1, \ldots, \alpha_n, b_1, \ldots b_m) = 0 \quad 1 \leq i \leq n$;
   iii)

$$\left| \begin{array}{ccc} \frac{\partial f_1}{\partial x_1} & \cdots & \frac{\partial f_1}{\partial x_n} \\ \frac{\partial f_n}{\partial x_1} & \cdots & \frac{\partial f_n}{\partial x_n} \end{array} \right| (\bar{\alpha}, \bar{b}) \neq 0.$$

Notes:

   a) For $\mathbf{R}, \mathbf{C}$ or $\mathbf{Z}_p$, the implicit function theorem gives $\bar{\alpha}$ as an isolated zero of the system

$$f_i(\bar{x}, \tilde{b}) = 0 \qquad 1 \leq i \leq n.$$

   b) In polynomial algebra the above idea would give a nonstandard, but equivalent, notion of *algebraic*.

2.5.   *E-algebraic Closure*

Definition 6 $CL_R^E(B)$ (the E-algebraic closure of B in R) is the set of all $\alpha \in R$, α E-algebraic over B.

Lemma 21 *(R a domain).* CL_R^E *is a closure operation, and* $CL_R^E(B)$ *is an* E-*subring of* R.

*Proof.*Simple Jacobian calculations. $\square$

2.6.   *CL^E and Elementary Substructures*

In analogy to a well-known result of Tarski for real-closed fields, Wilkie [22] proved the magnificent:

Theorem 22 *If B is a subset of* $\mathbf{R}$,

$$CL_{\mathbf{R}}^E(B) \prec \mathbf{R} \quad , \quad as \quad E - rings$$

Notes:

a) $CL_{\mathbf{C}}^E(\phi)$ is not an elementary substructure of $\mathbf{C}$, as may be seen by standard coding techniques.

b) I am confident that

$$CL_{\mathbf{Z}_p}^E(\phi) \prec \mathbf{Z}_p,$$

but do not have a complete proof.

2.7. *Finiteness and E-Algebraicity*

Consider systems

$$f_i(\bar{x}, \tilde{b}) = 0 \qquad\qquad 1 \leq i \leq n$$

as in Definition 5, where one is interested only in nonsingular zeros.

In $\mathbf{C}$, for fixed $\tilde{b}$, there may be infinitely many such zeros. This cannot happen in $\mathbf{R}$, as Hovanskii [9] saw, opening the way to Wilkie's work.

Theorem 23 *There is a recursive function β such that for all $n, m, b_1, \ldots, b_m$ in $\mathbf{R}$ and $f_1, \ldots, f_n \in \mathbf{Z}[x_1, \ldots x_n, y_1, \ldots y_m]^E$ the system*

$$f_i(\bar{x}, \tilde{b}) = 0 \qquad\qquad 1 \leq i \leq n$$

has $\leq \beta(f_1, \cdots, f_n)$ nonsingular solutions in $\mathbf{R}$.

Note: A similar bound β, not known to be recursive, exists for $\mathbf{Z}_p$.

2.8. *A Bound on Multiplicity.* I conclude this section with a different kind of uniformity, of remarkable generality. The proof needs only Hardy's idea (Theorem 16).

Theorem 24 *There is a recursive function μ such that if*

$$f(x_1, \ldots, x_n, y_1, \ldots, y_m) \in \mathbf{Z}[x_1, \ldots, x_n, y_1, \ldots y_m]^E,$$

and K is any E-field of characteristic 0, and $b_1, \ldots, b_m \in K$, then either

a) $f(\bar{x}, \tilde{b}) = 0$ \qquad *qua element of* \qquad $K[\bar{x}]^E$, *or*

b) $\exists k_1, \ldots k_n$ \qquad *with* \qquad $\sum k_i \leq \mu(f)$ *such that*

$$\left(\frac{\partial^{k_1}}{\partial x_1} \cdots \frac{\partial^{k_n}}{\partial x_n} f \right)(\bar{0}, \tilde{b}) \neq 0.$$

3   Geometry and Real Exponential Algebra

3.1. Hovanskii's Theorem 22, as he himself saw [9], can be combined with Morse Theory to give other Finiteness Theorems. Some of the ideas go back to Milnor [16], and are very clearly axiomatized in Tougeron's [21].

Theorem 25 *There is a recursive function c such that for all n, m,*

$$f_1, \ldots, f_k \in \mathbf{Z}[x_1, \ldots, x_n, y_1, \ldots, y_m]^E$$

and $b_1, \ldots b_m \in \mathbf{R}$, the zero set in $\mathbf{R}^n$ of the system

$$f_i(\bar{x}, \tilde{b}) = 0 \qquad\qquad 1 \leq i \leq k$$

has $\leq c(f_1, \ldots f_k)$ connected components.

Notes

    a) A nonrecursive version of the uniformity can be obtained by the general theory of *o*-minimality [22], using [9]. Since I will discuss this in the write-up of the 1993 Gödel Lecture, I do not elaborate this here.

    b) I would dearly love to have a p-adic formulation of the above.

3.2. There is a remarkable connection, discovered by Tougeron [21], and stressed in my Gödel Lecture, between Theorem 24 and *Noetherian phenomena*.

One should note first that neither $\mathbf{C}[x]^E$ nor $\mathbf{R}[x]^E$ is noetherian for E-ideals, by considering the E-ideal I generated by all

$$E(\frac{x}{n}) - 1$$

for $n \geq 1$.

For $\mathbf{C}$ the situation is even worse. For any E-ring R one can define the "exponential Zariski topology" on R^n by taking as basic closed sets the zero sets of elements of $R[x_1, \ldots, x_n]^E$. Then $\mathbf{C}$, under the exponential Zariski topology, is not a noetherian space, since the sets $2\pi i k! \mathbf{Z} (k = 1, 2, \ldots)$ form a strictly descending sequence of closed sets.

However(!)

Theorem 26 $\mathbf{R}^n$ *is a noetherian space for the exponential Zariski topology.*

Proof. See [21]. $\square$

Notes

a) In my Gödel Lecture I stressed how extraordinary this is. Sontag and I [15] applied it to learning theory.

b) The result is an instance of something much more general about Zariski topologies in *o*-minimal theories, as I remarked in my Gödel lecture. In recent work van den Dries [3] has made remarkable applications to model-completeness.

c) The analogue of Theorem 25 for $\mathbf{Z}_p$ is a triviality, by compactness.

4 Freedom at Infinity

4.1. Suppose $R \to S$ is an embedding of real-closed fields, and $\alpha \in S$. Then the type of α over R is determined by the cut α makes in R (this is just *o*-minimality). In particular, if $\alpha > R$ the type of α over R is unique, and one may reasonably call it the type at $+\infty$ over R.

Note that in particular $R(\alpha)$ is a pure transcendental extension $R(x)$ of R. Moreover, if we only assume R is ordered (and not necessarily real- closed) $R(x)$ has a unique order extending that on R and putting $x > R$.

The above results have rather nontrivial exponential analogues.

Firstly:

Theorem 27 *The real exponential field* $\mathbf{R}$ *is o-minimal.*

Proof. See [22]. □

It follows that if K is an ordered E-field, $K \equiv \mathbf{R}$, then there is a unique type at $+\infty$ over K. Indeed, if α is any realization of this type, K and α generate a unique ordered E-field. How to characterize the underlying E-field? In the algebraic case earlier discussed, $K(\alpha) \cong K(x)$, the rational function-field. Now we want the exponential analogue.

In the unpublished [12] the general construction $K(X)^E$ is given, where K is an E-field, and X any set. This is intended as the exponential analogue of $K(X)$. There are, however, subtleties in giving $K(X)^E$ as a universal construction, and I must refer to [12] for their discussion and resolution.

The essential point in the present context is that if $K \equiv \mathbf{R}$ and α realizes the type at $+\infty$, then K and α generate an E-field K-isomorphic to $K(x)^E$. The details are in [12], and were inspired by the work of the Berlin group under Dahn and Wolter [2, 23].

So far the point is that the type at $+\infty$ is *free*, as far as quantifier-free algebraic formulas are concerned.

What about $<$? Here there is a deep result, the Rigidity Theorem, due to Dahn [2] with improvements by van den Dries and me [6]. This is also to be interpreted in terms of "freedom at infinity". To state the results, one must isolate the *Taylor axioms*, which are of independent interest for real exponential algebra.

Initial Taylor Axioms. If $n \geq 1$ and $|y| \leq \frac{1}{2}$ then

$$|E(y) - \sum_{j=0}^{n} y^j/j \mid \leq |y|^{n+1}.$$

Final Taylor Axioms. If $n \geq 1$ and $y \neq 0$ then $E(y) \neq 1 + y^n/n!$

Note that the latter (for $n = 1$) imply that E is monotone increasing, and (less obviously) that E is differentiable, with derivative E. See [2] or [12].

Clearly the Taylor Axioms are $\forall_1$ and are true in **R**. Note too that the Taylor axioms imply, for a model K, that $e(= E(1))$ makes in Q the same cut as the real e, so e is transcendental over Q (and if **R** $\models SC$, the E-subring of K generated by ϕ is $\mathbf{Z}[\phi]^E$). ·

The Rigidity Theorem is:

Theorem 28 *If K is an ordered E-field satisfying the Taylor Axioms, $K(x)^E$ has a unique ordering $>$ (extending that of K) which satisfies the Taylor Axioms and has $x > K$.*

For a proof under stronger assumptions on K, see [2] and [12]. The proof in [11] adapts (using exponential power series) to give the above.

There are many ingredients in the proof, notably Rosenlicht's [19] combination of differential algebra and valuation theory, giving an abstract version of L'Hôpital's Rule at $+\infty$.

Theorem 28 implies a generalization of a very important result of Wolter [23], on the "last root problem". To motivate this, I go back to the theory of ordered fields, and consider a polynomial $f(x) = x^n + ax^{n-1} + \ldots + a_n$ in $K[x]$. K ordered. Then a well-known calculation shows that if α is a root of f in any ordered extension of K, then

$$|\alpha| \leq \max(1, \sum |a_i|).$$

Go now to the more general case of $g(x) = a_o x^n + a_, x^{n-1} + \ldots + g_n$. One deduces that there are rational functions $\rho_1(a_o, \ldots, a_n), \ldots, \rho_k(a_o, \ldots, a_n)$ such that unless $g \equiv 0$ then for any root α of g there exists i such that $\rho_i(a_o, \ldots, a_n)$ is defined and $|\alpha| \leq |\rho_i(a_0, \ldots a_n)|$. It is easy to see that the appeal to rational functions, and not merely polynomial functions, is necessary.

To formulate Wolter's result in the terms of this paper, one needs to refer to $Q(x_1, \ldots, x_n)^E$, the exponential - rational version of $\mathbf{Z}[x_1, \ldots, x_n]^E$. There is a straightforward construction of this in [12]. Note that elements of $Q(x_1, \ldots, x_n)^E$ define n-ary partial functions on any characteristic zero E-field.

Theorem 29 *There is a recursive map which to each $f \in Q[x, x_1, \ldots, x_n]^E$ assigns a finite set $\{g_1, \ldots g_k\} \in Q(x_1, \ldots, x_n)^E$, such that for all ordered E-fields K satisfying Taylor's Axioms, and all $b_1, \ldots, b_n \in K$, either*

i) $f(x, \bar{b}) \equiv 0$ *on K or*
ii) for any $\alpha \in K$ with $f(\alpha, \bar{b}) = 0$ there exists i such that

$$|\alpha| \le |g_i(\bar{b})|.$$

This can be deduced from the Rigidity Theorem by compactness. See [12]. *Note.* It is unknown whether the Taylor Axioms imply that $f(x, \bar{b}) = 0$ has only finitely many roots, if $f(x, \bar{b}) \not\equiv 0$. If one assumes a Rolle Scheme, one can prove (uniform, effective) Finite Theorems of this type.

4.2. The ideas of 4.1 suggest the following basic problem:

PROBLEM (Assume *SC*). If K is an ordered field satisfying Taylor's Axioms, can K be embedded in an ordered E-field $L \equiv \mathbf{R}$? That is, do the Taylor Axioms give the universal theory of $\mathbf{R}$, assuming *SC*?

4.3. *LE Power Series, and Another Kind of Freedom.*

The work of the Berlin group brought into focus a natural construction which has an extremely rich theory. This involves an elaboration of the ideas of 1.6.6, taking account also of $<$.

Start with an ordered E-field K satisfying Taylor's Axioms. Let Γ be any ordered abelian group, with a distinguished positive element 1. Consider $K((t^\Gamma))$ and $K[[t^\Gamma]]$ as in 1.6. Let $t = t^1, x = t^{-1}$. There is a standard way to make $K((t^\Gamma))$ an ordered field [7], namely

$$\sum k_\gamma t^\gamma > 0$$

if $k_{\gamma_o} > 0$ where γ_o is the least γ with $k_\gamma \ne 0$..

Then t is a positive element infinitesimal relative to K, and x is a positive element infinite relative to K

$K[[t^\Gamma]]$ has a natural E satisfying the Initial Taylor Axioms. Also

$$K((t^\Gamma)) = K[t^\Gamma] \oplus \Delta,$$

where Δ is the additive subgroup of elements of negative support. Δ has thus a natural order. How to define E on Δ to preserve Taylor Axioms? First extend

$$K((t^\Gamma)) \qquad to \qquad K((t^\Gamma))((t^\Delta)), i.e.$$

$$K((t^{\Gamma \oplus \Delta})) \qquad where \qquad \Gamma \oplus \Delta$$

has the anti-lexiocographic order (1.6.6, Lemma 5), thereby getting a new ordered field. Bearing in mind the Final Taylor Axioms, one now puts

$$E(\delta) = t^{-\delta} \quad , for \quad \delta \in \Delta,$$

and it works!

If one starts with $\Gamma = Q$, and repeats the above ω times one gets $K((t))^E$, which is a model of the Taylor Axioms.

The next step is the decisive one. Suppose in addition K *has logarithms*, i.e. every positive element of K is an exponential. This will certainly not transmit to $K((t))^E$, and indeed t will not be an exponential. However, there is an order embedding (over K) of E-fields

$$K((t))^E \to K((t))^E$$

sending t to $E(t)$. This gives a "construction of $\log(t)$". If we iterate this construction ω times, and take the limit, we get $K((t))^{LE}$, which has logarithms and satisfies the Taylor Axioms, and is a field of power series. It turns out to have a very rich structure, including a natural derivation $\partial/\partial t$ (and a $\partial/\partial x$). See [12] for background, and fine structure.

Most remarkably, using work of Ressayre [17] one can show:

Theorem 30 $\mathbf{R} \prec \mathbf{R}((t))^{LE}$.

The interpretation involving freeness is that if $F(x,y) \in \mathbf{R}[x,y]^E$ and $\mathbf{R} \models (\forall x \geq x_o)(\exists y)F(x,y) = 0$, then for x at $+\infty$ you get a formal solution y in $\mathbf{R}((t))^{LE}$.

5   Decidability and E-algebraic Numbers

5.1.   At the outset I stressed that SC affects the laws of exponentiation. At the other end of the logical spectrum, Wilkie and I [11] showed that if SC is *true* the real exponential field is decidable. There are two main components:

a) Finding a recursive set of true sentences, about restricted exponentiation, which is model-complete (no use is made of SC);

b) Using SC to embed the E-algebraic elements of $\mathbf{R}$ into every model of a suitable theory as in (a).

5.2.   As a byproduct we (and also van den Dries and Miller) show that if SC then π is not definable in the real exponential field.

5.3. I conclude with a challenging problem, which seems untouched despite the great progress made on real exponentiation.

Many variable last root problem.
Find informative bounds, in terms of $y_1, \ldots, y_r$, for the nonsingular zeros $\bar{x}$ of systems

$$f_i(x_1, \ldots, x_n, , y_1, \ldots, y_r) = 0 \qquad 1 \leq i \leq n.$$

This is known [11] to be linked to SC.

References

[1] B.F. Caviness and M.J. Prelle, *A note on algebraic independence of logarithmic and exponential constants*, SPGSAM Bull. **12** (1978).

[2] B.I. Dahn, *The limit behaviour of exponential terms*, Fund. Math. **124** (1984), 169–186.

[3] L.P.D. van den Dries, *Proof of an assertion of Gabrielov*, Manuscript, Urbana 1993.

[4] _____, *Tame topology and o-minimal structures*, Monograph(s), in preparation.

[5] _____, *Exponential polynomials, and exponential functions*, Pacific J. Math. **113** (1984), 51–66.

[6] L.P.D. van den Dries and A.J. Macintyre, *Unpublished notes on Dahn's work and LE series*.

[7] L.P.D. van den Dries, A.J. Macintyre, and D. Marker, *The elementary theory of restricted analytic fields with exponentiation*, to appear in Ann. of Math.

[8] G.H. Hardy, *Orders of infinity*, Cambridge, 1910.

[9] A. Hovanskii, *On a class of systems of transcendental equations*, Soviet Math. Dokl. **22** (1980), 762–765.

[10] F.W. Lawvere, *Some thoughts on the future of category theory*, Lecture Notes in Mathematics **1488**, Springer.

[11] A. Macintyre and A.J. Wilkie, *On the decidability of the real exponential field*, to appear in Oddifreddi's volume for Kreisel's 70th birthday.

[12] A.J. Macintyre, *Lecture Notes on Exponentiation*, Urbana, 1985, unpublished.

[13] _____, *Model completeness*, Handbook of Mathematical Logic (J. Barwise, ed.), North-Holland, 1977, pp. 139–180.

[14] _____, *Schanuel's Conjecture and free exponential rings*, Ann. Pure Appl. Logic **51** (1991), 241–246.

[15] A.J. Macintyre and E. Sontag, *Finiteness results for sigmoidal "neural" networks*, Proceedings of the 1993 IEEE STOC Symposium.

[16] J. Milnor, *On the Betti numbers of real varieties*, Proc. Amer. Math. Soc. **15** (1964), 275–280.

[17] J.P. Ressayre, *Integer parts of real closed exponential fields*, Arithmetic, Proof Theory and Computational Complexity (P. Clote and J. Krajicek, eds.), O.U.P., 1993.

[18] D. Richardson, *A solution of the identity problem for integral exponential functions*, Z. Math. Logik Grundlag. Math. **15** (1969), 333–340.

[19] M. Rosenlicht, *The rank of a Hardy field*, Trans. Amer. Math. Soc. **280** (1983), 659–671.

[20] S. Schanuel, *Negative sets have Euler characteristic and dimension*, Lecture Notes in Mathematics **1488** , Springer.

[21] J.C. Tougeron, *sur certaines algèbres de fonctions analytiques*, Séminaire sur la géometrie algèbrique reélle Publ. Math. Paris VII **24** (1986), 35–121.

[22] A.J. Wilkie, *Some model completeness results for expansions of the ordered field of reals by Pfaffian functions*, to appear in J. Amer. Math. Soc.

[23] H. Wolter, *On the "problem of the last root" for exponential terms*, Proceedings of the First Easter Conference on Model Theory 1983, Seminarberecht, vol. 49, Humboldt University, 1983, pp. 144–152.

MATHEMATICAL INSTITUTE, 24-29 ST GILES, OXFORD OX1 3LB, UK
    e-mail address: `ajm@vax.ox.ac.uk`

An algebraic version of categorical equivalence for varieties and more general algebraic categories

Ralph McKenzie[*]

Abstract

We characterize in a relatively simple algebraic fashion the functors establishing categorical equivalence between two equational classes of algebras. Our results will actually be proved to hold for categorical equivalences between any two classes $\mathcal{K}$ and $\mathcal{L}$ of finitary algebras where each of $\mathcal{K}$ and $\mathcal{L}$ contains the finitely generated free algebras of the variety it generates, and is hereditary in the sense that it contains all subalgebras of its algebras. They can easily be extended to classes of infinitary algebras of bounded signature, where $\mathcal{K}$ and $\mathcal{L}$ are assumed to be hereditary and to contain all their free algebras of ranks up to a common bound on the ranks of their operations.

The reader of this paper should possess a working knowledge of elementary category theory and elementary universal algebra. What is needed can be learned from the texts by S. Mac Lane [23] and R. McKenzie, G. McNulty, W. Taylor [29] (especially §§3.6, 4.1, 4.10–4.12). All classes $\mathcal{K}$ of algebras considered in this paper are assumed to consist of similar algebras (having the same finitary type). If $\mathcal{K}$ is such a class, then $\mathsf{HSP}\,\mathcal{K}$ denotes the class of all homomorphic images of subalgebras of (Cartesian) products of systems of algebras from $\mathcal{K}$. It is the smallest **variety** that includes $\mathcal{K}$ as a subclass. We use $\mathbf{F}_n(\mathcal{K})$ to denote the free algebra of rank n in $\mathsf{HSP}\,\mathcal{K}$. (This algebra is determined up to isomorphism.) We use $\mathsf{S}\,\mathcal{K}$ to denote the class of all algebras isomorphic to subalgebras of algebras in $\mathcal{K}$ and $\mathsf{SF}\,\mathcal{K}$ to denote $\mathsf{S}\,\mathcal{K}'$ where

$$\mathcal{K}' = \mathcal{K} \cup \{\mathbf{F}_n(\mathcal{K}) : \ 1 \leq n < \omega\}.$$

---

[*]Research supported by NSF Grants DMS 89 04014 and DMS 94 03187

Classes $\mathcal{K}$ that satisfy $\mathsf{SF}\,\mathcal{K} = \mathcal{K}$ will be called SF-classes.

Every class $\mathcal{K}$ can be considered to be a category: The "objects" of this category are the algebras in $\mathcal{K}$, while the "maps" of the category are the homomorphisms between algebras. We denote this category also by $\mathcal{K}$ and for $\mathbf{A}, \mathbf{B} \in \mathcal{K}$ we write $\mathcal{K}(\mathbf{A}, \mathbf{B})$ for the set of all homomorphisms from $\mathbf{A}$ into $\mathbf{B}$.

Two categories C and D are said to be *equivalent* if there are functors $F : C \to D$ and $G : D \to C$ such that the composite functors $F \circ G$ and $G \circ F$ are naturally isomorphic to the identities on D and C respectively. When $\mathcal{K}$ and $\mathcal{L}$ are classes of algebras, we write $\mathcal{K} \equiv_c \mathcal{L}$ to denote that as categories they are equivalent (and we say that they are *categorically equivalent*). We note that a functor $\varphi : \mathcal{K} \to \mathcal{L}$ is (part of) a category equivalence between the categories if and only iff φ is a bijection between $\mathcal{K}(\mathbf{A}, \mathbf{B})$ and $\mathcal{L}(\varphi(\mathbf{A}), \varphi(\mathbf{B}))$ for all $\mathbf{A}, \mathbf{B} \in \mathcal{K}$ and for all $\mathbf{U} \in \mathcal{L}$ there is some $\mathbf{A} \in \mathcal{K}$ with $\varphi(\mathbf{A}) \cong \mathbf{U}$ in $\mathcal{L}$. Given two algebras $\mathbf{A}$ and $\mathbf{U}$ of possibly very different types, we shall say that they are *categorically equivalent*, and write $\mathbf{A} \equiv_c \mathbf{U}$, iff we have $\varphi(\mathbf{A}) = \mathbf{U}$ for some functor $\varphi : \mathsf{HSP}(\mathbf{A}) \to \mathsf{HSP}(\mathbf{U})$ which is a category equivalence.

In this paper we set out to characterize, algebraically, in as concrete a fashion as may be possible, those functors which are category equivalences between varieties, and those pairs of varieties (and pairs of algebras) which are categorically equivalent. The results we obtain turn out to require only the closure under subalgebras and the presence of finitely generated free algebras in a class, and so we obtain also a characterization of category equivalences between SF-classes. Also it turns out that two algebras $\mathbf{A}$ and $\mathbf{U}$ are categorically equivalent, as defined, iff $\varphi(\mathbf{A}) = \mathbf{U}$ for some category equivalence $\varphi : \mathsf{SP}(\mathbf{A}) \to \mathsf{SP}(\mathbf{U})$ between the prevarieties generated by these algebras. Such a functor always has a unique extension to a category equivalence between $\mathsf{HSP}(\mathbf{A})$ and $\mathsf{HSP}(\mathbf{U})$.

The results obtained here were already known in many special cases. We can recall, for example, Morita's theorem, which provides concrete conditions on two rings with unit which are necessary and sufficient in order that their varieties of left unitary modules be equivalent as categories. B. Banaschewski [3] solved the special case of unary algebras. Other instances were proved using the tools of duality theory, as in [8], [9], [21], [31] and [34]. A result of Hu [18] suggests that categorical equivalence can be a useful concept for algebraic classification: if $\mathbf{B}$ denotes the two-element Boolean algebra, then $\mathbf{B}/\equiv_c$ consists of all the finite primal algebras. Since an early version of this current paper first circulated in 1992, several authors have employed our characterization in studies of category equivalence among finite algebras—see [5], [12], [16].

A different equivalence relation on SF-classes, much stronger than category equivalence, will be of importance in this paper. We say that two SF-classes $\mathcal{K}$ and $\mathcal{L}$ are *equivalent*, written $\mathcal{K} \equiv \mathcal{L}$, provided that there is a functor $\mu :$

$\mathcal{K} \to \mathcal{L}$ which is a category equivalence and satisfies $\mu(\mathbf{F}_n(\mathcal{K})) = \mathbf{F}_n(\mathcal{L})$ for all positive integers n, where $\mathbf{F}_n(\mathcal{K})$ is the free algebra of rank n in $\mathsf{HSP}(\mathcal{K})$. Such a functor μ will be called an *equivalence*. When $\mathcal{K}$ and $\mathcal{L}$ are varieties, many equivalent definitions of this concept of equivalence are known. For instance, two varieties $\mathcal{K}$ and $\mathcal{L}$ are equivalent iff their algebraic theories are isomorphic (category theory), or iff their clones are isomorphic (universal algebra), or finally, iff there is a category equivalence $\varphi : \mathcal{K} \to \mathcal{L}$ such that φ commutes with the forgetful functors to the category of sets—i.e., $\varphi(\mathbf{A})$ and $\mathbf{A}$ have the same universe and $\mathcal{K}(\mathbf{A}, \mathbf{B}) = \mathcal{L}(\varphi(\mathbf{A}), \varphi(\mathbf{B}))$ for all $\mathbf{A}, \mathbf{B} \in \mathcal{K}$. (Such an φ is, indeed, an equivalence.)

We say that algebras $\mathbf{A}$ and $\mathbf{B}$ are *equivalent*, and write $\mathbf{A} \equiv \mathbf{B}$, provided that we have $\mu(\mathbf{A}) = \mathbf{B}$ for some equivalence $\mu : \mathsf{HSP}(\mathbf{A}) \to \mathsf{HSP}(\mathbf{B})$. It is well known that $\mathbf{A} \equiv \mathbf{B}$ iff $\mathbf{A}$ is isomorphic to an algebra $\mathbf{C}$ such that $\mathbf{C}$ and $\mathbf{B}$ have the same universe and exactly the same clone of term operations—i.e., every operation of $\mathbf{C}$ is a composition of operations of $\mathbf{B}$ and every operation of $\mathbf{B}$ is a composition of operations of $\mathbf{C}$.

We shall say that two category equivalence functors $\varphi_1 : \mathcal{K} \to \mathcal{L}_1$ and $\varphi_2 : \mathcal{K} \to \mathcal{L}_2$ are *equivalent* provided that there is an equivalence $\mu : \mathcal{L}_1 \to \mathcal{L}_2$ such that $\mu \circ \varphi_1 = \varphi_2$.

The main results of this paper can now be summarized.

(I) With the above notion of equivalence between category equivalence functors sourced from a SF-class $\mathcal{K}$, we shall see that every category equivalence $\varphi : \mathcal{K} \to \mathcal{L}$ where $\mathcal{L}$ is any SF-class is determined up to equivalence by the algebra $\mathbf{P}$ in $\mathcal{K}$ such that $\varphi(\mathbf{P})$ is a free algebra $\mathbf{F}_1(\mathcal{L})$ in $\mathcal{L}$. In otherwords, if we have category equivalence functors $\varphi_1 : \mathcal{K} \to \mathcal{L}_1$ and $\varphi_2 : \mathcal{K} \to \mathcal{L}_2$, and if $\varphi_1(\mathbf{P}) = \mathbf{F}_1(\mathcal{L}_1)$ and $\varphi_2(\mathbf{P}) = \mathbf{F}_1(\mathcal{L}_2)$ for some algebra $\mathbf{P}$, then φ_1 and φ_2 are equivalent. The algebra $\mathbf{P}$ corresponding to $\mathbf{F}_1(\mathcal{L})$ under a category equivalence $\mathcal{K} \to \mathcal{L}$ must be finitely generated (as an algebra) and projective in $\mathcal{K}$—i.e., $\mathbf{P}$ is a retract of some $\mathbf{F}_n(\mathcal{K})$. Moreover, $\mathbf{P}$ must be a co-generator in $\mathcal{K}$, meaning that every algebra in $\mathcal{K}$ is a homomorphic image of some co-power $\kappa \star \mathbf{P}$ of $\mathbf{P}$.

(II) Conversely, if $\mathbf{P}$ is any finitely generated projective co-generator in $\mathcal{K}$, then there exists a SF-class $\mathcal{L}$ and a category equivalence $\varphi : \mathcal{K} \to \mathcal{L}$ with $\varphi(\mathbf{P}) = \mathbf{F}_1(\mathcal{L})$.

(III) There is a very transparent algebraic construction which produces $\mathcal{L}$ and the category equivalence $\varphi : \mathcal{K} \to \mathcal{L}$ with $\varphi(\mathbf{P}) = \mathbf{F}_1(\mathcal{L})$. Actually, what is needed is not $\mathbf{P}$ itself, but a certain definition of it. There is a sequence $\bar{\sigma}(\bar{x})$ of n terms in n variables, defined in the language of $\mathcal{K}$, such that the equations $\bar{\sigma}(\bar{a}) = \bar{a}$ (this is a set of n equations) constitute a presentation for $\mathbf{P}$ as generated by elements $\bar{a} = \langle a_0, a_1, \dots, a_{n-1} \rangle$ and such that the equations

$\bar{\sigma}(\bar{\sigma}(\bar{x})) \approx \bar{\sigma}(\bar{x})$ are valid in $\mathcal{K}$. Then $\mathcal{L}$ and φ are constructed directly from the data $\bar{\sigma}(\bar{x})$.

(IV) The SF-classes categorically equivalent to $\mathcal{K}$ can be enumerated, up to equivalence, by enumerating all the finite sequences of terms over $\mathcal{K}$ which satisfy two elementary conditions. With such a $\bar{\sigma}(\bar{x})$ we construct an SF-class $\mathcal{K}^{[n]}(\bar{\sigma})$ of algebras of a different type and a specific category equivalence $\varphi : \mathcal{K} \to \mathcal{K}^{[n]}(\bar{\sigma})$. The class $\mathcal{K}^{[n]}(\bar{\sigma})$ is a variety iff $\mathcal{K}$ is a variety, and is closed under products, or homomorphisms, or ultraproducts iff $\mathcal{K}$ has the same closure property. Thus, as it turns out, any SF-class is categorically equivalent to $\mathcal{K}$ iff it is equivalent to some class $\mathcal{K}^{[n]}(\bar{\sigma})$.

The submission of this paper for publication was delayed for three years due to a concern that the results may be entirely unoriginal. Category theorists generally maintained that the subject of categorical equivalence between varieties of algebras was thoroughly explored a long time ago and all questions were resolved in the papers of J. R. Isbell [19] and P. Freyd [15]. I am convinced that this view is mistaken. In Isbell's paper we find on page 25 "the classification problem" stated with no solution offered. This is the problem: "Which algebraic theories have categorically equivalent categories of models?" Freyd's paper may have been obliquely addressed to this problem, but does not really get at it. I believe that the results we obtain, as summarized above, constitute an approximate solution to Isbell's classification problem that is algebraically useful and provides a tool, that was previously missing, which can be used to solve the classification problem in many special cases. However I also believe that Isbell's problem, properly understood, will turn out to be undecidable, in the same sense that it is undecidable whether two finitely presented groups given by generators and defining equations are isomorphic.

P. Freyd [15] shows that a category equivalence $F : \mathcal{V} \to \mathcal{W}$ between algebraic categories must be representable and can be viewed as

$$\mathbf{A} \mapsto \mathbf{A} \otimes \mathbf{Q}$$

for a certain $\mathbf{Q} \in \mathcal{W}$. He even shows how we might in principle go about constructing $\mathbf{A} \otimes \mathbf{Q}$ through generators and relations. His treatment is very abstract and doesn't seem to have been of much use up to now in algebraic investigations. However, I would agree that the results summarized under (I) and (II) in the preceeding paragraphs are at least implicit in both Isbell's and Freyd's papers. What I believe is missing in both papers are the realizations that a variety may carry substantially more information than is definable in categorical terms, that categorical equivalence becomes thereby a potentially useful tool for the classification of varieties, and that it is possible to characterize this relation among varieties in sufficiently concrete algebraic terms so that it becomes available as a tool for the algebraist.

Throughout the paper, we assume that no algebra has nullary operations. Our arguments are made a bit cleaner by this convention. There is no loss of generality here since nullary operations, or constants, can always be replaced by unary operations with singleton range.

1 Examples

Assume that $\varphi : \mathcal{V} \to \mathcal{W}$ is a category equivalence between two varieties. We shall later verify the known fact, mentioned above, that φ carries the finitely generated free algebras of $\mathcal{V}$ to finitely generated projective algebras that are co-generators in $\mathcal{W}$. When $\varphi(\mathbf{F}_1(\mathcal{V})) = \mathbf{F}_n(\mathcal{W})$, $\mathcal{V}$ is equivalent to the variety $\mathcal{W}^{[n]}$ constructed in Example 1.

Example 1: Let $\mathbf{A}$ be an algebra and n be a positive integer. By $\mathbf{A}^{[n]}$ we denote the algebra whose universe is A^n and whose basic operations are, for each positive r and for each sequence $t = \langle t_0(\bar{x}), \ldots, t_{n-1}(\bar{x}) \rangle$ of rn-ary terms in the language of $\mathbf{A}$, the r-ary operation $m_t : (A^n)^r \to A^n$ whose component maps $A^{rn} \to A$ are the term operations $t_i^{\mathbf{A}}$ of $\mathbf{A}$. We call this algebra the $[n]$th *matrix power* of $\mathbf{A}$. It is well-known that if $\mathcal{W}$ is any variety of algebras, then the class $\mathcal{W}^{[n]}$, consisting of all algebras isomorphic to $\mathbf{A}^{[n]}$ for some $\mathbf{A} \in \mathcal{W}$, is a variety. Moreover the correspondence $\mathbf{A} \mapsto \mathbf{A}^{[n]}$ defines a functor between these varieties that is a category equivalence, $\mathcal{W} \equiv_c \mathcal{W}^{[n]}$. For any class $\mathcal{K}$ of algebras we have $\mathcal{K} \equiv_c \mathcal{K}^{[n]}$; moreover, $\mathcal{K}$ is an **SF**-class iff $\mathcal{K}^{[n]}$ is an **SF**-class. [This construction was certainly known to F. E. J. Linton during the 1960's (see also W. Felsher [14] and W. Neumann [30]); has been used, e.g., in W. Taylor [36] and [37]; and has found several non-trivial applications in the recent universal algebraic literature, such as in D. Hobby, R. McKenzie [17].]

We remark that the functor $[n] : \mathcal{W} \to \mathcal{W}^{[n]}$ carries $\mathbf{F}_1(\mathcal{W})$ to an algebra that is not free in $\mathcal{W}^{[n]}$ unless $n = 1$ or $\mathbf{F}_1(\mathcal{W}) \cong \mathbf{F}_{rn}(\mathcal{W})$ for some $r \geq 1$.

Example 2: Let $\mathcal{S}$ be the variety of sets and $\mathcal{W}$ be the variety of semigroups with one additional operation s, defined by the equations $x \cdot x \approx x$, $x \cdot (y \cdot z) \approx x \cdot z \approx (x \cdot y) \cdot z$, $s(x \cdot y) \approx s(y) \cdot s(x)$, $ss(x) \approx x$. Then $\mathcal{S}$ and $\mathcal{W}$ are categorically equivalent. In fact, $\mathcal{W}$ is equivalent to $\mathcal{S}^{[2]}$. These varieties cannot be equivalent since every member of $\mathcal{W}$ has cardinality a perfect square while $\mathcal{S}$ has a two-element member.

Example 3: J. Jezek [20] proved that if $\mathcal{W}$ is the variety of all algebras of some fixed type of operations, then a variety $\mathcal{V}$ is categorically equivalent to $\mathcal{W}$ iff $\mathcal{V}$ is equivalent to some finite matrix power $\mathcal{W}^{[n]}$. (Actually, his result is not formulated in precisely this way, but his argument proves this result.) The only special property of $\mathcal{W}$ needed for this conclusion is the property that every projective algebra in $\mathcal{W}$ is a free algebra in $\mathcal{W}$. This result, for $\mathcal{W}$ the variety of sets, was proved earlier in G. C. Wraith [38].

Example 4: By a *primal algebra* is meant a finite algebra of more than one element such that every finitary operation of at least one argument on the universe of the algebra is a term operation of the algebra. For example, the variety $\mathcal{B}$ of Boolean algebras is generated by a two-element primal algebra. In Theorem 6.5 we give a new proof of T. K. Hu's theorem [18] that a variety $\mathcal{V}$ is categorically equivalent to $\mathcal{B}$ iff $\mathcal{V}$ is generated by some primal algebra. On the other hand, a variety is equivalent to $\mathcal{B}$ iff it is generated by a two-element primal algebra.

2   The mother functors

Besides the concept of the nth matrix power variety $\mathcal{V}^{[n]}$, our characterization of category equivalences between varieties uses one further concept which we now introduce.

Definition 2.1 Let $\mathbf{A}$ be any algebra and $\sigma(x)$ be any unary term in the language of $\mathbf{A}$. By $\mathbf{A}(\sigma)$ we denote the algebra whose universe is $\sigma(A)$ (the range of $\sigma^{\mathbf{A}}$) and whose operations are, for every positive integer r and r-ary term $t(\bar{x})$ in the language of $\mathbf{A}$, the operation t_{σ} defined on $\sigma(A)$ by placing $t_{\sigma}(\bar{x}) = \sigma(t(\bar{x}))$. If $\mathcal{K}$ is any class of similar algebras and $\sigma(x)$ is a term of one variable in the language of $\mathcal{K}$, then we define $\mathcal{K}(\sigma)$ to be the class of algebras isomorphic to $\mathbf{A}(\sigma)$ for some $\mathbf{A} \in \mathcal{K}$.

We remark that this construction is used extensively in the structure theory of locally finite algebras presented in D. Hobby, R. McKenzie [17], where $\sigma(x)$ is not required to be a term, but can be any polynomial of an algebra.

Theorem 2.1 *Let $\mathcal{K}$ be an SF-class and $\sigma(x)$ be one of its unary terms such that $\mathcal{K} \models \sigma(\sigma(x)) \approx \sigma(x)$. Then $\mathcal{K}(\sigma)$ is an SF-class and it is a variety if $\mathcal{K}$ is.*

PROOF. First we show that $\mathcal{K}(\sigma)$ includes the finitely generated free algebras in the variety it generates. Let $\mathbf{F} = \mathbf{F}_{\mathcal{K}}(n)$ be the free algebra in $\mathcal{K}$ generated by $x_1, \ldots, x_n$. Let $\mathbf{B}$ be the subalgebra of $\mathbf{F}$ generated by $\sigma(x_1), \ldots, \sigma(x_n)$. Then obviously $\mathbf{B}(\sigma)$ is generated by the elements $y_i = \sigma(x_i)$. We claim that $\mathbf{B}(\sigma)$ has the universal mapping property in $\mathcal{K}(\sigma)$ relative to these generators. Indeed, let $\mathbf{C}(\sigma) \in \mathcal{K}(\sigma)$, $\mathbf{C} \in \mathcal{K}$, and let $c_1, \ldots, c_n \in \mathbf{C}(\sigma)$. Let $f : \mathbf{F} \to \mathbf{C}$ be the homomorphism with $f(x_i) = c_i$. Then $f(y_i) = \sigma(c_i) = c_i$ and so, $f|_{\mathbf{B}(\sigma)}$ is a homomorphism $\mathbf{B}(\sigma) \to \mathbf{C}(\sigma)$ that takes y_i to c_i.

Next, we show that $\mathcal{K}(\sigma)$ is closed under formation of subalgebras. Let $\mathbf{B}(\sigma) \in \mathcal{K}(\sigma)$, $\mathbf{B} \in \mathcal{K}$, and let $\mathbf{S}$ be one of its subalgebras. Let $\mathbf{B}'$ be the subalgebra of $\mathbf{B}$ generated by the set S (the universe of $\mathbf{S}$). Then it is easy to see that $\mathbf{B}'(\sigma)$ is identical with $\mathbf{S}$.

Having shown that $\mathcal{K}(\sigma)$ is an SF-class, let us now assume that $\mathcal{K}$ is a variety. We must show that $\mathcal{K}(\sigma)$, like $\mathcal{K}$, is closed under the formation of direct products and homomorphic images. For products, this is trivial and the argument is left to the reader.

To see that $\mathcal{K}(\sigma)$ is closed under homomorphisms, let $\mathbf{A} \in \mathcal{K}$ and $\pi : \mathbf{A}(\sigma) \to \mathbf{C}$ be a surjective homomorphism. Put $S = \sigma(A)$ and take $\mathbf{B}$ to be the subalgebra of $\mathbf{A}$ generated by S, so that $\mathbf{A}(\sigma) = \mathbf{B}(\sigma)$. Then let $\theta = \{(x, y) \in S^2 : \pi(x) = \pi(y)\}$, (the kernel of π); and define β as the congruence on $\mathbf{B}$ generated by the set θ. We claim that $\theta = \beta \cap S^2$. From the claim, it will follow that there is an injective map $\gamma : C \to B/\beta$ such that $\gamma\pi(x) = x/\beta$ for $x \in S$. This map obviously is an isomorphism of algebras, $\gamma : \mathbf{C} \cong (\mathbf{B}/\beta)(\sigma)$. Thus it will follow that $\mathbf{C} \in \mathcal{K}(\sigma)$, once the claim is proved.

To prove the claim, we employ Malcev's characterization of generated congruences: β is the least equivalence relation over B that contains all the pairs $(g(x), g(y))$ where $(x, y) \in \theta$ and g is any unary polynomial operation of $\mathbf{B}$. Now let g be a unary polynomial operation of $\mathbf{B}$. Then, since $\mathbf{B}$ is generated by $S = \sigma(B)$, the function $\sigma g|_S$ is a unary polynomial operation of the algebra $\mathbf{B}(\sigma)$; hence $(\sigma g(x), \sigma g(y))$ belongs to θ when (x, y) is in θ. From these facts, and Malcev's characterization, it follows that β is a subset of the equivalence relation $\sigma^{-1}(\theta)$ over B. Then if $(x, y) \in \beta \cap S^2$, we have $(x, y) = (\sigma(x), \sigma(y)) \in \theta$ since $(x, y) \in \sigma^{-1}(\theta)$. Thus $\beta \cap S^2 \subseteq \theta \subseteq \beta \cap S^2$ and the claimed equality is demonstrated. $\bullet$

Definition 2.2 Let $\mathcal{K}$ be a class of similar algebras and $\sigma(x)$ be a unary term of $\mathcal{K}$. We say that $\sigma(x)$ is *invertible* in $\mathcal{K}$ if there exist unary terms $t_0(x), \ldots, t_{n-1}(x)$ (for some positive integer n) and an n-ary term $t(x_0, \ldots, x_{n-1})$ of $\mathcal{K}$ such that

$$\mathcal{K} \models x \approx t(\sigma(t_0(x)), \ldots, \sigma(t_{n-1}(x))).$$

We say that $\sigma(x)$ is *idempotent* in $\mathcal{K}$ if $\mathcal{K} \models \sigma(\sigma(x)) \approx \sigma(x)$.

Theorem 2.2 *Suppose that $\mathcal{K}$ is a class of algebras closed under isomorphisms and $\sigma(x)$ is a unary term that is invertible and idempotent in $\mathcal{K}$. For $\mathbf{A}, \mathbf{B} \in \mathcal{K}$ and $\alpha \in \mathcal{K}(\mathbf{A}, \mathbf{B})$ put $\varphi(\mathbf{A}) = \mathbf{A}(\sigma)$ and $\varphi(\alpha) = \alpha|_{\sigma(A)}$.*

(i) φ is a category equivalence between $\mathcal{K}$ and $\mathcal{K}(\sigma)$.

(ii) For any homomorphism $\alpha \in \mathcal{K}(\mathbf{A}, \mathbf{B})$, $\varphi(\alpha) \in \mathcal{K}(\sigma)(\mathbf{A}(\sigma), \mathbf{B}(\sigma))$ is injective (surjective) iff α has the same property.

(iii) $\mathcal{K}$ is closed under subalgebras, or homomorphic images, or products iff $\mathcal{K}(\sigma)$ has the same property. Thus $\mathcal{K}$ is a variety iff $\mathcal{K}(\sigma)$ is a variety.

(iv) $\mathcal{K}$ is an SF-class iff $\mathcal{K}(\sigma)$ is an SF-class.

PROOF. We can assume that σ is invertible in $\mathcal{K}$ via terms $t, t_0, \ldots, t_{n-1}$. Let $\mathbf{A}, \mathbf{B} \in \mathcal{K}$. We first prove (i) and (ii). To do this, we demonstrate first that $\varphi : \mathcal{K}(\mathbf{A}, \mathbf{B}) \to \mathcal{K}(\sigma)(\mathbf{A}(\sigma), \mathbf{B}(\sigma))$ is a bijection and then that where $\alpha \in \mathcal{K}(\mathbf{A}, \mathbf{B})$ we have that α is injective iff $\alpha|_{\sigma(A)}$ is injective, and that $\alpha(A) = B$ iff $\alpha(\sigma(A)) = \sigma(B)$.

That φ is injective on maps follows easily from the invertibility equation for σ. Now suppose that $\beta \in \mathcal{K}(\sigma)(\mathbf{A}(\sigma), \mathbf{B}(\sigma))$. Define $\hat{\beta} : A \to B$ by

$$\hat{\beta}(x) = t(\beta(\sigma t_0(x)), \ldots, \beta(\sigma t_{n-1}(x))) \,.$$

We proceed to show that $\hat{\beta}$ is the (unique) homomorphism $\rho \in \mathcal{K}(\mathbf{A}, \mathbf{B})$ such that $\rho|_{\sigma(A)} = \beta$. To see that $\hat{\beta}$ is a homomorphism, let f be any (say m-ary) operation symbol of $\mathcal{K}$; let $a_0, \ldots, a_{m-1} \in A$. For $i < n$, we have the mn-ary term

$$s^{(i)}(x_0^0, \ldots, x_{n-1}^{m-1}) = t_i f(t(x_0^0, \ldots, x_{n-1}^0), \ldots, t(x_0^{m-1}, \ldots, x_{n-1}^{m-1}))$$

and the corresponding term $s_\sigma^{(i)}$ which induces a basic operation of the algebra $\mathbf{A}(\sigma)$. Since β is a homomorphism, then

$$\beta(s_\sigma^{(i)}(\sigma t_0(a_0), \ldots, \sigma t_{n-1}(a_{m-1}))) = s_\sigma^{(i)}(\beta \sigma t_0(a_0), \ldots, \beta \sigma t_{n-1}(a_{m-1})) \,.$$

This equation can also be read as

$$\beta \sigma t_i f(a_0, \ldots, a_{m-1}) = \sigma t_i f(\hat{\beta}(a_0), \ldots, \hat{\beta}(a_{m-1})) \,.$$

It implies, in combination with the invertibility equation for σ, that

$$\hat{\beta}(f(a_0, \ldots, a_{m-1})) = f(\hat{\beta}(a_0), \ldots, \hat{\beta}(a_{m-1})) \,,$$

which is the condition for $\hat{\beta}$ to be a homomorphism.

To see that $\hat{\beta}|_{\sigma(A)} = \beta$, let $a \in A$ with $\sigma(a) = a$. Now

$$\begin{aligned}
\hat{\beta}(a) &= t(\beta \sigma t_0(a), \ldots, \beta \sigma t_{n-1}(a)) \\
&= t(\sigma t_0 \beta(a), \ldots, \sigma t_{n-1} \beta(a)) \\
&= \beta(a)
\end{aligned}$$

since $\sigma t_i|_{\sigma(A)}$ is one of the basic operations of $\mathbf{A}(\sigma)$ for each $i < n$.

Again, let $\alpha \in \mathcal{K}(\mathbf{A}, \mathbf{B})$. It is clear that if α is injective or surjective then $\varphi(\alpha)$ has the same property. To argue the converse, suppose first that α is not injective, say we have $\alpha(a_1) = \alpha(a_2)$ where $a_1 \neq a_2$. There must exist $i < n$ with $c_1 = \sigma t_i(a_1) \neq \sigma t_i(a_2) = c_2$ (by invertibility). Obviously, $\alpha(c_1) = \alpha(c_2)$ since α is a homomorphism. This shows that $\alpha|_{\sigma(A)}$ is not injective. Now suppose that $\alpha|_{\sigma(A)}$ is surjective. Let $b \in B$ be arbitrary. Choose $a_i \in \sigma(A)$ with $\alpha(a_i) = \sigma t_i(b)$ for $0 \leq i < n$, and put $a = t(a_0, \ldots, a_{n-1})$. Clearly, we have $\alpha(a) = b$.

Now we consider (iii). To see that $\mathcal{K}$ is closed under homomorphisms iff $\mathcal{K}(\sigma)$ is, we first note that for $\mathbf{A} \in \mathcal{K}$, every congruence relation on $\mathbf{A}(\sigma)$ is the restriction of a unique congruence of $\mathbf{A}$. (The proof of this parallels the proof, above, that every homomorphism between $\mathbf{A}(\sigma)$ and $\mathbf{B}(\sigma)$ is the restriction of a unique homomorphism between $\mathbf{A}$ and $\mathbf{B}$.) Now let $\mathbf{A} \in \mathcal{K}$ and let θ and $\hat{\theta}$ be a congruence of $\mathbf{A}(\sigma)$ and its extension to $\mathbf{A}$. Observe that $\mathbf{A}(\sigma)/\theta \cong (\mathbf{A}/\hat{\theta})(\sigma)$. Thus if $\mathbf{A}/\hat{\theta}$ belongs to $\mathcal{K}$, then $\mathbf{A}(\sigma)/\theta$ belongs to $\mathcal{K}(\sigma)$. Conversely, if $\mathbf{A}(\sigma)/\theta$ belongs to $\mathcal{K}(\sigma)$, let $\mathbf{A}(\sigma)/\theta \cong \mathbf{B}(\sigma)$ with $\mathbf{B} \in \mathcal{K}$. Working in the larger class $\mathcal{K}'$ which is the variety generated by $\mathcal{K}$, and using two results already proved, we have that the isomorphism between $(\mathbf{A}/\hat{\theta})(\sigma)$ (isomorphic to $\mathbf{A}(\sigma)/\theta$) and $\mathbf{B}(\sigma)$ extends to an isomorphism between $\mathbf{A}/\hat{\theta}$ and $\mathbf{B}$—thus $\mathbf{A}/\hat{\theta}$ belongs to $\mathcal{K}$.

The truth of our assertion that $\mathcal{K}$ is closed under subalgebras iff $\mathcal{K}(\sigma)$ is closed under subalgebras follows by a similar argument based on the easily proved fact that for $\mathbf{A} \in \mathcal{K}$, every subuniverse of $\mathbf{A}(\sigma)$ is the intersection with $\sigma(A)$ of a unique subuniverse of $\mathbf{A}$.

The proof that $\mathcal{K}(\sigma)$ is closed under direct products iff $\mathcal{K}$ has this property follows the same pattern; and we leave it for the reader to supply.

Finally, for (iv), all that remains to be proved, in view of Theorem 2.1, is that $\mathcal{K}$ must contain its finitely generated free algebras, provided that $\mathcal{K}(\sigma)$ is an SF-class. Let us now assume that $\mathcal{K}(\sigma)$ is an SF-class. Note that if $\mathbf{F} = \mathbf{F}_{\mathcal{K}(\sigma)}(mn)$ is freely generated in $\mathcal{K}(\sigma)$ by $y_0, \ldots, y_{mn-1}$ and $\mathbf{F} = \mathbf{A}(\sigma)$ with $\mathbf{A} \in \mathcal{K}$, then the subalgebra of $\mathbf{A}$ generated by $x_0, \ldots, x_{m-1}$ where $x_i = t(y_{in}, \ldots, y_{(i+1)n-1})$, is freely generated in $\mathcal{K}$ by these elements. (This subalgebra belongs to $\mathcal{K}$, as we have seen, since $\mathcal{K}(\sigma)$ is closed under subalgebras.) Thus ends our proof of the theorem. ●

We now formulate the (better known) version of Theorem 2.2 for the matrix power construction.

Theorem 2.3 *Let $\mathcal{K}$ be a class of algebras and n be a positive integer. Let $[n] : \mathcal{K} \to \mathcal{K}^{[n]}$ be the functor defined in Example 1.*

(i) $[n]$ is a category equivalence between $\mathcal{K}$ and $\mathcal{K}^{[n]}$.

(ii) For any homomorphism $\alpha \in \mathcal{K}(\mathbf{A}, \mathbf{B})$, $\alpha^{[n]} : \mathbf{A}^{[n]} \to \mathbf{B}^{[n]}$ is injective, or surjective, iff α has the same property.

(iii) $\mathcal{K}$ is closed under subalgebras, or homomorphic images, or products iff $\mathcal{K}^{[n]}$ has the same property. Thus $\mathcal{K}$ is a variety iff $\mathcal{K}^{[n]}$ is a variety.

(iv) $\mathcal{K}$ is an SF-class iff $\mathcal{K}^{[n]}$ is an SF-class.

PROOF. Since most parts of the theorem are well-known and all parts are easy to prove, we leave this proof for the reader to construct. For proving (iv), observe that when $\mathbf{A} \in \mathcal{K}$, the elements $\langle a_0^0, \ldots, a_{n-1}^0 \rangle, \ldots, \langle a_0^{k-1}, \ldots, a_{n-1}^{k-1} \rangle$ freely generate $\mathbf{A}^{[n]}$ in $\mathcal{K}^{[n]}$ iff $a_0^0, \ldots, a_{n-1}^{k-1}$ freely generate $\mathbf{A}$ in $\mathcal{K}$. In particular, if $\mathbf{A}$ is kn-freely generated in $\mathcal{K}$, then $\mathbf{A}^{[n]}$ is k-freely generated in $\mathcal{K}^{[n]}$.   •

Consider the relation that holds between classes $\mathcal{K}$ and $\mathcal{L}$ just in case there exists a positive integer n and a unary term σ in the language of $\mathcal{K}^{[n]}$ which is idempotent and invertible in $\mathcal{K}^{[n]}$, such that $\mathcal{L}$ is equivalent with $\mathcal{K}^{[n]}(\sigma)$. The next three Remarks outline an elementary proof that this relation is reflexive, symmetric and transitive. We shall see in §§3–5 that for SF-classes, this relation is identical to the relation of categorical equivalence.

First observe that $\mathcal{K} \equiv \mathcal{L}$ implies $\mathcal{K}^{[n]} \equiv \mathcal{L}^{[n]}$; also, $\mathcal{K} \equiv \mathcal{L}$ implies that for every unary term σ idempotent and invertible in $\mathcal{K}$, there corresponds (via the equivalence) a unary term τ idempotent and invertible in $\mathcal{L}$ such that $\mathcal{K}(\sigma) \equiv \mathcal{L}(\tau)$.

Remark 1: Let $\mathcal{K}$ be any class of algebras and $\mathcal{L} = \mathcal{K}^{[n]}$ (for some $n > 0$). Define $\sigma(x)$ to be the term $\langle x_0, \ldots, x_0 \rangle$ in the language of $\mathcal{L}$, i.e.,

$$\sigma^{\mathbf{A}^{[n]}}(\langle a_0, a_1, \ldots, a_{n-1} \rangle) = \langle a_0, \ldots, a_0 \rangle.$$

This $\sigma(x)$ is idempotent and invertible in $\mathcal{L}$; moreover, we have $\mathcal{L}(\sigma) \equiv \mathcal{K}$, i.e., $\mathcal{K} \equiv \mathcal{L}^{[1]}(\sigma)$.

Remark 2: Let $\mathcal{K}$ be any abstract class of algebras and $\mathcal{L} = \mathcal{K}(\sigma)$ where σ is idempotent and invertible in $\mathcal{K}$. Say

$$\mathcal{K} \models x \approx t(\sigma t_0(x), \ldots, \sigma t_{n-1}(x)).$$

There is an invertible idempotent term $\tau(x)$ for $\mathcal{L}^{[n]}$ such that $\mathcal{K} \equiv \mathcal{L}^{[n]}(\tau)$. In fact, we can take

$$\tau(\bar{x}) = \langle \sigma t_0 t(\bar{x}), \ldots, \sigma t_{n-1} t(\bar{x}) \rangle.$$

To see that $\mathcal{K} \equiv \mathcal{L}^{[n]}(\tau)$, the reader should work out the details of showing that for each $\mathbf{A} \in \mathcal{K}$, the map

$$\langle x_0, \ldots, x_{n-1} \rangle \mapsto t(x_0, \ldots, x_{n-1})$$

is a bijection from the universe of $(\mathbf{A}(\sigma))^{[n]}(\tau)$ onto the universe of $\mathbf{A}$ and is an isomorphism of the former algebra with an algebra equivalent to $\mathbf{A}$. The terms satisfying the invertibility equations for τ in $\mathcal{L}^{[n]}$ are not difficult to find. Here, the composition of the category equivalence functors

$$\mathcal{K} \to \mathcal{K}(\sigma) \to \mathcal{K}(\sigma)^{[n]} \to \mathcal{K}(\sigma)^{[n]}(\tau)$$

is equivalent, in the sense defined in the introduction, to the identity functor on $\mathcal{K}$.

Remark 3: Let $\mathcal{K}$ be any class of algebras, $\mathcal{L} = \mathcal{K}^{[m]}(\sigma)$, and $\mathcal{M} = \mathcal{L}^{[n]}(\tau)$ where σ and τ are invertible idempotent unary terms in the respective varieties $\mathcal{K}^{[m]}$ and $\mathcal{L}^{[n]}$. Then we have that $\mathcal{M}$ is equivalent to $\mathcal{K}^{[r]}(\gamma)$ where $r = nm$ and

$$\gamma(\langle x_j^i : i < n, j < m \rangle) = \langle t_v^u \rangle \quad \text{with}$$

$\langle t_v^u : v < m \rangle$ the uth component of

$$\tau(\langle \sigma(x_j^i : j < m) \rangle : i < n \rangle).$$

The idempotence of γ in $\mathcal{K}^{[r]}$ is easy to prove. Invertibility is messy but straightforward. To see that $\mathcal{K}^{[r]}(\gamma)$ and $\mathcal{M}$ are equivalent, show first that the natural bijection (for an algebra $\mathbf{A} \in \mathcal{K}$) between the sets A^{nm} and $(A^m)^n$ gives a set-bijection between $\mathbf{A}^{[r]}(\gamma)$ and $(\mathbf{A}^{[m]}(\sigma))^{[n]}(\tau)$.

Example 5: Returning to Example 4, let $\mathcal{B}$ be the variety of Boolean algebras and $\mathcal{P}$ be a variety generated by a three-element primal algebra $\mathbf{P}$. The algebra $\mathbf{P}$ has an invertible idempotent term $\sigma(x)$ such that $|\sigma(P)| = 2$. The algebra $\mathbf{P}(\sigma)$ is equivalent with a two-element Boolean algebra $\mathbf{B}$, since $\mathbf{P}$ is primal and therefore induces all possible operations on its subset $\sigma(P)$. Now $\mathcal{P} = \mathsf{HSP}(\mathbf{P})$ implies that $\mathcal{P}(\sigma) = \mathsf{HSP}(\mathbf{P}(\sigma))$ since the canonical functor $\mathcal{P} \to \mathcal{P}(\sigma)$ is a category equivalence. Hence

$$\mathcal{P}(\sigma) = \mathsf{HSP}(\mathbf{P}(\sigma)) \equiv \mathsf{HSP}(\mathbf{B}) = \mathcal{B};$$

and also $\mathbf{B} \equiv_c \mathbf{P}$ and $\mathcal{B} \equiv_c \mathcal{P}$.

Example 6: (Supplied by W. Blok.) Let $\mathcal{A}$ be the variety of all Abelian lattice-ordered groups. If $\mathbf{A} = \langle A, +, -, \vee, \wedge, 0 \rangle$ belongs to $\mathcal{A}$ then the term $\sigma(x) = x \vee 0$ is idempotent and invertible for $\mathbf{A}$; in fact, $\mathbf{A}$ satisfies $x \approx (x \vee 0) - (-x \vee 0)$. The algebra $\mathbf{A}(\sigma)$, the positive cone of $\mathbf{A}$, is equivalent with $\langle \sigma(A), +, \to, 0 \rangle$, where $x \to y$ is defined to be $(y - x) \vee 0$. The variety $\mathcal{A}(\sigma)$ is equivalent with the variety of *cancellative hoops*, or algebras $\langle H, +, \to, 0 \rangle$ such that $\langle H, +, 0 \rangle$ is a commutative monoid and these equations are satisfied:

(i) $x \to x \approx 0$;

(ii) $(x \to y) + x \approx (y \to x) + y$;

(iii) $(x + y) \to z \approx x \to (y \to z)$;

(iv) $y \to (y + x) \approx x$.

3 Categorical notions in varieties

In the preceding section, we examined two algebraic constructions which produce from an SF-class $\mathcal{K}$ a new SF-class $\mathcal{K}'$ along with a category equivalence $\varphi : \mathcal{K} \equiv_c \mathcal{K}'$. In the next two sections, we shall prove that every category equivalence between SF-classes is equivalent to one that arises by iterating these constructions (first form $\mathcal{K}^{[n]}$ for some n, then form $\mathcal{K}^{[n]}(\sigma)$ for some σ). In the proofs, we shall need to know that certain algebraic concepts for the algebras in an SF-class $\mathcal{K}$ are "categorical"; i.e., can be defined in language that refers only to the objects, the maps, and the composition of maps in the category $\mathcal{K}$. Categorical notions are guaranteed to be preserved by any category equivalence $\varphi : \mathcal{K} \equiv_c \mathcal{L}$, and that is all we need to know.

Of course, not every algebraic concept is categorical, else this paper would be pointless. A most significant example of an algebraic concept that is not categorical is the concept of an algebra being free (or freely generated) in a class.

The facts presented in this section can all be found in the literature. B. Davey and H. Werner [9] contains a large collection of examples of algebraic properties that are preserved by categorical equivalence.

For the next few paragraphs, all algebras mentioned are assumed to be drawn from one SF-class $\mathcal{K}$ which is fixed but arbitrary.

The concept of an *injective homomorphism* $\varphi \in \mathcal{K}(\mathbf{A}, \mathbf{B})$ is categorical; indeed φ is injective iff φ is a monomorphism in $\mathcal{K}$, i.e., for all $\mathbf{C} \in \mathcal{K}$ and $\alpha, \beta \in \mathcal{K}(\mathbf{C}, \mathbf{A})$, if $\varphi\alpha = \varphi\beta$ then $\alpha = \beta$. This requires only the presence of the free algebras in $\mathcal{K}$.

The concept of a *surjective homomorphism* $\varphi \in \mathcal{K}(\mathbf{A}, \mathbf{B})$ is categorical, although in general it is not equivalent with the categorical notion of an epimorphism. It turns out that φ is onto $\mathbf{B}$ iff whenever $\varphi = \delta\pi$ where $\delta \in \mathcal{K}(\mathbf{C}, \mathbf{B})$ is injective, then δ is an isomorphism, i.e., there exists $\delta' \in \mathcal{K}(\mathbf{B}, \mathbf{C})$ with $\delta\delta' = \mathrm{id}_B$ and $\delta'\delta = \mathrm{id}_C$. Indeed, if φ is onto $\mathbf{B}$ then $\varphi = \delta\pi$ implies δ is onto $\mathbf{B}$, so if δ is also injective then it is an isomorphism. Conversely, if φ is not onto $\mathbf{B}$, then it can be factored as $\varphi = \delta\pi$ where $\pi \in \mathcal{K}(\mathbf{A}, \varphi(\mathbf{A}))$ is surjective and δ is injective but not an isomorphism.

In category theory, there is a concept of "projective object" defined in terms of any class of epimorphisms. We need the algebraic notion of a projective algebra, defined in terms of surjective homomorphisms. An algebra $\mathbf{P}$ is *projective* in $\mathcal{K}$ iff $\mathbf{P} \in \mathcal{K}$ and whenever $\alpha \in \mathcal{K}(\mathbf{P}, \mathbf{A})$ and $\pi \in \mathcal{K}(\mathbf{C}, \mathbf{A})$ for some $\mathbf{A}, \mathbf{C}$ and π is surjective, then there exists some $\tau \in \mathcal{K}(\mathbf{P}, \mathbf{C})$ with $\alpha = \pi\tau$. Since the concept of a surjective homomorphism is categorical, so is the concept of a projective algebra.

For the purposes of our proof, a subalgebra $\mathbf{C}$ of $\mathbf{A}$ will be identified with any injective homomorphism $\varphi : \mathbf{S} \to \mathbf{A}$ whose range is $\mathbf{C}$; that is, "subalgebra of $\mathbf{A}$" will mean, interchangeably, any such homomorphism or its range. We write $\alpha \subseteq \mathbf{A}$ to denote that $\alpha \in \mathcal{K}(\mathbf{C}, \mathbf{A})$ for some $\mathbf{C}$ and is injective. For subalgebras $\alpha, \beta \subseteq \mathbf{A}$, say $\alpha \in \mathcal{K}(\mathbf{C}, \mathbf{A})$ and $\beta \in \mathcal{K}(\mathbf{D}, \mathbf{A})$, we write $\alpha \subseteq \beta$ to denote that the range of α is contained in the range of β. This is a categorical notion: $\alpha \subseteq \beta$ iff there is $\delta \in \mathcal{K}(\mathbf{C}, \mathbf{D})$ with $\alpha = \beta\delta$.

Theorem 3.1 *These notions are categorical in an* SF-*class* $\mathcal{K}$.

(1) $\alpha \subseteq \mathbf{A}$;

(2) $\alpha \in \mathcal{K}(\mathbf{A}, \mathbf{B})$ *is surjective;*

(3) $\alpha \subseteq \beta$ *(where* $\alpha, \beta \subseteq \mathbf{A}$*);*

(4) $\mathbf{P}$ *is a projective algebra in* $\mathcal{K}$*;*

(5) $\mathbf{A}$ *is finitely generated;*

(6) $\mathbf{A}$ *is a one-element algebra;*

(7) $\mathbf{A}$ *is a finite algebra;*

(8) $\mathbf{A}$ *is a countable algebra.*

PROOF. These facts can all be found in B. Banaschewski [2] or J. Jezek [20] or B. Davey and H. Werner [9]. We have already proved the theorem for properties (1)–(4).

For (5), just translate the condition that the largest element of the subalgebra lattice of $\mathbf{A}$ is compact into categorical language: $\mathbf{A}$ is finitely generated iff for every collection $\{\varphi_t : t \in T\}$ of subalgebras of $\mathbf{A}$, if every subalgebra $\varphi \subseteq \mathbf{A}$ such that $\varphi_t \subseteq \varphi$ for all t is a surjective homomorphism then there is a finite subcollection $\{\varphi_{t_0}, \ldots, \varphi_{t_{n-1}}\}$ such that every $\varphi \subseteq \mathbf{A}$ with $\varphi_{t_i} \subseteq \varphi$ for all $i \in \{0, \ldots, n-1\}$ is a surjective homomorphism.

For (6), note that $\mathbf{A}$ is a one-element algebra iff it is a terminal object in $\mathcal{K}$—$\mathcal{K}(\mathbf{C}, \mathbf{A})$ is a one-element set for every $\mathbf{C} \in \mathcal{K}$ (since $\mathcal{K}$ contains a free algebra $\mathbf{F}_1(\mathcal{K})$). For (7), observe that $\mathbf{A}$ is a finite algebra iff every finitely generated algebra in $\mathcal{K}$ admits only finitely many homomorphisms into $\mathbf{A}$. (Of course, $\mathcal{K}(\mathbf{F}, \mathbf{A})$ has the same cardinality as $\mathbf{A}$ if $\mathbf{F} = \mathbf{F}_1(\mathcal{K})$.) For (8), modify the condition formulated for (7) in an appropriate fashion.   •

We shall need some observations about projective algebras, co-products and co-powers in an SF-class $\mathcal{K}$.

The *co-product* in $\mathcal{K}$ of a system of algebras $\mathbf{B}_t \in \mathcal{K}$ $(t \in T)$ is an algebra $\mathbf{B} \in \mathcal{K}$ together with a system of maps $\iota_t \in \mathcal{K}(\mathbf{B}_t, \mathbf{B})$ $(t \in T)$ such that for every $\mathbf{C} \in \mathcal{K}$ and system of maps $\varphi_t \in \mathcal{K}(\mathbf{B}_t, \mathbf{C})$ $(t \in T)$ there is a unique $\varphi \in \mathcal{K}(\mathbf{B}, \mathbf{C})$ with $\varphi_t = \varphi \circ \iota_t$ for all $t \in T$. A co-product (also called "free product") of a system of algebras in $\mathcal{K}$ is determined up to isomorphism if it exists. Not all systems of algebras in $\mathcal{K}$ need have a co-product, but they do if $\mathcal{K}$ is a variety. Also, we can observe that where $\mathbf{F}_m(\mathcal{K})$ denotes the free algebra in $\mathcal{K}$ freely generated by $\{x_0, \ldots, x_{m-1}\}$, we have that $\mathbf{F}_m(\mathcal{K})$ is a co-product in $\mathcal{K}$ of m copies of $\mathbf{F}_1(\mathcal{K})$ with respect to the maps $\iota_i \in \mathcal{K}(\mathbf{F}_1(\mathcal{K}), \mathbf{F}_m(\mathcal{K}))$ $(i < m)$ where $\iota_i(x) = x_i$. We write

$$\mathbf{F}_m(\mathcal{K}) = m \star \mathbf{F}_1(\mathcal{K})$$

to denote this state of affairs and say that $\mathbf{F}_m(\mathcal{K})$ is the mth co-power of $\mathbf{F}_1(\mathcal{K})$. (We use x to denote the free generator of $\mathbf{F}_1(\mathcal{K})$.)

Note also that we have

$$\mathbf{F}_{nm}(\mathcal{K}) = n \star \mathbf{F}_m(\mathcal{K})$$

in $\mathcal{K}$ for any positive integers n, m.

The only co-products we shall have occasion to use are those in which all factors are identical. As above, we use the notation $n \star \mathbf{A}$ to denote the nth co-power of $\mathbf{A}$ in $\mathcal{K}$, if it exists. When $n \star \mathbf{A} = \mathbf{B}$ exists in $\mathcal{K}$, relative to the maps $\iota_0, \ldots, \iota_{n-1} \in \mathcal{K}(\mathbf{A}, \mathbf{B})$ then for any $\mathbf{C} \in \mathcal{K}$ and $\varphi_0, \ldots, \varphi_{n-1} \in \mathcal{K}(\mathbf{A}, \mathbf{C})$ we use $\varphi_0 \sqcup \cdots \sqcup \varphi_{n-1}$ to denote the homomorphism $\varphi : n \star \mathbf{A} \to \mathbf{C}$ satisfying $\varphi \circ \iota_i = \varphi_i$ for all $i < n$.

We adopt the usual categorical definition of a *product* in $\mathcal{K}$. The algebra $\mathbf{A} \in \mathcal{K}$ is a product of algebras $\mathbf{A}_0, \ldots, \mathbf{A}_{n-1}$ in $\mathcal{K}$ via maps $\pi_i \in \mathcal{K}(\mathbf{A}, \mathbf{A}_i)$ iff for any algebra $\mathbf{B}$ in $\mathcal{K}$ and homomorphisms $\varphi_i : \mathbf{B} \to \mathbf{A}_i$ $(i < n)$ there is a unique homomorphism $\varphi : \mathbf{B} \to \mathbf{A}$ with $\pi_i \varphi = \varphi_i$ for all $i < n$. We denote this unique φ by $\varphi_0 \sqcap \cdots \sqcap \varphi_{n-1}$. Of course, since we are only assuming that $\mathcal{K}$ is an SF-class, products may seldom exist in $\mathcal{K}$.

4 Presentations of projective algebras

We continue to assume that $\mathcal{K}$ is an SF-class. Our first observation is that any finitely generated projective algebra $\mathbf{P}$ in $\mathcal{K}$ has all of its finite co-powers $n \star \mathbf{P}$ existing in $\mathcal{K}$. We break this observation into two separate useful facts.

We say that an algebra $\mathbf{A}$ is a *retract* of an algebra $\mathbf{B}$ iff there exist a pair of homomorphisms $\iota : \mathbf{A} \to \mathbf{B}$ (injection) and $\pi : \mathbf{B} \to \mathbf{A}$ (retraction) such that $\mathrm{id}_A = \pi \circ \iota$.

Lemma 4.1 *A finitely generated algebra $\mathbf{A}$ in $\mathcal{K}$ is projective in $\mathcal{K}$ iff $\mathbf{A}$ is a retract of some finitely generated free algebra of $\mathcal{K}$.* •

Lemma 4.2 *Suppose that $\mathbf{A}$ is a retract of $\mathbf{B}$. If $n \star \mathbf{B}$ exists in $\mathcal{K}$ then $n \star \mathbf{A}$ exists in $\mathcal{K}$ and $n \star \mathbf{A}$ is a retract of $n \star \mathbf{B}$.*

In fact, let $\iota \in \mathcal{K}(\mathbf{A}, \mathbf{B})$ and $\pi \in \mathcal{K}(\mathbf{B}, \mathbf{A})$ with $\pi\iota = id_A$ and suppose that $\mathbf{D} = n \star \mathbf{B}$ in $\mathcal{K}$ relative to the system of homomorphisms $\{\iota_i : i < n\} \subseteq \mathcal{K}(\mathbf{B}, \mathbf{D})$. Then the subalgebra $\mathbf{C}$ of $\mathbf{D}$ generated by the union of the sets $\iota_i\iota(A)$ is a co-power $n \star \mathbf{A}$ relative to the system $\{\iota'_i : i < n\} \subseteq \mathcal{K}(\mathbf{A}, \mathbf{C})$, $\iota'_i = \iota_i\iota$. Moreover $\mathbf{C}$ is a retract of $\mathbf{D}$ via the identical injection and the retraction $\pi' = (\iota'_0\pi) \sqcup \cdots \sqcup (\iota'_{n-1}\pi)$; and for $i < n$ we have $\pi' \circ \iota_i = \iota'_i\pi$. •

Corollary 4.1 *If $\mathbf{P}$ is any finitely generated projective algebra in $\mathcal{K}$ then $n \star \mathbf{P}$ exists in $\mathcal{K}$ for every positive integer n.* •

Now let $\mathbf{P}$ be any finitely generated projective algebra in $\mathcal{K}$. This means that for a certain n, $\mathbf{P}$ is a retract of $\mathbf{F}_n(\mathcal{K})$ via an injection ι and retraction π. Since $\mathbf{P}$ is isomorphic with $\iota(\mathbf{P})$, a subalgebra of $\mathbf{F}_n(\mathcal{K})$, we can without loss of generality assume that $\mathbf{P} \subseteq \mathbf{F}_n(\mathcal{K})$ and that π is an endomorphism of $\mathbf{F}_n(\mathcal{K})$ whose range is $\mathbf{P}$ and we have $\pi^2 = \pi$. Letting $x_0, \ldots, x_{n-1}$ be the free generators of $\mathbf{F}_n(\mathcal{K})$, there are terms $\sigma_0(\bar{x}), \ldots, \sigma_{n-1}(\bar{x})$ such that $\pi(x_i) = \sigma_i(x_0, \ldots, x_{n-1})$. The condition $\pi^2 = \pi$ is clearly equivalent to the equations

$$\sigma_i(\sigma_0(\bar{x}), \ldots, \sigma_{n-1}(\bar{x})) = \sigma_i(\bar{x}) \ \text{ for } i \in \{0, \ldots, n-1\}.$$

In other words, where $\bar{\sigma} = \langle \sigma_0, \ldots, \sigma_{n-1} \rangle$, $\mathcal{K} \models \bar{\sigma}(\bar{\sigma}(\bar{x})) \approx \bar{\sigma}(\bar{x})$ and so, $\bar{\sigma}(\bar{x})$ is an idempotent unary term for $\mathcal{K}^{[n]}$ as this notion was defined in §2.

Thus every finitely generated projective algebra $\mathbf{P}$ in $\mathcal{K}$ is, for some n and some idempotent unary term $\bar{\sigma}(\bar{x})$ for $\mathcal{K}^{[n]}$, isomorphic to the subalgebra of $\mathbf{F}_n(\mathcal{K})$ generated by $\{\sigma_i(\bar{x}) : i < n\}$. It is also true that $\mathbf{P}$ is defined up to isomorphism as the algebra in $\mathcal{K}$ generated by elements $a_0, \ldots, a_{n-1}$ subject only to the equations

$$\sigma_i(a_0, \ldots, a_{n-1}) = a_i, \ \text{ for } 0 \leq i < n.$$

In brief, $\mathbf{P}$ is the algebra with the presentation $\bar{\sigma}(\bar{a}) = \bar{a}$.

It should also be obvious from these considerations that for every idempotent unary term $\bar{\sigma}(\bar{x})$ for $\mathcal{K}^{[n]}$, there exists in $\mathcal{K}$ the algebra with presentation $\bar{\sigma}(\bar{a}) = \bar{a}$ and that this is a finitely generated projective algebra in $\mathcal{K}$; in fact, it is isomorphic to a retract of $\mathbf{F}_n(\mathcal{K})$.

Definition 4.1 When a finitely generated projective algebra $\mathbf{P}$ of $\mathcal{K}$ and a term $\bar{\sigma}(\bar{x})$ are related as above, we say that they are *correlated*, or that $\mathbf{P}$ *is defined by* $\bar{\sigma}(\bar{x})$.

An algebra $\mathbf{A}$ will be called a *co-generator* in $\mathcal{K}$ iff $\mathbf{A}$ has all its finite co-powers in $\mathcal{K}$ and $\mathbf{F}_1(\mathcal{K})$ is a retract of some algebra $n \star \mathbf{A}$.

Note that if $\mathcal{K}$ is a variety then $\mathbf{A} \in \mathcal{K}$ is a co-generator in $\mathcal{K}$ iff every algebra in $\mathcal{K}$ is a homomorphic image of $\kappa \star \mathbf{A}$ (the co-power computed in $\mathcal{K}$) for some cardinal κ.

Lemma 4.3 *Let $\mathbf{P}$ be the finitely generated projective algebra in $\mathcal{K}$ defined by an idempotent term $\bar{\sigma}(\bar{x})$ for $\mathcal{K}^{[n]}$.*

(1) *$\bar{\sigma}(\bar{x})$ is invertible in the sense of Definition 2.2 iff $\mathbf{P}$ is a co-generator in $\mathcal{K}$.*

(2) *For every $\mathbf{A} \in \mathcal{K}$ the map $f \mapsto \langle f(\sigma_0(\bar{x})), \dots, f(\sigma_{n-1}(\bar{x})) \rangle$ from $\mathcal{K}(\mathbf{P}, \mathbf{A})$ to A^n (evaluation at the generators) is a bijection between $\mathcal{K}(\mathbf{P}, \mathbf{A})$ and $A^{[n]}(\sigma)$ and these bijections constitute a natural isomorphism between the set-valued functor $\mathcal{K}(\mathbf{P}, \_)$ on the one hand, and the functor which takes $\mathbf{A}$ to the universe of $\mathbf{A}^{[n]}(\sigma)$ on the other hand.*

PROOF. To prove (1), suppose first that $\bar{\sigma}$ is invertible. Thus there are unary $\mathcal{K}^{[n]}$ terms $\bar{t}^{(0)}(\bar{x}), \dots, \bar{t}^{(k-1)}(\bar{x})$ for some k and a k-ary $\mathcal{K}^{[n]}$ term $\bar{s}$ so that
$$\mathcal{K}^{[n]} \models \bar{x} \approx \bar{s}(\bar{\sigma}(\bar{t}^{(0)}(\bar{x})), \dots, \bar{\sigma}(\bar{t}^{(k-1)}(\bar{x}))) \,.$$
Let ι be the homomorphism from $\mathbf{F}_n(\mathcal{K})$ into $\mathbf{F}_{kn}(\mathcal{K})$ that sends
$$x_i \mapsto \bar{s}_i(\bar{\sigma}(x_0^0, \dots, x_{n-1}^0), \dots, \bar{\sigma}(x_0^{k-1}, \dots, x_{n-1}^{k-1})) \,.$$
Note that ι actually maps $\mathbf{F}_n(\mathcal{K})$ into $k \star \mathbf{P} \subseteq \mathbf{F}_{kn}(\mathcal{K})$. Let π be the homomorphism from $\mathbf{F}_{kn}(\mathcal{K})$ to $\mathbf{F}_n(\mathcal{K})$ that sends
$$x_i^j \mapsto t_i^{(j)}(x_0, \dots, x_{n-1}) \,.$$

The first displayed equation is equivalent to the equality $\pi \circ \iota = \mathrm{id}_{F_n}$. Thus $\mathbf{F}_n(\mathcal{K})$ is a retract of $k \star \mathbf{P}$ via the injection ι and the retraction $\pi|_{k \star P}$. Since $\mathbf{F}_1(\mathcal{K})$ is a retract of $\mathbf{F}_n(\mathcal{K})$, the desired result follows.

Conversely, suppose that $\mathbf{F}_1(\mathcal{K})$ is a retract of $k \star \mathbf{P}$ via maps ι and π. Thus we have a $\mathcal{K}$ term s of kn variables so that

$$\iota(x) = s(\bar{\sigma}(\bar{x}^{(0)}), \ldots, \bar{\sigma}(\bar{x}^{(k-1)})).$$

And we have $\mathcal{K}$ terms $t_i^{(j)}$ $(j < k, i < n)$ of one variable so that

$$\pi(\bar{\sigma}_i(\bar{x}^{(j)})) = t_i^{(j)}(x) \quad \text{for } j < k \text{ and } i < n.$$

Since $\pi \circ \iota = \mathrm{id}_{F_1}$ we have that

$$\mathcal{K} \models x \approx s(\bar{t}^{(0)}(x), \ldots, \bar{t}^{(k-1)}(x))$$

where $\bar{t}^{(j)}(x) = \langle t_0^{(j)}(x), \ldots, t_{n-1}^{(j)}(x) \rangle$; and, because π is a homomorphism,

$$\mathcal{K} \models \bar{t}^{(j)}(x) \approx \bar{\sigma}(\bar{t}^{(j)}(x)).$$

Now let $k' = kn$ and for $j' = jn + u < k'$ (with $j < k$, $u < n$), define a unary $\mathcal{K}^{[n]}$ term $\bar{t}'^{(j')}$ by setting, for $i < n$, $t_i'^{(j')}(x_0, \ldots, x_{n-1}) = t_i^{(j)}(x_u)$. Also define a $\mathcal{K}^{[n]}$ term $\bar{s}' = \langle s_0', \ldots, s_{n-1}' \rangle$ of k' variables by

$$s_i'(\bar{y}^{(0)}, \ldots, \bar{y}^{(k'-1)}) =$$
$$= s(y_0^{(i)}, \ldots, y_{n-1}^{(i)}, y_0^{(n+i)}, \ldots, y_{n-1}^{(n+i)}, \ldots, y_0^{(\{k-1\}n+i)}, \ldots, y_{n-1}^{(\{k-1\}n+i)}).$$

We leave it to the reader to check that we have defined $\mathcal{K}^{[n]}$ terms demonstrating the invertibility over $\mathcal{K}^{[n]}$ of $\bar{\sigma}$.

Now to prove (2), let $\mathbf{A} \in \mathcal{K}$ and consider the mapping $\mathrm{eval} : \mathcal{K}(\mathbf{P}, \mathbf{A}) \to A^n$ where $\mathrm{eval}(f) = \langle f(\sigma_0(\bar{x})), \ldots, f(\sigma_{n-1}(\bar{x})) \rangle$. It is clear that eval is injective, since the elements $\{ \sigma_i(\bar{x}) : i < n \}$ generate $\mathbf{P}$. Moreover, since $\bar{\sigma}(\bar{\sigma}(\bar{x})) = \bar{\sigma}(\bar{x})$ then for any homomorphism $f : \mathbf{P} \to \mathbf{A}$, $\bar{\sigma}(\mathrm{eval}(f)) = \mathrm{eval}(f)$—i.e., $\mathrm{eval}(f) \in A^{[n]}(\bar{\sigma})$. Next, to see that $\mathrm{eval} : \mathcal{K}(\mathbf{P}, \mathbf{A}) \to A^n(\bar{\sigma})$ is surjective, let $\bar{a} = \langle a_0, \ldots, a_{n-1} \rangle \in A^{[n]}(\bar{\sigma})$. Define $g \in \mathcal{K}(\mathbf{F}_n(\mathcal{K}), \mathbf{A})$ by setting $g(x_i) = a_i$ for $i < n$; then take $f = g|_P$. Since $\bar{\sigma}(\bar{a}) = \bar{a}$, we have

$$f(\sigma_i(\bar{x})) = \sigma_i(\bar{a}) = a_i$$

for $i < n$—i.e., $\mathrm{eval}(f) = \bar{a}$.

Finally, to see that eval is a natural isomorphism, let $\mathbf{A}, \mathbf{B} \in \mathcal{K}$ and $h \in \mathcal{K}(\mathbf{A}, \mathbf{B})$. Then for any $f \in \mathcal{K}(\mathbf{P}, \mathbf{A})$, we have

$$\mathrm{eval}(\mathcal{K}(\mathbf{P}, h)(f)) = \mathrm{eval}(hf) =$$
$$= \langle h(f(\sigma_0(\bar{x}))), \ldots, h(f(\sigma_{n-1}(\bar{x}))) \rangle = h^{[n]}(\bar{\sigma})(\mathrm{eval}(f)).$$

The next lemma is an easy consequence of Lemmas 4.1 and 4.2 and Definition 4.1.

Lemma 4.4 *An algebra* **A** *is a co-generator in an* SF*-class* $\mathcal{K}$ *iff every finitely generated projective algebra in* $\mathcal{K}$ *is a retract of a finite co-power of* **A**. *Hence the property of being a co-generator is invariant under category equivalences between* SF*-classes.*   •

5   Algebraic theories, clones, co-algebras and hom-functors

In order to see that every categorical equivalence between varieties can be obtained by composing first a functor $\mathcal{V} \mapsto \mathcal{V}^{[n]}$, second a functor $\mathcal{W} \mapsto \mathcal{W}(\sigma)$, and third, an equivalence, we need to recall several familiar concepts from the categorical and universal algebraic literatures.

The set of all term operations of an algebra **A** is a clone; that is, it is closed under composition and contains the trivial n-ary operations (the projections) for all $n > 0$. In fact, the clone of term operations of **A**, denoted Clo **A**, is that set of finitary operations generated under compositions from the set consisting of the basic operations of the algebra **A** together with the trivial operations.

The clone of **A** can be viewed as an enriched category—namely, the objects are the finite Cartesian powers $A = A^1, A^2, A^3, \ldots, A^n, \ldots$ of the universe of **A** and the morphisms are all the functions $f : A^k \to A^m$ such that the component maps $f_i : A^k \to A$ $(i < m)$ of f are k-ary term operations of the algebra **A**, and the enrichment consists in designating for each n a special object, namely A^n, and n special morphisms, namely the projection maps $\pi_i^n : A^n \to A$.

Abstractly, a *clone* is a category $\mathcal{C}$ together with a distinguished denumerable list of its objects $c(1), c(2), \ldots, c(n), \ldots$, and a distinguished list of its morphisms $p_{n,i}^{\mathcal{C}} \in \mathcal{C}(c(n), c(1))$ $(0 \leq i < n < \omega)$ such that every object in $\mathcal{C}$ is equal to some $c(n)$ and, moreover, $c(n)$ is the nth direct power of $c(1)$ relative to the morphisms $p_{n,0}^{\mathcal{C}}, \ldots, p_{n,n-1}^{\mathcal{C}}$ in $\mathcal{C}$. Clones are the objects in the category of clones. Morphisms in the category of clones are functors $F : \mathcal{C} \to \mathcal{D}$ such that $F(c(n)) = d(n)$ and $F(p_{n,i}^{\mathcal{C}}) = p_{n,i}^{\mathcal{D}}$ for all $0 \leq i < n < \omega$. An isomorphism of clones is an invertible morphism in the category of clones. Now, it is not hard to show that every clone (in this sense) is isomorphic to the clone of term operations of some algebra.

The dual category of a clone is called an *algebraic theory* (following F. W. Lawvere [24]). That is, an algebraic theory is a category $\mathcal{E}$ together with a distinguished denumerable list of its objects $e(1), e(2), \ldots, e(n), \ldots$, and a distinguished list of its morphisms $\iota_{n,i}^{\mathcal{E}} \in \mathcal{E}(e(1), e(n))$ $(0 \leq i < n < \omega)$ such that every object in $\mathcal{E}$ is equal to some $e(n)$ and, moreover, $e(n)$ is the nth co-power $n \star e(1)$ of $e(1)$ relative to the morphisms $\iota_{n,0}^{\mathcal{E}}, \ldots, \iota_{n,n-1}^{\mathcal{E}}$ in $\mathcal{E}$.

The concepts of algebraic theory (or clone) and of variety are intimately related. Every SF-class $\mathcal{K}$ has its algebraic theory $\mathrm{Th}(\mathcal{K})$ which is the full subcategory of $\mathcal{K}$ formed on the objects $\mathbf{F}_{\mathcal{K}}(n)$ with the distinguished morphisms $\iota_{n,i} : \mathbf{F}_{\mathcal{K}}(1) \to \mathbf{F}_{\mathcal{K}}(n)$ satisfying $\iota_{n,i}(x) = x_i$. (What universal algebraists call the *clone* of $\mathcal{K}$ is isomorphic to the dual category of $\mathrm{Th}(\mathcal{K})$.) And on the other hand, every algebraic theory $\mathcal{E}$ is isomorphic to the algebraic theory of a variety, which is determined up to equivalence by $\mathcal{E}$. In fact, we have the Lawvere category $\mathcal{S}^{\mathcal{E}}$ whose objects are the set-valued contravariant functors $\mathcal{E} \to \mathcal{S}$ (with $\mathcal{S}$ the category of sets) which take co-powers to direct powers, and whose morphisms are natural transformations. $\mathcal{S}^{\mathcal{E}}$ can be identified with a variety. In fact, each object $F : \mathcal{E} \to \mathcal{S}$ in $\mathcal{S}^{\mathcal{E}}$ gives an algebra with universe $F(e(1)) = A$ and operations the functions $F(\alpha) : A^n \to A$ with α ranging over $\mathcal{E}(e(1), e(n))$ for all n; and all these algebras constitute a variety the algebraic theory of which is isomorphic with $\mathcal{E}$. For an SF-class $\mathcal{K}$, the variety $\mathcal{S}^{\mathrm{Th}(\mathcal{K})}$ is equivalent to the variety generated by $\mathcal{K}$.

Two varieties $\mathcal{V}$ and $\mathcal{V}'$ are equivalent iff their algebraic theories $\mathcal{T}$ and $\mathcal{T}'$ are isomorphic in the category of algebraic theories. Isbell's classification problem, once again, is "for which pairs of algebraic theories $\mathcal{T}$ and $\mathcal{T}'$ are the categories (or varieties) $\mathcal{S}^{\mathcal{T}}$ and $\mathcal{S}^{\mathcal{T}'}$ categorically equivalent?"

Let $\mathcal{K}$ be any SF-class and $\mathbf{Q}$ an algebra in $\mathcal{K}$ whose finite co-powers exist in $\mathcal{K}$. We denote by $\mathrm{Th}_{\mathcal{K}}(\mathbf{Q})$ (or just $\mathrm{Th}(\mathbf{Q})$) the algebraic theory whose reduct as a category is just the full subcategory of $\mathcal{K}$ on the objects $1 \star \mathbf{Q}, 2 \star \mathbf{Q}, \ldots, n \star \mathbf{Q}, \ldots$. Assume now that $\mathcal{K}(\mathbf{Q}, \mathbf{A}) \neq \emptyset$ for every $\mathbf{A} \in \mathcal{K}$ (which will certainly be the case if $\mathbf{Q}$ is a retract of some $\mathbf{F}_{\mathcal{K}}(n)$). Every algebra $\mathbf{A} \in \mathcal{K}$ then induces a member of $\mathcal{S}^{\mathrm{Th}(\mathbf{Q})}$ via the hom-functor $\hom_{\mathcal{K}}(-, \mathbf{A}) = \mathcal{K}(-, \mathbf{A})$ restricted to $\mathrm{Th}(\mathbf{Q})$. (Since $\hom_{\mathcal{K}}(-, \mathbf{A})$ is a contravariant functor $\mathcal{K} \to \mathcal{S}$ which converts co-powers to direct powers in $\mathcal{S}$.) Thus we have a corresponding algebra, with universe $\mathcal{K}(\mathbf{Q}, \mathbf{A})$, having an n-ary operation for each $\alpha \in \mathcal{K}(\mathbf{Q}, n \star \mathbf{Q})$. We shall denote this algebra by $\mathbf{A}_{\mathbf{Q}}$. Also we put

$$\mathcal{K}_{\mathbf{Q}} = \{\mathbf{A}_{\mathbf{Q}} : \mathbf{A} \in \mathcal{K}\}.$$

Then the functor $\hom_{\mathcal{K}}(\mathbf{Q}, -) : \mathcal{K} \to \mathcal{S}$ can also be regarded as a functor

$$\hom_{\mathcal{K}}(\mathbf{Q}, -) : \mathcal{K} \to \mathcal{K}_{\mathbf{Q}}.$$

In case $\mathbf{Q} = \mathbf{F}_{\mathcal{K}}(1)$, the above functor is actually an equivalence between SF-classes, $\mathcal{K} \equiv \mathcal{K}_{\mathbf{Q}}$.

Now let $\varphi : \mathcal{L} \to \mathcal{K}$ be a covariant co-power preserving functor between SF-classes. Then $\mathbf{Q} = \varphi(\mathbf{F}_{\mathcal{L}}(1))$ is an algebra whose finite co-powers exist in $\mathcal{K}$; in fact, we have an algebraic theory

$$\mathcal{T}' = \varphi(\mathrm{Th}(\mathcal{L}))$$

with objects $\varphi(\mathbf{F}_{\mathcal{L}}(n)) = n \star \mathbf{Q}$ and morphisms all the maps $\varphi(\alpha)$, $\alpha \in \mathcal{L}(\mathbf{F}_{\mathcal{L}}(m), \mathbf{F}_{\mathcal{L}}(n))$. The full algebraic theory $\mathrm{Th}_{\mathcal{K}}(\mathbf{Q}) = \mathcal{T}$ has $\mathcal{T}'$ as a sub-theory. Hence $\mathcal{T}'$ induces, for every $\mathbf{A} \in \mathcal{K}$ such that the set $\mathcal{K}(\mathbf{Q}, \mathbf{A})$ is non-empty, an algebra on this hom-set. This algebra is a reduct of $\mathbf{A}_{\mathbf{Q}}$, and we denote it by $\mathbf{A}_{\varphi}$. Note that $\mathcal{K}(\mathbf{Q}, \mathbf{A})$ is non-empty if $\mathbf{A} = \varphi(\mathbf{B})$ for some $\mathbf{B} \in \mathcal{L}$, and when this occurs, the functor φ provides an algebraic homomorphism $\mathbf{B}_{\mathbf{F}_{\mathcal{L}}(1)} \to \mathbf{A}_{\varphi}$. Even if $\mathbf{A}$ is not in the range of φ, the algebra $\mathbf{A}_{\varphi}$, if non-empty, belongs to the variety of models of $\mathrm{Th}(\mathcal{L})$ which, by the preceding paragraph, is equivalent to the variety $\mathsf{HSP}(\mathcal{L})$ generated by $\mathcal{L}$. This situation is summarized by saying that $\mathbf{Q}$ is a $\mathsf{HSP}(\mathcal{L})$-*co-algebra* in $\mathcal{K}$ that happens to be induced by the functor φ.

Theorem 5.1 *Let $\varphi : \mathcal{K} \to \mathcal{L}$ be a category equivalence between SF-classes. Let $\mathbf{Q}$ be an algebra such that $\mathbf{F}_{\mathcal{L}}(1) = \varphi(\mathbf{Q})$. Then $\mathbf{Q}$ is a finitely generated projective co-generator in $\mathcal{K}$. Choose a category equivalence $\psi : \mathcal{L} \to \mathcal{K}$ with $\psi(\mathbf{F}_{\mathcal{L}}(1)) = \mathbf{Q}$ and $\varphi \circ \psi$ and $\psi \circ \varphi$ each naturally isomorphic to the identity functor on their respective domains. Then for $\mathbf{B} \in \mathcal{L}$, the homomorphism defined above is an isomorphism $\mathbf{B}_{\mathbf{F}_{\mathcal{L}}(1)} \cong \psi(\mathbf{B})_{\psi}$; and moreover, $\psi(\mathbf{B})_{\mathbf{Q}} \equiv \psi(\mathbf{B})_{\psi} \equiv \mathbf{B}$. The functor that takes $\mathbf{B}$ to $\psi(\mathbf{B})_{\mathbf{Q}}$ and $f \in \mathcal{L}(\mathbf{B}, \mathbf{C})$ to $\mathrm{hom}_{\mathcal{K}}(\mathbf{Q}, \psi(f))$ is an equivalence: $\mathcal{L} \equiv \mathcal{K}_{\mathbf{Q}}$.*

PROOF. Since $\mathbf{F}_{\mathcal{L}}(1)$ is a finitely generated projective co-generator in $\mathcal{L}$, then $\mathbf{Q}$ has these properties in $\mathcal{K}$ (by Theorem 3.1 and Lemma 4.4). Note that φ and ψ automatically preserve existing co-powers.

Since ψ is now a bijection on hom-sets, the categories $\mathrm{Th}_{\mathcal{K}}(\mathbf{Q})$ and $\psi(\mathrm{Th}(\mathcal{L}))$ are identical. Thus for $\mathbf{B}$ in $\mathcal{L}$, the algebras $\psi(\mathbf{B})_{\psi}$ and $\psi(\mathbf{B})_{\mathbf{Q}}$ have exactly the same operations and are equivalent by a trivial renaming of basic operations. The homomorphism $\mathbf{B}_{\mathbf{F}_{\mathcal{L}}(1)} \to \psi(\mathbf{B})_{\psi}$ mentioned above is the bijection induced by ψ between the sets $\mathcal{L}(\mathbf{F}_{\mathcal{L}}(1), \mathbf{B})$ and $\mathcal{K}(\mathbf{Q}, \psi(\mathbf{B}))$; it is an isomorphism. So we have

$$\mathbf{B} \equiv \mathbf{B}_{\mathbf{F}_{\mathcal{L}}(1)} \cong \psi(\mathbf{B})_{\psi} \equiv \psi(\mathbf{B})_{\mathbf{Q}}$$

and this equivalence works the same for all $\mathbf{B}$ in $\mathcal{K}$ at the level of terms, assigning to the term t in $\mathrm{Th}(\mathcal{L})$ the term $\psi(t)$ in $\mathrm{Th}_{\mathcal{K}}(\mathbf{Q})$. Since every algebra $\mathbf{A} \in \mathcal{K}$ is isomorphic to some $\psi(\mathbf{B})$, then $\mathbf{A}_{\mathbf{Q}}$ is isomorphic to $\psi(\mathbf{B})_{\mathbf{Q}}$. We have an equivalence of SF-classes, $\mathcal{L} \equiv \mathcal{K}_{\mathbf{Q}}$ and it very clearly reflects the fact that the algebraic theories of these two classes, $\mathrm{Th}(\mathcal{L})$ and $\mathrm{Th}_{\mathcal{K}}(\mathbf{Q})$, are isomorphic via ψ.   ●

The above theorem reveals that when $\varphi : \mathcal{K} \to \mathcal{L}$ is a category equivalence between SF-classes and $\varphi(\mathbf{Q}) = \mathbf{F}_{\mathcal{L}}(1)$, then φ is equivalent to the canonical functor $\mathcal{K} \to \mathcal{K}_{\mathbf{Q}}$. Therefore this functor must be a category equivalence. To complete this circle of ideas, we must now show that for every finitely generated projective co-generator $\mathbf{P}$ in $\mathcal{K}$, the functor $\mathcal{K} \to \mathcal{K}_{\mathbf{P}}$ is a category equivalence. To do this, we shall show that it is equivalent to a composition of the functors introduced in §2, and in the process we shall, of course, verify that $\mathcal{K}_{\mathbf{P}}$ is an SF-class.

Theorem 5.2 *Let $\mathcal{K}$ be an SF-class and $\mathbf{P}$ be a finitely generated projective algebra in $\mathcal{K}$, defined by the idempotent term $\bar{\sigma}(\bar{x})$ of $\mathcal{K}^{[n]}$. Then $\mathcal{K}_{\mathbf{P}}$ is an SF-class equivalent to $\mathcal{K}^{[n]}(\bar{\sigma})$ and the functors $\mathcal{K} \to \mathcal{K}_{\mathbf{P}}$ and $\mathcal{K} \to \mathcal{K}^{[n]}(\bar{\sigma})$ are equivalent. These equivalent functors are category equivalences iff $\mathbf{P}$ is a co-generator in $\mathcal{K}$.*

PROOF. We need first to complete the work begun in Lemma 4.3, by showing that the bijection

$$\mathrm{eval} : \mathcal{K}(\mathbf{P}, \mathbf{A}) \to \bar{\sigma}(A^n)$$

establishes an equivalence between the algebras $\mathbf{A}_{\mathbf{P}}$ and $\mathbf{A}^{[n]}(\bar{\sigma})$ for all $\mathbf{A} \in \mathcal{K}$. We shall write $\mathbf{F}(m)$ for $\mathbf{F}_{\mathcal{K}}(m)$, and we put

$$\langle x_0, \ldots, x_{n-1} \rangle = \bar{x} \quad \text{and}$$

$$\langle x_0^j, \ldots, x_{n-1}^j \rangle = \bar{x}^j$$

where $x_0, \ldots, x_{n-1}$ are the free generators of $\mathbf{F}(n)$ and x_i^j $(0 \le j < m$ and $0 \le i < n)$ are the free generators of $\mathbf{F}(mn)$. The algebra $\mathbf{P}$ is taken as the subalgebra of $\mathbf{F}(n)$ generated by the elements $\sigma_i(\bar{x})$ of the tuple $\bar{\sigma}(\bar{x})$; and $m \star \mathbf{P}$ is taken as the subalgebra of $\mathbf{F}(mn)$ generated by all elements of the tuples $\bar{\sigma}(\bar{x}^j)$, $(0 \le j < m)$. For $0 \le j < m$, the jth insertion of $\mathbf{P}$ into $m \star \mathbf{P}$ is the homomorphism that takes $\bar{\sigma}(\bar{x})$ to $\bar{\sigma}(\bar{x}^j)$. Elements of $m \star \mathbf{P}$ can be written in the form $t(\bar{\sigma}(\bar{x}^0), \ldots, \bar{\sigma}(\bar{x}^{m-1}))$ where t is an $\mathcal{K}$-term of mn variables.

If $f \in \mathcal{K}(\mathbf{P}, m \star \mathbf{P})$ then since $\bar{\sigma}(\bar{\sigma}(\bar{x})) = \bar{\sigma}(\bar{x})$ we have that the n-tuple $f(\bar{\sigma}(\bar{x}))$ of elements of $m \star \mathbf{P}$ takes the form

$$f(\bar{\sigma}(\bar{x})) = \bar{\sigma}(\bar{t}(\bar{\sigma}(\bar{x}^0), \ldots, \bar{\sigma}(\bar{x}^{m-1})))$$

where $\bar{t}$ is an n-tuple of $\mathcal{K}$-terms of mn variables. We denote this homomorphism by $f = f_{\bar{t}}$. The corresponding m-ary operation O_f of an algebra $\mathbf{A}_{\mathbf{P}}$ takes an m-tuple $\langle u_0, \ldots, u_{m-1} \rangle$ of members of $\mathcal{K}(\mathbf{P}, \mathbf{A})$ to the element

$$u = (u_0 \sqcup u_1 \sqcup \cdots \sqcup u_{m-1}) \circ f \in \mathcal{K}(\mathbf{P}, \mathbf{A}).$$

Thus

$$u(\bar{\sigma}(\bar{x})) = \bar{\sigma}(\bar{t}(u_0(\bar{\sigma}(\bar{x})), \ldots, u_{m-1}(\bar{\sigma}(\bar{x})))) \quad \text{i.e.,}$$

$$\text{eval}(u) = \bar{\sigma}(\bar{t}(\text{eval}(u_0), \ldots, \text{eval}(u_{m-1})))\,.$$

In other words,

$$\text{eval}(O_f(u_0, \ldots, u_{m-1}) = \bar{t}_{\bar{\sigma}}(\text{eval}(u_0), \ldots, \text{eval}(u_{m-1}))$$

where $\bar{t}_{\bar{\sigma}}$ is the m-ary operation of the algebra $\mathbf{A}^{[n]}(\bar{\sigma})$ induced by the $\mathcal{K}^{[n]}$ term $\bar{t}$ according to Definition 2.1.

The displayed formula, combined with Lemma 4.3 (2) establishes that eval is an equivalence between the classes $\mathcal{K}_{\mathbf{P}}$ and $\mathcal{K}^{[n]}(\bar{\sigma})$ and that the functor $\mathcal{K} \to \mathcal{K}^{[n]}(\bar{\sigma})$ is the composition of the functor $\mathcal{K} \to \mathcal{K}_{\mathbf{P}}$ with this equivalence. Thus the functors $\mathcal{K} \to \mathcal{K}_{\mathbf{P}}$ and $\mathcal{K} \to \mathcal{K}^{[n]}(\bar{\sigma})$ are equivalent. Now it follows from Theorems 2.1 and 2.3 that $\mathcal{K}^{[n]}(\bar{\sigma})$ is an SF-class, and the equivalent class $\mathcal{K}_{\mathbf{P}}$ must also have this property.

Now for the last assertion in the theorem, suppose first that $\mathbf{P}$ is a co-generator (equivalently, by Lemma 4.3, $\bar{\sigma}$ is invertible). Thus $\mathcal{K} \to \mathcal{K}^{[n]}(\bar{\sigma})$ is a category equivalence, by Theorems 2.2 and 2.3. It follows readily that the equivalent functor $\mathcal{K} \to \mathcal{K}_{\mathbf{P}}$ is a category equivalence.

Finally, assume that $\mathcal{K} \to \mathcal{K}_{\mathbf{P}}$ is a category equivalence. Let $\mathbf{B}$ be the subalgebra of $\mathbf{F}(1)$ generated by the union of all the subalgebras $f(\mathbf{P})$ where f ranges over all homomorphisms from $\mathbf{P}$ to $\mathbf{F}(1)$. Clearly, the algebras $\mathbf{F}(1)_{\mathbf{P}}$ and $\mathbf{B}_{\mathbf{P}}$ are identical. Therefore, $\mathbf{F}(1)$ and $\mathbf{B}$ are isomorphic algebras, since our functor is an equivalence. But this obviously implies that $\mathbf{B} = \mathbf{F}(1)$. Then we have homomorphisms $f_0, \ldots, f_{r-1} : \mathbf{P} \to \mathbf{F}(1)$ so that $\mathbf{F}(1)$ is generated by $f_0(P) \cup \cdots \cup f_{r-1}(P)$. Hence $f_0 \sqcup \cdots \sqcup f_{r-1}$ is a surjective homomorphism $r \star \mathbf{P} \to \mathbf{F}(1)$. Since $\mathbf{F}(1)$ is projective, it must be a retract of $r \star \mathbf{P}$. We have shown that $\mathbf{P}$ is a co-generator.   •

6   The results and some applications

Theorem 6.1 and Corollaries 6.1–6.3 summarize the results contained in Theorems 2.2, 2.3, 5.1, and 5.2.

Theorem 6.1 *Suppose that φ is a category equivalence between SF-classes $\mathcal{M}$ and $\mathcal{N}$. Let $\mathbf{P} \in \mathcal{M}$ be an algebra such that $\varphi(\mathbf{P})$ is a free algebra on one generator in $\mathcal{N}$. Let $\bar{\sigma}(\bar{x})$ be an invertible idempotent term for $\mathcal{M}^{[n]}$ which defines $\mathbf{P}$. The class $\mathcal{N}$ is equivalent to each of the classes $\mathcal{M}_{\mathbf{P}}$ and $\mathcal{M}^{[n]}(\bar{\sigma})$. For $\mathbf{A} \in \mathcal{M}$ we have $\varphi(\mathbf{A}) \equiv \mathbf{A}_{\mathbf{P}} \equiv \mathbf{A}^{[n]}(\bar{\sigma})$. The functor φ is equivalent to each of the functors $\mathcal{M} \to \mathcal{M}_{\mathbf{P}}$ and $\mathcal{M} \to \mathcal{M}^{[n]}(\bar{\sigma})$. All of these functors and equivalences extend to the varieties generated by the three classes.*   •

Corollary 6.1 *Let* **A** *and* **B** *be any algebras. Then* $\mathbf{A} \equiv_c \mathbf{B}$ *iff* $\mathbf{B} \equiv \mathbf{A}^{[n]}(\bar{\sigma})$ *for some positive integer* n *and invertible idempotent term* $\bar{\sigma}$ *for* $\mathbf{A}^{[n]}$.

Corollary 6.2 *Let* $\mathcal{M}$ *be any* SF-*class. The* SF-*classes categorically equivalent to* $\mathcal{M}$ *are precisely those classes equivalent to* $\mathcal{M}^{[n]}(\bar{\sigma})$ *for some positive integer* n *and some invertible idempotent term* $\bar{\sigma}$ *for* $\mathcal{M}^{[n]}$. •

Let $\mathcal{C} = \mathrm{Th}(\mathcal{V})$ be the algebraic theory of a variety $\mathcal{V}$ and $\bar{\sigma}(\bar{x})$ be an invertible idempotent term for $\mathcal{V}^{[n]}$. We can identify $\bar{\sigma}$ with an element s of $\mathcal{C}(c(n), c(n))$, and its invertibility can be determined in $\mathcal{C}$. Namely, $s \in \mathcal{C}(c(n), c(n))$ is "invertible" iff for some m there are $\tau \in \mathcal{C}(c(n), c(mn))$ and $\nu_0, \ldots, \nu_{m-1} \in \mathcal{C}(c(n), c(n))$ such that

$$(s\nu_0 \sqcup s\nu_1 \sqcup \cdots \sqcup s\nu_{m-1}) \circ \tau = 1_{c(n)} .$$

The algebraic theory of $\mathcal{V}^{[n]}(\bar{\sigma})$ can of course be constructed from $\mathcal{C}$ and $\bar{\sigma}$ without reference to the algebras in the varieties. The theory resulting from the construction (isomorphic to $\mathrm{Th}(\mathcal{V}^{[n]}(\bar{\sigma}))$) we denote by $\mathcal{C}^{[n]}(\bar{\sigma})$. Here now is our "approximate" solution to J. Isbell's classification problem.

Corollary 6.3 *Let* $\mathcal{C}$ *and* $\mathcal{C}'$ *be algebraic theories. The categories of models of these theories are categorically equivalent iff* $\mathcal{C}'$ *is isomorphic to* $\mathcal{C}^{[n]}(s)$ *for some positive integer* n *and some idempotent invertible element* $s \in \mathcal{C}(c(n), c(n))$.

Every algebra **A** in a variety $\mathcal{V}$ gives rise to an algebraic theory which is the full subcategory of $\mathcal{V}$ formed on the objects

$$1 \star \mathbf{A}, 2 \star \mathbf{A}, \ldots, n \star \mathbf{A}, \ldots$$

but it is unclear just in what way the Lawvere variety of models of this theory is related to $\mathcal{V}$. Thus one is led to several questions concerning possible extensions of the results proved above. The next theorem answers one of these questions.

Theorem 6.2 *Suppose that* $\mathcal{K}$ *is an* SF-*class and* $\mathbf{Q} \in \mathcal{K}$. *These statements are equivalent.*

(i) $\mathrm{hom}(\mathbf{Q}, -) : \mathcal{K} \to \mathcal{K}_{\mathbf{Q}}$ *is a category equivalence and* $\mathcal{K}_{\mathbf{Q}}$ *is an* SF-*class.*

(ii) $\mathbf{Q}$ *is a finitely generated projective co-generator in* $\mathcal{K}$.

PROOF. (ii) ⇒ (i) is by Theorems 2.2, 2.3 and 5.2.

(i) ⇒ (ii): Under the assumptions of (i), we have by Theorem 3.1 that $\varphi =$ hom$(\mathbf{Q}, -)$ both preserves and reflects all of the properties listed in Theorem 3.1. In particular, if $\alpha \in \mathcal{K}(\mathbf{A}, \mathbf{B})$ is surjective, then so is $\varphi(\alpha)$. This condition is equivalent to $\mathbf{Q}$ being projective. Thus $\mathbf{Q}$ is projective.

To see that $\mathbf{Q}$ is finitely generated, consider the map $1_{\mathbf{Q}} \in \varphi(\mathbf{Q})$. Since every morphism in $\mathcal{K}_{\mathbf{Q}}$ is of the form $\varphi(\gamma)$, it is trivial to see that for every $u \in \mathbf{R} \in \mathcal{K}_{\mathbf{Q}}$, there is a unique homomorphism $f \in \mathcal{K}_{\mathbf{Q}}(\varphi(\mathbf{Q}), \mathbf{R})$ with $f(1_{\mathbf{Q}}) = u$. This implies, among other things, that the subalgebra of $\varphi(\mathbf{Q})$ generated by $1_{\mathbf{Q}}$ is isomorphic to the free algebra $\mathbf{T} = \mathbf{F}_{\mathcal{K}_{\mathbf{Q}}}(1)$. Now we have $\mathbf{T} = \varphi(\mathbf{P})$ for some $\mathbf{P} \in \mathcal{K}$, and then we have a monomorphism $g \in \mathcal{K}(\mathbf{P}, \mathbf{Q})$ so that $\varphi(g) \in \mathcal{K}_{\mathbf{Q}}(\mathbf{T}, \varphi(\mathbf{Q}))$ is a monomorphism taking the free generator x in $\mathbf{T}$ to $1_{\mathbf{Q}}$. This means that $g \circ x = 1_{\mathbf{Q}}$. Thus g is surjective. Again by Theorem 3.2, the algebra $\mathbf{P}$ must be finitely generated; hence its surjective image $\mathbf{Q}$ is finitely generated.

Now it follows from the Theorem 5.2 that $\mathbf{Q}$ is a co-generator in $\mathcal{K}$.   •

In the following problems, $\mathcal{K}$ is assumed to be an SF-class.

Problem 1: Assuming only that $\mathcal{K}_{\mathbf{Q}}$ is closed under subalgebras, or contains its finitely generated free algebras, or both, what can be concluded about $\mathbf{Q}$? If hom$(\mathbf{Q}, -) : \mathcal{K} \to \mathcal{K}_{\mathbf{Q}}$ is a category equivalence, what does that tell us about $\mathbf{Q}$?

Problem 2: Let $\mathbf{A}$ and $\mathbf{B}$ be arbitrary algebras in $\mathcal{K}$. Under what conditions do we have $\mathcal{K}_{\mathbf{A}} \cong_c \mathcal{K}_{\mathbf{B}}$, or $\mathcal{K}_{\mathbf{A}} \equiv \mathcal{K}_{\mathbf{B}}$?

Problem 3: Suppose that $\mathbf{P}$ and $\mathbf{Q}$ are both finitely generated projective co-generators in $\mathcal{K}$. Characterize algebraically (or categorically) the relation that holds between $\mathbf{P}$ and $\mathbf{Q}$ when $\mathcal{K}_{\mathbf{P}} \equiv \mathcal{K}_{\mathbf{Q}}$.

We shall now re-prove a long list of results from the literature. In most cases, our proofs are shorter and more painless than the originals. Some of the results appear in a stronger form than was known before; and some—such as the second statement in Theorem 6.3, Corollary 6.4, Theorem 6.6, parts (2) and (3) of Theorem 6.8 and Theorem 6.9—may be new.

The first result is applicable to varieties of lattices.

Theorem 6.3 *Suppose that $\mathcal{V}$ is a variety such that $\mathbf{F}_{\mathcal{V}}(1)$ is a one-element algebra.*

(i) If $\mathcal{V} \equiv_c \mathcal{W}$ and $|\mathbf{F}_{\mathcal{W}}(1)| = 1$ then $\mathcal{V} \equiv \mathcal{W}$.

(ii) Let $\mathbf{P}$ and $\mathbf{Q}$ be finitely generated projective algebras in $\mathcal{V}$. Then $\mathcal{V}_{\mathbf{P}} \equiv \mathcal{V}_{\mathbf{Q}}$ iff there is an equivalence $\varphi : \mathcal{V} \equiv \mathcal{V}$ such that $\varphi(\mathbf{P}) = \mathbf{Q}$.

PROOF. For (i), suppose that φ is a category equivalence $\mathcal{V} \to \mathcal{W}$ where the one-generator free algebras in both varieties are one-element algebras. Then by Theorem 3.1, φ carries $\mathbf{F}_{\mathcal{V}}(1)$ to $\mathbf{F}_{\mathcal{W}}(1)$. Hence by Theorem 5.1, we have

$$\mathcal{W} \equiv \mathcal{V}^{[1]}(\sigma) \equiv \mathcal{V}^{[1]} \equiv \mathcal{V}$$

(where $\sigma(x) = x$). This proof is due to J. Jezek [20].

Now suppose that $|\mathbf{F}_{\mathcal{V}}(1)| = 1$ and $\mathbf{P}$ and $\mathbf{Q}$ are finitely generated projective algebras in $\mathcal{V}$. Trivially, they are co-generators. Assume that we have an equivalence $\varphi : \mathcal{V}_{\mathbf{P}} \equiv \mathcal{V}_{\mathbf{Q}}$. Let $\mathbf{T} = \mathbf{F}_{\mathcal{V}}(1)$. Notice that $\mathbf{T}_{\mathbf{P}}$ and $\mathbf{T}_{\mathbf{Q}}$ are one-element algebras. Thus we can assume that $\varphi(\mathbf{T}_{\mathbf{P}}) = \mathbf{T}_{\mathbf{Q}}$. Moreover, since $\mathbf{P}_{\mathbf{P}}$ is the one-generator free algebra in $\mathcal{V}_{\mathbf{P}}$ and likewise for $\mathbf{Q}_{\mathbf{Q}}$ in $\mathcal{V}_{\mathbf{Q}}$, we can assume that $\varphi(\mathbf{P}_{\mathbf{P}}) = \mathbf{Q}_{\mathbf{Q}}$. Now the inverse functor to $\mathbf{A} \mapsto \mathbf{A}_{\mathbf{P}}$ is equivalent to $\mathbf{B} \mapsto \mathbf{B}_{\mathbf{T}_{\mathbf{P}}}$, in particular, we have that

$$\mathbf{A} \equiv (\mathbf{A}_{\mathbf{P}})_{\mathbf{T}_{\mathbf{P}}}$$

for all $\mathbf{A} \in \mathcal{V}$. Likewise

$$(\mathbf{A}_{\mathbf{Q}})_{\mathbf{T}_{\mathbf{Q}}} \equiv \mathbf{A} .$$

Moreover, since $\varphi(\mathbf{T}_{\mathbf{P}}) = \mathbf{T}_{\mathbf{Q}}$, we have

$$(\mathbf{A}_{\mathbf{P}})_{\mathbf{T}_{\mathbf{P}}} \equiv \varphi(\mathbf{A}_{\mathbf{P}})_{\mathbf{T}_{\mathbf{Q}}} .$$

The composition of these three equivalences is an equivalence $\psi : \mathcal{V} \to \mathcal{V}$ such that $\varphi(\mathbf{A}_{\mathbf{P}}) = \psi(\mathbf{A})_{\mathbf{Q}}$. Moreover, the formulas

$$\psi(\mathbf{P})_{\mathbf{Q}} = \varphi(\mathbf{P}_{\mathbf{P}}) = \mathbf{Q}_{\mathbf{Q}}$$

entail that $\psi(\mathbf{P}) \cong \mathbf{Q}$.   •

The last theorem takes an especially simple form if $\mathcal{V}$ is a variety of lattices.

Corollary 6.4 *Suppose that $\mathcal{V}$ is a variety of lattices and $\mathbf{P}$ and $\mathbf{Q}$ are finitely generated projective algebras in $\mathcal{V}$. Then we have $\mathcal{V}_{\mathbf{P}} \equiv \mathcal{V}_{\mathbf{Q}}$ iff either $\mathbf{P}$ is isomorphic to $\mathbf{Q}$ or else $\mathcal{V}$ is self-dual and $\mathbf{P}$ is isomorphic to the dual of $\mathbf{Q}$.*

PROOF. It is well known (and easy to see) that in any non-identical equivalence between two varieties of lattices, the basic operations $\vee$, $\wedge$ must correspond to $\wedge$, $\vee$. Thus if $\mathcal{V}$ has a non-identical equivalence with itself, that equivalence must be obtained by switching $\vee$ and $\wedge$, and so $\mathcal{V}$ is self-dual. •

Theorem 6.4 *Let $\mathcal{W}$ be a variety in which every finitely generated projective algebra is free, such as the variety of all groups, of all Abelian groups, or of all algebras of some type. Then for any variety $\mathcal{V}$ we have that $\mathcal{W} \equiv_c \mathcal{V}$ iff $\mathcal{W}^{[n]} \equiv \mathcal{V}$ for some n.*

PROOF. Suppose that φ is a category equivalence $\mathcal{W} \to \mathcal{V}$ and $\varphi(\mathbf{P}) = \mathbf{F}_{\mathcal{V}}(1)$. Then $\mathbf{P} = \mathbf{F}_{\mathcal{W}}(n)$ for some n. By Theorem 5.2, $\mathcal{V} \equiv \mathcal{W}^{[n]}$. •

Problem 4: Characterize the varieties $\mathcal{V}$ having the property that $\mathcal{W} \equiv_c \mathcal{V}$ implies $\mathcal{W}^{[n]} \equiv \mathcal{V}$ for some n.

The next theorem is essentially due to T. K. Hu [18]. In its present form, it probably first appeared in B. A. Davey and H. Werner [9].

Theorem 6.5 *A variety is categorically equivalent with the variety of Boolean algebras iff it is generated by some primal algebra.*

PROOF. Let $\mathcal{B}$ be the variety of Boolean algebras and let $\mathbf{B}$ be the two-element Boolean algebra. $\mathbf{B}$ is a primal algebra and $\mathsf{HSP}(\mathbf{B}) = \mathcal{B}$. The argument of Example 5 will work, where $\mathbf{P}$ is any primal algebra, to show that $\mathbf{B} \equiv \mathbf{P}(\sigma)$ for some idempotent invertible term σ for $\mathbf{P}$; i.e., $\mathbf{B} \equiv_c \mathbf{P}$, implying $\mathcal{B} \equiv_c \mathsf{HSP}(\mathbf{P})$. Conversely, suppose that $\mathcal{B} \equiv_c \mathcal{V}$, so that we have $\mathcal{V} \equiv \mathcal{B}^{[n]}(\bar{\sigma})$. Since $\mathbf{B}$ is primal, so is $\mathbf{B}^{[n]}$. It is easy to see that the functor $\mathbf{A} \mapsto \mathbf{A}(\bar{\sigma})$ preserves primality. (See Example 5.) Thus $\mathbf{B}^{[n]}(\bar{\sigma})$ is primal. Since this latter algebra generates the variety $\mathcal{B}^{[n]}(\bar{\sigma})$, we are done. •

Let $\mathbf{X} = \langle X, \leq \rangle$ be an ordered set (i.e., $\leq$ is a partial ordering of X), and denote by $\mathbf{A}(\mathbf{X})$ an algebra whose clone of term operations is the category of all monotone (with respect to $\leq$) mappings between finite direct powers of $\mathbf{X}$. Write $\mathcal{V}(\mathbf{X})$ for the variety generated by $\mathbf{A}(\mathbf{X})$ (whose clone is this same clone of all monotone operations on $\mathbf{X}$). The next theorem was proved for finite ordered sets in [7].

Theorem 6.6 *Let $\mathbf{X}$ be any ordered set and $\mathcal{W}$ be any variety categorically equivalent to $\mathcal{V}(\mathbf{X})$. Then $\mathcal{W} \equiv \mathcal{V}(\mathbf{Y})$ for some ordered set $\mathbf{Y}$. For any ordered sets $\mathbf{Y}$ and $\mathbf{Z}$ we have that $\mathcal{V}(\mathbf{Y}) \equiv_c \mathcal{V}(\mathbf{Z})$ iff each of $\mathbf{Y}$ and $\mathbf{Z}$ is order-isomorphic or dual order-isomorphic to an order-retract of a finite direct power of the other.*

PROOF. Let $\mathcal{W} \equiv_c \mathcal{V}(\mathbf{X})$, say $\mathcal{W} = \mathcal{V}(\mathbf{X})^{[n]}(\bar{\sigma})$. As in the preceding proof, we have that $\mathcal{W}$ is generated by the algebra $\mathbf{A}(\mathbf{X})^{[n]}(\bar{\sigma})$. It is trivial to see that the clone of this algebra consists of all the monotone operations over the ordered set $\bar{\sigma}(\mathbf{X}^n)$ (whose order is inherited from $\mathbf{X}^n$). Thus we can take $\mathbf{Y} = \bar{\sigma}(\mathbf{X}^n)$ and the first statement in the theorem is proved.

Now suppose that we have $\mathcal{V}(\mathbf{Y}) \equiv_c \mathcal{V}(\mathbf{Z})$ where $\mathbf{Y}$ and $\mathbf{Z}$ are ordered sets. Note that in $\mathcal{V}(\mathbf{Z})$ the algebra $\mathbf{A}(\mathbf{Z})$ is, up to isomorphism, the only algebra which generates the variety and has no proper subalgebras; moreover, this is a categorical property. Thus it must be the case that $\mathbf{A}(\mathbf{Z}) \equiv \mathbf{A}(\mathbf{Y})^{[n]}(\bar{\sigma})$, i.e., $\mathbf{A}(\mathbf{Z}) \equiv \mathbf{A}(\bar{\sigma}(\mathbf{Y}^n))$. This amounts to the existence of a bijection μ between the ordered sets $\sigma(\mathbf{Y}^n)$ and $\mathbf{Z}$ which converts the monotone clone on the first ordered set into the monotone clone on the second set. This implies that μ is either an isomorphism or a dual-isomorphism between ordered sets. Thus $\mathbf{Z}$ is isomorphic or dually-isomorphic to $\sigma(\mathbf{Y}^n)$, which is a retract of $\mathbf{Y}^n$. The same argument gives that $\mathbf{Y}$ is isomorphic or dually isomorphic to a retract of some finite direct power of $\mathbf{Z}$.

Conversely, suppose that each of $\mathbf{Y}$ and $\mathbf{Z}$ is isomorphic or dually isomorphic to a retract of a finite power of the other. Since $\mathcal{V}(\mathbf{Y}) = \mathcal{V}(\mathbf{Y}^\partial)$, we can assume that $\mathbf{Y} \subseteq \mathbf{Z}^n$ and there is a monotone map φ from $\mathbf{Z}^n$ onto $\mathbf{Y}$ with $\varphi^2 = \varphi$. Moreover, we have an order-isomorphism or dual order-isomorphism γ from $\mathbf{Z}$ onto an ordered subset $\mathbf{X}$ of $\mathbf{Y}^m$ and a monotone retraction π of $\mathbf{Y}^m$ onto $\mathbf{X}$, $\pi^2 = \pi$.

First we show that we can assume that γ is an order-isomorphism. Suppose that it is a dual order-isomorphism. Then take γ' to be the product of $\gamma : \mathbf{Z} \to \mathbf{X} \subseteq \mathbf{Z}^{nm}$ and $\gamma^{(nm)} : \mathbf{Z}^{nm} \to \mathbf{Y}^{mnm}$. Here γ' is an order-isomorphism of $\mathbf{Z}$ with a subset $\mathbf{X}'$ of $\mathbf{Y}^{mnm}$. We get an order-retraction π' of $\mathbf{Y}^{mnm}$ onto $\mathbf{X}'$ by composing first, $\pi^{(nm)} : \mathbf{Y}^{mnm} \to \mathbf{X}^{nm}$ (surjective), then $(\gamma^{(nm)})^{-1} : \mathbf{X}^{nm} \to \mathbf{Z}^{nm}$ (dual-isomorphism), then $\varphi^{(m)} : \mathbf{Z}^{nm} \to \mathbf{Y}^m$ (surjective), then $\pi : \mathbf{Y}^m \to \mathbf{X}$ (surjective), and finally $\gamma^{(nm)} : \mathbf{X} \to \mathbf{X}'$.

Thus we now assume that γ is an order-isomorphism. Now one can verify that φ is an idempotent unary term operation for the algebra $\mathbf{A}(\mathbf{Z})^{[n]}$ and γ and π provide the terms to make it invertible. We have, of course, that $\mathbf{A}(\mathbf{Y}) \equiv \mathbf{A}(\mathbf{Z})^{[n]}(\varphi)$ and so $\mathcal{V}(\mathbf{Y})$ is equivalent to $\mathcal{V}(\mathbf{Z})^{[n]}(\varphi)$ and category equivalent to $\mathcal{V}(\mathbf{Z})$. $\bullet$

Corollary 6.5 *([8], [31]) Let $\mathcal{D}$ be the variety of distributive lattices with 0 and 1. Then $\mathcal{D}$ is equivalent to $\mathcal{V}(\mathbf{Z})$ where $\mathbf{Z}$ is the two-element chain. A variety $\mathcal{V}$ is categorically equivalent to $\mathcal{D}$ iff $\mathcal{V}$ is equivalent to $\mathcal{V}(\mathbf{X})$ for some finite lattice-ordered set $\mathbf{X}$.*

PROOF. It is well-known that $\mathcal{D}$ and $\mathcal{V}(\mathbf{Z})$ are equivalent. Also, it is fairly easy to check that every finite lattice-ordered set $\mathbf{X}$ of more than one element is a retract of a finite power $\mathbf{Z}^n$, and of course $\mathbf{Z}$ is a retract of $\mathbf{X}$. So this corollary follows from the preceding theorem.   •

Theorem 6.7 *A variety $\mathcal{V}$ is categorically equivalent to $_\mathbf{R}\mathcal{M}$ (unitary modules over the ring $\mathbf{R}$ with unit) iff $\mathcal{V} \equiv {}_\mathbf{S}\mathcal{M}$ for some ring $\mathbf{S}$ that is Morita-equivalent to $\mathbf{R}$.*

PROOF. Suppose that $\mathcal{V} \equiv_c {}_\mathbf{R}\mathcal{M}$, i.e., we have $\mathcal{V} \equiv {}_\mathbf{R}\mathcal{M}^{[n]}(\bar{\sigma})$. Now it is well-known that $_\mathbf{R}\mathcal{M}^{[n]} \equiv {}_\mathbf{T}\mathcal{M}$ where $\mathbf{T}$ is the ring of n-by-n matrices over $\mathbf{R}$. Then $\bar{\sigma}$ is just left scalar multiplication by an idempotent element $e = e^2 \in T$ such that there are elements $u_j, v_j \in T$ for $j < k$ with $\mathbf{T} \models 1 = \sum_{0 \leq j < k} u_i e v_i$ (this is invertibility). Clearly, $_\mathbf{T}\mathcal{M}(\bar{\sigma})$ is equivalent to $_\mathbf{S}\mathcal{M}$ where $\mathbf{S}$ is the subring eTe of $\mathbf{T}$ with unit element e.   •

Theorem 6.8 *Suppose that $\mathcal{V} \equiv_c \mathcal{W}$, where $\mathcal{V}$ and $\mathcal{W}$ are varieties. For each of these properties (P) we have $\mathcal{V} \models (P)$ iff $\mathcal{W} \models (P)$:*

(1) finitely generated;

(2) equivalent to a variety of finite type;

(3) equivalent to a finitely presented variety;

(4) residually small;

(5) residually $< \lambda$ (λ any infinite cardinal);

(6) locally finite;

(7) has type set equal to J (a subset of $\{\mathbf{1}, \mathbf{2}, \mathbf{3}, \mathbf{4}, \mathbf{5}\}$).

Remark. *Type set* in the statement of property (7) is a concept of tame congruence theory; see [17]. A special case of (7) was proved in [11].

PROOF. We assume that $\varphi : \mathcal{V} \equiv_c \mathcal{W}$, and also that $\mathcal{W} \equiv \mathcal{V}^{[n]}(\bar{\sigma})$. By Theorem 3.1, $\mathbf{A}$ is of cardinality $\leq \lambda$ (or $< \lambda$, or finitely generated) iff $\varphi(\mathbf{A})$ is. Also, $\mathbf{A}$ is subdirectly irreducible iff $\varphi(\mathbf{A})$ is subdirectly irreducible. Furthermore, a class $\mathcal{K} \subseteq \mathcal{V}$ generates $\mathcal{V}$ iff $\varphi(\mathcal{K})$ generates $\mathcal{W}$. From these facts, it follows that $\mathcal{V} \models (P)$ iff $\mathcal{W} \models (P)$ for each of the properties (P) = (1), (4), (5), (6).

For (2) and (3), it is well-known (and easy to prove) that $\mathcal{V}^{[n]}$ is a variety of finite type, or equivalent to a finitely presented variety, iff $\mathcal{V}$ is. Hence what remains to be shown here is that $\mathcal{V}(\sigma)$ is equivalent to a variety of finite type (or a finitely presentable variety) when $\mathcal{V}$ is of finite type (or finitely presentable), assuming that σ is an invertible idempotent term for $\mathcal{V}$.

Suppose first that $\mathcal{V}$ is of finite type and let $f_0, \ldots, f_{k-1}$ be the operation symbols of $\mathcal{V}$. Let $t(\bar{x}), t_0(x), \ldots, t_{m-1}(x)$ be the invertibility terms for σ. What must be shown is that the clone of $\mathcal{V}(\sigma)$ is finitely generated, i.e., that there is some finite set of terms of $\mathcal{V}(\sigma)$ such that, modulo the equations valid in $\mathcal{V}(\sigma)$, every one of its terms s_σ is equal to one built via composition from terms in this fixed set. Well, for $i < k$ and $j < m$, where f_i is n_i-ary, define

$$g_i = \sigma f_i(t(\bar{x}^0), t(\bar{x}^1), \ldots, t(\bar{x}^{n_i-1}));$$

$$h_{i,j} = \sigma t_j f_i(t(\bar{x}^0), t(\bar{x}^1), \ldots, t(\bar{x}^{n_i-1}));$$

two $n_i m$-ary terms of $\mathcal{V}(\sigma)$. We claim that the clone of $\mathcal{V}(\sigma)$ is generated by the set

$$\Gamma = \{g_i, h_{i,j}, \sigma t_j : i < k, j < m\}.$$

The proof is by induction on terms of $\mathcal{V}$ to show that for every such term s, there are terms μ and ν_j constructed by composing terms in Γ so that $\mathcal{V}(\sigma) \models s_\sigma \approx \mu$ and for all $j < n$, $\mathcal{V}(\sigma) \models (t_j s)_\sigma \approx \nu_j$. This proof is straightforward and we leave it to the reader.

Now suppose that $\mathcal{V}$ has operation symbols as above and is finitely presentable, say, is defined by the finite set Σ of equations. First observe that $\mathcal{V} \equiv \mathcal{V}(\sigma)^{[m]}(\bar{\tau})$ where $\bar{\tau}$, an invertible idempotent term for $\mathcal{V}(\sigma)^{[m]}$, is $\langle \sigma t_0 t(\bar{x}), \ldots, \sigma t_{m-1} t(\bar{x}) \rangle$. In fact, if $\mathbf{A} \in \mathcal{V}$ then the correspondence $\varphi : x \mapsto \langle \sigma t_0(x), \ldots, \sigma t_{m-1}(x) \rangle$ is an isomorphism between $\mathbf{A}$ and an algebra $\langle \bar{\tau}(\sigma(A^m)), F_0, \ldots, F_{k-1} \rangle = \mathbf{A}'$ in which

$$F_i(\bar{\tau}(\bar{x}^0), \ldots, \bar{\tau}(\bar{x}^{n_i-1})) = \langle \sigma t_0(s), \ldots, \sigma t_{m-1}(s) \rangle$$

with

$$s = f_i(t(\bar{\tau}(\bar{x}^0)), \ldots, t(\bar{\tau}(\bar{x}^{n_i-1}))).$$

The term operation induced by σ in $\mathbf{A}'$ can be expressed as

$$\sigma^{\mathbf{A}'}(\bar{x}) = \langle \sigma t_0 \sigma t(\bar{x}), \ldots, \sigma t_{m-1} \sigma t(\bar{x}) \rangle.$$

$\mathbf{A}(\sigma)$ is isomorphic to $\mathbf{A}'(\sigma)$ via $\varphi|_{\mathbf{A}(\sigma)}$ and the inverse of this map is also given by a term of $\mathbf{A}(\sigma)$, namely we have $\mathbf{A} \models \sigma(x) = \sigma t(\varphi(\sigma(x)))$, or

$$\mathbf{A}(\sigma) \models x = t_\sigma(t_{0\sigma}(x), \ldots, t_{m-1\sigma}(x)).$$

Now for $\mathbf{B} \in \mathcal{V}(\sigma)$, the fact that $\mathbf{B}^{[m]}(\bar{\tau}) \models \Sigma$ can be written as a finite set Σ' of equations in a set G of operation symbols corresponding to the finite set Γ of terms selected in the previous paragraph, which generate the clone of $\mathcal{V}(\sigma)$. Then Σ' together with the above-displayed equation, the equations in G expressing the next displayed equation above, and the equations in G which express that the bijection set up above is an isomorphism between $\mathbf{B}$ and $\mathbf{B}^{[m]}(\bar{\tau})(\sigma)$, constitutes a presentation Σ'' for the variety $\mathcal{V}(\sigma)$.

The assertion that isomorphism of varieties preserves property (7) is an unpublished result of the author. Since the proof requires the machinery of tame congruence theory, we omit it.   •

Theorem 6.9 *Suppose that $\mathcal{V}$ and $\mathcal{W}$ are locally finite varieties and let $\mathcal{V}_{\mathrm{fin}}$ and $\mathcal{W}_{\mathrm{fin}}$ denote the class of finite algebras in the respective varieties. Then $\mathcal{V} \equiv_c \mathcal{W}$ iff $\mathcal{V}_{\mathrm{fin}} \equiv_c \mathcal{W}_{\mathrm{fin}}$. Any categorical equivalence between $\mathcal{V}_{\mathrm{fin}}$ and $\mathcal{W}_{\mathrm{fin}}$ extends to a categorical equivalence between $\mathcal{V}$ and $\mathcal{W}$.*

PROOF. If $\varphi : \mathcal{V} \equiv_c \mathcal{W}$ then it follows from Theorem 3.1 that the restriction of φ to $\mathcal{V}_{\mathrm{fin}}$ is a categorical equivalence between $\mathcal{V}_{\mathrm{fin}}$ and $\mathcal{W}_{\mathrm{fin}}$. On the other hand, if $\varphi : \mathcal{V}_{\mathrm{fin}} \equiv_c \mathcal{W}_{\mathrm{fin}}$ then (since both classes contain their finitely generated free algebras) it follows from Theorem 6.1 that $\mathcal{W}_{\mathrm{fin}} \equiv \mathcal{V}_{\mathrm{fin}}^{[n]}(\bar{\sigma})$, that φ is equivalent to the functor $\mathbf{A} \mapsto \mathbf{A}^{[n]}(\bar{\sigma})$ on $\mathcal{V}_{\mathrm{fin}}$, and that φ extends to a category equivalence between $\mathcal{V}$ and $\mathcal{W}$.   •

Let $\mathcal{M}$ be any variety. We say that a variety $\mathcal{V}$ satisfies the *Mal'cev condition* presented by $\mathcal{M}$ iff there exists a mapping I from operation symbols of $\mathcal{M}$ to terms of $\mathcal{V}$ such that every algebra $\mathbf{A} \in \mathcal{V}$ gives rise to an algebra $\mathbf{A}^I \in \mathcal{M}$ by interpreting each operation symbol f of $\mathcal{M}$ as the term operation defined on $\mathbf{A}$ by the term $I(f)$. (An equivalent condition is that there exists some homomorphism from $\mathrm{Th}(\mathcal{M})$ to $\mathrm{Th}(\mathcal{V})$ in the category of algebraic theories.)

The Mal'cev condition presented by $\mathcal{M}$ is said to be *linear* if $\mathcal{M}$ is the class of all algebras obeying some set of equations each member of which is *linear*, i.e., has the form $\sigma \approx \tau$ where each of σ and τ either is a variable, or is a basic operation symbol applied to variables. B. A. Davey, H. Werner [9] proved the following result, and asked whether an elementary algebraic proof could be found. Does the proof below qualify under those criteria?

Theorem 6.10 *If $\mathcal{V} \equiv_c \mathcal{W}$ then any linear Mal'cev condition satisfied by $\mathcal{V}$ is satisfied by $\mathcal{W}$ and vice versa.*

PROOF. $\mathcal{V}^{[n]}$ satisfies any Mal'cev condition satisfied by $\mathcal{V}$, because the clone of $\mathcal{V}^{[n]}$ contains an isomorphic copy of the clone of $\mathcal{V}$. $\mathcal{V}(\sigma)$ satisfies any linear Mal'cev condition satisfied by $\mathcal{V}$. Indeed, if I interprets $\mathcal{M}$ in $\mathcal{V}$, just let $I'(f(\bar{x})) = I(f(\bar{x}))_\sigma$ for operation symbols f. Then trivially, for $\mathbf{A} \in \mathcal{V}$, $\mathbf{A}(\sigma)^{I'}$ satisfies any linear equation satisfied by $\mathbf{A}^I$.   •

By a *pre-primal* algebra we mean a finite algebra of more than one element whose clone of term operations is a maximal member of the lattice of all clones on the universe of the algebra. I. Rosenberg [33] has classified pre-primal algebras into six types. By a *pre-primal variety* we mean any variety that is generated by some pre-primal algebra. The next theorem is an unpublished is due to K. Denecke and O. Lüders [12].

Theorem 6.11 *Suppose that* V *and* W *are categorically equivalent varieties. Then* V *is pre-primal iff* W *is, and if pre-primal, then the Rosenberg types of* V *and* W *are the same.*   •

References

[1] B. Banaschewski, *Functors into categories of m-sets*, Abhandlungen aus dem Math. Sem. der Universität Hamburg **38** (1972), 49–64.

[2] ———, *On the category invariance of cardinalities in a prevariety*, Algebra Universalis **14** (1982), 267–268.

[3] ———, *On categories of algebras equivalent to a variety*, Algebra Universalis **16** (1983), 264–267.

[4] ———, *More on compact hausdorff spaces and finitary duality*, Canad. J. Math. **36** (1984), 1113–1118.

[5] C. Bergman and J. Berman, *Morita equivalence of almost-primal clones*, (preprint, 1994).

[6] F. Borceaux and E. Vitale, *A morita theorem for algebraic theories*, (preprint).

[7] B. A. Davey, R. W. Quackenbush, and D. Schweigert, *Monotone clones and the varieties they determine*, Order **7** (1990), 145–167.

[8] B. A. Davey and I. Rival, *Exponents of lattice-ordered algebras*, Algebra Universalis **14** (1982), 87–98.

[9] B. A. Davey and H. Werner, *Dualities and equivalences for varieties of algebras*, "Contributions to Lattice Theory" (Proc. Conf. Szeged, 1980), Coll. Math. Soc. János Bolyai, no. 33, North-Holland, pp. 101–275.

[10] K. Denecke, *Categorical equivalences of varieties generated by algebras* **A** *and* **A**′, Mathematical Problems in Computation Theory, Banach Center Publ. Vol. 21, Polish Scientific Publishers, Warsaw, 1988.

[11] ———, *Minimal algebras and category equivalence*, Beiträge Algebra Geom. **31** (1991), 121–141.

[12] K. Denecke and O. Lüders, *Category equivalences of clones*, preprint, October 1992.

[13] B. Elkins and J. Zilber, *Categories of actions and morita equivalences*, Rocky Mountain J. Math. **6** (1976), 199–225.

[14] W. Felscher, *Equational maps*, "Contributions to Mathematical Logic", North Holland, 1968, pp. 121–161.

[15] P. Freyd, *Algebra valued functors in general and tensor products in particular*, Colloq. Math. **14** (1966), 89–106.

[16] G. Gierz, *Morita equivalence of quasiprimal algebras and sheaves*, preprint 1994.

[17] D. Hobby and R. McKenzie, *The Structure of Finite Algebras*, Contemporary Mathematics, American Mathematics Society, Providence, Rhode Island, 1988.

[18] T. K. Hu, *Stone duality for primal algebra theory*, Math. Z. **110** (1969), 180–198.

[19] J. R. Isbell, *Subobjects, adequacy, completeness and categories of algebras*, Rozprawy Matematyczne **36** (1964), 3–33.

[20] J. Jezek, *A note on isomorphic varieties*, Comment. Math. Univ. Carolin. **23** (1982), 579–588.

[21] K. Keimel and H. Werner, *Stone duality for varieties generated by quasiprimal algebras*, Mem. Amer. Math. Soc. **148** (1974), 59–85.

[22] A. Knoebel, *Free algebras – are they categorically definable*, 1983, pp. 333–341.

[23] S. Mac Lane, *Categories for the Working Mathematician*, Springer-Verlag, 1971.

[24] F. W. Lawvere, *Functorial semantics of algebraic theories*, Proc. Nat. Acad. Sci. U.S.A. **50** (1963), 869–872.

[25] _____, *Some algebraic aspects in the context of functorial semantics of algebraic theories*, Reports of the Midwest Category Seminar II (S. Mac Lane, ed.), Springer Lecture Notes **61**, Springer Verlag, 1968.

[26] H. Lindner, *Morita equivalence of enriched categories*, Cahiers Topologie Géom. Différentielle **15** (1974).

[27] F. E. J. Linton, *Some aspects of equational categories*, Proc. Conf. Categorical Algebra, La Jolla, 1965, Springer-Verlag, 1966, pp. 84–94.

[28] _____, *Applied functorial semantics, I*, 1970, pp. 1–14.

[29] R. McKenzie, G. McNulty, and W. Taylor, *Algebras, Lattices, Varieties*, Wadsworth/Brooks Cole, Monterrey, 1987.

[30] W. D. Neumann, *Representing varieties of algebras by algebras*, J. Austral. Math. Soc. Ser. A **11** (1970), 1–8.

[31] R. W. Quackenbush, *Pseudovarieties of finite algebras isomorphic to bounded distributive lattices*, Discrete Math. **28** (1979), 189–192.

[32] R.W. Quackenbush, *A new proof of rosenberg's primal algebra characterization theorem*, Finite Algebra and Multiple-Valued Logic, Colloq. Math. soc. János Bolyai, Szeged, 1979, pp. 603–634.

[33] I. Rosenberg, *Über die funktionale vollständigkeit in den mehrwertigen logiken*, Rozpravy Československé Akad. Věd Řada Mat. Přírod. Věd **80** (1970), 3–93.

[34] K. Sokolnicki, *Characterizations of certain algebraic theories*, Bull. Acad. Pol. Sci. Ser. Sci. Math. **27** (1979), 247–254.

[35] W. Taylor, *Characterizing Malcev conditions*, Algebra Universalis **3** (1973), 351–397.

[36] ———, *The fine spectrum of a variety*, Algebra Universalis **5** (1975), 263–303.

[37] ———, *Equational logic*, "Contributions to Universal Algebra, Proc. Szeged 1975", Colloq. Math. Soc. János Bolyai **17**, 1977, pp. 465–501.

[38] G. C. Wraith, *Algebraic Theories*, Aarhus University Lecture Note Series 22, (revised edition, 1975).

DEPARTMENT OF MATHEMATICS, VANDERBILT UNIVERSITY, NASHVILLE, TN 37212
e-mail address: mckenzie@math.vanderbilt.edu

Boolean Universal Algebra

Alden F. Pixley

Abstract

We survey the area of universal algebra which, roughly speaking, studies varieties of algebras which share some important properties with the variety of Boolean algebras. Special attention is devoted to the theory of affine complete varieties. An algebra **A** is affine complete if each congruence compatible function on A is a polynomial of **A**. A variety is affine complete if each of its members is affine complete; the variety of Boolean algebras is the most important example. All known affine complete varieties of finite type are finitely generated and we discuss some of the properties these share with the variety of Boolean algebras.

1 Introduction: Boolean algebras and primality

In order to motivate our discussion we begin by recalling several properties of Boolean algebras. Though we will often introduce notation and terminology as we go along, as much as possible we shall follow the book of McKenzie, McNulty, and Taylor, [14]. The properties of Boolean algebras we focus on are the following.

> Representation. The variety $\mathcal{B}$ of Boolean algebras is the variety generated by $\mathbf{B}_2$, the two element Boolean algebra and, in fact consists of—up to isomorphism—all subdirect powers of $\mathbf{B}_2$, that is $\mathcal{B} = IP_S(\mathbf{B}_2)$.

This fact is due to Marshall Stone [18]. Now every variety V is generated by a single algebra (the free algebra $\mathbf{F}$ in V on ω free generators), so Stone's result is interesting in this respect since it tells us that B is *finitely* generated. But also the variety $V(\mathbf{F})$, generated by $\mathbf{F}$, in general has the form $V = HSP(\mathbf{F})$ by one of Birkhoff's basic theorems on varieties. In general one of the most

challenging problems in studying a variety is to understand the role played by
the H operator, i.e.: what is the nature of homomorphisms of subalgebras of
products. The Stone representation $IP_S(\mathbf{B}_2)$ is a simpler *coordinate* representation which is much more appealing and transparent because of the absence
of the H operator. The other basic theorem of Birkhoff asserts that a variety
is generated by its subdirectly irreducible members: each algebra is a subdirect product of subdirectly irreducible members of the variety. Thus Stone's
theorem asserts that $\mathbf{B}_2$ is the only subdirectly irreducible Boolean algebra.
This is in sharp contrast to the general situation: a finitely generated variety
can have arbitrarily many nonisomorphic subdirectly irreducible members of
arbitrary size.

> Duality. Also due to Marshall Stone [18], the Stone duality theorem
> asserts, in the language of category theory, that as a category, the
> variety of Boolean algebras is equivalent to the dual of the category
> of Boolean spaces.

Any variety is a category where the morphisms are just the homomorphisms
between algebras in the variety. The significance of Stone duality is that
algebraic problems about Boolean algebra can be translated into (topological)
problems about continuous functions, and visa versa.

> Primality. The two element Boolean algebra, $\mathbf{B}_2$ is *primal*, meaning
> that every operation on the universe B of $\mathbf{B}_2$ is a term function of
> $\mathbf{B}_2$.

This fact is often called the Fundamental theorem of switching circuit theory and has been the starting point for much work in logic and computer
science. Claude Shannon's work in the 1930's capitalized on this fact.

> Functional and affine completeness. The two element Boolean al-
> gebra is also *functionally complete*, meaning that every operation
> on B is a polynomial function of $\mathbf{B}_2$. The variety $\mathcal{B}$ is *affine com-
> plete*, meaning that for each Boolean algebra the polynomials are
> precisely the operations which are compatible with the congruences
> of the algebra.

The functional completeness of $\mathbf{B}_2$ is trivial, since the polynomial functions
are just those functions which are obtained by inserting constants in some of
the argument places of a term. Notice that an algebra is functionally complete
iff it is both simple and affine complete. Hence the affine completeness of the
entire variety of Boolean algebras generalizes the functional completeness of
the generator. This fact was discovered in 1962 by G. Grätzer [5].

[The terminology *affine complete* is due to Wille [19] and arose in his study of *congruence class geometries*, geometries which are constructed by taking the congruence classes of an algebra as points. In this setting the dilations of the geometry are just the congruence compatible functions of the original algebra. This geometric source plays no apparent role in any of the current work on affine complete algebras — which is perhaps an oversight. On the other hand current usage in universal algebras gives a totally unrelated meaning to *affine algebras* and the literature on affine algebras is extensive. Thus a new name for affine completeness may be in order. See Section 2 for a proposal.]

The first step in the development of Boolean universal algebra came with the recognition that each of the properties of Boolean algebras listed above depends really only upon the primality of the two element generator of the variety. Thus we have the following properties which parallel those above:

Representation. For any primal algebra $\mathbf{P}$, i.e.: finite algebra $\mathbf{P} = (P, F)$ with the property that all operations on P are term functions of $\mathbf{P}$, the variety $V(\mathbf{P})$ generated by $\mathbf{P}$ has the representation $V(\mathbf{P}) = IP_S(\mathbf{P})$, the same attractive coordinate representation as in the Boolean case.

This fact seems to have been first discovered by L. I. Wade in 1945 but was reproved, exploited, and generalized by A. L. Foster in a series of papers beginning in the 1950's.

Duality. In 1969 T. K. Hu [6] directly generalized the Stone duality and in fact showed that for any primal algebra $\mathbf{P}$, the variety $V(\mathbf{P})$ is equivalent as a category to the dual of the category of Boolean spaces.

Thus any two primal algebras generate varieties which are category equivalent to each other. Even more, we have that if $\mathbf{P}$ is any primal algebra then a variety W is category equivalent to the variety $V(\mathbf{P})$ iff W is the variety generated by some primal algebra. Put another way, the class of all algebras generating varieties category equivalent to any primal variety is just the class of all primal algebras.

Functional and affine completeness. If $\mathbf{P}$ is any primal algebra then of course it is *a fortiori* functionally complete, as in the Boolean case, but also Hu [7] proved in 1971 that for any primal algebra $\mathbf{P}$ the variety $V(\mathbf{P})$ is affine complete.

Hu established this result, (generalizing Grätzer's result for Boolean algebras) as an application of the generalized Stone duality: that is by describing the congruence compatible operations on an algebra $\mathbf{A} \in V(\mathbf{P})$ in terms of continuous functions on the corresponding Stone space of $\mathbf{A}$.

What is behind all of this? The answer was discovered in the early 1960s as an application of the theory of Mal'cev conditions. The first part of the answer concerns congruence permutability.

In general the join of two congruence relations of an algebra is the union of all their finite relation products. If two congruences commute (permute) under relation product, then their join is just their product. Mal'cev's theorem asserts that a variety V is congruence permutable (all pairs of congruences of members of V permute) iff V has a *Mal'cev term*, a term $p(x, y, z)$ of V such that

$$p(x, x, z) = z, \quad p(x, z, z) = x$$

are equations of V. (Thus congruence permutability was the first Mal'cev condition.)

If $\mathbf{A}_1$ and $\mathbf{A}_2$ are algebras then the subalgebras of their direct product can be quite arbitrary. In the case of congruence permutability the possibilities are considerably restricted. For $i = 1, 2$ we let $\pi_i : \mathbf{A}_1 \times \mathbf{A}_2 \to \mathbf{A}_i$ denote the projection mapping the pair (x_1, x_2) to x_i. Suppose $\mathbf{S}$ is a subalgebra of $\mathbf{A}_1 \times \mathbf{A}_2$ and let $S_i = \pi_i(S)$. For congruences $\phi_i \in \mathrm{Con}\mathbf{S}_i$ suppose

$$\sigma : \mathbf{S}_1/\phi_1 \to \mathbf{S}_2/\phi_2$$

is an isomorphism. Then we always have that

$$S < \{(a_1, a_2) \in S_1 \times S_2 : \sigma(a_1/\phi_1) = a_2/\phi_2\}.$$

If equality holds above we will say that $\mathbf{S}$ is a *rectangular* subalgebra of the product, since it is the union of the blocks $a_1/\phi_1 \times a_2/\phi_2$. It is an easy exercise to show that another characterization is that a subalgebra $\mathbf{S}$ is rectangular iff it has the property that for all pairs $(x, y), (x, v), (u, v), (u, y)$ in $\mathbf{A}_1 \times \mathbf{A}_2$,

$$(x, y), (x, v), (u, v) \in S \Rightarrow (u, y) \in S,$$

which means that if three vertices of a rectangle are in S then so is the fourth. This gives another easy way to picture a rectangular subalgebra. Notice that a congruence relation of an algebra $\mathbf{A}$ is just the universe of a *diagonal* rectangular subdirect product in $\mathbf{A} \times \mathbf{A}$. (Diagonal, means that the subdirect product contains $\{(x, x) : x \in A\}$.) The following result is due essentially to I. Fleischer [3] and explains the significance of congruence permutability for our purposes.

Theorem 1.1 *A variety is congruence permutable iff for all pairs of algebras* **A** *and* **B** *in* V, *every subalgebra of* $\mathbf{A} \times \mathbf{B}$ *is rectangular.*

As a simple application of this theorem observe that if both **A** and **B** are simple, then any irredundant subdirect product of them is their full direct product. This generalizes easily to more than two factors: any irredundant subdirect product of simple algebras in a congruence permutable variety is the full direct product. Now in the variety of Boolean algebras, if $\vee$ and $\wedge$ are lattice join and meet and $'$ is complementation, the term

$$p(x, y, z) = (x \vee y') \wedge (y' \vee z) \wedge (x \vee z)$$

is a Mal'cev term so this explains why every finite Boolean algebra is a finite direct power of $\mathbf{B}_2$.

The second part of the answer to what underlies the basic properties of primal algebras concerns congruence distributivity. By an important theorem of B. Jónsson [8] a variety is congruence distributive iff for some integer $n \geq 2$ there exist ternary terms $t_0, \ldots, t_n$ such that a certain set Δ_n of identities in the t_i holds in V. The terms t_0 and t_n are just the first and third projections respectively, and in case $n = 2$ the condition Δ_2 is just that there exists a single ternary majority term, i.e.: a term $m(x, y, z)$ such that

$$m(x, x, z) = x, \quad m(x, y, x) = x, \quad m(x, y, y) = y$$

are identities of V. Moreover for each n, Δ_{n+1} are deducible from Δ_n, so satisfying Δ_2 is the strongest way in which a variety can be congruence distributive. The most important example occurs in the variety of lattices which derives its congruence distributivity from the fact that the so called *median* term

$$m(x, y, z) = (x \vee y) \wedge (y \vee z) \wedge (x \vee z),$$

is a majority term for the variety. Notice that for Boolean algebras the Mal'cev term $p(x, y, z) = m(x, y', z)$.

Another reason Δ_2 is so important is that in a congruence permutable variety, i.e.: in the presence of a Mal'cev term, all of Δ_n collapse to Δ_2. In this case the variety is *arithmetical*, i.e.: is both congruence permutable and congruence distributive. Arithmeticity is thus a Mal'cev condition which can thus be defined by the Mal'cev identities together with Δ_2. Arithmeticity can also be defined by a single ternary term:

Theorem 1.2 ([16]) *A variety* V *is arithmetical iff there is a ternary term* $t(x, y, z)$ *such that*

$$t(x, x, y) = y, \quad t(x, y, x) = x, \quad t(x, y, y) = x$$

are identities of V.

The significance of congruence distributivity is first, that in a congruence distributive variety the congruences of a subdirect product of finitely many algebras are simply products of congruences on the factors (hence there are no *skew congruences*). An early result of the author, (appearing in [4], proof of Theorem 5.4), uses this observation to show that if the finite algebra **A** generates a congruence distributive variety, V then

$$V = IP_S HS(\mathbf{A}).$$

(A second fundamental theorem of Jónsson [8] generalizes this representation of V to the case where it is congruence distributive and generated by an arbitrary class of algebras.)

In the case of a primal algebra **P** we can find many terms $t(x, y, z)$ so that the identities of Theorem 1.2 hold in **P**. Since **P** has no proper subalgebras and is simple we obtain a direct proof of the representation of the variety generated by a primal algebra. One important specific way of choosing the term $t(x, y, z)$ to satisfy the identities defining arithmeticity is to choose t to be the ternary *discriminator* function on P:

$$t(x, y, z) = \begin{cases} z & \text{if } x = y, \\ x & \text{otherwise.} \end{cases}$$

Notice that the Mal'cev term $p(x, y, z)$ defined above for Boolean algebras is the discriminator on $\mathbf{B}_2$. Notice also that if t is the discriminator on a set then $m(x, y, z) = t(x, t(x, y, z), z)$ is a majority function on the set.

Using the results summarized so far we can obtain the following important characterization of primal algebras. (But also see the discussion following Theorem 1.5 below for a simple proof.)

Theorem 1.3 ([16]) *A finite algebra* **A** *is primal iff*
   a) **A** *has no proper subalgebras,*
   b) **A** *is rigid (i.e.: has only the trivial automorphism),*
   c) **A** *is simple, and*
   d) $V(\mathbf{A})$ *is arithmetical, i.e.:* **A** *has a term satisfying the identities of Theorem 1.2.*
Conditions c) and d) can be replaced by the single condition
   e) The discriminator is a term function of **A**.

Prior to this characterization theorem primality did not seem to be interesting from a purely algebraic point of view since the definition entailed none of the usual associated algebraic structures (subalgebras, automorphism group, congruence lattice) normally considered in algebra. Theorem 1.3 places primality at the center of primary algebraic considerations. Also, as we shall see, it suggests generalizations of primality.

It is significant that the ternary majority condition Δ_2 generalizes in a way which implies more than distributivity: We say that an $(n+1)$-ary function $u(x_0, \ldots, x_n)$ is a *near unanimity function* if each of the following *near unanimity identities* are satisfied by u.

$$u(x, \ldots, x, y, x, \ldots, x) = x.$$

(All but at most one of the variables are identified.)

The following theorems summarize important properties of near unanimity functions.

Theorem 1.4 (Mitschke [15]) *If a variety has a near unanimity term then the variety is congruence distributive. (Which Δ_n is satisfied depends upon the arity of the near unanimity term.)*

Theorem 1.5 ([1]) *For a variety V and integer $n \geq 2$ the following are equivalent:*

1) V has an $(n+1)$-ary near unanimity term.

2) If $\mathbf{A}$ is a subalgebra of a direct product of $m \geq n$ algebras of V, then $\mathbf{A}$ can be uniquely determined from the knowledge of its various n-fold projections into the product.

3) For any algebra $\mathbf{A}$ of V an m-ary partial operation f on A, with finite domain, has an interpolating term function iff all subalgebras of $\mathbf{A}^n$ are closed under f (where defined).

An important consequence of condition 2) of Theorem 1.5 is the following finite generation condition for clones on a finite set.

Corollary 1.1 *If A is a finite set and C is a clone on A (i.e.: a set of operations on A which contains the projections and is closed under composition) and if C contains a near unanimity function, then C is finitely generated.*

An important consequence of condition 3) of Theorem 1.5 results from taking $\mathbf{A}$ to be a finite algebra in a variety with a ternary majority ($n = 2$) term (e.g.: the variety of lattices with $m(x, y, z)$ the median term). In this case we conclude from condition 3) that an operation on A is a term function iff each subuniverse in $A \times A$ is closed under the operation. Thus if for example f is a

k-ary operation, f will be a term function iff for each subuniverse S in $A \times A$, and $k \times 2$-matrix

$$
\begin{pmatrix}
a_{1,1} & a_{1,2} \\
a_{2,1} & a_{2,2} \\
\cdots & \cdots \\
a_{k,1} & a_{k,2}
\end{pmatrix}
$$

of elements of A, if the rows of the matrix are in S then the pair which is the result of applying f to the columns of the matrix is also in S. (For $n > 2$ the same holds for subuniverses of $\mathbf{A}^n$ and $k \times n$ matrices.)

In summary, what lies behind the properties, considered above, of varieties generated by a primal algebra $\mathbf{P}$ and, in particular, the variety of Boolean algebras, are

> first, because of the presence of a majority term, an operation of the algebra is a term function iff it is compatible with the "subalgebras of the square" of the algebra,

and

> second, because of the presence of a Mal'cev term, the subalgebras of the square are all rectangular.

The conditions a), b), c) of Theorem 1.3 merely insure that, aside from the diagonal, $\mathbf{P} \times \mathbf{P}$ has no other proper subalgebras, so any function is a term function.

2 Generalizations; "almost" primal algebras

As we noted above, the term functions of any algebra $\mathbf{A}$ form a clone on the set A, denoted by Clo$\mathbf{A}$. Likewise the set of polynomials forms a (generally larger) clone denoted by Pol$\mathbf{A}$. The informal terminology "almost" primal algebras, (introduced by Bergman and Berman [2]) is intended to describe algebras $\mathbf{A}$ for which Clo$\mathbf{A}$ is precisely the set of all operations on A which are compatible with various derived structures of $\mathbf{A}$; for example the subalgebra lattice, Sub$\mathbf{A}$, congruence lattice Con$\mathbf{A}$, or automorphism group Aut$\mathbf{A}$ have been popular choices. Notice that each of these structures can be construed as sets of subuniverses of the square $\mathbf{A} \times \mathbf{A}$, i.e.: binary relations on A which are compatible with (i.e.: closed under) the operations of $\mathbf{A}$. (We identify Sub$\mathbf{A}$ with those subuniverses of the square contained in the diagonal, and Aut$\mathbf{A}$ with the subuniverses of the square consisting of the graphs of automorphisms.) More generally, an almost primal algebra is one in which Clo$\mathbf{A}$ is the subclone of the clone of all operations on A which preserve some (hopefully interesting) relations on A.

To clarify this we recall the Galois correspondence between clones and sets of relations on a set A. For a set R of relations on A,

$$\mathcal{F}R = F = \text{ all operations } f \text{ of } A \text{ for which}$$
$$(\forall \theta \in R)\{\vec{x}, \vec{y}, \ldots \in \theta \Rightarrow (f(x_1, y_1, \ldots), f(x_2, y_2, \ldots), \ldots) \in \theta\}$$

is the clone determined by R. For a set F of operations on A,

$$\mathcal{R}F = R = \text{ all relations } \theta \text{ on } A \text{ for which}$$
$$(\forall f \in F)\{\vec{x}, \vec{y}, \ldots \in \theta \Rightarrow (f(x_1, y_1, \ldots), f(x_2, y_2, \ldots), \ldots) \in \theta\}$$

is the set of relations determined by F. $\mathcal{F}R$ is always a clone and a set of operations F is a clone iff $\mathcal{F}\mathcal{R}F = F$. $\mathcal{R}F$ can also be described as the collection of all subuniverses of the various finite direct powers of the algebra (A, F).

Following Bergman and Berman [2], if X is a name which designates a set of (usually binary) relations on A, we say that an algebra $\mathbf{A}$ is X primal if X denotes the set of relations R on A and $\text{Clo}\mathbf{A} = \mathcal{F}R$. For example in the literature a *hemiprimal* algebra is an algebra in which the term functions are precisely operations which are compatible with the congruence relations of the algebra. Using the terminology of Bergman and Berman, just discussed, we would call these *congruence primal* algebras.

Perhaps the most popular of the almost primal algebras are the *quasiprimal* algebras, which in this terminology would be the finite "subalgebra isomorphism" primal algebras; specifically a finite algebra $\mathbf{A}$ is quasiprimal if $\text{Clo}\mathbf{A} = \mathcal{F}IS\mathbf{A}$ where $IS\mathbf{A}$ is the set of all those binary relations on A which are the graphs of isomorphisms between subalgebras of $\mathbf{A}$. (For example if $\mathbf{A}$ has no proper subalgebras then $IS\mathbf{A}$ is just $\text{Aut}\mathbf{A}$.) The following characterization is a direct consequence of the theorems of the preceding section. Note how it generalizes the characterization of primal algebras, Theorem 1.3.

Theorem 2.1 ([16]) *For a finite algebra $\mathbf{A}$ the following are equivalent:*

a) $\mathbf{A}$ is quasiprimal,

b) Each subalgebra of $\mathbf{A}$ is simple and $V(\mathbf{A})$ is arithmetical, i.e.: $\text{Clo}\mathbf{A}$ contains a function satisfying the identities of Theorem 1.2.

c) The discriminator is in $\text{Clo}\mathbf{A}$.

"Functional completeness", or "completeness" for short, and its generalizations, "almost (functionally) complete" algebras are then obtained by replacing $\text{Clo}\mathbf{A}$ by $\text{Pol}\mathbf{A}$ in the discussion above. Thus $\mathbf{A}$ is "complete" provided $\text{Pol}\mathbf{A}$

is the clone of all operations on A. Almost complete algebras, corresponding to almost primal algebras, result from taking $\text{Pol}A$ to be the subclone of the clone of all operations on A which preserve some prescribed relations on A. The most interesting of these are the affine complete algebras, i.e.: where $\text{Pol}A = \mathcal{F}\text{Con}A$, and are the $\text{Pol}A$ analogs of the hemiprimal (= congruence primal) algebras. Using the terminology of Bergman and Berman we would call them *congruence complete* rather than affine complete. As noted in Section 1, it is perhaps unfortunate that the terminology of the past 30 years has been "affine" complete instead of "congruence" complete. Throughout the remainder of this paper we will however continue to use the older terminology "affine complete" and "hemiprimal".

An intriguing development of this subject has been the that quite parallel characterizations for congruence primal algebras and congruence complete algebras exist. The first of these deals with the case where the algebra is simple and was originally discovered by Werner [20], but is an easy consequence of Theorem 1.3. It generalizes Theorem 2.1 characterizing quasiprimality which in turn generalizes Theorem 1.3 characterizing primality.

Theorem 2.2 *A finite algebra* $\mathbf{A}$ *is (functionally) complete iff*
   a) $\mathbf{A}$ *is simple, and*
   b) $\text{Pol}\mathbf{A}$ *contains a function satisfying the identities of Theorem 1.2.*
Conditions a) and b) can be replaced by the single condition
   c) The discriminator is in $\text{Pol}\mathbf{A}$.

PROOF.(Part) If the discriminator of $\mathbf{A}$ is a polynomial then it is a term function of $\mathbf{A}^{+}$, the algebra obtained by adding the elements of A as new nullary operations. Since this addition makes the algebra rigid and destroys all of its proper subalgebras, it follows from Theorem 1.3 that $\mathbf{A}^{+}$ is primal and we conclude that $\mathbf{A}$ is complete. The other parts of the proof are equally straightforward.

Theorem 2.2 is an intriguing generalization of the characterization of quasiprimality since it simply drops the requirement that the proper subalgebras be simple and replaces $\text{Clo}\mathbf{A}$ by the larger $\text{Pol}\mathbf{A}$. Thus the two theorems are almost parallel. This suggests that we find a parallel way of characterizing both hemiprimal affine complete algebras. Since every algebra is a reduct of a hemiprimal algebra (just replace the basic operations by all of the congruence compatible operations) there is little chance of doing this without some restriction. The clue is to observe that $\mathbf{A}$ is simple iff $\text{Con}\mathbf{A}$ is the two element chain and thus $\mathbf{A}$ is arithmetical. With this restriction we have the following parallel generalizations of the primality characterization (Theorem 1.3) and the functional completeness characterization (Theorem 2.2).

Theorem 2.3 ([17]) *A finite arithmetical algebra* **A** *is hemiprimal iff*
  a) **A** *has no proper subalgebras,*
  b) *For congruences* θ_1, θ_2 *of* **A** *any isomorphism*

$$\phi : \mathbf{A}/\theta_1 \to \mathbf{A}/\theta_2$$

is the identity, and
  c) Clo**A** *contains a function satisfying the identities of Theorem 1.2.*

Theorem 2.4 ([17]) *A finite arithmetical algebra is affine complete iff* Pol**A** *contains a function satisfying the identities of Theorem 1.2.*

These theorems can be proved as in [17] or by a more streamlined approach using Theorems 1.1 and 1.5.

Theorem 2.4 has an immediate corollary.

Corollary 2.1 *Any finite algebra in an arithmetical variety is affine complete.*

We also have the following finiteness criterion for determining when a finitely generated congruence distributive variety is affine complete.

Theorem 2.5 *If* V *is a congruence distributive variety generated by a finite algebra having no proper subalgebras, then* V *is affine complete if each finite algebra in* V *is affine complete.*

PROOF OF THEOREM 2.5 Suppose $V = V(\mathbf{A})$ where **A** is finite having no subalgebras. By congruence distributivity $V = \mathbf{IP}_S\mathbf{H}(\mathbf{A})$ so that if $\mathbf{B} \in V$, we may suppose

$$\mathbf{B} < \Pi\{\mathbf{A}_i : i \in I\}$$

where each $\mathbf{A}_i$ is $\mathbf{A}/\theta$ for some $\theta \in Con$ **A**. Let $f : B^k \to B$ be compatible with Con **B**. Then if $\mathbf{B} = (B, F)$ it follows that the algebra $\mathbf{B}^* = (B, F \cup \{f\})$ has the same universe and congruence lattice as **B**. Also, by compatibility, f naturally induces various functions on the subdirect factors of **B**, say f_i is the function induced on A_i. It follows from this that $\mathbf{B}^*$ is a subdirect product

$$\mathbf{B}^* < \Pi\{\mathbf{A}_i^* : i \in I\},$$

where $\mathbf{A}_i^*$ is obtained by adding f_i to the operations of $\mathbf{A}_i$. Since A is finite there are only finitely many non-isomorphic $\mathbf{A}_i^*$. Hence if we pick any $a \in B$, then the subalgebra $\mathbf{B}^*(a)$ of $\mathbf{B}^*$ generated by a is isomorphic to a subdirect product in $\Pi\{\mathbf{A}_i^* : i \in I_0\}$ for some finite subset $I_0 \subset I$, and hence $\mathbf{B}^*(a)$ is finite. Therefore the reduct $\bar{\mathbf{B}}$ obtained from it by discarding the operation f is a finite subalgebra of **B**. Also f (restricted) is in $\bar{B}^k \to \bar{B}$ and is compatible with Con $\bar{\mathbf{B}}$. Hence, by hypothesis, f (restricted) is a polynomial of $\bar{\mathbf{B}}$. Hence for some $(k+m)$–ary term t and $\mathbf{a} \in \bar{B}^m$, $f(\mathbf{x}) = t(\mathbf{x}, \mathbf{a})$ for all $\mathbf{x} \in \bar{B}^k$. But since $\mathbf{A}$ has no subalgebras, the projections of $\bar{B}$ are onto each of the factors A_i, for *all* $i \in I$. Hence for any $\mathbf{x} \in B^k$ and $i \in I$, choose $\bar{\mathbf{x}} \in \bar{B}^k$ with $\bar{\mathbf{x}}_i = \mathbf{x}_i$. Then

$$f(\mathbf{x})_i = f_i(\mathbf{x}_i) = f_i(\bar{\mathbf{x}}_i) = f(\bar{\mathbf{x}})_i = t(\bar{\mathbf{x}}, \mathbf{a})_i = t(\bar{\mathbf{x}}_i, \mathbf{a}_i) = t(\mathbf{x}_i, \mathbf{a}_i) = t(\mathbf{x}, \mathbf{a})_i$$

and hence f is a polynomial of $\mathbf{B}$. We conclude that V is affine complete.

From this theorem, combined with Corollary 2.1 we have the following immediate consequence which explains what is behind Grätzer's and Hu's theorems on the affine completeness of a variety generated by a primal algebra.

Corollary 2.2 ([11]) *If V is an arithmetical variety generated by a finite algebra having no proper subalgebras, then V is affine complete.*

Corollary 2.2 makes it easy to construct many affine complete varieties which naturally generalize the variety generated by a primal algebra. In particular, by Theorem 2.3 any finite arithmetical hemiprimal algebra generates an affine complete variety. On the other hand Theorem 2.5 does not give any such ready prescription for constructing an affine complete variety which is not arithmetical. The first recognized example was constructed by Kaarli and Pixley [11]. In that paper we exhibit a finitely generated affine complete variety which is not arithmetical but which is congruence distributive. (In [9] the construction is systematized so that it becomes easy to construct many other examples of affine complete varieties not of the type given by Corollary 2.2.)

It is worth noticing that there is no such thing as a nontrivial hemiprimal variety, i.e.: a variety in which each algebra $\mathbf{A}$ has the property that all Con$\mathbf{A}$ compatible operations are term functions. For if V were such a variety, and $\mathbf{A}$ a nontrivial member and if T were the set of all terms of V then $\mathbf{A}^{|T|}$ would be in V and hence all of its elements would be (constant) terms of V, and this is impossible on cardinality grounds.

Finally, before devoting the remainder of the paper to affine complete varieties, we should remark that a good deal of study has been and continues to be devoted to affine complete algebras *per se*. Since every algebra is a reduct of an affine complete algebra—even a hemiprimal algebra (just add all Con$\mathbf{A}$ compatible functions as new operations) it seems somewhat fruitless to attempt any kind of characterization. Perhaps for this reason a good deal of this research has been focused on describing the affine complete algebras in some specific axiomatic classes, usually varieties. On the other hand, in the present context at least, we understand *Boolean universal algebra* to be the study of varieties which naturally generalize the variety of Boolean algebras.

3 Affine complete varieties

In this section we turn around our approach: we start with an affine complete variety and discover what we can of its properties, selecting some proofs. First we consider the size of subdirectly irreducibles. The situation here is very different than in an arbitrary variety, even with some finiteness conditions (finite type or local finiteness), where, by an important theorem of Magari ([13]), the variety contains a (likely very large infinite) simple algebra. An easy counting argument shows that if $\mathbf{A}$ is any subdirectly irreducible affine complete algebra, and if the number of operations is a most countable, then $\mathbf{A}$ is finite. The following theorem says that in an affine complete variety this remains true independent of the number of operations. The proof shows that a variety which contains an infinite subdirectly irreducible affine complete algebra always contains another subdirectly irreducible member which is not affine complete.

Theorem 3.1 ([11]) *If V is affine complete then V is residually finite (i.e.: each subdirectly irreducible member of V is finite).*

PROOF. Let $\mathbf{A} = (A, F)$ be subdirectly irreducible in the affine complete variety V. Let μ be the monolith congruence, i.e.: the proper congruence which is the intersection of all proper congruences. Suppose that $\aleph_0 \leq | A |$, so that

$$| \text{Pol}\mathbf{A} | \leq | A | + | F | .$$

Let a, b be distinct but congruent mod μ. Define $f : A^3 \to A$ for all $u, v, x \in A$ by

$$
\begin{aligned}
f(u, u, a) &= a \\
f(u, v, x) &= b \text{ if } v \neq u \text{ or } x \neq a.
\end{aligned}
$$

Then since $(a, b) \in \mu$ and μ is contained in any proper congruence, it follows that f is Con$\mathbf{A}$–compatible. Hence for some integer k and $(k + 3)$–ary term $t(x, y, z, \mathbf{w})$,

$$f(x, y, z) = t(x, y, z, \mathbf{c})$$

for some $\mathbf{c} \in A^k$. Consider the first order sentence

$$\Phi := (\exists x, y)\{x \neq y \wedge (\exists \mathbf{c})(\forall u, v)[t(u, u, x, \mathbf{c}) = x \wedge t(u, u, y, \mathbf{c}) = y$$

$$\wedge (u \neq v \to t(u, v, x, \mathbf{c}) = t(u, v, y, \mathbf{c}))]\}.$$

Taking $x = a$ and $y = b$ we see that $\mathbf{A} \models \Phi$. Since $\aleph_0 \leq | A |$, by the upwards Löwenheim–Skolem Theorem, there is $\mathbf{B} \in V$ with

$$| F | + \aleph_0 \leq | B |$$

and such that $\mathbf{B} \models \Phi$.

Accordingly, there are distinct $x, y \in B$ and $\mathbf{c} \in B^k$ such that for any distinct $u, v \in B$,

$$x = t(u, u, x, \mathbf{c})\theta(u, v)t(u, v, x, \mathbf{c}) = t(u, v, y, \mathbf{c})\theta(u, v)t(u, u, y, \mathbf{c}) = y.$$

Hence the pair (x, y) is in any non–zero congruence of $\mathbf{B}$. Therefore $\mathbf{B}$ is subdirectly irreducible. But then, since $\mid F \mid + \aleph_0 \leq \mid B \mid$, we have

$$\mid \mathrm{Pol}\mathbf{B} \mid \leq \mid B \mid + \mid F \mid + \aleph_0 = \mid B \mid < \mid \{x, y\}^B \mid \leq \mid \mathrm{Pol}\mathbf{B} \mid,$$

a contradiction. Hence A must be finite.

Congruence, distributivity plays a critical role in affine complete varieties, at least in the locally finite case:

Theorem 3.2 *If V is a locally finite affine complete variety then*
a) V has a near unanimity term and hence is congruence distributive, and
b) V is n-permutable for some integer n.

The fact that V is congruence distributive and n-permutable is an unpublished result due to R. McKenzie, obtained as a consequence of tame congruence theory. The stronger fact, that V has a near unanimity term is very recent and is due to Kaarli [10]; the proof makes use of the underlying ideas of tame congruence theory but is more direct. n-permutability generalizes ordinary permutability (which is 2-permutability) and means that the join of a pair of congruences can always be expressed as their n-fold relation product. The significance of n permutability for congruence complete varieties is not yet understood. On the other hand the existence of a near unanimity term (and the consequent congruence distributivity) yields most of what we presently know about affine complete varieties. In fact the following problem is the most important open problem in this area.

Problem 3.1 *Is there a non congruence distributive affine complete variety?*

The problem is particularly interesting because from Theorems 3.1 and 3.2 we know that an affine complete variety is not only residually finite but also each finitely generated subvariety is congruence distributive. This leads naturally to the following more general version of Problem 3.1, which is independent of affine completeness:

Problem 3.2 *Is there a non congruence distributive variety which is both residually finite and has all of its finitely generated subvarieties congruence distributive?*

Hereafter we focus on congruence distributive affine complete varieties.

For brevity, if an operation $f : A^k \to A$ preserves, or is compatible with, all congruences of the algebra $\mathbf{A}$, let us say that f is Con$\mathbf{A}$ compatible. Notice that this is true iff for all $\vec{x}, \vec{y} \in A^k$, we have

$$(f(\vec{x}), f(\vec{y})) \in \theta(\vec{x}, \vec{y}).$$

($\theta(\vec{x}, \vec{y})$ is the least congruence which collapses the k pairs of corresponding components of $\vec{x}$ and $\vec{y}$.)

Now if $\mathbf{B}$ is an affine complete subalgebra of some arbitrary algebra $\mathbf{A}$, then each Con$\mathbf{B}$ compatible operation $f:B^k \to B$ has at least one Con$\mathbf{A}$ compatible extension to $A^k \to A$, namely the polynomial which agrees with f on B^k. If $V(\mathbf{A})$ is congruence distributive affine complete then this extension is unique:

Theorem 3.3 ([11]) *If V is a congruence distributive affine complete variety, $\mathbf{B}$ a subalgebra of $\mathbf{A} \in V$, and if $f:B^k \to B$ is Con$\mathbf{B}$ compatible, then there is exactly one Con$\mathbf{A}$ compatible extension of f to $A^k \to A$.*

PROOF(Sketch) For simplicity of notation suppose $k = 1$. We suppose the theorem is false, namely that f has extensions g and h and that $g(u) \neq h(u)$ for some $u \in A \setminus B$. Let $\mathbf{D}$ be the subdirect power of $\mathbf{A}$ with

$$D = \{\mathbf{a} = (a_1, a_2, ...) \in A^\omega : a_i \text{ are eventually constant and in } B\}.$$

Certainly $\mathbf{D} \in V$. Define $c:D \to D$ by

$$
\begin{aligned}
c(\mathbf{x})_n &= g(x_n) \ n \text{ odd}, \\
&= h(x_n) \ n \text{ even}.
\end{aligned}
$$

Since $\mathbf{x}$ is eventually constant in B, so is $c(\mathbf{x})$ and thus $c:D \to D$.

To show that c is Con$\mathbf{D}$ compatible requires that we show that

$$(c(\mathbf{x}), c(\mathbf{y})) \in \theta(\mathbf{x}, \mathbf{y})$$

for all $\mathbf{x}, \mathbf{y} \in D$, and this is readily done using the fact that $\mathbf{x}$ and $\mathbf{y}$ are eventually constant together with the fact (a consequence of congruence distributivity), that congruences on finite products are products of factor congruences. Therefore for some, say $(m+1)$–ary term, t,

$$c(\mathbf{x}) = t(\mathbf{x}, \mathbf{d}^1, \dots, \mathbf{d}^m)$$

where each $\mathbf{d}^i = (d_1^i, d_2^i, \dots)$ is eventually constant, say d^i, in B. Hence for n_0 sufficiently large and $n \geq n_0$

$$c(\mathbf{x})_n = t(x_n, d^1, \dots, d^m).$$

For such an n select $\mathbf{x} \in D$ with $x_n = x_{n+1} = u$. Then

$$g(u) = c(x_n) = t(x_n, d^1, \dots, d^m) = t(x_{n+1}, d^1, \dots, d^m) = c(x_{n+1}) = h(u),$$

a contradiction.

Theorem 3.3 has several important consequences which we list in the following four corollaries:

Corollary 3.1 *If $V(\mathbf{A})$ is a congruence distributive affine complete variety and $\mathbf{B}$ is a subalgebra of $\mathbf{A}$, then $V(\mathbf{B}) = V(\mathbf{A})$.*

PROOF. For if p and q are terms and $p = q$ is one of the defining equations of $V(\mathbf{B})$, then by Theorem 3.3 $p = q$ is also an equation of $\mathbf{A}$. Hence $V(\mathbf{B}) = V(\mathbf{A})$.

Notice that in Theorem 3.3, the subalgebra $\mathbf{B}$ of $\mathbf{A}$ could have only a single element. Hence we have

Corollary 3.2 *If $\mathbf{A}$ is a nontrivial algebra in a congruence distributive affine complete variety, then $\mathbf{A}$ has no trivial subalgebras.*

Corollary 3.3 *If V is a congruence distributive affine complete variety, then the subdirectly irreducible algebras in V have no proper subalgebras.*

PROOF. Suppose $\mathbf{B} < \mathbf{A}$ where $\mathbf{A}$ is subdirectly irreducible and $V(\mathbf{A})$ is congruence distributive affine complete. Corollary 3.1 asserts that $\mathbf{A} \in V(\mathbf{B}) = IP_S HS(\mathbf{B})$ by the finiteness of B. Hence the subdirect irreducibility of $\mathbf{A}$ asserts that $\mathbf{A} \in HS(\mathbf{B})$ so $B = A$.

Finally, for affine complete varieties the two properties, *locally finite* and *finitely generated* coincide, and either implies that the variety is essentially of finite type:

Corollary 3.4 *If V is a locally finite affine complete variety then*
  a) V is generated by a finite algebra having no proper subalgebras, and
  b) V is term equivalent to a variety of finite type.

PROOF. a) By local finiteness the free algebra $\mathbf{F}(1)$ in V on one free generator is finite and by Corollary 3.3 each subdirectly irreducible member of V is a homomorphic image of $\mathbf{F}(1)$, so there are only finitely many subdirectly irreducibles and hence V is finitely generated. Any minimal subalgebra of a generator will have no proper subalgebras and will also generate V by Corollary 3.1.

b) By Theorem 3.2 V has a near unanimity term and, by a) above, is generated by a finite algebra $\mathbf{A}$. Hence, by Corollary 1.1, Clo$\mathbf{A}$ is finitely generated so that $\mathbf{A}$ is term equivalent to an algebra of finite type. Since $V = V(\mathbf{A})$, V is term equivalent to a variety of finite type.

Not all congruence distributive affine complete varieties are locally finite. See [11] for an example having countably many operations. According to the following theorem, such an example cannot have only finitely many operations. Without assuming local finiteness Theorem 3.4 is the best result we presently have regarding the (finite) sizes of the subdirectly irreducibles in an affine complete variety.

Theorem 3.4 ([11]) *If V is congruence distributive affine complete and has at most countable type, then for some integer N, $\mid A \mid \le N$ for all subdirectly irreducible members of V, (i.e.: V is residually $\le N$ for some integer N).*

PROOF(Sketch) We know that each subdirectly irreducible is finite and has no proper subalgebras. If we suppose their sizes are unbounded then we can choose a sequence of them $\mathbf{A}_1, \mathbf{A}_2, \ldots$ with $\mid A_i \mid < \mid A_{i+1} \mid$ for all i. As in the proof of Theorem 3.3 we again use a combinatorial argument to obtain a contradiction.

For each $\mathbf{A}_i$ let μ_i be the monolith and choose $\mathbf{a} = (a_1, a_2, \ldots) \in \prod A_i$ such that $\mid a_i / \mu_i \mid \ge 2$ for all i. Take $\mathbf{B}$ to be the subalgebra of all elements in $\prod A_i$ which are eventually in the subalgebra generated by $\mathbf{a}$. $\mathbf{B}$ is a subdirect product since the $\mathbf{A}_i$ have no proper subalgebras. $\mathbf{B}$ is countable because V has at most countable type. Hence $\mid \text{Pol}\mathbf{B} \mid \le \aleph_0$. A contradiction is then obtained without difficulty by showing how to actually construct $2^{\aleph_0}$ Con$\mathbf{B}$ compatible functions $f : B \to B$. The construction depends in an apparently essential way on the countability of B.

Thus our next problem is the following:

Problem 3.3 *Is Theorem 3.4 true for varieties of uncountable type?*

Another general problem on congruence distributive varieties, independent of affine completeness, but which arose in considering Problem 3.3, is the following:

Problem 3.4 *If V is a congruence distributive variety of finite type and is residually finite, is V residually $\le N$ for some integer N?*

In any case if V has finite type then from Theorem 3.4 we have that the number of non-isomorphic subdirectly irreducible members of V is finite. Their direct product generates V and by Corollary 3.1 so does a minimal subalgebra of the product. Combining this with Corollary 3.4 we have the following conclusion.

Theorem 3.5 *For an affine complete variety V the following are equivalent:*
  a) V is locally finite,
  b) V is finitely generated,
  c) V is congruence distributive and is term equivalent to a variety of finite type.

Combining Theorem 3.5 with Corollary 2.2 we obtain the following theorem which (subject to the same finiteness conditions) settles the question of when an arithmetical variety is affine complete.

Theorem 3.6 *An arithmetical variety which is either of finite type or is locally finite is affine complete iff it is generated by a finite algebra having no proper subalgebras.*

Further refinements on the structure of arithmetical affine complete varieties appear in [12].

As indicated by Problem 3.1, for non arithmetical varieties we have no result so neat as Theorem 3.6 characterizing when a variety is affine complete. However, combining the results of this section with Theorem 2.5 we do have the following partial characterization result:

Theorem 3.7 *A locally finite variety V is affine complete iff*
  a) V has a near unanimity term (hence is congruence distributive),
  b) V is finitely generated by a finite algebra having no proper subalgebras (hence is essentially of finite type), and
  c) the finite members of V are affine complete.

Using more refined methods, Kaarli [10] has recently obtained much sharper results on the characterization problem, and in particular, has shown that condition *c)* in the theorem above can be replaced by the weaker condition:

  c') each algebra in V which has no proper subalgebras (and hence is finite) is affine complete.

This condition can obviously be replaced by the requirement that the free V algebra with one free generator together with all of its homomorphic images, be affine complete.

Finally, the preceding discussion throws some light on the question of arbitrary (not necessarily congruence distributive) affine complete varieties (Problem 3.2). Recall ([14]) that two algebras **A** and **B** are *term equivalent* if Clo(**A**) = Clo(**B**) and they have the same universe. **A** and **B** are *weakly isomorphic* if **A** is isomorphic to an algebra which is term equivalent to **B**. Suppose **A** is subdirectly irreducible in an arbitrary affine complete variety V. Then **A** is finite (Theorem 3.1) so the subvariety $V(\mathbf{A})$ of V has a near unanimity term hence **A** has no proper subalgebras and Clo(**A**) is finitely generated. It follows that for a given integer n there are at most countably many non-weakly isomorphic subdirectly irreducibles of size n in V and hence at most countably many non-weakly isomorphic subdirectly irreducible algebras in V altogether.

4 Categorical equivalence

In section 1 we discussed Hu's generalization of the Stone duality theory: if **P** is primal then $V(\mathbf{P})$ (considered as a category), is category equivalent to the dual of the category of Stone spaces. From this it follows that a variety W is category equivalent to $V(\mathbf{P})$ iff $W = V(\mathbf{P}')$ for some primal algebra **P**'. This remarkable fact of Boolean universal algebra has recently been generalized by Bergman and Berman [2] to the (arithmetical) varieties generated by several types of almost primal algebras. We briefly record some of their results here. Without going into detail, we note that their results are not obtained by developing an appropriate duality theory, but instead by using a powerful variety equivalence theorem due to McKenzie and which provides necessary and sufficient conditions for algebras **A** and **B** to generate category equivalent varieties. McKenzie's theorem is a far reaching extension of the Morita theorem (describing necessary and sufficient conditions on two rings with unit insuring that their varieties of unital modules be category equivalent).

We first adopt the notation and terminology of [2]. If **A** and **B** are algebras, not necessarily of the same type, we say that **A** and **B** are *categorically equivalent* if there is a functor F which is an equivalence from the category $V(\mathbf{A})$ to $V(\mathbf{B})$ such that $F(\mathbf{A}) = \mathbf{B}$. We write $\mathbf{A} \equiv_c \mathbf{B}$ to indicate this relationship. Since '$\equiv_c$' is an equivalence relation on the class of all algebras, Hu's theorem states that **P** is primal iff the equivalence class $\mathbf{P}/\equiv_c$ is the class of all primal algebras. The principal results of Bergman and Berman are characterizations of the equivalence class $\mathbf{A}/\equiv_c$ in case **A** is a finite subalgebra primal (also known as semiprimal), arithmetical congruence primal, (arithmetical hemiprimal), or automorphism primal algebra. In case **A** is subalgebra primal, $\mathbf{A} \equiv_c \mathbf{B}$ iff **B** is also a finite subalgebra primal and there is an isomorphism between their subalgebra lattices which preserves one element subalgebras. The characterization of $\mathbf{A}/\equiv_c$ for $\mathbf{A}$ automorphism primal is more complex, apparently due to the possible presence of subalgebras arising as fixed points of some automorphism. But if we consider only those $\mathbf{A}$ without such subalgebras, then $\mathbf{A} \equiv_c \mathbf{B}$ iff $\mathbf{B}$ is also automorphism primal and Aut$\mathbf{A}$ is isomorphic to Aut$\mathbf{B}$.

From the standpoint of our present interest in affine complete varieties the most interesting of their results is the following:

Theorem 4.1 ([2], Corollary 4.5) *Let $\mathbf{A}$ be a finite arithmetical hemiprimal algebra. Then for any algebra $\mathbf{B}$, $\mathbf{A} \equiv_c \mathbf{B}$ iff $\mathbf{B}$ is also a finite arithmetical hemiprimal and Con$\mathbf{A}$ is isomorphic to Con$\mathbf{B}$.*

Hu's theorem for primal algebras is just the special case where Con$\mathbf{A}$ is the two element lattice.

Now according to Theorem 3.6 the arithmetical affine complete varieties of finite type are precisely those which are generated by finite algebras $\mathbf{A}$ of finite type, having no proper subalgebras, for which $V(\mathbf{A})$ is arithmetical. Arithmetical hemiprimal algebras are thus a special case by Theorem 2.3. And according to Theorem 2.3 such an $\mathbf{A}$ fails to be hemiprimal only by virtue of possible nonidentity isomorphisms $\phi : \mathbf{A}/\theta_1 \to \mathbf{A}/\theta_2$ between quotients, i.e.: possible rectangular subuniverses of $\mathbf{A} \times \mathbf{A}$ which are not congruence relations because they fail to contain the diagonal.

Thus Theorem 4.1 asserts that for an arithmetical affine complete variety V with hemiprimal generator $\mathbf{A}$, V is category equivalent to a variety W iff W is also arithmetical affine complete generated by a hemiprimal algebra $\mathbf{B}$ and the lattices of rectangular diagonal subdirect products in $\mathbf{A} \times \mathbf{A}$ and $\mathbf{B} \times \mathbf{B}$ are isomorphic. This leads to the following problem:

Problem 4.1 *If V is an arithmetical affine complete variety generated by the finite algebra $\mathbf{A}$ having no proper subalgebras, find necessary and sufficient conditions on an algebra $\mathbf{B}$ which insure that $\mathbf{A} \equiv_c \mathbf{B}$.*

From the preceding discussion if $\mathbf{A}$ satisfies the hypotheses and $\mathbf{A} \equiv_c \mathbf{B}$ holds, then $\mathbf{B}$ is finite with no proper subalgebras, generates an arithmetical (affine complete) variety, and the lattices of (rectangular) subdirect products in $\mathbf{A} \times \mathbf{A}$ and in $\mathbf{B} \times \mathbf{B}$ are isomorphic. What additional information is needed to characterize $\mathbf{A} \equiv_c \mathbf{B}$ is not yet clear.

References

[1] Baker, K. A., Pixley, A. F., *Polynomial interpolation and the Chinese Remainder Theorem for algebraic systems*, Math. Z. **143** (1975), 165–174.

[2] Bergman, C., Berman, J., *Morita equivalence of almost-primal clones*, Preprint

[3] Fleischer, I., *A note on subdirect products*, Acta Math. Acad. Sci. Hungar. **6** (1955), 463–465.

[4] Foster, A. L., Pixley, A. F., *Semi-categorical algebras*, Math. Z. **85** (1964), 169–184.

[5] Grätzer, G., *On Boolean functions (Notes on lattice theory II)*, Rev. Roumaine Math. Pures Appl. **7** (1962), 693–697.

[6] Hu, T. K., *Stone duality for primal algebra theory*, Math. Z. **110** (1969), 180–198.

[7] Hu, T. K., *Characterization of algebraic functions in equational classes generated by independent primal algebras*, Algebra Universalis **1** (1971), 187–191.

[8] Jónsson, B., *Algebras whose congruence lattices are distributive*, Math. Scand. **21** (1967), 110–121.

[9] Kaarli, K., *On affine complete varieties generated by hemi-primal algebras with Boolean congruence lattices*, Tartu Ülikooli Toimetised **878** (1990), 23–38.

[10] Kaarli, K., *Locally finite affine complete varieties*, Preprint.

[11] Kaarli, K., Pixley, A. F., *Affine complete varieties*, Algebra Universalis **24** (1987), 74–90.

[12] Kaarli, K., Pixley, A. F., *Finite arithmetical affine complete algebras having no proper subalgebras*, Algebra Universalis, to appear.

[13] Magari, R., *Una dimostrazione del fatto che ogni varietà ammette algebre semplici*, Ann. Univ. Ferrara Sez. VII (NS) **14**, 1–4.

[14] McKenzie, R. N., McNulty, G. F., Tayler, W. F., Algebras, Lattices, Varieties, vol. I, Wadsworth and Brooks/Cole, Monterey, California, 1987.

[15] Mitschke, A., *Near unanimity identities and congruence distributivity in equational classes*, Algebra Universalis **4** (1974), 287–288.

[16] Pixley, A. F., *The ternary discriminator function in universal algebra*, Math. Ann. **191** (1971), 167–180.

[17] Pixley, A. F., *Completeness in arithmetical algebras*, Algebra Universalis **2** (1972), 179–196.

[18] Stone, M. H., *The theory of representations for Boolean algebras*, Trans. Amer. Math. Soc. **41** (1936), 37–111.

[19] Wille, R., Kongruezklassengeometrien, Lecture Notes in Math. **113** Berlin-Heidelberg-New York, Springer, 1970.

[20] Werner, H., *Eine Charakterisierung funktional vollständiger Algebren*, Arch. Math. (Basel) **21** (1970), 381–385.

DEPARTMENT OF MATHEMATICS, HARVEY MUDD COLLEGE, CLAREMONT, CA 91711

e-mail address: apixley@hmc.edu

Restructuring Mathematical Logic: An Approach Based on Peirce's Pragmatism

Rudolf Wille

Abstract

Restructuring mathematics is considered as the activity of reworking mathematical developments in order to integrate and to rationalize origins, connections, interpretations, and applications. Philosophically, this activity is mostly supported by Pragmatism as founded by Ch. S. Peirce. To activate more connections of mathematical logic to reality, an approach to restructuring is suggested which revitalizes the traditional paradigm of logic given by the three main functions of thinking: concepts, judgments, and conclusions. The described formalizations are based on ideas as they have been developed in formal concept analysis during the last fifteen years.

Contents

1   Restructuring Mathematics

"Restructuring Mathematics" designates a basic idea of activity which arose sixteen years ago in a seminar at the TH Darmstadt in which mathematicians met to work on a better understanding of the relationship of mathematics to the real world. A basic question was: *Why develop mathematics as we do today?* To find substantial answers, a broad spectrum of real world problems and theoretical views, including different philosophical positions, were discussed in

the seminar. The discussion was mostly influenced by the conception of *"Wissenschaftsdidaktik"* of Hartmut von Hentig, a well-known German scholar of education. Hentig has elaborated his ideas of "Wissenschaftsdidaktik" in his book *"Magician or Master? On the Unity of Science in the Process of Understanding"* [18], in which he widely discusses present problems of science and makes proposals on how to overcome those problems. Against the increasing specialization and instrumentalization, Hentig set up his charge to *"restructure"* scientific disciplines which he explains as follows:

> "Sciences have to examine their disciplinarity, and this means: to uncover their unconscious purposes, to declare their conscious purposes, to select and to adjust their means according to those purposes, to explain possible consequences comprehensibly and publicly, and to make accessible their ways of scientific finding and their results by the every-day language." ([18] p. 136f)

> "The restructuring of scientific disciplines within themselves becomes more and more necessary to make them more learnable, mutually available, and criticizable in more general surroundings, also beyond disciplinary competence. This restructuring may and must be performed by general patterns of perception, thought, and action of our civilization." ([18] p. 33f)

The first attempt to restructure a mathematical discipline was concerned with a course on *"Linear Algebra"* in the academic year 1978/79 (see [20]). Other attempts of restructuring followed in lattice theory, order theory, introductory algebra, axiomatic geometry, basics of statistics and calculus. All attempts are far from being complete, but they have been astonishingly fruitful and inspiring. Restructuring lattice theory has even created a new discipline called *"Formal Concept Analysis"* which is already used in many areas and has even reached the state of supporting the development of commercial products (cf. [21],[26],[10]).

The developed understanding of restructuring mathematics may be explained best with respect to mathematical education:

> "In the ideal case, a restructured development unfolds a mathematical theory in continuous relation with reality; in doing so, not only the necessary abstractions are embodied in suggestive concretizations, but the often unconscious purposes, limitations, and risks of applications of the theory are explained too. Therefore, restructuring is supposed to make a theory not only more communicable, learnable, and available, but also criticizable. The demand to build on general patterns of perception, thought, and action especially

means that the disciplinary communication should be performed in the every-day language as much as possible. For a student, the restructured development presents itself as a sequence of growing areas of content so that each by itself is a significant unit of learning which qualifies for further fruitful studies; in addition to this, the student is continuously urged to consider meaning and substance of the theory." ([20] p. 105)

With respect to research, the idea is that mathematics consists of a network of more or less autonomous subcultures which grow out of an immeasurable wealth of stimulations and motivations. Then, restructuring is understood as creatively reworking the produced developments in order to integrate and to rationalize origins, connections, interpretations, and applications. The aim is to reach a structured theory which unfolds the formal thoughts according to meaningful interpretations allowing a broad communication and critical discussion of the content, so that it becomes clear what are its effects and consequences beyond the theory. This means that the restructured theory is substantially determined by a pragmatic understanding, so that the theory is opened as much as possible to the surrounding which the theory might effect. This understanding suggests viewing restructuring in the wider conception of *"Allgemeine Wissenschaft"* which has been developed since 1987 (see [22],[27]).

2 Pragmatism and Discourse Philosophy

The charge to restructure scientific disciplines has a philosophical background which relies on a pragmatic understanding of science. The branch of philosophy mostly concerned in this understanding is *Pragmatism* founded by *Charles Sanders Peirce* (1839-1914) which today finds its convincing continuation by the *Discourse Philosophy* of *Karl-Otto Apel* and *Jürgen Habermas*. Peirce published ideas of his pragmatism first in his papers *"The Fixation of Belief"* (1877) and *"How to Make Our Ideas Clear"* (1878). The second paper contains the most influential *"Pragmatic Maxim"*:

> "Consider what effects, that might conceivably have practical bearings, we conceive the object of our conception to have. Then our conception of these effects is the whole of our conception of the object." ([15] 5.402)

The maxim claims that the significance of a conception consists of its effects. Therefore scientific concepts and theories have their final designation in all the effects which they may produce. This is the reason why pragmatism sees a tight connection between theory and practice. This also determines the epistemological understanding of reality: Only in an unlimited process of investigations reality is constituted and in each situation investigators have to build on common sense in treating problems of a necessarily restricted nature. Thus, scientific disciplines are not only determined by the originality of theoretical thoughts, they heavily rely on common sense, on views and purposes with respect to problems in reality.

In pragmatism, Peirce has founded ontology and logic on a new list of categories. After a careful examination of Kant's *"Transcendental Analytic"*, in particular of Kant's table of categories, Peirce derived his three *universal categories* which are part of each conception. In his *"Lectures on Pragmatism"* (1903), he describes his categories as follows [15] 5.66:

- Category the First is the Idea of that which is such as it is of anything else. That is to say, it is a *quality* of feeling.

- Category the Second is the Idea of that which is such as it is as being Second to some First, regardless of anything else, and in particular regardless of any Law, although it may conform to a law. That is to say, it is *Reaction* as an element of the phenomenon.

- Category the Third is the Idea of that which is such as it is as being a Third, or Medium, between a Second and a First. That is to say, it is *Representation* as an element of the phenomenon.

The category of firstness covers everything which does not need further determinations. The category of secondness is concerned with everything whose nature is to be related to some firstness. The category of thirdness comprises everything which mediates and interprets between firstness and secondness. Peirce has elaborated this foundation in his logic as a theory of signs. In this way he also became the founder of semiotics.

Rationality of argumentation is a central theme of pragmatism and discourse philosophy. In general, philosophers are far from reaching an agreement how to understand reasoning and most of them doubt an objective *"foundation of reasoning"*. Karl-Otto Apel discusses this theme in the pragmatic tradition:

> "In view of this problematic situation [of rational argumentation] it is more obvious not to give up reasoning at all, but to break with the concept of reasoning which is orientated by the pattern of logic-mathematical proofs. In accordance with a new foundation of critical rationalism, Kant's question of transcendental reasoning has to be taken up again as the question about the normative conditions of the possibility of discursive communication and understanding (and therewith discursive criticism too). Reasoning then

appears primarily not as deduction of propositions out of propositions within an objectivizable system of propositions in which one has always abstracted from the actual pragmatic dimension of argumentation, but as answering of why-questions of all sorts within the scope of argumentative discourse." ([4] p. 19)

Pragmatism and discourse philosophy locates the foundation of reasoning within the intersubjective community of communication and argumentation. Only the process of discourse and understanding leads to comprehensive states of rationality which properly respects the pragmatic dimension of reality. This does not exclude logic-mathematical proofs, but they can only be part of a broader argumentative discourse. With his transformation of Kant's transcendental philosophy to intersubjectivity, Peirce has given a convincing philosophical foundation for the constitutive role of the intersubjective community of communication and argumentation (cf. [3]). This does not only apply to epistemology and knowledge, but also to intersubjective ethics which Apel has elaborated in his important treatise *"The Apriori of the Community of Communication and the Foundations of Ethics"* [2].

3 Mathematical Logic

From the viewpoint of Peirce's pragmatism, restructuring mathematical logic means that, first of all, its purposes and effects have to be clarified. *Mathematical Logic*, understood as the modern version of formal logic, is concerned with the formulation of expressions, the validity of propositions, and the correctness of conclusions on the base of mathematically defined languages. In the introductory article to the *"Handbook of Mathematical Logic"*, Jon Barwise emphasizes the role of mathematical logic within mathematics:

"Mathematical logic adds a new dimension to this science by paying attention to the language used in mathematics, to the ways abstract objects are defined, and to the laws of logic which govern us as we reason about these objects. The logician undertakes this study with the hope of understanding the phenomena of mathematical experience and eventually contributing to mathematics, both in terms of important results that arise out of the subject itself (Gödel's Second Incompleteness Theorem is the most famous example) and in terms of applications to other branches of mathematics." ([5] p. 6)

But mathematical logic does not only contribute to mathematics, it is increasingly used by other disciplines too, especially by computer science. In particular, *Artificial Intelligence* has been shaped to a great extent by the paradigm of mathematical logic and, as a consequence, artificial intelligence is stimulating research in mathematical logic with growing influence. Thus, the actual applications make it necessary to rethink traditionally formulated purposes of mathematical logic and to analyze the effects of results of logic research. Serious critics of developments in artificial intelligence must be considered with respect to the role of logic within those developments. A basic question is: How much does mathematical logic support further mechanizations of human thinking so that the cognitive autonomy of human beings becomes endangered? (cf. [17])

It has to be emphasized that ethics cannot be avoided in discussing the fundamental understanding of logic and its mathematization. Based on Peirce's pragmatism, Karl-Otto Apel has even shown that, in the sense of transcendental philosophy, ethics is necessary as a condition of the possibility of logic; he argues:

> "The logical validity of arguments cannot be examined without presuming, in principle, a community of thinkers who are qualified for intersubjective understanding and consensus formation. [...] With the real community of argumentation, the logical justification of our thinking also presupposes the act on a fundamental moral norm [...]: In the community of communication the mutual acknowledgment of all members as equal partners of discussion is assumed." ([2] p. 399f)

Concerning mathematical logic, the human communication and argumentation must be particularly considered since logic-mathematical methods cannot grasp realities without an eventually serious loss of content. Therefore mathematizations have always to keep the connections to their origins so that the consequences of mathematical treatments may be rationally analyzed and interpreted in human communication. The connections of logic to reality have been narrowed since Frege's turn to predicate logic, the leading paradigm of mathematical logic today. Thus, restructuring has to establish a broader understanding of mathematical logic, in particular, by elaborating the pragmatic dimension.

For activating real communication and argumentation, it seems to be most important to build enough bridges from the logic-mathematical theory to reality. One way to do this is to revitalize the traditional paradigm of logic given by "the three essential main functions of thinking - *concepts, judgments, and conclusions*" ([9] p. 6). This would particularly bring in the epistemological view which is essential for an appropriate understanding of mathematics. Peirce has already included this view in his definitions of mathematics: "the one by its method, that of drawing necessary conclusions; the other by its aim and subject matter, as study of hypothetical states of things" ([15] 4.236). An appropriate formalization of concepts may serve as the most elementary connection between mathematical logic and reality. Of course, mathematical logic uses already several references and abstractions for a concept, for instance: sign, symbol, term, expression, predicate, set, class etc. But they mostly reflect a too restricted understanding of a concept; in particular, a tight connection between the extension and the intension as constitutive components of a concept is usually missing. If the formalization of concepts has already deficits for human communication and argumentation, judgments as combinations of concepts cannot be formalized appropriately, which also has negative consequences for the treatment of conclusions. Thus, a formal theory of concepts including the pragmatic view will be basic for a reality-oriented mathematical logic. An approach starting from such basic theory should allow us to "restructure" a large body of the rich mathematical theory of logic so that its purposes, interpretations, and effects may be available for discussion and understanding in the human community of communication.

4 Formalization of Concepts

There is a long philosophical tradition concerned with understanding a *concept* as constituted by its extension and its intension (cf. [7], [19]); the *extension* consists of all objects belonging to the concept while the *intension* comprises all attributes valid for those objects. These two components represent the fundamental complementarity of particularity and generality as it is present in the relationship of reality and thought (cf. [14]). A formalization of concepts in the pragmatic view should grasp both components to allow enough references for human communication and argumentation. Such an approach, which has been proven successful in many applications, has been developed in *Formal Concept Analysis* (see [21],[26],[6]). It starts with the primitive notion of a *formal context* which is defined as a triple (G, M, I) where G and M are sets while I is a binary relation between G and M, i.e. $I \subseteq G \times M$; the elements of G and M are called *objects* (in German: *Gegenstände*) and *attributes* (in German: *Merkmale*), respectively, and gIm, i.e. $(g, m) \in I$, is read: the object g *has* the attribute m. For constructing formal concepts within a given formal context (G, M, I), the following *derivation operators* are defined for $X \subseteq G$ and $Y \subseteq M$:

$$X \mapsto X' := \{m \in M \mid gIm \text{ for all } g \in X\},$$
$$Y \mapsto Y' := \{g \in G \mid gIm \text{ for all } m \in Y\}.$$

A *formal concept* of a formal context (G, M, I) is defined as a pair (A, B) with $A \subseteq G$, $B \subseteq M$, $A = B'$, and $B = A'$; A and B are called the *extent* and the *intent* of the concept (A, B), respectively. A general method of constructing formal concepts uses the derivation operators to obtain for $X \subseteq G$ and $Y \subseteq M$ the formal concepts (X'', X') and (Y', Y''). The formalization of the *subconcept-superconcept-relation* is defined by

$$(A_1, B_1) \leq (A_2, B_2) :\Longleftrightarrow A_1 \subseteq A_2 \ (\Longleftrightarrow B_1 \supseteq B_2).$$

The set of all formal concepts of (G, M, I) together with the defined order relation is denoted by $\underline{\mathfrak{B}}(G, M, I)$. For a restructured approach to mathematical logic based on a formalization of concepts, it is important to clarify the structure formed by the formal concepts. This is achieved for $\underline{\mathfrak{B}}(G, M, I)$ by the following theorem (see [21]):

The Basic Theorem of Dyadic Concept Analysis. *Let (G, M, I) be a formal context. Then $\underline{\mathfrak{B}}(G, M, I)$ is a complete lattice, called the* concept lattice *of (G, M, I), for which infima and suprema can be described as follows:*

$$\bigwedge_{t \in T} (A_t, B_t) = (\bigcap_{t \in T} A_t, (\bigcup_{t \in T} B_t)''), \quad \bigvee_{t \in T} (A_t, B_t) = ((\bigcup_{t \in T} A_t)'', \bigcap_{t \in T} B_t).$$

In general, a complete lattice L is isomorphic to $\underline{\mathfrak{B}}(G, M, I)$ if and only if there exist mappings $\tilde{\gamma} : G \longrightarrow L$ and $\tilde{\mu} : M \longrightarrow L$ such that $\tilde{\gamma}G$ is supremum-dense *in L (i.e. $L = \{\bigvee X \mid X \subseteq \tilde{\gamma}G\}$), $\tilde{\mu}G$ is* infimum-dense *in L (i.e. $L = \{\bigwedge X \mid X \subseteq \tilde{\mu}M\}$), and $gIm \Longleftrightarrow \tilde{\gamma}g \leq \tilde{\mu}m$ for $g \in G$ and $m \in M$; in particular, $L \cong \underline{\mathfrak{B}}(L, L, \leq)$.*

By the Basic Theorem, the ordered sets $\underline{\mathfrak{B}}(G, M, I)$ are exactly the complete lattices up to isomorphism. Since concept lattices can be effectively represented by *labelled line diagrams* (at least in the finite case), they have proved to be as successful tools for communication and argumentation (cf. [24],[25]). If one attaches the name of each object g to the represented concept $\gamma g := (\{g\}'', \{g\}')$ and the name of each attribute m to the represented concept $\mu m := (\{m\}', \{m\}'')$ then extensions, intensions, and the underlying context can be recovered so that various bridges to the original data may be activated for discussing the content. Experiences have shown that, in many cases, the specific meanings of the relationships given by I should also have formal elements of reference. This together with Peirce's ontological view of his three universal categories has recently led to a triadic approach to formal concept analysis (see [12]).

The triadic formalization of concepts is based on the notion of a *triadic context* defined as a quadruple (G, M, B, Y) where G, M, and B are sets and Y

is a ternary relation between G, M, and B, i.e. $Y \subseteq G \times M \times B$; the elements of G, M, and B are called *objects*, *attributes*, and *conditions*, respectively (cf. [11]), and $(g, m, b) \in Y$ is read: the object g *has* the attribute m *under* the condition b (the relational notation $b(g, m)$ may be also used for $(g, m, b) \in Y$). The conditions in B may include relations, mediations, interpretations, evaluations, meanings, reasons etc. concerning connections between objects and attributes. For defining derivation operators it is convenient to use K_1, K_2, K_3 for G, M, B, which may also indicate that K_i corresponds to Peirce's *i-th category*. A triadic context $\mathbb{K} := (K_1, K_2, K_3, Y)$ gives rise to numerous *dyadic contexts*; in particular, let us define $\mathbb{K}^{(1)} := (K_1, K_2 \times K_3, Y^{(1)})$, $\mathbb{K}^{(2)} := (K_2, K_1 \times K_3, Y^{(2)})$, $\mathbb{K}^{(3)} := (K_3, K_1 \times K_2, Y^{(3)})$ where

$$gY^{(1)}(m, b) :\Longleftrightarrow mY^{(2)}(g, b) :\Longleftrightarrow bY^{(3)}(g, m) :\Longleftrightarrow (g, m, b) \in Y$$

and, for $\{i, j, k\} = \{1, 2, 3\}$ and $A_k \subseteq K_k$, $\mathbb{K}_{A_k}^{ij} := (K_i, K_j, Y_{A_k}^{ij})$ where

$$(a_i, a_j) \in Y_{A_k}^{ij} :\Longleftrightarrow a_i, a_j, a_k \text{ are related by } Y \text{ for all } a_k \in A_k.$$

The *dyadic derivations* of subsets X are denoted by $X^{(i)}$ in $\mathbb{K}^{(i)}$ and by $X^{(i,j,A_k)}$ in $\mathbb{K}_{A_k}^{ij}$. A *triadic concept* of $\mathbb{K}$ is defined as a triple (A_1, A_2, A_3) with $A_i \subseteq K_i$ for $i = 1, 2, 3$ and $A_i = (A_j \times A_k)^{(i)}$ for $\{i, j, k\} = \{1, 2, 3\}$ with $j < k$; A_1, A_2, and A_3 are called the *extent*, the *intent*, and the *modus* of the triadic concept (A_1, A_2, A_3), respectively. The triadic concepts may be considered as elementary units of thought, because they are exactly the maximal triples (A_1, A_2, A_3) with $A_1 \times A_2 \times A_3 \subseteq Y$ ("dyadic" concepts are exactly the maximal pairs (A_1, A_2) with $A_1 \times A_2 \subseteq I$). Triadic concepts may be generated by two sets: for instance, $X_1 \subseteq K_1$ and $X_3 \subseteq K_3$ give rise to the triadic concept (A_1, A_2, A_3) with $A_2 := X_1^{(1,2,X_3)}$, $A_1 := A_2^{(1,2,X_3)}$, and $A_3 := (A_1 \times A_2)^{(3)}$ which, under the conditions of X_3, yields the smallest extent containing X_1.

The triadic concepts are naturally structured by the quasiorders $\lesssim_i$ with

$$(A_1, A_2, A_3) \lesssim_i (B_1, B_2, B_3) :\Longleftrightarrow A_i \subseteq B_i \quad (i = 1, 2, 3).$$

To understand the structure $\underline{\mathfrak{T}}(\mathbb{K})$ of all triadic concepts of a triadic context $\mathbb{K}$ together with those quasiorders, some further definitions are necessary: A *triordered set* is defined as a relational structure $(S, \lesssim_1, \lesssim_2, \lesssim_3)$ for which the $\lesssim_i$ are quasiorders on S such that $\lesssim_i \cap \lesssim_j \subseteq \gtrsim_k$ for $\{i, j, k\} = \{1, 2, 3\}$ and $\sim_1 \cap \sim_2 \cap \sim_3 = id_S$ where $\sim_i := \lesssim_i \cap \gtrsim_i$ $(i = 1, 2, 3)$. For $X_i, X_k \subseteq S$, an element u of S is called an *ik-bound* of (X_i, X_k) if $u \gtrsim_i x$ for all $x \in X_i$ and $u \gtrsim_k x$ for all $x \in X_k$. An *ik*-bound u of (X_i, X_k) is an *ik-limit* of (X_i, X_k) if $u \gtrsim_j v$ for all *ik*-bounds v of (X_i, X_k) and an *ik*-limit u is the *ik-join* of (X_i, X_k) if $u \lesssim_k w$ for all $w \in S$ with $w \sim_j u$ and $w \gtrsim_k x$ for all

$x \in X_k$ ($\{i,j,k\} = \{1,2,3\}$). There exists at most one ik-join of (X_i, X_k) in $(S, \lesssim_1, \lesssim_2, \lesssim_3)$ which is denoted by $\nabla_{ik}(X_i, X_k)$. A *complete trilattice* is defined as a triordered set $(S, \lesssim_1, \lesssim_2, \lesssim_3)$ in which the ik-joins exists for all $i \neq k$ in $\{1,2,3\}$ and all pairs of subsets of S. Examples of complete trilattices which are of particular interest to logic are the *Boolean trilattices* based on an arbitrary set S: such a trilattice consists of all triples (A_1, A_2, A_3) of subsets of S with $A_1 \cap A_2 \cap A_3 = \emptyset$ and $A_i \cup A_j = S$ for $i \neq j$ in $\{1,2,3\}$, the quasiorders are defined as in the case of triadic concepts.

It can be shown that the relational structures $\underline{\mathfrak{T}}(\mathbb{K})$ are exactly the complete trilattices up to isomorphism. This is a consequence of the triadic version of the Basic Theorem for which some further notions shall be introduced. For a complete trilattice $\underline{S} := (S, \lesssim_1, \lesssim_2, \lesssim_3)$, the set of all order filters of $(S, \lesssim_i)$ is denoted by $\mathcal{F}_i(\underline{S})$ ($i = 1,2,3$). A subset $\mathcal{X}$ of $\mathcal{F}_i(\underline{S})$ is called *i-dense* if each principal filter of $(S, \lesssim_i)$ is the intersection of some order filters from $\mathcal{X}$. For $l \in \{1,2,3\}$ and $\mathcal{X} \subseteq \underline{\mathfrak{T}}(\mathbb{K})$, let $\bigcup_l \mathcal{X} := \bigcup\{X_l \mid (X_1, X_2, X_3) \in \mathcal{X}\}$. Now, the theorem can be formulated (see [28]):

The Basic Theorem of Triadic Concept Analysis.
Let $\mathbb{K} := (K_1, K_2, K_3, Y)$ be a triadic context. Then $\underline{\mathfrak{T}}(\mathbb{K})$ is a complete trilattice, called the concept trilattice *of $\mathbb{K}$, for which the ik-joins can be described as follows ($\{i,j,k\} = \{1,2,3\}$):*

$$\nabla_{ik}(\mathcal{X}_i, \mathcal{X}_k) = (A_1, A_2, A_3) \text{ with } A_j := (\textstyle\bigcup_i \mathcal{X}_i)^{(i,j, \bigcup_k \mathcal{X}_k)}, \; A_i := A_j^{(i,j, \bigcup_k \mathcal{X}_k)},$$
$$\text{and } A_k := (A_i \times A_j)^{(k)} \text{ if } i < j \text{ or } A_k := (A_j \times A_i)^{(k)} \text{ if } j < i.$$

In general, a complete trilattice $\underline{L} := (L, \lesssim_1, \lesssim_2, \lesssim_3)$ is isomorphic to $\underline{\mathfrak{T}}(\mathbb{K})$ if and only if there exist mappings $\tilde{\kappa}_i : K_i \longrightarrow \mathcal{F}_i(\underline{L})$ ($i = 1,2,3$) such that $\tilde{\kappa}_i(K_i)$ is i-dense and $A_1 \times A_2 \times A_3 \subseteq Y \iff \bigcap_{i=1}^3 \bigcap_{a_i \in A_i} \tilde{\kappa}_i(a_i) \neq \emptyset$ for all $A_1 \subseteq K_1$, $A_2 \subseteq K_2$, and $A_3 \subseteq K_3$; in particular, $\underline{L} \cong \underline{\mathfrak{T}}(L, L, L, Y_L)$ where

$$(x_1, x_2, x_3) \in Y_L :\iff \exists u \in L \forall i \in \{1,2,3\} : x_i \lesssim_i u.$$

Concept trilattices may be also visualized by diagrams for communication and argumentation. But labelling may become more complicated, since the components of the triadic concepts do not necessarily form closure systems (as in the dyadic case). Further research is needed to see how concept trilattices can be used for human discourse.

5   Judgments and Conclusions

In philosophy and traditional logic, *judgments* are central for the theme of cognition and knowledge. This designation becomes apparent in the following quotation from an actual lexicon of philosophy:

> "*Judgment*, in the sense of logic, is an act of affirmation or negation in which two concepts (subject and predicate) are mutually set in relation. [...] Each judgment internally and inseparably contains a relation to embodiments of possible subjects of knowledge, of possible facts, and of necessary conditions." ([16] p. 679f)

The ontological and epistemological nature of judgments are basic for the mediation of thought and reality. Therefore the replacement of judgments by *formal propositions* in mathematical logic has been a dramatic change because it has weakened the ontological ties of logic; in particular, the understanding of concepts within judgments, concerning their role of joining thought with reality, disappeared in formal logic. It is an important task of restructuring mathematical logic to recover judgments in their traditional understanding and to integrate them by appropriate formalizations. In view of Peirce's pragmatism, restructuring must take into consideration that the validity of judgments can only be confirmed by an argumentative discourse in the intersubjective community of communication. This again suggests the contextual approach of formalization as it has been already used for mathematizing concepts and their relationships.

In this paper, only an approach based on formal contexts as defined in Section 4 can be outlined. *Irreducible formal judgments* within a formal context (G, M, I) are introduced as relationships

$$\mathfrak{a} \leq \mathfrak{b} \quad \text{and} \quad \mathfrak{a} \not\leq \mathfrak{b}$$

between formal concepts $\mathfrak{a}$ and $\mathfrak{b}$ of (G, M, I) which might be true or false. Other elementary types of relationships can be expressed by irreducible judgments, namely:

⟨ the object g has the attribute m ⟩ by $\gamma g \leq \mu m$,
⟨ the object g is an instance of the concept $\mathfrak{a}$ ⟩ by $\gamma g \leq \mathfrak{a}$, and
⟨ the attribute m belongs to the concept $\mathfrak{a}$ ⟩ by $\mathfrak{a} \leq \mu m$ (cf. [13]).

As in propositional logic, a *formal judgment logic* can be developed - starting from the irreducible formal judgments of a given formal context - by recursively defining the *formal judgments* using the propositional connectives $\wedge$, $\vee$, $\neg$, and $\rightarrow$. The *semantics* of a formal judgment logic is not only based on truth values, but on the relationships in the concept lattice of the underlying formal context too; in particular, a formal judgment is either true or false. An advantage is that logical connectives may be turned into lattice operations or relations as the following equivalences of formal judgments show:

$$(a \leq b) \wedge (a \leq c) \iff a \leq (b \wedge c),$$
$$(a \leq b) \wedge (b \leq c) \iff (a \vee b) \leq c,$$
$$(a \not\leq b) \vee (a \leq c) \iff a \not\leq (b \wedge c),$$
$$(a \not\leq c) \vee (b \not\leq c) \iff (a \vee b) \not\leq c,$$
$$\neg(a \leq b) \iff a \not\leq b,$$
$$\neg(a \not\leq b) \iff a \leq b,$$
$$(a \leq b) \rightarrow (a \not\leq c) \iff a \not\leq (b \wedge c),$$
$$(a \leq c) \rightarrow (b \not\leq c) \iff (a \vee b) \not\leq c.$$

Of course, a language of *judgment formulas* should also be elaborated, but experiences have shown that in many applications it is enough to use a formal judgment logic directly. Just the aim to keep the ties to reality as close as possible underlines the necessity of arguing in tight connection with the given data context.

An extension of formal judgment logic has been suggested by a formal treatment of inference and acquisition of conceptual knowledge (see [23],[13],[26], [1],[29]). The basic idea of this approach is to distinguish between the present and the potential knowledge. The potential conceptual knowledge has to be specified by a *conceptual universe* which is defined to be a formal context $\mathbb{U} := (G_{\mathbb{U}}, M_{\mathbb{U}}, I_{\mathbb{U}})$. The model of the present knowledge consists of conceptual knowledge which can be derived from the conceptual universe; in particular, the formal judgment logic of the conceptual universe is basic for the representation of the present knowledge. To enrich the possibilities of expressing conceptual knowledge, the notion of formal concepts has been generalized to the notion of semiconcepts, which yields richer algebraic structures also including operations of negation (s. [13]). A *semiconcept* of a formal context (G, M, I) is defined as a pair (A, B) with $A \subseteq G$ and $B \subseteq M$ such that $A = B'$ or $B = A'$. The semiconcepts are ordered by

$$(A_1, B_1) \sqsubseteq (A_2, B_2) :\iff A_1 \subseteq A_2 \text{ and } B_1 \supseteq B_2.$$

This order does not yield a lattice structure in general, but there are natural generalizations of the meet and join operation of $\underline{\mathfrak{B}}(G, M, I)$ and, in addition, order-reversing operations of negation:

$$(A_1, B_1) \sqcap (A_2, B_2) := (A_1 \cap A_2, (A_1 \cap A_2)'),$$
$$(A_1, B_1) \sqcup (A_2, B_2) := ((B_1 \cap B_2)', B_1 \cap B_2),$$
$$\neg(A, B) := (G \setminus A, (G \setminus A)'),$$
$$\lrcorner(A, B) := ((M \setminus B)', M \setminus B).$$

The constants $(\emptyset, M)$ and $(G, \emptyset)$ are considered as nullary operations $\bot$ and $\top$, respectively. The set of all semiconcepts of (G, M, I) together with the operations $\sqcap$, $\sqcup$, $\neg$, $\lrcorner$, $\bot$, and $\top$ is called the *algebra of semiconcepts* of (G, M, I) and denoted by $\mathfrak{H}(G, M, I)$. It is covered by two canonical Boolean algebras $(\{(A, A') \mid A \subseteq G\}, \sqcap, \neg, \bot)$ and $(\{(B', B) \mid B \subseteq M\}, \sqcup, \lrcorner, \top)$ which have the concept lattice $\mathfrak{B}(G, M, I)$ as intersection.

The algebra of semiconcepts of the conceptual universe $\mathbb{U}$ allows us to represent inferences of conceptual knowledge by algebraic inequalities. Thus, procedures concluding new knowledge from present knowledge can be treated as *word problems* according to the algebra $\mathfrak{H}(\mathbb{U})$. It is helpful to know that the word problems of algebras of semiconcepts are solvable (see [8]), but fast algorithms to perform the solution of those word problems are still missing. In general, an effective theory of *formal conclusions* must be developed for formal judgment logics (and their potential triadic extensions). In view of Peirce's pragmatism, this theory should also be designed to join thought and reality and to support human communication and argumentation.

Let us finally remark that the presented ideas of restructuring mathematical logic should only be understood as a first indication how one may restructure formal logic in view of pragmatism and discourse philosophy. Other attempts and approaches should be performed. Of course, a restructured development should capture as much as possible of the rich culture of modern mathematical logic. In particular, it is desirable that restructuring creates a pragmatic version of the language and theory of first order logic.

References

[1] U. Andelfinger, *Begriffliche Wissenssysteme aus pragmatisch-semiotischer Sicht*, Begriffliche Wissensverarbeitung: Grundfragen und Aufgaben (R. Wille and M. Zickwolff, eds.), B.I.-Wissenschaftsverlag, Mannheim, 1994, pp. 153–172.

[2] K.-O. Apel, *Das Apriori der Kommunikationsgemeinschaft und die Grundlagen der Ethik*, Transformation der Philosophie. Band 2: Das Apriori der Kommunikationsgemeinschaft, Suhrkamp Taschenbuch Wissenschaft **165**, Frankfurt, 1976.

[3] _____, *Von Kant zu Peirce: Die semiotische Transformation der Transzendentalen Logik*, Transformation der Philosophie. Band 2: Das Apriori der Kommunikationsgemeinschaft, Suhrkamp Taschenbuch Wissenschaft **165**, Frankfurt, 1976.

[4] _____, *Begründung*, Handlexikon der Wissenschaftstheorie (H. Seiffert and G. Radnitzky, eds.), Ehrenwirth, München, 1989, pp. 14–19.

[5] J. Barwise, *An introduction to predicate logic*, Handbook of mathematical logic (J. Barwise, ed.), North Holland, Amsterdam, 1977, pp. 5–46.

[6] B. Ganter and R. Wille, *Formale Begriffsanalyse: Mathematische Grundlagen*, (in preparation).

[7] N. Hartmann, *Aristoteles und das Problem des Begriffs*, Abh. Preuss. Akad. Wiss. Jg. 1939. Phil. –hist. Kl. Nr. 5, Verlag Akad. Wiss., Berlin, 1939.

[8] C. Herrmann, P. Luksch, M. Skorsky, and R. Wille, *Algebras of semiconcepts and double Boolean algebras*, (in preparation).

[9] I. Kant, *Logic*, Dover, New York, 1988.

[10] W. Kollewe, M. Skorsky, F. Vogt, and R. Wille, *TOSCANA - ein Werkzeug zur begrifflichen Analyse und Erkundung von Daten*, (R. Wille and M. Zickwolff, eds.), Begriffliche Wissensverarbeitung: Grundfragen und Aufgaben, B.I.-Wissenschaftsverlag, Mannheim, 1994, pp. 267–288.

[11] S. Krolak Schwerdt, P. Orlik, and B. Ganter, *TRIPAT: a model for analyzing three-mode binary data*, Information systems and data analysis (H.-H. Bock, W. Lenski, and M.M. Richter, eds.), Springer, Berlin-Heidelberg, 1994, pp. 298–307.

[12] F. Lehmann and R. Wille, *A triadic approach to formal concept analysis*, (in preparation).

[13] P. Luksch and R. Wille, *A mathematical model for conceptual knowledge systems*, Classification, data analysis, and knowledge organization (H.-H. Bock and P. Ihm, eds.), Springer, Berlin-Heidelberg, 1991, pp. 156–162.

[14] M. Otte, *Das Formale, das Soziale und das Subjektive: eine Einführung in die Philosophie und Didaktik der Mathematik*, Suhrkamp Taschenbuch Wissenschaft **1106**, Suhrkamp, Frankfurt, 1994.

[15] Ch.S. Peirce, *Collected Papers*, Harvard Univ. Press, Cambridge, 1931–35.

[16] H. Schmidt, *Philosophisches Wörterbuch*, 19. Aufl., Kröner, Stuttgart, 1974.

[17] G. Unseld, *Maschinenintelligenz oder Menschenphantasie? Ein Plädoyer für den Ausstieg aus unserer technisch-wissenschaftlichen Kultur*, Suhrkamp-Taschenbuch Wissenschaft **987**, Suhrkamp, Frankfurt, 1992.

[18] H. von Hentig, *Magier oder Magister? über die Einheit der Wissenschaft im Verständigungsprozess*, Suhrkamp Taschenbuch **207**, Suhrkamp, Frankfurt, 1974.

[19] M. Weitz, *Theories of concepts: A history of the major philosophical tradition*, Routledge, London, 1988.

[20] R. Wille, *Versuch der Restrukturierung von Mathematik am Beispiel der Grundvorlesung "Lineare Algebra"*, Beiträge zum Mathematikunterricht 1981 (B. Artmann, ed.), Schroedel, Hannover, 1981, pp. 102–112.

[21] _____, *Restructuring lattice theory: an approach based on hierarchies of concepts*, Ordered sets (I. Rival, ed.), Reidel, Dordrecht-Boston, 1982, pp. 445–470.

[22] _____, *Allgemeine Wissenschaft als Wissenschaft für die Allgemeinheit*, Verantwortung in der Wissenschaft (H. Böhme and H.-J. Gamm, eds.), TH Darmstadt, 1988, pp. 159–176.

[23] _____, *Knowledge acquisition by methods of formal concept analysis*, Data analysis, learning symbolic and numeric knowledge (E. Diday, ed.), Nova Science Publ., New York-Budapest, 1989, pp. 365–380.

[24] _____, *Lattices in data analysis: how to draw them with a computer*, Algorithms and order (I. Rival, ed.), Kluwer, Dordrecht-Boston, 1989, pp. 33–58.

[25] _____, *Begriffliche Datensysteme als Werkzeug der Wissenskommunikation*, Mensch und Maschine - Informationelle Schnittstellen der Kommunikation (H. H. Zimmermann, H.-D. Luckhardt, and A. Schulz, eds.), Universitätsverlag Konstanz, Konstanz, 1992, pp. 63–73.

[26] _____, *Concept lattices and conceptual knowledge systems*, Computers & Mathematics with Applications **23** (1992), 493–515.

[27] _____, *Plädoyer für eine philosophische Grundlegung der Begrifflichen Wissensverarbeitung*, Begriffliche Wissensverarbeitung: Grundfragen und Aufgaben (R. Wille and M. Zickwolff, eds.), B.I.-Wissenschaftsverlag, Mannheim, 1994, pp. 11–25.

[28] _____, *The basic theorem of triadic concept analysis*, 1995, pp. 149–159.

[29] M. Zickwolff, *Zur Rolle der Formalen Begriffsanalyse in der Wissensakquisition*, Begriffliche Wissensverarbeitung: Grundfragen und Aufgaben (R. Wille and M. Zickwolff, eds.), B.I.-Wissenschaftsverlag, Mannheim, 1994, pp. 173–189.

FACHBEREICH MATHEMATIK, TECHNISCHE HOCHSCHULE DARMSTADT, D-6100 DARMSTADT, GERMANY

e-mail address: WILLE@MATHEMATIK.TH-DARMSTADT.DE

The development of research in Algebra in Italy from 1850 to 1940 [*]

Guido Zappa

We choose 1850 as the starting point for our exposition of the development of algebra in Italy. That year immediately follows the dramatic events of 1848-1849, which from one side ended up with a failure on the political-military ground, but on the other side have leaded to the ripening, at least in the social circles more open to novelties, of the consciousness of the belonging of the Italians to one and the same country and of the need of unify themselves into one and the same state. The mathematicians who in the following years will contribute more to the scientific growth, all had been marked by the fight for independence. Thus Francesco Brioschi had participated to the Five Days revolt in Milan, Enrico Betti had fought in the ranks of the battalion of the Tuscan Universities in the battle of Curtatone and Montanara, Giuseppe Battaglini had been persecuted by the Bourbon government in Naples.

Moreover, in the decade 1850-1860 took place the development of the scientific activity of the two most prominent exponents among italian algebraists at that time: Betti and Brioschi.

Finally, precisely in 1850, appeared the first issue of the journal "Annali di Scienze Matematiche Fisiche e Naturali", edited by the abbot Barnaba Tortolini. This was the first italian periodical devoted exclusively to mathematics and physics; later it will become "Annali di Matematica Pura ed Applicata", that is still one of our best mathematical journals.

For obvious reasons of time and space, our review cannot contain all the algebraic contributions of the italian mathematicians of the period in consideration; we will have to limit ourselves to the most prominent authors and the more meaningful results.

---

[*]This lecture was originally given in French; the english translation was cured by the Editors of the Proceedings.

1   Enrico Betti and Galois theory.

The publication, in 1846, in the "Journal of Liouville" of the celebrated memoir on the resolution of algebraic equations, written by Evariste Galois before his tragic death in 1832, awakened the interest of the young Tuscan mathematician Enrico Betti (1823 - 1892).

After graduating in Pisa in 1846, Betti published in 1851 a memoir entitled "Sopra la risolubilità per radicali delle equazioni algebriche di grado primo" (On the solvability by radicals of algebraic equations of prime degree), where it is proved that an irreducible equation of prime degree is solvable by radicals if and only if its roots are expressible as rational functions of any two of them. The necessity of the condition has been already proven by Abel, while Galois had remarked that it is also sufficient, giving a hint of the proof.

Betti returned on the problem of resolution by radical of algebraic equations, on the path drawn by Galois, in several successive papers, among which is prominent the memoir "Sulla risoluzione delle equazioni algebriche" (On the resolution of algebraic equations) in 1852. In its first part, he expounds and develops the elements of the theory of permutation groups contained in Galois writings. As it is well-known, these were very concise and sometimes difficult to interpret. In them, in particular, the term "group" is used with different meanings. Betti tries to clarify the matter; nevertheless also his treatment is not totally clear. Among other things, he introduces for the first time (except for a hint already in Galois) the concept of "quotient group" of a permutation group G over a normal subgroup N. The quotient group is introduced as a group of permutation induced on the set of the right cosets of N in G, by right multiplication with elements of G.

In the second part of that memoir, he proves the theorems on solvability by radicals of algebraic equations, given in a sketchy way by Galois.

2   The solution of 5th and 6th degree equations. The theory of algebraic forms and Francesco Brioschi.

A problem that stimulated the mathematicians at that time was to find solution formulas of the general equation of degree 5. Those formulas, of course, could not involve only rational operations and root extractions, but also functions of other kind. Betti explored this path, but afterwards he gave up, since he had been preceded by Hermite and Kronecker who, independently, expressed the desired solution by elliptic functions. Brioschi too gave remarkable contributions to this problem. We also owe to him the determination of solution formulas for the general equation of 6th degree via iperelliptic functions.

Betti gave up before long from dealing with algebraic problems, being drawn by questions of mathematical physics, but in his teachings at Pisa University and at the Scuola Normale Superiore - of which he was the principal - he contributed in keeping alive the interest in algebra.

Brioschi (1824 - 1873) instead went on all his life in dealing with algebraic problems, besides analysis and applications. He graduated in Pavia in 1845. He was a prominent mathematician both as a scientist and as an organizer. He was, for a certain period, General Secretary at the Ministry of Education, and as such he thoroughly engaged himself for the scientific and didactic growth of italian universities. In 1865 he established the Polytechnic in Milan, of which he was the principal and professor of hydraulics up to his death.

Brioschi helped in spreading in Italy some theories which had arisen in other countries, adding to them very valuable contributions. We owe to him one of the first treatise on determinants. Many of his papers dealt with the theory of algebraic forms, of invariants and covariants. This theory was founded by Cayley, who took inspiration from a work of Boole, and initially was developed by Cayley himself and Sylvester. Brioschi became an authoritative part in the framework of researchers in this field, producing an impressive number of major results, up to those of scholars of other countries. We cannot quote his main results, due to the complexity of the theory. Rather, let us emphasize the distinctive features of his scientific work, very well described by Beltrami in his commemoration before the Accademia dei Lincei. Beltrami maintains that "Brioschi si è sempre mantenuto fedele all'indirizzo prettamente classico, [....], versando a piene mani in ogni sua ricerca il tesoro inesauribile delle sue formule agili e penetranti e conseguendo fama indisputata d'insuperabile virtuosità nel maneggio di ogni più raffinato strumento dell'arte analitica" [1].

Regarding the theory of invariants, it is to be remembered also the name of Francesco Faà di Bruno (1825 - 1868), professor in Turin, who gave a few valuable contributions to this field of studies and published a treatise on binary forms, which was very praised by the foremost scholars in the field and was translated into German with an integration by Max Noether.

The study of algebraic forms has also a geometric interest, therefore several italian geometers produced noticeable contributions to the subject.

---

[1] "Brioschi constantly abode by the very classical stream [....], shedding by handfuls in each of his works the endless treasure of his nimble and acute formulas, thus gaining unquestioned reputation of unsurpassable virtuosity in the handling of all most refined tools of the analytical art."

3   The second generation of algebraists after the Unity of Italy.   Alfredo Capelli, Giovanni Frattini and Luigi Bianchi.

Coming to more recent times, the most representative algebraist who first appears is Alfredo Capelli (1855 - 1910). Born in Milan, he graduated in Rome in 1877, under the supervision of Giovanni Battaglini. The latter had formerly been professor at Naples University, where he started the "Giornale di Matematiche" which after his death became "Giornale di Matematiche di Battaglini". This journal too helped considerably with the spreading of the results of our scholars in the last century.

In 1870 appeared in France the famous "Traité des substitutions" of Camille Jordan. It supplied the first systematic exposition of the theory of permutation groups and of the resolution of algebraic equations according to Galois.

In 1875 - 1876 Battaglini lectured in Rome on permutation groups, following the guidelines of Jordan, and advised the young Capelli to tackle problems on groups in his thesis. From this thesis sprang a huge memoir, appeared in "Giornale di Matematiche" in 1878. There, among other things, Capelli discovers anew, with original methods, the famous theorem of Sylow, that he did not know, and gives remarkable properties of p-groups. In 1884 he published a memoir of the Accademia dei Lincei, again dedicated to groups, proving that any non nilpotent normal subgroup of a finite group admits a proper supplement and drawing several properties of nilpotent groups. But the main field of research of Capelli is the theory of algebraic forms, to which he contributed with a great number of works. He based his theory on the notion of "polar", and this allowed him to proceed in a more unitary way. We owe to Capelli a well-known formulation, particularly expressive, of the theorem which gives the conditions for the existence of solutions of systems of linear equations (known as "Rouché-Capelli" or "Kronecker-Capelli" theorem). From 1881 he was professor at the University of Palermo and from 1886 at the University of Naples, where he worked intensively as a teacher. He published also a treatise of "Istituzioni di Analisi Algebrica" and contributed to keep alive in Italy the interest in Algebra.

To the quoted memory by Capelli on groups, is linked the main result of Giovanni Frattini (1852 - 1925). Teacher in a Secondary Technical School in Rome, he published several works in group theory and arithmetic, but his main result is the definition and the investigation of that particular subgroup of a (finite) group which today we just call the Frattini-subgroup. Usually it is defined as the intersection of all maximal subgroups, but Frattini defined it as the subgroup composed by those elements which do not belong to any minimal system of generators. Among other things he showed that this subgroup is nilpotent, using an argument taken from Capelli's memoir. The Frattini-subgroup plays an important role in the modern foundation of Group theory.

Albeit he dealt with algebraic problems only marginally, Luigi Bianchi (1856 - 1928) was active in keeping alive in Italy the interest for algebra. Born in Parma, he was a student at the Scuola Normale Superiore in Pisa and then professor at Pisa University. His reputation is due principally to his researches in Differential Geometry and Number Theory. As regards algebra we have a juvenile work (on the trail of Betti, his master) on the Lagrange resolvent of an equation of prime degree, solvable by radicals, and a series of notes on finite algebraic extensions of the rational field. He published several excellent treatises on Differential Geometry, Number Theory and Complex Functions. As far as algebra is concerned, particularly important is a volume on the theory of finite permutation groups and of algebraic equations according to Galois, sprang from a course he gave unceasingly at the Scuola Normale Superiore, that was carried on by Giovanni Ricci after his death. It is a book of rather classic style, but extraordinarily clear, which helped very much in preserving the interest in group theory in that period, when the attention payed by italian mathematicians was weakening.

Also Giuseppe Peano (1858 - 1932), the famous mathematician from Piedmont, had a relevant influence on the development of algebra. In his early years he made noticeable contributions to the theory of algebraic forms. Then, with his so-called geometric calculus, he set the foundations of vector calculus. Finally, because of his well-known investigations on the logic foundation of mathematics, he can be considered a pioneer in algebraic logic.

4 The new algebraic school in Sicily: Francesco Gerbaldi, Giuseppe Bagnera, Michele Cipolla.

Some italian geometers dealt with problems relative to finite groups of collineations. Among them, we point out Francesco Gerbaldi (1858 - 1934). In 1890 he became professor of Geometry at the University of Palermo and there he carried out a steady activity in teaching, addressing several students towards problems in geometry and algebra. One of those, Giuseppe Bagnera (1865 - 1927), mathematician of deep insight, published among other things some works on groups. Among those the one stands out, in which a complete classification of all groups of order p^5, p prime, is obtained.

Also Michele Cipolla (1880 -1947) grew up at Gerbaldi's school. He graduated in Palermo in 1902 and became professor at Catania University in 1911, then he was back in Palermo in 1923. He is a remarkable figure of algebraist and started a valid school, to which some damage was maybe caused by an excessive isolation from the international frame. Besides his remarkable results in Number Theory, Cipolla is remembered for having conceived and developed the theory of the so-called "fundamental subgroups" of a finite group. By a "fundamental system" of a group we mean a subset composed exactly of those elements which share one and the same normalizer. To every fundamental system one can associate a "fundamental subgroup", namely the common normalizer of the elements of the system.

Cipolla analyzed thoroughly the structure that consists of the fundamental subgroups and the fundamental systems of a group. He and others, among which Giuseppe Mignosi, his colleague in Palermo, and Vincenzo Amato, his student in Catania, determined the fundamental systems and subgroups of several classes of groups. This is a very polished chapter, even if it hardly fits in the general development of group theory.

Among Gerbaldi's students, should be mentioned also Fortunato Bucca, who died in Palermo in 1899 when he was only 24, after having obtained some smart results in Galois theory.

5 Changes in the mathematical stage in Italy and abroad at the beginning of the new century. Other italian algebraists in this period.

Between the end of the last century and the World War I, a notable change in the position of italian algebraists took place. Previously, chiefly because of Betti, Brioschi and Capelli, Italy was fully in the main line of progress of algebra in the world. But by the end of the 1800 there are fundamental changes in mathematics. Algebraic geometry was born, consisting essentially of the study of algebraic forms and varieties with geometric tools. Italy, by the works of Guido Castelnuovo, Federigo Enriques and Francesco Severi, rises to prominence in this field. This naturally yielded a weakening of the interest for "proper" algebra in Italy, because of the use by the mentioned masters and their students of essentially geometric methods. There are still some valuable algebraists, but they start mainly local schools, like the mentioned sicilian school, scarcely connected with the themes that were prevalent in other countries.

Some researchers of this period in Linear Algebra should be mentioned; among them, in Pisa, Onorato Niccoletti (1872 - 1929), who dealt with Hermitian matrices, and his student Francesco Cecioni (1884 - 1968), author of compelling memoirs in matrix equations; in Naples Ernesto Pascal (1865 - 1940), to whom we owe a praised volume on determinants.

In the meanwhile, abroad, Algebra was undergoing a deep transformation. What came to be known as Modern Algebra, or Abstract Algebra, was coming out. An important stage in this change is the celebrated memoir by Steinitz on field extensions, while the full performance of the new trend is in the works of Artin and Emmy Noether and in the treatise "Modern Algebra" by van der Wärden. Remarkable contributions in this direction are due to some american scholars, like Dickson and Wedderburn, who studied the systems of hypercomplex numbers, afterward called "algebras".

At the root of the new theories stays the axiomatic definition of groups, rings, fields and other similar structures; one leaves out of consideration the nature of the elements which constitutes these structures, while one deduces the general properties of the structures from the axioms. Subsequently these properties are studied in the single concrete cases.

6 Gaetano Scorza, Riemann matrices and the theory of Algebras.

The new trend asserts itself in Italy by the work of Gaetano Scorza (1876 - 1939). This illustrious mathematician from Calabria was a student in Pisa at the Scuola Normale Superiore, and initially addresses his research to algebraic geometry, following the path of his master Eugenio Bertini. In 1912 he was called as a professor of Geometry at the University of Cagliari, and subsequently went to the Universities of Parma, Catania, Naples and Rome.

In a certain stage of his research, Scorza became interested in problems regarding abelian functions. Given a complex algebraic curve of genus p, one can associate to it p independent complex functions (called abelian functions), each having $2p$ independent complex periods. The values of these $2p$ periods for the p functions form a $p \times 2p$ matrix, whose entries are linked with a special inequality. A matrix of this kind was called by Scorza a "Riemann matrix". Such matrices stem out too from other problems in algebraic geometry. From this fact Scorza was leaded to investigate Riemann matrices, regardless the nature of the problems from which they stem out. To each Riemann matrix he associates a rational algebra, i.e. finite dimensional ring over the field of rational numbers. The properties of the various Riemann matrices (hence, in the case of abelian functions, the properties of the algebraic curves from which they come) correspond to properties of the algebras associated to them. These are always unitary and semisimple (i.e. without proper nilpotent ideals) and sometimes simple (i.e. without proper ideals) and sometimes even primitive (i.e. without zero-divisors).

The general theory of Riemann matrices was exposed by Scorza in a celebrated memoir appeared in 1916 in the "Rendiconti del Circolo Matematico di Palermo", and continued afterwards in 1921 in a second memoir on the same journal. In the first part of the latter one, he presents the foundation of the general theory of Algebras, which was almost unknown in Italy. A more thorough and exhaustive presentation was obtained with the publication, still in 1921, of Scorza's volume entitled: "Corpi Numerici ed Algebre", a very praised work which largely contributed to the rebirth in our country of a more lively interest in Algebra in his more modern aspects. Form those years on Scorza's scientific activity was devoted almost exclusively to algebras, obtaining important results, up to the bets foreign scholars. Among other things we owe to him the complete classification of the algebras of the 3rd and 4th order, on any field.

In 1923 the already quoted mathematician Cecioni, who had been deeply interested to researches on algebra by reading Scorza's works, published a valuable memoir where he thoroughly investigated a particular class of primitive algebras. Moreover, soon a group of Scorza's students arose,and they gave remarkable contributions to the theory of algebras; among those, excel Nicolo' Spampinato, professor in Catania and then in Naples , and Salvatore Cherubino, professor in Pisa. Particularly remarkable are their elaborations aimed at extending the theory of analytical functions to functions defined on an algebra.

7   Other algebraic researches between the two World Wars. Concluding remarks.

Scorza was also interested in other algebraic problems. First, he gave attention to the algebras connected to finite groups. Given a field **K** and a finite group **G**, linear combinations of elements of **G** with coefficients in **K**, where operations of addition and multiplication are defined in a natural way, constitute an algebra over **K**, called the "algebra su **K** legata a **G**". Contributions to the knowledge of such algebras where produced by Scorza and by his student Giuseppe Fichera. Scorza, in this period , became interested also in Group Theory, giving substantial contribution to the theory of fundamental subgroups of Cipolla. Also remarkable is a simple, but exceedingly elegant, paper where he characterizes those groups which are a set-theoretical union of three proper subgroups.

Among the algebraic works in this period which do not belong to Scorza's school, three memoirs by Luigi Onofri stand out. They appeared on the "Annali di Matematica" in 1925,1928,and 1929 and concerned permutation groups on a countable set. An earlier investigation on this subject had already been made by Giulio Andreoli in 1915.

On the whole, one notes in this period a certain isolation of italian algebraists with regard to other countries. This is due in part to the fact that they wrote mainly in Italian. But the chief causes are deeper. The period 1920-1940 is characterized by the prevailing of nationalisms, very evident in totalitarian countries, but somehow present in democratic countries too. Even at the international Congresses of mathematicians , one witnessed by no means casual and very revealing absences of great mathematicians from this or that country.

Scorza was, anyway, among italian algebraists the one who was most engaged in updating our researches. He devoted the last years of his life to prepare a treatise on abstract groups, to be published only in 1942, three years after his death. It is a treatise of very modern setting, which fits well side by side to the one, appeared a few years before, by Zassenhaus. In both the treatises a general theory of groups, finite or infinite, is developed, thus outdoing the traditional position focused on finite groups.

We put the upper bound of our survey to 1940. In 1939 Gaetano Scorza died, and World War II had broken out; Italy went into war in 1940. This was the end of an age in which, as we said above, isolation was prevailing. In the meanwhile, in other countries Algebra had undergone great advances. The synthetical methods of the italian school of algebraic geometry had given grand results, but they were no more enough to face the more and more complex problems which came up. It was necessary to go back to the algebraic basis, and to elaborate new tools. Hence the development of commutative Algebra. Then the rise of new areas in mathematics, like Logic and Computer Science, called for the creation of new algebraic tools. After World War II, in the renewed expansive mood of mathematics, in our country too, Algebra blossomed up.

VIA QUINTINO SELLA 45, I-50136, FIRENZE, ITALY

A Criterion to Decide the Semantic Matching Problem

G. Aguzzi U. Modigliani

Abstract

A class of theories presented by constructor-based convergent rewrite systems with decidable matching problem is characterized by means of some semantic requirements on the rewrite system. Such requirements are satisfied whenever in any reduction each recursive call to a function occurs into a term in a position such that the term is P increasing at that position, P being a property of terms like depth, size, and so on. The P increase property enables a sorting of the equations generated during the execution of the adopted matching procedure according to a well-founded ordering.

1   Introduction

Semantic matching is generally considered a simpler version of semantic unification w.r.t. a given equational theory (see however [4] for a more detailed discussion). While the latter process generates a basis set of unifiers, substitutions which make two terms equal in the given theory, the former restricts the substitutions to apply only to the term on the left (the pattern) of the given equation (goal) $s \to^? t$.

Semantic matching has potential applications in pattern-directed languages. It is, therefore, of some importance to find suitable classes of equational theories for which the semantic matching problem is decidable. In fact, in the general case, both semantic unification and matching are undecidable problems, even when the theory is presented as a finite and convergent set of rewrite rules [3, 9, 6]. Despite such undecidability, nontrivial classes of equational theories presented by convergent term rewriting systems possess a decidable matching problem. But, for some special classes of theories, for instance associativity and commutativity, even semantic unification is decidable.

Following the lines yet expressed in our previous works [2, 1] and the terse exposition given in [7], we characterize a class of theories presented by constructor-based convergent rewrite systems with decidable matching problem by means of some semantic requirements on the rewrite system. Such requirements are satisfied whenever in any reduction each recursive call to a function occurs into a term in a position such that the term is P increasing at that position, P being a property of terms like depth, size, and so on. The P increase property enables us to sort the equations generated during the execution of the matching procedure, presented below, according to a well-founded ordering.

In our previous papers [2, 1] a decidability condition for the matching problem was described in terms of increase of a particular property of terms, namely increase of the length of terms like naturals and lists, and the way to prove such an increase was also technically different and less powerful w.r.t. the way adopted in this paper.

We consider the class of theories presented by constructor-based rewrite systems because our method is directly applicable to such rewrite systems, while our increase criterion seems difficult to be proved in the general case of non constructor-based rewrite systems. On the other hand, even if in [7] such a restriction is not assumed, it can be shown that their depth non-decrease sufficient conditions are not actually sufficient in the general case.

We show that the characterization given in [7] can be refined thus establishing the termination of the matching procedure for a class of rewrite systems wider than that characterized there.

Section 2 contains a brief expositions of basic notions and notations adopted in the paper. The matching procedure for which a termination criterion will be given in section 4 is presented in section 3 along with some of its main properties. Section 4 is devoted to the introduction and development of the conditions for a term being depth non-decreasing (depth increasing) at a given position w.r.t. a term rewriting system. These conditions are discussed referring to the depth of a term, but they can be also appropriately applied to other suitable properties. Finally, the main criterion assuring the decidability of the matching problem is reported. Section 5 illustrates some final remarks and conclusions.

2   Basic notions and notations

In the following we adopt the standard concepts, notations and definitions as given in [6, 11].

We briefly review the relevant basic notions and notations for equational theories and rewrite systems.

Terms are constructed from a given set of function symbols and variables. Letters l, p, q, r, s, and t will usually denote terms, if not otherwise stated, while variables are denoted by x, y, w, and z. A *ground* term is one containing no variables. A term is said to be *linear* in a variable x if x occurs only once in it. For example. the term $s(x) * z + y$ is linear in all its three variables.

The *size* of a ground term is the number of function symbols it has, whereas its *depth* is the length of the longest path in its tree representation. A *substitution* is a mapping from variables to terms. We use lower case Greek letters θ, σ, etc. to denote substitutions, and write them out as $\{x_1 \mapsto t_1, \ldots, x_n \mapsto t_n\}$.

A (ground) term t *matches* a pattern (term) s in an (equational) theory E if $E \models s\sigma = t$ for some substitution σ. For example, $s(s(0))$ matches $x + y$ with the substitution $\{x \mapsto s(s(0)), y \mapsto 0\}$ in the theory $\{x + 0 = x\}$.

A term s *unifies* with a term t in a theory E if $E \models s\sigma = t\sigma$ for some substitution σ. We say that a substitution σ is *at least as general* as a substitution μ if there exists a substitution θ such that μ and the composition of σ and θ give equal terms (equal in E), for each variable in the problem. For example, a most general unifier of $x + y$ and $u + v$ is the substitution $\{x \mapsto u, y \mapsto v\}$. Semantic unification is the process of finding all such substitutions.

An *equation* is an unordered pair of terms written in the form $s = t$. The possible variables occurring in an equation are understood as being universally quantified.

A *rewrite rule* is an oriented equation between terms, written $l \rightarrow r$ such that the set of variables occurring in r is contained in that of l. A *rewrite system* is a set of such rules. A rewrite rule is *left linear* if its left hand side is linear for all the variables. A rewrite rule is said to be *variable-preserving* (*non-erasing*) if all the variables in its left hand side also appear in its right hand side term.

A function symbol f is said to be a *defined function* w.r.t. a rewrite system R if there exists a rule in R with f as the topmost symbol of its left hand side. This rule is called f-*rule*. If there is no such rule, then f is called a *constructor*. In a *constructor-based* rewrite system each rewrite rule has the left hand side term of the form $f(p_1, \cdots, p_n)$ where each p_i is a term constituted only by variables and/or constructors. Then, terms in $\mathcal{T}(\mathcal{F} \cup \mathcal{C}, \mathcal{X})$ are constructed by means of elements of $\mathcal{F}$, the set of defined function symbols, elements of $\mathcal{C}$, the set of constructors, and variables of $\mathcal{X}$.

An f-*term* is a term with f as topmost operator. A term t is called a *flat* term if $t = f(t_1, \ldots, t_n)$, $f \in \mathcal{F}$ and, for each i, $t_i \in \mathcal{T}(\mathcal{C}, \mathcal{X})$, $1 \leq i \leq n$. Moreover we call *flattening of* t the term s obtained from t where each of its proper g-subterms, $g \in \mathcal{F}$, has been replaced, in a consistent manner, by a fresh variable. Then, for example, the flattening of $+(*(x, y), *(x, y))$ is $+(z, z)$. In the following we usually indicate with $\bar{t}$ an instance of the term t.

For a given system R, the *rewrite* relation $\rightarrow$ replaces an instance $l\sigma$ of a left hand side l by the corresponding instance $r\sigma$ of the right hand side r. Unlike equations, replacements are not allowed in the reverse direction. We write $s \rightarrow t$ if s rewrites to t in one step; $s \rightarrow^* t$ if t is *derivable* from s, that is, if s rewrites to t in zero or more steps; $s \downarrow t$ if s and t *join*, that is, if $s \rightarrow^* w$ and $t \rightarrow^* w$ for some term w. A term s is said to be *irreducible*, or in *normal form*, if there is no term t such that $s \rightarrow t$. We write $s \rightarrow^! t$ if $s \rightarrow^* t$ and t is in normal form.

In the sequel, we will refer to rules of the rewrite system by means of labels, so that a labelled rule is $h) \ l \rightarrow r$.

Moreover, we write $t \underset{r}{\rightarrow} s$ if the rule labeled with r has been applied to t to get s. In general, if $t \underset{r_1}{\rightarrow} \cdots \underset{r_n}{\rightarrow} s$ is a derivation we call $r_1 \cdots r_n$, by abuse of language, a derivation too, and also write $t \underset{r_1 \cdots r_n}{\rightarrow} s$. Given a rewrite system R, the *surreduction* relation $\rightsquigarrow$ holds between terms t and s, and we write $t \underset{h,\sigma}{\rightsquigarrow} s$ iff $s = t\sigma[r\sigma]_p$ and a rule $l \rightarrow r \in R$, labeled with h, and position p in t exist such that $t|_p\sigma = l\sigma$. Moreover, if $t \underset{r_1,\sigma_1}{\rightsquigarrow} \cdots \underset{r_n,\sigma_n}{\rightsquigarrow} s$ is a surreduction derivation we simply write $t \underset{r_1 \cdots r_n,\sigma_1 \cdots \sigma_n}{\rightsquigarrow} s$.

All the matching problems we consider are of the form $s \rightarrow^? N$ meaning: find a substitution (*solution*) σ such that $s\sigma$ has normal form N (we will frequently use N to stand for a term in normal form). A solution is *irreducible* if each of the terms substituted for the variables is irreducible.

A rewrite relation is *terminating* if there is no infinite chain of rewrites: $t_1 \rightarrow t_2 \rightarrow \cdots \rightarrow t_k \rightarrow \cdots$. A rewrite relation is *ground confluent* if, whenever two ground terms s and t are derivable from a term u, then a term v is derivable from both s and t. That is, if $u \rightarrow^* s$ and $u \rightarrow^* t$, then $s \rightarrow^* v$ and $t \rightarrow^* v$ for some term v.

A rewrite system that is both ground confluent and terminating is said to be *ground convergent*; in the following, whenever we say "convergent", we mean "ground convergent". Convergent rewrite systems are important for the following reason: if R is a convergent rewrite system and E is the underlying equational theory (E is R with each rule taken as an equation), then $E \models s = t$ (for ground terms s and t) iff $s \downarrow t$ in R. In the rest of the paper we assume any rewrite system (ground) convergent and constructor-based.

3 The matching procedure

We are going to present a termination property of a method for solving the semantic matching problem as described, in more details, in [2, 1, 7]. Such a method is a simplified version of the generally complete system for unification as given in [8, 6, 10]. The goal we consider has the form $s \rightarrow^? N$ where N is a ground normal form w.r.t. the theory presented by a constructor-based convergent rewrite system. Solving this goal means: find a substitution σ such that $s\sigma$ has normal form N.

The method consists in transforming the given equation until all the obtained equations are of the form $x \mapsto N$, where x is a variable and N is a normal form, provided that whenever the same variable appears on the left in more than one equation the same term appears on the right. A finite set $\{x_1 \mapsto t_1, \ldots, x_n \mapsto t_n\}$ satisfying the above condition represents a substitution. A complete set of solutions to the equation $s \rightarrow^? N$ is obtained by nondeterministically applying the following transformation rules to the initial set $G = \{s \rightarrow^? N\}$.

Eliminate	$\{x \rightarrow^? t\} \cup G$ $\Rightarrow$ $\{x \mapsto t\} \cup G\{x \mapsto t\}$
Decompose	$\{f(s_1, \ldots, s_n) \rightarrow^? f(t_1, \ldots, t_n)\} \cup G$ $\Rightarrow$ $\{s_1 \rightarrow^? t_1, \ldots, s_n \rightarrow^? t_n\} \cup G$
Mutate	$\{f(s_1, \ldots, s_n) \rightarrow^? t\} \cup G$ $\Rightarrow$ $\{r \rightarrow^? t, s_1 \rightarrow^? l_1, \ldots, s_n \rightarrow^? l_n\} \cup G$ where $f(l_1, \ldots, l_n) \rightarrow r$ is a renamed rule in R

We remark that even if the application of the **Mutate** rule may introduce equations with a non-ground right hand side, due to the adopted selection rule expressed below, the selected equation to be transformed has always a ground right hand side.

The following completeness result holds:

Theorem 3.1 *(Completeness [7])*
Let R be either a variable-preserving or a left-linear convergent rewrite system. If the goal $s \rightarrow^? N$ has a solution θ, then there is a derivation of the form

$$\{s \xrightarrow{?} N\} \xRightarrow{!} \mu,$$

such that μ is a substitution at least as general as θ.

Our matching procedure applies at each step one of the above transformations to a selected equation according to a given selection rule. The adopted
selection rule applies **Eliminate** whenever possible, otherwise the first subgoal
obtained by **Mutate** is the selected one. Such a selection rule preserves the
completeness property [1, 7]. The above procedure can be shown to be terminating under suitable conditions satisfied by the rewrite systems expressing
the theory.

Others selection rules can be adopted. But, having in mind the termination of the matching procedure, the above presented is the best one since
the selected equation is always either syntactically solvable or with a ground
right-hand side.

In order to present such conditions we need to introduce the following
concepts and definitions.

Definition 3.2 (Suitable Property [7])
A *suitable property* P is a measure (like *depth*, *size*, etc.) associated with
ground terms, along with a well-founded total ordering $>$ which compares
values of P, such that P is strictly larger, under $>$, for terms than for its
subterms.

Definition 3.3 (Non-Decreasing [7])
Let P be a suitable property. A function symbol f is defined to be *non-
decreasing* (w.r.t. P) iff whenever $f(s_1, \ldots, s_n) \to^! N$, where N and s_i,
$i \in \{1, \cdots, n\}$, are in ground normal form, $P(s_i) \leq P(N)$, $i \in \{1, \cdots, n\}$.

By means of these concepts it is possible to give a sufficient condition for
the termination of the matching procedure.

Theorem 3.4 *(Theorem 3.6 in [7])*
*Let R be a convergent variable-preserving term rewriting system, and P some
suitable property. If*

- *all right hand sides of rules in R are either variables, or have a constructor
at the top-level, and*

- *all right hand sides are such that no defined function is nested below any
function decreasing with respect to P,*

then the semantic matching problem is decidable for R.

This theorem poses both semantic and syntactic restrictions on the rules.
The semantic one imposes that defined function can appear only below non-
decreasing functions in the right hand side of the rules. Unfortunately, the non-
decrease property is undecidable. It is then essential to give sufficient criteria

implying the non-decrease property. For instance, in [7] the following simple criterion is given: if every variable occurs below at least the same number of constructors on the right hand side, as on the left, then the corresponding function is depth non-decreasing. This criterion together with theorem 3.4, permits to guarantee the decidability of the matching problem for the following system R_{M_1} giving the definition of addition and multiplication functions, the latter defined only on pairs of natural numbers with the second different from 0 (see [7]):

$$+(x, 0) \to x$$
$$+(x, s(y)) \to s(+(x, y))$$
$$*(x, s(0)) \to x$$
$$*(s(x), s(y)) \to s(+(x, *(s(x), y)))$$

On the other hand, such conditions are not sufficient to guarantee the termination of the matching procedure for the following system R_{M_2}, which reduces terms to the same normal form computed by the previous one.

$$a_1) \ +(x, 0) \to x$$
$$a_2) \ +(x, s(y)) \to s(+(x, y))$$
$$m_1) \ *(x, s(0)) \to x$$
$$m_2) \ *(s(x), s(y)) \to +(s(x), *(s(x), y))$$

In fact, the right hand side of rule m_2) does not have a constructor at the top level. Nevertheless, it can be shown that the semantic matching problem is decidable for such system. The main purpose of this work is then to show that systems like the previous ones satisfy a more general condition.

The characterization we are going to present is a refinement of that expressed in theorem 3.4 for a twofold reason:

- it generalizes the possibility of having a constructor at the top-level and defined functions below non-decreasing function symbols, with the following scheme: in each rule a recursive call can occur only in a *P increasing* term at the call position. Roughly speaking, a P increasing term at a position p is such that any its ground instance has a normal form greater than that of the corresponding instance of the subterm at position p, referring to a suitable property P. Observe that this property is similar to the non-decrease property. In fact, if f is non-decreasing then the term $f(x_1, \ldots, x_n)$ is P non-decreasing w.r.t. any of its parameter positions.

- remarking that the P increase property of terms is also undecidable in general, a more powerful sufficient criterion has been given for testing the increase property of a term considering the depth property.

The interest of such an investigation also arises from the consideration that given a rewrite system not fulfilling the conditions of theorem 3.4 we don't know how to build (if it is possible!) an equivalent rewrite system satisfying those conditions. The rewrite systems R_{M_1} and R_{M_2} show that in the case of multiplication the above transformation is possible. On the other hand, if you consider the usual specification of exponential not satisfying the conditions of theorem 3.4 we were not able to find an equivalent rewrite system satisfying those conditions.

4   Depth non-decrease (increase) property and main theorems

Let us consider the previous system R_{M_2}.

$$a_1) \ + (x, 0) \to x$$
$$a_2) \ + (x, s(y)) \to s(+(x, y))$$
$$m_1) \ *(x, s(0)) \to x$$
$$m_2) \ *(s(x), s(y)) \to +(s(x), *(s(x), y))$$

It has a decidable matching problem and it is a simple example clearly showing the main reasons for such a decidability property. The essential features of its rules are:

- variable preservingness;

- *depth increase* of right hand members at each recursive call position.

While the first point is also requested by conditions of theorem 3.4, the second point is the one we are going to present.

Let us give the general definition of the increase property of a term.

Definition 4.1 (*P* increase)
A term t is *P non-decreasing (P increasing)* at a position $p \neq \lambda$ w.r.t. a TRS R iff for each of its ground instance $\bar{t}$ such that $\bar{t} \to^! n$ and $\bar{t}|_p \to^! m$, then $P(n) \geq P(m)$ $(P(n) > P(m))$ for a suitable property P.

The depth non-decrease (increase) property of a term t w.r.t. one of its positions is undecidable as it can be shown in an analogous way as done in [7] for the case of the depth non-decrease property of a function.

As yet remarked, our depth non-decrease property of a term is similar to the depth non-decrease property of a function but it differs from that property in the following respects:

- it is a property of a term and not of a function;

- the inequality is requested just for one specified subterm and not globally over each parameter of the function that is at the top-level.

Remark that according to this definition any constructor term is P increasing w.r.t. any of its proper subterms, since p is suitable and a constructor is not a defined function symbol.

Example 4.2 In the following system defining the addition over natural numbers

$$a_1) \ +(x, 0) \to x$$
$$a_2) \ +(x, s(y)) \to s(+(x, y))$$

the function $+$ is depth non-decreasing, i.e. the depth of the normal form of any ground instance of $+(x, y)$ is greater than or equal to that of the corresponding normal forms of x and y. Moreover, we can establish that the term $+(s(x), y)$ is depth increasing w.r.t. y, i.e. the depth of the normal form of any its ground instance is greater than that of the corresponding normal form of y.

The criterion we are going to present establishes the depth non-decrease (increase) property of a term. In order to give such criterion we need to introduce some new concepts.

Let us give a brief outline and motivation for the notions we are going to present. First of all we will discuss the concept of *sets of terminal paths and of repeatable paths* of a term t w.r.t. a given TRS R as a means to characterize as much closely as possible the form of any derivation of any instance of t. This will lead us to automatically derive the so called *refined grammar* from the basic *derivation grammar* for t in R. Let us explain with an example these concepts. Consider the following rewrite system defining the addition over naturals artificially using mutual recursion:

$$a_1) \ +(x, 0) \to x$$
$$a_2) \ +(x, s(y)) \to \oplus(x, y)$$
$$a_3) \ \oplus(x, y) \to s(+(x, y))$$

The set of terminal paths for the term $+(x, s(y))$ is $\{a_2 a_3 a_1\}$. It is constituted by the label sequence (derivation) corresponding to the reduction

$$+(x, s(0)) \underset{a_2}{\to} \oplus(x, 0) \underset{a_3}{\to} s(+(x, 0)) \underset{a_1}{\to} s(x)$$

i.e. the shortest reduction leading to a normal form such that no subpath has been repeated in it.

The set of repeatable paths is $\{a_2 a_3\}$ meaning that a recursive call to the function $+$ will occur, in any derivation, after the application of the pair of rules a_2, a_3. In fact, by surreducing $+(x, s(y))$ with rules a_2 and then a_3 we get

$$+(x, s(y)) \underset{a_2 a_3, \{y \mapsto y1\}}{\rightsquigarrow} s(+(x, y1))$$

i.e.

$$+(x, s(y))\{y \mapsto y1\} = +(x, s(y1)) \underset{a_2 a_3}{\rightarrow} s(+(x, y1))$$

On the other hand, by considering only the rules of the rewrite system we can know which are the shortest reductions leading to a normal form such that no subpath has been repeated (basic terminal cases), and we can collect this information in the production set P of a context-free grammar whose language will be the superset of rule label sequences corresponding to the derivation of any instance of the generic term $+(w, z)$. In the above example we have that

$$< + > ::= a_1$$

belong to P, meaning that $+(w, z) \underset{a_1, z \mapsto 0}{\rightsquigarrow} w$. But from the given rewrite rules we also know that other, longer derivations, for suitable instances of $+(w, z)$ are of the form indicated by the productions

$$< + > ::= a_2 < \oplus >$$
$$< \oplus > ::= a_3 < + >$$

All these productions constitute the production set P of the derivation grammar $G(+)$ for the generic term $+(w, z)$.

Using the information obtained by sets of terminal and repeatable paths we know that a recursive call to $+$ occurs after the application of rules a_2 and a_3, then we can unfold the production defining $< \oplus >$ into the right-hand side of the production defining $< + >$, thus getting the complete *refined* set of productions for $+(x, s(y))$:

$$< + > ::= a_2 a_3 a_1$$
$$< + > ::= a_2 a_3 < + >$$

The fact that we can know in advance which are the actual repeatable paths greatly reduces the number of all possible unfoldings that could have been applied in the general case. On the other hand, we will show that the definitions of sets of terminal and repeatable paths, based on the technique of simulating the construction of a complete surreduction tree for the term t, lead to an actual algorithm for constructing them.

These sets represent a very powerful information for proving inductive properties of a term t: knowing the form of any derivation, namely which are the repeated subpaths in any computation, enables us to easily show, by structural induction on the derivations, inductive properties of t with respect to the given rewrite system. This is achieved by only imposing suitable conditions on the generic term obtained by surreducing t with a repeatable path, in order to assure the maintenance of the desired property, yet tested on the terminal cases.

In the following, when we use the relation $\succ_{emb}$ we refer to the extended homeomorphic embedding relation, that is the usual embedding [6] where $x \succ_{emb} y$, $x, y \in \mathcal{X}$, holds too. This definition is similar but more general than *homeomorphic variable embedding relation* as defined in [5].

The first concept we are going to introduce to achieve our goal is that of *sets of terminal paths and repeatable paths* of a flat term t. As before pointed out, by means of these sets we can have some important semantic information about possible reductions of instances of t. These sets are constructed by considering the minimal (i.e. shortest) initial derivations of suitable instances of t. These derivations are obtained by simulating the construction of the complete surreduction tree of t, and cutting each branch at those nodes that are embedding an ancestor node when the rule sequence applied to get the ancestor has been repeated to get the nodes we are cutting. In this way, we maintain only the minimal initial surreduction derivation and also have attached to each such derivation the important information about the repeated rule sequence we could have by proceeding in that branch. Such a rule sequence is then called *repeatable* in the next definition. Thus, referring to the rewrite system of example 4.2, where the usual definition of addition is given, the surreduction tree for $+(x, y)$ is

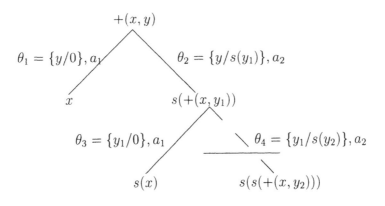

Then, the minimal initial derivations are

1) $+ (x, 0) \underset{a_1}{\rightarrow} x$

2) $+ (x, s(0)) \underset{a_2}{\rightarrow} s(+(x, 0)) \underset{a_1}{\rightarrow} s(x)$

the first derivation will be adorned by the information that its label sequence a_1 is a terminal path, while the second derivation will be adorned by the information that a_2 is a repeatable rule in possible (longer) derivations of $+(x, y)$ instances.

Definition 4.3 (Sets of terminal paths and repeatable paths)
Let $t = f(t_1, \ldots, t_n)$, $f \in \mathcal{F}$, be a flat term. Let us consider the following surreduction derivation $t = t_1 \leadsto_{r_1, \theta_1} t_2 \leadsto \cdots \leadsto_{r_{i-1}, \theta_{i-1}} t_i$, $i \geq 1$.

- If t_i is in normal form and for each $1 < j \leq i$ at most one m, $1 \leq m < j$, exists with $t_j \succ_{emb} t_m$ then the set T-*paths* constituted by the elements $\pi = r_1 \ldots r_i$ is called *set of terminal paths for t*.

- Otherwise, if $t_i \leadsto_{r_i, \theta_i} t_{i+1}$ and there exists an index $1 < j < i$ such that $t_{i+1} \succ_{emb} t_j$, and, letting $\lambda_1 = r_1 \ldots r_{j-1}$ and $\lambda_2 = r_1 \ldots r_{j-1} r_j \ldots r_i$, there exists a suffix σ of both λ_1 and λ_2, then the set R-*paths* constituted by the elements $\rho = r_j \ldots r_i$ is called *set of repeatable paths for t*.

Proposition 4.4
For any flat term t an algorithm for computing its sets of terminal and repeatable paths does exist.

Proof. If these sets would not be computable, then it would exist an infinite derivation ρ such that two indexes i, j, $i < j$, $t_j \succ_{emb} t_i$ do not exist, where t_i, t_j are obtained by applying the same rule. This is an absurd. In fact, if this would happen we could consider, due to the Tree theorem [12], an infinite subderivation with each term embedded into another and obtained by always applying different rules implying the infiniteness of the rewrite system R. □
QED

Looking at the above definition, one can argue that if the considered term is $f(x_1, \ldots, x_n)$ and the function f is recursively defined in the given rewrite system, some repeatable path starting with the label of an f-rule has to be found when computing the set R-paths. But this is not always the case, as shown in the following example. Consider the term $f(x)$ and the rewrite system:

$$h_1)\ half(0) \rightarrow 0$$
$$h_2)\ half(s(s(x))) \rightarrow s(half(x))$$
$$f_1)\ f(0) \rightarrow 0$$
$$f_2)\ f(s(x)) \rightarrow s(f(half(s(x))))$$

The R-paths set of the term $f(x)$ is $\{h_2\}$. However, we can easily recognize that f is recursively defined and no repeatable path starting with the label of an f-rule has been found. This is due to the fact that before reapplying the rule f_2, which is a recursive one, a normalizing reduction of a $half$-term has to be performed. This suggests that if we substitute the parameter $half(s(x))$ into the call to f in the right-hand side of the rule f_2 and compute the R-paths set for the new system, we find the repeatable path f_2. In fact, we are interested in collecting only those repeatable paths starting with an f-rule label whenever it is possible. The above idea has been formally described into the following recursive procedure computing the *refined set of repeatable paths* containing only paths starting with f-rule labels if it is possible.

We are now in a position to give the following definition of refined set of repeatable paths.

Definition 4.5 (Refined set of repeatable paths)
Let $t = f(t_1, \ldots, t_n)$, $f \in \mathcal{F}$, be a flat term and R be a term rewriting system, and R-*paths* be the set of repeatable paths for t. The *refined set RP of repeatable paths* of t w.r.t. R is recursively defined as follows:

- if R-*paths* does not contain any ρ whose first label does not refer to an f-rule, then $RP=R$-*paths*;

- otherwise,

    - let R-*paths*$= P_f \cup P_{\neq f}$, with P_f the set of paths whose first label refers to an f-rule and $P_{\neq f}$ the set of paths whose first label does not refer to an f-rule,

    - let R' be the rewrite system obtained from R by replacing into the right side of its rules each g-term with a fresh variable, for each function g defined by the rule referred to by the first label of some $\rho \in P_{\neq f}$,

    - let RP' be the recursively computed refined set of repeatable paths of t w.r.t. R',

    then $RP = P_f \cup RP'$.

Then, a simple but important consequence of the above definitions of set of repeatable paths and refined set of repeatable paths RP of a flat term $t = f(t_1, \ldots, t_n)$ is that for any $\rho \in RP$ there exists a term s, $t \underset{\rho}{\rightsquigarrow} s$, such that s contains f-subterms.

In order to establish our criterion for the validity of the depth non-decrease property of a term t, we consider the syntactic form of the derivation of any instance of t according to the given rewrite system. As before remarked, this leads us to the definition of a context-free grammar whose language (necessarily!) is a superset of the language constituted by the rule label sequences labeling the actual derivations of instances of t.

For example, if we consider the term $t = +(x, y)$ we soon understand that any such label sequence is of the form, referring to the rewrite system of example 4.2, a_1 or $a_2 \cdots a_2 a_1$, reflecting the fact that an instance of t is reduced either by applying rule a_1 or by first repeatedly applying a_2, and then rule a_1. This can be formalized into the following productions defining such derivations:

$$1) \; <+> ::= a_1$$
$$2) \; <+> ::= a_2 <+>$$

In general, given a flat term $t = f(t_1, \ldots, t_n)$, this task is accomplished in two steps: the first defines such a derivation grammar for the generic term $t = f(x_1, \ldots, x_n)$, $x_i \in \mathcal{X}$; the second, leading to a refinement of the previous grammar, takes into account the information extracted from the sets of terminal and repeatable paths for the term t. This is important since the grammar obtained at the second step defines a language which more closely captures the form of any actual derivation of instances of t. For example, some derivations of the first language are excluded in the second as illustrated below.

Then, the main idea of our criterion for deciding the depth non-decrease (increase) property of a term t is to check the desired property on the cases presented in the set of terminal paths and, if these satisfy it, to go on by considering any further derivation, whose structure is given by the grammar, and imposing suitable conditions on terms obtained by surreducing t by means of a repeatable path in order to guarantee the maintenance of the property checked on the set of terminal paths.

Definition 4.6 (Derivation grammar)
Given a defined function symbol f and a rewrite system R, the following context-free grammar $G(f)$ is called the *derivation grammar* of f: $G(f) = <T, N, P, A>$ where

- $T = \{h \mid h)\ l \to r \in R\}$;

- $N = \{<g> \mid g \text{ is a defined function symbol in } R\}$;

- $P = \left\{ \begin{array}{c} <f> ::= l_i <g_{j_1}> \cdots <g_{j_k}> \mid l_i)\ f(t_1, \ldots, t_n) \to r \in R, \\ g_1, \cdots, g_k \text{ are the defined function symbols of } r, \text{ and} \\ <g_{j_1}> \cdots <g_{j_k}> \text{ is any permutation of } <g_1> \cdots <g_k> \end{array} \right\} \cup Q$
  where Q contains the productions of the derivation grammar recursively defined for each g_{j_h} (avoiding loops in the case of recursively defined function symbols);

- $A = <f>$.

Property 4.7 The label sequence ρ of the reduction of any instance of a given flat term $t = f(t_1, \ldots, t_n)$ belongs to $L_{G(f)}$.

Proof. The proof is easily carried on by induction on the structure of the label sequence. □ QED

Considering that RP is the refined set of repeatable paths, we can refine the above grammar $G(f)$ by selectively unfolding its productions according to such repeatable paths found in RP. Let us illustrate such a refinement by means of the following example.

Example 4.8 Let R be the rewrite system defining the Ackermann function

$$ac_1)\ ack(0, x) \rightarrow s(x)$$
$$ac_2)\ ack(s(x), 0) \rightarrow ack(x, s(0))$$
$$ac_3)\ ack(s(x), s(y)) \rightarrow ack(x, ack(s(x), y))$$

the derivation grammar of ack is $G(ack) = <T, N, P, A>$ where:

$$
\begin{aligned}
T = \{ &\ ac1, ac2, ac3\}, \\
N = \{ &\ <ack>\}, \\
P = \{ &\ 1)\ <ack> ::= ac1 \\
&\ 2)\ <ack> ::= ac2 <ack> \\
&\ 3)\ <ack> ::= ac3 <ack><ack>\} \\
A = &\ <ack>
\end{aligned}
$$

The sets of terminal and refined repeatable paths, *T-paths* and RP respectively, for $t = ack(s(x), y)$ are

$$
\begin{aligned}
T\text{-}paths = \{ &\ ac2\ ac1, \\
&\ ac3\ ac2\ ac1\ ac1, \\
&\ ac3\ ac2\ ac1\ ac1, \\
&\ ac2\ ac3\ ac2\ ac1\ ac1\}
\end{aligned}
$$

$$
\begin{aligned}
RP = \{ &\ ac3, \\
&\ ac3\ ac2, \\
&\ ac3\ ac1, \\
&\ ac2\ ac3\}
\end{aligned}
$$

By unfolding the third production in the second, we get 2') $<ack> ::= ac2ac3 <ack><ack>$ and by unfolding the second production in the third,

we get $3'$) $< ack > ::= ac3ac2 < ack >< ack >$. By unfolding the first in the
third we get $3''$) $< ack > ::= ac3ac1 < ack >$. By unfolding the first in the
second we get $1'$) $< ack > ::= ac2ac1$. We can finally consider the following set
of productions $\{1'), 2'), 3'), 3''), 3)\}$. The productions in this set cover all the
computed repeatable paths; moreover, the grammar constituted by this set of
productions is a refinement of the previous one in the sense that productions
$1), 2)$ are no more present: $2)$ has been replaced by $1'$) and $2'$), whereas $1)$ has
been deleted since it does not correspond to the reduction of any instance of
the term $ack(s(x), y)$ as shown by T-paths and RP sets. The label sequences
$ac1$ and $ac2ac2 \cdots$ do not belong to the language $L_{G_r(f)}$.

An algorithm to get the *refined grammar* $G_r(f)$ starting from $G(f)$ for a
term $t = f(t_1, \ldots, t_n)$ can be formally defined as follows.

Definition 4.9 (Refined grammar)
Given the flat term $t = f(t_1, \ldots, t_n)$, $f \in \mathcal{F}$, the rewrite system R, let $G(f) =$
$< T, N, P, A >$ be the derivation grammar of f, RP be the refined set of repeatable paths for t, T-paths the set of terminal paths, and $P_{<f>}$ be the subset of
P constituted by the productions whose left member is $< f >$ and whose right
member contains non terminals. Let $P_{T\text{-paths}} = \{< f > ::= \pi \mid \pi \in T\text{-paths}\}$.
The *refined set PR of productions for t* is defined as follows. First, let PR_0
be the set of productions for t whose left hand side is $< f >$ obtained in the
following way ($l_1, \ldots, l_k$ standing for rule labels):

- $PR_0 = P_{<f>} \cup P_{T\text{-paths}}$ if $RP = \emptyset$; otherwise

- $< f > ::= l_1\mathtt{nt} \in PR_0$ if $< f > ::= l_1\mathtt{nt} \in P_{<f>}$, $l_1 \in RP$, $\mathtt{nt} \in N^+$;

- $< f > ::= l_1 \ldots l_k\mathtt{nt} \in PR_0$, $\mathtt{nt} \in N^+$ and $k > 1$, if $l_1 \ldots l_k \in RP$ and $\mathtt{nt}$
  has been obtained by unfolding each production of the form $< g > ::=$
  $l_i\mathtt{nt_{g'}} \in P$, $\mathtt{nt_g} \in N^+$, into the production $< f > ::= l_1 \ldots l_{i-1} < g > \mathtt{nt_i}$
  recursively obtained by starting with a production of the form $< f > ::=$
  $l_1\mathtt{nt_1} \in P_{<f>}$, $\mathtt{nt_i}, \mathtt{nt_1} \in N^+$.

Then, let PR_1 be the set of productions for t whose left hand side is different
from $< f >$ obtained in the following way:

- PR_1 is the union of the refined sets of productions of each flattened g-
  subterm, $g \in \mathcal{F}$, $g \neq f$, of the term s such that $t \leadsto_\rho s$, for any ρ occurring
  in a production of PR_0.

Finally, $PR = PR_0 \cup PR_1$. The *refined grammar for t* is $G_r(f) = < T, N, PR, A >$

Property 4.10 The label sequence ρ of the reduction of any instance of a
given flat term $t = f(t_1, \ldots, t_n)$ belongs to $L_{G_r(f)}$.

In order to establish our next criterion on the depth non-decrease or depth increase property of a flat term w.r.t. a rewrite system R, we introduce the following notion of *depth increase degree*. For example, given the rewrite system R_{M_1} defining the addition, and the term $t = +(x, s(y))$, the depth increase degree of t at position 1 is at least one. In fact, for any ground normal form instance of x and y, the difference between the depth of the corresponding normal form of t and that of the instance of x is at least one. For the same reason, the term $+(x, y)$ has depth increase degree at position 1 equal to zero.

We call the term $t = f(g_1, \ldots, g_i, \ldots, g_n)$, with all g_i in ground normal form, *d-depth increasing at position i*, $d \geq 0$, iff $t \to^! N$ and $depth(N) - depth(g_i) \geq d$.

From this definition and the subterm property of *depth* it follows that a constructor term $t = c(g_1, \ldots, g_n)$, $c \in C$, is 1-depth increasing at any of its parameter positions. Then, a 0-depth increasing term at a given position is depth non-decreasing at that position, while a d-depth increasing term, $d > 0$, is depth increasing.

We can now give the following notion of *minimal estimated depth (m.e.d.)*, $\delta(o, t, R)$, *of an occurrence o in a term t w.r.t. a rewrite system R* defined as follows:

$$\delta(o, t, R) = \sum_{\alpha.m \text{ prefix of } o, \ m \in \mathbb{N}} d_\alpha$$

where $t|_\alpha$ is d_α-depth increasing at position m. Remark that, when some $t|_\alpha$ is such that its d_α is less than zero the m.e.d. is undefined.

In the following $\delta(o, t, R)$ will usually refer to the occurrence o of a given variable in t. Then, for example, if $t = +(x, s(y))$, $\delta(1, t, R_{M_2}) = d_\lambda = 1$ since the term $t|_\lambda = +(x, s(y))$ is 1-depth increasing at position 1.

Finally, let x be a variable, t be a term, we indicate with $\Delta(x, t, R)$ the maximum among m.e.d.'s of any occurrence of x in t, i.e.

$$\Delta(x, t, R) = \max_{o:t|_o = x} \delta(o, t, R)$$

For example, let t be the term $s(x)$, then x occurs at position 1 in t and the m.e.d. of the occurrence 1 of x in t is $\delta(1, t, R) = 1$, for any rewrite system R where s is a constructor, while $\delta(1, f(x), R)$ will be d, if $d \geq 0$, according to the (provable) depth increase degree of the defined function symbol f w.r.t. R, or undefined otherwise. The main theorem reported below gives a criterion to establish the depth increase degree of a term.

The essential idea for proving a term $t = f(t_1, \ldots, t_i, \ldots, t_n)$ being d-depth increasing, $d \geq 0$, at position i is the following:

- check the desired property on the basis cases expressed by the terminal production of $G_r(f)$; if this test succeeds, then we can assure the desired property if

- for every repeatable path ρ occurring into the recursive productions for $< f >$, such that $f(\hat{t}_1, \ldots, \hat{t}_i, \ldots, \hat{t}_n) = t\theta \xrightarrow{\rho} s$, the maximal depth at which each variable of $\hat{t}_i$ occurs in $\hat{t}_i$, is maintained or increased for each occurrence of that variable in s.

Observe that the second condition is just a generalization of the criterion given in [7] that tests whether each variable occurring in the left-hand side of the rules occurs on the right below as much constructors as in the left.

Referring to the position o of a variable into the term q, the quantity $\delta(o, q, R)$ is the estimated depth at which that variable lies in q. As its definition states, this number is computed by considering the depth increasing degree at position m of each superterm occurring at α such that $\alpha.m$ is a prefix of the occurrence of the considered variable. Thus, for example, let us consider $t = fib(x)$ and the rewrite system R:

$$
\begin{aligned}
&a_1) \ +(x, 0) \to x \\
&a_2) \ +(x, s(y)) \to s(+(x, y)) \\
&fib_1) \ fib(0) \to s(0) \\
&fib_2) \ fib(s(0)) \to s(s(0)) \\
&fib_3) \ fib(s(s(x))) \to +(fib(s(x)), fib(x))
\end{aligned}
$$

We will show that t is 1-depth increasing at position 1. The cases corresponding to terminal productions in $G_r(fib)$ verify our thesis. Thus, we can assume as inductive hypothesis that t is 1-depth increasing. Any recursive production for $< fib >$ in $G_r(fib)$ is of the form (denoting with $perm(a_1 \ldots a_k)$ any permutation of $a_1 \ldots a_k$):

$$< fib > ::= fib_3 \ perm(< fib >< fib >< + >)$$

with repeatable path fib_3 such that $\bar{t} = fib(s(s(x))) \xrightarrow{fib_3} +(fib(s(x)), fib(x)) = q$. Then $\Delta(x, s(s(x)), R) = 2$. Considering the occurrence 1.1.1 of x in q, the estimated depth of this occurrence is given by the addition of

$$
\begin{aligned}
d_{1.1} &= \text{degree of } s(x) \text{ at position 1,} \\
d_1 &= \text{degree of } fib(z) \text{ at position 1,} \\
d_\lambda &= \text{degree of } +(z, fib(x)) \text{ at position 1.}
\end{aligned}
$$

Now, $d_{1.1} = 1$ since s is a constructor, $d_1 = 1$ as inductive hypothesis, and d_λ will be computed by taking the minimum between the degrees of $+(z, s(0))$ and of $+(z, s(s(0)))$ at position 1. These two terms are obtained by replacing into $+(z, fib(x))$ the term $fib(x)$ with $s(0)$ and $s(s(0))$, corresponding to the basic reductions $fib(0) \xrightarrow{fib_1} s(0)$ and $fib(s(0)) \xrightarrow{fib_2} s(s(0))$ where both $fib(0)$ and

$fib(s(0))$ are matched by the term $fib(x)$. Hence, the minimum degree is 1 given by the degree of $+(z, s(0))$. Thus, the minimal estimated depth at which x lies in q is 3. Analogously, by considering the occurrence 2.1 of x in q we have that its minimal estimated depth is 3. Due to this, the previous second condition is satisfied. Hence, we have proven that t is 1-depth increasing at position 1.

Theorem 4.11 *(Main Theorem)*
Let $t = f(t_1, \ldots, t_i, \ldots, t_n)$ be a flat term, R a rewrite system, and $G_r(f) = \; < T, N, P, A >$ the refined grammar for t in R, then t is d-depth increasing at position i in R, if:

a1) *for each non-recursive production in P of the form $< f > ::= \rho$ such that $t \leadsto_{\rho, \theta} s$ (i.e. $\bar{t} = t\theta \xrightarrow{\rho} s$), then*

$$depth(s) - depth(\bar{t}_i) \geq d \tag{1}$$

a2) *for each non-recursive production in P of the form $< f > ::= \rho N^+$ such that $t \leadsto_{\rho, \theta} s$ (i.e. $\bar{t} = t\theta \xrightarrow{\rho} s$), then, letting $Var(\bar{t}_i) = \{x_1, \ldots, x_k\}$, for each j, $1 \leq j \leq k$, and for each occurrence o of x_j in s*

$$\delta(o, s, R) - \Delta(x_j, \bar{t}_i, R) \geq d \tag{2}$$

b) *for each recursive production in P of the form $< f > ::= \rho N^+$ such that $t \leadsto_{\rho, \theta} s$ (i.e. $\bar{t} = t\theta \xrightarrow{\rho} s$), then, letting $Var(\bar{t}_i) = \{x_1, \ldots, x_k\}$, for each j, $1 \leq j \leq k$, and for each occurrence o of x_j in s*

$$\Delta(x_j, \bar{t}_i, R) \leq \min_r \; \delta(o, s, R) \tag{3}$$

where $\delta(o, s, R)$ is computed in the following way, recalling that

$$\delta(o, s, R) = \sum_{\alpha.m \text{ prefix of } o, \; m \in \mathbf{N}} d_\alpha$$

- *if $s|_\alpha$ is an f-term, then $d_\alpha = d$*
- *if $s|_\alpha = h(h_1, \ldots, h_m, \ldots, h_k)$, $h \neq f$, then d_α is the depth increase degree at position m of the term $h(\hat{h}_1, \ldots, z, \ldots, \hat{h}_k)$ where z is a fresh variable and $\hat{h}_j = h_j$ if none f-subterm occurs in h_j, or $\hat{h}_j$ is obtained from h_j by recursively replacing any of its f-subterms, q, (proceeding in an innermost way) with a term r such that the following condition (C) is satisfied.*

  A non recursive production $< f > ::= \rho_1 N^*$ exists and $t\sigma \xrightarrow{\rho_1} r$ and q matches $t\sigma$. $\tag{C}$

Proof. First of all, we remark that for each label sequence ρ in the considered productions there exists a pair of terms $\bar{t}$, s such that $\bar{t} \underset{\rho}{\to} s$, where $\bar{t}$ is an instance of t. In fact, ρ is a label sequence belonging to either the terminal paths' set or repeatable paths' set of t. Hence, $\bar{t} = t\theta_1 \cdots \theta_k$, where $t \underset{\rho,\theta_1\cdots\theta_k}{\rightsquigarrow} s$. The proof is carried on by induction on the structure of the derivation for t. The base cases are those illustrated in the non recursive productions of $G_r(f)$ for which conditions a1) and a2) assure the requested property. In the inductive case, since the considered derivation must be of the form indicated by a recursive production, we consider recursive productions of the refined grammar. Condition (3) assures the maintenance of the property holding for the base cases. In fact, any variable of $\bar{t}_i$ is assured to be, into the term s, always under suitable operators such that its m.e.d. is greater than or equal to the maximal depth at which it lies in $\bar{t}_i$. This holds since condition (3) is verified by assuming the depth increase degree of f-subterms as d; while for h-subterms, $h \neq f$, as the minimum among the depth increase degrees (guaranteed by this theorem) of terms obtained by replacing f-subterms with the term r as defined in condition (3).   □ QED

As a simple consequence of this theorem we have that a flat term t is depth non-decreasing (depth increasing) at position i if it is 0-depth increasing (d-depth increasing, $d > 0$) at position i.

The previous property permits to give the following criterion for deciding the depth increase property of any term at any position.

Property 4.12 (Depth non-decrease (depth increase) criterion for generic terms)
A term $t = f(t_1, \ldots, t_n)$ is depth non-decreasing (depth increasing) at a position p, w.r.t. a rewrite system R if the term $aux(x_1, \ldots, x_k)$ is depth non-decreasing (depth increasing) at position 1 w.r.t. the system $R \cup \{h\}$ $aux(x_1, \ldots, x_k) \to \bar{t}\}$, where $\bar{t} = t[x_1]_p$, $x_1 \notin \mathcal{V}ar(t)$, $\{x_1, \ldots, x_k\} = \mathcal{V}ar(\bar{t})$ and aux is a new defined function symbol.

Example 4.13 Referring to the system R_{M_2} the term $t = s(+(x, *(s(x), y)))$ is depth increasing at position 1.2 since $aux(x_1, x)$ is depth increasing at position 1 w.r.t. the system $R_{M_2} \cup \{aux(x_1, x) \to s(+(x, x_1))\}$.

Let us now give the fundamental decidability criterion for the semantic matching problem.

Theorem 4.14 *(Decidability criterion for semantic matching)*
Let R be a constructor-based, variable-preserving, and convergent term rewriting system, without mutual recursion functions, and P some suitable property.
If

- *for each rule $f(p_1, \ldots, p_n) \to r$ such that $f(t_1, \ldots, t_n)$ occurs in r, and for each occurrence p of a recursive call in r the term $r[x]_p$ is P increasing at position p, where x is a new variable.*

then the semantic matching problem is decidable for R.

Proof.(Proof Sketch) The proof of this theorem is based on analogous arguments to the ones used in the proof of theorem 3.6 in [7]. A first induction is made w.r.t. the ordering $\succ$ to show that all solutions to any goal of the form $t \to^? N$, where t is a term whose unique defined function symbol is the top-level one, can be computed in finite time. The ordering $\succ$ on equations is defined as follows: $s_1 \to^? N_1 \succ s_2 \to^? N_2$ iff either $P(N_1) > P(N_2)$ or $P(N_1) = P(N_2)$ and s_2 is a strict subterm of s_1. A second induction is made on the structure of any left hand side of goals. $\square$ QED

An analogous theorem can be stated even for rewrite systems with mutually recursive functions. In fact, considering each defined function f, the generic term $t = f(x_1, \ldots, x_n)$, and the corresponding refined grammar $G_r(f)$, the condition expressed in the above theorem must be fulfilled between each pair of the form $f(p_1, \ldots, p_n)$ and r such that $f(p_1, \ldots, p_n) \underset{\rho}{\to}^? r$, ρ being a repeatable path occurring in a recursive production of $G_r(f)$.

We remark that, considering the following system

$$
\begin{aligned}
a_1) &\ +(x, 0) \to x \\
a_2) &\ +(x, s(y)) \to s(+(x, y)) \\
m_1) &\ *(x, s(0)) \to x \\
m_2') &\ *(x, s(y)) \to +(x, *(x, y))
\end{aligned}
$$

the right member of rule m_2') violates the condition of the theorem when P is the depth of a term, since the term $+(x, w)$ is not depth increasing w.r.t. w. So, when solving the goal $*(x, y) \to^? 0$, we have to solve infinitely many times a subgoal which is a variant of it. This happens because when solving the subgoal $+(x_1, *(x_1, y_1)) \to^? 0$ by applying **Mutate** using rule a_1, we have again to solve the goal $*(x_1, y_1) \to^? 0$. This fact leads us to believe that the property P increase be indeed necessary. This agrees with the conclusions of the example 3.5 in [7]. In fact, the system reported there does not satisfy the condition of depth increase.

By means of this theorem we can establish that the previous rewrite systems R_{M_1} and R_{M_2}, as well as many others (like append, reverse, etc.), have a decidable matching problem. In fact, as shown in example 4.13, the term $s(+(x, *(s(x), y)))$ is depth increasing at the recursive call position 1.2 and the term $+(s(x), *(s(x), y))$ is depth increasing at the recursive call position 2. In fact, in these cases the depth increase criterion is applicable.

5   Conclusion

As pointed out in [7], the above theorem can also be applied in the case where we have variable dropping rules provided that the rewrite system be left-linear and if r is the right hand side of a rule where any function which has variable dropping rule occurs, then any goal of the form $r \rightarrow^? N$ has only ground solutions. Finally, we remark that we are trying to extend our criterion in order to establish the P increase property for a larger class of rewrite systems. On the other hand, we are also studying the application of the (analogous) P decrease criterion for directly deciding the termination of general rewrite systems.

References

[1] G. Aguzzi, U. Modigliani, and M. C. Verri, *A Universal Termination Condition for Solving Goals in Equational Languages*, Proceedings of the Second International Workshop on Conditional Term Rewriting Systems, CTRS '90, LNCS no. 516 (S. Kaplan and M. Okada, eds.), Springer-Verlag, Berlin, 1991, pp. 418–423.

[2] G. Aguzzi and M.C. Verri, *On the termination of an algorithm for unification in equational theories*, Proceedings of IEEE TENCON '89, IEEE, 1989, pp. 1034–1039.

[3] A. Bockmayr, *A note on a canonical theory with undecidable unification and matching problem*, JAR **3** (1987), 379–381.

[4] H. J. Bürckert, *Matching – A Special Case of Unification?*, Unification (C. Kirchner, ed.), Academic Press, San Diego, CA, 1990, pp. 125–138.

[5] H. Chen, J. Hsiang, and H.-C. Kong, *On Finite Representations of Infinite Sequences of Terms*, Proceedings of the Second International Workshop on Conditional Term Rewriting Systems, CTRS '90, LNCS no. 516 (S. Kaplan and M. Okada, eds.), Springer-Verlag, Berlin, 1991, pp. 100–114.

[6] N. Dershowitz and J. P. Jouannaud, *Rewrite systems*, Handbook of Theoretical Computer Science, volume B: Formal Models and Semantics, chapter 6 (J. van Leeuwen, ed.), Elsevier, Amsterdam, and The Mit Press, Cambridge, 1990.

[7] N. Dershowitz, S. Mitra, and G. Sivakumar, *Decidable matching for convergent systems*, Proceedings of the 11th International Conference on Automated Deduction, Saratoga Springs, New York, USA, June 15–18, LNAI no. 607 (Deepak Kapur, ed.), Springer-Verlag, 1992, pp. 589–602.

[8] N. Dershowitz and G. Sivakumar, *Solving goals in equational languages*, Proceedings of the First International Workshop on Conditional Term Rewriting Systems, LNCS no. 308, Springer-Verlag, Berlin, 1987, pp. 45–55.

[9] S. Heilbrunner and S. Holldobler, *The undecidability of the unification and matching problem for canonical theories*, Acta Informatica **24** (1987), 157–171.

[10] J. P. Jouannaud and C. Kirchner, *Solving equations in abstract algebras: A rule-based survey of unification*, Computational Logic: Essays in Honor of Alan Robinson, chapter 8 (J. L. Lassez and G. Plotkin, eds.), The Mit Press, Cambridge, MA, 1991.

[11] J. W. Klop, *Term rewriting systems*, Handbook of Logic in Computer Science, vol. 2, Background: Computational Structures (S. Abramski, D. M. Gabbay, and T. S. E. Maiubaum, eds.), Oxford University Press, Oxford, 1992, pp. 1–116.

[12] J.B. Kruskal, *Well-quasi-ordering, the tree theorem, and Vazsonyi's conjecture*, Trans. Amer. Math. Soc. **95** (1960), 210–225.

DIPARTIMENTO DI SISTEMI E INFORMATICA, UNIVERSITÀ DI FIRENZE, VIA C. LOMBROSO, 6/17 – I–50134 FIRENZE, ITALY,
    e-mail address: Gianni.Aguzzi@dsi2.ing.unifi.it
    e-mail address: Umberto.Modigliani@dsi2.ing.unifi.it

Remarks On Magari Algebras of PA and $I\Delta_0+$EXP [*]

Lev Beklemishev

Abstract

We give a simple example of a formula of propositional second order provability logic that holds for the interpretation w.r.t. PA, but does not hold w.r.t. $I\Delta_0+$EXP, and establish some related results on Magari algebras of the two theories.

The concept of the *diagonalizable algebra* $M(T)$ of a theory T was introduced by R.Magari [4][1]. $M(T)$ is, essentially, the Lindenbaum sentence algebra of T endowed by the unary operator $\Box_T$ of provability in T. Well-known results of R.Solovay imply that the equational theory of $M(T)$, for any reasonable (i.e. recursively enumerable, sound, and containing a sufficiently strong fragment of arithmetic) theory T, is axiomatized by the Gödel-Löb logic GL, and that the universal and existential theories of this algebra are decidable.

Recently, Magari algebras have been extensively studied by V.Shavrukov [6, 7, 8]. The following two results of his are of a particular interest:

1. For any reasonable theory T, the first order theory of $M(T)$ is undecidable (in fact, not even arithmetical) [8].

2. If a theory U contains the uniform Σ_1^0-reflection principle for a theory T, then $M(T)$ and $M(U)$ are not isomorphic [6].

Later (cf [9]) he improved this result by showing that, actually, the algebras $M(T)$ and $M(U)$ are not elementary equivalent, that is, their first order theories do not coincide. It follows that there exists a formula of propositional second

---

[*]Research supported by the Russian Foundation for Fundamental Reasearch (project 93-011-16015) and by the International Science Foundation (Grant No. NFQ000). I am particularly grateful to the organizers of Magari conference whose support allowed me to participate in it.

[1]In the following we shall call such algebras *Magari algebras*.

order provability logic that holds for the interpretation w.r.t. the theory T, but not w.r.t. U.

In this note we give a different example of such a formula for two particular theories: PA (Peano Arithmetic) and $I\Delta_0+$EXP (PA with the induction schema restricted to Δ_0-formulas, together with an axiom asserting the totality of exponentiation function). Our example has a disadvantage that it has an occurrence of a quantifier under the scope of modality. On the other hand, the overall number of propositional quantifiers used, as well as the number of quantifierchanges for this example is much smaller.

Recall that the *uniform Σ_1^0-reflection principle* $\mathrm{RFN}_{\Sigma_1^0}(T)$ for a theory T is the schema

$$\forall x\,(\Box_T\sigma(x) \to \sigma(x)),$$

for all arithmetical Σ_1^0-formulas $\sigma(x)$. It is well-known that, due to the existence of partial truthdefinitions, $\mathrm{RFN}_{\Sigma_1^0}(T)$ is equivalent to a single Π_2^0-sentence over theories T containing $I\Delta_0+$EXP.

For a reasonable theory T, let $\mathrm{M}'(T)$ denote the Magari algebra of T with a constant $\mathrm{RFN}_{\Sigma_1^0}$ for the uniform Σ_1^0-reflection principle for T added to its signature.

Theorem 1. *The algebras* $\mathrm{M}'(\mathrm{PA})$ *and* $\mathrm{M}'(I\Delta_0+\mathrm{EXP})$ *are not elementary equivalent. More precisely, there is a first order formula Q which holds in* $\mathrm{M}'(\mathrm{PA})$ *but fails in* $\mathrm{M}'(I\Delta_0+\mathrm{EXP})$ *such that Q has the quantifier prefix $\forall\forall\exists$, and the constant* $\mathrm{RFN}_{\Sigma_1^0}$ *occurs in Q only once.*

The example provided by Theorem 1 is optimal w.r.t. the number of quantifier alternations. This follows from our next theorem, which is a direct corollary of the results in [2].

Theorem 2. *The universal theories of* $\mathrm{M}'(\mathrm{PA})$ *and* $\mathrm{M}'(I\Delta_0+\mathrm{EXP})$ *are decidable and actually coincide.*

At present, we do not know, if the constant $\mathrm{RFN}_{\Sigma_1^0}$ is first order definable in the language of $\mathrm{M}(T)$. Moreover, it is unknown, if there are any other definable elements in, say, $\mathrm{M}(\mathrm{PA})$ except for those of its $\bot$-generated subalgebra. (Here and below $\bot$ denotes the logical constant "falsum" and the corresponding element of Magari algebras.) However, an easy lemma given below shows that $\mathrm{RFN}_{\Sigma_1^0}(T)$ is definable in *propositional second order provability logic* $\mathrm{PL}^2(T)$ of the theory T.

Recall that the language of $\mathrm{PL}^2(T)$ is obtained from that of GL by adding the quantifiers $\forall$ and $\exists$ with the following formula formation rule: if $A(p)$ is a formula and p is a propositional variable occurring free in $A(p)$ only under the scope of $\Box$, then $\forall p\, A(p)$ and $\exists p\, A(p)$ are formulas. (Free and bound occurrences of variables are defined in a standard manner. A formula is *closed* if it has no free occurrences.)

Formulas of this language are naturally translated into the language of arithmetic, so that the modality $\Box$ is interpreted as the provability predicate for a given theory T, and propositional variables range over (Gödel numbers of) T-sentences. $\mathrm{PL}^2(T)$ is the set of all closed formulas true under this interpretation in the standard model of arithmetic. It is not difficult to see that the elementary theory of $\mathrm{M}(T)$ can be identified with the fragment of $\mathrm{PL}^2(T)$ consisting of the formulas with no occurrences of quantifiers inside the scopes of modal operators.

Lemma 1. *For any reasonable theory T,*

$$T \vdash \mathrm{RFN}_{\Sigma_1^0}(T) \leftrightarrow (\neg\Box_T\bot \wedge \forall\alpha \in Sent_T\, (\Box_T\Box_T\alpha \to \Box_T\alpha)),$$

where $Sent_T$ represents the set of (Gödel numbers) of T-sentences in T.

Proof. The implication $(\to)$ follows from the fact that $\Box_T x$ is a Σ_1^0-formula. To derive $(\leftarrow)$ notice that, by a free-variable version of Goldfarb's Principle, for any Σ_1^0-formula $\sigma(x)$ there is a formula $\alpha(x)$ such that

$$T \vdash \forall x\, (\Box_T\alpha(x) \leftrightarrow \sigma(x) \vee \Box_T\bot). \tag{1}$$

Assume $\neg\Box_T\bot \wedge \forall\alpha \in Sent_T\, (\Box_T\Box_T\alpha \to \Box_T\alpha)$ and reason inside T:

For all n, if $\Box_T\sigma(\bar{n})$ then $\Box_T\Box_T\alpha(\bar{n})$ by (1), hence $\Box_T\alpha(\bar{n})$, because $\alpha(\bar{n})$ is a T-sentence. From $\Box_T\alpha(\bar{n})$ we infer $\sigma(\bar{n}) \vee \Box_T\bot$ using (1) once again, ergo $\sigma(\bar{n})$.

So, we have proved $\forall x\, (\Box_T\sigma(x) \to \sigma(x))$ for every Σ_1^0-formula $\sigma(x)$, **q.e.d.**

Corollary 1. *There is a formula of second order propositional provability logic which holds for the interpretation w.r.t. PA but not w.r.t. IΔ_0+EXP, and which has altogether 4 occurrences of quantifiers, of them 1 under the scope of modality.*

Now we turn to the proof of Theorem 1. This theorem is an almost immediate corollary of Theorem 3 formulated below, which is also of some independent interest.

Let $exp(x)$ denote the function 2^x; EXP, as usual, stands for the axiom $\forall x \exists y\, (exp(x) = y)$. *Superexponentiation* function $s(x)$ is defined recursively as $s(0) = 1$, $s(x+1) = exp(s(x))$; for $n \geq 0$, let $s^n(x)$ denote the n-th iterate of

$s(x)$. We assume all these functions to be defined by their Δ_0-graphs, so that all their elementary properties, including monotonicity and recursive definition clauses, except for the totality, are provable in $I\Delta_0+\Omega_1$ (cf. [10]). Further, let $s^n(x)!$ denote the formula $\exists y\ s^n(x) = y$, and let SUPEXP be the sentence $\forall x \exists y\ (s(x) = y)$ asserting the totality of superexponentiation.

Extensions of a theory T by iterated consistency assertions are defined as follows:

$$(T)_0 \rightleftharpoons T; \quad (T)_{n+1} \rightleftharpoons (T)_n + \mathrm{Con}((T)_n); \quad (T)_\omega \rightleftharpoons \bigcup_{n<\omega} (T)_n.$$

Here and below $\mathrm{Con}(U)$ denotes the consistency statement for a theory U.

Theorem 3. $I\Delta_0+\mathrm{SUPEXP}$ *is* Π_1^0-*conservative over* $(I\Delta_0+\mathrm{EXP})_\omega$.

Proof. Theorems of a similar character can be found in the literature (cf. e.g. [5]), although we have not encountered this particular one. For our present needs we adapt an easy model-theoretic proof of a characterization of Π_1^0-consequences of $I\Delta_0+\mathrm{EXP}$ over $I\Delta_0+\Omega_1$ given by A.Wilkie and J.Paris. We follow their proof very closely, therefore we omit some technical details carefully treated in [10].

Lemma 2. *Let* T *be a* Π_1^0-*axiomatized theory, and let* $\phi(x)$ *be a* $\Delta_0(exp)$-*formula. Then,*

$$T + \mathrm{SUPEXP} \vdash \forall x\ \phi(x) \quad \Rightarrow \quad \exists n \in \mathbf{N}\ \ T + \Omega_1 \vdash \forall x\ (s^n(x)! \to \phi(x)).$$

Proof. Suppose that, on the contrary, for all $n \in \mathbf{N}$

$$T + \Omega_1 \not\vdash \forall x\ (s^n(x)! \to \phi(x)).$$

Add to the language of arithmetic a new constant symbol c, and let T_1 denote the theory axiomatized over $T+\Omega_1$ by the axiom $\neg\phi(c)$ together with $\{s^n(c)! : n \in \mathbf{N}\}$. Compactness Theorem implies that T_1 has a model. Indeed, if M_n is a model of $T + \Omega_1$ such that

$$M_n \not\models \forall x\ (s^n(x)! \to \phi(x)),$$

and c is such that

$$M_n \models s^n(c)! \wedge \neg\phi(c),$$

then by the obvious properties of $s^n(x)$ provable in $I\Delta_0+\Omega_1$ one also has

$$M_n \models s^m(c)!,$$

for all $m < n$. Therefore, with the growth of n, M_n gets to be a model of an arbitrarily large fragment of T_1.

Now let M be a model of T_1; define

$$M^* \rightleftharpoons \{a \in M \ : \ \exists n \in \mathbf{N} \ M \models (a \leq s^n(c))\}.$$

Since T is Π_0^1-axiomatized and M^* is an initial segment of M, M^* is a model of T. M is closed under superexponentiation, because

$$
\begin{aligned}
a \in M^* \ &\Rightarrow \ a \leq s^n(c) \ \text{for some } n \in \mathbf{N}, \\
&\Rightarrow \ s(a) \leq s^{n+1}(c) \ \text{by provable monotonicity of } s, \\
&\Rightarrow \ s(a) \in M^*.
\end{aligned}
$$

It follows that $M^* \models \text{SUPEXP}$. Finally, $M^* \not\models \phi(c)$, because $\phi(x) \in \Delta_0(exp)$ and $M^* \models \text{EXP}$. This contradicts the assumption that

$$T + \text{SUPEXP} \vdash \forall x \ \phi(x), \qquad\qquad \textbf{q.e.d.}$$

Lemma 3. *Let T be a reasonable theory extending* IΔ₀+EXP, *and let $\phi(x)$ be a Δ_0-formula, possibly containing other parameters than just x. Then,*

$$T \vdash \forall x \ (s(x)! \to \phi(x)) \quad \Rightarrow \quad \text{IΔ}_0\text{+EXP+Con}(T) \vdash \forall x \ \phi(x).$$

Proof. Clearly,

$$\text{IΔ}_0\text{+EXP} \vdash s(0)! \wedge \forall x \ (s(x)! \to s(x+1)!).$$

By shortening techniques, there is a cut $I(x)$ in IΔ₀+EXP, closed under $+$, $\cdot$ and ω_1, such that IΔ₀+EXP $\vdash \forall x \ (I(x) \to s(x)!)$. Hence, if $T \vdash \forall x \ (s(x)! \to \phi(x))$, then $T \vdash \forall x \ (I(x) \to \phi(x))$, and thus $T \vdash \forall x \ (I(x) \to \phi^I(x))$, because $\phi(x) \in \Delta_0$ (cf. [10], Lemma 7.2).

Since $I(x)$ is closed under ω_1, this means that $I(x)$ defines a relative interpretation of the theory IΔ₀+Ω_1+$\forall x \ \phi(x)$ in T. It follows that

$$\text{IΔ}_0\text{+}\Omega_1 \vdash \text{Con}(T) \to \text{Con}(\text{IΔ}_0\text{+}\Omega_1\text{+}\forall x \ \phi(x)). \qquad\qquad (2)$$

On the other hand, by provable Σ_1^0-completeness in IΔ₀+EXP we have

$$
\begin{aligned}
\text{IΔ}_0\text{+EXP} \vdash \neg\forall x \ \phi(x) \ &\to \ \Box_{\text{IΔ}_0\text{+}\Omega_1} \neg\forall x \ \phi(x) \\
&\to \ \neg\text{Con}(\text{IΔ}_0\text{+}\Omega_1\text{+}\forall x \ \phi(x)). \qquad (3)
\end{aligned}
$$

From (2) and (3) one concludes that

$$\text{IΔ}_0\text{+EXP} \vdash \text{Con}(T) \to \forall x \ \phi(x), \qquad\qquad \textbf{q.e.d.}$$

Proof of Theorem 2. By metamathematical induction on n we show that, for all $\phi(x) \in \Delta_0$,

$$I\Delta_0 + EXP \vdash \forall x\, (s^n(x)! \to \phi(x)) \quad \Rightarrow \quad (I\Delta_0 + EXP)_n \vdash \forall x\, \phi(x).$$

This, together with Lemma 2, immediately yields the result.

BASIS of induction is trivial.

INDUCTION STEP: Suppose

$$I\Delta_0 + EXP \vdash \forall x\, (s^{n+1}(x)! \to \phi(x)),$$

then clearly

$$I\Delta_0 + EXP \vdash \forall x, u\, (s^n(u)! \to (\phi(x) \vee s(x) \neq u)).$$

The induction hypothesis implies

$$
\begin{aligned}
(I\Delta_0 + EXP)_n \quad & \vdash \quad \forall x, u\, (\phi(x) \vee s(x) \neq u) \\
& \vdash \quad \forall x\, (\exists u\, (s(x) = u) \to \phi(x)).
\end{aligned}
$$

By Lemma 3 we obtain

$$I\Delta_0 + EXP + Con((I\Delta_0 + EXP)_n) \vdash \forall x\, \phi(x), \qquad\qquad \textbf{q.e.d.}$$

Corollary 2. *The theory* $I\Delta_0 + EXP + RFN_{\Sigma_1^0}(I\Delta_0 + EXP)$ *proves the same arithmetical* Π_1^0-*sentences as* $(I\Delta_0 + EXP)_\omega$.

Proof. Essentially by the results of A.Wilkie and J.Paris (cf. [10], Lemmas 8.10, 8.18) $I\Delta_0 + SUPEXP$ contains $RFN_{\Sigma_1^0}(I\Delta_0 + EXP)$.[2] So, by Theorem 3, Π_1^0-consequences of $I\Delta_0 + EXP + RFN_{\Sigma_1^0}(I\Delta_0 + EXP)$ are contained in $(I\Delta_0 + EXP)_\omega$.

For the converse inclusion, show by an obvious metamathematical induction on n that, for all $n \in \mathbf{N}$,

$$I\Delta_0 + EXP + RFN_{\Sigma_1^0}(I\Delta_0 + EXP) \vdash (I\Delta_0 + EXP)_n, \qquad\qquad \textbf{q.e.d.}$$

Notice that the analog of Corollary 2 for PA does not hold: the metamathematical induction above can be transformed into a mathematical one, proving

$$PA + RFN_{\Sigma_1^0}(PA) \vdash Con((PA)_\omega).$$

---

[2]A.Visser proved that SUPEXP is actually equivalent to $RFN_{\Sigma_1^0}(I\Delta_0 + EXP)$ over $I\Delta_0 + \Omega_1$.

Remark 1. Similar conservativity results for Σ_1^0-reflection hold over Primitive Recursive Arithmetic PRA and over extensions of PRA by iterated uniform Σ_1^0-reflection principles. These results have been obtained by U.Schmerl [5] using syntactic methods. However, it is also not difficult to check that the proof of Corollary 2 given above works for these theories instead of $I\Delta_0$+EXP as well, along with all the rest of the results of this paper.

Proof of Theorem 1. In the following we shall write Magari algebraic formulas using proof theoretic notation, i.e., we only omit subscripts and any other references to particular theories. Let $\alpha \in P$ denote the following Magari algebraic formula:

$$\alpha \in P \quad \Leftrightarrow \quad \exists x \, (\vdash \alpha \leftrightarrow \Diamond x),$$

where, as usual, $\Diamond$ abbreviates $\neg\Box\neg$. By Goldfarb's Principle, an arithmetic sentence is in P iff it is (provably equivalent) to a Π_1^0-formula and is stronger than consistency.

We claim that the following statement holds in $M'(PA)$, but does not hold in $M'(I\Delta_0$+EXP):

$$\exists \alpha \, [\, \vdash \alpha \to \neg\Box\bot \quad \& \quad \forall \pi \in P \, (\vdash \alpha \to \pi \; \Rightarrow \; \vdash \alpha \to \Diamond\pi \,) \tag{4}$$

$$\& \quad \exists \pi \in P \, (\vdash \mathrm{RFN}_{\Sigma_1^0} \to \pi \; \& \; \nvdash \alpha \to \pi \,) \,]. \tag{5}$$

First of all, it is easy to see that the above formula does not hold in $M'(I\Delta_0$+EXP). Indeed, if a sentence $\pi \in P$ satisfies

$$I\Delta_0\text{+EXP} \; \vdash \; \mathrm{RFN}_{\Sigma_1^0}(I\Delta_0\text{+EXP}) \to \pi,$$

then, by Theorem 3, there holds

$$I\Delta_0\text{+EXP} \; \vdash \; \neg\Box_{I\Delta_0\text{+EXP}}^n\bot \to \pi,$$

for some $n \in \mathbf{N}$. On the other hand, if

$$\vdash \alpha \to \neg\Box\bot \quad \& \quad \forall \pi \in P \, (\vdash \alpha \to \pi \; \Rightarrow \; \vdash \alpha \to \Diamond\pi \,),$$

then, by metamathematical induction on n, $\vdash \alpha \to \neg\Box^n\bot$ for all $n \in \mathbf{N}$. Therefore one also has

$$I\Delta_0\text{+EXP} \vdash \alpha \to \pi,$$

which contradicts (5).

To show that our formula holds in $M'(PA)$ it is sufficient to construct an arithmetical sentence α satisfying the following two claims:

1. For every Σ_1^0-sentence σ, $PA + \alpha \vdash \Box_{PA}\sigma \to \sigma$, that is, $PA+\alpha$ contains the local Σ_1^0-reflection principle $\mathrm{Rfn}_{\Sigma_1^0}(PA)$ for PA;

2. PA+α $\nvdash$ Con($(PA)_\omega$).

Claim 1, obviously, ensures the validity of (4), and Claim 2, with Con($(PA)_\omega$) playing the role of π, validates (5).

The required sentence α is constructed by methods of P.Lindström (cf. [3]). We also need the following result, proved easily with the aid of modal logic and contained, e.g., in [1].

Lemma 4. *Let T be an axiomatized theory containing* $I\Delta_0$+EXP *such that* $(T)_\omega$ *is consistent. Then* $T + \mathrm{Rfn}_{\Sigma_1^0}(T)$ *is consistent as well.*

Taking the theory PA+$\neg$Con($(PA)_\omega$) for T in the above lemma we conclude that the theory

$$U \rightleftharpoons \mathrm{PA} + \neg\mathrm{Con}((\mathrm{PA})_\omega) + \mathrm{Rfn}_{\Sigma_1^0}(\mathrm{PA})$$

is consistent (because $\mathrm{Rfn}_{\Sigma_1^0}(T)$ is even stronger that $\mathrm{Rfn}_{\Sigma_1^0}(\mathrm{PA})$). Notice that U is axiomatized over PA by a set of sentences having bounded arithmetical complexity. Therefore, by Theorem 4 of [3], there exists an arithmetic sentence α such that PA+α is consistent and contains U. This readily shows that α satisfies Claims 1 and 2, **q.e.d.**

References

[1] L.D. Beklemishev, *On the classification of propositional provability logics*, Math. USSR-Izv. **35** (1990), 247–275 (Russian).

[2] _____, *Bimodal logics for extensions of arithmetical theories*, Prepublication Series Logic and Computer Science 7, Department of Math. Logic, Steklov Math. Institute, Moscow, Nov. 1992.

[3] P. Lindström, *On partially conservative sentences and interpretability*, Proc. Amer. Math. Soc. **91** (1984), no. 3, 436–443.

[4] R. Magari, *The diagonalizable algebras (the algebraization of the theories which express Theor.:II)*, Boll. Un. Mat. Ital. **12** (1975), 117–125.

[5] U.R. Schmerl, *A fine structure generated by reflection formulas over Primitive Recursive Arithmetic*, Logic Colloquium '78 (Amsterdam) (M. Boffa, D. van Dalen, and K. McAloon, eds.), North-Holland, 1979, pp. 335–350.

[6] V.Yu. Shavrukov, *A note on the diagonalizable algebras of PA and ZF*, Ann. Pure Appl. Logic **61** (1993), 161–173.

[7] _____, *Subalgebras of diagonalizable algebras of theories containing arithmetic*, Dissertationes Math. (Rozprawy Mat.), vol. 323, 1993.

[8] _____, *Undecidability in diagonalizable algebras*, ILLC prepublication series ML–93–13, University of Amsterdam, 1993.

[9] _____, *Undecidability in diagonalizable algebras*, Typeset manuscript, Apr. 1994.

[10] A. Wilkie and J. Paris, *On the scheme of induction for bounded arithmetic formulas*, Ann. Pure Appl. Logic **35** (1987), 261–302.

STEKLOV MATHEMATICAL INSTITUTE, VAVILOVA 42, 117966 MOSCOW, RUSSIA,
e-mail address: lev@bekl.mian.su

Undecidability in Weak Membership Theories

Dorella Bellè   Franco Parlamento [*]

Abstract

Let φ range over the collection of conjunctions of equalities with one inequality in the language with the constant $\emptyset$ for the empty set and a binary operation $\mathbf{w}$ which, applied to any two sets x and y yields the result of adding y as an element to x. The satisfiability of sentences of the form $\forall\varphi$ with respect to the theory NW, having the obvious axioms for $\emptyset$ and $\mathbf{w}$, is undecidable. The satisfiability of sentences of the form $\exists\forall\varphi$ with respect to the theory obtained by adding to NW the following equational consequences of the extensionality axiom: (E_1) $x\mathbf{w}y\mathbf{w}z = x\mathbf{w}z\mathbf{w}y$ and $(E_2)x\mathbf{w}y\mathbf{w}y = x\mathbf{w}y$, is undecidable. General conditions on codings of Turing Machine computations into an equational language, which entail undecidability, are specified.

1   Introduction

Investigations into the decision problem for "small axiomatic fragments of set theory", to use Tarski's wording in [17], date back, at least, to [16] (see also the ensuing [3]), which stated the interpretability of Robinson's Arithmetic Q into the theory having the axioms (N) $\forall x(x \notin \emptyset)$ and (W) $\forall x\forall y\forall z(x \in \mathbf{w}(y, z) \leftrightarrow x \in y \lor x = z)$ as well as the Extensionality Axiom (E) $\forall x\forall y(\forall z(z \in x \leftrightarrow z \in y) \to x = y)$ to be henceforth denoted with NWE.

More recently such theories have attracted attention from both the foundational and the computational point of view.

Under the stimulus of foundational issues the interpretability of Q into NWE has been improved to obtain the interpretability of Q into NW (see [9]).

---

[*]This work has been supported by funds MURST 60% and 40% of Italy.

The purpose of developing a theorem prover devoted to deal with set-theoretic notions [2], as well as efforts aimed at obtaining extensions of logic programming "with sets" [5] have been motivating reasons leading to investigating the theory NWL obtained by adding to the language of NW the binary function symbol ℓ and the axiom

$$(L): \qquad \forall x \forall y \forall z (z \in x \, \ell \, y \leftrightarrow z \in x \wedge z \neq y)$$

[11], [10], [12]. See [13] for an overview.

In this note we will deal mainly with the theory NW and one of its extensions with the purpose of establishing undecidability results to complement the decidability results in [1].

Although the interpretability of Q in NW implies the essential undecidability of NW (cf. also [18]), it is of little help to assess precisely to which class of sentences, classified accordingly to the quantificational prefix, the undecidability of NW applies to. The obvious reduction, through the deduction theorem, to the decision problem for pure logic alone, extensively investigated in [6] and [8], also does not yield any useful information in that respect.

In this note we will show that:

(a) the satisfiability, with respect to NW of sentences of the form $\forall x \varphi$ where φ is a conjunction of equalities with one inequality in the language with the constant $\emptyset$ and the binary function symbol $\mathbf{w}$, is undecidable;

(b) (a) fails to hold when the following two equational consequences E_1 and E_2 of the Extensionality Axiom: (E_1) $x\mathbf{w}y\mathbf{w}z = x\mathbf{w}z\mathbf{w}y$ and (E_2) $x\mathbf{w}y\mathbf{w}y = x\mathbf{w}y$ are added to NW (we are using the operator $\mathbf{w}$ infix and left associative);

(c) the satisfiability, with respect to $NW + E_1 + E_2$ of sentences of the form $\exists y \forall x \varphi$ where φ is a conjunction of equalities with one inequality in the language with the constant $\emptyset$ and the binary function symbol $\mathbf{w}$, is undecidable.

2   Pure Logic

Abstracting from the specific coding used in [7], we note that, to establish undecidability for an equational language L, it suffices to define a 1-1 map $\hat{\ }$ from a countable alphabet Σ to the set T_x of terms of the language L in a variable x such that:

for all $a, b, c \in \Sigma$, $t \in T_x$, if $\hat{a}(x/t) \sqsubseteq \widehat{bc}$ then $a = b$ and $t = \hat{c}$ or $a = c$ and $t = x$

where $\sqsubseteq$ denotes the subterm relation. As a matter of fact a map of this kind can be extended to a monomorphism $\hat{\ }$ from the monoid of strings $(\Sigma^*, \epsilon, \circ)$, with the concatenation operation $\circ$, to $(T_x, x, /)$ with the operation $/$ of simultaneous substitution in x. Furthermore the following property holds:

For all $v, w \in \Sigma^*$, $w \neq \epsilon$, $t \in T_x$, if $\hat{w}(x/t) \sqsubseteq \hat{v}$ then there are $z, z' \in \Sigma^*$ such that $v = z'wz$ and $\hat{z} = t$.

Using this property we can show that the halting problem for Turing Machines can be reduced to the derivability problem of one identity from a set of identities.

We will refer to a (deterministic) Turing Machine M as represented by a set of 5-tuple of the form (q, s, q', s', m), where $q, q' \in Q$ are states and $s, s' \in \Sigma$ are tape symbols, meaning that if M is in the state q and it is scanning s then it replaces s by s', moves to the left if m is L or to the right if m is R and enters state q'. q_0 and q_f denote the initial and final state and b the blank symbol. A configuration of M is a finite string of symbols from M with exactly one occurrence of a state symbol. Two configurations which differ only for a number of b's, at the beginning or the end will be considered as equivalent. The transition relation $\Rightarrow_M$ is defined by

$$
\begin{aligned}
waqsw' &\Rightarrow_M wq'as'w' &&\text{if } (q, s, q', s', L) \in M \\
wqsw' &\Rightarrow_M ws'q'w' &&\text{if } (q, s, q', s', R) \in M
\end{aligned}
$$

where w and w' are strings of tape symbols of M. By $\Leftrightarrow_M^*$ we denote the equivalence relation generated by $\Rightarrow_M$.

To the transition relation of M we make correspond the (smallest) set of equations S_M such that:

1. for every $(q, s, q', s', L) \in M$ and tape symbol a, $\widehat{aqs} = \widehat{q'as'} \in S_M$;

2. for every $(q, s, q', s', R) \in M$ $\widehat{qs} = \widehat{s'q'} \in S_M$;

3. $\hat{B} = \widehat{Bb}, \hat{E} = \widehat{bE} \in S_M$

Finally every configuration vqw of M is coded with the term $B\widehat{vqw}E$.

PROPOSITION 2.1 $vqw \Rightarrow_M v'q_f w'$ *iff* $S_M \vdash B\widehat{vqw}E = B\widehat{v'q_f w'}E$.

Proof: $\Rightarrow$ is straightforward.

$\Leftarrow$ From the assumption that $S_M \vdash B\widehat{vqw}E = B\widehat{v'q_f w'}E$, by the completeness of the term rewriting version of Birkhoff's calculus, it follows that

$$B\widehat{vqw}E \leftrightarrow^* B\widehat{v'q_f w'}E$$

namely that there are terms $h_1, \ldots, h_{n-1}$ such that:

$$(***) \qquad B\widehat{vqw}E \leftrightarrow h_1 \leftrightarrow \cdots \leftrightarrow h_{n-1} \leftrightarrow B\widehat{v'q_f w'}E$$

where $h_i \leftrightarrow h_{i+1}$ means that either h_j can be obtained from h_i or h_i can be obtained from h_j by applying one of the rewrite rule provided by S_M.

Due to the property of the map $\hat{\cdot}$, it is straightforward to see that for each $1 \leq i \leq n-1$ h_i codes an instantaneous configuration of M and $(***)$ can be seen as the way in which the Thue-process associated to M leads from vqw to $v'q_f w'$. Since q_f is a final state, by Post's lemma [4] $(***)$ can be shortened (if needed) to obtain a derivation of $v'q_f w'$ from vqw in the semi-Thue process associated with M and therefore an halting computation of M starting from the configuration vqw. $\square$

As an immediate consequence of Prop. 2.1, we have the following result:

PROPOSITION 2.2 *A Turing Machine M halts with empty tape when started on the empty tape iff $S_M \vdash B\widehat{q_0}E = B\widehat{q_f}E$.*

Therefore the satisfiability of a conjunction of identities, i.e. sentences of the form $\forall x(r = t)$, with a negation of an identity, in a language which permits the coding $\hat{\cdot}$ is undecidable.

A coding $\hat{\cdot}$ fulfilling the required condition can be immediately obtained if the language L contains infinitely unary function symbols since it suffices to let $\hat{a} = f_a(x)$, with the assumption that for $a \neq b$, $f_a \neq f_b$ [7]. Actually it is enough to assume that the language contains two unary function symbols f_0, f_1. In fact, if $\{a_n : n \in \omega \setminus \{0\}\}$ is an enumeration without repetitions of Σ, it suffices to let

$$\hat{a}_n = f_0(f_1^n(f_0(x)))$$

where f_1^n indicates the application of f_1 iterated n times, and $f_i(t)$ stands for $f_i(x/t)$, to obtain a map $\hat{\cdot}$ fulfilling the required condition, as it is easy to check. To have in L a single binary function symbol $\mathbf{w}$ is also enough, as it is seen by letting $f_0(x)$ denote $\mathbf{w}(x, x)$ and $f_1(x)$ denote $\mathbf{w}(x, \mathbf{w}(x, x))$ and defining $\hat{a}_n$ as in the previous case.

If c is a constant which does not appear in S_M, $S_M \vdash B\widehat{q_0}E = B\widehat{q_f}E$ iff $S_M \vdash B\widehat{q_0}E(c) = B\widehat{q_f}E(c)$. Therefore, if the language contains a constant which is not used in the coding, we can refer the previous undecidability result to sentences of the form $\forall x \varphi$, where φ is a conjunction of equalities with one inequality (which involves the constant c). In the sequel we will make use of a coding which uses, besides a binary function symbol $\mathbf{w}$ also the constant $\emptyset$. In this case, to obtain the undecidability result in the latter form, requires a stronger condition on the coding.

Letting H be the Herbrand Universe over L, $\hat{\cdot} : \Sigma \to T_x$ is required to satisfy the following condition:

$\forall a, b \in \Sigma, \forall t, t' \in T_x \cup H$ if $\hat{a}(x/t) \sqsubseteq \hat{b}(x/t')$ then $a = b$ and $t = t'$ or $\hat{a}(x/t) \sqsubseteq t'$

This condition, which generalizes the previous one, as it is seen by taking $t' = \hat{c}$, ensures that

for all $v, w \in \Sigma^*$, $w \neq \epsilon$, $t \in H$, c constant: if $\hat{w}(x/t) \sqsubseteq \hat{v}(x/c)$ then there are $z, z' \in \Sigma^*$ such that $v = z'wz$ and $\hat{z}(x/c) = t$.

Using this property, following the proof of Prop. 2.1 it is immediate to obtain:

PROPOSITION 2.3 $vqw \Rightarrow_M v'q_f w'$ *iff* $S_M \vdash B\widehat{vqw}E(c) = B\widehat{v'q_f w'}E(c)$.

PROPOSITION 2.4 *A Turing Machine M halts with empty tape when started on the empty tape iff* $S_M \vdash \widehat{Bq_0}E(c) = \widehat{Bq_f}E(c)$.

Therefore the satisfiability of sentences of the form $\forall x \varphi$ where φ is a conjunction of equalities with one inequality in a language which permits the coding $\hat{\cdot}$ is undecidable.

3 NW

If we consider the previous satisfiability problem with respect to a given theory T, we can still use the result of the preceding paragraph, provided that we are able to produce a map $\hat{\cdot}$ such that the following further condition holds:

$$T + S_M \vdash \widehat{Bq_0}E(c) = \widehat{Bq_f}E(c) \quad \text{iff} \quad S_M \vdash \widehat{Bq_0}E(c) = \widehat{Bq_f}E(c)$$

As far as NW is concerned, consider the map defined as follows. Let $\{a_n : n \in \omega \setminus \{0\}\}$ be an enumeration without repetitions of $\Sigma \cup Q \cup \{B, E\}$ and let 0 and 1 stand for $\emptyset$ and $\emptyset w\emptyset$ respectively. Then to every symbol a_n in $\Sigma \cup Q \cup \{B, E\}$ we associate the term

$$\hat{a}_n(x) = x\mathbf{w}0 \underbrace{\mathbf{w}1 \ldots \mathbf{w}1}_{n} \mathbf{w}0.$$

On the ground of that map, which satisfy the required conditions, we can state the following:

PROPOSITION 3.1 *Let t_1, t_2 be ground terms, then*

$$NW + S_M \vdash t_1 = t_2 \quad \text{iff} \quad S_M \vdash t_1 = t_2$$

Proof : Let H be the Herbrand universe over $\emptyset, \mathbf{w}$, and let $\in^H$ be the natural interpretation (under W) of $\in$ in M. Letting $\mathcal{M}/_\sim$ denote the quotient of $\mathcal{M} = (H, \emptyset, \mathbf{w}, \in^H)$ with respect to the equivalence relation $\sim$ defined on H as

$$t_1 \sim t_2 \quad \text{iff} \quad S_M \vdash t_1 = t_2$$

we clearly obtain a model of S_M and N.

By induction on the height of a derivation in Birkhoff calculus of $s_1 \mathbf{w} t_1 \mathbf{w} \ldots \mathbf{w} t_n = s_2 \mathbf{w} k_1 \mathbf{w} \ldots \mathbf{w} k_m$ from S_M, it is rather straightforward to prove that the following condition holds:

$$(*) \quad \begin{array}{l} \text{if } S_M \vdash s_1 \mathbf{w} t_1 \mathbf{w} \ldots \mathbf{w} t_n = s_2 \mathbf{w} k_1 \mathbf{w} \ldots \mathbf{w} k_m \text{ then } s_1 = s_2 \text{ and} \\ \quad \forall 1 \leq i \leq n \ \exists j \ 1 \leq j \leq m \ \ S_M \vdash t_i = k_j \\ \text{and conversely } \forall 1 \leq j \leq m \ \exists i \ 1 \leq i \leq n \ \ S_M \vdash t_i = k_j. \end{array}$$

Using $(*)$ it is easy to verify that $\mathcal{M}/_\sim$ is also a model of W.

Obviously the (ground) equalities true in $\mathcal{M}/_\sim$ are exactly those derivable (in the equational calculus) from S_M. Therefore we have established that, for $t_1, t_2 \in H$,

$$NW + S_M \vdash t_1 = t_2 \quad \text{iff} \quad S_M \vdash t_1 = t_2$$

In particular

$$(**) \quad \begin{array}{l} \text{if } NW + S_M \vdash B\widehat{vqw}E(\emptyset) = B\widehat{v'q'w'}E(\emptyset) \\ \text{then } S_M \vdash B\widehat{vqw}E(\emptyset) = B\widehat{v'q'w'}E(\emptyset) \end{array}$$

$\square$

Since a term which encodes an instantaneous configuration is ground, from Prop. 2.3 and Prop. 3.1 it follows:

COROLLARY 3.1 *The satisfiability with respect to NW of sentences of the form $\forall x \varphi$ where φ is a conjunction of equalities with one inequality in the language $\emptyset, \mathbf{w}$ is undecidable.*

4   NW+$\mathbf{E_1 E_2}$

The previous proof of Prop. 2.3 shows that NW has no effect at all as far as the derivability of equations from S_M is concerned. In some sense the effect of NW on $=$ is too weak to matter. Adding axioms, like the Extensionality Axiom, which have a stronger impact on $=$, may change completely the situation. In fact for $NW + E_1 E_2$ the following decidability result holds:

PROPOSITION 4.1 *The satisfiability problem for purely universal sentences in $\emptyset, \mathbf{w} =, \in$ with respect to $NW + E_1 E_2$ is decidable.*

Proof: It is straightforward to see that $NW + E_1E_2$ has only one normal model which is isomorphic to the structure $\mathcal{HF}$ of hereditarily finite sets with the empty set, the membership relation and $x\mathbf{w}y$ interpreted as $x \cup \{y\}$. From that it follows that a $\forall^*$-sentence $\forall x_1 \ldots \forall x_n \varphi$ (φ an open formula in $\emptyset, \mathbf{w} =, \in$) is unsatisfiable with respect to $NW + E_1E_2$ iff $\mathcal{HF} \models \exists x_1 \ldots \exists x_n \neg\varphi$.

That provides a reduction of our problem to the problem of establishing whether a given $\exists^*$-sentence with matrix in $\emptyset, \mathbf{w} =, \in$ is true in $\mathcal{HF}$; a problem which is known to be decidable (see [1]). $\square$

Prop. 4.1 show that for $NW + E_1E_2$ undecidability must depend on higher logical complexity than that offered by universal quantification. The following proposition tells us precisely what is the increment in logical complexity which is responsible for the reappearance of undecidability.

THEOREM 4.1 *The satisfiability problem for sentences of the form $\exists y \forall x \varphi$ where φ is a conjunction of equalities with one inequality in $\emptyset, \mathbf{w}$ with respect to $NW + E_1E_2$ is undecidable.*

Proof: We add to $\emptyset, \mathbf{w}$ a new constant k and then proceed to coding strings of symbols in the following way. Letting $\overline{0} = \emptyset$ and $\overline{n+1} = \overline{n}\mathbf{w}\overline{n}$, for every symbol a_n in $\Sigma \cup Q \cup \{B, E\}$ we let

$$\hat{a}_n(x) \equiv k\mathbf{w}\overline{n}\mathbf{w}x$$

Every configuration vqw of M is coded with the ground term $\widehat{BvqwE}(k)$.

Clearly the map $\hat{\ }$ just defined satisfies the stronger condition stated in section 2. which is needed for Prop. 2.4.

To see that $NW + E_1E_2$ has no effect on the derivability of equations from S_M we first show how to build a model of $NW + E_1E_2$ in which the ground equations true are precisely those derivable from $E_1E_2 + S_M$ and then, in the case of terms encoding instantaneous configurations, that also the equations E_1E_2 can be discharged.

Let H be the Herbrand Universe over $\emptyset, \mathbf{w}, k$ and let $\in^H$ be obtained from the natural interpretation of $\in$ in H by adding $\{(t, k) : t \in H\}$ and then closing under W. Note that in this way k behaves as a universal set in H.

Let $\mathcal{N}$ be the quotient of $(H', \emptyset, \mathbf{w}, \in^{H'})$ with respect to the equivalence relation $\sim$ defined as

$$t_1 \sim t_2 \quad \text{iff} \quad S_M, E_1, E_2 \vdash t_1 = t_2.$$

Clearly $\mathcal{N}$ is a model of S_M and $N + E_1E_2$. The crucial point is to verify that $\mathcal{N} \models W$. That can be verified on the ground of the following fact:

if $S_M \vdash s_1\mathbf{w}t_1\mathbf{w}\ldots\mathbf{w}t_n = s_2\mathbf{w}k_1\mathbf{w}\ldots\mathbf{w}k_m$ then $s_1 = s_2$ and, if $s_1 \neq k$, J
$$\forall 1 \leq i \leq n \ \exists j 1 \leq j \leq m \ \ S_M \vdash t_i = k_j$$
and conversely $\forall 1 \leq j \leq m \ \exists i 1 \leq i \leq n \ \ S_M \vdash t_i = k_j$.

For the second part, consider the function $Canon$ defined as follows, on the ground of a total order $<$ of the terms in H:

$$
\begin{aligned}
Canon(g_i) &= g_i \quad \text{for } 0 \leq i \leq k \\
Canon(g_h\mathbf{w}r_1\mathbf{w}\ldots\mathbf{w}r_l) &= g_h\mathbf{w}Canon(r_{i_1})\mathbf{w}\ldots\mathbf{w}Canon(r_{i_n})
\end{aligned}
$$

where $Canon\{r_1,\ldots,r_l\} = Canon\{r_{i_1},\ldots,r_{i_n}\}$ and, for $1 \leq j < k \leq n$, $Canon(r_{i_j}) < Canon(r_{i_k})$.

It is easy to see that $Canon(t_1) = Canon(t_2)$ iff $E_1, E_2 \vdash t_1 = t_2$.

Let t be a term of the form $s\mathbf{w}t_1\mathbf{w}\ldots\mathbf{w}t_n$ where s is a variable or a constant symbol; then s is called *seed* of t, while each t_i is an *element* of t.

Now we extend the language with a new constant $\underline{k}$ and consider the rewriting system $\underline{R}$ which has all the rewriting rules that come from E_1, the equations in S_M, the equations obtained by substituting $\underline{k}$ in place of k in S_M and furthermore the rewriting rules obtained from $E_2 : x\mathbf{w}y\mathbf{w}y = x\mathbf{w}y$ by replacing y with a term that has seed different from $\underline{k}$, and the rewriting rules obtained from $x\mathbf{w}\underline{y}\mathbf{w}y = x\mathbf{w}y$ or $x\mathbf{w}y\mathbf{w}\underline{y} = x\mathbf{w}y$, by replacing y with a term t that has seed $\underline{k}$ and y with the term obtained by replacing all occurrences in t of $\underline{k}$ with k. Let A be the set of terms that satisfy the following properties:

> the seed of the term is $\underline{k}$; every subterm having $\underline{k}$ as seed has no elements or otherwise it has exactly one element with seed $\underline{k}$, at least one element with seed $\emptyset$ and all the elements with seed $\emptyset$ have the same canonical representation.

Observe that the application of any rewriting rule of $\underline{R}$ to a term in A yields a term in A.

On A we define inductively on the height of terms the following map:

$$m(\underline{k}\mathbf{w}t_1\mathbf{w}\ldots\mathbf{w}t_n) = \underline{k}\mathbf{w}h\mathbf{w}m(t_i)$$

where h is the canonical representation of the elements which have $\emptyset$ as seed and t_i is the unique term with seed $\underline{k}$.

From the assumption that $E_1, E_2 + S_M \vdash \hat{w}(k) = \hat{w'}(k)$, where w, w' are strings, by the completeness of the term rewriting version of Birkhoff's calculus, it follows that

$$\hat{w}(k) \to^* \hat{w'}(k)$$

namely that there are terms $h_1,\ldots,h_{n-1}$ such that:

$$\hat{w}(k) = h_0 \to h_1 \to \cdots \to h_{n-1} \to h_{n+1} = \hat{w'}(k)$$

where $h_i \to h_{i+1}$ means that either h_{i+1} can be obtained from h_i by applying one of the rewrite rule provided by E_1, E_2 or by S_M.

If we replace all occurrences of k in $\hat{w}(k)$ with $\underline{k}$ we obtain a term $\underline{h}_0$ in A and furthermore we have that $m(\underline{h}_0) = h_0$. Furthermore some of the occurrences of k in $h_1, \ldots, h_{n-1}, h_n$ can be marked to obtain a derivation $h_0 \to \underline{h}_1 \to \cdots \to \underline{h}_{n-1} \to \underline{h}_{n+1}$ in $\underline{R}$. Since A is closed for application of rules in $\underline{R}$, $\underline{h}_1, \ldots, \underline{h}_{n-1}, \underline{h}_n$ are all in A.

It is easy to verify that if the rule which leads from $\underline{h}_i$ to $\underline{h}_{i+1}$ comes from E_1 or (a variant of) E_2 or it comes from equations in S_M in which k is not marked then $m(\underline{h}_i) = m(\underline{h}_{i+1})$.

If the rule comes from an equation in S_M in which k is marked we have that a (possibly different) instance of the same equation and in the same direction can be applied to $m(\underline{h}_i)$ obtaining $m(\underline{h}_{i+1})$.

For example if we are replacing the subterm $\underline{k}\mathbf{w}\overline{n}_q\mathbf{w}(\underline{k}\mathbf{w}\overline{n}_s\mathbf{w}t)$ in $\underline{h}_i$ with the subterm $\underline{k}\mathbf{w}\overline{n}_{s'}\mathbf{w}(\underline{k}\mathbf{w}\overline{n}_{q'}\mathbf{w}t)$ (in $\underline{h}_{i+1}$) then, due to the definition of m, $k\mathbf{w}\overline{n}_q\mathbf{w}(\underline{k}\mathbf{w}\overline{n}_s\mathbf{w}m(t))$ is in $m(\underline{h}_i)$ and if we replace it in $m(\underline{h}_i)$ by $k\mathbf{w}\overline{n}_{s'}\mathbf{w}(\underline{k}\mathbf{w}\overline{n}_{q'}\mathbf{w}m(t))$ we obtain $m(\underline{h}_{i+1})$.

Hence, we can produce a derivation of $m(\underline{h}_{n+1})$ from h_0 using only rewriting rules derived from equations in S_M.

But, as for $\underline{h}_0$, $m(\underline{h}_{n+1}) = h_{n+1}$ and we can conclude that there exists a derivation from $\hat{w}(k)$ to $\hat{w}'(k)$ that uses only rewriting rules that come from S_M.

We have thus established that:

$$E_1, E_2 + S_M \vdash \hat{w}(k) = \hat{w'}(k) \text{ iff } S_M \vdash \hat{w}(k) = \hat{w}'(k)$$

and the proof of Theor. 4.1 is concluded. $\square$

The proof of Theor. 4.1 relies essentially on the lack in $NW + E_1 E_2$ of the Regularity Axiom, to make it possible the presence of a universal set in $\mathcal{N}$, as well as on the absence of the Extensionality Axiom, otherwise all the identities corresponding to Turing Machine instructions (independently of the specific Turing Machine we are dealing with) would be trivially satisfied in $\mathcal{N}$. That leaves open the problem to extend Theor. 4.1 to theories which adopt these axioms.

On the other hand, both Theor. 2.4 and Theor.4.1 can be extended when the language is enriched by adding the function symbol ℓ and axiom L is adopted.

Any bound on the level of nesting of $\mathbf{w}$ in φ would make the satisfiability problem in Theor.2.4 and Theor.4.1 trivially decidable, since then there would be only finitely many non equivalent sentences to test. For undecidability to be possible in that case, arbitrarily long lists of quantifiers must be allowed.

Concerning the resulting problem, we point out that the satisfiability of $\exists^*\forall$ sentences in the language $=, \emptyset, \in$ with respect to NWL and $NWL + E_1E_2$ as well as the satisfiability of $\exists^*\forall\forall$ in the language $=, \emptyset, \in$ with respect to NWL have been proved to be decidable in [11] and [14]. Finally we add that the essential undecidability of the theory $NWL + E_1E_2$ with respect to sentences of the form $\exists^*\forall^*$ in the language $=, \emptyset, \in, \mathbf{w}$ is known to follow from results in [15], which shows how to encode Turing machine's computations by using prenex formulae involving only restricted universal quantification, in the language with equality, the membership relation $\in$ and $\emptyset, \mathbf{w}$.

References

[1] D. Bellè and F. Parlamento, *Decidability and Completeness of Open Formulae in Membership Theories*, Technical Report 25/93, Universitá di Udine (1993).

[2] D. Cantone, A. Ferro, and E.G Omodeo, *Computable Set Theory. Vol. 1*, Int. Series of Monographs on Computer Science, Oxford University Press, 1989.

[3] G. E. Collins and J. D. Halpern, *On the Interpretability of Arithmetic in Set Theory*, Notre Dame J. Formal Logic **XI(4)** (1970), 477–483.

[4] M.D. Davis and E.J. Weyuker, *Computability, Complexity, and Languages*, Academic Press, New York, 1983.

[5] A. Dovier and G. Rossi, *Embedding extensional finite sets in clp*, Proceedings of the 4th International Logic Programming Symposium (D. Miller, ed.), 1993.

[6] B. Dreben and W.D. Goldfarb, *The Decision Problem. Solvable Classes of Quantificational Formulas*, Addison Wensley, 1979.

[7] C. Fermüller and G. Salzer, *Ordered Paramodulation and Resolution as Decision Procedure*, Lecture Notes in Artificial Intelligence, 698, Springer Verlag, 1993.

[8] H.R. Lewis, *Unsolvable Classes of Quantificational Formulas*, Addison Wensley, 1979.

[9] A. Mancini and F. Montagna, *A Minimal Predicative Set Theory*, Notre Dame J. Formal Logic, to appear.

[10] E. Omodeo, F. Parlamento, and A. Policriti, *A contribution to the automated treatment of membership theories*, Atti del Convegno Nazionale sulla Programmazione Logica, 1990.

[11] _____, *Decidability of $\exists^*\forall$-sentences in Membership Theories*, Technical Report 6, Università di Udine, Dip. di Matematica e Informatica (1992).

[12] _____, *A derived algorithm for evaluating ϵ-expressions over sets*, J. Symb. Comput. **15** (1993), 673–704.

[13] F. Parlamento, *Il problema della decisione per teorie deboli dell'appartenenza*, Atti del XV Incontro di Logica Matematica - Camerino, 1992.

[14] F. Parlamento and A. Policriti, *The satisfiability problem for the class* ∃*∀∀ *in a set-theoretic setting*, Technical Report 13/92, Universitá di Udine (1992).

[15] _____ , *Undecidability Results for Restricted Universally Quantified Formulae of Set Theory*, Comm. Pure Appl. Math. **XLVI(1)** (1993), 57–73.

[16] A. Tarski and W. Szmielew, *Mutual Interpretability of Some Essentially Undecidable Theories*, Proc. of Intl. Cong. of Mathematicians, Cambridge, Cambrige University Press, 1950, volume 1.

[17] A. Tarsky, A. Mostowsky, and R.M. Robinson, *Undecidable Theories*, North Holland, 1953.

[18] R.L. Vaught, *On a theorem of Cobham concerning undecidable theories*, Proceedings of the 1960 International Congress on Logic, Methodology and Philosophy of Science, Stanford University Press, 1962, pp. 14–25.

DIPARTIMENTO DI MATEMATICA E INFORMATICA, UNIVERSITÀ DI UDINE, VIA ZANON 6,, 33100 UDINE, ITALY

e-mail address: belle@dimi.uniud.it

e-mail address: Franco.Parlamento@uniud.it

Infinite λ-calculus and non-sensible models [*]

Alessandro Berarducci

Abstract

We define a model of $\lambda\beta$-calculus which is similar to the model of Böhm trees, but it does not identify all the unsolvable lambda-terms. The role of the unsolvable terms is taken by a much smaller class of terms which we call mute. Mute terms are those zero terms which are not β-convertible to a zero term applied to something else. We prove that it is consistent with the $\lambda\beta$-calculus to simultaneously equate all the mute terms to a fixed arbitrary closed term. This allows us to strengthen some results of Jacopini and Venturini Zilli concerning easy λ-terms. Our results depend on an infinitary version of λ-calculus. We set the foundations for such a calculus, which might turn out to be a useful tool for the study of non-sensible models of λ-calculus.

Dedicated to the memory of Roberto Magari

1 Introduction

Our aim is to define a new model of λ-calculus in which two λ-terms are identified if they have the same "asymptotic behaviour", namely they approach the same limit by repeated β-reductions. Such an idea is already present in the notion of "Böhm tree" (see [5, 1]) but it is not fully exploited in the sense that the asymptotic behaviour of the unsolvable terms is completely ignored. The Böhm tree of a λ-term, is a kind of infinite unfolding of the λ-term with respect to β-reduction. Consider for instance the Turing fixed point combinator $\mathbf{Y_t} \equiv QQ$ where $Q \equiv \lambda x, y.y(xx)$ (as usual $\equiv$ between λ-terms

---

[*]Work partially supported by the Esprit project "Gentzen" and by the research projects 60% and 40% of the Italian Ministero dell' Università e della Ricerca Scientifica e Tecnologica. Presented at the meeting "Common foundations of logic and functional programming" held in Torino, Feb. 1994, and at the conference in honor of Roberto Magari, Siena, April 1994.

is syntactic identity up to renaming of bound variables, whilst $=$ denotes β-convertibility). The Böhm tree $BT(\mathbf{Y_t})$ is obtained as a limit of the ω-sequence of β-reductions $\mathbf{Y_t} \to \lambda y.y(QQ) \to \lambda y.y(y(QQ)) \to \ldots$, namely $BT(\mathbf{Y_t})$ is the "completely unfolded" infinite λ-term $\lambda y.y(y(y(\ldots)))$. Böhm trees are more often depicted as trees, but we prefer to think of them as infinite λ-terms. For the reader's convenience we recall the formal definition of Böhm tree in section 2. Since the Böhm tree of an unsolvable λ-term is defined to be $\bot$ (where $\bot$ is a special symbol staying for "bottom", or "undefined", or "empty"), Böhm trees do not distinguish among unsolvable terms.

The importance of Böhm trees both for the proof-theory and for the semantics of lambda-calculus needs not be stressed. Let us just recall that two λ-terms are equal in the Plotkin model $P\omega$ if and only if they have the same Böhm tree (see [9, 1]). Moreover two λ-terms are equal in the Scott model D_∞ if and only if they have the same "$\beta\eta$-Böhm tree" (see [9, 20, 1]).

The fact that the Böhm tree of an unsolvable term is $\bot$ can be a drawback: it means that Böhm trees give no information on the inner structure of the unsolvable terms. It has been argued that the unsolvable terms correspond to the notion of "undefined" in λ-calculus (see [1]), and therefore one does not need to look inside them. However some recent papers [11, 10, 4] have suggested that some unsolvable terms can have an operational meaning and therefore one should take as undefined elements a smaller set of terms than the unsolvables, for instance the *zero terms* (Statman) or the *easy terms* [12, 13, 14, 21]. Using a result of Visser [19] Statman proved (see [2]) that one can take any Π_1 set of terms closed under β-conversion to represent the undefined value of a partial recursive function. Topological models of λ-calculus where not all the unsolvable are identified are studied in [8, 18].

With these motivations in mind we extend non-trivially the notion of Böhm tree to a large class of unsolvable terms. We do so by introducing what we call the *infinite $\beta\bot$-normal form* of a λ-term. The definition of the infinite $\beta\bot$-normal form is very natural: we just apply the idea of infinite unfolding also to the unsolvable terms. Consider for instance the unsolvable term $\Omega_3 \equiv \omega_3\omega_3$ where $\omega_3 \equiv \lambda x.xxx$. We have the ω-sequence of β-reductions

$$\Omega_3 \to \Omega_3\omega_3 \to \Omega_3\omega_3\omega_3 \to \ldots$$

such a sequence "converges" to the infinite β-reduction $\Omega_3 \to_\infty NF^\infty(\Omega_3)$, where $NF^\infty(\Omega_3)$ is the unique infinite term satisfying the syntactic identity $NF^\infty(\Omega_3) \equiv NF^\infty(\Omega_3)\omega_3$. This can be depicted as a tree with binary application nodes:

$$NF^\infty(\Omega_3) \equiv$$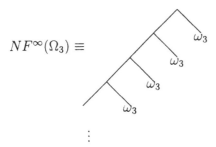

$$\vdots$$

The notion of convergence here used is convergence with respect to the usual topology of infinite trees plus the stronger requirement that the depth of the redexes reduced in an infinite β-reduction must go to infinity (such a notion is called strong-convergence in [16] in the context of term-rewriting systems for first order infinite terms). Unlike what happens for finite lambda-terms, the above example shows that there are terms, like $NF^\infty(\Omega_3)$, which are in normal form (as they have no β-redexes) and yet they begin neither with a λ-abstraction nor with a variable.

The main result of this paper is that the infinite $\beta\bot$-normal forms are, like the Böhm trees, a model of the $\lambda\beta$-calculus. Such a model is not *sensible*, i.e. not all the unsolvable are identified. What plays the role of the unsolvable terms is a much smaller class of terms which we call *mute*. We recall that a *zero term* is a term which cannot by reduced to an abstraction term $\lambda x.T$. Mute terms are then defined as those zero terms which cannot be reduced to a variable or to a zero term applied to some other term. If A is mute we set by convention $NF^\infty(A) = \bot$. For instance $\Omega \equiv (\lambda x.xx)(\lambda x.xx)$ is mute, whilst Ω_3 is a zero term but it is not mute. The idea is that mute terms have a totally undefined operational behaviour.

To prove our main result the crucial lemma is to show that infinite $\beta\bot$-normal forms behave well under substitutions, i.e. we must show:

$$NF^\infty(A[x := B]) = NF^\infty(NF^\infty(A)[x := NF^\infty(B)]) \qquad (*)$$

A similar problem arises if one deals with Böhm trees, and in that context it is usually handled by using finite approximations to infinite trees and proving some kind of continuity theorem. Here we propose a different approach. While in the usual treatment of Böhm trees one deals exclusively with infinite λ-terms which are in normal form, we take the more liberal view of considering a full fledged infinite λ-calculus in which we allow arbitrary infinite λ-terms and infinite reductions among them. This extra freedom allows us to state and prove some general results about infinite λ-calculus of which $(*)$ is an immediate consequence. More precisely we formulate two versions of infinite λ-calculus, an infinite $\lambda\beta$-calculus and an infinite $\lambda\beta\bot$-calculus. The infinite $\lambda\beta$-calculus is not Church-Rosser and not normalizing (with respect to infinite

β-reductions), but if a finite term has an infinite β-normal form, then it is unique. The infinite $\beta\bot$-calculus behaves better: it is Church-Rosser, at least for reductions starting from a finite term, and normalizing. So a finite term A has one and only one $\beta\bot$-normal form, which coincides with what we have called $NF^\infty(A)$. The equality $(*)$ follows at once since the two sides coincide with the (unique) infinite $\beta\bot$-normal form of $A[x := B]$.

In our approach finite approximations are not explicitly used but they are somehow implicit in the notion of infinite reduction.

In section 14 we prove, that mute terms are much more "undefined" than the unsolvable terms in the sense that it is consistent with the $\lambda\beta$-calculus to simultaneously equate all mute terms to an arbitrary closed term (not necessarily mute). In particular each mute term is *easy*, i.e. it can be consistently equated to every closed term. The proof consists in defining a suitable Church-Rosser extension of λ-calculus. The method of proving consistency results via Church-Rosser extensions was used by Mitschke (see [1] Section 15.3, [17]) and by Intrigila [11] in a more complex situation. The method is improved in [4] where the use of infinite Böhm trees is introduced. A general overview is given in [3].

The fact that all mute terms can be simultaneously equated to an arbitrary closed fixed term is a very strong property which is not shared by the class of the easy terms, or even by the smaller class of the closed recurrent zero terms: for instance $\Omega \equiv (\lambda x.xx)(\lambda x.xx)$ and $\Omega\mathbf{I}$ (where $\mathbf{I} \equiv \lambda x.x$) are two closed recurrent zero terms which cannot be simultaneously equated to $\lambda xy.x$.

B. Intrigila [11] showed that the class of the easy terms is not semantically stable in the sense that if we equate all the closed easy terms to Ω, then the rules of the $\lambda\beta$-calculus force us to equate an easy term to a non-easy one. Our main result shows instead that the class of all mute terms is stable, in the sense that it is possible to have a model in which all mute terms are identified without having equalities between mute and non-mute terms.

We finish the paper by comparing mute terms with another class of previously studied easy terms, namely closed recurrent zero terms studied in [13] (A is recurrent if every reduct of A can be reduced to A). Neither class is included in the other but every closed recurrent zero term is a *strong zero term of finite degree* (see section 15), namely it is β-convertible to a term of the form $AM_1 \ldots M_n$ with A mute. We show that every strong zero term of finite degree is easy. (This is an immediate consequence of the fact that mute terms are easy and that if A is easy, then AM is also easy.) We thus obtain a strengthening of the fact that closed recurrent zero terms are easy.

By our results the problem of which zero terms are easy, is reduced to the case of the strong zero terms of infinite degree. Here both possibilities can happen and some difficult problems remain: for instance it is still not known whether $\mathbf{Y_t}\Omega_3$ is easy (see [13, 14, 4]). We hope that our results will shed some light on the elusive nature of easy-terms.

We finish this introduction with the following two remarks, which point out the limitations of the infinite $\beta\perp$-normal forms. First it is clear that there is no "Böhm out technique" for the infinite $\beta\perp$-normal forms. Secondly note that while the unsolvable terms form a Π_1 set, the mute terms form prima facie a more complicated set, namely a Π_2 set (complete?). Moreover, unlike the case of Böhm trees, there seems to be no general algorithm to compute higher and higher finite approximations to an infinite $\beta\perp$-normal form, although in many special cases, for instance the case of Ω_3, this is possible.

For related work in the area of term rewriting systems and infinite first order terms see the last section.

Notation: A *context* $C[\ ]$ is a term containing some occurrences of a special constant "hole". $C[B]$ is the term obtained by replacing all the occurrences of the holes with the term B. With the notation $A[x := B]$ we indicate as usual the result of substituting all the free occurrences of x in A with B after renaming the bound variables of A in such a way that the free variables of B do not become bound in $A[x := B]$. The difference between substitutions and contexts is that the free variables of B might become bound in $C[B]$ but not in $A[x := B]$.

A *notion of reduction* is an arbitrary binary relation on lambda-terms. We consider also notion of reductions ρ different from β-reduction. $\rightarrow_\rho$ denotes one-step ρ-reductions, i.e. the closure of ρ under substitutions and contexts (see [1] Definition 3.1.5, p. 51). $\Rightarrow_\rho$ is the reflexive and transitive closure of $\rightarrow_\rho$ and $\rightarrow_{=\rho}$ is the reflexive closure of $\rightarrow_\rho$. If we omit subscripts we mean β-reduction. The sign $\equiv$ between terms stands for syntactical identity up to renaming of bound variables (α-conversion). The sign $=$ between two terms stands for provable equality in some theory (in most of the cases $=$ is β-convertibility).

Added in proofs: An infinitary version of λ-calculus has been independently introduced by Kennaway, Klop, Sleep and Van de Vries in a recent manuscript [15]. Our approach is different since: 1) we do not equate all the unsolvable closed terms; 2) we allow infinite terms of the form $(((\ldots)A_2)A_1)A_0$ (infinitely many parenthesis). In this paper the Church-Rosser theorem for infinite $\beta\perp$-reductions is only proved for the class of those infinite terms B which arise from finite terms, in the sense that there is an infinite reduction $A \rightarrow_{\beta\perp\infty} B$ with A a finite term. Later investigations in collaboration with B. Intrigila showed that the Church-Rosser property holds even without this restriction (see also [3]).

2   Infinite λ-terms and normal forms

We identify λ-terms with their parsing trees, so we write λ-terms either in linear form or in tree form. This is convenient when we consider infinite λ-terms. An infinite λ-term is defined as a finite or infinite rooted tree such that each leaf is labeled by a variable and the inner nodes are either binary "application nodes", or unary "abstraction nodes", in which case they have a label of the form λx where x is a variable. For later purposes we expand the language with a constant $\perp$ which can then appear as a label of a leaf. Unless otherwise stated "term" means "infinite λ-term" possibly containing some occurrences of $\perp$. Finite λ-terms are special cases of infinite λ-terms. We have seen examples of infinite λ-terms in the introduction, for instance $BT(\mathbf{Y_t})$ or $NF^\infty(\Omega_3)$. When we write terms in linear form we follow the usual convention of left-associativity, thus ABC is $(AB)C$ etc. This corresponds to a tree whose root has left-son AB and right-son C. The term $\lambda x.A$ is identified with the tree with root λx having the son A.

The Böhm tree of a (finite) λ-term A is an infinite λ-term $BT(A)$ (possibly containing some occurrences of $\perp$), defined as follows.

- if there is a (multistep) β-reduction of the form

  $$A \Rightarrow \lambda x_1, \ldots, x_n.x_i A_1 \ldots A_k,$$

  then

  $$BT(A) \equiv \lambda x_1, \ldots, x_n.x_i BT(A_1) \ldots BT(A_k).$$

- if A is *unsolvable*, i.e. there is no reduction of the above form, then $BT(A) = \perp$.

Notice that a finite term in normal form coincides with its Böhm tree. It can be shown that the map $A \mapsto BT(A)$ is single valued, namely it does not depend on the non-deterministic choice of the β-reductions in the definition of $BT(A)$.

Our way of depicting Böhm trees is different from the usual one, for instance we write $\lambda x, y.xy$ as

This change of notation is justified by the extension of the notion of Böhm tree that we want to do. We recall that a *zero term* is a term which cannot

be reduced (by a multiple β-reduction) to an abstraction term, i.e. to a term of the form $\lambda x.A$ (it then follows that a zero term is not β-convertible to an abstraction term). Note that a variable is a zero term. The key property of zero terms is that if A is a zero term, M is an arbitrary term, and $AM \Rightarrow B$, then B has the form $A'M'$ with $A \Rightarrow A'$ and $M \Rightarrow M'$. The infinite $\beta\bot$-normal form $NF^\infty(A)$ of a term A is defined as follows.

- If $A \Rightarrow x$ for some variable x, then $NF^\infty(A) \equiv x$.

- If $A \Rightarrow \lambda x.B$ for some B, then $NF^\infty(A) \equiv \lambda x.NF^\infty(B)$.

- If $A \Rightarrow BC$ where B is a zero term, then $NF^\infty(A) \equiv NF^\infty(B)NF^\infty(C)$.

- In the remaining cases we say that A is *mute* and we set $NF^\infty(A) \equiv \bot$.

Thus a term is mute if and only if it is a zero term and it is not β-reducible to a variable or to a zero term applied to some other term.

The fact that $A \mapsto NF^\infty(A)$ is single valued will follow from the Church-Rosser theorem for infinite $\beta\bot$-reductions (starting from a finite term) to be proved later.

Mute terms can be characterized as follows. We say that a term is in *top normal form* (or is a top normal form) if it is either a variable, or an abstraction term $\lambda x.T$, or a term of the form BC where B is a zero term. We say that A *has a top normal form* if there is a β-reduction $A \Rightarrow A'$ with A' in top normal form. It follows from the definitions that a term A is mute iff it has no top normal form. (If we work in ordinary $\lambda\beta$-calculus a term is in top normal form if and only if it can never be reduced to a β-redex, however this characterization fails if we extend the language with a new constant $\bot$. According to our definition $\bot$ is mute.)

The name top normal form is justified by the remark that any β-reduct of a term in top normal form is in top normal form. Since to be a zero term is an undecidable property, to *be* in top normal form is also undecidable (and to *have* a top normal form is even more complex). The intuitive reason for the undecidability is that, unlike what happens for head normal forms (see [1]), in order to recognize if a term is in top normal form, it does not suffice to look at the first level of its tree-representation, it is sometimes necessary to look at the whole term.

Example 2.1 We have already seen in the introduction the infinite normal form of Ω_3. Now let $\mathbf{Y}_t\Omega_3$ be the Curry fixed point of Ω_3. Then $\mathbf{Y}_t\Omega_3 \rightarrow \Omega_3(\mathbf{Y}_t\Omega_3)$ and it can be shown that $NF^\infty(\mathbf{Y}_t\Omega_3) \equiv NF^\infty(\Omega_3)NF^\infty(\mathbf{Y}\Omega_3)$. In tree-form this means:

$NF^\infty(\mathbf{Y}\Omega_3) \equiv$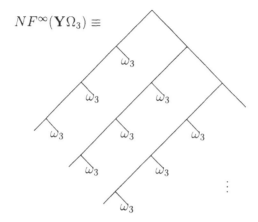

It is easy to see that the infinite normal form of a term is "finer" than its Böhm tree, in the sense that $NF^\infty(A)$ can be obtained from $BT(A)$ by replacing all the $\perp$'s in $BT(A)$ by suitable terms (possibly containing other $\perp$'s). This depends on the fact that every term in head-normal form is in top normal form but not conversely, for instance Ω_3 has no head-normal form.

Notice that an infinite $\beta\perp$-normal form can have infinitely many initial abstractions, consider for instance $NF^\infty(K^\infty)$ where K^∞ is the Turing fixed point of $K \equiv \lambda x, y.x$. Then $NF^\infty(K^\infty) \equiv \lambda x_1, x_2, x_3, x_4, \ldots$. On the other hand since K^∞ is unsolvable, the Böhm tree of K^∞ is $\perp$.

3   Infinite $\lambda\beta$-calculus

So far the notion of infinite $\beta\perp$-normal form that we have defined in the previous section is not associated to any notion of reduction. Moreover we have defined $NF^\infty(A)$ only for a finite term A. In the sequel of this paper we will fill these gaps by defining an infinite $\lambda\beta$-calculus (in this section) and an infinite $\lambda\beta\perp$-calculus (in later sections).

First notice that β-reduction $\rightarrow$ is defined for infinite lambda terms in exactly the same way as for finite ones, namely $(\lambda x.A)B \rightarrow A[x := B]$ (as usual $\rightarrow$ is closed under substitutions and contexts). Having defined β-reduction we can extend the notion of zero term, mute term, etc. to the infinite terms. Therefore the definition of $NF^\infty(A)$ makes sense also for infinite terms.

To obtain the infinite $\lambda\beta$-calculus we will define a notion of infinite β-reduction $\rightarrow_\infty$.

Definition 3.1 Given two terms A and B, we say $A \equiv_n B$ if A and B coincide up to the n-th level of their tree-representation. More precisely:

1. $A \equiv_0 B$ iff A and B have the same root.

2. $A \equiv_{n+1} B$ iff A and B have the same root and each immediate subterm of A is in relation $\equiv_n$ to the corresponding immediate subterm of B.

If n is negative we make the convention that $A \equiv_n B$ holds for every A and B.

Note that the root of a term, determines in particular whether the term is an application term (of the form BC), or an abstraction term (of the form $\lambda x.U$), or a variable. In the last two cases the root specifies also which is the corresponding variable.

As an example we have: $AB \not\equiv_0 \lambda x.U \equiv_0 \lambda x.V \not\equiv_0 \lambda y.Q$ where x and y are different variables. Note that $A \equiv B$ iff $A \equiv_n B$ for every n.

Since terms are trees, they possess a natural topology and a notion of limit.

Definition 3.2 Let $< A_n \mid n \in \omega >$ be a sequence of terms. We say $\lim < A_n > \equiv A$ if A is a term, and $\forall k \exists n \forall m \geq n \ A_m \equiv_k A$.

Definition 3.3 The *depth* of a specific occurrence of a subterm A in B, is defined as the length of the path connecting the root of B to the root of A in the tree-representation of B.

Example 3.4 The depth of A in A is 0. The depth of A in $\lambda x.A$ is 1. The depth of A in $\lambda x.AB$ and in $\lambda x.BA$ is 2. The depth of A in $\lambda x.ABC$ is 3.

Definition 3.5 The depth of a β-reduction $A \to B$ is defined as the depth of the redex being contracted. The depth of a multistep β-reduction $A \Rightarrow B$ is the minimum of the depths of the one-step β-reductions of which the multistep reduction is composed. The empty reduction has infinite depth.

Remark 3.6 If the reduction $A \Rightarrow B$ has depth n, then $A \equiv_{n-1} B$.

We define infinite β-reduction $A \to_\infty B$ as follows.

Definition 3.7 Let $\sigma: A_0 \Rightarrow A_1 \Rightarrow A_2 \Rightarrow A_3 \Rightarrow \ldots$ be an infinite sequence of β-reductions such that the depth of the reduction $A_i \Rightarrow A_{i+1}$ tends to infinity with i. Then by the previous remark $B \equiv \lim < A_i >$ exists and we set by definition $A_0 \to_\infty B$ via the reduction σ. We also say that the ω-sequence σ *converges*.

So an ω-sequence of reductions is an infinite reduction if and only if it converges. Note that by our convention empty-reductions have infinite depth. It follows that $A \Rightarrow B$ implies $A \to_\infty B$.

Definition 3.8 The depth of an infinite reduction $A \to_\infty B$ given by $A \equiv A_0 \Rightarrow A_1 \Rightarrow A_2 \Rightarrow \ldots \to_\infty B$, is defined as the minimum of the depths of the reductions $A_i \Rightarrow A_{i+1}$.

Remark 3.9 Let $\omega \equiv \lambda x.xx$. The infinite sequence of β-reductions $\omega\omega \to \omega\omega \to \omega\omega \to \ldots$ does not converge because the depth of redexes does not go to infinity. However we still have $\omega\omega \to_\infty \omega\omega$ (with another reduction) because empty reductions have infinite depth .

4   Residuals under infinite β-reductions

The residuals of a subterm under a (finite) β-reduction can be defined as in the case of finite lambda-terms. A simple way of doing this is to introduce labels. We do this in some detail in order to extend it to infinite reductions. We make a distinction between "residuals" and "extended residuals". The latter will be used in section 11.

Labeled terms are terms in which some (possibly all) subterm occurrences have a label (we can take the natural numbers as set of labels). Since terms are trees and subterms are subtrees, we can consider a labeled term as a tree in which some nodes are labeled by a natural number. Substitutions are defined for labeled terms as follows: $A[x := B]$ is obtained by erasing all the labels attached to the free occurrences of x in A and then replacing all such free occurrences with the term B (with all its labels). So if a free occurrence of x has a label, then it looses its label in the substitution $x[x := B]$. (This ensures that there are no overlappings of labels when a substitution $A[x := B]$ occurs.)

Having defined $A[x := B]$ for labeled terms we can define β-reduction for labeled terms as usual: $(\lambda x.A)B \to A[x := B]$. We can depict this in tree-form:

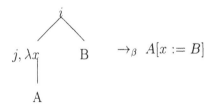

Note that the labels i and j are lost during the reduction. Clearly if B has a label, then $A[x := B]$ has as many copies of that label as the number of free occurrences of x in A (which loose their labels).

We now define the *residuals* of a specific occurrence of a subterm $A \subset B$ under a β-reduction $B \to C$. This is done as follows: give a label n to the given occurrence of A in B (and to nothing else). Perform the β-reduction $B \to C$ and look for the set of all subterms of C with label n. These subterms (actually subterm occurrences) are the residuals of A. From the above picture we see that if a redex is contracted, it has no residuals.

We now extend these notions to infinite reductions. If $A \to_\infty B$ via the sequence of reductions $A \equiv A_0 \to A_1 \to A_2 \to \ldots$, and if some subterms of A are labeled, then the labeling of A induces a unique labeling of the A_i's. Moreover since the depth of the reductions $A_i \to A_{i+1}$ tends to infinity, there is a unique labeling of B such that the A_i's converge to B as labeled terms. It thus make sense to define the residuals of a subterm of A under an infinite β-reduction.

Extended residuals are defined exactly as residuals except that we add the further clause that if $A \to A'$, then A' is an extended residual of A. A precise definition can be obtained by modifying the above picture in such a way that the label i is not lost but it is adjoined to the root of $A[x := B]$ in addition to the other labels possibly present. In this process the labels can cumulate, so unlike what happens for residuals, two subterms can have the same extended residual.

5 Failure of the Church-Rosser property for infinite $\lambda\beta$-calculus

The next example shows that the Church-Rosser property fails for infinite β-reductions.

Example 5.1 Let $I \equiv \lambda x.x$, $\omega \equiv \lambda x.xx$ and $Q \equiv \lambda x.I(xx)$. We have the reductions

$$
\begin{array}{ccccccc}
QQ & \to & I(QQ) & \to & I(I(QQ)) & \to_\infty & I(I(I(I(\ldots)))) \\
\Downarrow & & \Downarrow & & \Downarrow & & \\
\omega\omega & \to & \omega\omega & \to & \omega\omega & \ldots & ?
\end{array}
$$

but there is no reduction, whether finite or infinite, from $I(I(I(I(\ldots))))$ (infinitely many I's) to $\omega\omega$.

The term QQ responsible for the failure of the Church-Rosser property is mute. Later we will see that mute terms are the only responsible for the failure of the Church-Rosser property in the sense that if we add a rule sending all the mute terms to $\bot$, then the Church-Rosser property for infinite reductions is restored (at least if we start from a finite term).

6 Projections of β-reductions

For finite λ-calculus given two reductions σ and ρ starting from the same term, there is a well known manner to define in a canonical way a reduction

σ/ρ called the *projection* of σ over ρ. We investigate up to what extent this can be extended to infinite $\lambda\beta$-calculus. We will see that there are limitations. We start by recalling the finite case.

Definition 6.1 Let A be a finite term. Consider two β-reductions $\sigma: A \to B$. and $\rho: A \to C$ and suppose that Δ is the redex contracted by σ. Define $\sigma/\rho: C \Rightarrow D$ as the multistep β-reduction obtained by reducing from left to right all the residuals of Δ under ρ.

Proposition 6.2 *If A is a finite term and $B \overset{\rho}{\leftarrow} A \overset{\sigma}{\to} C$, then ρ/σ and σ/ρ end up in the same term D, so that we have the following diagram*

$$
\begin{array}{ccc}
A & \overset{\rho}{\to} & B \\
\sigma \downarrow & & \Downarrow \sigma/\rho \\
C & \overset{\rho/\sigma}{\Rightarrow} & D
\end{array}
$$

Proof. It is instructive to recall the proof. If the two redexes being reduced in $B \leftarrow A \to C$ are disjoint or coincide, the result is obvious. If they are nested one inside the other, then. after removing the outer context, we are in one of the following two cases.

Case 1.

$$
\begin{array}{ccc}
(\lambda x.A)B & \to & A[x := B] \\
\downarrow & & \Downarrow \\
(\lambda x.A)B' & \to & A[x := B']
\end{array}
$$

where the vertical reductions are induced by a given β-reduction $B \to B'$.

Case 2.

$$
\begin{array}{ccc}
(\lambda x.A)B & \to & A[x := B] \\
\downarrow & & \downarrow \\
(\lambda x.A')B & \to & A'[x := B]
\end{array}
$$

where the vertical reductions are induced by a given β-reduction $A \to A'$.
QED

Definition 6.3 The diagram appearing in the statement of the above proposition is called the *elementary* diagram determined by σ and ρ, and the reduction σ/ρ is called the *projection* of σ over ρ (similarly for ρ/σ).

Note that an elementary diagram can split on at most one side, i.e. either σ/ρ or ρ/σ is a one-step or empty β-reduction. To define the projection of a multistep reduction, we need to recall the notion of development.

Definition 6.4 A finite or infinite sequence of reductions $A \to A_1 \to A_2 \to \ldots$ is called a *development* if it reduces only residuals of redexes occurring in A.

Theorem 6.5 *(see [1]) If A is finite, then all developments starting from A are finite.*

Proposition 6.6 *(see [1]) If A is finite, and $B \overset{\rho}{\Leftarrow} A \overset{\sigma}{\Rightarrow} C$ are multistep β-reductions, we can define in a canonical way the projections σ/ρ and ρ/σ so that we have:*

$$
\begin{array}{ccc}
A & \overset{\rho}{\Rightarrow} & B \\
\sigma \Downarrow & & \Downarrow \sigma/\rho \\
C & \overset{\rho/\sigma}{\Rightarrow} & D
\end{array}
$$

Proof. (Sketch) The projection σ/ρ is defined by putting many elementary diagrams side to side until they "fill the rectangle" and yield the desired reductions. However since elementary diagrams can split, we need finiteness of developments to see that this process terminates after finitely many steps. QED

Since the Church-Rosser property fails for infinite $\lambda\beta$-calculus there is no hope to define σ/ρ for arbitrary infinite β-reductions σ and ρ. However in some special cases this can be done.

Definition 6.7 If A is an infinite term and $B \overset{\rho}{\Leftarrow} A \overset{\sigma}{\to} C$ are single step reductions, we define σ/ρ as the possibly infinite β-reduction which is obtained by reducing in some fixed order the residuals under ρ of the redex Δ contracted in $A \overset{\sigma}{\to} C$ (the precise order is not important: the residuals are all disjoint and we can reduce them in any order obtaining an infinite sequence of reductions in which the depth tends to infinity).

The following example shows that σ/ρ can be infinite even if σ and ρ are finite.

Example 6.8 Let $T \equiv \lambda x.x(x(x \ldots))$ (infinitely many x's) and let Δ be a β-redex with contractum Δ'. Then we have the diagram

$$
\begin{array}{ccc}
T\Delta & \overset{\sigma}{\to} & T\Delta' \\
\rho \downarrow & & \downarrow \rho/\sigma \\
\Delta(\Delta(\Delta \ldots)) & \overset{\sigma/\rho}{\to}_\infty & \Delta'(\Delta'(\Delta' \ldots))
\end{array}
$$

It is easy to see that σ/ρ is well defined as an infinite β-reduction in the sense that the depth of the redexes which are reduced tends to infinity.

Another special case in which we can define σ/ρ is the following.

Definition 6.9 If A_0 is a finite term and σ is an ω-sequence of multistep β-reductions $A_0 \Rightarrow A_1 \Rightarrow A_2 \Rightarrow \ldots$, and ρ is a multistep β-reduction $A_0 \Rightarrow B_0$, we define the ω-sequence of reductions $\sigma/\rho\colon B_0 \Rightarrow B_1 \Rightarrow B_2 \Rightarrow \ldots$ in such a way that each $B_i \Rightarrow B_{i+1}$ is the projection of $A_i \Rightarrow A_{i+1}$ under ρ. We thus obtain

$$\sigma\colon \quad A_0 \;\Rightarrow\; A_1 \;\Rightarrow\; A_2 \;\Rightarrow\; \ldots$$
$$\Downarrow \qquad \Downarrow \qquad \Downarrow$$
$$\sigma/\rho\colon \; B_0 \;\Rightarrow\; B_1 \;\Rightarrow\; B_2 \;\Rightarrow\; \ldots$$

It might happen that σ converges to an infinite reduction $\sigma\colon A \to_\infty \lim < A_i >$, while the ω-sequence σ/ρ does not converge to an infinite reduction (as the depth of $B_i \Rightarrow B_{i+1}$ might not go to infinity). This is exactly what happens in the example showing the failure of the Church-Rosser property. However if σ/ρ *does converge* to an infinite reduction, then $\sigma/\rho\colon B_0 \to_\infty \lim < B_i >$ is called the *projection of σ over ρ*.

7   Depth of redexes

All terms in this section are assumed to be finite. We investigate how the relation $\equiv_n$ behaves in connection with β-reduction. The general flavor of the results of this section is that the depth of a subterm cannot decrease too much under a finite β-reduction (although it can increase by an arbitrary amount). The next proposition shows that deep subterms have deep residuals.

Proposition 7.1 *If T' is a residual of T in a β-reduction $A \to B$, then the depth of T' in B is greater or equal than the depth of T in A minus 2.*

Proof. Clear from the tree-representation of β-reduction:

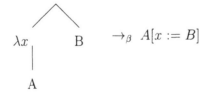

In the above picture the root of A goes up from depth 2 on the left to depth 0 on the right (provided that A is not a variable in which case it has no residuals). A case analysis shows that all the other subterms do not decrease in depth by more than two. This clearly holds even if the reduction takes place inside a context. QED

Note that the depth of B can increase by an arbitrary amount under the above depicted β-reduction (although it cannot decrease by more than 1).

Definition 7.2 Given a term A a *position* in A is a finite sequence of ternary digits which tell us whether to go down, left or right in the tree-representation of A. Two reductions $A \rightarrow A'$ and $B \rightarrow B'$ are called *similar* if they are obtained by reducing two redexes Δ in A and Δ' in B which occur at the same position.

Note that similar reductions have the same depth.

Proposition 7.3 *If $A \equiv_n B$ and the two reductions $A \rightarrow A'$ and $B \rightarrow B'$ are similar, then $A' \equiv_{n-2} B'$.*

Proof. By Proposition 7.1. QED

We now consider what happens to the depth of a subterm in an elementary diagram.

Proposition 7.4 *Consider two β-reductions $B \xleftarrow{\rho} A \xrightarrow{\sigma} C$. We have:*

1. *If σ is strictly deeper than ρ, then ρ and ρ/σ are similar.*

2. *The depth of ρ/σ is greater or equal than the depth of ρ minus 2.*

3. *If σ and ρ have depth $\geq n$, then also the reductions σ/ρ and ρ/σ have depth $\geq n$.*

Proof. By a simple case analysis. QED

8   Transitivity of infinite β-reductions

We prove that if $A \rightarrow_\infty B \rightarrow_\infty C$, then $A \rightarrow_\infty C$. We begin with a preliminary result.

Lemma 8.1 *If $A \rightarrow_\infty B$ and $B \rightarrow C$, then $A \rightarrow_\infty C$.*

Proof. Let $A \rightarrow_\infty B$ be given by the sequence of β-reductions $A \equiv B_0 \rightarrow B_1 \rightarrow B_2 \rightarrow \ldots \rightarrow_\infty B$. This means that the depth of the reductions $\sigma_i\colon B_i \rightarrow B_{i+1}$ tends to infinity as $i \rightarrow \infty$, and $\lim < B_i > \equiv B$. If we choose n big enough we can write:

$$
\begin{array}{ccccccccc}
A & \Rightarrow & B_n & \rightarrow & B_{n+1} & \rightarrow & B_{n+2} & \rightarrow & \cdots & \rightarrow_\infty & B \\
& & \downarrow & & \downarrow & & \downarrow & & & & \downarrow \\
& & C_n & \rightarrow_\infty & C_{n+1} & \rightarrow_\infty & C_{n+2} & \rightarrow_\infty & \cdots & & C
\end{array}
$$

where all the vertical reductions are similar in the sense of Definition 7.2 and the horizontal ones, with the exception of $A \Rightarrow B_n$, are deeper than the vertical ones. It suffices to show that $C_n \to_\infty C$. It is clear that the reductions $C_i \to C_{i+1}$ have a depth tending to infinity with i (since those of the upper row do), but the resulting "reduction" from C_n to C has length $\omega \cdot \omega$ (at most) instead of ω. If A is finite, then all the reductions from C_i to C_{i+1} are finite (for $i \geq n$), and we have $C_n \to_\infty C$. If A is infinite, we have to reorder the $\omega \cdot \omega$-sequence of reductions in order to get an equivalent converging sequence of length ω. This can be done as follows. Let $\Delta \equiv (\lambda x.S)T$ be the redex reduced in $B_n \to C_n$. Since for $i \geq n$ the reductions $B_i \to B_{i+1}$ are deeper than the reductions $B_i \to C_i$, Δ has one and only one residual $\Delta_i \equiv (\lambda x.S_i)T_i$ in each B_i and C_i is obtained from B_i by replacing Δ_i with its contractum $S_i[x := T_i]$. The reason why the reduction $C_i \to_\infty C_{i+1}$ can be infinite is that, although T_{i+1} is certainly obtained from T_i by a one-step or empty β-reduction, there might be infinitely many occurrences of T_i inside $S_i[x := T_i]$. However an important point to notice is that the various reductions $T_i \to_{=\beta} T_{i+1}$ (for the various i's and the various occurrences of T_i) do not affect and are not affected in any way by the surrounding context in the sense that:

1) nothing is substituted inside the T_i's by the outer context (since the reductions $B_i \to B_{i+1}$ have a depth bigger than the one at which Δ_i occurs);

2) No T_i is the first half of a redex which is reduced somewhere in the above diagram (i.e. in the above diagram there are no reductions of redexes of the form $(\lambda x.U)V$ where $\lambda x.U$ is one of the T_i's).

It follows that the reductions affecting the various occurrences of the T_i's and the outer context are independent of each other and they can be performed in any order. All these reductions can then be reordered in a converging ω-sequence yielding the desired reduction $C_n \to_\infty C$. QED

Theorem 8.2 *If $A \to_\infty B \to_\infty C$, then $A \to_\infty C$.*

Proof. Let $A \to_\infty B$ be given by $A \equiv A_0 \to A_1 \to A_2 \to \ldots \to_\infty B$, and let $B \to_\infty C$ be given by $B \equiv B_0 \to B_1 \to B_2 \to \ldots \to_\infty C$. By a repeated application of the proof of the previous lemma we can construct a diagram of the form

$$
\begin{array}{ccccccc}
A & \Rightarrow & D_0 & \dots & & \to_\infty & B_0 \\
& & \downarrow & & & & \downarrow \\
& & C_1 & \Rightarrow & D_1 \dots & \to_\infty & B_1 \\
& & & & \downarrow & & \downarrow \\
& & & & C_2 \Rightarrow D_2 & \to_\infty & B_2 \\
& & & & \downarrow & & \downarrow \\
& & & & & C_3 \to_\infty & B_3 \\
& & & & & \vdots & \downarrow_\infty \\
& & & & & & C
\end{array}
$$

where each vertical reduction $D_i \to C_{i+1}$ is similar to $B_i \to B_{i+1}$ and the depth of the reductions $C_i \Rightarrow D_i \to_\infty B_i$ is bigger or equal than the depth of $B_{i-1} \to B_i$ minus two. It follows that the depth of the reductions $C_i \Rightarrow D_i$ and $D_i \to C_{i+1}$ tends to infinity with i and therefore $A \to_\infty \lim < C_i > \equiv \lim < B_i > \equiv C$. QED

9   Unicity of infinite β-normal forms

If a (possibly infinite) term A has no β-redexes, we say that A *is in β-normal form*, written $A \in NF$. We say that A *has an infinite β-normal form* if there is an infinite β-reduction $A \to_\infty A'$ with A' in β-normal form. Some terms, for instance the mute terms, do not have an infinite β-normal form. In this section we prove that if a finite term A has an infinite β-normal form, then it has a unique infinite β-normal form.

The next proposition shows that we can bound the depth of a development just looking at the starting term. The proof is easy and left to the reader.

Proposition 9.1 *If A is a term and $\mathcal{F}$ is a set of redex occurrences in A each of which has depth $\geq n$, then every development $A \Rightarrow B$ which reduces only the residuals of the redexes in $\mathcal{F}$ has depth $\geq n$.*

Lemma 9.2 *If A is a finite term and $C \leftarrow A \to_\infty B$ with $B \in NF$, then $C \to_\infty B$.*

Proof. Since A is a finite term, we can use Definition 6.9 to construct a diagram of the form

$$
\begin{array}{ccccccccc}
A & \equiv & B_0 & \to & B_1 & \to & B_2 & \to & \dots & \to_\infty & B \\
& & \downarrow & & \Downarrow & & \Downarrow & & \\
C & \equiv & C_0 & \Rightarrow & C_1 & \Rightarrow & C_2 & \Rightarrow & \dots
\end{array}
$$

where each subdiagram

$$B_i \quad \rightarrow \quad B_{i+1}$$
$$\Downarrow \qquad \quad \Downarrow$$
$$C_i \quad \Rightarrow \quad C_{i+1}$$

is defined by taking projections. All the vertical reductions in this diagram are developments since they contract only the residuals of the redex Δ contracted in $A \rightarrow C$. Since $B \equiv \lim < B_i >$ is a normal form (hence it has no redexes), the minimal depth of the residuals of Δ in B_i tends to infinity with i, and therefore also the depth of the reduction $B_i \Rightarrow C_i$ tends to infinity by Proposition 9.1. But then $\lim < C_i >$ exists and it is equal to $\lim < B_i >$, i.e. $\lim < C_i > \equiv B$.

Since the depths of $B_i \Rightarrow B_{i+1}$ and $B_i \Rightarrow C_i$ tend to infinity with i, by Proposition 7.4 also the depth of the reductions $C_i \Rightarrow C_{i+1}$ tends to infinity, hence $C_0 \rightarrow_\infty B$. QED

The notion of zero term has been defined in terms of finite β-reductions. The next lemma and the more general result of Lemma 10.5, show that zero terms are well behaved with respect to infinite reductions.

Lemma 9.3 *If A is finite and $A \rightarrow_\infty B \in NF$, then A is a zero term if and only if B is a zero term.*

Proof. If A is not a zero term, then there is an abstraction term $\lambda x.T$ such that $A \Rightarrow \lambda x.T$. Since $B \in NF$ by the previous lemma $\lambda x.T \rightarrow_\infty B$. But then B must be of the form $\lambda x.T'$ and therefore B is not a zero term.

Conversely if B is not a zero term, then being a normal form it must be an abstraction term. Since $A \rightarrow_\infty B$, in this reduction there is an intermediate finite step $A \Rightarrow A' \rightarrow_\infty B$ where A' coincides with B up to depth 1. Hence A' is an abstraction term and A is not a zero term. QED

Lemma 9.4 *Let B be a normal form and for every i let $B_i \rightarrow_\infty B$ be a reduction of depth $\geq i$ with B_i a finite term. If for each i we have a reduction $B_i \rightarrow_\infty C_i$, then the depth of the reductions $B_i \rightarrow_\infty C_i$ tends to infinity with i.*

Proof. It is enough to show that the depth of $B_i \rightarrow_\infty C_i$ is greater than 0 if i is big enough, for then we can apply the same reasoning to the subterms of B_i, B and C_i.

Case 1. For some i, B_i is a variable or an abstraction term $\lambda x.U$. Then the result is clear: every multistep β-reduction starting from B_i has depth > 0 (if we work in the expanded language with $\bot$ we can treat $\bot$ as a free variable, since so far we have not introduced any special rule concerning $\bot$).

Case 2. B_i is an application term, say $B_i \equiv U_i V_i$. We can assume $i > 0$, so the depth of $B_i \to_\infty B$ is greater than 0. It then follows that B is of the form UV with $U_i \to_\infty U$ and $V_i \to_\infty V$. Clearly U is a zero term and a normal form, otherwise UV cannot be a normal form. Since $U_i \to_\infty U$, by Lemma 9.3 U_i is also a zero term. It follows that any reduction starting from $U_i V_i$ has depth > 0. QED

Theorem 9.5 *If A is finite, $B_\infty \leftarrow A \to_\infty C$ and B, C are normal forms, then $B \equiv C$.*

Proof. Let $A \to_\infty B$ be given by $A \equiv B_0 \to B_1 \to B_2 \to \ldots \to_\infty B$. Since C is a normal form, by Lemma 9.2 the reduction $B_0 \to_\infty C$ induces reductions $B_n \to_\infty C$ for every n and we can construct a diagram

$$
\begin{array}{ccccccccc}
B_0 & \to & B_1 & \to & B_2 & \to & \ldots & \to_\infty & B \\
\downarrow_\infty & & \downarrow_\infty & & \downarrow_\infty & & & & \\
C & \equiv & C & \equiv & C & \equiv & \ldots & &
\end{array}
$$

In order to show that $B \equiv C$ it suffices to show that the depth of the reductions $B_i \to_\infty C$ tends to infinity. Here we use the fact that B is a normal form and Lemma 9.4. QED

10   Head reductions

We recall the well known notion of head-redex and internal redex and we extend it to infinite terms. An important warning is that an infinite term can have a redex without having a left-most redex.

If a term T is of the form $T \equiv \lambda x_1, \ldots, x_k.\Delta M_1 \ldots M_n$ where $k, n \geq 0$ and Δ has the form $(\lambda x.A)B$, then we say that the displayed occurrence of Δ in T is the *head-redex* of T. An *internal redex* is a redex which is not the head-redex. An *head-reduction* is a β-reduction $A \Rightarrow_h B$ where only head redexes are reduced. An *internal reduction* is a β-reduction $A \Rightarrow_i B$ where only internal redexes are reduced.

Infinite internal reductions $A \to_{i\infty} B$ are defined in the obvious way and we will see that if $A \to_{i\infty} B$, then A is a top normal form if and only if B is such.

The main result of this section is that if a (possibly infinite) term has a top normal form, then it can be reduced to a top normal form by an head reduction. For finite terms this would be an easy consequence of the standardization theorem of Curry and Feys that says that internal reductions can always be postponed, but for infinite terms the problem is more delicate because if we try to postpone a one-step internal reduction, then we might generate an infinite internal reduction (see Example 6.8).

Let us recall the situation for finite terms:

Theorem 10.1 *If A is finite, then every β-reduction $A \Rightarrow B$ can be factored as an head-reduction followed by an internal reduction.*

Proof. See [1] Lemma 11.4.6. p. 299. QED

The next lemma says in particular that we can postpone a single internal reduction after an head reduction at the expense of getting an infinite internal reduction. The result holds for infinite terms. Note that if $U \rightarrow_i V$, then U has an head redex if and only if V has an head redex.

Lemma 10.2 *If $U \rightarrow_i V$ and U has an head redex, then by taking projections we can construct a diagram of the form*

$$
\begin{array}{ccc}
U & \rightarrow_i & V \\
\downarrow h & & \downarrow h \\
V' & \rightarrow_{i\infty} & W
\end{array}
$$

Proof. Define $V' \rightarrow_{i\infty} W$ as the converging sequence of reductions obtained by reducing in some fixed order the residuals of the redex Δ contracted in $U \rightarrow_i V$. Example 6.8 gives an idea of the general situation. QED

Lemma 10.3 *Any diagram $C \ _h\!\leftarrow A \rightarrow_\infty B$ can be extended to a diagram $C \rightarrow_\infty B \ _{=h}\!\leftarrow B$.*

Proof. Let $A \equiv A_0$ and $C \equiv C_0$. The result follows by constructing the diagram:

$$
\begin{array}{ccccccccc}
A_0 & \rightarrow & A_1 & \rightarrow & A_2 & \rightarrow & \ldots & \rightarrow_\infty & B \\
\downarrow h & & \downarrow_{=h} & & \downarrow_{=h} & & & & \downarrow_{=h} \\
C_0 & \rightarrow_\infty & C_1 & \rightarrow_\infty & C_2 & \rightarrow_\infty & \ldots & \rightarrow_\infty & D
\end{array}
$$

where each square is defined by taking projections and $B_i \rightarrow_{=h} C_i$ is the empty reduction if and only if the sequence $B_0 \rightarrow \ldots \rightarrow B_i$ contains some head reductions. Reasoning as in Lemma 8.1 we get $C_0 \rightarrow_\infty D$. QED

We can now extend to infinite terms a result which is well known for finite terms:

Theorem 10.4 *If A is not a zero term, then there exists an head-reduction of the form $A \Rightarrow_h \lambda x.T$.*

Proof. Given a reduction of the form $\sigma\colon A \Rightarrow \lambda x.S$, we prove by induction on its length that there exists a term T and an head-reduction $A \Rightarrow_h \lambda x.T$ such that the length of $A \Rightarrow_h \lambda x.T$ is less or equal than the length of $A \Rightarrow \lambda x.S$.

If σ consists of a (possibly empty) head-reduction $A \Rightarrow_h B$ followed by a non-empty internal reduction $B \Rightarrow_i \lambda x.S$, then B has the form $\lambda x.T$ for some T and therefore we can take ρ to be $A \Rightarrow_h B$.

Otherwise σ can be factored as $A \Rightarrow U \rightarrow_i V \rightarrow_h W \Rightarrow_i \lambda x.S$. Since V has an head redex and $U \rightarrow_i V$ is an internal reduction, U has an head-redex. Hence there is V' and an head reduction $U \rightarrow_h V'$. By Lemma 10.2 we can write $A \Rightarrow U \rightarrow_h V' \rightarrow_{i\infty} W \Rightarrow_i \lambda x.S$. But then V' has the form $\lambda x.S'$ for some S' and therefore we can apply the induction hypothesis to $A \Rightarrow U \rightarrow_h V'$. QED

We can now strengthen Lemma 9.3 as follows:

Lemma 10.5 *If $A \rightarrow_\infty B$ is an infinite β-reduction, then A is a zero term if and only if B is a zero term.*

Proof. If $B \Rightarrow \lambda x.T$, then $A \rightarrow_\infty \lambda x.T$ by transitivity, hence there is T' such that $A \Rightarrow \lambda x.T'$. This shows that if B is not a zero term, then A is not a zero term.

Conversely if A is not a zero term, then by Theorem 10.4 there is an head-reduction of the form $A \Rightarrow_h \lambda x.S$. By Lemma 10.3 there is a term D and reductions $B \Rightarrow_h D$ and $\lambda x.S \rightarrow_\infty D$. Hence D has the form $\lambda x.S'$ and B is not a zero term. QED

Corollary 10.6 *If $A \rightarrow_{i\infty} B$, then A is a top normal form if and only if B is a top normal form.*

Proof. The only non-trivial case is when A and B are application terms. But then we can write $A \equiv UV$ and $B \equiv U'V'$ with $U \rightarrow_\infty U'$ and $V \rightarrow_\infty V'$. By Lemma 10.5 U is a zero term if and only if U' is such. Hence A is in top normal form if and only if B is in top normal form. QED

We can now prove:

Theorem 10.7 *If a term A has a top normal form, then there exists an head-reduction $A \Rightarrow_h T$ with T in top normal form.*

Proof. Given a reduction $\sigma\colon A \Rightarrow B$ with B in top normal form, we prove by induction on the its length that there exists an head-reduction $\rho\colon A \Rightarrow T$

with T in top normal form and the length of $A \Rightarrow T$ less or equal than the length of $A \Rightarrow B$.

If σ consists of a (possibly empty) head-reduction $A \Rightarrow_h B'$ followed by a non-empty internal reduction $B' \Rightarrow_i B$, then by Corollary 10.6 B' is a top normal form and therefore we can take ρ to be $A \Rightarrow_h B'$.

Otherwise σ can be decomposed as $A \Rightarrow U \to_i V \to_h W \Rightarrow_i B$. Since V has an head redex, U has an head-redex. Hence there is V' and an head reduction $U \to_h V'$. By Lemma 10.2 we can write $A \Rightarrow U \to_h V' \to_{i\infty} W \Rightarrow_i B$. By Corollary 10.6 (applied twice), V' is a top normal form and therefore we can apply the induction hypothesis to $A \Rightarrow U \to_h V'$. QED

11   Substitution instances of zero terms

All terms and contexts in this section are possibly infinite terms and contexts in the expanded language with $\bot$. A substitution instance of a zero term is not necessarily a zero term, take for instance a zero term of the form $xT_1 \ldots T_k$ where x is a variable. In this section we characterize those terms A such that every substitution instance of A is a zero term and we prove some related results.

Definition 11.1 We say that M is *not active* in the β-reduction $C[M] \Rightarrow T$ if no redex contracted in this reduction is of the form $(\lambda x.A)B$ where $\lambda x.A$ is an extended residual of M. We say that M is *not touched* in the β-reduction $C[M] \Rightarrow T$ if no redex contracted in this reduction is of the form $(\lambda x.A)B$ where $\lambda x.A$ is a subterm of an extended residual of M.

Note that "not touched" implies "not active". If a term is not touched in a reduction, then its residuals coincide with its extended residuals and they are just substitution instances of the term itself.

If M is not active in a β-reduction we can replace everywhere M and its "maximal" extended residuals by a free variable x still getting a valid reduction, where a maximal extended residual is an extended residual which is not properly included in any other one. Thus we have:

Lemma 11.2 *If $C[M] \Rightarrow T[M_1, \ldots, M_n]$ where $M_1, \ldots, M_n$ are all the maximal extended residuals of M and M is not active, then $C[x] \Rightarrow T[x, \ldots, x]$ where x is free in $C[x]$.*

The above lemma holds a fortiori if M is not touched.

Definition 11.3 We say that M *goes to the head* in a reduction $C[M] \Rightarrow T$, if the given reduction contains an intermediate step of the form $C[M] \Rightarrow \lambda y_1 \ldots y_n.M'Q_1 \ldots Q_k \Rightarrow T$ where $n, k \geq 0$ and M' is an extended residual of M.

Remark 11.4   1. In any head reduction $C[M] \Rightarrow_h T$, M and its maximal extended residuals can be replaced by a free variable for as long as M does not go to the head.

2. From 1. it follows that if M goes to the head in $C[M] \Rightarrow_h T$, then there exists a reduction of the form $C[x] \Rightarrow_h \lambda y_1 \ldots y_n.xT_1 \ldots T_k$ where x is free in $C[x]$ and $n, k \geq 0$.

Lemma 11.5 *Let M be an arbitrary term and let x be free in $A[x]$. We have:*

1. If $A[M]$ is a zero term, so is $A[x]$.

2. If $A[M]$ is a top normal form, so is $A[x]$.

3. If $A[M]$ has a top normal form, so does $A[x]$.

Proof. If $A[x]$ is not a zero term there exists a reduction of the form $A[x] \Rightarrow \lambda y.S$. Substituting M for x in this reduction we obtain a reduction of the form $A[M] \Rightarrow \lambda y.S'$. This proves 1. For point 2. the only non trivial case is when the top normal form $A[M]$ is an application $U[M]V[M]$ with $U[M]$ a zero term. By point 1. $U[x]$ is a zero term, hence $A[x] \equiv U[x]V[x]$ is a top normal form. To prove point 3. we use Theorem 10.7. Suppose that there is a reduction $A[M] \Rightarrow T$ with T a top normal form. We can assume that $A[M] \Rightarrow T$ is an head reduction. If M goes to the head in this reduction, then by Remark 11.4 there is a reduction of the form $A[x] \Rightarrow xT_1 \ldots T_k$, hence $A[x]$ has a top normal form. If M does not go to the head, we can replace M and all its maximal extended residuals by x getting again a top normal form of $A[x]$ (by point 2.). QED

Corollary 11.6 *If $A[x]$ is mute, so is $A[M]$ for every M (x free in $A[x]$).*

We would like to prove some partial converses of the above results. We need restrictions on M that will ensure that M behaves like a free variable in any reduction.

Definition 11.7 A term A is a *strong zero term*, if it is a zero term and there is no reduction of the form $A \Rightarrow xT_1 \ldots T_k$ where x is a variable and $k \geq 0$.

Note that any mute term is a strong zero term. In particular $\perp$ is a strong zero term.

Lemma 11.8 *If x is free in $A[x]$ and $A[x]$ is a strong zero term, then $A[B]$ is a strong zero term for every B.*

Proof. First we prove that $A[B]$ is a zero term. If it is not, then by Theorem 10.4 there exists an head reduction of the form $A[B] \Rightarrow_h \lambda y.S$. If B goes to the head in this reduction, then $A[x] \Rightarrow_h \lambda y_1 \dots y_n.x T_1 \dots T_k$ for some $n, k \geq 0$ and some $T_1, \dots, T_k$. This is absurd since $A[x]$ is a strong zero term. Otherwise B is not active in the given head reduction and therefore its maximal extended residuals can be replaced by x yielding $A[x] \Rightarrow_h \lambda y.S'$ for some S', which is again a contradiction since $A[x]$ is a zero term.

It remains to show that $A[B]$ is a strong zero term. If it is not then there exists an head reduction of the form $A[B] \Rightarrow_h y S_1 \dots S_k$. If B goes to the head we reach a contradiction as above. Otherwise B is not touched and we can replace its maximal extended residuals by x getting a reduction from the term $A[x]$ to a term beginning with a variable, which is absurd since $A[x]$ is a strong zero term. QED

Corollary 11.9 *A term A is a strong zero term if and only if every substitution instance of A is a zero term.*

Proof. By the above lemma if A is a strong zero term, then any substitution instance of A is a strong zero term. Conversely assume that every substitution instance of A is a zero term. Then certainly A itself is a zero term. If it is not a strong zero term, there is a reduction of the form $A \Rightarrow x T_1 \dots T_k$. By substituting x in this reduction by $U_1^{k+1} \equiv \lambda y_1, \dots, y_{k+1}.y_{k+1}$, we obtain $A[x := U_1^{k+1}] \Rightarrow \lambda y_{k+1}.y_{k+1}$ contradicting the fact that any substitution instance of A is a zero term. QED

Corollary 11.10   *1. Any extended residual of a strong zero term is a strong zero term.*

2. A strong zero term is not active in any β-reduction.

Proof. 2. follows from 1. To prove 1. notice that any β-reduct and any substitution instance of a strong zero term is a strong zero term. Hence every extended residual U' of a strong zero term U is a strong zero term. QED

We can now prove a converse to Lemma 11.5

Lemma 11.11 *Let M be a strong zero-term and let x be free in $A[x]$. Suppose $A[x] \neq_\beta x$. We have:*

1. *$A[x]$ is a zero-term if and only if $A[M]$ is a zero-term.*

2. *$A[x]$ is a top normal form if and only if $A[M]$ is a top normal form.*

3. *$A[x]$ has a top normal form if and only if $A[M]$ has a top normal form.*

Proof. We have already proved one direction of the double implication for an arbitrary term M. Let us prove the other direction.

1. If $A[M]$ is not a zero-term, there is a reduction of the form $A[M] \Rightarrow \lambda y.C[M_1, \ldots, M_n]$ where $M_1, \ldots, M_n$ are all the occurrences of maximal extended residuals of M ($\lambda y.C[M_1, \ldots, M_n]$ cannot be an extended residual of M because M is a strong zero term). Replacing all the maximal extended residuals of M by x we obtain a reduction $A[x] \Rightarrow \lambda y.C[x, \ldots, x]$, thus $A[x]$ is not a zero-term.

2. The only interesting case is when $A[x]$ has the form $U[x]V[x]$ and therefore $A[M] \equiv U[M]V[M]$. By the part 1. $U[x]$ is a zero term if and only if $U[M]$ is such. Hence $A[x]$ is a top normal form if and only if $A[M]$ is such.

3. Suppose $A[x] \Rightarrow T[x]$ where $T[x]$ is a top normal form. Then $A[M] \Rightarrow T[M']$ where M' is a substitution instance of M, hence a strong zero term (by Corollary 11.9). By assumption $T[x] \neq_\beta x$. The result now follows from part 2. QED

Corollary 11.12 *If M_1 and M_2 are mute, then $A[M_1]$ is mute if and only if $A[M_2]$ is mute.*

Proof. If $A[M_1]$ is mute, then $A[x] =_\beta x$ or $A[x]$ is mute. In both cases $A[M_2]$ is mute. QED

12   Infinite $\lambda\beta\bot$-calculus

We define a $\bot$-*redex* as a mute term different from $\bot$. We define $\bot$-reduction $\rightarrow_\bot$ as the reduction generated (under substitutions and contexts) by $A \rightarrow_\bot \bot$ for every $\bot$-redex A. Since mute terms are closed under substitutions, a generic $\bot$-reduction has the form $C[A] \rightarrow_\bot C[\bot]$ where A is a mute term different form $\bot$. A term is a $\bot$-*normal form* is it has not $\bot$-redexes and it is a β-*normal form* if it has no β-redexes. Since every $\bot$-redex contains a β-redex, β-normal forms are also $\bot$-normal forms and we can simply speak about normal forms without further qualifications. Note that $\bot\bot$ and $\lambda x. \bot$ are normal forms.

$\beta\bot$-reduction is defined in the obvious way: $A \rightarrow_{\beta\bot} B$ if and only if either $A \rightarrow_\beta B$ or $A \rightarrow_\bot B$. The infinite reductions $\rightarrow_{\infty\bot}$ and $\rightarrow_{\infty\beta\bot}$ are defined in a completely similar way than infinite β-reductions $\rightarrow_\infty$ (also called $\rightarrow_{\beta\infty}$).

The notion of *residual* can be extended to $\bot$-reductions by defining $\bot$-reductions with labels in the following way:

$$(A)^n \rightarrow_\bot \bot$$

where A is a mute term with label n (the above reduction can take place inside a context). So A and all its subterms have no residuals in the reduction $A \rightarrow_\bot \bot$.

We want to define σ/ρ and ρ/σ for $\beta\bot$-reductions. For one step reductions we can give the following definition.

Definition 12.1 If $\rho\colon A \rightarrow_{\beta\bot} B$ and $\sigma\colon A \rightarrow_{\beta\bot} C$ are one-step $\beta\bot$-reductions, $\rho/\sigma\colon C \rightarrow_{\beta\bot\infty} D$ is the reduction which reduces all the residuals of σ in C in some fixed order (they are disjoint and if there are infinitely many of them, their depth must tend to infinity).

Lemma 12.2 *If σ and ρ are one-step $\beta\bot$-reductions starting from the same term, then σ/ρ and ρ/σ end up in the same term.*

Proof. We need three facts:
1) mute terms are closed under substitutions;
2) a mute term cannot be of the form $\lambda x.T$.
3) if U and $C[U]$ are mute, then $C[\bot]$ is mute (by Corollary 11.12).
Fact 1) is used to construct the following diagram (where U is the $\bot$-redex):

$$\begin{array}{ccc} (\lambda x.A[U])B & \rightarrow_\beta & A[U][x := B] \\ \downarrow_\bot & & \downarrow_\bot \\ (\lambda x.A[\bot])B & \rightarrow_\beta & A[\bot][x := B] \end{array}$$

Fact 2) is used to ensure that the reduction of a $\bot$-redex U does not destroy those β-redexes that are not contained in U, and so it allows us to postpone a β-reduction after a $\bot$-reduction when this is necessary to construct the appropriate diagram.

Fact 3) allows us to construct the following diagram (where U is the inner $\bot$-redex and $C[U]$ is the outer M-redex):

$$\begin{array}{ccc} C[U] & \rightarrow_\bot & C[\bot] \\ \downarrow_\bot & & \downarrow_\bot \\ \bot & \equiv & \bot \end{array}$$

Form these special cases the result easily follows. QED

To extend these notions to multistep reductions we need the notion of $\beta\bot$-development.

Definition 12.3 A $\beta\perp$-*development* is a finite or infinite sequence of $\beta\perp$-reductions $A_0 \to_{\beta\perp} A_1 \to_{\beta\perp}\to \ldots$ which reduce only the residuals of some redexes of A.

As in the case of β-reduction it can be shown that if A is a finite term, then every $\beta\perp$-development $A \equiv A_0 \to_{\beta\perp} A_1 \to_{\beta\perp}\to \ldots$ is finite ($\perp$-reduction is a very simple collapsing reduction, so it does not pose any problem for the finiteness of developments).

It follows that if we start from a finite term, we can define σ/ρ also for multistep $\beta\perp$-reductions σ and ρ as we did for β-reductions.

Lemma 12.4 *If A is finite and $\rho\colon A \Rightarrow_{\beta\perp} B$ and $\sigma\colon A \Rightarrow_{\beta\perp} C$ are multistep $\beta\perp$-reductions, there is a canonical way of defining a term D and two multistep $\beta\perp$-reductions $\rho/\sigma\colon C \Rightarrow_{\beta\perp} D$ and $\sigma/\rho\colon B \Rightarrow_{\beta\perp} D$.*

In particular $\beta\perp$-reduction for finite terms is Church-Rosser. For infinite terms there might be problems to define σ/ρ as in the case of β-reduction.

Theorem 12.5 $\to_{\beta\perp\infty}$ *is transitive.*

Proof. Completely analogous to the proof of the transitivity of $\to_\infty$. QED

The next theorem says that a finite term has at most one infinite $\beta\perp$-normal form:

Theorem 12.6 *If A is finite, $B \,_{\beta\perp\infty}\!\leftarrow A \to_{\beta\perp\infty} C$ and B,C are normal forms, then $B \equiv C$.*

Proof. Exactly as the proof of Theorem 9.5. In fact in that theorem we used only few properties of β-reductions which continue to hold for $\beta\perp$-reduction: namely the fact that we can define projections σ/ρ for reductions among finite terms, and the fact that the depth of residuals under one-step reductions decreases by 2 at most (see section 3). Such properties clearly hold for $\beta\perp$-reductions. QED

Unlike the case of infinite $\lambda\beta$-calculus, for infinite $\lambda\beta\perp$-calculus we also have existence of infinite normal forms.

Theorem 12.7 $NF^\infty(A)$ *is a normal form of A with respect to infinite $\beta\perp$-reduction.*

Proof. It is clear from the definitions that $NF^\infty(A)$ is a normal form. It is also clear that by "unfolding" the definition of $NF^\infty(A)$ we obtain an infinite $\beta\perp$-reduction $A \twoheadrightarrow_{\beta\perp\infty} NF^\infty(A)$. QED

Thus for infinite $\beta\perp$-calculus we have both existence and unicity of infinite normal forms of finite terms.

Corollary 12.8 *The Church-Rosser theorem for infinite $\beta\perp$-calculus holds, provided we restrict to finitely generated terms (where B is finitely generated if there is a finite term A and a reduction $A \twoheadrightarrow_{\beta\perp\infty} B$)[1].*

Proof. The set of finitely generated terms is closed under $\twoheadrightarrow_{\beta\perp\infty}$ by transitivity of $\twoheadrightarrow_{\beta\perp\infty}$. The Church-Rosser theorem follows at once from the existence and the unicity of infinite normal forms of finite terms. QED

13   A new model of lambda-calculus

We recall that a model of λ-calculus can be defined as a pair $(X, \cdot)$ where X is a non-empty set and $\cdot$ is a binary operation on X together with a semantic map $(A, \rho) \mapsto [[A]]_\rho$ which associates to every (finite) lambda-term A and every map $\rho: V \to X$, where V includes the free variables of A, an element $[[A]]_\rho \in X$ in such a way that the following conditions are satisfied:

1. $[[x]]_\rho = \rho(x)$ if x is a variable.

2. $[[(AB)]]_\rho = [[A]]_\rho \cdot [[B]]_\rho$.

3. $[[\lambda x.A]]_\rho \cdot b = [[A]]_{\rho[x:=b]}$.

4. If for all $d \in X$ we have $[[A]]_{\rho[x:=d]} = [[B]]_{\rho[x:=d]}$, then $[[\lambda x.A]]_\rho = [[\lambda x.B]]_\rho$.

5. $[[A]]_\rho = [[A]]_\tau$ if ρ and τ agree on the free variables of A.

6. $[[A]]_\rho = [[B]]_\rho$ if A and B differ only by a renaming of bound variables.

Definition 13.1 We define a model of lambda-calculus as follows. Take X to be the set of all normal forms of (finite) lambda-terms with respect to $\twoheadrightarrow_{\beta\perp\infty}$. For $A, B \in X$ define $A \cdot B$ as $NF^\infty(AB)$. This is well defined by Corollary 12.8. Finally define $[[A]]_\rho$ as the infinite normal form of the term A_ρ obtained by performing the substitution ρ on the term A.

---

[1]Added in proofs: later research with B. Intrigila has shown that the restriction to finitely generated terms is not necessary.

Theorem 13.2 *The triple* $(X, \cdot, [[\ ]])$ *defined above, is a model of lambda-calculus.*

Proof. The crucial condition to verify is the third. It suffices to show that for every two finite terms A and B, $NF^\infty(A[x := B]) \equiv NF^\infty(NF^\infty(A)[x := NF^\infty(B)])$. This follows by the unicity of infinite normal forms for finitely generated terms, the transitivity of $\rightarrow_{\beta\perp\infty}$, and the fact that $A[x := B] \rightarrow_{\beta\perp\infty} NF^\infty(A)[x := NF^\infty(B)]$. QED

We call $NF^\infty(\Lambda)$ the model defined above. This model identifies all the mute terms and more generally all the terms with the same infinite $\beta\perp$-normal form. For instance let A be a (finite) normal form and let X be a (finite) λ-term such that $X \Rightarrow XA$. Then the infinite $\beta \perp$-normal form of X is uniquely determined by $NF^\infty(X) \equiv (((\ldots)A)A)A$ (infinitely many A's). It follows that all the terms X such that $X \Rightarrow XA$ are identified in the model $NF^\infty(\Lambda)$. On the other hand B. Intrigila remarked, in a private communication, that the model $NF^\infty(\Lambda)$ does not equate all the terms with $X = XA$. In fact one can construct many such terms which are in (finite) normal form (see [6]). This is an example of how the model discriminates between β-reduction and β-convertibility.

14   It is consistent to equate all mute terms to an arbitrary closed term

All terms and reductions in this section are finite. We prove that it is consistent with the $\lambda\beta$-calculus to simultaneously equate all the mute terms to a fixed arbitrary closed term M, i.e. the theory $\lambda\beta + \{M = U \mid U$ is mute $\}$ is consistent. We prove this by defining a Church-Rosser notion of reduction which sends all the mute terms to M. The main lemma is the following.

Lemma 14.1 *If U is mute and $C[\ ]$ is a context such that $C[U]$ is mute, then either $C[x] \Rightarrow x$ where x is a variable not in $C[\ ]$, or for every closed term N, $C[N]$ is mute.*

Proof. If $C[N]$ is not mute, then there exists an head reduction $C[N] \Rightarrow_h Q[N_1, \ldots, N_n]$ where $Q[N_1, \ldots, N_n]$ is a top normal form and we have displayed all the occurrences of maximal extended residuals $N_1, \ldots, N_n$ of N. If in this reduction N goes to the head, then there are terms $T_1, \ldots, T_k$ $(n \geq 0)$ and an head-reduction $C[x] \Rightarrow_h xT_1 \ldots T_k$ where x is a variable. It follows that $C[U] \Rightarrow_h U'T_1^* \ldots T_k^*$ where $U', T_1^*, \ldots, T_k^*$ are substitution instances of

$U, T_1, \ldots, T_k$. U' is mute by Corollary 11.6. Since $C[U]$ and U' are mute, $k = 0$. Thus $C[x] \Rightarrow x$.

On the other hand if N does not go to the head in the head reduction $C[N] \Rightarrow_h Q[N_1, \ldots, N_n]$, then N and all its extended residuals can be replaced by x yielding $C[x] \Rightarrow Q[x, \ldots, x]$. It follows that $C[U] \Rightarrow Q[U_1, \ldots, U_n]$ where each U_i is a substitution instance of U. Since $Q[N_1, \ldots, N_n]$ is a top normal form by Lemma 11.5 $Q[x_1, \ldots, x_n]$ is a top normal form. Since $U_1, \ldots, U_n$ are strong zero terms, by Lemma 11.11 also $Q[U_1, \ldots, U_n]$ is a top normal form. This is absurd because $C[U] \Rightarrow Q[U_1, \ldots, U_n]$ and $C[U]$ is mute. QED

Definition 14.2 Fix an arbitrary closed term M. Define M-reduction by: $C[U] \to_M C[M]$ for every mute term U and every context $C[\ ]$ with exactly one hole. We say that U *is the M-redex* of the given M-reduction.

Since mute terms are closed under substitutions and β-reductions, every substitution instance of an M-redex is an M-redex. The main difficulty with M-reduction is that there seems to be no sensible way of defining projections σ/ρ for M-reductions. The next lemma gives a partial substitute.

Lemma 14.3 *Any diagram $C \;_M\!\!\leftarrow A \to_M B$ can be extended to a diagram of one of the following forms:*

$$
\begin{array}{ccccccc}
A & \to_M & B & \quad & A & \to_M & B \\
\downarrow_M & & \Downarrow_\beta & \quad & \downarrow_M & & \\
C & \equiv & D & \quad & C & \Rightarrow_\beta & D
\end{array}
\qquad
\begin{array}{ccc}
A & \to_M & B \\
\downarrow_M & & \downarrow_{=M} \\
C & \to_{=M} & D
\end{array}
$$

Proof. Note that the first two diagrams are one the transpose of the other. If the two M-redexes are disjoint or coincide the proof is trivial (and we obtain the third kind of diagram). Consider the case of nested M-redexes U and $C[U]$. By symmetry we can assume that $C[U]$ is the one reduced in $A \to_M B$. Since U and $C[U]$ are mute, by Lemma 14.1 either $C[x] \Rightarrow x$, where x is a variable not in $C[\ ]$, or $C[M]$ is mute. In the first case $C[M] \Rightarrow_\beta M$ and we obtain the following diagram:

$$
\begin{array}{ccc}
C[U] & \to_M & C[M] \\
\downarrow_M & & \Downarrow_\beta \\
M & \equiv & M
\end{array}
$$

The result follows by writing $A \equiv A_1[C[U]]$ and inserting the four terms of the above diagram inside the context $A_1[\ ]$.

In the second case $C[M]$ is mute and we obtain an instance of the third kind of diagram:

$$
\begin{array}{ccc}
C[U] & \to_M & C[M] \\
\downarrow_M & & \downarrow_M \\
M & \equiv & M
\end{array}
$$

QED

Definition 14.4 The notion of *residual* is defined for M-reductions $C[U] \to_M C[M]$ by stipulating that if $H[U]$ is a subterm of $C[M]$ properly containing (the given occurrence of) U, then $H[M]$ is the residual of $H[U]$ under the given M-reduction. U itself has no residuals.

Lemma 14.5 *Any diagram* $C \, _M{\leftarrow} \, A \to_\beta B$ *can be extended to a diagram of the form:*

$$
\begin{array}{ccc}
A & \to_\beta & B \\
\downarrow_M & & \Downarrow_M \\
C & \to_{=\beta} & D
\end{array}
$$

Proof. Let U be the M-redex reduced by $A \to_M C$. Since mute terms are closed under substitutions, every residual of an M-redex under a β-reduction is an M-redex. So we can define $B \Rightarrow_M D$ as the multistep β-reduction which reduces all the residuals of U from left to right. Since U is a zero term, it is not the first part of a β-redex. It follows that M-reductions do not destroy β-redexes, unless the β-redex is contained in the M-redex. Thus either $C \equiv D$ or we can define $C \to_\beta D$ as the one-step β-reduction which reduces the residual of the β-redex of $A \to_\beta B$ under the M-reduction $A \to_M C$. QED

Corollary 14.6 $\Rightarrow_{\beta M}$ *satisfies the weak Church-Rosser property, i.e. given reductions* $C \, _{\beta M}{\leftarrow} \, A \to_{\beta M} B$, *there is a term* D *and reductions* $C \, _{\beta M}{\Rightarrow} D \Leftarrow_{\beta M} B$.

From what we have proved so far the Church-Rosser property for $\Rightarrow_{\beta M}$ does not follow because a priori one can imagine diagrams such as:

$$
\begin{array}{ccccccccccc}
\cdot & \to_M & \cdot & \to_M & \cdot & \to_M & \cdot \\
\downarrow_M & & \downarrow_\beta & & \downarrow_\beta & & \downarrow_\beta \\
\cdot & \equiv & \cdot & \to_M & \cdot & \Rightarrow_M & \cdot \\
\downarrow_M & & \downarrow_M & & \equiv & & \equiv \\
\cdot & \equiv & \cdot & \Rightarrow_\beta & \cdot & \Rightarrow_M & \cdot \\
\downarrow_M & & \downarrow_M & & \Downarrow_M & & \\
\cdot & \equiv & \cdot & \Rightarrow_\beta & \cdot & & \cdot
\end{array}
$$

Fig.1

This would cause troubles because to complete the diagram we must find a common reduct for the two multi-steps M-reduction on the lower-right corner, and these two M-reductions are, a priori, longer than those we started with.

To avoid the occurrence of such diagrams we need the following.

Definition 14.7 We say that a β-reduction $\sigma: A \Rightarrow_\beta B$ is *collapsing*, written $\sigma: A \to_c B$, if there exist two contexts $C[\ ]$ and $H[\ ]$ with exactly one hole, and a closed term N, such that $A \equiv C[H[N]]$, $B \equiv C[N]$, and the β-reduction $\sigma: C[H[N]] \Rightarrow C[N]$ is induced by a β-reduction $H[x] \Rightarrow_\beta x$ where x is not in $H[\ ]$.

We say that $H[N]$ (or better the pair N, $H[\ ]$) is the *c-redex* of the reduction $\sigma: A \to_c B$.

As the notation $A \to_c B$ suggests, we are going to treat $\to_c$ as a one-step reduction, even if it actually consists of several β-reductions. When we write $\Rightarrow_c$ we mean a sequence of $(\to_c)$-reductions (possibly with different $H[\ ]$'s). The motivation for introducing collapsing reductions is the following:

Remark 14.8 The β-reductions mentioned in Lemma 14.3 are collapsing.

Remark 14.9 If in the definition of collapsing reductions we allow $H[\ ]$ to have several holes, we obtain an equivalent definition (but $C[\ ]$ is always assumed to have exactly one hole). In fact even if $H[\ ]$ has several holes, in the reduction $H[x] \Rightarrow x$ only one specific occurrence of x has a residual so we can redefine $H[\ ]$ in such a way that only that occurrence is placed inside the hole.

The above remark will be repeatedly used (without explicit note) in the following way: in order to prove a result of the form: "if some reduction σ is collapsing, then some other reduction σ' is also collapsing", we can assume for σ the one-hole definition, and for σ' the several-holes definition.

Note that, unlike arbitrary β-reductions, collapsing reductions do not duplicate subterms (each subterm has zero or one residual under a collapsing reduction). As a consequence we have:

Lemma 14.10 *Any diagram $C \twoheadleftarrow A \to_\beta B$, can be extended to a diagram of the form:*

$$
\begin{array}{ccc}
A & \to_\beta & B \\
\downarrow c & & \Downarrow c \\
C & \to_{=\beta} & D
\end{array}
$$

(The key point to observe is that $C \to_{=\beta} D$ is a one-step or empty reduction.)

Proof. Let $\Delta \equiv (\lambda y.S)T$ be the β-redex contracted in $A \rightarrow_\beta B$, and let $H[N]$ be the c-redex of $A \rightarrow_c C$. So we have $H[x] \Rightarrow x$ for some variable x not in $H[\ ]$ and we can write $A \equiv A_1[H[N]] \rightarrow_c A_1[N] \equiv C$ for some context $A_1[\ ]$ with exactly one hole.

Case 1. Suppose that $\Delta \subset N$ (here we are implicitly using the assumption that the contexts in the definition of collapsing reduction have exactly one hole: so it is clear which occurrence of N we refer to). By contracting Δ we obtain a reduction $N \rightarrow_\beta N'$. Thus we can write:

$$
\begin{array}{ccc}
A \equiv A_1[H[N]] & \rightarrow_\beta & A_1[H[N']] \\
\downarrow_c & & \downarrow_c \\
A_1[N] & \rightarrow_\beta & A_1[N']
\end{array}
$$

which gives the desired result.

Case 2. Suppose that $N \subset \Delta \subset H[N]$ and $N \not\equiv \Delta$.

N cannot be the "$\lambda x.S$" part of the redex $\Delta \equiv (\lambda y.S)T$, because otherwise $H[N] \equiv H_1[NT]$ for some context $H_1[\ ]$ such that $H_1[xT] \Rightarrow x$, which is impossible since T cannot be erased without erasing x as well. So N is contained either in S or in T. Suppose $N \subset T$. Then $\Delta \equiv (\lambda y.S)T_1[N]$ and $H[N] \equiv H_1[(\lambda y.S)T_1[N]]$ for some contexts $T_1[\ ]$ and $H_1[\ ]$ (with one hole) such that $H_1[(\lambda y.S)T_1[x]] \Rightarrow x$. It then follows (form the unicity of normal forms for β-reduction) that $H_1[S[y := T_1[x]]] \Rightarrow x$. Hence, by Remark 14.9, $H_1[S[y := T_1[N]]] \rightarrow_c N$ and we have:

$$
\begin{array}{ccc}
A \equiv A_1[H_1[(\lambda y.S)T_1[N]]] & \rightarrow_\beta & A_1[H_1[S[y := T_1[N]]]] \\
\downarrow_c & & \downarrow_c \\
A_1[N] & \rightarrow_\beta & A_1[N]
\end{array}
$$

The case in which $N \subset S$ is similar.

Case 3. Suppose that $\Delta \subset H[N]$ and Δ is disjoint form N. Then $\Delta \subset H[\ ]$ and by contracting Δ we obtain a context $H'[\ ]$ with $H'[x] \Rightarrow x$. The desired result follows.

Case 4. Suppose that $H[N] \subset \Delta$ and $\Delta \not\equiv H[N]$. Since $H[x] \Rightarrow x$, if $H[N] \equiv \lambda y.S$, then the only possibility is that $H[N] \equiv N \equiv \lambda y.S$. In this case $A \rightarrow_c C$ is the empty reduction and there is nothing to prove.

So we can assume that $H[N]$ is contained either in S or in T. Suppose $H[N] \subset T$. Then $\Delta \equiv (\lambda y.S)T_1[H[N]]$ for some context $T_1[\ ]$ with one hole. So we can write:

$$
\begin{array}{ccc}
A \equiv A_1[(\lambda y.S)T_1[H[N]]] & \rightarrow_\beta & A_1[S[y := T_1[H[N]]]] \\
\downarrow_c & & \Downarrow_c \\
A_1[(\lambda y.S)T_1[N]] & \rightarrow_\beta & A_1[S[y := T_1[N]]]
\end{array}
$$

This time the vertical reduction on the right is $\Rightarrow_c$ rather than $\rightarrow_c$ because there are several occurrences of $H[\ ]$ being erased.

Case 5. If Δ and $H[N]$ are disjoint the result is trivial. QED

Lemma 14.11 $\Rightarrow_\beta$ *and* $\Rightarrow_M$ *commute, i.e any diagram of the form* $C \underset{M}{\Leftarrow} A \Rightarrow_\beta B$ *can be extended to a diagram of the form:*

$$
\begin{array}{ccc}
A & \Rightarrow_\beta & B \\
\Downarrow_M & & \Downarrow_M \\
C & \Rightarrow_\beta & D
\end{array}
$$

Proof. By induction on the length of $A \Rightarrow_\beta B$ and Lemma 14.5. QED

Corollary 14.12 $\Rightarrow_\beta$ *and* $\Rightarrow_{\beta M}$ *commute.*

Proof. Clear form the Church-Rosser property of $\Rightarrow_\beta$ and the previous lemma. QED

Lemma 14.13 *Any diagram* $C \underset{M}{\leftarrow} A \rightarrow_c B$ *can be extended to a diagram of the form:*

$$
\begin{array}{ccc}
A & \rightarrow_c & B \\
\downarrow_M & & \downarrow_{=M} \\
C & \rightarrow_{=c} & D
\end{array}
$$

Proof. Let U be the M-redex contracted in $A \rightarrow_M C$ and let $H[N]$ be the c-redex of $A \rightarrow_c B$.

Case 1. Suppose $U \subset N$. Then we can reason as in Case 1 of Lemma 14.10 with U instead of Δ.

Case 2. Suppose that $N \subset U \subset H[N]$ with $N \not\equiv U$. We can then write $U \equiv U_1[N]$ and $H[N] \equiv H_1[U_1[N]]$ for some contexts $U_1[\ ]$ and $H_1[\ ]$ with one hole, such that $H_1[U_1[x]] \Rightarrow x$, where x is a fresh variable. Since $U_1[N]$ is mute, it is in particular a zero term. Hence $U_1[x]$ is also a zero term. But then the only possibility to have $H_1[U_1[x]] \Rightarrow x$ is that $U_1[x] \Rightarrow x$ and $H_1[x] \Rightarrow x$. It follows that N is mute because $U_1[N] \Rightarrow N$. Thus we have:

$$
\begin{array}{ccc}
H_1[U_1[N]] & \rightarrow_c & H_1[N] \\
\downarrow_M & & \downarrow_M \\
H_1[M] & \equiv & H_1[M]
\end{array}
$$

and the result follows.

Case 3. Suppose that $U \subset H[N]$ and U is disjoint from N. Then we can write $H[N] \equiv H_1[U, N]$ for some context $H_1[\ ,\ ]$ such that $H_1[U, x] \Rightarrow x$.

Since U is mute, it behaves as a free variable in any reduction and therefore it cannot give any contribution to this reduction. Hence $H_1[M,x] \Rightarrow x$ and $H_1[M,N] \to_c N$. Thus we have:

$$H_1[U,N] \quad \to_c \quad N$$
$$\downarrow_M \qquad\qquad \equiv$$
$$H_1[M,N] \quad \to_c \quad N$$

and the result follows.

Case 4. Suppose $H[N] \subset U$ and $H[N] \not\equiv U$. We can write $U \equiv U_1[H[N]]$. Since $U_1[H[N]]$ is mute and $U_1[H[N]] \Rightarrow U_1[N]$, $U_1[N]$ is mute. Hence we have:

$$U_1[H[N]] \quad \to_c \quad U_1[N]$$
$$\downarrow_M \qquad\qquad \downarrow_M$$
$$M \qquad \equiv \qquad M$$

and we are done. QED

Lemma 14.14 *Any diagram $C \underset{c}{\Leftarrow} A \Rightarrow_{\beta M} B$ can be extended to a diagram of the form:*

$$A \quad \Rightarrow_{\beta M} \quad B$$
$$\Downarrow_c \qquad\qquad \Downarrow_c$$
$$C \quad \Rightarrow_{\beta M} \quad D$$

where the length of $C \Rightarrow_{\beta M} D$ is less or equal than the length of $A \Rightarrow_{\beta M} B$.

Proof. By induction on the length of $A \Rightarrow_{\beta M} B$ by Lemma 14.13 and Lemma 14.10. QED

Lemma 14.15 *Any diagram $C_M \Leftarrow A \Rightarrow_{\beta M} B$ can be extended to a diagram of the form:*

$$A \quad \Rightarrow_{\beta M} \quad B$$
$$\Downarrow_M \qquad\qquad \Downarrow_{\beta M}$$
$$C \quad \Rightarrow_{\beta M} \quad D$$

Proof. By induction on the length of $A \Rightarrow_{\beta M} B$. If the first step of the reduction $A \Rightarrow_{\beta M} B$ is a β-reduction we can use Lemma 14.11 and then we apply the induction hypothesis. If the first reduction of $A \Rightarrow_{\beta M} B$ is an M-reduction, then there are three cases corresponding to the three diagrams of Lemma 14.3. In the first case (using Remark 14.8) we can write:

$$
\begin{array}{ccccc}
A & \to_M & B' & \Rightarrow_{\beta M} & B \\
\downarrow_M & & \downarrow_c & & \Downarrow_c \\
C' & \equiv & C' & \Rightarrow_{\beta M} & E \\
\Downarrow_M & & \Downarrow_M & & \\
C & \equiv & C & &
\end{array}
$$

where $C' \Rightarrow_{\beta M} E$ is obtained using Lemma 14.14 and has length less or equal than the length of $B' \Rightarrow_{\beta M} B$. We can now find a common reduct to the diagram $C_M \Leftarrow C' \Rightarrow_{\beta M} E$ by applying the induction hypothesis.

The other two cases are similar. QED

Lemma 14.16 $\Rightarrow_{\beta M}$ *satisfies the Church-Rosser property.*

Proof. By Lemma 14.15 and Corollary 14.12, $\Rightarrow_{\beta M}$ commutes both with $\Rightarrow_\beta$ and with $\Rightarrow_M$, hence with $\Rightarrow_{\beta M}$. QED

Theorem 14.17 *For every closed term M, the theory $T = \lambda\beta + \{M = U | U$ is mute $\}$ is consistent.*

Proof. If T derives a contradiction, say $T \vdash 0 = 1$ (where 0 and 1 are the Church numerals for zero and one), then $0 =_{\beta M} 1$ where $=_{\beta M}$ is the transitive closure of $\Rightarrow_{\beta M}$. This is absurd since $\Rightarrow_{\beta M}$ is Church-Rosser and 0 and 1 are normal forms. QED

15   Zero terms of finite degree are easy

All terms and reductions in this section are finite. We recall that an *easy* term is a term U such that for every closed term M the theory $\lambda\beta + \{M = U\}$ is consistent. In the previous section we have shown that the class of mute terms has a much stronger property: all mute terms can be simultaneously equated to an arbitrary closed term. In this section we prove that all strong zero term (in particular all closed zero terms) of finite "degree" are easy, thus strengthening some results of [13].

Definition 15.1 Let U be a strong zero term. We say that U has degree 0 if it is mute (i.e. it is not β-convertible to VM with V a zero term). We say that U has degree $n + 1$ if it is β-convertible to a term of the form VM where V is a strong zero term of degree n. We say that U has infinite degree in the remaining cases.

Lemma 15.2 *([13, 14]) If U is easy, then for every M, UM is easy.*

Proof. If $UM = Q$ is inconsistent, then so is $U = \mathbf{K}Q$ where $\mathbf{K} \equiv \lambda x, y.x$. QED

Since mute terms are easy it follows:

Theorem 15.3 *All strong zero terms of finite degree are easy.*

This is a strengthening of a theorem of [13] which says that all recurrent closed zero terms are easy, where "recurrent" is defined as follows:

Definition 15.4 A term A is *recurrent* if whenever $A \Rightarrow B$, there is a reduction $B \Rightarrow A$.

Lemma 15.5 *Every recurrent closed zero term A is a strong zero terms of finite degree (hence it is easy).*

Proof. Let k be the number of subterms of A. If A is not of finite degree, there is $n > k$ and a reduction $A \Rightarrow BT_1 \ldots T_n$ where B is a zero term and each $T_i \in \Lambda_0$. But then every reduct of $BT_1 \ldots T_n$ has the form $B'Q_1 \ldots Q_n$ so it has at least n subterms, and therefore cannot coincide with A, contradicting the fact that A is recurrent. QED

The problem of classifying the closed zero terms which are easy has thus been reduced to the case of those of infinite degree. Ω_3 is an example of a zero term of infinite degree which is not easy. In [11] there is an example of an easy term of infinite degree. The fixed point $\mathbf{Y}_t\Omega_3$ has infinite order and it is not known to be easy (see [13, 14]). In [4] it is shown that $\mathbf{Y}_t\Omega_3$ can be consistently equated to every closed normal form.

16   Related work

Infinite reductions and unicity of normal forms are considered in [7] and [16] in the context of term rewriting systems for infinite first order terms. I was not initially aware of this fact, which was pointed out to me by M. Venturini Zilli. In particular the notion of infinite β-reduction, based on the assumption that in an infinite β-reduction the depth or redexes tends to infinity, correspond exactly to the notion of strongly converging reduction in [16]. The use of residues under an infinite reduction can also be found there. It should be noted however that we work with infinite terms with lambda-abstractions, while [7] and [16] work with infinite first order terms, i.e. they do not allow binding of variables. The notion of non-top-terminating in [7] resembles very closely the notion of mute. In [16] we find the idea that such terms, which are there called terms without head-normal form, are meaningless from an operational point of view. This idea is there formalized in a Church-Rosser theorem "up to hypercollapsing terms" (Theorem 7.4). In [16] we also find a counterexample to the infinite Church-Rosser property for combinatory logic. In [7] there is a theorem about the unicity of $\perp$-normal forms for $\perp$-converging semi-$\perp$-confluent rewrite systems (Theorem 8). This is reminiscent of our theorem on uniqueness of infinite β-normal forms. However λ-calculus seems to be neither $\perp$-converging nor semi-$\perp$-confluent if we extend the notions in the natural way.

Acknowledgements. The original motivation for this research comes from the previous paper with Benedetto Intrigila [4]. In that paper we left open the question, still not solved, of whether $\mathbf{Y_t}\Omega_3$ is easy. During our attempts to solve the problem Intrigila suggested to expand the language of λ-calculus with a new constant δ together with the rule $\delta \to \delta\omega_3$. The idea was that the operational behaviour of Ω_3 is fully described by δ. The infinite normal form $NF^\infty(\Omega_3)$ defined in this paper achieves a similar effect without expanding the language. I am grateful to Marisa Venturini Zilli for introducing me to the study of infinite first order terms, for pointing out the references [7, 16], and for a careful reading of a preliminary version of this paper. My dearest thanks go to Corrado Böhm without whose teachings over the years this paper would not exist.

References

[1] H.P. Barendregt, *The Lambda Calculus*, Studies in Logic **103**, North-Holland, Amsterdam, 1984.

[2] ———, *Representing 'undefined' in lambda calculus*, Theoretical pearls, J. Functional Programming **2** (1992), no. 3, 367–374.

[3] A. Berarducci and B. Intrigila, *Church-Rosser λ-theories, infinite λ-calculus and consistency problems*, manuscript.

[4] ———, *Some new results on easy lambda-terms*, Theoret. Comput. Sci. **121** (1993), 71–88.

[5] C. Böhm, *Alcune proprietà delle forme $\beta - \eta$-normali nel $\lambda - k$-calcolo*, Pubblicazioni dell'Istituto per le Applicazioni del Calcolo, n. 696, Roma, 1968.

[6] C. Böhm and B. Intrigila, *The Ant-Lion paradigm for strong normalization*, Information and computation **114** (1994), 30–49.

[7] N. Dershowitz, S. Kaplan, and D. A. Plaisted, *Rewrite, Rewrite, Rewrite, Rewrite, Rewrite, ...*, Theoret. Comput. Sci. **83** (1991).

[8] F. Honsell and S. Ronchi della Rocca, *An approximation theorem for topological lambda models and the topological incompleteness of lambda calculus*, J. Comput. System Sci. **45** (1992), 49–75.

[9] J.M.E. Hyland, *A syntactic characterization of the equality in some models of the λ-calculus*, J. London Math. Soc. (2) **12** (1976), 361–370.

[10] B. Intrigila, *Some results on numeral systems in λ-calculus*, to appear in the Notre Dame Journal of Formal Logic.

[11] _____, *A problem on easy terms in λ-calculus*, Fundamenta Informaticae **XV.1** (1991).

[12] G. Jacopini, *A condition for identifying two elements of whatever model of combinatory logic*, Springer Lecture Notes in Computer Science **37** (C. Böhm, ed.), 1975.

[13] G. Jacopini and M. Venturini Zilli, *Equating for Recurrent Terms of λ-calculus and Combinatory Logic*, Quaderni dello IAC, serie III (1978), no. 85, Roma, 3–14.

[14] _____, *Easy terms in the lambda-calculus*, Fundamenta Informaticae **VIII.2** (1985).

[15] R. Kennaway, J. W. Klop, R. Sleep, and F-J de Vries, *The infinite lambda calculus λ∞*, manuscript (1994).

[16] _____, *Transfinite reductions in orthogonal term rewriting systems*, preprint 1993, to appear in Information and Computation.

[17] G. Mitschke, *λ-Kalkül, δ-conversion und axiomatische Rekursionstheorie*, Preprint Nr. 274, Technische Hochschule, Darmstadt Fachbereit Mathematik, 1976.

[18] D. Park, *The y-combinator in Scott's lambda calculus models*, Theory of computation, Report n. 13, University of Warwick, Dept. of Comput. Sci., 1976.

[19] A. Visser, *Numerations λ-calculus and arithmetic*, To H.B. Curry: Essays on Combinatory Logic, Lambda Calculus and Formalism (J.P. Seldin and J.R.Hindley, eds.), Academic Press, London, 1980, pp. 259–284.

[20] C.P. Wadsworth, *The relation between computational and denotational properties for Scott's d_∞ models of the lambda-calculus*, SIAM J. Comput. **5** (1976), 488–521.

[21] C. Zylberajch, *Sintaxe et semantique de la facilitè en lambda calcul*, These de Doctorat, Universitè de Paris VII, 1991.

DIPARTIMENTO DI MATEMATICA, UNIVERSITÀ DI PISA, VIA BUONARROTI 2, 56127 PISA, ITALY
e-mail address: berardu@dm.unipi.it

A Computer Study of 3-Element Groupoids

Joel Berman   Stanley Burris

Abstract

In [2] it is noted that current computing power may be inadequate to answer some basic questions about a three-element algebra, e.g., does it generate a Mal'cev variety? or a congruence distributive variety? With this challenge in mind a study of *all* three-element groupoids was undertaken.

1   The study

Two-element groupoids are well known (just think of the logical connectives "and", "or", etc.), but 3-element groupoids offer a much bigger challenge. There are 19,683 groupoid tables on $\{0, 1, 2\}$; and up to isomorphism there are 3,330 such tables. We started by making a catalog of these groupoids by selecting from each isomorphism class (of groupoids on $\{0, 1, 2\}$) the lexicographically first member, treating each table as the 9 letter word:

row1 row2 row3.

Such a catalog of isomorphism representatives took only a couple of seconds to generate, and was stored in a 3,330-by-9 array.

In initial work on the 3-element groupoids the 3,330 isomorphism representatives were used and thirteen properties were analyzed, namely the properties on page 394 except numbers 3 and 4, together with the cardinalities of the free algebras on 0, 1 and 2 generators in *some* of the varieties generated by these groupoids. (For 0 generators the number of constant unary term functions was used.)

Using this information the pre-order given by $\mathbf{A} \leq \mathbf{B}$ *iff the clone of* $\mathbf{A}$ *is a subset of the clone of some isomorphic copy of* $\mathbf{B}$ was determined.[1] The

---

[1] In [7] Anne Fearnley has carried out a detailed study of the clones on three-elements which are of the form Pol(ρ) for ρ a unary or binary relation. This study does not bear directly on our work, but it is certainly a valuable companion.

induced equivalence relation, called *clone equivalence*, had 411 equivalence classes (actually the classes were determined before the ordering $\leq$). Using representatives of these 411 classes the cardinalities of the free algebras on 0,1 and 2 generators in *all* the varieties generated by these groupoids were determined. This completed the analysis of the 13 properties mentioned above. Then, with the help of the partial ordering $\leq$ on the 411 clone equivalence representatives, the groupoids that generate congruence distributive or congruence modular varieties were determined. And finally the set of types of each of the 411 groupoids was determined. These three additional properties gave a total of 16 properties analyzed; the results are presented on pages 406–408.

Many properties, such as Mal'cev conditions and the sixteen properties considered, are invariant within each clone equivalence class. Consequently it was decided to make the presentation of data more compact by giving information in terms of the 411 clone equivalence representatives.

Now let us look in more detail at the sequence of steps followed. The main tool to analyze a given groupoid $\mathbf{A}$ was the computation of various subuniverses $S(\mathbf{A}, n, X)$ of powers $\mathbf{A}^n$ of $\mathbf{A}$, generated by suitable X — this tool for computer analysis was pioneered in the paper [5] of Berman and Wolk. From a programming point of view it was easiest to simply fix n as the largest power needed, and for smaller powers the coordinates considered were restricted.

Because of the theoretical work in [2] mentioned above the first project selected was to determine which $\mathbf{A}$'s had a Mal'cev term. Thus we concentrated on $S(\mathbf{A}, 15, X)$ where X consisted of the three 15-tuples:

$$
\begin{array}{lccccccccccccccc}
px & 0 & 0 & 0 & 0 & 1 & 1 & 1 & 1 & 2 & 2 & 2 & 2 & 0 & 1 & 2 \\
py & 0 & 0 & 1 & 2 & 0 & 1 & 1 & 2 & 0 & 1 & 2 & 2 & 0 & 1 & 2 \\
pz & 1 & 2 & 1 & 2 & 0 & 0 & 2 & 2 & 0 & 1 & 0 & 1 & 0 & 1 & 2 \; ;
\end{array}
$$

$\mathbf{A}$ had a Mal'cev term iff the 15-tuple

$$
\begin{array}{lccccccccccccccc}
pm & 1 & 2 & 0 & 0 & 1 & 0 & 2 & 1 & 2 & 2 & 0 & 1 & 0 & 1 & 2
\end{array}
$$

was in $S(\mathbf{A}, 15, X)$. The method for computing $S(\mathbf{A}, n, X)$ was a straightforward application of the groupoid operation to pairs of previously found elements, proceeding in successive sweeps, each sweep yielding terms with tree depth one greater than the previous sweep. The program made about 85 million applications of the groupoid operation per hour (on a Sun Sparc II).

Early attempts to find the Mal'cev terms soon led to the realization that attempting to compute the full $S(\mathbf{A}, 15, X)$ for each of the 3,330 isomorphism representatives to determine if a Mal'cev term existed would likely take too long. This led to the development of shortcuts via the multiphase attack described next.

Phase 1: When **A** does not have a Mal'cev term one can often show this by working with a small subset of the 15 coordinates, by showing that the restriction of pm to this subset does not lie in the closure under the groupoid operation of the restrictions of px, py, pz to this same subset. This was the first phase of the attack, to try several small subsets of the 15 coordinates in hopes of proving there is no Mal'cev term. If the coordinates are labelled 0 through 14 then the following choices were tried in the first phase: (0 1 2), (1 2 3),..., (9 10 11), (0 2 4 5 12 13), (1 3 8 10 12 14), (6 7 9 11 13 14), (evens), (odds). The sets of coordinates (0 2 4 5 12 13), (1 3 8 10 12 14), and (6 7 9 11 13 14) were used to determine if there is a Mal'cev term on $\{0, 1\}$, $\{0, 2\}$, and $\{1, 2\}$ respectively.

An important time saving observation was that if the Cayley table of **A** is the transpose of the Cayley table of **B**, i.e., $x \cdot_{\mathbf{A}} y = y \cdot_{\mathbf{B}} x$, then both or neither have Mal'cev terms. A subroutine to check if a new table to be considered was actually a representative of a transpose of a previous one was incorporated. Since there are 1,596 pairs related by transposes in the catalog of 3,330 isomorphism representatives there is indeed a considerable amount of work saved. For the cases which were not isomorphism representatives of transposes of previous ones, the test in Phase 1 usually required only a second or two per groupoid; and in this phase 1,792 of the groupoids were identified as not having Mal'cev terms.

Phase 2: If the attempt to refute the existence of a Mal'cev term for a groupoid **A** by using few coordinates failed then a quick proof of the existence of a Mal'cev term was sought by looking for

- a binary term $b(x, y)$ and unary terms $u_1(x), u_2(x)$ such that the unary terms have 2-element ranges and the equation $b(u_1(x), u_2(x)) \approx x$ holds in the groupoid.

Such a binary term $b(x, y)$ is said to be *invertible*. Suppose such a term exists. As **A** has passed Phase 1, there must exist terms $m_1(x, y, z)$ and $m_2(x, y, z)$ which give Mal'cev terms on the ranges of u_1, resp. u_2. Then it follows that

$$b(m_1(u_1(x), u_1(y), u_1(z)), m_2(u_2(x), u_2(y), u_2(z)))$$

is a Mal'cev term for **A**. This test, along with the test for transposes, identified 1,474 of the groupoids as having Mal'cev terms.

Phase 3: This left 64 of the 3,330 groupoids to be cataloged. For the final phase we turned to the generation of $S(\mathbf{A}, 15, X)$. In all but 7 of the remaining cases pm was found in the generated subuniverse.

The last seven cases (591, 746, 774, 951; and transposes 978, 1487, 1512) generate the same clone. This was discovered after trying a direct computation on #591 lasting over 60 hours, without completion or resolution of the existence of a Mal'cev term. Then it was found that the groupoid operation of #599 is a term function of #591; and #599 had a Mal'cev term. Thus all seven groupoids had a Mal'cev term.

With this a complete classification of which 3-element groupoids have Mal'cev terms[2] was obtained. In total only a few hours of CPU time were needed to run this part of the program.

Having given a fairly detailed account of the part of the project devoted to Mal'cev terms we will now simply outline some of the ideas behind the programs used for the study of the other properties, and for the determination of $\leq$.

- The test for a 3-element groupoid to be *quasiprimal* comes from [8], namely *it must be hereditarily simple*[3], *have a Mal'cev term, and nontrivial subalgebras must be nonabelian*; thus a 3-element groupoid is quasiprimal iff it satisfies 7, 10 and 11 on page 394.

- The following result from Berman and McKenzie [4] was used to find the *Abelian* and *strongly Abelian* groupoids (by Abelian, respectively strongly Abelian, we mean the condition TC, respectively TC*, holds for all terms):

  Let $\mathbf{A}$ *be an algebra. Let* T *be the subuniverse of* $\mathbf{A}^4$ *generated by*
  $$\{(a,a,b,b) \mid a,b \in A\} \cup \{(a,b,a,b) \mid a,b \in A\}$$
  and let S *be the subuniverse of* $\mathbf{A}^4$ *generated by*
  $$\{(a,b,a,b)|a,b \in A\} \cup \{(a,b,c,c) \mid a,b,c \in A\}.$$
  (i) $\mathbf{A}$ *is Abelian if and only if whenever* $(a,a,b,c) \in T$ *or* $(a,b,a,c) \in T$, *then* $b = c$.
  (ii) $\mathbf{A}$ *is strongly Abelian if and only if* $(a,a,b,c) \in S$ *implies* $b = c$.

- *Affine* is given by 7 and 8 on page 394.

- Those *generating decidable varieties* were determined as follows. For a 3-element groupoid $\mathbf{A}$ the McKenzie and Valeriote theorem says that $V(\mathbf{A})$ is decidable iff one of the following holds:

  - $\mathbf{A}$ is quasiprimal;

---

[2]In a previous version of this study we worked with *near* Mal'cev terms, defined as terms which act like Mal'cev terms provided $x \neq y$ or $y \neq z$. This allowed us to reduce the number of coordinates to 12 in the above work. However it was erroneously claimed that such terms would guarantee the existence of a Mal'cev term. We, and independently Ralph McKenzie, discovered this flawed reasoning. It turns out that 74 of the groupoids have near Mal'cev terms, but not Mal'cev terms (e.g., #565).

[3]Of course simple 3-element algebras are hereditarily simple.

- **A** is affine and the associated variety of modules is decidable;

- **A** is strongly Abelian, essentially unary, and the monoid of the free algebra on one generator is linear.

By inspecting the 5 affine algebras, namely the groupoids with numbers

$$2124 \quad 2155 \quad 2302 \quad 2346 \quad 2934,$$

one quickly sees that the associated rings (of idempotent binary term functions) have size 3, and hence the associated rings are $\mathbf{Z}_3$. Thus the varieties of modules associated with the affine algebras are actually vector spaces over a finite field, and hence they are decidable.

And one also easily checks that each of the 13 strongly Abelian algebras, namely the groupoids with numbers

$$1 \quad 14 \quad 27 \quad 275 \quad 366 \quad 394 \quad 1045 \quad 2029 \quad 2243 \quad 2466 \quad 3161 \quad 3242 \quad 3302,$$

is essentially unary[4], and that the monoids of the free algebras on one generator are 1-generated, hence linear.

Thus a 3-element groupoid generates a decidable variety iff it is quasiprimal or affine or strongly Abelian.

- For the pre-order $\leq$ defined on page 389 note that if $\mathbf{A} \leq \mathbf{B}$ and $\mathbf{B} \leq \mathbf{A}$ then the clone determined by $\mathbf{A}$ is the same as the clone of some isomorphic copy of $\mathbf{B}$; if this is the case $\mathbf{A}$ and $\mathbf{B}$ are said to be *clone equivalent*, written $\mathbf{A} \sim \mathbf{B}$. A simple straightforward algorithm to determine if $\mathbf{A} \leq \mathbf{B}$ would be to generate $F(2)$, the elements of the free algebra on two generators, for the variety generated by $\mathbf{B}$ and check if the operation of any of the groupoids on $\{0, 1, 2\}$ isomorphic to $\mathbf{A}$ appears in $F(2)$. To do this for the more than 10 million pairs $(\mathbf{A}, \mathbf{B})$ would likely have required a prohibitive amount of time.

So an alternative strategy was adopted. Before determining the pre-order $\leq$ the clone equivalence relation $\sim$ was determined. It was noted that the thirteen properties first studied were invariant under $\sim$; these properties were used to obtain an upper bound to $\sim$ with 132 equivalence classes. Then using time-limited calculations to obtain some of the $F(2)$'s a lower bound to $\sim$ was established with 440 equivalence classes. Next representatives from each of these 440 classes were selected, and using the induced equivalence relation from the 132 classes, certain $F(2)$'s were calculated to refine these induced equivalence classes. The final result was the collection of 411 clone equivalence classes. Then, taking a representative from each of the 411 classes (the lexicographically first elements) we returned to calculating appropriate $F(2)$'s to determine the partial order $\leq$ on the 411 representatives.

---

[4]Actually any 3-element strongly Abelian algebra is essentially unary by 0.17iii of [9].

- The *free spectra* $|F(0)|$, $|F(1)|$, $|F(2)|$ of the varieties generated by the 411 clone equivalence representatives were determined by straightforward computations of the closure of suitable generators in $\mathbf{A}^3$ and $\mathbf{A}^9$.

- Using the partial ordering the 411 clone equivalence representatives were searched for those which *generate congruence distributive varieties* as follows. First the 12 representatives *with a majority term* were determined (see the figure on page 409), starting with the 10 quasiprimal representatives and considering subcovers. Next the subcovers of these 12 were examined to see which generated a congruence distributive variety; and this procedure was iterated if necessary. As it turns out, each of the groupoids so encountered which did not generate a congruence distributive variety had either a **1**, **2** or **5** in its set of types, or had a two-element subalgebra which did not generate a congruence distributive variety. For the others a $S(\mathbf{A}, 21, X)$ program was used to search for Jónsson terms. Whenever a new element was generated the program tried to find Jónsson terms incorporating a term corresponding to the new element.

- A similar approach was used for *congruence modularity*, starting with the classification of the congruence distributive and Mal'cev cases. Again the typeset of $\mathbf{A}$ and the two-element subalgebras of $\mathbf{A}$ sufficed to eliminate the negative cases; and a $S(\mathbf{A}, 21, X)$ program to search for Gumm terms made the verifications in the positive cases.

The Jónsson and Gumm terms presented on page 407 are the simplest possible in the sense that the number is minimal in each case, and within that constraint the tree depth is smallest possible; and the Mal'cev terms have minimal tree depth.

The following table lists twelve of the properties considered, and the number of groupoids with each property (out of 19,683), the number up to isomorphism (out of 3,330), and the number up to clone equivalence (out of 411):

PROPERTY	NUMBER OF GROUPOIDS	NUMBER OF GROUPOIDS up to isomorphism	NUMBER OF GROUPOIDS up to clone equivalence
1. generates a decidable variety	8,914	1,503	20
2. is quasiprimal	8,851	1,485	10
3. generates a congruence distributive variety	12,199	2,050	57
4. generates a congruence modular variety	13,117	2,207	76
5. is affine	12	5	3
6. is strongly Abelian	51	13	7
7. has a Mal'cev term	9,145	1,538	20
8. is Abelian	117	27	16
9. has an invertible binary term	11,442	1,907	32
10. Abelian subalgebras are trivial	14,259	2,399	156
11. is simple	16,009	2,693	191
12. is rigid (i.e., trivial automorphism group)	19,422	3,237	372

The detailed data on the 411 clone equivalence representatives is presented in several tables. First there is a table of sixteen properties, followed by a summary of this table. There is a picture of an upper segment of this poset given by the 20 representatives with a Mal'cev term; and also one for the 12 representatives with a majority term. To relate the 411 clone equivalence representatives to the 3,330 isomorphism representatives the clone equivalence class of each of these 411 is presented. Instead of the ordering $\leq$ a table of covering elements is given, namely for each of the 411 representatives the sub-covers and covers are listed. Next comes a ranking of the 411 representatives by the length of the maximal chain from the smallest element. At the end is a catalog of the 3,330 isomorphism representatives with the numbering used here. There are two types of entries, namely consider

1432 002 210 121 #534 and #**1433** 002 210 200

The first gives the groupoid table for isomorphism representative #1432 and says that its clone equivalence representative is #534. The second gives the groupoid table for isomorphism representative #1433 and says that it is one of the 411 clone equivalence representatives. The tables are abbreviated — the

usual table for #1432 would look like

$$
\begin{array}{c|ccc}
 & 0 & 1 & 2 \\
\hline
0 & 0 & 0 & 2 \\
1 & 2 & 1 & 0 \\
2 & 1 & 2 & 1 \\
\end{array}
$$

ACKNOWLEDGEMENTS. We are indebted to Ross Willard for his interest and comments throughout the project. The second author thanks NSERC for support of this research.

2   Some properties

The next pages discuss the following 16 items for each of the 411 clone equivalence representatives:

- **Types** — the set of types realized in the groupoid (in the sense of tame congruence theory).

- **Dec.** — the variety generated by the groupoid has a decidable first order theory.

- **QP** — the groupoid is quasiprimal.

- **CD** — the groupoid generates a congruence distributive variety.

- **CM** — the groupoid generates a congruence modular variety.

- **AF** — the groupoid is affine.

- **SA** — the groupoid is strongly Abelian.

- **MT** — the groupoid has a Mal'cev term.

- **AB** — the groupoid is Abelian.

- **IT** — the groupoid has an invertible binary term $b(x, y)$.

- **HNA** — nontrivial subalgebras of the groupoid are not Abelian.

- **S** — the groupoid is simple.

- **R** — the groupoid is rigid.

- $|F(0)|$ — the number of constant unary term functions.

- $|F(1)|$ — the size of the free algebra on 1 generator in the variety generated by the groupoid.

- $|F(2)|$ — the size of the free algebra on 2 generators in the variety generated by the groupoid.

	Types	Dec.	QP	CD	CM	AF	SA	MT	AB	IT	HNA	S	R	$|F(0)|$	$|F(1)|$	$|F(2)|$
1	1	•					•		•					1	2	3
2	1												•	1	3	6
3	1 5												•	0	2	5
4	1												•	1	2	5
5	1												•	1	3	8
6	1 5												•	0	2	7
8	5											•	•	1	4	17
9	1 5												•	0	2	7
10	1												•	1	3	12
11	1												•	1	3	12
12	1 5												•	0	2	11
13	1												•	1	3	7
14	1	•					•		•				•	1	3	5
15	1 5												•	0	2	7
16	3											•	•	1	4	53
18	5											•	•	0	2	13
19	1 3												•	1	3	16
21	1												•	0	2	6
22	3											•	•	1	3	24
24	3											•	•	0	2	8
25	1 3												•	1	3	8
26	1 3												•	1	4	17
27	1	•					•		•				•	0	2	4
30	1												•	1	2	6
31	1												•	1	3	9
32	1 5												•	0	2	8
33	5											•		1	2	5
34	5											•	•	1	4	27
35	5											•	•	0	2	11
36	1												•	1	3	18
37	1												•	1	3	18
38	1 5												•	0	2	15
39	1												•	1	3	12
43	3											•	•	1	4	70
44	4											•	•	0	2	20
45	3											•	•	1	3	24
46	3											•	•	1	4	65
47	3											•	•	0	2	10
48	3											•	•	1	3	36
49	3											•	•	1	4	71
50	3											•	•	0	2	14
51	1 3												•	1	3	12
52	1 3												•	1	4	23
53	1												•	0	2	6
59	5											•	•	1	4	28

	Types	Dec.	QP	CD	CM	AF	SA	MT	AB	IT	HNA	S	R	$|F(0)|$	$|F(1)|$	$|F(2)|$
60	1 5												•	0	2	8
61	3											•	•	1	4	55
63	3											•	•	0	2	29
65	5											•	•	1	4	23
66	4											•	•	0	2	19
67	3											•	•	1	4	137
69	4											•	•	0	2	29
70	1 3												•	1	3	26
72	1												•	0	2	10
73	3											•	•	1	3	50
75	3											•	•	0	2	18
78	1												•	0	2	6
79	1 5										•		•	0	3	11
80	5										•			0	1	3
81	1 5										•		•	0	3	13
82	5										•		•	0	1	5
83	5										•	•	•	0	3	24
85	1 3										•		•	0	3	30
87	3 5										•		•	0	1	14
88	1 5										•		•	0	3	19
89	1 5										•		•	0	3	17
90	5										•		•	0	1	10
91	3										•	•	•	0	3	102
93	3										•	•	•	0	1	34
94	3 5										•		•	0	2	18
96	1 5												•	0	1	6
97	3										•	•	•	0	2	30
99	3											•	•	0	1	10
100	3 5										•		•	0	2	10
101	3 5										•		•	0	3	42
102	1 5												•	0	1	4
104	1 5												•	0	2	5
105	5										•		•	0	1	3
106	3 5										•		•	0	2	10
107	1 5													0	1	4
111	1 5										•		•	0	3	22
112	1 5												•	0	2	7
113	5										•		•	0	1	5
115	3										•	•	•	0	2	26
116	4											•	•	0	1	8
117	3										•	•	•	0	2	34
119	3											•	•	0	1	10
120	3										•	•	•	0	2	44
121	3											•	•	0	2	9
122	3											•	•	0	1	4

	Types	Dec.	QP	CD	CM	AF	SA	MT	AB	IT	HNA	S	R	$|F(0)|$	$|F(1)|$	$|F(2)|$
123	3 5										•		•	0	2	18
124	3 5										•		•	0	2	14
125	1 5												•	0	1	4
129	2 5												•	0	2	8
130	3										•	•	•	0	3	70
132	3										•	•	•	0	1	38
134	5										•	•	•	0	2	16
135	4											•	•	0	1	10
136	3										•	•	•	0	3	141
137	3											•	•	0	2	24
138	4										•	•	•	0	1	7
139	3 5										•		•	0	2	28
141	1 5												•	0	1	10
142	3										•	•	•	0	2	52
143	3										•	•	•	0	2	34
144	3											•	•	0	1	6
147	1 5													0	1	4
148	5										•	•		1	3	9
149	5										•	•	•	1	7	57
151	3			•	•						•	•	•	1	7	241
153	3			•	•					•	•	•	•	0	3	459
155	5										•	•	•	1	7	49
157	3			•	•						•	•	•	1	7	313
160	1 3										•		•	1	4	29
161	3			•	•				•		•	•	•	1	9	1377
162	1												•	0	3	12
163	3			•	•				•		•	•	•	1	6	480
165	3											•	•	0	3	132
166	1 3										•		•	1	4	31
168	1												•	0	3	8
169	3 5										•		•	0	4	24
170	1 5													0	2	8
171	3			•	•						•	•	•	1	9	497
175	5										•	•	•	0	4	56
176	3				•							•	•	0	2	68
178	3											•	•	0	2	68
179	3			•	•						•	•	•	0	2	70
180	3			•	•						•	•	•	1	4	64
182	3											•	•	0	3	60
183	3			•	•					•	•	•	•	1	6	594
184	3			•	•						•	•	•	0	4	272
185	3											•	•	0	2	24

	Types	Dec.	QP	CD	CM	AF	SA	MT	AB	IT	HNA	S	R	$\|F(0)\|$	$\|F(1)\|$	$\|F(2)\|$
186	3			•	•						•	•	•	1	4	52
188	5											•	•	0	2	12
194	5											•	•	0	2	16
195	3			•	•						•	•	•	0	2	102
198	4											•	•	0	2	13
199	1 3										•		•	1	5	114
201	1												•	0	3	22
203	3											•	•	0	2	36
204	3											•	•	0	2	32
207	1 3										•		•	1	3	8
209	3 5										•		•	0	2	60
213	3			•	•						•	•	•	1	6	408
215	3			•	•						•	•	•	0	2	136
216	1 3										•			1	2	7
218	3											•	•	0	2	24
219	3			•	•						•	•	•	1	2	40
221	3											•	•	0	2	48
222	3			•	•						•		•	1	2	14
223	3			•	•						•		•	1	3	48
224	1 3												•	0	2	16
235	1 3												•	0	2	16
239	3			•	•						•			1	2	6
241	1 3												•	0	2	8
244	3			•	•						•	•	•	0	2	160
250	3			•	•						•	•	•	0	2	198
252	4											•	•	0	2	18
253	3			•	•						•	•		1	2	18
255	3											•	•	0	2	72
257	1												•	0	3	26
258	1												•	0	3	28
259	1 5												•	0	1	18
260	1												•	0	3	10
261	1												•	0	3	12
262	1 5												•	0	1	10
263	3											•	•	0	3	90
265	3											•	•	0	1	30
266	3											•	•	0	3	90
267	4											•		0	1	4
268	3											•	•	0	3	54
269	3											•	•	0	1	10
270	1 3												•	0	3	36
271	1												•	0	1	4

	Types	Dec.	QP	CD	CM	AF	SA	MT	AB	IT	HNA	S	R	$|F(0)|$	$|F(1)|$	$|F(2)|$
272	1												•	0	3	14
273	1 5													0	1	6
274	3											•	•	0	3	54
275	1	•					•		•					0	1	2
278	3											•	•	0	1	44
280	1 3												•	0	2	20
281	1												•	0	1	6
282	3											•	•	0	3	162
283	1 2												•	0	2	16
284	1 5												•	0	1	6
286	1 3												•	0	2	24
287	1													0	1	4
298	3			•	•						•	•		1	3	45
305	1 3										•		•	0	4	128
306	1 2												•	0	2	32
308	1												•	0	2	20
309	1 3										•		•	0	2	32
311	1													0	2	16
316	1												•	0	2	12
317	1 3										•		•	0	2	48
320	1												•	0	2	6
321	1													0	2	8
322	1 3										•			1	3	13
341	3			•	•						•	•		1	3	25
347	3			•	•						•	•		0	1	15
349	3			•	•						•	•	•	0	1	153
353	1												•	0	2	10
354	1 3										•		•	0	1	10
356	1													0	1	4
359	1													0	1	8
366	1	•					•		•					0	2	4
376	1												•	1	3	13
377	1												•	1	3	13
378	1 5												•	0	2	15
379	1												•	1	3	14
380	1												•	1	3	14
381	1 5												•	0	2	14
382	3											•	•	1	4	134
384	3											•	•	0	2	46
385	3											•	•	1	4	107
387	3											•	•	0	2	37
388	3											•	•	1	4	170

	Types	Dec.	QP	CD	CM	AF	SA	MT	AB	IT	HNA	S	R	$|F(0)|$	$|F(1)|$	$|F(2)|$
390	3											•	•	0	2	58
391	1 3												•	1	4	35
405	1								•				•	1	3	6
406	1												•	1	3	6
407	1 5												•	0	2	8
410	4											•	•	0	2	27
417	1 3												•	1	4	23
434	3											•	•	1	4	164
436	4											•	•	0	2	33
437	3											•	•	1	4	245
439	3											•	•	0	2	83
454	1 3										•		•	0	3	31
455	1 3										•		•	0	3	31
456	3 5										•		•	0	1	15
457	1 3										•		•	0	3	33
458	1 3										•		•	0	3	33
459	3 5										•		•	0	1	17
460	3										•	•	•	0	3	138
462	3										•	•	•	0	1	46
463	3										•	•	•	0	3	174
465	3										•	•	•	0	1	58
469	3										•		•	0	3	78
483	1 5										•		•	0	3	23
484	1 5												•	0	2	8
485	5										•		•	0	1	6
487	3										•	•	•	0	2	36
488	4											•	•	0	1	12
493	3											•	•	0	2	12
494	3										•	•	•	0	1	10
495	3 5										•		•	0	3	51
496	3 5										•		•	0	2	20
512	3										•	•	•	0	3	168
513	3											•	•	0	2	28
514	4								•				•		0	1	5
515	3								•				•	•	0	3	249
517	3								•				•	•	0	1	83
519	3											•	•	•	0	2	58
520	3												•	•	0	1	20
522	3												•	•	0	2	40
532	3			•	•							•	•	•	1	9	849
534	3	•	•	•	•			•		•	•	•	•	0	3	2187
538	3	•	•	•	•			•		•	•	•	•	1	9	6561

	Types	Dec.	QP	CD	CM	AF	SA	MT	AB	IT	HNA	S	R	$|F(0)|$	$|F(1)|$	$|F(2)|$
562	4			•	•						•	•	•	0	4	82
563	3				•			•				•	•	0	2	324
565	3											•	•	0	2	324
566	3			•	•						•	•	•	0	2	486
571	3	•	•	•	•			•			•	•	•	0	4	1296
600	1 3										•		•	1	4	49
602	3 5										•		•	0	2	100
603	1 3										•		•	1	5	138
604	1 3										•		•	1	5	140
606	3			•	•						•	•	•	1	4	104
608	3			•	•						•	•	•	0	2	208
609	1 3										•		•	1	5	136
612	1 3										•		•	1	5	140
613	1 3										•		•	1	5	144
615	3			•	•						•	•	•	1	6	624
618	3			•	•						•	•	•	1	6	768
620	3			•	•						•	•	•	0	2	256
624	3			•	•						•		•	1	3	72
629	1 3										•		•	1	5	140
630	1 3										•		•	1	5	144
632	3			•	•						•	•	•	1	4	78
638	1 3										•		•	1	5	145
639	1 3										•		•	1	5	154
652	3			•	•					•	•	•	•	1	6	1008
654	3			•	•						•	•	•	0	2	336
658	3			•	•					•	•	•	•	1	6	1458
677	1												•	0	3	18
678	1								•				•	0	3	10
679	1 5												•	0	1	10
680	1												•	0	3	34
681	1 5												•	0	1	10
682	3											•	•	0	3	108
684	3											•	•	0	1	60
687	3											•	•	0	3	180
690	1 3												•	0	3	36
691	3											•	•	0	3	336
693	3											•	•	0	1	112
695	3											•	•	0	2	80
696	3											•	•	0	1	24
697	3											•	•	0	3	486
698	3											•	•	0	2	72
704	3											•	•	0	2	48

	Types	Dec.	QP	CD	CM	AF	SA	MT	AB	IT	HNA	S	R	$\|F(0)\|$	$\|F(1)\|$	$\|F(2)\|$
705	4											•	•	0	1	14
707	3											•	•	0	2	48
710	3											•	•	0	1	162
712	3											•	•	0	2	72
755	3			•	•						•	•	•	0	4	896
756	3				•							•	•	0	2	224
758	3											•	•	0	2	224
780	3											•	•	0	2	144
792	3			•	•					•	•	•	•	1	7	409
870	3	•	•	•	•			•			•	•	•	0	1	729
885	3	•	•	•	•			•			•	•		0	1	27
898	3	•	•	•	•			•			•	•		0	1	9
984	3											•	•	0	2	36
1012	1 2												•	1	3	18
1014	3											•	•	1	3	34
1038	3											•	•	1	3	38
1040	3											•	•	0	2	13
1065	1 2								•				•	1	3	6
1066	1 2												•	1	4	12
1084	2 5												•	0	2	20
1086	3											•	•	0	2	36
1107	3											•	•	0	2	40
1108	3										•	•		0	1	3
1132	2 5												•	0	2	8
1133	2 5												•	0	2	13
1151	1 2												•	1	5	130
1153	3				•					•		•	•	1	6	672
1176	3				•			•		•		•	•	1	6	972
1200	1 2												•	1	5	34
1202	3			•	•					•	•	•	•	1	4	164
1205	3			•	•					•	•	•	•	1	4	240
1219	3				•					•		•	•	1	4	160
1221	3				•					•		•	•	1	4	216
1225	3			•	•					•	•		•	1	4	68
1227	1 3												•	0	2	20
1231	3			•	•					•	•		•	1	4	96
1233	1 3												•	0	2	32
1242	2 3				•					•			•	1	4	64
1249	3			•	•					•	•		•	1	6	432
1268	2 3				•					•			•	1	4	40
1269	2 3				•					•	•		•	1	6	252
1271	1 3												•	0	3	66

	Types	Dec.	QP	CD	CM	AF	SA	MT	AB	IT	HNA	S	R	$\lvert F(0)\rvert$	$\lvert F(1)\rvert$	$\lvert F(2)\rvert$
1277	1 3												•	0	3	144
1281	1 2				•								•	0	2	12
1321	2 3				•					•			•	1	6	288
1433	2 5												•	0	2	68
1437	2 5												•	0	2	20
1481	1 2								•				•	1	5	18
1700	1 2								•				•	1	3	6
1708	1 2								•				•	1	3	14
1791	3 5										•		•	0	4	264
1793	1 5												•	0	2	28
1799	1 5												•	0	2	52
1818	1 5												•	0	2	20
1829	1 2									•			•	1	5	49
1837	1 2									•			•	1	4	15
1962	1 5												•	0	2	12
2088	1 2													1	2	6
2090	3				•			•				•	•	1	2	36
2102	2				•			•					•	1	2	9
2104	3				•			•			•	•		1	2	12
2116	2				•			•					•	1	3	42
2124	2	•			•	•		•	•		•			1	3	9
2135	1 5													0	2	20
2144	1 2									•				1	3	21
2159	3	•	•	•	•			•			•	•		1	3	81
2171	1 5													0	2	12
2346	2	•			•	•		•	•			•		0	1	3
2353	1 3									•			•	2	5	18
2354	1 3									•			•	2	9	514
2357	3			•	•				•	•	•	•	•	2	1 2	2688
2369	3			•	•				•	•			•	2	12	1152
2393	3	•	•	•	•			•	•	•	•	•	•	2	12	3888
2407	3	•	•	•	•			•	•	•	•	•	•	3	27	19683
2428	2 3				•			•	•	•			•	2	12	672
2430	1 2												•	0	6	68
2436	3										•	•	•	0	6	972
2460	1												•	0	5	34
2461	1												•	0	5	38
2462	1												•	0	5	34
2463	1								•				•	0	5	18
2464	3										•	•	•	0	6	672
2466	1	•					•		•				•	0	3	6
2467	1												•	0	5	14

	Types	Dec.	QP	CD	CM	AF	SA	MT	AB	IT	HNA	S	R	$|F(0)|$	$|F(1)|$	$|F(2)|$
2472	1												•	0	3	14
2476	1 3												•	0	6	132
2478	1												•	0	5	74
2479	1												•	0	5	78
2480	1												•	0	5	66
2483	1												•	0	5	26
2486	1 3												•	0	6	288
2487	3											•	•	0	6	432
2493	3											•	•	0	6	108
2529	3	•	•	•	•			•			•	•		0	3	27
2539	1 3												•	0	6	108
2545	1 3												•	0	6	72
2552	1									•			•	0	6	52
2558	1									•			•	0	5	18
2636	1 3												•	0	6	36
2654	1 3								•	•			•	2	6	56
2686	1 3								•	•			•	2	6	72
2698	1 3									•			•	3	10	83
2702	1 3								•	•			•	3	12	207
2739	1												•	0	5	78
2799	1 3								•	•			•	3	15	333
2803	1 3								•	•			•	3	15	525
2934	2	•			•	•		•	•			•		0	3	9
3242	1	•				•			•			•		0	3	6

3   Some Information on the Poset of the 411 Clone Equivalence Representatives

DECIDABLE : 1 14 27 275 366 534 538 571 870 885 898 2124 2159 2346 2393 2407 2466 2529 2934 3242

number = 20

QUASIPRIMAL : 534 538 571 870 885 898 2159 2393 2407 2529

number = 10
minimals: 571 885 898
maximal non quasiprimal cases: 161 532 658 792 1176 2124 2357 2428 2436 2803 2934

MAL'CEV : 534 538 563 571 870 885 898 1176 2090 2102 2104 2116 2124 2159 2346 2393 2407 2428 2529 2934

number = 20
minimals (with Mal'cev terms):

563: $((xy)(z((xz)(xz))))$
$\quad (((z(x(zz)))((z(xx))((xx)(xy))))(((z(xx))((xx)(zz)))(((xx)(zz))((yz)(zz)))))$
2102: $(x(yz))(z(x(xz)))$
2104: $(xy)z$
2346: $y(xz)$
maximal non Mal'cev cases: 161 532 658 792 2357 2436 2803 3242

$\boxed{\text{CD}}$: 151 153 157 161 163 171 179 180 183 184 186 195 213 215 219 222 223 239 244 250 253 298 341 347 349 532 534 538 562 566 571 606 608 615 618 620 624 632 652 654 658 755 792 870 885 898 1202 1205 1225 1231 1249 2159 2357 2369 2393 2407 2529

number = 57
minimals (with Jónsson terms):
179: $p1 = (x(((xx)y)z))(((xz)y)(x(xz)))$,
$\quad p2 = ((z((zz)x))(xx))(((z((zz)y))((z(xx))((yx)x)))((((zz)x)z)(((zx)(zy))(xy))))$
186: $p1 = x((xx)(((xx)(xy))(xz)))$, $p2 = (x((xx)z))((xz)((xy)(xx)))$,
$\quad p3 = (z((xx)z))((zx)(y(zz)))$, $p4 = z((zz)((yy)(xx)))$
215: $p1 = (x((xy)((xy)(zx))))(((xy)(yx))((z(yx))(zx)))$,
$\quad p2 = (z((zx)((zy)(zz))))(((zz)(xx))((yx)y))$
239: $p1 = x((xy)z)$, $p2 = z(yx)$
347: $p1 = (x(yx))((x(zx))(z(yx)))$, $p2 = (z(yz))((yz)x)$
562: $p1 = (xy)(z((xx)(yy)))$
898: $p1 = (((xy)x)((zx)((yx)(zx))))$
maximal non CD cases: 149 155 175 515 1176 1791 2124 2428 2436 2803 2934

$\boxed{\text{CM}}$: 151 153 157 161 163 171 176 179 180 183 184 186 195 213 215 219 222 223 239 244 250 253 298 341 347 349 532 534 538 562 563 566 571 606 608 615 618 620 624 632 652 654 658 755 756 792 870 885 898 1153 1176 1202 1205 1219 1221 1225 1231 1242 1249 1268 1269 1321 2090 2102 2104 2116 2124 2159 2346 2357 2369 2393 2407 2428 2529 2934
number = 76
minimals (with Gumm terms if not CD or Mal'cev):
176: $p1 = x$, $p2 = ((xx)(xy))((x(xz))((xz)(yx)))$,
$\quad p3 = (((zz)x)((zz)(yy)))((zx)(((zx)(zy))(xx)))$
179 186 215 239 347 562
756: $p1 = x$, $p2 = (x(yz))(yz)$, $p3 = (z(xz))((xz)(xy))$
1242: $p1 = x$, $p2 = x((x(yz))(z(yx)))$, $p3 = (z(xy))((yx)(xy))$
1268: $p1 = x$, $p2 = x((x(yz))(z(yx)))$, $p3 = z(yx)$
2102 2104 2346
maximal non CM cases: 149 155 175 305 437 515 1791 2354 2436 2803 3242

$\boxed{\text{AFFINE}}$: 2124 2346 2934

number = 3
minimals: 2346
maximals: 2124 2934

$\boxed{\text{STRONGLY ABELIAN}}$: 1 14 27 275 366 2466 3242

number = 7

maximals: 14 366 2466 3242
minimal non strongly abelian cases: 2 3 4 13 21 33 53 78 80 102 104 105 107 122 125 147 168
267 271 287 320 356 405 406 678 1065 1108 1700 2346

ABELIAN : 1 14 27 275 366 405 678 1065 1481 1700 2124 2346 2463 2466 2934 3242

number = 16
maximals: 1481 2124 2463 2934
minimal non Abelian cases: 2 3 4 13 21 33 53 78 80 102 104 105 107 122 125 147 168 267 271
287 320 356 406 1108

INVERTIBLE : 153 161 163 183 534 538 652 658 792 1153 1176 1202 1205 1219 1221 1225
1231 1242 1249 1268 1269 1321 2357 2369 2393 2407 2428 2654 2686 2702 2799 2803

number = 32
minimals (with binary/unary pairs):

    153: b(x,y) = xy, u1 = xx, u2 = x(xx)
    163: b(x,y) = yx, u1 = (xx)x, u2 = x((xx)x)
    1225: b(x,y) = xy, u1 = x(xx), u2 = (xx)x
    1242: b(x,y) = yx, u1 = (xx)x, u2 = x((xx)x)
    1268: b(x,y) = yx, u1 = (xx)x, u2 = x((xx)x)
    2654: b(x,y) = yx, u1 = xx, u2 = x(xx)
maximal non invertible cases: 157 437 515 532 571 618 870 1829 2090 2116 2159 2354 2436
2529 2552 2698

SIMPLE : 8 16 18 22 24 33 34 35 42 43 44 45 46 47 48 49 50 59 61 63 65 66 67 69 73 75 83
91 93 97 99 115 116 117 119 120 121 122 130 132 134 135 136 137 138 142 143 144 148 149
151 153 155 157 161 163 165 171 175 176 178 179 180 182 183 184 185 186 188 194 195 198
203 204 213 215 218 219 221 244 250 252 253 255 263 265 266 267 268 269 274 278 282 298
341 347 349 382 384 385 387 388 390 410 434 436 437 439 460 462 463 465 487 488 493 494
512 513 514 515 517 519 520 522 532 534 538 562 563 565 566 571 606 608 615 618 620 632
652 654 658 682 684 687 691 693 695 696 697 698 704 705 707 710 712 755 756 758 780 792
870 885 898 984 1014 1038 1040 1086 1107 1108 1153 1176 1202 1205 1219 1221 2090 2104
2124 2159 2346 2357 2393 2407 2436 2464 2487 2493 2529 2934 3242

number = 191
minimals: 8 18 24 33 35 47 83 99 116 119 121 122 134 138 144 186 188 194 198 267 514 984
1108 2346 3242
maximal non simple cases: 169 2369 2428 2803

NON-RIGID : 1 33 80 107 147 148 170 216 239 253 267 273 275 287 298 311 321 322 341
347 356 359 366 514 681 885 898 1108 2088 2104 2124 2135 2144 2159 2171 2346 2529 2934
3242

number = 39
maximals = 2159 2529
minimal rigid cases: 4 14 27 82 102 105 122 125 271

4 Two pictures

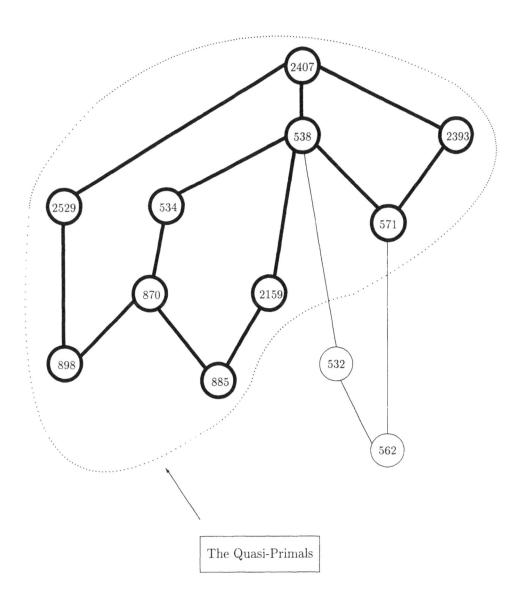

Figure 1. Clone equivalence representatives with a majority term.

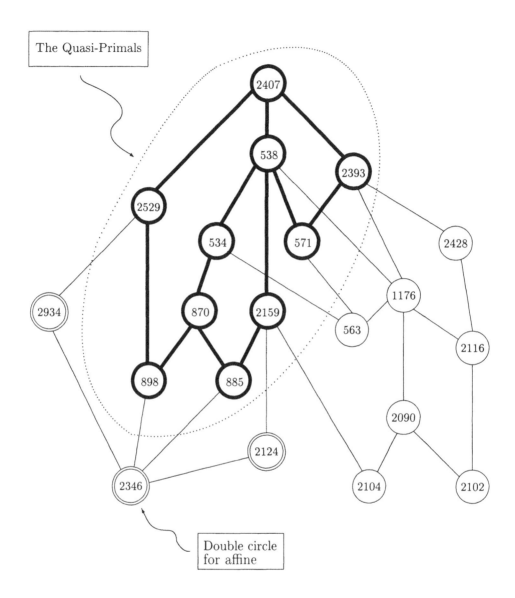

The Quasi-Primals

Double circle
for affine

Figure 2.   Clone equivalence representatives with a Mal'cev term.

5 Clone Equivalence Classes

This is a listing of the clone equivalence classes among the 3,330 representatives of isomorphism types. Two groupoids are clone equivalent if some isomorphic copy of the first generates the same clone as the second. A boxed number denotes the beginning of an equivalence class. There are 411 such classes below: the first (boxed) element of each is the clone equivalence representative of that class.

1 **2** **3** **4** 7 **5** 28 **6** 29 **8** 54 **9** 55 **10** 367 **11** 368
12 369 **13** 393 **14** 394 **15** 395 **16** 17 20 23 419 420 996 1020 **18** 421
19 995 **21** 997 **22** 1019 **24** 1021 **25** 1043 **26** 1044 **27** 1045
30 **31** **32** **33** 56 **34** 57 **35** 58 **36** 370 **37** 371 **38** 372
39 396 **40** 397 **41** 398 **42** 422 **43** 423 **44** 424 **45** 998 **46** 999
47 1000 **48** 1022 **49** 1023 **50** 1024 **51** 1046 **52** 1047 **53** 1048
59 **60** **61** 62 64 373 374 399 **63** 375 **65** 400 **66** 401 **67** 68 71 74 77 425 426 1002 1026 1050 **69** 427 **70** 76 1001 1049 **72** 1003 **73** 1025 **75** 1027 **78** 1051 **79** **80** **81** 103 **82** 84 **83** 126 **85** 86 108 109 445 446 448 449 **87** 110 447 450 **88** 471 **89** 472 **90** 473 **91** 92 95 98 114 118 497 498 500 1068 1071 1092 **93** 499 **94** 1067 **96** 1069 **97** 1091 **99** 1093 **100** 1111 **101** 1112 **102** 1113 **104** **105** **106** 127 **107** 128 **111** 474 **112** 475 **113** 476 **115** 501 **116** 502 **117** 1070 **119** 1072 **120** 1094 **121** 1064 **122** 1095 **123** 1114 **124** 1115 **125** 1116 **129** **130** 131 133 451 452 477 **132** 453 **134** 478 **135** 479 **136** 140 503 1074 **137** 146 504 1118 **138** 505 **139** 145 1073 1117 **141** 1075 **142** 1096 **143** 1097 **144** 1098 **147** 1119 **148** **149** 150 **151** 152 154 190 191 193 323 325 326 339 342 362 523 524 529 530 549 555 2173 2176 2177 2191 2195 2228 **153** 156 159 173 192 291 294 297 303 314 324 327 330 332 335 338 340 343 345 346 348 350 352 357 525 528 531 551 577 1190 1337 1404 1407 1515 1629 1632 1651 1673 1888 1889 1896 1897 1903 1904 2070 2138 2158 2164 **155** 550 **157** 158 575 576 **160** 1134 **161** 164 167 177 181 196 200 289 290 292 293 295 296 299 300 301 302 304 307 312 313 315 318 328 329 331 333 334 337 344 363 578 581 1135 1138 1141 1158 1181 1627 1628 1630 1631 1633 1634 1649 1650 1652 1655 1671 1672 1674 1677 2037 2038 2174 2175 2192 2193 2209 2211 2212 2227 **162** 1136 **163** 202 242 243 245 246 1157 1163 1533 1534 1555 1556 **165** 276 277 279 1159 1329 1598 1599 **166** 1180 **168** 1182 **169** **170** 189 **171** 172 174 526 527 552 **175** 553 **176** 554 **178** 187 579 1184 **179** 580 **180** 1137 **182** 1139 **183** 248 249 251 254 256 1160 1577 1578 1977 1996 2014 **184** 197 206 582 1161 1187 **185** 1162 **186** 1183 **188** 1185 **194** 556 **195** 557 **198** 583 **199** 205 208 210 211 225 226 228 229 1140 1186 1528 1530 1531 1549 1550 1552 1553 **201** 1142 **203** 1164 **204** 1165 **207** 1527 **209** 212 227 230 1529 1532 1551 1554 **213** 214 217 220 231 232 234 237 1571 1572 1574 1575 1972 1974 1991 1993 **215** 233 1573 1576 **216**

1971 218 1617 219 236 1976 1992 221 238 1624 1994 222 1973 223
240 2010 2012 224 1609 235 1975 239 2011 241 2013 244 247 1340
1518 250 1193 252 1510 253 1995 255 1332 257 1593 258 1594
259 1595 260 1507 261 1396 262 1441 263 264 1605 1606 265
1607 266 1616 267 1618 268 1399 269 1444 270 2028 271 1610
272 1596 273 1597 274 1608 275 2029 278 1452 280 1250 281
1272 282 285 1611 1623 283 288 1612 2030 284 1125 286 1253 287
1274 298 336 2036 2210 305 310 319 364 1653 1678 2052 2242 306 360 1654
2108 308 1675 309 355 1676 2017 311 365 2067 2257 316 1656 317
358 1258 1261 320 1057 321 2053 322 2172 341 361 2194 2226 347
1898 349 351 1449 1890 353 1054 354 1122 356 2032 359 1278
366 2243 376 377 378 379 402 380 403 381 404 382 383
408 409 428 429 431 432 384 430 385 386 411 412 1004 1005 1007 1008 387
413 1006 1009 388 389 414 415 1028 1029 1031 1032 390 416 1030 1033 391
392 1052 1053 405 406 407 410 433 417 418 1055 1056 434 435
436 437 438 440 441 443 444 1010 1011 1013 1015 1018 1034 1035 1037 1039 1042
1058 1059 1061 1063 439 442 775 979 454 455 456 457 480 458
481 459 482 460 461 486 506 507 509 462 508 463 464 466 467 489 490
492 1076 1077 1079 1080 1099 1100 1102 465 468 491 1078 1081 1101 469 470 1120
1121 483 484 485 487 510 488 511 493 1016 494 1088
495 1123 496 1124 512 513 514 515 516 518 521 1082 1083 1085
1087 1090 1103 1106 1126 1129 517 907 519 1104 1110 1131 520 699 522
1127 532 533 535 536 558 559 561 534 537 540 543 546 560 569 586 594 718 721
724 727 730 733 736 739 744 753 761 769 778 786 794 797 800 803 806 809 812 815 820 823
826 829 832 837 840 845 848 851 854 857 862 865 868 869 871 872 874 875 877 878 880 881
883 884 886 887 889 890 892 893 894 896 897 899 900 902 905 908 909 911 914 916 917 918
921 925 928 931 934 937 944 949 958 973 982 1145 1148 1168 1293 1296 1315 1360 1363 1382
1385 1417 1420 1423 1425 1426 1428 1429 1431 1432 1434 1436 1438 1440 1447 1448 1450
1453 1456 1471 1474 1493 1637 1640 1659 1681 1759 1762 1767 1770 1781 1789 1802 1810
1823 1826 1831 1834 1845 1848 1853 1866 1869 1874 1885 1886 1891 1892 1893 1894 1899
1900 1901 1902 1905 1906 1907 1909 1910 1915 1920 1940 1960 2041 2056 2119 2129 2146
2152 2162 2165 2180 2198 2215 2231 2246 2272 2285 2307 2332 538 539 541 542 544
545 547 548 564 567 568 570 573 584 585 587 590 592 593 595 598 716 717 719 720 722 723
725 726 728 729 731 732 734 735 737 738 740 741 742 743 745 748 751 752 754 757 759 760
762 765 767 768 770 773 776 777 779 782 784 785 787 790 795 796 798 799 801 802 804 805
807 808 810 811 813 814 816 817 821 822 824 825 827 828 830 831 833 834 835 836 838 839
841 842 846 847 849 850 852 853 855 856 858 859 860 861 863 864 866 867 926 927 929 930
932 933 935 936 938 939 940 941 942 943 945 946 950 953 956 957 959 962 964 965 966 969
974 977 980 981 983 986 988 989 990 993 1143 1144 1146 1147 1149 1150 1152 1154 1156
1166 1167 1169 1172 1175 1188 1189 1191 1194 1197 1283 1286 1289 1291 1292 1294 1295
1297 1298 1300 1302 1304 1306 1313 1314 1316 1319 1322 1328 1335 1336 1338 1341 1344
1350 1353 1356 1358 1359 1361 1362 1364 1365 1367 1369 1371 1373 1376 1379 1380 1381
1383 1384 1386 1387 1389 1391 1393 1395 1398 1401 1402 1403 1405 1406 1408 1409 1411
1413 1415 1462 1465 1468 1469 1470 1472 1473 1475 1476 1478 1480 1482 1484 1491 1492

1494 1497 1500 1506 1513 1514 1516 1519 1522 1635 1636 1638 1639 1641 1642 1643 1644
1645 1646 1647 1648 1657 1658 1660 1663 1665 1666 1667 1669 1679 1680 1682 1684 1686
1687 1688 1690 1757 1758 1760 1761 1763 1764 1765 1766 1768 1769 1771 1772 1773 1774
1775 1776 1777 1778 1779 1780 1782 1785 1787 1788 1790 1792 1794 1795 1796 1798 1800
1801 1803 1806 1808 1809 1811 1813 1815 1816 1817 1819 1821 1822 1824 1825 1827 1828
1830 1832 1833 1835 1836 1838 1839 1840 1841 1842 1843 1844 1846 1847 1849 1850 1851
1852 1854 1855 1856 1857 1858 1859 1860 1861 1862 1863 1864 1865 1867 1868 1870 1871
1872 1873 1875 1876 1877 1878 1879 1880 1881 1882 1883 1884 1911 1912 1913 1914 1916
1917 1918 1919 1921 1922 1923 1924 1925 1926 1927 1928 1929 1930 1931 1932 1933 1936
1938 1939 1941 1943 1945 1946 1947 1949 1951 1952 1953 1956 1958 1959 1961 1963 1965
1966 1967 1969 2039 2040 2042 2043 2044 2045 2047 2049 2051 2054 2055 2057 2059 2062
2068 2069 2071 2073 2076 2117 2118 2120 2121 2122 2123 2125 2126 2127 2128 2130 2132
2136 2137 2139 2141 2145 2147 2148 2149 2150 2151 2153 2154 2156 2157 2160 2161 2167
2168 2178 2179 2181 2182 2183 2184 2185 2186 2187 2188 2189 2190 2196 2197 2199 2200
2201 2202 2203 2204 2205 2206 2207 2208 2213 2214 2216 2217 2218 2219 2220 2221 2222
2223 2224 2225 2229 2230 2232 2233 2234 2235 2236 2237 2238 2239 2240 2241 2244 2245
2247 2249 2251 2252 2253 2255 2258 2259 2260 2262 2264 2265 2266 2268 2271 2273 2274
2275 2276 2278 2279 2280 2281 2282 2283 2284 2286 2287 2288 2289 2290 2291 2292 2294
2295 2296 2297 2298 2299 2300 2301 2303 2304 2305 2306 2308 2309 2310 2311 2313 2314
2315 2316 2317 2319 2323 2324 2325 2327 2331 2333 2334 2335 2336 2337 2338 2339 2341
2342 2344 2345 2348 2349 | 562 | | 563 | 572 747 906 952 1171 1446 | 565 | 574 588
596 749 771 954 975 1173 1179 1192 1199 1312 1331 1490 1509 | 566 | 589 597 660 671 674
750 772 903 955 976 1443 | 571 | 591 599 746 774 951 978 1170 1177 1195 1201 1309 1334
1487 1512 | 600 | 601 626 627 1349 1394 1461 1505 | 602 | 605 611 614 628 631 637 640
1416 1422 1439 1445 1537 1540 1559 1562 | 603 | 1355 | 604 | 1467 | 606 | 607 1352
1464 | 608 | 617 634 643 1419 1442 1581 1584 | 609 | 610 635 636 1535 1536 1538 1539
| 612 | 1557 | 613 | 1558 | 615 | 616 641 642 1579 1580 1582 1583 | 618 | 619 621 622
644 645 647 648 1978 1979 1981 1982 1997 1998 2000 2001 | 620 | 623 646 649 1805 1955
1980 1999 | 624 | 625 650 651 2015 2016 2018 2019 | 629 | 1400 | 630 | 1511 | 632 |
633 1397 1508 | 638 | 1560 | 639 | 1561 | 652 | 653 655 656 661 662 664 665 1203 1206
1211 1212 1214 1215 1226 1229 1234 1235 1237 1238 1282 1288 1327 1333 1372 1378 1483
1489 1541 1542 1543 1544 1545 1546 1563 1564 1565 1566 1567 1568 | 654 | 657 663 666
781 789 912 919 985 992 1213 1216 1236 1239 1451 1458 | 658 | 659 667 668 669 670 672
673 675 676 1209 1217 1218 1220 1222 1224 1232 1240 1241 1243 1245 1247 1255 1262 1263
1265 1267 1285 1308 1330 1375 1486 1585 1586 1587 1588 1589 1590 1591 1592 1983 1984
1986 1988 1990 2002 2003 2005 2007 2009 2020 2021 2023 2025 | 677 | 1351 | 678 | 1463
| 679 | 1418 | 680 | 685 1357 1600 | 681 | 686 1424 1601 | 682 | 683 1354 1466 | 684 |
689 1421 1615 | 687 | 688 1613 1614 | 690 | 2031 | 691 | 692 694 700 701 703 1284 1290
1374 1485 1602 1603 | 693 | 702 913 920 | 695 | 1210 | 696 | 713 | 697 | 706 708 709
711 714 1279 1287 1377 1619 1621 1622 1625 2033 | 698 | 1276 1310 1488 | 704 | 1204
| 705 | | 707 | 715 1280 1620 1626 2034 | 710 | 904 | 712 | 1207 1230 1275 | 755 |
763 783 791 960 967 987 994 1317 1324 1342 1348 1495 1502 1520 1526 1661 1668 1685 1691
1783 1807 1934 1957 2058 2064 2074 2080 2248 2254 2263 2269 | 756 | 764 915 922 961
968 1318 1435 1454 1457 1460 1496 1662 1738 1741 1746 1784 1935 2083 2096 | 758 | 766
963 970 1320 1326 1498 1504 1683 1689 1804 1954 2072 2078 2261 2267 | 780 | 788 991

1339 1346 1524 1664 1670 1786 1937 2060 2066 [792] 793 818 819 843 844 923 924 947
948 971 972 [870] 873 876 879 882 888 891 895 901 910 1427 1430 1887 1895 [885]
2163 [898] 1908 [984] 1517 2250 2256 [1012] 1017 1060 [1014] 1036 1041 1062
[1038] [1040] [1065] [1066] [1084] 1089 1128 [1086] 1105 1109 1130
[1107] [1108] [1132] [1133] [1151] 1155 1196 1366 1410 1414 1477 1521 1525
1692 1693 1703 1704 1714 1715 1717 1718 1722 1723 1725 1726 [1153] 1174 1178 1198
1299 1303 1343 1347 1368 1388 1392 1412 1479 1499 1503 1523 1698 1699 1706 1707 1709
1710 1711 1720 1721 1728 1729 1730 1731 1732 1733 1736 1737 <u>1739 1740</u> 1744 1745 1747
1748 2046 2050 2075 2079 2082 2084 2085 2094 2095 2097 2098 [1176] 1301 1323 1345
1390 1501 1742 1743 1749 1750 1751 1752 1753 1754 1755 1756 <u>2061 2063</u> 2065 2086 2087
2089 2091 2093 2099 2100 2101 2103 2105 2111 2112 2113 2115 [1200] 1370 1695 1701
[1202] [1205] 1208 1228 1254 [1219] 1223 1264 1985 1989 2022 [1221] 1244 1266
2004 2006 2008 [1225] 1248 [1227] 1270 [1231] 1251 [1233] 1273 [1242] 1246
1987 2024 [1249] 1252 1256 1257 1259 1260 1305 1311 1547 1548 1569 1570 [1268]
2026 [1269] 2027 [1271] 1307 [1277] 1604 [1281] 2035 [1321] 1325 1712 1713
1734 1735 2048 2077 2106 2107 2109 2110 [1433] 1455 1459 1694 1705 1716 1719 1724
1727 [1437] 1697 1702 [1481] 1696 [1700] [1708] 2081 [1791] 1797 1814
1820 1942 1948 1964 1970 2131 2134 2142 2169 2318 2321 2328 2350 [1793] 1812 1950
1968 [1799] 1944 2140 2326 [1818] 2133 [1829] 2270 [1837] [1962] 2320
[2088] [2090] 2092 [2102] [2104] 2114 [2116] [2124] 2155 2302 [2135]
2143 2170 2329 [2144] 2277 2330 [2159] 2166 2293 2312 2340 2343 2347 [2171]
2322 2351 2352 [2346] [2353] 2650 [2354] 2355 2356 2359 2360 2361 2362 2371
2372 2373 2374 2377 2378 <u>2379 2380</u> 2643 2644 2645 2646 2649 2651 2652 2659 2660 2661
2662 2665 2666 2667 2668 [2357] 2358 2363 2364 2365 2366 2367 2368 2375 2376 2381
2382 2383 2384 2385 2386 2389 2390 2391 2392 2395 2396 2397 2398 2647 2648 2653 2655
2656 2657 2658 2663 2664 2669 2670 2671 2672 2673 2674 2675 2676 2677 2678 2681 2683
<u>2684</u> 2869 2870 2871 2872 2875 2876 2877 2878 2883 2884 2885 2886 2889 2890 2891 2892
[2369] 2370 2387 2388 2541 2620 2626 2638 2897 2898 2899 2900 2903 2904 2905 2906
[2393] 2394 2399 2400 2401 2402 2403 2404 2405 2406 2434 2440 2454 2535 2614 2632
2679 2680 2685 2687 2688 2689 2690 2710 2716 2722 2728 2782 2860 2873 <u>2874 2879</u> 2880
2881 2882 2887 2888 2893 2894 2895 2896 2901 2902 2907 2908 2909 2910 [2407] 2408
2409 2410 2411 2412 2413 2414 2415 2416 2417 2418 2419 2420 2421 2422 2423 2424 2425
2426 2427 2429 2431 2432 2433 2435 2437 2438 2439 2441 2443 2444 2445 2447 2449 2450
2451 2453 2455 2456 2457 2459 2504 2505 2506 2507 2508 2509 2510 2511 2512 2513 2514
2515 2516 2517 2518 2519 2520 2521 2522 2523 2524 2525 2526 2527 2528 2530 2531 2532
2534 2536 2537 2538 2540 2542 2543 2544 2546 2547 2548 2549 2550 2551 2553 2554 2555
2556 2557 2559 2560 2561 2562 2563 2564 2565 2566 2567 2568 2569 2570 2571 2572 2573
2574 2575 2576 2577 2578 2579 2580 2581 2582 2583 2584 2585 2586 2587 2588 2589 2590
2591 2592 2593 2594 2595 2596 2597 2598 2599 2600 2601 2602 2603 2604 2605 2606 2607
2608 2609 2610 2611 2612 2613 2615 2617 2618 2619 2621 2623 2624 2625 2627 2628 2629
2631 2633 2634 2635 2637 2639 2640 2641 2691 2692 2693 2694 2695 2696 2697 2699 2700
2701 2703 2704 2705 2706 2707 2708 2709 2711 2713 2714 2715 2717 2719 2720 2721 2723
2724 2725 2727 2729 2731 2733 2734 2735 2736 2761 2762 2763 2764 2765 2766 2767 2768
2769 2770 2771 2772 2773 2774 2775 2776 2777 2778 2779 2781 2783 2784 2785 2787 2788
2789 2790 2792 2793 2794 2795 2796 2797 2798 2800 2801 2802 2804 2805 2806 2807 2808

2809 2810 2811 2812 2813 2814 2815 2816 2817 2818 2819 2820 2821 2822 2823 2824 2825
2826 2827 2828 2829 2831 2832 2833 2834 2835 2836 2837 2838 2839 2840 2841 2842 2843
2844 2846 2847 2848 2849 2850 2851 2852 2853 2855 2856 2857 2859 2861 2863 2865 2866
2867 2911 2912 2913 2914 2915 2916 2917 2918 2919 2920 2921 2922 2923 2924 2925 2926
2927 2928 2930 2931 2932 2933 2935 2936 2937 2939 2940 2941 2942 2944 2945 2953 2954
2955 2956 2957 2958 2959 2960 2961 2962 2963 2964 2965 2966 2967 2968 2970 2971 2972
2973 2975 2976 2978 2979 2980 2981 2982 2983 2984 2985 2986 2987 2988 2989 2990 2991
2992 2993 2994 2995 2996 2997 2998 2999 3000 3001 3002 3003 3004 3005 3006 3007 3008
3009 3010 3011 3012 3013 3014 3015 3016 3017 3018 3019 3020 3021 3022 3023 3024 3025
3026 3027 3028 3029 3031 3032 3033 3034 3036 3037 3069 3070 3071 3072 3073 3074 3075
3076 3077 3078 3079 3080 3081 3082 3083 3085 3086 3087 3088 3090 3091 3092 3093 3094
3095 3096 3097 3098 3099 3100 3101 3102 3103 3104 3105 3107 3108 3109 3110 3111 3112
3113 3114 3115 3116 3117 3118 3119 3120 3121 3122 3123 3124 3125 3126 3127 3128 3129
3130 3131 3132 3133 3134 3135 3136 3137 3138 3139 3140 3141 3142 3143 3144 3145 3146
3147 3148 3149 3150 3151 3152 3153 3155 3156 3157 3159 3160 3176 3177 3178 3179 3180
3181 3182 3184 3185 3186 3187 3188 3190 3191 3192 3193 3194 3195 3196 3197 3198 3199
3200 3201 3202 3203 3204 3205 3206 3207 3208 3210 3211 3213 3214 3215 3216 3217 3218
3219 3220 3221 3222 3223 3224 3225 3226 3227 3228 3229 3230 3231 3232 3233 3234 3235
3236 3237 3238 3239 3240 3241 3247 3248 3249 3250 3251 3252 3253 3254 3255 3257 3258
3259 3260 3261 3262 3263 3264 3265 3266 3267 3268 3269 3271 3272 3273 3274 3275 3276
3277 3278 3279 3280 3281 3282 3283 3284 3285 3286 3287 3288 3289 3290 3291 3292 3293
3294 3295 3296 3297 3298 3299 3300 3301 3303 3304 3305 3306 3307 3308 3309 3310 3311
3312 3313 3314 3315 3316 3317 3318 3319 3320 3322 3323 3325 3326 3327 3329 | 2428 |
2448 | 2430 | 2446 | 2436 | 2442 2452 2458 2491 2492 2497 2498 2499 2500 2501 2502
2503 2533 2616 2630 2712 2718 2726 2732 2737 2750 2751 2756 2757 2758 2759 2760 2780
2858 2929 2938 2943 2946 2950 2952 2969 3030 3061 3066 3067 3068 3084 3089 3154 3174
3175 3246 | 2460 | 2468 2469 3039 3046 3047 | 2461 | 3040 | 2462 | 3051 | 2463 |
3052 | 2464 | 2465 2470 2471 2473 2474 2475 2481 2482 2484 2485 2741 2742 2744 2745
3057 3058 3059 3060 3062 3063 3064 3065 3166 3167 3168 3169 3170 3171 3172 | 2466 |
3161 | 2467 | 3162 | 2472 | 3173 | 2476 | 2477 3189 3243 | 2478 | 2738 3041 3044
| 2479 | 3042 | 2480 | 2740 3053 3055 | 2483 | 2743 3163 3164 | 2486 | 2622 3244 3245
| 2487 | 2488 2489 2490 2495 2496 2746 2747 2748 2749 2754 2755 2786 2791 2868 2974
2977 3038 3043 3048 3049 3050 3054 3056 | 2493 | 2494 2752 2753 2864 3035 3158 3165
| 2529 | 2854 3106 3256 3328 3330 | 2539 | 2642 2947 2948 | 2545 | 2949 | 2552 | 3212
| 2558 | 3183 | 2636 | 2951 | 2654 | 2682 | 2686 | 2698 | 2830 2862 | 2702 | 2730
| 2739 | 3045 | 2799 | 2845 3209 3321 | 2803 | 3270 3324 | 2934 | 3242 | 3302

6 Covers and Subcovers in the Poset of 411 Clone Equivalence Representatives

In the following the subcovers of a boxed element are to its left, the covers are to its right.

275 | 1 | 4 14 25 33 207 216 239 1065 1700 2088 2124

14 $\boxed{2}$ 5 8 11 26 39 1200
27 $\boxed{3}$ 6 9 15 25 100 1132
1 $\boxed{4}$ 8 11 19 30 39 222 2654
2 $\boxed{5}$ 10 31 34 40 59
3 $\boxed{6}$ 8 12 18 22 32 35 41 83 94 2654
2 4 6 $\boxed{8}$ 16 34 59 65
3 $\boxed{9}$ 18 19 35 60 79 123
5 13 $\boxed{10}$ 16 36 37 61 160 376 379
2 4 13 $\boxed{11}$ 16 36 37 61 160 377 380
6 15 $\boxed{12}$ 16 38 63 381
14 $\boxed{13}$ 10 11 26 39 1200
1 $\boxed{14}$ 2 13 405 406
3 $\boxed{15}$ 12 18 26 41 407
8 10 11 12 18 22 26 $\boxed{16}$ 43 46
6 9 15 $\boxed{18}$ 16 44 66
4 9 21 25 $\boxed{19}$ 22 45 70 160 1225 1268
27 $\boxed{21}$ 19 24 47 72 218 235 260 1281
6 19 24 $\boxed{22}$ 16 48 73 1202
21 $\boxed{24}$ 22 50 63 75 704
1 3 $\boxed{25}$ 19 26 51 2353
2 13 15 25 $\boxed{26}$ 16 52 417
275 $\boxed{27}$ 3 21 53 78 104 168 241 320 678 1065 1700 2466
4 $\boxed{30}$ 43 59 70 180 219 223 1012 2102 2686
5 $\boxed{31}$ 43 180 199 223 1151 2116 2803
6 $\boxed{32}$ 44 139 1084 2686
1 $\boxed{33}$ 34 59 65 73 148 219 253 1014 1202 2104
5 8 33 35 $\boxed{34}$ 43 61 149
6 9 $\boxed{35}$ 34 44 59 66 73 117 1086
10 11 $\boxed{36}$ 67 199 223 382 600 1151 2116 2803
10 11 $\boxed{37}$ 67 199 223 382 600 1151 2116 2803
12 41 $\boxed{38}$ 67 209 384 1433 2803
2 4 13 $\boxed{39}$ 42 199 223 606 1151 2116 2702
5 $\boxed{40}$ 43 199 223 606 1151 2116 2702
6 15 $\boxed{41}$ 38 44 2702
39 43 $\boxed{42}$ 67 213 382
16 30 31 34 40 44 $\boxed{43}$ 42
18 32 35 41 $\boxed{44}$ 43 69 215 410
19 47 51 $\boxed{45}$ 46 48 61 73 180 1202
16 45 52 $\boxed{46}$ 49 157 385
21 53 $\boxed{47}$ 45 50 63 75 182 704
22 45 50 $\boxed{48}$ 49 1205 1219
46 48 $\boxed{49}$ 67 213 382

24 47 $\boxed{50}$ 48 221 384 707 1038

25 53 $\boxed{51}$ 45 52 186 1225 1242 2654

26 51 $\boxed{52}$ 46 391 2702

27 $\boxed{53}$ 47 51 65 66 198 224 283 353 1708 2472

5 8 30 33 35 $\boxed{59}$ 67 171

9 $\boxed{60}$ 69 70 111 120 180 562 1107

10 11 34 45 65 $\boxed{61}$ 67 151 163 382 385

12 24 47 66 $\boxed{63}$ 67 157 244 384 387

8 33 53 $\boxed{65}$ 61 155

18 35 53 $\boxed{66}$ 63 69 410

36 37 38 42 49 59 61 63 69 73 $\boxed{67}$ 183 434

44 60 66 78 $\boxed{69}$ 67 250 436

19 30 60 72 $\boxed{70}$ 73 199 1231

21 78 $\boxed{72}$ 70 75 201 252 257 1012 1233

22 33 35 45 70 75 $\boxed{73}$ 67 163 1205 1221

24 47 72 $\boxed{75}$ 73 165 255 439 695 698 1014

27 $\boxed{78}$ 69 72 185 188 198 272

9 104 $\boxed{79}$ 81

275 $\boxed{80}$ 82 100 148 239 347

79 $\boxed{81}$ 83 89

80 $\boxed{82}$ 83 90 94 123 149 186 222

6 81 82 107 $\boxed{83}$ 91 130

111 $\boxed{85}$ 91 130 199 454 455

90 $\boxed{87}$ 93 209 223 456

89 $\boxed{88}$ 111

81 112 $\boxed{89}$ 88

82 113 $\boxed{90}$ 87 101

83 85 93 101 115 120 $\boxed{91}$ 136 213 460

87 99 116 119 122 $\boxed{93}$ 91 132 215

6 82 96 100 $\boxed{94}$ 97 117 139 1225

102 $\boxed{96}$ 94 99 119 141 186 218 235 262

94 99 123 $\boxed{97}$ 120 142 157 1202

96 125 $\boxed{99}$ 93 97 221 278 704 1086

3 80 102 $\boxed{100}$ 94 123

90 111 123 124 $\boxed{101}$ 91 495

275 $\boxed{102}$ 96 100 155 241 354 679 1132 1962

27 $\boxed{104}$ 79 106 112 129 198 207

275 $\boxed{105}$ 106 113 120 138 139 180 219

104 105 107 $\boxed{106}$ 115 134 169

275 $\boxed{107}$ 83 106 116 117 129 170 216 347

60 88 $\boxed{111}$ 85 101 483

104 $\boxed{112}$ 89 121 124 134 484

105 | 113 | 90 124 485

106 116 121 122 124 | 115 | 91 143 179 487

107 125 | 116 | 93 115 135 203

35 94 107 119 123 | 117 | 120 130 142 151 215 1202

96 125 | 119 | 93 117 180 182 221 265 704 1086

60 97 105 117 122 | 120 | 91 250 1205

112 | 121 | 115 137 493

275 | 122 | 93 115 120 221 494 520 1107

9 82 100 125 | 123 | 97 101 117 1225

112 113 125 | 124 | 101 115 496

275 | 125 | 99 116 119 123 124 134 144 188 194 224 269 284 984

104 107 | 129 | 137 169

83 85 117 134 | 130 | 136 163 171 460

93 135 | 132 | 136 244 349 462

106 112 125 | 134 | 130 143 175 487

116 | 135 | 132 137 143 178 488

91 130 132 137 138 142 143 | 136 | 153 183 512

121 129 135 144 | 137 | 136 176 513

105 147 | 138 | 136 179 250 349 517

32 94 105 141 | 139 | 142 171 209 1231

96 147 | 141 | 139 182 255 259 517 1084 1233

97 117 139 144 | 142 | 136 244 1205

115 134 135 144 | 143 | 136 195 519

125 | 144 | 137 142 143 178 185 204 255 278 520 696 1086

275 | 147 | 138 141 188 273 514

33 80 366 | 148 | 149 155 341

34 82 148 | 149 | 151

61 117 149 155 194 218 222 235 341 | 151 | 157 171

136 176 195 250 282 349 | 153 | 161 534

65 102 148 | 155 | 151

46 63 97 151 | 157 | 161 792

10 11 19 | 160 | 166

153 157 171 183 184 298 | 161 | 538

260 | 162 | 182 199 268 270 1151 2461

61 73 130 165 213 244 | 163 | 183 652

75 204 263 266 270 278 | 165 | 163 282 691

160 | 166 | 186 199 600

27 | 168 | 201 257 270 274 677 2467

106 129 170 | 169 | 175

107 366 | 170 | 169 194 322

59 130 139 151 175 180 182 199 | 171 | 161 532

134 169 188 194 | 175 | 171 184 562

137 185 198 306 $\boxed{176}$ 153 184 563
135 144 203 308 311 $\boxed{178}$ 184 565
115 138 188 198 309 $\boxed{179}$ 195
30 31 45 60 105 119 186 $\boxed{180}$ 171 213 606
47 119 141 162 188 201 224 258 261 $\boxed{182}$ 171 263 682
67 136 163 250 253 282 $\boxed{183}$ 161 658
175 176 178 195 305 $\boxed{184}$ 161 571
78 144 283 $\boxed{185}$ 176 282 698
51 82 96 166 224 $\boxed{186}$ 180 632
78 125 147 $\boxed{188}$ 175 179 182 204 252 274
125 170 $\boxed{194}$ 151 175 203
143 179 204 317 $\boxed{195}$ 153 184 566
53 78 104 320 $\boxed{198}$ 176 179 439
31 36 37 39 40 70 85 162 166 201 207 258 261 $\boxed{199}$ 171 213 609
72 168 260 $\boxed{201}$ 182 199 268 1151 1277 2739
116 194 316 321 $\boxed{203}$ 178
144 188 280 $\boxed{204}$ 165 195 695
1 104 $\boxed{207}$ 199 2353
38 87 139 259 $\boxed{209}$ 215 602
42 49 91 180 199 215 219 223 263 266 270 $\boxed{213}$ 163 615
44 93 117 209 221 265 $\boxed{215}$ 213 244 608
1 107 $\boxed{216}$ 219 253 322 1202 2654
21 96 224 267 $\boxed{218}$ 151 221 266 632 682 704
30 33 105 216 222 269 $\boxed{219}$ 213 1205
50 99 119 122 218 235 269 $\boxed{221}$ 215 255 712
4 82 239 271 $\boxed{222}$ 151 219 223 1225
30 31 36 37 39 40 87 222 $\boxed{223}$ 213 624
53 125 241 271 $\boxed{224}$ 182 186 218 268 274 280 309
21 96 241 271 $\boxed{235}$ 151 221 270 606 682 1227
1 80 $\boxed{239}$ 222 253 341
27 102 $\boxed{241}$ 224 235 690 2636
63 132 142 215 278 $\boxed{244}$ 163 250 654
69 120 138 244 255 284 $\boxed{250}$ 153 183 566
72 188 $\boxed{252}$ 255 268 562 695
33 216 239 267 287 $\boxed{253}$ 183 298
75 141 144 221 252 287 $\boxed{255}$ 250 282 1221
72 168 260 272 $\boxed{257}$ 258 2478
257 $\boxed{258}$ 182 199 268 1151 1277 2479
141 262 273 $\boxed{259}$ 209 265 266 1277 1433 1799
21 $\boxed{260}$ 162 201 257 261
260 $\boxed{261}$ 182 199 268 270 1151 2460
96 $\boxed{262}$ 259 270 1793

182 265 274 $\boxed{263}$ 165 213 687

119 259 271 $\boxed{265}$ 215 263 278 684

218 259 268 269 274 $\boxed{266}$ 165 213 687

275 $\boxed{267}$ 218 253 269 341 347 705 2104

162 201 224 252 258 261 $\boxed{268}$ 266 682

125 267 271 $\boxed{269}$ 219 221 266 278 684 696 2090

162 168 235 261 262 $\boxed{270}$ 165 213 687 1271 2539

275 $\boxed{271}$ 222 224 235 265 269 281 316 984 2102

78 $\boxed{272}$ 257 274 680 2483

147 $\boxed{273}$ 259 274 681 2171

168 188 224 272 273 $\boxed{274}$ 263 266 2493

$\boxed{275}$ 1 27 80 102 105 107 122 125 147 267 271 287 356 366 1108 2346 3242

99 144 265 269 281 $\boxed{278}$ 165 244 349 693

224 281 $\boxed{280}$ 204 286 704 1225 1268

271 $\boxed{281}$ 278 280 283 308 705 1227 1281 1708 2472

165 185 255 284 286 $\boxed{282}$ 153 183 697

53 281 287 $\boxed{283}$ 185 306 707 1242

125 $\boxed{284}$ 250 282 354 496 520 1133

280 287 $\boxed{286}$ 282 317 712 1231

275 $\boxed{287}$ 253 255 283 286 359 696 1233 2088

253 311 341 347 $\boxed{298}$ 161 2159

306 308 311 317 $\boxed{305}$ 184 755 2369 2428

283 353 359 $\boxed{306}$ 176 305 756 1321

281 316 $\boxed{308}$ 178 305 758 1829 2430 2476 2552

224 353 354 356 $\boxed{309}$ 179 317 620

321 359 $\boxed{311}$ 178 298 305 758 2144 2486

271 366 $\boxed{316}$ 203 308 780 2539

286 309 359 $\boxed{317}$ 195 305 654 1249 1269

27 $\boxed{320}$ 198 353 417 1066 1837 2558

356 366 $\boxed{321}$ 203 311 780 2545

170 216 $\boxed{322}$ 341 2799

148 239 267 322 $\boxed{341}$ 151 298

80 107 267 359 $\boxed{347}$ 298 349 885

132 138 278 347 354 $\boxed{349}$ 153 870

53 320 $\boxed{353}$ 306 309 387 391 1829 2552

102 284 $\boxed{354}$ 309 349 465 469

275 $\boxed{356}$ 309 321 359 624 690 2116

287 356 $\boxed{359}$ 306 311 317 347 693 1277

275 $\boxed{366}$ 148 170 316 321 1962 2124 2171

10 $\boxed{376}$ 382 600 624 1151 2116 2803

11 $\boxed{377}$ 382 600 624 1151 2116 2803

381 $\boxed{378}$ 384 602 1433 2803

10 $\boxed{379}$ 382 600 624 1151 2116 2799

11 $\boxed{380}$ 382 600 624 1151 2116 2799

12 407 $\boxed{381}$ 378 2799

36 37 42 49 61 376 377 379 380 384 405 406 $\boxed{382}$ 388 434 615

38 50 63 378 410 $\boxed{384}$ 382 390 608

46 61 387 391 $\boxed{385}$ 388 792

63 353 $\boxed{387}$ 385 390

382 385 390 $\boxed{388}$ 437 618 1153

384 387 $\boxed{390}$ 388 439 620 756

52 353 417 $\boxed{391}$ 385 1249 1321 2799

14 $\boxed{405}$ 382 606 624 638 1066 1481 2116

14 $\boxed{406}$ 382 606 624 639 1200 2116 2698

15 $\boxed{407}$ 381 410 1437 2698

44 66 407 $\boxed{410}$ 384 436

26 320 $\boxed{417}$ 391 2698

67 382 436 $\boxed{434}$ 437

69 410 $\boxed{436}$ 434 439

388 434 439 1038 1066 $\boxed{437}$ 658 1176

75 198 390 436 1040 $\boxed{439}$ 437 563 566

85 $\boxed{454}$ 457

85 $\boxed{455}$ 458

87 $\boxed{456}$ 459

454 $\boxed{457}$ 460 469 532 1269 2354

455 $\boxed{458}$ 460 469 532 1269 2354

456 $\boxed{459}$ 462 469 602 624

91 130 457 458 462 483 487 $\boxed{460}$ 463 512 615

132 459 485 488 $\boxed{462}$ 460 465 608

460 465 469 493 $\boxed{463}$ 515 618

354 462 494 $\boxed{465}$ 463 517 620

354 457 458 459 495 $\boxed{469}$ 463 1249

111 $\boxed{483}$ 460 495 532 1269 2354

112 $\boxed{484}$ 487 493 496,513 562 1133 2354

113 $\boxed{485}$ 462 487 494 496 532 602 624

115 134 484 485 488 $\boxed{487}$ 460 519

135 $\boxed{488}$ 462 487 513 520

121 484 $\boxed{493}$ 463 1040

122 485 $\boxed{494}$ 465 519

101 483 496 $\boxed{495}$ 469

124 284 484 485 $\boxed{496}$ 495 519

136 460 513 514 $\boxed{512}$ 515

137 484 488 $\boxed{513}$ 512 522

147 $\boxed{514}$ 512 517 885

463 512 517 519 522 1107 $\boxed{515}$ 534 658
138 141 465 514 520 1108 $\boxed{517}$ 515 566 870
143 487 494 496 520 1040 1108 $\boxed{519}$ 515 566
122 144 284 488 $\boxed{520}$ 517 519 522 710
513 520 1040 1133 $\boxed{522}$ 515 563
171 457 458 483 485 562 606 639 682 1084 1151 $\boxed{532}$ 538
153 515 563 566 697 870 $\boxed{534}$ 538
161 532 534 571 658 792 1176 1829 2159 $\boxed{538}$ 2407
60 175 252 484 $\boxed{562}$ 532 571
176 439 522 698 710 756 1107 $\boxed{563}$ 534 571 1176
178 710 758 $\boxed{565}$ 571 2436
195 250 439 517 519 654 695 710 $\boxed{566}$ 534 571 658
184 562 563 565 566 755 $\boxed{571}$ 538 2393
36 37 166 376 377 379 380 $\boxed{600}$ 609 632
209 378 459 485 679 681 $\boxed{602}$ 608 1249 1791
609 $\boxed{603}$ 604 612 629
603 $\boxed{604}$ 613 630
39 40 180 235 405 406 632 $\boxed{606}$ 532 615
215 384 462 602 684 $\boxed{608}$ 615 620
199 600 $\boxed{609}$ 603
603 $\boxed{612}$ 613 638
604 612 $\boxed{613}$ 639
213 382 460 606 608 639 687 $\boxed{615}$ 618
388 463 615 620 624 690 $\boxed{618}$ 652
309 390 465 608 $\boxed{620}$ 618 654
223 356 376 377 379 380 405 406 459 485 $\boxed{624}$ 618 1249
603 $\boxed{629}$ 630 638
604 629 $\boxed{630}$ 639
186 218 600 $\boxed{632}$ 606
405 612 629 $\boxed{638}$ 639
406 613 630 638 $\boxed{639}$ 532 615 1249 1269 2354
163 618 654 691 1205 1249 $\boxed{652}$ 658 2357
244 317 620 693 712 $\boxed{654}$ 566 652 755
183 437 515 566 652 697 1221 1269 $\boxed{658}$ 538 2393
168 678 $\boxed{677}$ 680 690 1200 2462
27 $\boxed{678}$ 677 1481 2463
102 $\boxed{679}$ 602 684 690 1437 1818
272 677 $\boxed{680}$ 682 1151 1269 1277 2480
273 $\boxed{681}$ 602 684 885 1277 1433 2135
182 218 235 268 680 $\boxed{682}$ 532 687
265 269 679 681 $\boxed{684}$ 608 687 693 780
263 266 270 682 684 $\boxed{687}$ 615 691 2487

241 356 677 679 $\boxed{690}$ 618 1277 2545
165 687 693 695 1277 $\boxed{691}$ 652 697 1153 2464
278 359 684 696 $\boxed{693}$ 654 691 710 756 758
75 204 252 696 704 1233 $\boxed{695}$ 566 691 1205 1219
144 269 287 705 $\boxed{696}$ 693 695 707 712
282 691 698 710 712 $\boxed{697}$ 534 658 1176 2436
75 185 707 $\boxed{698}$ 563 697 1221
24 47 99 119 218 280 705 1227 $\boxed{704}$ 695 712 1202
267 281 $\boxed{705}$ 696 704
50 283 696 1281 $\boxed{707}$ 698 756 1219
520 693 $\boxed{710}$ 563 565 566 697 870
221 286 696 704 $\boxed{712}$ 654 697 1205 1221
305 654 756 758 1791 $\boxed{755}$ 571 2357
306 390 693 707 1086 1433 $\boxed{756}$ 563 755 1153
308 311 693 780 $\boxed{758}$ 565 755 2464
316 321 684 984 1799 1818 2135 $\boxed{780}$ 758 2487
157 385 1202 $\boxed{792}$ 538
349 517 710 885 898 $\boxed{870}$ 534
347 514 681 2346 $\boxed{885}$ 870 2159
1108 2346 $\boxed{898}$ 870 2529
125 271 1962 2171 $\boxed{984}$ 780 2493
30 72 1065 $\boxed{1012}$ 1014 1151 1242
33 75 1012 $\boxed{1014}$ 1038 1219
50 1014 $\boxed{1038}$ 437 1221
493 $\boxed{1040}$ 439 519 522
1 27 $\boxed{1065}$ 1012 1066 1268 1481 2353
320 405 1065 $\boxed{1066}$ 437 1269 2698
32 141 1132 $\boxed{1084}$ 532 1086 1242 1433
35 99 119 144 1084 $\boxed{1086}$ 756 1107 1219
60 122 1086 $\boxed{1107}$ 515 563 1221
275 $\boxed{1108}$ 517 519 898
3 102 $\boxed{1132}$ 1084 1268 1437
284 484 $\boxed{1133}$ 522 1269
31 36 37 39 40 162 201 258 261 376 377 379 380 680 1012 1200 $\boxed{1151}$ 532 1321 2354
388 691 756 1219 1321 1708 $\boxed{1153}$ 1176 2357
437 563 697 1153 1221 2090 2116 $\boxed{1176}$ 538 2393
2 13 406 677 1481 $\boxed{1200}$ 1151
22 33 45 97 117 216 704 1225 $\boxed{1202}$ 792 1205
48 73 120 142 219 695 712 1202 1231 $\boxed{1205}$ 652
48 695 707 1014 1086 1242 1268 $\boxed{1219}$ 1153 1221
73 255 698 712 1038 1107 1219 $\boxed{1221}$ 658 1176
19 51 94 123 222 280 1227 $\boxed{1225}$ 1202 1231

235 281 | 1227 | 704 1225 1233 1271 2654
70 139 286 1225 1233 | 1231 | 1205 1249
72 141 287 1227 | 1233 | 695 1231 1242 1277 2686
51 283 1012 1084 1233 | 1242 | 1219 1321
317 391 469 602 624 639 1231 1277 | 1249 | 652 2369
19 280 1065 1132 1281 | 1268 | 1219 1269
317 457 458 483 639 680 1066 1133 1268 | 1269 | 658 2428
270 1227 | 1271 | 1277 2476 2702
201 258 259 359 680 681 690 1233 1271 | 1277 | 691 1249 1321 2486
21 281 | 1281 | 707 1268 2430 2654
306 391 1151 1242 1277 1433 | 1321 | 1153 2369
38 259 378 681 1084 1437 | 1433 | 756 1321 1791
407 679 1132 | 1437 | 1433
405 678 1065 1700 | 1481 | 1200
1 27 | 1700 | 1481 1708 1837 2353
53 281 1700 | 1708 | 1153 1829 2654
602 1433 1799 1818 2135 | 1791 | 755 2369
262 1962 | 1793 | 1799 2539
259 1793 2171 | 1799 | 780 1791 2486
679 1962 | 1818 | 780 1791 2545
308 353 1708 1837 | 1829 | 538 2799
320 1700 | 1837 | 1829 2698
102 366 | 1962 | 984 1793 1818 2636
1 287 | 2088 | 2104 2144 2686
269 2102 2104 | 2090 | 1176
30 271 | 2102 | 2090 2116
33 267 2088 | 2104 | 2090 2159
31 36 37 39 40 356 376 377 379 380 405 406 2102 | 2116 | 1176 2428
1 366 2346 | 2124 | 2159
681 2171 | 2135 | 780 1791 2159 2486
311 2088 | 2144 | 2159 2803
298 885 2104 2124 2135 2144 | 2159 | 538
273 366 | 2171 | 984 1799 2135
275 | 2346 | 885 898 2124 2934
25 207 1065 1700 2466 | 2353 | 2354 2654 2698
457 458 483 484 639 1151 2353 2461 2479 2480 2739 | 2354 | 2369 2428
652 755 1153 2369 2464 2654 | 2357 | 2393
305 1249 1321 1791 2354 2486 | 2369 | 2357
571 658 1176 2357 2428 2436 2686 | 2393 | 2407
538 2393 2529 2803 | 2407 |
305 1269 2116 2354 2430 | 2428 | 2393
308 1281 2461 | 2430 | 2428 2436 2799

565 697 2430 2464 $\boxed{\textbf{2436}}$ 2393
261 2466 $\boxed{\textbf{2460}}$ 2461 2478
162 2460 $\boxed{\textbf{2461}}$ 2354 2430 2539
677 2463 2467 $\boxed{\textbf{2462}}$ 2480 2545
678 2466 $\boxed{\textbf{2463}}$ 2462
691 758 2472 2486 2487 $\boxed{\textbf{2464}}$ 2357 2436
27 $\boxed{\textbf{2466}}$ 2353 2460 2463 2467 2472 2483 2558
168 2466 $\boxed{\textbf{2467}}$ 2462 2478 2636
53 281 2466 $\boxed{\textbf{2472}}$ 2464 2552 2654
308 1271 2539 $\boxed{\textbf{2476}}$ 2486 2799
257 2460 2467 2483 $\boxed{\textbf{2478}}$ 2479 2739
258 2478 $\boxed{\textbf{2479}}$ 2354 2486 2487
680 2462 2483 $\boxed{\textbf{2480}}$ 2354 2486 2487
272 2466 $\boxed{\textbf{2483}}$ 2478 2480 2493
311 1277 1799 2135 2476 2479 2480 2545 2739 $\boxed{\textbf{2486}}$ 2369 2464 2803
687 780 2479 2480 2493 2539 2545 2739 $\boxed{\textbf{2487}}$ 2464
274 984 2483 2636 $\boxed{\textbf{2493}}$ 2487
898 2934 $\boxed{\textbf{2529}}$ 2407
270 316 1793 2461 2636 $\boxed{\textbf{2539}}$ 2476 2487
321 690 1818 2462 2636 $\boxed{\textbf{2545}}$ 2486 2487
308 353 2472 2558 $\boxed{\textbf{2552}}$ 2799
320 2466 $\boxed{\textbf{2558}}$ 2552 2698
241 1962 2467 $\boxed{\textbf{2636}}$ 2493 2539 2545
4 6 51 216 1227 1281 1708 2353 2472 $\boxed{\textbf{2654}}$ 2357 2686 2702
30 32 1233 2088 2654 $\boxed{\textbf{2686}}$ 2393 2803
406 407 417 1066 1837 2353 2558 $\boxed{\textbf{2698}}$ 2702
39 40 41 52 1271 2654 2698 $\boxed{\textbf{2702}}$ 2799
201 2478 $\boxed{\textbf{2739}}$ 2354 2486 2487
322 379 380 381 391 1829 2430 2476 2552 2702 $\boxed{\textbf{2799}}$ 2803
31 36 37 38 376 377 378 2144 2486 2686 2799 $\boxed{\textbf{2803}}$ 2407
2346 3242 $\boxed{\textbf{2934}}$ 2529
275 $\boxed{\textbf{3242}}$ 2934

7 Measuring the longest chain from the bottom

height = 0: 275

height = 1: 1 27 80 102 105 107 122 125 147 267 271 287 356 366 1108 2346 3242

height = 2: 3 4 14 21 33 53 78 82 96 104 113 116 138 144 168 170 216 239 241 269 273 281

284 316 320 321 359 514 678 679 898 1065 1700 1962 2088 2124 2466 2934

height = 3: 2 6 9 13 15 24 25 30 47 72 90 99 100 106 112 119 129 135 141 148 188 194 198
207 222 224 235 253 260 262 272 283 308 311 322 347 353 354 405 406 485 677 681 705 1132
1281 1708 1818 1837 2104 2171 2463 2467 2472 2529 2558

height = 4: 5 8 11 12 18 19 26 32 35 39 41 50 51 60 75 79 87 94 121 123 124 134 162 169
185 201 203 218 219 252 257 259 261 274 280 306 309 341 407 484 488 494 680 690 696 885
984 1012 1066 1227 1481 1793 1829 2102 2135 2144 2353 2462 2483 2552 2636

height = 5: 10 22 31 34 38 40 44 45 52 59 65 66 70 81 93 97 115 117 137 139 175 178 204
221 258 265 270 286 298 377 380 381 417 456 493 496 520 704 707 1014 1084 1133 1200 1225
1233 1268 1437 1799 2090 2460 2480 2493 2545 2654

height = 6: 16 36 37 48 61 63 69 73 83 89 120 132 142 143 149 155 160 176 179 182 209 255
268 278 317 376 378 379 391 410 459 487 513 562 684 695 698 712 1038 1040 1086 1202 1231
1242 1271 2159 2461 2478 2686 2698

height = 7: 43 46 88 151 166 195 215 223 263 266 305 349 384 387 436 462 519 522 602 682
693 780 1107 1151 1205 1219 1277 1433 2116 2430 2479 2539 2702 2739

height = 8: 42 49 111 157 165 184 186 244 385 390 465 600 608 624 687 710 758 1221 1321
1791 2476

height = 9: 67 85 101 180 250 282 382 439 483 517 565 620 632 691 756 792 2486 2487 2799

height = 10: 91 130 199 388 434 454 455 495 563 606 654 697 870 2464 2803

height = 11: 136 171 213 437 457 458 566 609 755 1153 2436

height = 12: 153 163 460 469 571 603 1176

height = 13: 183 463 512 604 612 629

height = 14: 161 515 613 630 638

height = 15: 534 639

height = 16: 532 615 1249 1269 2354

height = 17: 618 2369 2428

height = 18: 652

height = 19: 658 2357

height = 20: 538 2393
height = 21: 2407

8 The Catalog of Isomorphism Representatives of 3-element groupoids

The following pages give the 3,330 representatives of the isomorphism types of three - element groupoids. The number with a # in front is the number of the clone equivalence representative of the table being considered.

#1 000 000 000	#2 000 000 001	#3 000 000 002	#4 000 000 010
#5 000 000 011	#6 000 000 012	7 000 000 020 #4	#8 000 000 021
#9 000 000 022	#10 000 000 100	#11 000 000 101	#12 000 000 102
#13 000 000 110	#14 000 000 111	#15 000 000 112	#16 000 000 120
17 000 000 121 #16	#18 000 000 122	#19 000 000 200	20 000 000 201 #16
#21 000 000 202	#22 000 000 210	23 000 000 211 #16	#24 000 000 212
#25 000 000 220	#26 000 000 221	#27 000 000 222	28 000 001 001 #5
29 000 001 002 #6	#30 000 001 010	#31 000 001 011	#32 000 001 012
#33 000 001 020	#34 000 001 021	#35 000 001 022	#36 000 001 100
#37 000 001 101	#38 000 001 102	#39 000 001 110	#40 000 001 111
#41 000 001 112	#42 000 001 120	#43 000 001 121	#44 000 001 122
#45 000 001 200	#46 000 001 201	#47 000 001 202	#48 000 001 210
#49 000 001 211	#50 000 001 212	#51 000 001 220	#52 000 001 221
#53 000 001 222	54 000 002 001 #8	55 000 002 002 #9	56 000 002 010 #33
57 000 002 011 #34	58 000 002 012 #35	#59 000 002 021	#60 000 002 022
#61 000 002 100	62 000 002 101 #61	#63 000 002 102	64 000 002 110 #61
#65 000 002 111	#66 000 002 112	#67 000 002 120	68 000 002 121 #67
#69 000 002 122	#70 000 002 200	71 000 002 201 #67	#72 000 002 202
#73 000 002 210	74 000 002 211 #67	#75 000 002 212	76 000 002 220 #70
77 000 002 221 #67	#78 000 002 222	#79 000 010 001	#80 000 010 002
#81 000 010 011	#82 000 010 012	#83 000 010 021	84 000 010 022 #82
#85 000 010 100	86 000 010 101 #85	#87 000 010 102	#88 000 010 110
#89 000 010 111	#90 000 010 112	#91 000 010 120	92 000 010 121 #91
#93 000 010 122	#94 000 010 200	95 000 010 201 #91	#96 000 010 202
#97 000 010 210	98 000 010 211 #91	#99 000 010 212	#100 000 010 220
#101 000 010 221	#102 000 010 222	103 000 011 001 #81	#104 000 011 011
#105 000 011 012	#106 000 011 021	#107 000 011 022	108 000 011 100 #85
109 000 011 101 #85	110 000 011 102 #87	#111 000 011 110	#112 000 011 111
#113 000 011 112	114 000 011 120 #91	#115 000 011 121	#116 000 011 122
#117 000 011 200	118 000 011 201 #91	#119 000 011 202	#120 000 011 210
#121 000 011 211	#122 000 011 212	#123 000 011 220	#124 000 011 221
#125 000 011 222	126 000 012 001 #83	127 000 012 011 #106	128 000 012 012 #107
#129 000 012 021	#130 000 012 100	131 000 012 101 #130	#132 000 012 102
133 000 012 110 #130	#134 000 012 111	#135 000 012 112	#136 000 012 120
#137 000 012 121	#138 000 012 122	#139 000 012 200	140 000 012 201 #136
#141 000 012 202	#142 000 012 210	#143 000 012 211	#144 000 012 212
145 000 012 220 #139	146 000 012 221 #137	#147 000 012 222	#148 000 020 001
#149 000 020 011	150 000 020 021 #149	#151 000 020 100	152 000 020 101 #151
#153 000 020 102	154 000 020 110 #151	#155 000 020 111	156 000 020 112 #153
#157 000 020 120	158 000 020 121 #157	159 000 020 122 #153	#160 000 020 200
#161 000 020 201	#162 000 020 202	#163 000 020 210	164 000 020 211 #161
#165 000 020 212	#166 000 020 220	167 000 020 221 #161	#168 000 020 222
#169 000 021 011	#170 000 021 021	#171 000 021 100	172 000 021 101 #171
173 000 021 102 #153	174 000 021 110 #171	#175 000 021 111	#176 000 021 112
177 000 021 120 #161	#178 000 021 121	#179 000 021 122	#180 000 021 200
181 000 021 201 #161	#182 000 021 202	#183 000 021 210	#184 000 021 211
#185 000 021 212	#186 000 021 220	187 000 021 221 #178	#188 000 021 222
189 000 022 011 #170	190 000 022 100 #151	191 000 022 101 #151	192 000 022 102 #153
193 000 022 110 #151	#194 000 022 111	#195 000 022 112	196 000 022 120 #161
197 000 022 121 #184	#198 000 022 122	#199 000 022 200	200 000 022 201 #161
#201 000 022 202	202 000 022 210 #163	#203 000 022 211	#204 000 022 212
205 000 022 220 #199	206 000 022 221 #184	#207 000 100 100	208 000 100 101 #199
#209 000 100 102	210 000 100 110 #199	211 000 100 111 #199	212 000 100 112 #209
#213 000 100 120	214 000 100 121 #213	#215 000 100 122	#216 000 100 200
217 000 100 201 #213	#218 000 100 202	#219 000 100 210	220 000 100 211 #213
#221 000 100 212	#222 000 100 220	#223 000 100 221	#224 000 100 222

225 000 101 100 #199 | 226 000 101 101 #199 | 227 000 101 102 #209 | 228 000 101 110 #199
229 000 101 111 #199 | 230 000 101 112 #209 | 231 000 101 120 #213 | 232 000 101 121 #213
233 000 101 122 #215 | 234 000 101 201 #213 | #235 000 101 202 | 236 000 101 210 #219
237 000 101 211 #213 | 238 000 101 212 #221 | #239 000 101 220 | 240 000 101 221 #223
#241 000 101 222 | 242 000 102 100 #163 | 243 000 102 101 #163 | #244 000 102 102
245 000 102 110 #163 | 246 000 102 111 #163 | 247 000 102 112 #244 | 248 000 102 120 #183
249 000 102 121 #183 | #250 000 102 122 | 251 000 102 201 #183 | #252 000 102 202
#253 000 102 210 | 254 000 102 211 #183 | #255 000 102 212 | 256 000 102 221 #183
#257 000 110 100 | #258 000 110 101 | #259 000 110 102 | #260 000 110 110
#261 000 110 111 | #262 000 110 112 | #263 000 110 120 | 264 000 110 121 #263
#265 000 110 122 | #266 000 110 201 | #267 000 110 202 | #268 000 110 211
#269 000 110 212 | #270 000 110 221 | #271 000 110 222 | #272 000 111 100
#273 000 111 102 | #274 000 111 120 | #275 000 111 222 | 276 000 112 100 #165
277 000 112 101 #165 | #278 000 112 102 | 279 000 112 110 #165 | #280 000 112 111
#281 000 112 112 | #282 000 112 120 | #283 000 112 121 | #284 000 112 122
285 000 112 201 #282 | #286 000 112 211 | #287 000 112 212 | 288 000 112 221 #283
289 000 120 100 #161 | 290 000 120 101 #161 | 291 000 120 102 #153 | 292 000 120 110 #161
293 000 120 111 #161 | 294 000 120 112 #153 | 295 000 120 120 #161 | 296 000 120 121 #161
297 000 120 122 #153 | #298 000 120 201 | 299 000 120 211 #161 | 300 000 120 221 #161
301 000 121 100 #161 | 302 000 121 101 #161 | 303 000 121 102 #153 | 304 000 121 110 #161
#305 000 121 111 | #306 000 121 112 | 307 000 121 120 #161 | #308 000 121 121
#309 000 121 122 | 310 000 121 211 #305 | #311 000 121 221 | 312 000 122 100 #161
313 000 122 101 #161 | 314 000 122 102 #153 | 315 000 122 110 #161 | #316 000 122 111
#317 000 122 112 | 318 000 122 120 #161 | 319 000 122 121 #305 | #320 000 122 122
#321 000 122 211 | #322 000 200 100 | 323 000 200 101 #151 | 324 000 200 102 #153
325 000 200 110 #151 | 326 000 200 111 #151 | 327 000 200 112 #153 | 328 000 200 120 #161
329 000 200 121 #161 | 330 000 200 122 #153 | 331 000 201 101 #161 | 332 000 201 102 #153
333 000 201 110 #161 | 334 000 201 111 #161 | 335 000 201 112 #153 | 336 000 201 120 #298
337 000 201 121 #161 | 338 000 201 122 #153 | 339 000 202 101 #151 | 340 000 202 102 #153
#341 000 202 110 | 342 000 202 111 #151 | 343 000 202 112 #153 | 344 000 202 121 #161
345 000 202 122 #153 | 346 000 210 101 #153 | #347 000 210 102 | 348 000 210 111 #153
#349 000 210 112 | 350 000 210 121 #153 | 351 000 210 122 #349 | 352 000 211 101 #153
#353 000 211 111 | #354 000 211 112 | 355 000 211 121 #309 | #356 000 211 122
357 000 212 101 #153 | 358 000 212 111 #317 | #359 000 212 112 | 360 000 212 221 #306
361 000 220 101 #341 | 362 000 220 111 #151 | 363 000 220 121 #161 | 364 000 221 111 #305
365 000 221 121 #311 | #366 000 222 111 | 367 001 000 000 #10 | 368 001 000 001 #11
369 001 000 002 #12 | 370 001 000 010 #36 | 371 001 000 011 #37 | 372 001 000 012 #38
373 001 000 020 #61 | 374 001 000 021 #61 | 375 001 000 022 #63 | #376 001 000 100
#377 001 000 101 | #378 001 000 102 | #379 001 000 110 | #380 001 000 111
#381 001 000 112 | #382 001 000 120 | 383 001 000 121 #382 | #384 001 000 122
#385 001 000 200 | 386 001 000 201 #385 | #387 001 000 202 | #388 001 000 210
389 001 000 211 #388 | 390 001 000 212 | 391 001 000 220 | 392 001 000 221 #391
393 001 001 000 #13 | 394 001 001 001 #14 | 395 001 001 002 #15 | 396 001 001 010 #39
397 001 001 011 #40 | 398 001 001 012 #41 | 399 001 001 020 #61 | 400 001 001 021 #65
401 001 001 022 #66 | 402 001 001 100 #379 | 403 001 001 101 #380 | 404 001 001 102 #381
#405 001 001 110 | #406 001 001 111 | #407 001 001 112 | 408 001 001 120 #382
409 001 001 121 #382 | #410 001 001 122 | 411 001 001 200 #385 | 412 001 001 201 #385
413 001 001 202 #387 | 414 001 001 210 #388 | 415 001 001 211 #388 | 416 001 001 212 #390
#417 001 001 220 | 418 001 001 221 #417 | 419 001 002 000 #16 | 420 001 002 001 #16
421 001 002 002 #18 | 422 001 002 010 #42 | 423 001 002 011 #43 | 424 001 002 012 #44
425 001 002 020 #67 | 426 001 002 021 #67 | 427 001 002 022 #69 | 428 001 002 100 #382
429 001 002 101 #382 | 430 001 002 102 #384 | 431 001 002 110 #382 | 432 001 002 111 #382
433 001 002 112 #410 | #434 001 002 120 | 435 001 002 121 #434 | #436 001 002 122
#437 001 002 200 | 438 001 002 201 #437 | 439 001 002 202 | 440 001 002 210 #437
441 001 002 211 #437 | 442 001 002 212 #439 | 443 001 002 220 #437 | 444 001 002 221 #437
445 001 010 000 #85 | 446 001 010 001 #85 | 447 001 010 002 #87 | 448 001 010 010 #85
449 001 010 011 #85 | 450 001 010 012 #87 | 451 001 010 020 #130 | 452 001 010 021 #130
453 001 010 022 #132 | #454 001 010 100 | #455 001 010 101 | #456 001 010 102
#457 001 010 110 | #458 001 010 111 | #459 001 010 112 | #460 001 010 120
461 001 010 121 #460 | #462 001 010 122 | #463 001 010 200 | 464 001 010 201 #463
465 001 010 202 | 466 001 010 210 #463 | 467 001 010 211 #463 | 468 001 010 212 #465
#469 001 010 220 | 470 001 010 221 #469 | 471 001 011 000 #88 | 472 001 011 001 #89
473 001 011 002 #90 | 474 001 011 010 #111 | 475 001 011 011 #112 | 476 001 011 012 #113
477 001 011 020 #130 | 478 001 011 021 #134 | 479 001 011 022 #135 | 480 001 011 100 #457
481 001 011 101 #458 | 482 001 011 102 #459 | #483 001 011 110 | 484 001 011 111
#485 001 011 112 | 486 001 011 120 #460 | #487 001 011 121 | #488 001 011 122
489 001 011 200 #463 | 490 001 011 201 #463 | 491 001 011 202 #465 | 492 001 011 210 #463
#493 001 011 211 | #494 001 011 212 | #495 001 011 220 | #496 001 011 221
497 001 012 000 #91 | 498 001 012 001 #91 | 499 001 012 002 #93 | 500 001 012 010 #91
501 001 012 011 #115 | 502 001 012 012 #116 | 503 001 012 020 #136 | 504 001 012 021 #137

505 001 012 022 #138	**506** 001 012 100 #460	**507** 001 012 101 #460	**508** 001 012 102 #462
509 001 012 110 #460	**510** 001 012 111 #487	**511** 001 012 112 #488	#**512** 001 012 120
#**513** 001 012 121	#**514** 001 012 122	#**515** 001 012 200	**516** 001 012 201 #515
#**517** 001 012 202	**518** 001 012 210 #515	#**519** 001 012 211	#**520** 001 012 212
521 001 012 220 #515	#**522** 001 012 221	**523** 001 020 000 #151	**524** 001 020 001 #151
525 001 020 002 #153	**526** 001 020 010 #171	**527** 001 020 011 #171	**528** 001 020 012 #153
529 001 020 020 #151	**530** 001 020 021 #151	**531** 001 020 022 #153	#**532** 001 020 100
533 001 020 101 #532	#**534** 001 020 102	**535** 001 020 110 #532	**536** 001 020 111 #532
537 001 020 112 #534	#**538** 001 020 120	**539** 001 020 121 #538	**540** 001 020 122 #534
541 001 020 200 #538	**542** 001 020 201 #538	**543** 001 020 202 #534	**544** 001 020 210 #538
545 001 020 211 #538	**546** 001 020 212 #534	**547** 001 020 220 #538	**548** 001 020 221 #538
549 001 021 000 #151	**550** 001 021 001 #155	**551** 001 021 002 #153	**552** 001 021 010 #171
553 001 021 011 #175	**554** 001 021 012 #176	**555** 001 021 020 #151	**556** 001 021 021 #194
557 001 021 022 #195	**558** 001 021 100 #532	**559** 001 021 101 #532	**560** 001 021 102 #534
561 001 021 110 #532	#**562** 001 021 111	#**563** 001 021 112	**564** 001 021 120 #538
#**565** 001 021 121	#**566** 001 021 122	**567** 001 021 200 #538	**568** 001 021 201 #538
569 001 021 202 #534	**570** 001 021 210 #538	#**571** 001 021 211	**572** 001 021 212 #563
573 001 021 220 #538	**574** 001 021 221 #565	**575** 001 022 000 #157	**576** 001 022 001 #157
577 001 022 002 #153	**578** 001 022 010 #161	**579** 001 022 011 #178	**580** 001 022 012 #179
581 001 022 020 #161	**582** 001 022 021 #184	**583** 001 022 022 #198	**584** 001 022 100 #538
585 001 022 101 #538	**586** 001 022 102 #534	**587** 001 022 110 #538	**588** 001 022 111 #565
589 001 022 112 #566	**590** 001 022 120 #538	**591** 001 022 121 #571	**592** 001 022 200 #538
593 001 022 201 #538	**594** 001 022 202 #534	**595** 001 022 210 #538	**596** 001 022 211 #565
597 001 022 212 #566	**598** 001 022 220 #538	**599** 001 022 221 #571	#**600** 001 100 000
601 001 100 001 #600	#**602** 001 100 002	#**603** 001 100 010	#**604** 001 100 011
605 001 100 012 #602	#**606** 001 100 020	**607** 001 100 021 #606	#**608** 001 100 022
#**609** 001 100 100	**610** 001 100 101 #609	**611** 001 100 102 #602	#**612** 001 100 110
#**613** 001 100 111	**614** 001 100 112 #602	#**615** 001 100 120	**616** 001 100 121 #615
617 001 100 122 #608	#**618** 001 100 200	**619** 001 100 201 #618	#**620** 001 100 202
621 001 100 210 #618	**622** 001 100 211 #618	**623** 001 100 212 #620	#**624** 001 100 220
625 001 100 221 #624	**626** 001 101 000 #600	**627** 001 101 001 #600	**628** 001 101 002 #602
#**629** 001 101 010	#**630** 001 101 011	**631** 001 101 012 #602	#**632** 001 101 020
633 001 101 021 #632	**634** 001 101 022 #608	**635** 001 101 100 #609	**636** 001 101 101 #609
637 001 101 102 #602	#**638** 001 101 110	#**639** 001 101 111	**640** 001 101 112 #602
641 001 101 120 #615	**642** 001 101 121 #615	**643** 001 101 122 #608	**644** 001 101 200 #618
645 001 101 201 #618	**646** 001 101 202 #620	**647** 001 101 210 #618	**648** 001 101 211 #618
649 001 101 212 #620	**650** 001 101 220 #624	**651** 001 101 221 #624	#**652** 001 102 000
653 001 102 001 #652	#**654** 001 102 002	**655** 001 102 010 #652	**656** 001 102 011 #652
657 001 102 012 #654	#**658** 001 102 020	**659** 001 102 021 #658	**660** 001 102 022 #566
661 001 102 100 #652	**662** 001 102 101 #652	**663** 001 102 102 #654	**664** 001 102 110 #652
665 001 102 111 #652	**666** 001 102 112 #654	**667** 001 102 120 #658	**668** 001 102 121 #658
669 001 102 200 #658	**670** 001 102 201 #658	**671** 001 102 202 #566	**672** 001 102 210 #658
673 001 102 211 #658	**674** 001 102 212 #566	**675** 001 102 220 #658	**676** 001 102 221 #658
#**677** 001 110 000	#**678** 001 110 001	#**679** 001 110 002	#**680** 001 110 010
#**681** 001 110 012	#**682** 001 110 020	**683** 001 110 021 #682	#**684** 001 110 022
685 001 110 102 #680	**686** 001 110 110 #681	#**687** 001 110 120	**688** 001 110 121 #687
689 001 110 122 #684	#**690** 001 110 220	#**691** 001 112 000	**692** 001 112 001 #691
#**693** 001 112 002	**694** 001 112 010 #691	#**695** 001 112 011	#**696** 001 112 012
#**697** 001 112 020	#**698** 001 112 021	**699** 001 112 022 #520	**700** 001 112 100 #697
701 001 112 101 #691	**702** 001 112 102 #693	**703** 001 112 110 #691	#**704** 001 112 111
#**705** 001 112 112	**706** 001 112 120 #697	#**707** 001 112 121	**708** 001 112 200 #697
709 001 112 201 #697	#**710** 001 112 202	**711** 001 112 210 #697	#**712** 001 112 211
713 001 112 212 #696	**714** 001 112 220 #697	**715** 001 112 221 #707	**716** 001 120 000 #538
717 001 120 001 #538	**718** 001 120 002 #534	**719** 001 120 010 #538	**720** 001 120 011 #538
721 001 120 012 #534	**722** 001 120 020 #538	**723** 001 120 021 #538	**724** 001 120 022 #534
725 001 120 100 #538	**726** 001 120 101 #538	**727** 001 120 102 #534	**728** 001 120 110 #538
729 001 120 111 #538	**730** 001 120 112 #534	**731** 001 120 120 #538	**732** 001 120 121 #538
733 001 120 122 #534	**734** 001 120 200 #538	**735** 001 120 201 #538	**736** 001 120 202 #534
737 001 120 210 #538	**738** 001 120 211 #538	**739** 001 120 212 #534	**740** 001 120 220 #538
741 001 120 221 #538	**742** 001 121 000 #538	**743** 001 121 001 #538	**744** 001 121 002 #534
745 001 121 010 #538	**746** 001 121 011 #571	**747** 001 121 012 #563	**748** 001 121 020 #538
749 001 121 021 #565	**750** 001 121 022 #566	**751** 001 121 100 #538	**752** 001 121 101 #538
753 001 121 102 #534	**754** 001 121 110 #538	#**755** 001 121 111	#**756** 001 121 112
757 001 121 120 #538	#**758** 001 121 121	**759** 001 121 200 #538	**760** 001 121 201 #538
761 001 121 202 #534	**762** 001 121 210 #538	**763** 001 121 211 #755	**764** 001 121 212 #756
765 001 121 220 #538	**766** 001 121 221 #758	**767** 001 122 000 #538	**768** 001 122 001 #538
769 001 122 002 #534	**770** 001 122 010 #538	**771** 001 122 011 #565	**772** 001 122 012 #566
773 001 122 020 #538	**774** 001 122 021 #571	**775** 001 122 022 #439	**776** 001 122 100 #538
777 001 122 101 #538	**778** 001 122 102 #534	**779** 001 122 110 #538	#**780** 001 122 111
781 001 122 112 #654	**782** 001 122 120 #538	**783** 001 122 121 #755	**784** 001 122 200 #538

785 001 122 201 #538	**786** 001 122 202 #534	**787** 001 122 210 #538	**788** 001 122 211 #780
789 001 122 212 #654	**790** 001 122 220 #538	**791** 001 122 221 #755	**#792** 001 200 000
793 001 200 001 #792	**794** 001 200 002 #534	**795** 001 200 010 #538	**796** 001 200 011 #538
797 001 200 012 #534	**798** 001 200 020 #538	**799** 001 200 021 #538	**800** 001 200 022 #534
801 001 200 100 #538	**802** 001 200 101 #538	**803** 001 200 102 #534	**804** 001 200 110 #538
805 001 200 111 #538	**806** 001 200 112 #534	**807** 001 200 120 #538	**808** 001 200 121 #538
809 001 200 122 #534	**810** 001 200 200 #538	**811** 001 200 201 #538	**812** 001 200 202 #534
813 001 200 210 #538	**814** 001 200 211 #538	**815** 001 200 212 #534	**816** 001 200 220 #538
817 001 200 221 #538	**818** 001 201 000 #792	**819** 001 201 001 #792	**820** 001 201 002 #534
821 001 201 010 #538	**822** 001 201 011 #538	**823** 001 201 012 #534	**824** 001 201 020 #538
825 001 201 021 #538	**826** 001 201 022 #534	**827** 001 201 100 #538	**828** 001 201 101 #538
829 001 201 102 #534	**830** 001 201 110 #538	**831** 001 201 111 #538	**832** 001 201 112 #534
833 001 201 120 #538	**834** 001 201 121 #538	**835** 001 201 200 #538	**836** 001 201 201 #538
837 001 201 202 #534	**838** 001 201 210 #538	**839** 001 201 211 #538	**840** 001 201 212 #534
841 001 201 220 #538	**842** 001 201 221 #538	**843** 001 202 000 #792	**844** 001 202 001 #792
845 001 202 002 #534	**846** 001 202 010 #538	**847** 001 202 011 #538	**848** 001 202 012 #534
849 001 202 020 #538	**850** 001 202 021 #538	**851** 001 202 022 #534	**852** 001 202 100 #538
853 001 202 101 #538	**854** 001 202 102 #534	**855** 001 202 110 #538	**856** 001 202 111 #538
857 001 202 112 #534	**858** 001 202 120 #538	**859** 001 202 121 #538	**860** 001 202 200 #538
861 001 202 201 #538	**862** 001 202 202 #534	**863** 001 202 210 #538	**864** 001 202 211 #538
865 001 202 212 #534	**866** 001 202 220 #538	**867** 001 202 221 #538	**868** 001 210 000 #534
869 001 210 001 #534	**#870** 001 210 002	**871** 001 210 010 #534	**872** 001 210 011 #534
873 001 210 012 #870	**874** 001 210 020 #534	**875** 001 210 021 #534	**876** 001 210 022 #870
877 001 210 100 #534	**878** 001 210 101 #534	**879** 001 210 102 #870	**880** 001 210 110 #534
881 001 210 111 #534	**882** 001 210 112 #870	**883** 001 210 120 #534	**884** 001 210 121 #534
#885 001 210 122	**886** 001 210 200 #534	**887** 001 210 201 #534	**888** 001 210 202 #870
889 001 210 210 #534	**890** 001 210 211 #534	**891** 001 210 212 #870	**892** 001 210 220 #534
893 001 210 221 #534	**894** 001 211 000 #534	**895** 001 211 002 #870	**896** 001 211 100 #534
897 001 211 200 #534	**#898** 001 211 202	**899** 001 212 000 #534	**900** 001 212 001 #534
901 001 212 002 #870	**902** 001 212 010 #534	**903** 001 212 011 #566	**904** 001 212 012 #710
905 001 212 020 #534	**906** 001 212 021 #563	**907** 001 212 022 #517	**908** 001 212 100 #534
909 001 212 101 #534	**910** 001 212 102 #870	**911** 001 212 110 #534	**912** 001 212 111 #654
913 001 212 112 #693	**914** 001 212 120 #534	**915** 001 212 121 #756	**916** 001 212 200 #534
917 001 212 201 #534	**918** 001 212 210 #534	**919** 001 212 211 #654	**920** 001 212 212 #693
921 001 212 220 #534	**922** 001 212 221 #756	**923** 001 220 000 #792	**924** 001 220 001 #792
925 001 220 002 #534	**926** 001 220 010 #538	**927** 001 220 011 #538	**928** 001 220 012 #534
929 001 220 020 #538	**930** 001 220 021 #538	**931** 001 220 022 #534	**932** 001 220 100 #538
933 001 220 101 #538	**934** 001 220 102 #534	**935** 001 220 110 #538	**936** 001 220 111 #538
937 001 220 112 #534	**938** 001 220 120 #538	**939** 001 220 121 #538	**940** 001 220 200 #538
941 001 220 201 #538	**942** 001 220 210 #538	**943** 001 220 211 #538	**944** 001 220 212 #534
945 001 220 220 #538	**946** 001 220 221 #538	**947** 001 221 000 #792	**948** 001 221 001 #792
949 001 221 002 #534	**950** 001 221 010 #538	**951** 001 221 011 #571	**952** 001 221 012 #563
953 001 221 020 #538	**954** 001 221 021 #565	**955** 001 221 022 #566	**956** 001 221 100 #538
957 001 221 101 #538	**958** 001 221 102 #534	**959** 001 221 110 #538	**960** 001 221 111 #755
961 001 221 112 #756	**962** 001 221 120 #538	**963** 001 221 121 #758	**964** 001 221 200 #538
965 001 221 201 #538	**966** 001 221 210 #538	**967** 001 221 211 #755	**968** 001 221 212 #756
969 001 221 220 #538	**970** 001 221 221 #758	**971** 001 222 000 #792	**972** 001 222 001 #792
973 001 222 002 #534	**974** 001 222 010 #538	**975** 001 222 011 #565	**976** 001 222 012 #566
977 001 222 020 #538	**978** 001 222 021 #571	**979** 001 222 022 #439	**980** 001 222 100 #538
981 001 222 101 #538	**982** 001 222 102 #534	**983** 001 222 110 #538	**#984** 001 222 111
985 001 222 112 #654	**986** 001 222 120 #538	**987** 001 222 121 #755	**988** 001 222 200 #538
989 001 222 201 #538	**990** 001 222 210 #538	**991** 001 222 211 #780	**992** 001 222 212 #654
993 001 222 220 #538	**994** 001 222 221 #755	**995** 002 000 000 #19	**996** 002 000 001 #16
997 002 000 002 #21	**998** 002 000 010 #45	**999** 002 000 011 #46	**1000** 002 000 012 #47
1001 002 000 020 #70	**1002** 002 000 021 #67	**1003** 002 000 022 #72	**1004** 002 000 100 #385
1005 002 000 101 #385	**1006** 002 000 102 #387	**1007** 002 000 110 #385	**1008** 002 000 111 #385
1009 002 000 112 #387	**1010** 002 000 120 #437	**1011** 002 000 121 #437	**#1012** 002 000 200
1013 002 000 201 #437	**#1014** 002 000 210	**1015** 002 000 211 #437	**1016** 002 000 212 #493
1017 002 000 220 #1012	**1018** 002 000 221 #437	**1019** 002 001 000 #22	**1020** 002 001 001 #16
1021 002 001 002 #24	**1022** 002 001 010 #48	**1023** 002 001 011 #49	**1024** 002 001 012 #50
1025 002 001 020 #73	**1026** 002 001 021 #67	**1027** 002 001 022 #75	**1028** 002 001 100 #388
1029 002 001 101 #388	**1030** 002 001 102 #390	**1031** 002 001 110 #388	**1032** 002 001 111 #388
1033 002 001 112 #390	**1034** 002 001 120 #437	**1035** 002 001 121 #437	**1036** 002 001 200 #1014
1037 002 001 201 #437	**#1038** 002 001 210	**1039** 002 001 211 #437	**#1040** 002 001 212
1041 002 001 220 #1014	**1042** 002 001 221 #437	**1043** 002 002 000 #25	**1044** 002 002 001 #26
1045 002 002 002 #27	**1046** 002 002 010 #51	**1047** 002 002 011 #52	**1048** 002 002 012 #53
1049 002 002 020 #70	**1050** 002 002 021 #67	**1051** 002 002 022 #78	**1052** 002 002 100 #391
1053 002 002 101 #391	**1054** 002 002 102 #353	**1055** 002 002 110 #417	**1056** 002 002 111 #417

1057 002 002 112 #320	1058 002 002 120 #437	1059 002 002 121 #437	1060 002 002 200 #1012
1061 002 002 201 #437	1062 002 002 210 #1014	1063 002 002 211 #437	1064 002 002 212 #121
#1065 002 002 220	#1066 002 002 221	1067 002 010 000 #94	1068 002 010 001 #91
1069 002 010 002 #96	1070 002 010 010 #117	1071 002 010 011 #91	1072 002 010 012 #119
1073 002 010 020 #139	1074 002 010 021 #136	1075 002 010 022 #141	1076 002 010 100 #463
1077 002 010 101 #463	1078 002 010 102 #465	1079 002 010 110 #463	1080 002 010 111 #463
1081 002 010 112 #465	1082 002 010 120 #515	1083 002 010 121 #515	#1084 002 010 200
1085 002 010 201 #515	#1086 002 010 210	1087 002 010 211 #515	1088 002 010 212 #494
1089 002 010 220 #1084	1090 002 010 221 #515	1091 002 011 000 #97	1092 002 011 001 #91
1093 002 011 002 #99	1094 002 011 010 #120	1095 002 011 012 #122	1096 002 011 020 #142
1097 002 011 021 #143	1098 002 011 022 #144	1099 002 011 100 #463	1100 002 011 101 #463
1101 002 011 102 #465	1102 002 011 110 #463	1103 002 011 120 #515	1104 002 011 121 #519
1105 002 011 200 #1086	1106 002 011 201 #515	#1107 002 011 210	#1108 002 011 212
1109 002 011 220 #1086	1110 002 011 221 #519	1111 002 012 000 #100	1112 002 012 001 #101
1113 002 012 002 #102	1114 002 012 010 #123	1115 002 012 011 #124	1116 002 012 012 #125
1117 002 012 020 #139	1118 002 012 021 #137	1119 002 012 022 #147	1120 002 012 100 #469
1121 002 012 101 #469	1122 002 012 102 #354	1123 002 012 110 #495	1124 002 012 111 #496
1125 002 012 112 #284	1126 002 012 120 #515	1127 002 012 121 #522	1128 002 012 200 #1084
1129 002 012 201 #515	1130 002 012 210 #1086	1131 002 012 211 #519	#1132 002 012 220
#1133 002 012 221	1134 002 020 000 #160	1135 002 020 001 #161	1136 002 020 002 #162
1137 002 020 010 #180	1138 002 020 011 #161	1139 002 020 012 #182	1140 002 020 020 #199
1141 002 020 021 #161	1142 002 020 022 #201	1143 002 020 100 #538	1144 002 020 101 #538
1145 002 020 102 #534	1146 002 020 110 #538	1147 002 020 111 #538	1148 002 020 112 #534
1149 002 020 120 #538	1150 002 020 121 #538	#1151 002 020 200	1152 002 020 201 #538
#1153 002 020 210	1154 002 020 211 #538	1155 002 020 220 #1151	1156 002 020 221 #538
1157 002 021 000 #163	1158 002 021 001 #161	1159 002 021 002 #165	1160 002 021 010 #183
1161 002 021 011 #184	1162 002 021 012 #185	1163 002 021 020 #163	1164 002 021 021 #203
1165 002 021 022 #204	1166 002 021 100 #538	1167 002 021 101 #538	1168 002 021 102 #534
1169 002 021 110 #538	1170 002 021 111 #571	1171 002 021 112 #563	1172 002 021 120 #538
1173 002 021 121 #565	1174 002 021 200 #1153	1175 002 021 201 #538	#1176 002 021 210
1177 002 021 211 #571	1178 002 021 220 #1153	1179 002 021 221 #565	1180 002 022 000 #166
1181 002 022 001 #161	1182 002 022 002 #168	1183 002 022 010 #186	1184 002 022 011 #178
1185 002 022 012 #188	1186 002 022 020 #199	1187 002 022 021 #184	1188 002 022 022 #100
1189 002 022 101 #538	1190 002 022 102 #153	1191 002 022 110 #538	1192 002 022 111 #565
1193 002 022 112 #250	1194 002 022 120 #538	1195 002 022 121 #571	1196 002 022 200 #1151
1197 002 022 201 #538	1198 002 022 210 #1153	1199 002 022 211 #565	#1200 002 022 220
1201 002 022 221 #571	#1202 002 100 000	1203 002 100 001 #652	1204 002 100 002 #704
#1205 002 100 010	1206 002 100 011 #652	1207 002 100 012 #712	1208 002 100 020 #1205
1209 002 100 021 #658	1210 002 100 022 #695	1211 002 100 100 #652	1212 002 100 101 #652
1213 002 100 102 #654	1214 002 100 110 #652	1215 002 100 111 #652	1216 002 100 112 #654
1217 002 100 120 #658	1218 002 100 121 #658	#1219 002 100 200	1220 002 100 201 #658
#1221 002 100 210	1222 002 100 211 #658	1223 002 100 220 #1219	1224 002 100 221 #658
#1225 002 101 000	1226 002 101 001 #652	#1227 002 101 002	1228 002 101 010 #1205
1229 002 101 011 #652	1230 002 101 012 #712	#1231 002 101 020	1232 002 101 021 #658
#1233 002 101 022	1234 002 101 100 #652	1235 002 101 101 #652	1236 002 101 102 #654
1237 002 101 110 #652	1238 002 101 111 #652	1239 002 101 112 #654	1240 002 101 120 #658
1241 002 101 121 #658	#1242 002 101 200	1243 002 101 201 #658	1244 002 101 210 #1221
1245 002 101 211 #658	1246 002 101 220 #1242	1247 002 101 221 #658	1248 002 102 000 #1225
#1249 002 102 001	1250 002 102 002 #280	1251 002 102 010 #1231	1252 002 102 011 #1249
1253 002 102 012 #286	1254 002 102 020 #1205	1255 002 102 021 #658	1256 002 102 100 #1249
1257 002 102 101 #1249	1258 002 102 102 #317	1259 002 102 110 #1249	1260 002 102 111 #1249
1261 002 102 112 #317	1262 002 102 120 #658	1263 002 102 121 #658	1264 002 102 200 #1219
1265 002 102 201 #658	1266 002 102 210 #1221	1267 002 102 211 #658	#1268 002 102 220
#1269 002 102 221	1270 002 112 000 #1227	#1271 002 112 001	1272 002 112 002 #281
1273 002 112 010 #1233	1274 002 112 012 #287	1275 002 112 020 #712	1276 002 112 021 #698
#1277 002 112 100	1278 002 112 110 #359	1279 002 112 120 #697	1280 002 112 121 #707
#1281 002 112 220	1282 002 120 000 #652	1283 002 120 001 #538	1284 002 120 002 #691
1285 002 120 010 #658	1286 002 120 011 #538	1287 002 120 012 #697	1288 002 120 020 #652
1289 002 120 021 #538	1290 002 120 022 #691	1291 002 120 100 #538	1292 002 120 101 #538
1293 002 120 102 #534	1294 002 120 110 #538	1295 002 120 111 #538	1296 002 120 112 #534
1297 002 120 120 #538	1298 002 120 121 #538	1299 002 120 200 #1153	1300 002 120 201 #538
1301 002 120 210 #1176	1302 002 120 211 #538	1303 002 120 220 #1153	1304 002 120 221 #538
1305 002 121 000 #1249	1306 002 121 001 #538	1307 002 121 002 #1271	1308 002 121 010 #658
1309 002 121 011 #571	1310 002 121 012 #698	1311 002 121 020 #1249	1312 002 121 021 #565
1313 002 121 100 #538	1314 002 121 101 #538	1315 002 121 102 #534	1316 002 121 110 #538
1317 002 121 111 #755	1318 002 121 112 #756	1319 002 121 120 #538	1320 002 121 121 #758
#1321 002 121 200	1322 002 121 201 #538	1323 002 121 210 #1176	1324 002 121 211 #755
1325 002 121 220 #1321	1326 002 121 221 #758	1327 002 122 000 #652	1328 002 122 001 #538
1329 002 122 002 #165	1330 002 122 010 #658	1331 002 122 011 #565	1332 002 122 012 #255
1333 002 122 020 #652	1334 002 122 021 #571	1335 002 122 100 #538	1336 002 122 101 #538

1337 002 122 102 #153	**1338** 002 122 110 #538	**1339** 002 122 111 #780	**1340** 002 122 112 #244
1341 002 122 120 #538	**1342** 002 122 121 #755	**1343** 002 122 200 #1153	**1344** 002 122 201 #538
1345 002 122 210 #1176	**1346** 002 122 211 #780	**1347** 002 122 220 #1153	**1348** 002 122 221 #755
1349 002 200 000 #600	**1350** 002 200 001 #538	**1351** 002 200 002 #677	**1352** 002 200 010 #606
1353 002 200 011 #538	**1354** 002 200 012 #682	**1355** 002 200 020 #603	**1356** 002 200 021 #538
1357 002 200 022 #680	**1358** 002 200 100 #538	**1359** 002 200 101 #538	**1360** 002 200 102 #534
1361 002 200 110 #538	**1362** 002 200 111 #538	**1363** 002 200 112 #534	**1364** 002 200 120 #538
1365 002 200 121 #538	**1366** 002 200 200 #1151	**1367** 002 200 201 #538	**1368** 002 200 210 #1153
1369 002 200 211 #538	**1370** 002 200 220 #1200	**1371** 002 200 221 #538	**1372** 002 201 000 #652
1373 002 201 001 #538	**1374** 002 201 002 #691	**1375** 002 201 010 #658	**1376** 002 201 011 #538
1377 002 201 012 #697	**1378** 002 201 020 #652	**1379** 002 201 021 #538	**1380** 002 201 100 #538
1381 002 201 101 #538	**1382** 002 201 102 #534	**1383** 002 201 110 #538	**1384** 002 201 111 #538
1385 002 201 112 #534	**1386** 002 201 120 #538	**1387** 002 201 121 #538	**1388** 002 201 200 #1153
1389 002 201 201 #538	**1390** 002 201 210 #1176	**1391** 002 201 211 #538	**1392** 002 201 220 #1153
1393 002 201 221 #538	**1394** 002 202 000 #600	**1395** 002 202 001 #538	**1396** 002 202 002 #261
1397 002 202 010 #632	**1398** 002 202 011 #538	**1399** 002 202 012 #268	**1400** 002 202 020 #629
1401 002 202 021 #538	**1402** 002 202 100 #538	**1403** 002 202 101 #538	**1404** 002 202 102 #153
1405 002 202 110 #538	**1406** 002 202 111 #538	**1407** 002 202 112 #153	**1408** 002 202 120 #538
1409 002 202 121 #538	**1410** 002 202 200 #1151	**1411** 002 202 201 #538	**1412** 002 202 210 #1153
1413 002 202 211 #538	**1414** 002 202 220 #1151	**1415** 002 202 221 #538	**1416** 002 210 000 #602
1417 002 210 001 #534	**1418** 002 210 002 #679	**1419** 002 210 010 #608	**1420** 002 210 011 #534
1421 002 210 012 #684	**1422** 002 210 020 #602	**1423** 002 210 021 #534	**1424** 002 210 022 #681
1425 002 210 100 #534	**1426** 002 210 101 #534	**1427** 002 210 102 #870	**1428** 002 210 110 #534
1429 002 210 111 #534	**1430** 002 210 112 #870	**1431** 002 210 120 #534	**1432** 002 210 121 #534
#**1433** 002 210 200	**1434** 002 210 201 #534	**1435** 002 210 210 #756	**1436** 002 210 211 #534
#**1437** 002 210 220	**1438** 002 210 221 #534	**1439** 002 212 000 #602	**1440** 002 212 001 #534
1441 002 212 002 #262	**1442** 002 212 010 #608	**1443** 002 212 011 #566	**1444** 002 212 012 #269
1445 002 212 020 #602	**1446** 002 212 021 #563	**1447** 002 212 100 #534	**1448** 002 212 101 #534
1449 002 212 102 #349	**1450** 002 212 110 #534	**1451** 002 212 111 #654	**1452** 002 212 112 #278
1453 002 212 120 #534	**1454** 002 212 121 #756	**1455** 002 212 200 #1433	**1456** 002 212 201 #534
1457 002 212 210 #756	**1458** 002 212 211 #654	**1459** 002 212 220 #1433	**1460** 002 212 221 #756
1461 002 220 000 #600	**1462** 002 220 001 #538	**1463** 002 220 002 #678	**1464** 002 220 010 #606
1465 002 220 011 #538	**1466** 002 220 012 #682	**1467** 002 220 020 #604	**1468** 002 220 021 #538
1469 002 220 100 #538	**1470** 002 220 101 #538	**1471** 002 220 102 #534	**1472** 002 220 110 #538
1473 002 220 111 #538	**1474** 002 220 112 #534	**1475** 002 220 120 #538	**1476** 002 220 121 #538
1477 002 220 200 #1151	**1478** 002 220 201 #538	**1479** 002 220 210 #1153	**1480** 002 220 211 #538
#**1481** 002 220 220	**1482** 002 220 221 #538	**1483** 002 221 000 #652	**1484** 002 221 001 #538
1485 002 221 002 #691	**1486** 002 221 010 #658	**1487** 002 221 011 #571	**1488** 002 221 012 #698
1489 002 221 020 #652	**1490** 002 221 021 #565	**1491** 002 221 100 #538	**1492** 002 221 101 #538
1493 002 221 102 #534	**1494** 002 221 110 #538	**1495** 002 221 111 #755	**1496** 002 221 112 #756
1497 002 221 120 #538	**1498** 002 221 121 #758	**1499** 002 221 200 #1153	**1500** 002 221 201 #538
1501 002 221 210 #1176	**1502** 002 221 211 #755	**1503** 002 221 220 #1153	**1504** 002 221 221 #758
1505 002 222 000 #600	**1506** 002 222 001 #538	**1507** 002 222 002 #260	**1508** 002 222 010 #632
1509 002 222 011 #565	**1510** 002 222 012 #252	**1511** 002 222 020 #630	**1512** 002 222 021 #571
1513 002 222 100 #538	**1514** 002 222 101 #538	**1515** 002 222 102 #153	**1516** 002 222 110 #538
1517 002 222 111 #984	**1518** 002 222 112 #244	**1519** 002 222 120 #538	**1520** 002 222 121 #755
1521 002 222 200 #1151	**1522** 002 222 201 #538	**1523** 002 222 210 #1153	**1524** 002 222 211 #780
1525 002 222 220 #1151	**1526** 002 222 221 #755	**1527** 011 000 000 #207	**1528** 011 000 001 #199
1529 011 000 002 #209	**1530** 011 000 010 #199	**1531** 011 000 011 #199	**1532** 011 000 012 #209
1533 011 000 020 #163	**1534** 011 000 021 #163	**1535** 011 000 100 #609	**1536** 011 000 101 #609
1537 011 000 102 #602	**1538** 011 000 110 #609	**1539** 011 000 111 #609	**1540** 011 000 112 #602
1541 011 000 120 #652	**1542** 011 000 121 #652	**1543** 011 000 200 #652	**1544** 011 000 201 #652
1545 011 000 210 #652	**1546** 011 000 211 #652	**1547** 011 000 220 #1249	**1548** 011 000 221 #1249
1549 011 001 000 #199	**1550** 011 001 001 #199	**1551** 011 001 002 #209	**1552** 011 001 010 #199
1553 011 001 011 #199	**1554** 011 001 012 #209	**1555** 011 001 020 #163	**1556** 011 001 021 #163
1557 011 001 100 #612	**1558** 011 001 101 #613	**1559** 011 001 102 #602	**1560** 011 001 110 #638
1561 011 001 111 #639	**1562** 011 001 112 #602	**1563** 011 001 120 #652	**1564** 011 001 121 #652
1565 011 001 200 #652	**1566** 011 001 201 #652	**1567** 011 001 210 #652	**1568** 011 001 211 #652
1569 011 001 220 #1249	**1570** 011 001 221 #1249	**1571** 011 002 000 #213	**1572** 011 002 001 #213
1573 011 002 002 #215	**1574** 011 002 010 #213	**1575** 011 002 011 #213	**1576** 011 002 012 #215
1577 011 002 020 #183	**1578** 011 002 021 #183	**1579** 011 002 100 #615	**1580** 011 002 101 #615
1581 011 002 102 #608	**1582** 011 002 110 #615	**1583** 011 002 111 #615	**1584** 011 002 112 #608
1585 011 002 120 #658	**1586** 011 002 121 #658	**1587** 011 002 200 #658	**1588** 011 002 201 #658
1589 011 002 210 #658	**1590** 011 002 211 #658	**1591** 011 002 220 #658	**1592** 011 002 221 #658
1593 011 010 000 #257	**1594** 011 010 001 #258	**1595** 011 010 002 #259	**1596** 011 010 010 #272
1597 011 010 012 #273	**1598** 011 010 020 #165	**1599** 011 010 021 #165	**1600** 011 010 100 #680
1601 011 010 102 #681	**1602** 011 010 120 #691	**1603** 011 010 121 #691	**1604** 011 010 220 #1277
1605 011 012 000 #263	**1606** 011 012 001 #263	**1607** 011 012 002 #265	**1608** 011 012 010 #274
1609 011 012 011 #224	**1610** 011 012 012 #271	**1611** 011 012 020 #282	**1612** 011 012 021 #283
1613 011 012 100 #687	**1614** 011 012 101 #687	**1615** 011 012 102 #684	**1616** 011 012 110 #266

1617 011 012 111 #218	1618 011 012 112 #267	1619 011 012 120 #697	1620 011 012 121 #707
1621 011 012 200 #697	1622 011 012 201 #697	1623 011 012 210 #282	1624 011 012 211 #221
1625 011 012 220 #697	1626 011 012 221 #707	1627 011 020 000 #161	1628 011 020 001 #161
1629 011 020 002 #153	1630 011 020 010 #161	1631 011 020 011 #161	1632 011 020 012 #153
1633 011 020 020 #161	1634 011 020 021 #161	1635 011 020 100 #538	1636 011 020 101 #538
1637 011 020 102 #534	1638 011 020 110 #538	1639 011 020 111 #538	1640 011 020 112 #534
1641 011 020 120 #538	1642 011 020 121 #538	1643 011 020 200 #538	1644 011 020 201 #538
1645 011 020 210 #538	1646 011 020 211 #538	1647 011 020 220 #538	1648 011 020 221 #538
1649 011 021 000 #161	1650 011 021 001 #161	1651 011 021 002 #153	1652 011 021 010 #161
1653 011 021 011 #305	1654 011 021 012 #306	1655 011 021 020 #161	1656 011 021 021 #316
1657 011 021 100 #538	1658 011 021 101 #538	1659 011 021 102 #534	1660 011 021 110 #538
1661 011 021 111 #755	1662 011 021 112 #756	1663 011 021 120 #538	1664 011 021 121 #780
1665 011 021 200 #538	1666 011 021 201 #538	1667 011 021 210 #538	1668 011 021 211 #755
1669 011 021 220 #538	1670 011 021 221 #780	1671 011 022 000 #161	1672 011 022 001 #161
1673 011 022 002 #153	1674 011 022 010 #161	1675 011 022 011 #308	1676 011 022 012 #309
1677 011 022 020 #161	1678 011 022 021 #305	1679 011 022 100 #538	1680 011 022 101 #538
1681 011 022 102 #534	1682 011 022 110 #538	1683 011 022 111 #758	1684 011 022 120 #538
1685 011 022 121 #755	1686 011 022 200 #538	1687 011 022 201 #538	1688 011 022 210 #538
1689 011 022 211 #758	1690 011 022 220 #538	1691 011 022 221 #755	1692 011 100 000 #1151
1693 011 100 001 #1151	1694 011 100 002 #1433	1695 011 100 010 #1200	1696 011 100 011 #1481
1697 011 100 012 #1437	1698 011 100 020 #1153	1699 011 100 021 #1153	#1700 011 100 100
1701 011 100 101 #1200	1702 011 100 102 #1437	1703 011 100 110 #1151	1704 011 100 111 #1151
1705 011 100 112 #1433	1706 011 100 120 #1153	1707 011 100 121 #1153	#1708 011 100 200
1709 011 100 201 #1153	1710 011 100 210 #1153	1711 011 100 211 #1153	1712 011 100 220 #1321
1713 011 100 221 #1321	1714 011 101 000 #1151	1715 011 101 001 #1151	1716 011 101 002 #1433
1717 011 101 010 #1151	1718 011 101 011 #1151	1719 011 101 012 #1433	1720 011 101 020 #1153
1721 011 101 021 #1153	1722 011 101 100 #1151	1723 011 101 101 #1151	1724 011 101 102 #1433
1725 011 101 110 #1151	1726 011 101 111 #1151	1727 011 101 112 #1433	1728 011 101 120 #1153
1729 011 101 121 #1153	1730 011 101 200 #1153	1731 011 101 201 #1153	1732 011 101 210 #1153
1733 011 101 211 #1153	1734 011 101 220 #1321	1735 011 101 221 #1321	1736 011 102 000 #1153
1737 011 102 001 #1153	1738 011 102 002 #756	1739 011 102 010 #1153	1740 011 102 011 #1153
1741 011 102 012 #756	1742 011 102 020 #1176	1743 011 102 021 #1176	1744 011 102 100 #1153
1745 011 102 101 #1153	1746 011 102 102 #756	1747 011 102 110 #1153	1748 011 102 111 #1153
1749 011 102 120 #1176	1750 011 102 121 #1176	1751 011 102 200 #1176	1752 011 102 201 #1176
1753 011 102 210 #1176	1754 011 102 211 #1176	1755 011 102 220 #1176	1756 011 102 221 #1176
1757 011 120 000 #538	1758 011 120 001 #538	1759 011 120 002 #534	1760 011 120 010 #538
1761 011 120 011 #538	1762 011 120 012 #534	1763 011 120 020 #538	1764 011 120 021 #538
1765 011 120 100 #538	1766 011 120 101 #538	1767 011 120 102 #534	1768 011 120 110 #538
1769 011 120 111 #538	1770 011 120 112 #534	1771 011 120 120 #538	1772 011 120 121 #538
1773 011 120 200 #538	1774 011 120 201 #538	1775 011 120 210 #538	1776 011 120 211 #538
1777 011 120 220 #538	1778 011 120 221 #538	1779 011 121 000 #538	1780 011 121 001 #538
1781 011 121 002 #534	1782 011 121 010 #538	1783 011 121 011 #755	1784 011 121 012 #756
1785 011 121 020 #538	1786 011 121 021 #780	1787 011 121 100 #538	1788 011 121 101 #538
1789 011 121 102 #534	1790 011 121 110 #538	#1791 011 121 111	1792 011 121 120 #538
#1793 011 121 121	1794 011 121 200 #538	1795 011 121 201 #538	1796 011 121 210 #538
1797 011 121 211 #1791	1798 011 121 220 #538	#1799 011 121 221	1800 011 122 000 #538
1801 011 122 001 #538	1802 011 122 002 #534	1803 011 122 010 #538	1804 011 122 011 #758
1805 011 122 012 #620	1806 011 122 020 #538	1807 011 122 021 #755	1808 011 122 100 #538
1809 011 122 101 #538	1810 011 122 102 #538	1811 011 122 110 #538	1812 011 122 111 #1793
1813 011 122 120 #538	1814 011 122 121 #1791	1815 011 122 200 #538	1816 011 122 201 #538
1817 011 122 210 #538	#1818 011 122 211	1819 011 122 220 #538	1820 011 122 221 #1791
1821 011 200 000 #538	1822 011 200 001 #538	1823 011 200 002 #534	1824 011 200 010 #538
1825 011 200 011 #538	1826 011 200 012 #534	1827 011 200 020 #538	1828 011 200 021 #538
#1829 011 200 100	1830 011 200 101 #538	1831 011 200 102 #534	1832 011 200 110 #538
1833 011 200 111 #538	1834 011 200 112 #534	1835 011 200 120 #538	1836 011 200 121 #538
#1837 011 200 200	1838 011 200 201 #538	1839 011 200 210 #538	1840 011 200 211 #538
1841 011 200 220 #538	1842 011 200 221 #538	1843 011 201 000 #538	1844 011 201 001 #538
1845 011 201 002 #534	1846 011 201 010 #538	1847 011 201 011 #538	1848 011 201 012 #534
1849 011 201 020 #538	1850 011 201 021 #538	1851 011 201 100 #538	1852 011 201 101 #538
1853 011 201 102 #534	1854 011 201 110 #538	1855 011 201 111 #538	1856 011 201 120 #538
1857 011 201 121 #538	1858 011 201 200 #538	1859 011 201 201 #538	1860 011 201 210 #538
1861 011 201 211 #538	1862 011 201 220 #538	1863 011 201 221 #538	1864 011 202 000 #538
1865 011 202 001 #538	1866 011 202 002 #534	1867 011 202 010 #538	1868 011 202 011 #538
1869 011 202 012 #534	1870 011 202 020 #538	1871 011 202 021 #538	1872 011 202 100 #538
1873 011 202 101 #538	1874 011 202 102 #534	1875 011 202 110 #538	1876 011 202 111 #538
1877 011 202 120 #538	1878 011 202 121 #538	1879 011 202 200 #538	1880 011 202 201 #538
1881 011 202 210 #538	1882 011 202 211 #538	1883 011 202 220 #538	1884 011 202 221 #538
1885 011 210 000 #534	1886 011 210 001 #534	1887 011 210 002 #870	1888 011 210 010 #153
1889 011 210 011 #153	1890 011 210 012 #349	1891 011 210 020 #534	1892 011 210 021 #534
1893 011 210 100 #534	1894 011 210 101 #534	1895 011 210 102 #870	1896 011 210 110 #153

1897 011 210 111 #153	**1898** 011 210 112 #347	**1899** 011 210 120 #534	**1900** 011 210 121 #534
1901 011 210 200 #534	**1902** 011 210 201 #534	**1903** 011 210 210 #153	**1904** 011 210 211 #153
1905 011 210 220 #534	**1906** 011 210 221 #534	**1907** 011 212 000 #534	**1908** 011 212 002 #898
1909 011 212 100 #534	**1910** 011 212 200 #534	**1911** 011 220 000 #538	**1912** 011 220 001 #538
1913 011 220 010 #538	**1914** 011 220 011 #538	**1915** 011 220 012 #534	**1916** 011 220 020 #538
1917 011 220 021 #538	**1918** 011 220 100 #538	**1919** 011 220 101 #538	**1920** 011 220 102 #534
1921 011 220 110 #538	**1922** 011 220 111 #538	**1923** 011 220 120 #538	**1924** 011 220 121 #538
1925 011 220 200 #538	**1926** 011 220 201 #538	**1927** 011 220 210 #538	**1928** 011 220 211 #538
1929 011 220 220 #538	**1930** 011 220 221 #538	**1931** 011 221 000 #538	**1932** 011 221 001 #538
1933 011 221 010 #538	**1934** 011 221 011 #755	**1935** 011 221 012 #756	**1936** 011 221 020 #538
1937 011 221 021 #780	**1938** 011 221 100 #538	**1939** 011 221 101 #538	**1940** 011 221 102 #534
1941 011 221 110 #538	**1942** 011 221 111 #1791	**1943** 011 221 120 #538	**1944** 011 221 121 #1799
1945 011 221 200 #538	**1946** 011 221 201 #538	**1947** 011 221 210 #538	**1948** 011 221 211 #1791
1949 011 221 220 #538	**1950** 011 221 221 #1793	**1951** 011 222 000 #538	**1952** 011 222 001 #538
1953 011 222 010 #538	**1954** 011 222 011 #758	**1955** 011 222 012 #620	**1956** 011 222 020 #538
1957 011 222 021 #755	**1958** 011 222 100 #538	**1959** 011 222 101 #538	**1960** 011 222 102 #534
1961 011 222 110 #538	#**1962** 011 222 111	**1963** 011 222 120 #538	**1964** 011 222 121 #1791
1965 011 222 200 #538	**1966** 011 222 201 #538	**1967** 011 222 210 #538	**1968** 011 222 211 #1793
1969 011 222 220 #538	**1970** 011 222 221 #1791	**1971** 012 000 000 #216	**1972** 012 000 001 #213
1973 012 000 010 #222	**1974** 012 000 011 #213	**1975** 012 000 012 #235	**1976** 012 000 020 #219
1977 012 000 021 #183	**1978** 012 000 100 #618	**1979** 012 000 101 #618	**1980** 012 000 102 #620
1981 012 000 110 #618	**1982** 012 000 111 #618	**1983** 012 000 120 #658	**1984** 012 000 121 #658
1985 012 000 200 #1219	**1986** 012 000 201 #658	**1987** 012 000 210 #1242	**1988** 012 000 211 #658
1989 012 000 220 #1219	**1990** 012 000 221 #658	**1991** 012 001 001 #213	**1992** 012 001 010 #219
1993 012 001 011 #213	**1994** 012 001 012 #221	**1995** 012 001 020 #253	**1996** 012 001 021 #183
1997 012 001 100 #618	**1998** 012 001 101 #618	**1999** 012 001 102 #620	**2000** 012 001 110 #618
2001 012 001 111 #618	**2002** 012 001 120 #658	**2003** 012 001 121 #658	**2004** 012 001 200 #1221
2005 012 001 201 #658	**2006** 012 001 210 #1221	**2007** 012 001 211 #658	**2008** 012 001 220 #1221
2009 012 001 221 #658	**2010** 012 002 001 #223	**2011** 012 002 010 #239	**2012** 012 002 011 #223
2013 012 002 012 #241	**2014** 012 002 021 #183	**2015** 012 002 100 #624	**2016** 012 002 101 #624
2017 012 002 102 #309	**2018** 012 002 110 #624	**2019** 012 002 111 #624	**2020** 012 002 120 #658
2021 012 002 121 #658	**2022** 012 002 200 #1219	**2023** 012 002 201 #658	**2024** 012 002 210 #1242
2025 012 002 211 #658	**2026** 012 002 220 #1268	**2027** 012 002 221 #1269	**2028** 012 012 001 #270
2029 012 012 012 #275	**2030** 012 012 021 #283	**2031** 012 012 100 #690	**2032** 012 012 102 #356
2033 012 012 120 #697	**2034** 012 012 121 #707	**2035** 012 012 220 #1281	**2036** 012 020 001 #298
2037 012 020 011 #161	**2038** 012 020 021 #161	**2039** 012 020 100 #538	**2040** 012 020 101 #538
2041 012 020 102 #534	**2042** 012 020 110 #538	**2043** 012 020 111 #538	**2044** 012 020 120 #538
2045 012 020 121 #538	**2046** 012 020 200 #1153	**2047** 012 020 201 #538	**2048** 012 020 210 #1321
2049 012 020 211 #538	**2050** 012 020 220 #1153	**2051** 012 020 221 #538	**2052** 012 021 011 #305
2053 012 021 021 #321	**2054** 012 021 100 #538	**2055** 012 021 101 #538	**2056** 012 021 102 #534
2057 012 021 110 #538	**2058** 012 021 111 #755	**2059** 012 021 120 #538	**2060** 012 021 121 #780
2061 012 021 200 #1176	**2062** 012 021 201 #538	**2063** 012 021 210 #1176	**2064** 012 021 211 #755
2065 012 021 220 #1176	**2066** 012 021 221 #780	**2067** 012 022 011 #311	**2068** 012 022 100 #538
2069 012 022 101 #538	**2070** 012 022 102 #153	**2071** 012 022 110 #538	**2072** 012 022 111 #758
2073 012 022 120 #538	**2074** 012 022 121 #755	**2075** 012 022 200 #1153	**2076** 012 022 201 #538
2077 012 022 210 #1321	**2078** 012 022 211 #758	**2079** 012 022 220 #1153	**2080** 012 022 221 #755
2081 012 100 100 #1708	**2082** 012 100 101 #1153	**2083** 012 100 102 #756	**2084** 012 100 110 #1153
2085 012 100 111 #1153	**2086** 012 100 120 #1176	**2087** 012 100 121 #1176	#**2088** 012 100 200
2089 012 100 201 #1176	#**2090** 012 100 210	**2091** 012 100 211 #1176	**2092** 012 100 220 #2090
2093 012 100 221 #1176	**2094** 012 101 100 #1153	**2095** 012 101 101 #1153	**2096** 012 101 102 #756
2097 012 101 110 #1153	**2098** 012 101 111 #1153	**2099** 012 101 120 #1176	**2100** 012 101 121 #1176
2101 012 101 201 #1176	#**2102** 012 101 210	**2103** 012 101 211 #1176	#**2104** 012 101 220
2105 012 101 221 #1176	**2106** 012 102 100 #1321	**2107** 012 102 101 #1321	**2108** 012 102 102 #306
2109 012 102 110 #1321	**2110** 012 102 111 #1321	**2111** 012 102 120 #1176	**2112** 012 102 121 #1176
2113 012 102 201 #1176	**2114** 012 102 210 #2104	**2115** 012 102 211 #1176	#**2116** 012 102 221
2117 012 120 100 #538	**2118** 012 120 101 #538	**2119** 012 120 102 #534	**2120** 012 120 110 #538
2121 012 120 111 #538	**2122** 012 120 120 #538	**2123** 012 120 121 #538	#**2124** 012 120 201
2125 012 120 211 #538	**2126** 012 120 221 #538	**2127** 012 121 100 #538	**2128** 012 121 101 #538
2129 012 121 102 #534	**2130** 012 121 110 #538	**2131** 012 121 111 #1791	**2132** 012 121 120 #538
2133 012 121 121 #1818	**2134** 012 121 211 #1791	#**2135** 012 121 221	**2136** 012 122 100 #538
2137 012 122 101 #538	**2138** 012 122 102 #153	**2139** 012 122 110 #538	**2140** 012 122 111 #1799
2141 012 122 120 #538	**2142** 012 122 121 #1791	**2143** 012 122 211 #2135	#**2144** 012 200 100
2145 012 200 101 #538	**2146** 012 200 102 #534	**2147** 012 200 110 #538	**2148** 012 200 111 #538
2149 012 200 120 #538	**2150** 012 200 121 #538	**2151** 012 201 101 #538	**2152** 012 201 102 #534
2153 012 201 110 #538	**2154** 012 201 111 #538	**2155** 012 201 120 #2124	**2156** 012 201 121 #538
2157 012 202 101 #538	**2158** 012 202 102 #153	#**2159** 012 202 110	**2160** 012 202 111 #538
2161 012 202 121 #538	**2162** 012 210 101 #534	**2163** 012 210 102 #885	**2164** 012 210 111 #153
2165 012 210 121 #534	**2166** 012 220 101 #2159	**2167** 012 220 111 #538	**2168** 012 220 121 #1793
2169 012 221 111 #1791	**2170** 012 221 121 #2135	#**2171** 012 222 111	**2172** 021 000 000 #322
2173 021 000 001 #151	**2174** 021 000 010 #161	**2175** 021 000 011 #161	**2176** 021 000 020 #151

2177 021 000 021 #151 | 2178 021 000 100 #538 | 2179 021 000 101 #538 | 2180 021 000 102 #534
2181 021 000 110 #538 | 2182 021 000 111 #538 | 2183 021 000 120 #538 | 2184 021 000 121 #538
2185 021 000 200 #538 | 2186 021 000 201 #538 | 2187 021 000 210 #538 | 2188 021 000 211 #538
2189 021 000 220 #538 | 2190 021 000 221 #538 | 2191 021 001 001 #151 | 2192 021 001 010 #161
2193 021 001 011 #161 | 2194 021 001 020 #341 | 2195 021 001 021 #151 | 2196 021 001 100 #538
2197 021 001 101 #538 | 2198 021 001 102 #534 | 2199 021 001 110 #538 | 2200 021 001 111 #538
2201 021 001 120 #538 | 2202 021 001 121 #538 | 2203 021 001 200 #538 | 2204 021 001 201 #538
2205 021 001 210 #538 | 2206 021 001 211 #538 | 2207 021 001 220 #538 | 2208 021 001 221 #538
2209 021 002 001 #161 | 2210 021 002 010 #298 | 2211 021 002 011 #161 | 2212 021 002 021 #161
2213 021 002 100 #538 | 2214 021 002 101 #538 | 2215 021 002 102 #534 | 2216 021 002 110 #538
2217 021 002 111 #538 | 2218 021 002 120 #538 | 2219 021 002 121 #538 | 2220 021 002 200 #538
2221 021 002 201 #538 | 2222 021 002 210 #538 | 2223 021 002 211 #538 | 2224 021 002 220 #538
2225 021 002 221 #538 | 2226 021 020 001 #341 | 2227 021 020 011 #161 | 2228 021 020 021 #151
2229 021 020 100 #538 | 2230 021 020 101 #538 | 2231 021 020 102 #534 | 2232 021 020 110 #538
2233 021 020 111 #538 | 2234 021 020 120 #538 | 2235 021 020 121 #538 | 2236 021 020 200 #538
2237 021 020 201 #538 | 2238 021 020 210 #538 | 2239 021 020 211 #538 | 2240 021 020 220 #538
2241 021 020 221 #538 | 2242 021 021 011 #305 | 2243 021 021 021 #366 | 2244 021 021 100 #538
2245 021 021 101 #538 | 2246 021 021 102 #534 | 2247 021 021 110 #538 | 2248 021 021 111 #755
2249 021 021 120 #538 | 2250 021 021 121 #984 | 2251 021 021 200 #538 | 2252 021 021 201 #538
2253 021 021 210 #538 | 2254 021 021 211 #755 | 2255 021 021 220 #538 | 2256 021 021 221 #984
2257 021 022 011 #311 | 2258 021 022 100 #538 | 2259 021 022 101 #538 | 2260 021 022 110 #538
2261 021 022 111 #758 | 2262 021 022 120 #538 | 2263 021 022 121 #755 | 2264 021 022 200 #538
2265 021 022 201 #538 | 2266 021 022 210 #538 | 2267 021 022 211 #758 | 2268 021 022 220 #538
2269 021 022 221 #755 | 2270 021 100 100 #1829 | 2271 021 100 101 #538 | 2272 021 100 102 #534
2273 021 100 110 #538 | 2274 021 100 111 #538 | 2275 021 100 120 #538 | 2276 021 100 121 #538
2277 021 100 200 #2144 | 2278 021 100 201 #538 | 2279 021 100 210 #538 | 2280 021 100 211 #538
2281 021 100 220 #538 | 2282 021 100 221 #538 | 2283 021 101 100 #538 | 2284 021 101 101 #538
2285 021 101 102 #534 | 2286 021 101 110 #538 | 2287 021 101 111 #538 | 2288 021 101 120 #538
2289 021 101 121 #538 | 2290 021 101 201 #538 | 2291 021 101 210 #538 | 2292 021 101 211 #538
2293 021 101 220 #2159 | 2294 021 101 221 #538 | 2295 021 102 100 #538 | 2296 021 102 101 #538
2297 021 102 110 #538 | 2298 021 102 111 #538 | 2299 021 102 120 #538 | 2300 021 102 121 #538
2301 021 102 201 #538 | 2302 021 102 210 #2124 | 2303 021 102 211 #538 | 2304 021 102 221 #538
2305 021 120 100 #538 | 2306 021 120 101 #538 | 2307 021 120 102 #534 | 2308 021 120 110 #538
2309 021 120 111 #538 | 2310 021 120 120 #538 | 2311 021 120 121 #538 | 2312 021 120 201 #2159
2313 021 120 211 #538 | 2314 021 120 221 #538 | 2315 021 121 100 #538 | 2316 021 121 101 #538
2317 021 121 110 #538 | 2318 021 121 111 #1791 | 2319 021 121 120 #538 | 2320 021 121 121 #1962
2321 021 121 211 #1791 | 2322 021 121 221 #2171 | 2323 021 122 100 #538 | 2324 021 122 101 #538
2325 021 122 110 #538 | 2326 021 122 111 #1799 | 2327 021 122 120 #538 | 2328 021 122 121 #1791
2329 021 122 211 #2135 | 2330 021 200 100 #2144 | 2331 021 200 101 #538 | 2332 021 200 102 #534
2333 021 200 110 #538 | 2334 021 200 111 #538 | 2335 021 200 120 #538 | 2336 021 200 121 #538
2337 021 201 101 #538 | 2338 021 201 110 #538 | 2339 021 201 111 #538 | 2340 021 201 120 #2159
2341 021 201 121 #538 | 2342 021 202 101 #538 | 2343 021 202 110 #2159 | 2344 021 202 111 #538
2345 021 202 121 #538 | #2346 021 210 102 | 2347 021 220 101 #2159 | 2348 021 220 111 #538
2349 021 220 121 #538 | 2350 021 221 111 #1791 | 2351 021 221 121 #2171 | 2352 021 222 111 #2171
#2353 100 000 000 | #2354 100 000 001 | 2355 100 000 010 #2354 | 2356 100 000 011 #2354
#2357 100 000 020 | 2358 100 000 021 #2357 | 2359 100 000 100 #2354 | 2360 100 000 101 #2354
2361 100 000 110 #2354 | 2362 100 000 111 #2354 | 2363 100 000 120 #2357 | 2364 100 000 121 #2357
2365 100 000 200 #2357 | 2366 100 000 201 #2357 | 2367 100 000 210 #2357 | 2368 100 000 211 #2357
#2369 100 000 220 | 2370 100 000 221 #2369 | 2371 100 001 000 #2354 | 2372 100 001 001 #2354
2373 100 001 010 #2354 | 2374 100 001 011 #2354 | 2375 100 001 020 #2357 | 2376 100 001 021 #2357
2377 100 001 100 #2354 | 2378 100 001 101 #2354 | 2379 100 001 110 #2354 | 2380 100 001 111 #2354
2381 100 001 120 #2357 | 2382 100 001 121 #2357 | 2383 100 001 200 #2357 | 2384 100 001 201 #2357
2385 100 001 210 #2357 | 2386 100 001 211 #2357 | 2387 100 001 220 #2369 | 2388 100 001 221 #2369
2389 100 002 000 #2357 | 2390 100 002 001 #2357 | 2391 100 002 010 #2357 | 2392 100 002 011 #2357
#2393 100 002 020 | 2394 100 002 021 #2393 | 2395 100 002 100 #2357 | 2396 100 002 101 #2357
2397 100 002 110 #2357 | 2398 100 002 111 #2357 | 2399 100 002 120 #2393 | 2400 100 002 121 #2393
2401 100 002 200 #2393 | 2402 100 002 201 #2393 | 2403 100 002 210 #2393 | 2404 100 002 211 #2393
2405 100 002 220 #2393 | 2406 100 002 221 #2393 | #2407 100 020 000 | 2408 100 020 001 #2407
2409 100 020 010 #2407 | 2410 100 020 011 #2407 | 2411 100 020 020 #2407 | 2412 100 020 021 #2407
2413 100 020 100 #2407 | 2414 100 020 101 #2407 | 2415 100 020 110 #2407 | 2416 100 020 111 #2407
2417 100 020 120 #2407 | 2418 100 020 121 #2407 | 2419 100 020 200 #2407 | 2420 100 020 201 #2407
2421 100 020 210 #2407 | 2422 100 020 211 #2407 | 2423 100 020 220 #2407 | 2424 100 020 221 #2407
2425 100 021 000 #2407 | 2426 100 021 001 #2407 | 2427 100 021 010 #2407 | #2428 100 021 011
2429 100 021 020 #2407 | #2430 100 021 021 | 2431 100 021 100 #2407 | 2432 100 021 101 #2407
2433 100 021 110 #2407 | 2434 100 021 111 #2393 | 2435 100 021 120 #2407 | #2436 100 021 121
2437 100 021 200 #2407 | 2438 100 021 201 #2407 | 2439 100 021 210 #2407 | 2440 100 021 211 #2393
2441 100 021 220 #2407 | 2442 100 021 221 #2436 | 2443 100 022 000 #2407 | 2444 100 022 001 #2407
2445 100 022 010 #2407 | 2446 100 022 011 #2430 | 2447 100 022 020 #2407 | 2448 100 022 021 #2428
2449 100 022 100 #2407 | 2450 100 022 101 #2407 | 2451 100 022 110 #2407 | 2452 100 022 111 #2436
2453 100 022 120 #2407 | 2454 100 022 121 #2393 | 2455 100 022 200 #2407 | 2456 100 022 201 #2407

2457 100 022 210 #2407	**2458** 100 022 211 #2436	**2459** 100 022 220 #2407	#**2460** 100 100 000
#**2461** 100 100 001	#**2462** 100 100 010	#**2463** 100 100 011	#**2464** 100 100 020
2465 100 100 021 #2464	#**2466** 100 100 100	#**2467** 100 100 101	**2468** 100 100 110 #2460
2469 100 100 111 #2460	**2470** 100 100 120 #2464	**2471** 100 100 121 #2464	#**2472** 100 100 200
2473 100 100 201 #2464	**2474** 100 100 210 #2464	**2475** 100 100 211 #2464	#**2476** 100 100 220
2477 100 100 221 #2476	#**2478** 100 101 000	#**2479** 100 101 001	#**2480** 100 101 010
2481 100 101 020 #2464	**2482** 100 101 021 #2464	#**2483** 100 101 100	**2484** 100 101 120 #2464
2485 100 101 121 #2464	#**2486** 100 101 220	#**2487** 100 102 000	**2488** 100 102 001 #2487
2489 100 102 010 #2487	**2490** 100 102 011 #2487	**2491** 100 102 020 #2436	**2492** 100 102 021 #2436
#**2493** 100 102 100	**2494** 100 102 101 #2493	**2495** 100 102 110 #2487	**2496** 100 102 111 #2487
2497 100 102 120 #2436	**2498** 100 102 121 #2436	**2499** 100 102 200 #2436	**2500** 100 102 201 #2436
2501 100 102 210 #2436	**2502** 100 102 211 #2436	**2503** 100 102 220 #2436	**2504** 100 120 000 #2407
2505 100 120 001 #2407	**2506** 100 120 010 #2407	**2507** 100 120 011 #2407	**2508** 100 120 020 #2407
2509 100 120 021 #2407	**2510** 100 120 100 #2407	**2511** 100 120 101 #2407	**2512** 100 120 110 #2407
2513 100 120 111 #2407	**2514** 100 120 120 #2407	**2515** 100 120 121 #2407	**2516** 100 120 200 #2407
2517 100 120 201 #2407	**2518** 100 120 210 #2407	**2519** 100 120 211 #2407	**2520** 100 120 220 #2407
2521 100 120 221 #2407	**2522** 100 121 000 #2407	**2523** 100 121 001 #2407	**2524** 100 121 020 #2407
2525 100 121 100 #2407	**2526** 100 121 101 #2407	**2527** 100 121 200 #2407	**2528** 100 121 201 #2407
#**2529** 100 121 220	**2530** 100 122 000 #2407	**2531** 100 122 001 #2407	**2532** 100 122 010 #2407
2533 100 122 011 #2436	**2534** 100 122 020 #2407	**2535** 100 122 021 #2393	**2536** 100 122 100 #2407
2537 100 122 101 #2407	**2538** 100 122 110 #2407	#**2539** 100 122 111	**2540** 100 122 120 #2407
2541 100 122 121 #2369	**2542** 100 122 200 #2407	**2543** 100 122 201 #2407	**2544** 100 122 210 #2407
#**2545** 100 122 211	**2546** 100 200 000 #2407	**2547** 100 200 001 #2407	**2548** 100 200 010 #2407
2549 100 200 011 #2407	**2550** 100 200 020 #2407	**2551** 100 200 021 #2407	#**2552** 100 200 100
2553 100 200 101 #2407	**2554** 100 200 110 #2407	**2555** 100 200 111 #2407	**2556** 100 200 120 #2407
2557 100 200 121 #2407	#**2558** 100 200 200	**2559** 100 200 201 #2407	**2560** 100 200 210 #2407
2561 100 200 211 #2407	**2562** 100 200 221 #2407	**2563** 100 201 000 #2407	**2564** 100 201 001 #2407
2565 100 201 010 #2407	**2566** 100 201 011 #2407	**2567** 100 201 020 #2407	**2568** 100 201 021 #2407
2569 100 201 100 #2407	**2570** 100 201 101 #2407	**2571** 100 201 110 #2407	**2572** 100 201 111 #2407
2573 100 201 120 #2407	**2574** 100 201 121 #2407	**2575** 100 201 200 #2407	**2576** 100 201 201 #2407
2577 100 201 210 #2407	**2578** 100 201 211 #2407	**2579** 100 202 000 #2407	**2580** 100 202 001 #2407
2581 100 202 010 #2407	**2582** 100 202 011 #2407	**2583** 100 202 020 #2407	**2584** 100 202 021 #2407
2585 100 202 100 #2407	**2586** 100 202 101 #2407	**2587** 100 202 110 #2407	**2588** 100 202 111 #2407
2589 100 202 120 #2407	**2590** 100 202 121 #2407	**2591** 100 202 200 #2407	**2592** 100 202 201 #2407
2593 100 202 210 #2407	**2594** 100 202 211 #2407	**2595** 100 220 000 #2407	**2596** 100 220 001 #2407
2597 100 220 010 #2407	**2598** 100 220 011 #2407	**2599** 100 220 020 #2407	**2600** 100 220 021 #2407
2601 100 220 100 #2407	**2602** 100 220 101 #2407	**2603** 100 220 110 #2407	**2604** 100 220 111 #2407
2605 100 220 120 #2407	**2606** 100 220 121 #2407	**2607** 100 220 200 #2407	**2608** 100 220 201 #2407
2609 100 220 210 #2407	**2610** 100 220 211 #2407	**2611** 100 221 000 #2407	**2612** 100 221 001 #2407
2613 100 221 010 #2407	**2614** 100 221 011 #2393	**2615** 100 221 020 #2407	**2616** 100 221 021 #2436
2617 100 221 100 #2407	**2618** 100 221 101 #2407	**2619** 100 221 110 #2407	**2620** 100 221 111 #2369
2621 100 221 120 #2407	**2622** 100 221 121 #2486	**2623** 100 221 200 #2407	**2624** 100 221 201 #2407
2625 100 221 210 #2407	**2626** 100 221 211 #2369	**2627** 100 222 000 #2407	**2628** 100 222 001 #2407
2629 100 222 010 #2407	**2630** 100 222 011 #2436	**2631** 100 222 020 #2407	**2632** 100 222 021 #2393
2633 100 222 100 #2407	**2634** 100 222 101 #2407	**2635** 100 222 110 #2407	#**2636** 100 222 111
2637 100 222 120 #2407	**2638** 100 222 121 #2369	**2639** 100 222 200 #2407	**2640** 100 222 210 #2407
2641 100 222 210 #2407	**2642** 100 222 211 #2539	**2643** 101 000 000 #2354	**2644** 101 000 001 #2354
2645 101 000 010 #2354	**2646** 101 000 011 #2354	**2647** 101 000 020 #2357	**2648** 101 000 021 #2357
2649 101 000 100 #2354	**2650** 101 000 110 #2353	**2651** 101 000 110 #2354	**2652** 101 000 111 #2354
2653 101 000 120 #2357	#**2654** 101 000 121	**2655** 101 000 200 #2357	**2656** 101 000 201 #2357
2657 101 000 210 #2357	**2658** 101 000 211 #2357	**2659** 101 001 000 #2354	**2660** 101 001 001 #2354
2661 101 001 010 #2354	**2662** 101 001 011 #2354	**2663** 101 001 020 #2357	**2664** 101 001 021 #2357
2665 101 001 100 #2354	**2666** 101 001 101 #2354	**2667** 101 001 110 #2354	**2668** 101 001 111 #2354
2669 101 001 120 #2357	**2670** 101 001 121 #2357	**2671** 101 001 200 #2357	**2672** 101 001 201 #2357
2673 101 001 210 #2357	**2674** 101 001 211 #2357	**2675** 101 002 000 #2357	**2676** 101 002 001 #2357
2677 101 002 010 #2357	**2678** 101 002 011 #2357	**2679** 101 002 020 #2393	**2680** 101 002 021 #2393
2681 101 002 100 #2357	**2682** 101 002 101 #2654	**2683** 101 002 110 #2357	**2684** 101 002 111 #2357
2685 101 002 120 #2393	#**2686** 101 002 121	**2687** 101 002 200 #2393	**2688** 101 002 201 #2393
2689 101 002 210 #2393	**2690** 101 002 211 #2393	**2691** 101 020 000 #2407	**2692** 101 020 001 #2407
2693 101 020 010 #2407	**2694** 101 020 011 #2407	**2695** 101 020 020 #2407	**2696** 101 020 021 #2407
2697 101 020 100 #2407	#**2698** 101 020 101	**2699** 101 020 110 #2407	**2700** 101 020 111 #2407
2701 101 020 120 #2407	#**2702** 101 020 121	**2703** 101 020 200 #2407	**2704** 101 020 201 #2407
2705 101 020 210 #2407	**2706** 101 020 211 #2407	**2707** 101 021 000 #2407	**2708** 101 021 001 #2407
2709 101 021 010 #2407	**2710** 101 021 011 #2393	**2711** 101 021 020 #2407	**2712** 101 021 021 #2436
2713 101 021 100 #2407	**2714** 101 021 110 #2407	**2715** 101 021 110 #2407	**2716** 101 021 111 #2393
2717 101 021 120 #2407	**2718** 101 021 121 #2436	**2719** 101 021 200 #2407	**2720** 101 021 201 #2407
2721 101 021 210 #2407	**2722** 101 021 211 #2393	**2723** 101 022 000 #2407	**2724** 101 022 001 #2407
2725 101 022 010 #2407	**2726** 101 022 011 #2436	**2727** 101 022 020 #2407	**2728** 101 022 021 #2393
2729 101 022 100 #2407	**2730** 101 022 101 #2702	**2731** 101 022 110 #2407	**2732** 101 022 111 #2436
2733 101 022 120 #2407	**2734** 101 022 200 #2407	**2735** 101 022 201 #2407	**2736** 101 022 210 #2407

2737 101 022 211 #2436	2738 101 100 000 #2478	#2739 101 100 001	2740 101 100 010 #2480
2741 101 100 020 #2464	2742 101 100 021 #2464	2743 101 100 100 #2483	2744 101 100 120 #2464
2745 101 100 121 #2464	2746 101 102 000 #2487	2747 101 102 001 #2487	2748 101 102 010 #2487
2749 101 102 011 #2487	2750 101 102 020 #2436	2751 101 102 021 #2436	2752 101 102 100 #2493
2753 101 102 101 #2493	2754 101 102 110 #2487	2755 101 102 111 #2487	2756 101 102 120 #2436
2757 101 102 200 #2436	2758 101 102 201 #2436	2759 101 102 210 #2436	2760 101 102 211 #2436
2761 101 120 000 #2407	2762 101 120 001 #2407	2763 101 120 010 #2407	2764 101 120 011 #2407
2765 101 120 020 #2407	2766 101 120 021 #2407	2767 101 120 100 #2407	2768 101 120 101 #2407
2769 101 120 110 #2407	2770 101 120 111 #2407	2771 101 120 120 #2407	2772 101 120 121 #2407
2773 101 120 200 #2407	2774 101 120 201 #2407	2775 101 120 210 #2407	2776 101 120 211 #2407
2777 101 122 000 #2407	2778 101 122 001 #2407	2779 101 122 010 #2407	2780 101 122 011 #2436
2781 101 122 020 #2407	2782 101 122 021 #2393	2783 101 122 100 #2407	2784 101 122 101 #2407
2785 101 122 110 #2407	2786 101 122 111 #2487	2787 101 122 120 #2407	2788 101 122 200 #2407
2789 101 122 201 #2407	2790 101 122 210 #2407	2791 101 122 211 #2487	2792 101 200 000 #2407
2793 101 200 001 #2407	2794 101 200 010 #2407	2795 101 200 011 #2407	2796 101 200 020 #2407
2797 101 200 021 #2407	2798 101 200 100 #2407	#2799 101 200 101	2800 101 200 110 #2407
2801 101 200 111 #2407	2802 101 200 120 #2407	#2803 101 200 121	2804 101 200 200 #2407
2805 101 200 201 #2407	2806 101 200 210 #2407	2807 101 200 211 #2407	2808 101 201 000 #2407
2809 101 201 001 #2407	2810 101 201 010 #2407	2811 101 201 011 #2407	2812 101 201 020 #2407
2813 101 201 021 #2407	2814 101 201 100 #2407	2815 101 201 101 #2407	2816 101 201 110 #2407
2817 101 201 111 #2407	2818 101 201 120 #2407	2819 101 201 200 #2407	2820 101 201 201 #2407
2821 101 201 210 #2407	2822 101 201 211 #2407	2823 101 202 000 #2407	2824 101 202 001 #2407
2825 101 202 010 #2407	2826 101 202 011 #2407	2827 101 202 020 #2407	2828 101 202 021 #2407
2829 101 202 100 #2407	2830 101 202 101 #2698	2831 101 202 110 #2407	2832 101 202 111 #2407
2833 101 202 120 #2407	2834 101 202 200 #2407	2835 101 202 201 #2407	2836 101 202 210 #2407
2837 101 202 211 #2407	2838 101 220 000 #2407	2839 101 220 001 #2407	2840 101 220 010 #2407
2841 101 220 011 #2407	2842 101 220 020 #2407	2843 101 220 021 #2407	2844 101 220 100 #2407
2845 101 220 101 #2799	2846 101 220 110 #2407	2847 101 220 111 #2407	2848 101 220 120 #2407
2849 101 220 200 #2407	2850 101 220 201 #2407	2851 101 220 210 #2407	2852 101 220 211 #2407
2853 101 221 000 #2407	2854 101 221 200 #2529	2855 101 222 000 #2407	2856 101 222 001 #2407
2857 101 222 010 #2407	2858 101 222 011 #2436	2859 101 222 020 #2407	2860 101 222 021 #2393
2861 101 222 100 #2407	2862 101 222 101 #2698	2863 101 222 110 #2407	2864 101 222 111 #2493
2865 101 222 120 #2407	2866 101 222 201 #2407	2867 101 222 210 #2407	2868 101 222 211 #2487
2869 102 000 000 #2357	2870 102 000 001 #2357	2871 102 000 010 #2357	2872 102 000 011 #2357
2873 102 000 020 #2393	2874 102 000 021 #2393	2875 102 000 100 #2357	2876 102 000 101 #2357
2877 102 000 110 #2357	2878 102 000 111 #2357	2879 102 000 120 #2393	2880 102 000 201 #2393
2881 102 000 210 #2393	2882 102 000 211 #2393	2883 102 001 000 #2357	2884 102 001 001 #2357
2885 102 001 010 #2357	2886 102 001 011 #2357	2887 102 001 020 #2393	2888 102 001 021 #2393
2889 102 001 100 #2357	2890 102 001 101 #2357	2891 102 001 110 #2357	2892 102 001 111 #2357
2893 102 001 120 #2393	2894 102 001 201 #2393	2895 102 001 210 #2393	2896 102 001 211 #2393
2897 102 002 000 #2369	2898 102 002 001 #2369	2899 102 002 010 #2369	2900 102 002 011 #2369
2901 102 002 020 #2393	2902 102 002 021 #2393	2903 102 002 100 #2369	2904 102 002 101 #2369
2905 102 002 110 #2369	2906 102 002 111 #2369	2907 102 002 120 #2393	2908 102 002 201 #2393
2909 102 002 210 #2393	2910 102 002 211 #2393	2911 102 020 000 #2407	2912 102 020 001 #2407
2913 102 020 010 #2407	2914 102 020 011 #2407	2915 102 020 020 #2407	2916 102 020 021 #2407
2917 102 020 100 #2407	2918 102 020 101 #2407	2919 102 020 110 #2407	2920 102 020 111 #2407
2921 102 020 120 #2407	2922 102 020 201 #2407	2923 102 020 210 #2407	2924 102 020 211 #2407
2925 102 021 000 #2407	2926 102 021 001 #2407	2927 102 021 010 #2407	2928 102 021 020 #2407
2929 102 021 021 #2436	2930 102 021 100 #2407	2931 102 021 101 #2407	2932 102 021 120 #2407
2933 102 021 201 #2407	#2934 102 021 210	2935 102 022 000 #2407	2936 102 022 001 #2407
2937 102 022 010 #2407	2938 102 022 011 #2436	2939 102 022 020 #2407	2940 102 022 100 #2407
2941 102 022 101 #2407	2942 102 022 110 #2407	2943 102 022 111 #2436	2944 102 022 120 #2407
2945 102 022 201 #2407	2946 102 022 211 #2436	2947 102 102 000 #2539	2948 102 102 001 #2539
2949 102 102 010 #2545	2950 102 102 020 #2436	2951 102 102 100 #2636	2952 102 102 120 #2436
2953 102 120 000 #2407	2954 102 120 001 #2407	2955 102 120 010 #2407	2956 102 120 011 #2407
2957 102 120 020 #2407	2958 102 120 021 #2407	2959 102 120 100 #2407	2960 102 120 101 #2407
2961 102 120 110 #2407	2962 102 120 111 #2407	2963 102 120 120 #2407	2964 102 120 201 #2407
2965 102 120 211 #2407	2966 102 122 000 #2407	2967 102 122 001 #2407	2968 102 122 010 #2407
2969 102 122 011 #2436	2970 102 122 020 #2407	2971 102 122 100 #2407	2972 102 122 101 #2407
2973 102 122 110 #2407	2974 102 122 111 #2487	2975 102 122 120 #2407	2976 102 122 201 #2407
2977 102 122 211 #2487	2978 102 200 000 #2407	2979 102 200 001 #2407	2980 102 200 010 #2407
2981 102 200 011 #2407	2982 102 200 020 #2407	2983 102 200 021 #2407	2984 102 200 100 #2407
2985 102 200 101 #2407	2986 102 200 110 #2407	2987 102 200 111 #2407	2988 102 200 120 #2407
2989 102 200 201 #2407	2990 102 201 000 #2407	2991 102 201 010 #2407	2992 102 201 020 #2407
2993 102 201 010 #2407	2994 102 201 011 #2407	2995 102 201 020 #2407	2996 102 201 100 #2407
2997 102 201 101 #2407	2998 102 201 110 #2407	2999 102 201 111 #2407	3000 102 201 120 #2407
3001 102 201 201 #2407	3002 102 201 211 #2407	3003 102 202 000 #2407	3004 102 202 001 #2407
3005 102 202 010 #2407	3006 102 202 011 #2407	3007 102 202 020 #2407	3008 102 202 100 #2407
3009 102 202 101 #2407	3010 102 202 110 #2407	3011 102 202 111 #2407	3012 102 202 120 #2407
3013 102 202 201 #2407	3014 102 202 211 #2407	3015 102 220 000 #2407	3016 102 220 001 #2407

3017 102 220 010 #2407	**3018** 102 220 011 #2407	**3019** 102 220 020 #2407	**3020** 102 220 100 #2407
3021 102 220 101 #2407	**3022** 102 220 110 #2407	**3023** 102 220 111 #2407	**3024** 102 220 120 #2407
3025 102 220 201 #2407	**3026** 102 220 211 #2407	**3027** 102 222 000 #2407	**3028** 102 222 001 #2407
3029 102 222 010 #2407	**3030** 102 222 011 #2436	**3031** 102 222 020 #2407	**3032** 102 222 100 #2407
3033 102 222 101 #2407	**3034** 102 222 110 #2407	**3035** 102 222 111 #2493	**3036** 102 222 120 #2407
3037 102 222 201 #2407	**3038** 102 222 211 #2487	**3039** 110 000 000 #2460	**3040** 110 000 001 #2461
3041 110 000 010 #2478	**3042** 110 000 011 #2479	**3043** 110 000 020 #2487	**3044** 110 000 100 #2478
3045 110 000 101 #2739	**3046** 110 000 110 #2460	**3047** 110 000 111 #2460	**3048** 110 000 120 #2487
3049 110 000 201 #2487	**3050** 110 000 211 #2487	**3051** 110 001 000 #2462	**3052** 110 001 001 #2463
3053 110 001 010 #2480	**3054** 110 001 020 #2487	**3055** 110 001 100 #2480	**3056** 110 001 120 #2487
3057 110 002 000 #2464	**3058** 110 002 001 #2464	**3059** 110 002 010 #2464	**3060** 110 002 011 #2464
3061 110 002 020 #2436	**3062** 110 002 100 #2464	**3063** 110 002 101 #2464	**3064** 110 002 110 #2464
3065 110 002 111 #2464	**3066** 110 002 120 #2436	**3067** 110 002 201 #2436	**3068** 110 002 211 #2436
3069 110 020 000 #2407	**3070** 110 020 001 #2407	**3071** 110 020 010 #2407	**3072** 110 020 011 #2407
3073 110 020 020 #2407	**3074** 110 020 100 #2407	**3075** 110 020 101 #2407	**3076** 110 020 110 #2407
3077 110 020 111 #2407	**3078** 110 020 120 #2407	**3079** 110 020 201 #2407	**3080** 110 020 211 #2407
3081 110 022 000 #2407	**3082** 110 022 001 #2407	**3083** 110 022 010 #2407	**3084** 110 022 011 #2436
3085 110 022 020 #2407	**3086** 110 022 100 #2407	**3087** 110 022 101 #2407	**3088** 110 022 110 #2407
3089 110 022 111 #2436	**3090** 110 022 120 #2407	**3091** 110 022 201 #2407	**3092** 110 120 000 #2407
3093 110 120 001 #2407	**3094** 110 120 010 #2407	**3095** 110 120 011 #2407	**3096** 110 120 020 #2407
3097 110 120 100 #2407	**3098** 110 120 101 #2407	**3099** 110 120 110 #2407	**3100** 110 120 111 #2407
3101 110 120 120 #2407	**3102** 110 120 201 #2407	**3103** 110 120 211 #2407	**3104** 110 122 000 #2407
3105 110 122 001 #2407	**3106** 110 122 020 #2529	**3107** 110 122 100 #2407	**3108** 110 122 101 #2407
3109 110 122 201 #2407	**3110** 110 200 000 #2407	**3111** 110 200 001 #2407	**3112** 110 200 010 #2407
3113 110 200 011 #2407	**3114** 110 200 100 #2407	**3115** 110 200 101 #2407	**3116** 110 200 110 #2407
3117 110 200 111 #2407	**3118** 110 200 120 #2407	**3119** 110 200 201 #2407	**3120** 110 200 211 #2407
3121 110 201 000 #2407	**3122** 110 201 001 #2407	**3123** 110 201 010 #2407	**3124** 110 201 011 #2407
3125 110 201 100 #2407	**3126** 110 201 101 #2407	**3127** 110 201 110 #2407	**3128** 110 201 111 #2407
3129 110 201 120 #2407	**3130** 110 201 201 #2407	**3131** 110 202 000 #2407	**3132** 110 202 001 #2407
3133 110 202 010 #2407	**3134** 110 202 011 #2407	**3135** 110 202 100 #2407	**3136** 110 202 101 #2407
3137 110 202 110 #2407	**3138** 110 202 111 #2407	**3139** 110 202 120 #2407	**3140** 110 202 201 #2407
3141 110 220 000 #2407	**3142** 110 220 001 #2407	**3143** 110 220 010 #2407	**3144** 110 220 011 #2407
3145 110 220 100 #2407	**3146** 110 220 101 #2407	**3147** 110 220 110 #2407	**3148** 110 220 111 #2407
3149 110 220 120 #2407	**3150** 110 220 201 #2407	**3151** 110 222 000 #2407	**3152** 110 222 001 #2407
3153 110 222 010 #2407	**3154** 110 222 011 #2436	**3155** 110 222 100 #2407	**3156** 110 222 101 #2407
3157 110 222 110 #2407	**3158** 110 222 111 #2493	**3159** 110 222 120 #2407	**3160** 110 222 201 #2407
3161 111 000 001 #2466	**3162** 111 000 001 #2467	**3163** 111 000 020 #2483	**3164** 111 000 100 #2483
3165 111 000 120 #2493	**3166** 111 002 000 #2464	**3167** 111 002 001 #2464	**3168** 111 002 010 #2464
3169 111 002 011 #2464	**3170** 111 002 100 #2464	**3171** 111 002 101 #2464	**3172** 111 002 110 #2464
3173 111 002 111 #2472	**3174** 111 002 120 #2436	**3175** 111 002 201 #2436	**3176** 111 020 000 #2407
3177 111 020 001 #2407	**3178** 111 020 010 #2407	**3179** 111 020 011 #2407	**3180** 111 020 100 #2407
3181 111 020 101 #2407	**3182** 111 020 110 #2407	**3183** 111 020 111 #2558	**3184** 111 020 120 #2407
3185 111 020 201 #2407	**3186** 111 022 000 #2407	**3187** 111 022 001 #2407	**3188** 111 022 010 #2407
3189 111 022 011 #2476	**3190** 111 022 100 #2407	**3191** 111 022 101 #2407	**3192** 111 022 110 #2407
3193 111 022 120 #2407	**3194** 111 022 201 #2407	**3195** 111 120 000 #2407	**3196** 111 120 001 #2407
3197 111 120 010 #2407	**3198** 111 120 011 #2407	**3199** 111 120 100 #2407	**3200** 111 120 101 #2407
3201 111 120 110 #2407	**3202** 111 120 111 #2407	**3203** 111 120 120 #2407	**3204** 111 120 201 #2407
3205 111 200 000 #2407	**3206** 111 200 001 #2407	**3207** 111 200 010 #2407	**3208** 111 200 011 #2407
3209 111 200 100 #2799	**3210** 111 200 101 #2407	**3211** 111 200 110 #2407	**3212** 111 200 111 #2552
3213 111 200 120 #2407	**3214** 111 200 201 #2407	**3215** 111 201 000 #2407	**3216** 111 201 001 #2407
3217 111 201 010 #2407	**3218** 111 201 011 #2407	**3219** 111 201 100 #2407	**3220** 111 201 101 #2407
3221 111 201 110 #2407	**3222** 111 201 120 #2407	**3223** 111 201 201 #2407	**3224** 111 202 000 #2407
3225 111 202 001 #2407	**3226** 111 202 010 #2407	**3227** 111 202 011 #2407	**3228** 111 202 100 #2407
3229 111 202 101 #2407	**3230** 111 202 110 #2407	**3231** 111 202 120 #2407	**3232** 111 202 201 #2407
3233 111 220 000 #2407	**3234** 111 220 001 #2407	**3235** 111 220 010 #2407	**3236** 111 220 011 #2407
3237 111 220 100 #2407	**3238** 111 220 101 #2407	**3239** 111 220 110 #2407	**3240** 111 220 120 #2407
3241 111 220 201 #2407	**#3242** 111 222 000	**3243** 112 002 001 #2476	**3244** 112 002 010 #2486
3245 112 002 100 #2486	**3246** 112 002 120 #2436	**3247** 112 020 001 #2407	**3248** 112 020 010 #2407
3249 112 020 011 #2407	**3250** 112 020 100 #2407	**3251** 112 020 101 #2407	**3252** 112 020 110 #2407
3253 112 020 120 #2407	**3254** 112 020 201 #2407	**3255** 112 022 001 #2407	**3256** 112 022 010 #2529
3257 112 022 100 #2407	**3258** 112 022 101 #2407	**3259** 112 022 120 #2407	**3260** 112 022 201 #2407
3261 112 120 001 #2407	**3262** 112 120 011 #2407	**3263** 112 120 100 #2407	**3264** 112 120 101 #2407
3265 112 120 110 #2407	**3266** 112 120 120 #2407	**3267** 112 120 201 #2407	**3268** 112 200 001 #2407
3269 112 200 011 #2407	**3270** 112 200 100 #2803	**3271** 112 200 101 #2407	**3272** 112 200 110 #2407
3273 112 200 120 #2407	**3274** 112 200 201 #2407	**3275** 112 201 001 #2407	**3276** 112 201 100 #2407
3277 112 201 101 #2407	**3278** 112 201 110 #2407	**3279** 112 201 120 #2407	**3280** 112 201 201 #2407
3281 112 202 001 #2407	**3282** 112 202 100 #2407	**3283** 112 202 101 #2407	**3284** 112 202 110 #2407
3285 112 202 120 #2407	**3286** 112 202 201 #2407	**3287** 112 220 001 #2407	**3288** 112 220 100 #2407
3289 112 220 101 #2407	**3290** 112 220 110 #2407	**3291** 112 220 120 #2407	**3292** 112 220 201 #2407
3293 120 020 001 #2407	**3294** 120 020 100 #2407	**3295** 120 020 101 #2407	**3296** 120 020 110 #2407

3297 120 020 120 #2407	**3298** 120 020 201 #2407	**3299** 120 120 001 #2407	**3300** 120 120 100 #2407
3301 120 120 101 #2407	**3302** 120 120 120 #3242	**3303** 120 120 201 #2407	**3304** 120 200 001 #2407
3305 120 200 100 #2407	**3306** 120 200 101 #2407	**3307** 120 200 110 #2407	**3308** 120 200 201 #2407
3309 120 201 001 #2407	**3310** 120 201 100 #2407	**3311** 120 202 001 #2407	**3312** 120 202 100 #2407
3313 120 202 101 #2407	**3314** 120 202 110 #2407	**3315** 120 220 001 #2407	**3316** 120 220 100 #2407
3317 120 220 101 #2407	**3318** 120 220 110 #2407	**3319** 121 020 001 #2407	**3320** 121 020 100 #2407
3321 121 020 101 #2799	**3322** 121 020 110 #2407	**3323** 121 200 001 #2407	**3324** 121 200 100 #2803
3325 121 202 001 #2407	**3326** 121 202 100 #2407	**3327** 121 202 110 #2407	**3328** 121 220 100 #2529
3329 122 020 001 #2407	**3330** 122 020 110 #2529		

References

[1] J. Berman, *Free spectra of 3-element algebras*, Proceedings of "Universal Algebra and Lattice Theory", Puebla 1982, Lecture Notes in Mathematics **1004**, 1983, pp. 10–53.

[2] J. Berman and S. Burris, *Algebras which generate decidable varieties*, in preparation.

[3] J. Berman, E.W. Kiss, P. Pröhle, and Á. Szendrei, *The set of types of a finitely generated variety*, Discrete Math. **112** (1993), 1–20.

[4] J. Berman and R. McKenzie, *Clones satisfying the term condition*, Discrete Math. **52** (1984), 7–29.

[5] J. Berman and B. Wolk, *Free lattices in some small varieties*, Algebra Universalis **10** (1980), 269–289.

[6] S. Burris and H.P. Sankappanavar, *A Course in Universal Algebra*, Springer Verlag, 1981.

[7] A. Fearnley, *Les clones sur trois elements de la forme pol(ρ) où ρ est une relation unaire où binaire*, Master's thesis, Université de Montréal, 1992.

[8] R. Freese, R. McKenzie, G. McNulty, and W. Taylor, *Algebras, Lattices, Varieties Vol. II*, Wadsworth and Brooks/Cole, in preparation.

[9] R. McKenzie and M. Valeriote, *The Structure of Decidable Locally Finite Varieties*, Birkhauser, 1989.

DEPARTMENT OF MATHEMATICS, STATISTICS, AND COMPUTER SCIENCE, UNIVERSITY OF ILLINOIS AT CHICAGO, 851 S. MORGAN, CHICAGO, ILLINOIS, USA 60607-7045

DEPARTMENT OF PURE MATHEMATICS, UNIVERSITY OF WATERLOO, WATERLOO, ONTARIO, CANADA N2L 3G1
   e-mail address: snburris@thoralf.uwaterloo.ca

Ideal properties in congruences

Ivan Chajda

Abstract

The concept of an "ideal" in universal algebra was introduced by
A. Ursini around 1970. If I is an ideal of an algebra $\mathbf{A}$ and $p = q$
is a congruence identity, we say that $\mathbf{A}$ satisfies the ideal congruence
identity $p = q$ if $[I]_p = \{x : \langle x, i \rangle \in p$ for some $i \in I\} = [I]_q$. Mal'cev
type conditions for ideal congruence permutability and ideal congruence
distributivity are given.

The concept of an "ideal" in universal algebras was introduced by A. Ursini
[6] and was intensively investigated by H.P. Gumm and A. Ursini [5].

Let us recall some basic definitions. Let $\mathcal{K}$ be a class of algebras of
type τ. If all algebras have a constant 0 which is either a nullary operation or an equationally definable constant, then we say that $\mathcal{K}$ is a class
with 0. An $(n + m) - ary$ term p of type τ is called an **ideal term in**
$y_1, \ldots, y_m$ if $p(x_1, \ldots, x_n, 0, \ldots, 0) = 0$ is an identity of $\mathcal{K}$. A non-empty
subset I of an algebra $\mathbf{A} \in \mathcal{K}$ is an **ideal** of $\mathbf{A}$ if for every ideal term
$p(x_1, \ldots, x_n, y_1, \ldots, y_m)$ in $y_1, \ldots, y_m$ and for all $a_1, \ldots, a_n \in A$, $i_1, \ldots, i_m \in I$,
we have $p(a_1, \ldots, a_n, i_1, \ldots, i_m) \in I$, see [6], [5]. If C is a subset of an algebra
$\mathbf{A}$ and R is a binary relation on A, set

$$[C]_R = \{x \in A : \langle x, c \rangle \in R \text{ for some } c \in C\}.$$

G. Gratzer [4] introduced the following concept: an algebra $\mathbf{A}$ has **weakly
associative congruences** if for every subalgebra $\mathbf{B}$ of $\mathbf{A}$ and any $\theta, \varphi \in$
$\mathrm{Con}(\mathbf{A})$

$$[[B]_\theta]_\varphi = [[B]_\varphi]_\theta.$$

An example of such an algebra is given in [1] and it is also proven that for a
variety $\mathcal{V}$ this condition is equivalent to congruence permutability of $\mathcal{V}$. We
can modify this concept replacing subalgebras with ideals:

Definition 1 An algebra $\mathbf{A}$ with 0 has **ideal permutable congruences** if
$[I]_{\theta \circ \varphi} = [I]_{\varphi \circ \theta}$ for all $\theta, \varphi \in \mathrm{Con}(\mathbf{A})$ and for every ideal I of $\mathbf{A}$. A variety $\mathcal{V}$
with 0 has **ideal permutable congruences** if every $\mathbf{A} \in \mathcal{V}$ has this property.

431

The following special concept was introduced independently in [1] and [5].

Definition 2 An algebra **A** wit 0 is **congruence permutable at** 0 if $[0]_{\theta \circ \varphi} = [0]_{\varphi \circ \theta}$ for all $\theta, \varphi \in \mathrm{Con}(\mathbf{A})$. A variety $\mathcal{V}$ with 0 is **congruence permutable at** 0 if every $\mathbf{A} \in \mathcal{V}$ has this property.

Since $\{0\}$ is always an ideal of **A**, this concept of permutability at 0 is a special case of the former one. Varieties which are congruence permutable at 0 have been characterized by a Mal'cev condition in [1], [3], [5].

Evidently the set I(**A**) of all ideals of an algebra **A** with 0 forms a complete lattice with respect to set inclusion. Therefore for any subset M of A, there exists the least ideal of **A** containing M; denote it by $I(M)$ and call it the **ideal generated by** M. If M is a finite subset, say $\{b_1, \ldots, b_k\}$, then $I(M)$ will be denoted by $I(b_1, \ldots, b_k)$. The following Lemma is obvious (see also [5]):

Lemma 1 *Let $\mathbf{A}$ be an algebra with 0 and let $b \in A$. Then*

$$I(b) = \{p(a_1, \ldots, a_n, b) : p(x_1, \ldots, x_n, y) \text{ an ideal term in } y, \ a_1, \ldots, a_n \in A\}.$$

The following result was first mentioned by J. Duda (unpublished).

Theorem 1 *Let $\mathcal{V}$ be a variety with 0. The following conditions are equivalent:*

(1) $\mathcal{V}$ has ideal permutable congruences;

(2) $\mathcal{V}$ is congruence permutable at 0.

Proof. (1) $\Longrightarrow$ (2) is trivial, since $\{0\}$ is always an ideal of **A** for every $\mathbf{A} \in \mathcal{V}$.

(2) $\Longrightarrow$ (1). Let $\mathcal{V}$ be a variety with 0 which is congruence permutable at 0. By Theorem 1 in [1] (2) is equivalent to the existence of a binary term $d(x, y)$ satisfying the identities

$$d(x, x) = 0 \qquad\qquad d(x, 0) = x.$$

Now, let $\mathbf{A} \in \mathcal{V}$, let I be an ideal of **A** and let $\theta, \varphi \in \mathrm{Con}(\mathbf{A})$. Suppose, e.g., that $a \in [I]_{\theta \circ \varphi}$ for some $a \in A$. Then there exist elements $b \in A$ and $i \in I$ with

$$\langle a, b \rangle \in \theta \qquad \text{and} \qquad \langle b, i \rangle \in \varphi.$$

Set $p(x, y, z) = d(x, d(y, z))$ and $q(x, y) = d(x, d(x, y))$. Clearly $q(x, y)$ is an ideal term in y and

$$p(x, y, y) = d(x, d(y, y)) = d(x, 0) = x.$$

Thus $\langle b, i \rangle \in \varphi$ implies that

$$\langle a, p(a, b, i) \rangle = \langle p(a, b, b), p(a, b, i) \rangle \in \varphi$$

and $\langle a, b \rangle \in \theta$ implies that

$$\langle p(a, b, i), q(a, i) \rangle = \langle p(a, b, i), , p(a, a, i) \rangle \in \theta.$$

But $q(a, i) \in I$, giving $a \in [I]_{\varphi \circ \theta}$. $\qquad \square$

For some examples of varieties that are congruence permutable at 0 see [1], [2] and [3].

We are going to deal with another important congruence property, namely distributivity

Definition 3 And algebra **A** with 0 is **ideal congruence distributive** if

$$[I]_{\theta \wedge (\varphi \vee \psi)} = [I]_{(\theta \wedge \varphi) \vee (\theta \wedge \psi)}$$

for every ideal I of **A** and all $\theta, \varphi, \psi \in \mathrm{Con}(\mathbf{A})$. A variety $\mathcal{V}$ with 0 is **ideal congruence distributive** if every $\mathbf{A} \in \mathcal{V}$ has ideal distributive congruences.

Theorem 2 *For a variety $\mathcal{V}$ with 0 the following conditions are equivalent:*

(1) $\mathcal{V}$ us ideal congruence distributive;

(2) there exist an integer n and ternary terms $p_0(x, y, z), \ldots, p_n(x, y, z)$ such that

$$\begin{aligned}
&p_0(x, y, z) = x, \qquad p_n(x, y, 0) = 0 \\
&p_i(x, y, x) = x \ \text{ for each } \ i = 0, \ldots, n \\
&p_i(x, x, z) = p_{i+1}(x, x, z) \quad \text{for } i \text{ even} \\
&p_i(x, z, z) = p_{i+1}(x, z, z) \quad \text{for } i \text{ odd.}
\end{aligned}$$

Proof. (1) $\implies$ (2) Let $\mathbf{F}_{\mathcal{V}}(x, y, z)$ be the free algebra of $\mathcal{V}$ on x, y, z and $I = I(z)$ the ideal of $\mathbf{F}_{\mathcal{V}}(x, y, z)$ generated by the set $\{z\}$. Let $\theta = \vartheta(x, z)$, $\varphi = \vartheta(x, y)$, $\psi = \vartheta(y, z)$ and set $\Gamma = \theta \wedge (\varphi \vee \psi)$, $\Omega = (\theta \wedge \varphi) \vee (\theta \wedge \psi)$. Since $\langle x, z \rangle \in \Gamma$ and $z \in I$ we have $x \in [I]_{\Gamma}$. By (1), this gives $x \in I_{\Omega}$, i.e. there exists an element $w \in I(z)$ with $\langle x, w \rangle \in \Omega$. Thus there exists elements $c_0, \ldots, c_n \in \mathbf{F}_{\mathcal{V}}(x, y, z)$ such that $x = c_0, c_n = w$ and

$$\begin{aligned}
&\langle c_i, c_{i+1} \rangle \in \theta \quad \text{for } i = 0, \ldots, n - 1 \\
&\langle c_i, c_{i+1} \rangle \in \varphi \quad \text{for } i \text{ even} \\
&\langle c_i, c_{i+1} \rangle \in \psi \quad \text{for } i \text{ odd.}
\end{aligned}$$

Hence there exists ternary terms $p_0(x, y, z), \ldots, p_n(x, y, z)$, $(i = 0, \ldots, n)$ with $c_i = p_i(x, y, z)$ satisfying

$$x = p_0(x, y, z) \qquad p_n(x, y, z) \in I(z)$$

and all the remaining identities of (2) except for $p_n(x, y, 0) = 0$. We now prove the last one. Since $p_n(x, y, z) \in I(z)$, by Lemma 1, there exists an ideal term $t(x_1, \ldots, x_n, u)$ in u such that $p(x, y, z) = t(x_1, \ldots, x_n, z)$ for some $x_1, \ldots, x_n \in \mathbf{F}_\mathcal{V}(x, y, z)$. Since $\mathbf{F}_\mathcal{V}(x, y, z)$ has exactly three free generators, x, y, z, we can take $n = 3$ and $x_1 = x, x_2 = y, x_3 = z$. Thus $p_n(x, y, z) = t(x, y, z, z)$ and, since t is an ideal term $t(x, y, z, 0) = 0$. Hence $p_n(x, y, 0) = t(x, y, 0, 0) = 0$.

(2) $\Longrightarrow$ (1). Let $\mathbf{A} \in \mathcal{V}$ and let $\theta, \varphi, \psi \in \mathrm{Con}(\mathbf{A})$. Let I be an ideal of $\mathbf{A}$ and let $a \in [I]_\Gamma$ for $\Gamma = \theta \wedge (\varphi \vee \psi)$. Then $\langle a, v \rangle \in \Gamma$ for some $v \in I$, i.e. $\langle a, v \rangle \in \theta$ and there exists $b_0, \ldots, b_m \in A$ with $a = b_0$, $b_m = v$ and

$$\langle b_j, b_{j+1} \rangle \in \varphi \quad \text{for } i \text{ even}$$
$$\langle b_j, b_{j+1} \rangle \in \psi \quad \text{for } i \text{ odd}.$$

Using (1) we obtain

$$\langle a, p_i(a, b_j, v) \rangle = \langle p_i(a, b_j, a), p_i(a, b_j, v) \rangle \in \theta$$

for all i, j. Moreover

$$p_i(a, a, v) \, \varphi \, p_i(a, b_1, v) \, \psi \, p_i(a, b_2, v) \, \psi \, \ldots \, p_i(a, v, v),$$

thus also

$$p_i(a, a, v) \, \theta \wedge \varphi \, p_i(a, b_1, v) \, \theta \wedge \psi \, p_i(a, b_2, v) \, \ldots \, p_i(a, v, v).$$

Hence $\langle p_i(a, a, v), p_i(a, v, v) \rangle \in \Omega$ for $\Omega = (\theta \wedge \varphi) \vee (\theta \wedge \psi)$ and every $i = 0, \ldots, n$. We conclude that

$$a = p_0(a, a, v) = p_1(a, a, v) \, \Omega \, p_1(a, v, v) = p_2(a, v, v) \, \Omega \, \ldots p_n(a, x, v)$$

where $x = v$ for i even and $x = a$ for i odd. Since $p_n(x, y, 0) = 0$, p_n is an ideal term in the last variable, thus $v \in I$ implies also that $p_n(a, x, v) \in I$, whence $a \in [I]_\Omega$. $\qquad\qquad\qquad\qquad\qquad\qquad\qquad\qquad\qquad\qquad\qquad\qquad\qquad\square$

Remark 1 By Jónsson's Lemma, every congruence distributive variety with 0 is, via Theorem 2, also ideal congruence distributive. On the other hand it was shown by G. Czédli that, for $I = \{0\}$, the congruence identity of Definition 3 is not equivalent to its dual in the general case (e.g., in the case of semilattices with 0).

Example 1 The variety $\mathcal{P}$ of all relatively pseudocomplemented semilattices is a variety of type $(2, 2, 0)$, with operations $\wedge, *, 0$, satisfying the semilattice identities for $\wedge$ and, moreover,

$$y * x \wedge x = x \qquad z * z \wedge x = x$$
$$x * z \wedge x = x \wedge z \qquad x \wedge 0 = 0.$$

Put $n = 3$ and $p_0(x, y, z) = p_1(x, y, z) = x$, $p_2(x, y, z) = y * z \wedge x$, and $p_3(x, y, z) = x \wedge z$. It is easy to show that $\mathcal{P}$ is ideal congruence distributive.

Example 2 The variety $\mathcal{S}$ of skew lattices with 0 is a variety of type $(2, 2, 0)$ with operations denoted by $\vee, \wedge, 0$ satisfying the following identities:

$$x \wedge x = x \qquad x \vee x = x \qquad x \wedge 0 = 0$$
$$x \wedge (y \vee x) = x \qquad x \wedge (x \vee y) = x.$$

The operations $\vee$ and $\wedge$ need not be associative or commutative in general. Put $n = 2$ and

$$p_0(x, y, z) = x, p_1(x, y, z) = x \wedge (z \vee y), p_2(x, y, z) = x \wedge z.$$

It is easy to verify (2) of Theorem 2, thus $\mathcal{S}$ is ideal congruence distributive.

The concept of ideal permutable congruences can be generalized in the following way: and algebra $\mathbf{A}$ with 0 has **ideal 3-permutable congruences** if for every ideal I of $\mathbf{A}$ and every $\theta, \varphi \in \mathrm{Con}(\mathbf{A})$ the following holds

$$[I]_{\theta \circ \varphi \circ \theta} = [I]_{\varphi \circ \theta \circ \varphi},$$

which is clearly equivalent to $[I]_{\theta \vee \varphi} = [I]_{\theta \circ \varphi \circ \theta}$. Recall that $\mathbf{A}$ is **congruence 3-permutable** if $\theta \circ \varphi \circ \theta = \varphi \circ \theta \circ \varphi$ for every $\theta, \varphi \in \mathrm{Con}(\mathbf{A})$.

It was first shown by B. Jónsson that, if $\mathbf{A}$ is congruence 3-permutable, then $\mathrm{Con}(\mathbf{A})$ is a modular lattice. A modification of this statement can be proven also for ideal 3-permutable congruences:

Theorem 3 *Let $\mathbf{A}$ be an algebra with 0 having ideal 3-permutable congruences. Let $\theta, \varphi, \psi \in \mathrm{Con}(\mathbf{A})$ with $\varphi \subseteq \theta$. If I is an ideal of $\mathbf{A}$ satisfying $I \subseteq [0]_\theta$, then*

$$[I]_{\theta \wedge (\varphi \vee \psi)} = [I]_{\varphi \vee (\theta \wedge \psi)}.$$

Proof. Let $x \in [I]_{\theta \wedge (\varphi \vee \psi)}$ for some $x \in A$. Then there exists $i \in I$ with $\langle x, i \rangle \in \theta$ and $\langle x, i \rangle \in \varphi \vee \psi$, i.e., $x \in [I]_\theta$ and $x \in [I]_{\varphi \vee \psi}$. By the assumption, $[I]_{\varphi \vee \psi} = [I]_{\varphi \circ \psi \circ \varphi}$, i.e.,

$$\langle x, p \rangle \in \varphi \quad \langle p, q \rangle \in \psi \quad \langle q, j \rangle \in \varphi \qquad (*)$$

for some $p, q, \in A$ and $j \in I$. By using that $\varphi \subseteq \theta$, we obtain $\langle x, p \rangle \in \theta$, $\langle j, q \rangle \in \theta$. However, $\langle x, i \rangle \in \theta$ and $i, j \in I \subseteq [0]_\theta$, whence $\langle i, j \rangle \in \theta$. We conclude that $\langle p, q \rangle \in \theta$. Thus $\langle p, q, \rangle \in \theta \wedge \varphi$ and $(*)$ implies that

$$\langle x, j \rangle \in \varphi \vee (\theta \wedge \psi),$$

i.e., $x \in [I]_{\varphi \vee (\theta \wedge \psi)}$. The converse inclusion is trivial. $\square$

Remark 2 The condition $I \subseteq [0]_\theta$ in the assumption of Theorem 3 is not necessary. It is trivial that for $\theta = \omega$, the least congruence in $\mathrm{Con}(\mathbf{A})$, the conclusion holds for every ideal of $\mathbf{A}$.

References

[1] I. Chajda, *A localization of some congruence conditions in varieties with nullary operations*, Ann. Univ. Sci. Budapest. Eötvös Sect. Math. **30** (1987), 17–23.

[2] ———, *Modifications of congruence permutability*, Acta Sci. Math. (Szeged) **53** (1989), 225–232.

[3] J. Duda, *Arithmeticity at 0*, Czechoslovak Math. J. **37** (1989), 197–206.

[4] G. Gratzer, *On the Jordan-Hölder theorem for universal algebras*, Publ. Math. Inst. Hungar. Acad. Sci. Ser. A **3** (1963), 397–406.

[5] H.P. Gumm and A. Ursini, *Ideals in universal algebra*, Algebra Universalis **19** (1984), 45–54.

[6] A. Ursini, *Sulle varietà di algebre con una buona teoria degli ideali*, Boll. Un. Mat. Ital. **6** (1972), 90–95.

DEPARTMENT OF ALGEBRA AND GEOMETRY, PALACKÝ UNIVERSITY, TOMKOVA 40, 77900 OLOMOUC, CZECH REPUBLIK
        e-mail address: CHAJDA@RISC.UPOL.CZ

Dualisability in general
and endodualisability in particular

Brian A. Davey [*]

Abstract

Sufficient conditions are given for the dualisability of a finite algebra $\underline{\mathbf{M}}$ in the quasivariety generated by a dualisable algebra $\underline{\mathbf{D}}$. The dual structure on $\underline{\mathbf{M}}$ may be obtained by adding a suitable subset of $\operatorname{End}\underline{\mathbf{M}}$ to the structure inherited from $\underline{\mathbf{D}}$. The results show that every finite abelian group is dualisable and that as dual structure we may take the original group operations along with the endomorphism monoid. Similar results follow for finite distributive lattices, Boolean algebras, semilattices and vector spaces, for example. The results are applied to yield large classes of endodualisable algebras. For example, it is proved that a finite vector space over a finite field is endodualisable if and only if it doesn't have dimension 1, that every finite boolean algebra and every finite algebra in the class of Heyting algebras generated by a three-element chain is endodualisable, while a finite Stone algebra is endodualisable if and only if it is boolean or its dense set is nonboolean. We also obtain the result of Davey, Haviar and Priestley that a distributive lattice is endodualisable if and only if it is nonboolean. Endodualisable algebras are of interest because they are necessarily endoprimal, that is, their term functions are just the finitary maps which preserve the action of the endomorphism monoid.

Keywords. natural duality, dualisability, endodualisability, endoprimality, abelian group, distributive lattice, semilattice, Stone algebra, Heyting algebra. 1991 *Math. Subj. Class.* 08A02, 08B05.

1   Two examples

Abelian groups of exponent 4.   The variety of abelian groups generated by a four-element cyclic group $\underline{\mathbf{C}}_4 = \langle C_4; \cdot, {}^{-1}, 1 \rangle$ is the class $\mathcal{A}_4$ of abelian groups

---

[*]This research was supported by a grant from the Australian Research Council.

of exponent 4. Since the only subdirectly irreducible groups in $\mathcal{A}_4$ are the two- and four-element cyclic groups, $\mathcal{A}_4$ equals the class $\mathbb{ISP}\,\underline{C}_4$ of all isomorphic copies of subgroups of powers of $\underline{C}_4$. The Pontryagin dual of an abelian group $\mathbf{A} \in \mathcal{A}_4$ is the compact topological abelian group $D(\mathbf{A}) := \mathcal{A}_4(\mathbf{A}, \underline{C}_4)$ of all homomorphisms from $\mathbf{A}$ into $\underline{C}_4$ equipped with the pointwise group operations and topology. Thus, if we define $\underline{C}_4 := \langle C_4; \cdot, ^{-1}, 1, \mathcal{T} \rangle$, where $\mathcal{T}$ is the discrete topology, then $D(\mathbf{A})$ is a closed subgroup of $\underline{C}_4^A$.

For each $\mathbf{A} \in \mathcal{A}_4$ and each $a \in A$. there is an evaluation map

$$e_a : \mathcal{A}_4(\mathbf{A}, \underline{C}_4) \to \underline{C}_4 \text{ given by } e_a(x) := x(a) \text{ for all } x \in \mathcal{A}_4(\mathbf{A}, \underline{C}_4).$$

Stripped of the categorical "general nonsense", the algebraic half of the Pontryagin duality for $\mathcal{A}_4$ says precisely that, for each $\mathbf{A} \in \mathcal{A}_4$, the map $e : a \mapsto e_a$ is an isomorphism of $\mathbf{A}$ onto the group of all continuous homomorphisms from $D(\mathbf{A})$ to $\underline{C}_4$.

(To dress this up for the more categorically minded, we proceed as follows. Let $\mathcal{X}_4$ be the class of all compact topological abelian groups of exponent 4. While it is a little more difficult to prove than its algebraic counterpart, it is nevertheless true that $\mathcal{X}_4$ equals the class $\mathbb{IS}_c\mathbb{P}\,\underline{C}_4$ of all isomorphic copies of closed subgroups of powers of $\underline{C}_4$ (see [12] ,pp.153–154). The Pontryagin dual of a compact topological group $\mathbf{X} \in \mathcal{X}_4$ is the subgroup $E(\mathbf{X}) := \mathcal{X}_4(\mathbf{X}, \underline{C}_4)$ of $\underline{C}_4^X$ consisting of all continuous homomorphisms from $\mathbf{X}$ into $\underline{C}_4$. Pontryagin duality states that the categories $\mathcal{A}_4$ and $\mathcal{X}_4$ are dually equivalent via the natural hom-functors $D : \mathcal{A}_4 \to \mathcal{X}_4$ and $E : \mathcal{X}_4 \to \mathcal{A}_4$. The map $e : \mathbf{A} \to ED(\mathbf{A})$, introduced above, is one coordinate of the natural transformation, from the identity functor on $\mathcal{A}_4$ to the functor ED, which serves as the unit of the dual adjunction resulting from this dual equivalence.)

The same argument applies to any finite cyclic group. But what happens if we start with a noncyclic abelian group, say $\underline{M} = \mathbf{C}_2 \times \mathbf{C}_4$? First note that $\mathcal{A}_4 = \mathbb{ISP}\,\underline{M}$ so we are attempting to find a new duality for the class of abelian groups of exponent 4. By analogy with the $\underline{C}_4$ case, we could define $\underline{M}_0 := \langle C_2 \times C_4; \cdot, ^{-1}, 1, \mathcal{T} \rangle$ and then define the dual of $\mathbf{A} \in \mathcal{A}_4$ to be the closed subgroup $D(\mathbf{A}) := \mathcal{A}_4(\mathbf{A}, \underline{M})$ of $\underline{M}_0^A$. This fails immediately: there are (continuous) homomorphisms from $D(\widetilde{C}_2) \cong \mathbf{C}_2 \times \mathbf{C}_2$ into $\mathbf{C}_2 \times \mathbf{C}_4$ other than the two evaluation maps. The situation can be remedied by adding extra structure to $\underline{M}_0$.

Let a and b be generators of $\mathbf{C}_2$ and $\mathbf{C}_4$ respectively and let α and γ be the endomorphisms of $\mathbf{C}_2 \times \mathbf{C}_4$ determined by $\alpha(a, 1) := (1, 1)$, $\alpha(1, b) := (1, b)$ and $\gamma(a, 1) := (1, b^2)$, $\gamma(1, b) := (1, 1)$. Now define $\underline{M} := \langle C_2 \times C_4; \alpha, \gamma, \cdot, ^{-1}, 1, \mathcal{T} \rangle$. It is easily seen that $D(\mathbf{A}) := \mathcal{A}_4(\mathbf{A}, \underline{M})$ is closed under both α and γ and so is a substructure of $\underline{M}^A$ for all $\mathbf{A} \in \mathcal{A}_4$. It is an immediate consequence of Theorem 3.3 (see 5.1) that the map $e : a \mapsto e_a$ is an isomorphism of $\mathbf{A}$

onto the group of all morphisms (that is, continuous group homomorphisms which preserve α and γ) from $D(\mathbf{A})$ to $\underset{\sim}{\mathbf{M}}$. Again, this representation theorem lifts to a categorical statement: define $\underset{\sim}{\mathcal{X}} := \mathbb{IS}_c\mathbb{P}\,\underset{\sim}{\mathbf{M}}$, then the hom-functors $D : \mathcal{A}_4 \to \mathcal{X}$ and $E : \mathcal{X} \to \mathcal{A}_4$, given by homming into $\underset{\sim}{\mathbf{M}}$ and $\mathbf{M}$ respectively, yield a dual adjunction between $\mathcal{A}_4$ and $\mathcal{X}$ such that the unit of the adjunction, $e : \mathrm{id}_{\mathcal{A}_4} \to ED$, is a natural isomorphism.

Distributive lattices. The variety $\mathcal{D}$ of distributive lattices is generated by the two-element lattice $\mathbf{D} = \langle \{0,1\}; \vee, \wedge \rangle$. The Priestley duality for $\mathcal{D}$ is obtained by taking $\underset{\sim}{\mathbf{D}} = \langle \{0,1\}; 0, 1, \leq_D, \mathcal{T} \rangle$, where $\leq_D = \{(0,0),(0,1),(1,1)\}$ is the order relation on the two-element chain. Then the dual of $\mathbf{A} \in \mathcal{D}$ is the bounded, ordered, compact topological space $D(\mathbf{A}) := \mathcal{D}(\mathbf{A}, \underset{\sim}{\mathbf{D}}) \subseteq \underset{\sim}{\mathbf{D}}^A$. Again, for each $\mathbf{A} \in \mathcal{D}$ and each $a \in A$, there is an evaluation map

$$e_a : \mathcal{D}(\mathbf{A}, \underset{\sim}{\mathbf{D}}) \to \underset{\sim}{\mathbf{D}} \text{ given by } e_a(x) := x(a) \text{ for all } x \in \mathcal{D}(\mathbf{A}, \underset{\sim}{\mathbf{D}}).$$

The algebraic half of the Priestley duality asserts that the map $e : a \mapsto e_a$ is an isomorphism of $\mathbf{A}$ onto the lattice of all morphisms (that is, continuous order-preserving maps which preserve 0 and 1) from $D(\mathbf{A})$ to $\mathbf{D}$.

Now consider the three-element chain $\underset{\sim}{\mathbf{M}} = \langle \{0, d, 1\}; \vee, \wedge \rangle$. Define endomorphisms α and γ of $\mathbf{M}$ by $\alpha(0) = \gamma(0) = 0$, $\alpha(1) = \gamma(1) = 1$ and $\alpha(d) = 1$ while $\gamma(d) = 0$. It follows at once from Theorem 3.1 that if we take $\underset{\sim}{\mathbf{M}} = \langle \{0, d, 1\}; \alpha, \gamma, 0, 1, \leq_D, \mathcal{T} \rangle$ and define $D(\mathbf{A}) := \mathcal{D}(\mathbf{A}, \underset{\sim}{\mathbf{M}}) \subseteq \underset{\sim}{\mathbf{M}}^A$ for each $\mathbf{A} \in \mathcal{D}$, then the map $e : a \mapsto e_a$ is an isomorphism of $\mathbf{A}$ onto the lattice of all morphisms (that is, continuous maps which preserve α, γ, 0, 1 and $\leq_D$) from $D(\mathbf{A})$ to $\mathbf{M}$. By Theorem 3.3, without disturbing the duality, we can replace the order $\leq_D$ on the sublattice $\underset{\sim}{\mathbf{D}}$ of $\mathbf{M}$ by the order $\leq_M$ on $\mathbf{M}$ itself. In fact, by a result of Davey, Haviar and Priestley [6], since $\mathbf{M}$ is nonboolean we can delete the order completely and still preserve the duality! Thus $\underset{\sim}{\mathbf{M}}$ is endodualisable (see 5.4 below).

As Theorems 3.1 and 3.3 show, the important common features of these two examples are that α and γ map $\mathbf{M}$ into the algebra which carried the original duality ($\underline{\mathbf{C}}_4$ in the abelian group case and $\mathbf{D}$ in the distributive lattice case), that α is a retraction of $\mathbf{M}$ onto the original algebra and that α and γ together separate the points of M.

These examples and several others will be discussed in more generality in Section 5.

2 The dualisability problem

One of the most fundamental, tantalizing and frustrating problems in the theory of natural dualities is:

(∗) *Characterize those finite algebras which are dualisable.*

Since this problem seems to be particularly difficult, it has been natural to consider it in two halves.

(∗₁) *Give sufficient conditions for a finite algebra to be dualisable, thereby extending the list of algebras which are known to be dualisable.*

(∗₂) *Give necessary conditions for a finite algebra to be dualisable, thereby extending the list of algebras which are known to be non-dualisable.*

One of the best examples of (∗₁) is the NU Duality Theorem (see [12], pp.139–142) which states that every finite algebra $\underline{M}$ which possesses a $(k+1)$-ary near unanimity term is dualisable and, moreover, the set of all subalgebras of $\underline{M}^k$ yields a duality on the quasivariety $\mathbb{ISP}\,\underline{M}$ generated by $\underline{M}$. Almost the only example of (∗₂) is the main result of Davey, Heindorf and McKenzie [8] which gives a partial converse to the NU Duality Theorem and states that if $\underline{M}$ generates a congruence-distributive variety but does not possess a near unanimity term, then $\underline{M}$ is not dualisable.

For the fundamentals of the theory of natural dualities the reader is referred to the author's survey [5] or to the soon-to-be-published monograph Clark and Davey [1]. We shall give only a very brief reminder of what it means for a finite algebra $\underline{M}$ to be dualisable.

Let $\underline{M}$ be a finite algebra and let $\underset{\sim}{M} = \langle M; G, H, R, \mathcal{T} \rangle$ where $\mathcal{T}$ is the discrete topology, G is a family of total operations on M each of which is a homomorphism from a finite power of $\underline{M}$ to $\underline{M}$, while H is a family of partial operations on M each of which is a homomorphism from a subalgebra of a finite power of $\underline{M}$ to $\underline{M}$, and R is a family of finitary relations each of which is (the universe of) a subalgebra of a finite power of $\underline{M}$. For each algebra $\mathbf{A}$ in the quasivariety $\mathcal{A} := \mathbb{ISP}\,\underline{M}$ generated by $\underline{M}$ define $D(\mathbf{A})$ to be the set $\mathcal{A}(\mathbf{A}, \underline{M})$ of all homomorphisms from $\mathbf{A}$ to $\underline{M}$ and extend the topology and the structure on $\underset{\sim}{M}$ pointwise to $D(\mathbf{A})$. If, for every $\mathbf{A} \in \mathcal{A}$, the evaluation maps

$$e_a : D(\mathbf{A}) \to M, \text{ given by } e_a(x) = x(a) \text{ for all } x \in D(\mathbf{A}),$$

for $a \in A$, are the only continuous structure preserving maps from $D(\mathbf{A})$ to $\underset{\sim}{M}$, then we say that $\underset{\sim}{M}$ **yields a duality on** $\mathcal{A}$ or, if we are more concerned with the algebra $\underline{M}$ rather than the quasivariety it generates, then we say that $\underset{\sim}{M}$ **dualises** $\underline{M}$. If there is some choice of the structure $\underset{\sim}{M}$ such that $\underset{\sim}{M}$ dualises $\underline{M}$, we say that $\underline{M}$ **admits a duality** or that $\underline{M}$ is **dualisable**. The uninitiated should consult [5] to see these concepts in their proper context. We shall be particularly interested in the situation where $\underset{\sim}{M} := \langle M; \operatorname{End}\underline{M}, \mathcal{T} \rangle$ dualises $\underline{M}$ in which case we shall say that $\underline{M}$ is **endodualisable**. The interest in endodualisable algebras stems from the fact that an endodualisable algebra is necessarily **endoprimal**, that is, for all $n \in \mathbb{N}$, the n-ary term functions on $\underline{\mathbf{M}}$ are the maps from M^n to M which commute with the endomorphisms of $\underline{\mathbf{M}}$. When this holds for a fixed $n \in \mathbb{N}$, we say that $\underline{\mathbf{M}}$ is n-**endoprimal**.

The theorems presented in Section 3 grew out of the author's desire to develop some general theory from which it would follow easily that every finite abelian group is dualisable. Since every finite abelian group belongs to the quasivariety generated by the finite cyclic group $\mathbf{C}_m$, for some m, and since $\mathbf{C}_m$ is dualisable (see [12], pp.151–154 or [5]), it is natural to consider the following problem.

$(*_3)$ *Give conditions on a finite dualisable algebra $\underline{\mathbf{D}}$ under which each finite algebra $\underline{\mathbf{M}}$ in the quasivariety generated by $\underline{\mathbf{D}}$ is dualisable.*

In fact, for every finite abelian group $\mathbf{A}$ there is a finite cyclic group $\mathbf{C}_m$ such that $\mathbf{A}$ and $\mathbf{C}_m$ generate the same quasivariety. Hence we can specialise $(*_3)$ as follows.

$(*_4)$ *Give conditions on a finite dualisable algebra $\underline{\mathbf{D}}$ under which each finite algebra $\underline{\mathbf{M}}$, which generates the same quasivariety as $\underline{\mathbf{D}}$, is dualisable.*

The results in the next section address Problem $(*_4)$.

3   Dualisability in general

Throughout this section we shall consider a finite algebra $\underline{\mathbf{M}}$ belonging to the quasivariety generated by a finite dualisable algebra $\underline{\mathbf{D}}$. To simplify the notation in our first theorem, instead of assuming that $\underline{\mathbf{D}}$ is a retract of $\underline{\mathbf{M}}$ via a retraction $\alpha : \underline{\mathbf{M}} \to \underline{\mathbf{D}}$ and coretraction $\beta : \underline{\mathbf{D}} \to \underline{\mathbf{M}}$ with $\alpha \circ \beta = \mathrm{id}_{\underline{\mathbf{D}}}$, we shall assume that $\underline{\mathbf{D}}$ is a subalgebra of $\underline{\mathbf{M}}$ and that β is the inclusion map.

While our first theorem and its corollary are more general than our second theorem, unlike the second theorem they require us to replace a reflexive relation on D by an irreflexive relation on M and a total operation on D by a partial operation on M.

Let $\underline{\mathbf{D}}$ be a subalgebra of $\underline{\mathbf{M}}$ and let r be an n-ary algebraic operation on $\underline{\mathbf{D}}$. Since $\underline{\mathbf{r}} \leq \underline{\mathbf{D}}^n$ and $\underline{\mathbf{D}}^n \leq \underline{\mathbf{M}}^n$ we may view r as an n-ary algebraic relation on $\underline{\mathbf{M}}$. For emphasis, we shall denote this relation *on* $\underline{\mathbf{M}}$ by r_D. Similarly, if g is a total (or partial) operation on D then g_D denotes g regarded as a *partial* operation on M. Let $R_D := \{\, r_D \mid r \in R \,\}$, $G_D := \{\, g_D \mid g \in G \,\}$, and $H_D := \{\, h_D \mid h \in H \,\}$.

Theorem 3.1 *Let* $\underset{\sim}{\mathbf{M}}$ *be a finite algebra in* $\mathfrak{D} := \mathbb{ISP}\,\underset{\sim}{\mathbf{D}}$, *let* $\mathbf{D}$ *be a subalgebra of* $\underset{\sim}{\mathbf{M}}$ *and assume that* $\underset{\sim}{\mathbf{D}}$ *is a retract of* $\underset{\sim}{\mathbf{M}}$. *If* $\underset{\sim}{\mathbf{D}}$ *is dualisable, then* $\underset{\sim}{\mathbf{M}}$ *is dualisable. More specifically, assume that* $\underset{\sim}{\mathbf{D}} = \langle D; G, H, R, \mathcal{T} \rangle$ *yields a duality on* $\mathfrak{D}$ *and that* $\alpha : \underset{\sim}{\mathbf{M}} \to \underset{\sim}{\mathbf{D}}$ *is a retraction. If* Γ *is a subset of* $\operatorname{End}\underset{\sim}{\mathbf{M}}$ *such that* $\alpha \in \Gamma$, *each* $\gamma \in \Gamma$ *maps into* D *and* Γ *separates the points of* M, *then* $\underset{\sim}{\mathbf{M}} := \langle M; \Gamma, G_D \cup H_D, R_D, \mathcal{T} \rangle$ *yields a duality on* $\mathbb{ISP}\,\underset{\sim}{\mathbf{M}} = \mathfrak{D}$.

Proof. Let $\mathbf{A} \in \mathfrak{D}$ and let $u : D(\mathbf{A}) \to \underset{\sim}{\mathbf{M}}$ be an $\underset{\sim}{\mathbf{M}}$-morphism, where $D(\mathbf{A}) := \mathfrak{D}(\mathbf{A}, \underset{\sim}{\mathbf{M}}) \leq \underset{\sim}{\mathbf{M}}^A$. Since $\underset{\sim}{\mathbf{D}} \leq \underset{\sim}{\mathbf{M}}$, the set $H(\mathbf{A}) := \mathfrak{D}(\mathbf{A}, \underset{\sim}{\mathbf{D}})$ is a subset of $D(\mathbf{A})$. If $x \in H(\mathbf{A})$, then, since α is a retraction of $\underset{\sim}{\mathbf{M}}$ onto $\underset{\sim}{\mathbf{D}}$, we have $x = \alpha \circ x$ and hence, since u preserves α,

$$u(x) = u(\alpha \circ x) = u(\alpha(x)) = \alpha(u(x)) \in D.$$

Thus $u{\restriction}_{H(\mathbf{A})} : H(\mathbf{A}) \to \underset{\sim}{\mathbf{D}}$ is a $\underset{\sim}{\mathbf{D}}$-morphism. Since, by assumption, $\underset{\sim}{\mathbf{D}}$ yields a duality on $\mathfrak{D}$, it follows that there exists $a \in A$ such that $u(x) = x(a)$ for all $x \in H(\mathbf{A})$. Since Γ separates the points of M, in order to show that $u(x) = x(a)$ for all $x \in D(\mathbf{A})$ it suffices to show that $\gamma(u(x)) = \gamma(x(a))$ for all $\gamma \in \Gamma$. Let $x \in D(\mathbf{A})$ and let $\gamma \in \Gamma$. Then

$$
\begin{aligned}
\gamma(u(x)) &= u(\gamma(x)) && \text{as } u \text{ preserves } \gamma \\
&= u(\gamma \circ x) && \text{by the definition of } \gamma(x) \\
&= (\gamma \circ x)(a) && \text{as } \gamma \circ x \in H(A) \\
&= \gamma(x(a)),
\end{aligned}
$$

as required. □

Note that $\underset{\sim}{\mathbf{M}} \in \mathbb{ISP}\,\underset{\sim}{\mathbf{D}}$ if and only if the homomorphisms from $\underset{\sim}{\mathbf{M}}$ into $\underset{\sim}{\mathbf{D}}$ separate the points of M. Thus, provided $\underset{\sim}{\mathbf{D}}$ is a retract of $\underset{\sim}{\mathbf{M}}$, we can always choose Γ in the theorem above to be $\mathfrak{D}(\underset{\sim}{\mathbf{M}}, \underset{\sim}{\mathbf{D}})$. Since we can add extra endomorphisms without destroying the duality, we obtain the following corollary.

Corollary 3.2 *Let* $\underset{\sim}{\mathbf{M}}$ *be a finite algebra in* $\mathfrak{D} := \mathbb{ISP}\,\underset{\sim}{\mathbf{D}}$ *such that* $\underset{\sim}{\mathbf{D}}$ *is a subalgebra of* $\underset{\sim}{\mathbf{M}}$ *and assume that* $\underset{\sim}{\mathbf{D}}$ *is a retract of* $\underset{\sim}{\mathbf{M}}$. *If* $\underset{\sim}{\mathbf{D}} = \langle D; G, H, R, \mathcal{T} \rangle$ *yields a duality on* $\mathfrak{D}$, *then* $\underset{\sim}{\mathbf{M}} := \langle M; \operatorname{End}\underset{\sim}{\mathbf{M}}, G_D \cup H_D, R_D, \mathcal{T} \rangle$ *yields a duality on* $\mathbb{ISP}\,\underset{\sim}{\mathbf{M}} = \mathfrak{D}$.

It follows at once from Theorem 3.1 and Corollary 3.2 that in many familiar quasivarieties every finite algebra is dualisable. This is fine at a theoretical level, but may be inconvenient in practice since, for example, a total operation in the type of $\underset{\sim}{\mathbf{D}}$ is replaced by a partial operation in the type of $\underset{\sim}{\mathbf{M}}$. At the cost of the loss of a little generality, we can guarantee that the type of $\underset{\sim}{\mathbf{M}}$ is just the type of $\underset{\sim}{\mathbf{D}}$ with the addition of the total operations in Γ.

If $\underset{\sim}{\mathbf{M}}$ is a subalgebra of $\mathbf{D}^k$, then as well as the relation r_D we could also consider the pointwise extension, r_M, of the relation r on D to an n-ary relation on M:

$$((a_{11}, \ldots, a_{1k}), \ldots, (a_{n1}, \ldots, a_{nk})) \in r_M$$
$$\Longleftrightarrow (a_{1i}, \ldots, a_{ni}) \in r \text{ for } i = 1, \ldots, k.$$

For example, let $\leq$ be the order on the two-element lattice, $\mathbf{D}$, let $\underset{\sim}{\mathbf{M}}$ be a sublattice of $\mathbf{D}^k$ and identify D with the subset $\{(0, \ldots, 0), (1, \ldots, 1)\}$ of D^k. Then $\leq_D$ is the order on the two-element sublattice $\mathbf{D}$ of $\underset{\sim}{\mathbf{M}}$ while $\leq_M$ is the order relation of the lattice $\underset{\sim}{\mathbf{M}}$ itself. If $\underset{\sim}{\mathbf{M}}$ is a subalgebra of $\mathbf{D}^k$ and moreover M is a substructure of $\mathbf{D}^k$ then, by the definition of the operations on $\mathbf{D}^k$, each total operation g in $\underset{\sim}{G}$ has a pointwise extension to a total operation, g_M, on M and, similarly, each partial operation h in H has a pointwise extension to a partial operation, h_M, on M. Since these are the usual definitions of the relations and (partial) operations on a subset of a power of $\underset{\sim}{\mathbf{D}}$, we shall often write r, g and h instead of r_M, g_M and h_M.

Theorem 3.3 *Let $\underset{\sim}{\mathbf{D}}$ be a finite algebra, let $\underset{\sim}{\mathbf{M}}$ be a subalgebra of $\mathbf{D}^k$ and assume that $\mathbf{D} = \langle D; G, H, R, \mathcal{T} \rangle$ yields a duality on $\mathcal{D} := \mathbb{ISP}\,\mathbf{D}$. Assume that $\alpha : \underset{\sim}{\mathbf{M}} \to \underset{\sim}{\mathbf{D}}$ and $\beta : \underset{\sim}{\mathbf{D}} \to \underset{\sim}{\mathbf{M}}$ are homomorphisms satisfying $\alpha \circ \beta = \mathrm{id}_{\mathbf{D}}$ and that Λ is a subset of $\mathcal{D}(\underset{\sim}{\mathbf{M}}, \underset{\sim}{\mathbf{D}})$ which contains α and separates the points of M. Define $\Gamma := \{\, \beta \circ \lambda \mid \lambda \in \Lambda \,\} \subseteq \mathrm{End}\,\underset{\sim}{\mathbf{M}}$. If M is a substructure of $\mathbf{D}^k$ and $\beta : D \to M$ is an embedding with respect to the structure on $\underset{\sim}{\mathbf{D}}$, then $\underset{\sim}{\mathbf{M}} := \langle M; \Gamma \cup G, H, R, \mathcal{T} \rangle$ (and therefore $\underset{\sim}{\mathbf{M}}' := \langle M; \mathrm{End}\,\underset{\sim}{\mathbf{M}} \cup G, H, R, \mathcal{T} \rangle$) yields a duality on $\mathbb{ISP}\,\underset{\sim}{\mathbf{M}} = \mathcal{D}$.*

Proof. Since M is a substructure of $\mathbf{D}^k$ and $\beta : D \to M$ is an embedding with respect to the structure on $\underset{\sim}{\mathbf{D}}$, it follows that the $\underset{\sim}{\mathbf{D}}$-structure induced on $\beta(D)$ from $\underset{\sim}{\mathbf{D}}$ by the map β coincides with the $\underset{\sim}{\mathbf{D}}$-structure induced on $\beta(D)$ from $\mathbf{D}^k$. Thus we may identify D with $\beta(D)$ and apply Theorem 3.1. It now suffices to show that for all $\mathbf{A} \in \mathcal{D}$ each continuous map $u : D(\mathbf{A}) \to M$ which preserves the endomorphisms in Γ and preserves the relations r_M, the operations g_M and the partial operations h_M necessarily preserves r_D, g_D and h_D. Since u preserves $\beta \circ \alpha \in \Gamma$, it follows that u preserves the fixpoint set D of $\beta \circ \alpha$. Thus u preserves $r_D = r_M \cap D^n$ and, similarly, u preserves g_D and h_D. $\square$

Note that in Theorem 3.3 we may choose Λ to consist of α along with the k projections, whence the structure on $\underset{\sim}{\mathbf{M}}$ may be obtained from the structure on $\underset{\sim}{\mathbf{D}}$ by adding at most $k + 1$ endomorphisms of $\underset{\sim}{\mathbf{M}}$.

Remark 3.4 One of the most straightforward ways to ensure that the assumptions of Theorem 3.3 are met is to let $\underline{\mathbf{M}}$ be a diagonal subalgebra of $\underline{\mathbf{D}}^k$, take β to be the natural map, $d \mapsto (d, \dots, d)$, and then take, for the operations in $G \cup H$, maps which are simultaneously homomorphisms and term functions (or restrictions of term functions in the case of partial maps). Any projection will serve for the retraction α. Provided $\underset{\sim}{\mathbf{D}}$ yields a duality on $\mathcal{D}$, the assumptions of the theorem will be satisfied by taking Γ to consist of the k projections. This applies, for example, in the case of abelian groups, vector spaces over a finite field, semilattices and Stone algebras.

Theorem 3.1 was first proved as an application of the Piggyback Duality Theorem (see [13], [14] and [9]): since $\underline{\mathbf{M}}$ belongs to $\mathcal{D}$ and we have a duality for $\mathcal{D}$, we can try to derive a duality for $\mathbb{ISP}\,\underline{\mathbf{M}}$ by riding piggyback on the given duality for $\mathcal{D}$. In order to apply the most general version of the Piggyback Duality Theorem, we need to know that there is a subset Ω of $\mathcal{D}(\underline{\mathbf{M}}, \underline{\mathbf{D}})$ such that for all $\mathbf{A} \in \mathbb{ISP}\,\underline{\mathbf{M}}$ the maps $\Phi_\alpha : \mathcal{D}(\mathbf{A}, \underline{\mathbf{M}}) \to \mathcal{D}(\mathbf{A}, \underline{\mathbf{D}})$, given by $x \mapsto \alpha \circ x$, are jointly surjective, that is, for all $y \in \mathcal{D}(\mathbf{A}, \underline{\mathbf{D}})$ there exists $x \in \mathcal{D}(\mathbf{A}, \underline{\mathbf{M}})$ and $\alpha \in \Omega$ such that $\Phi_\alpha(x) = y$. Our final lemma shows that this condition is equivalent to the condition assumed in Theorem 3.1, namely that $\underline{\mathbf{D}}$ is a retract of $\underline{\mathbf{M}}$.

Lemma 3.5 *Assume that* $\underline{\mathbf{M}}$ *is a finite algebra which generates the quasi-variety* $\mathcal{D}$. *Then the following are equivalent:*

(a) *the maps* $\Phi_\alpha : \mathcal{D}(\mathbf{A}, \underline{\mathbf{M}}) \to \mathcal{D}(\mathbf{A}, \underline{\mathbf{D}})$ *are jointly surjective for all* $\mathbf{A} \in \mathcal{D}$;

(b) *there exists* $\alpha \in \Omega$ *such that* $\Phi_\alpha : \mathcal{D}(\mathbf{A}, \underline{\mathbf{M}}) \to \mathcal{D}(\mathbf{A}, \underline{\mathbf{D}})$ *is surjective for all* $\mathbf{A} \in \mathcal{D}$;

(c) *some* $\alpha \in \Omega$ *is a retraction of* $\underline{\mathbf{M}}$ *onto* $\underline{\mathbf{D}}$.

Proof. Assume (a). Since $\mathrm{id}_{\underline{\mathbf{D}}} \in \mathcal{D}(\underline{\mathbf{D}}, \underline{\mathbf{D}})$, there is some $\alpha \in \Omega$ and some $\beta \in \mathcal{D}(\underline{\mathbf{D}}, \underline{\mathbf{M}})$ such that $\Phi_\alpha(\beta) = \mathrm{id}_{\underline{\mathbf{D}}}$, that is, $\alpha \circ \beta = \mathrm{id}_{\underline{\mathbf{D}}}$ whence α is a retraction. Hence (c) holds.

Now assume (c). Say $\alpha \circ \beta = \mathrm{id}_{\underline{\mathbf{D}}}$ for some $\beta \in \mathcal{D}(\underline{\mathbf{D}}, \underline{\mathbf{M}})$. Let $\mathbf{A} \in \mathcal{D}$ and let $y \in \mathcal{D}(\mathbf{A}, \underline{\mathbf{D}})$. Define $x \in \mathcal{D}(\mathbf{A}, \underline{\mathbf{M}})$ by $x := \beta \circ y$. Then $\Phi_\alpha(x) = \alpha \circ \beta \circ y = y$, whence Φ_α is surjective. Thus (b) holds. This completes the proof since (b) $\Rightarrow$ (a) is trivial. $\square$

4 Endodualisability in particular

We now wish to apply the results of the previous section to the study of endodualisable and almost endodualisable algebras. A finite algebra $\mathbf{D}$ is called *n-almost endodualisable* if $\underset{\sim}{\mathbf{D}} = \langle D; G, s, \mathcal{T} \rangle$ dualises $\mathbf{D}$ for some subset G of $\operatorname{End} \mathbf{D}$ and some n-ary algebraic relation s. If the arity of s is unimportant, we shall refer to $\mathbf{D}$ as *almost endodualisable*. It is proved in Davey and Priestley [10] that every finite Heyting algebra of the form $\mathbf{2}^n \oplus \mathbf{1}$ is 2-almost endodualisable. Assume that $\underset{\sim}{\mathbf{D}} = \langle D; d_1, \ldots, d_\ell, s, \mathcal{T} \rangle$ dualises $\mathbf{D}$ where $d_1, \ldots, d_\ell$ are constants and s is some n-ary algebraic relation on $\mathbf{D}$. Since each d_i may be replaced by the corresponding constant endomorphism, by Theorem 3.1 every finite algebra $\mathbf{M}$ in $\mathbb{ISP}\,\mathbf{D}$ which has $\mathbf{D}$ as a retract is n-almost endodualisable. For example, it follows immediately from Theorem 3.1 that every finite distributive lattice is 2-almost endodualisable: see 5.4 below. The results presented below combine to give useful sufficient conditions for the endodualisablity of a finite algebra in the quasivariety generated by an (n-almost) endodualisable algebra. Throughout this discussion we shall work with an n-ary algebraic relation s, but the results apply equally well to an algebraic operation g or partial operation h—simply take s to be graph of the (partial) operation.

It is clear that if $\mathbf{D}$ is (n-almost) endodualisable and the assumptions of Theorem 3.3 hold, then the algebra $\mathbf{M}$ is also (n-almost) endodualisable. By imposing injectivity conditions on $\mathbf{D}$ we can also use Theorem 3.1 to lift (n-almost) endodualisability of $\mathbf{D}$ up to the algebra $\mathbf{M}$. Recall that an algebra $\mathbf{A} \in \mathcal{D}$ is *injective* in $\mathcal{D}$ if for all algebras $\mathbf{B} \in \mathcal{D}$ every homomorphism from a subalgebra of $\mathbf{B}$ into $\mathbf{A}$ extends to a homomorphism from $\mathbf{B}$ into $\mathbf{A}$.

Theorem 4.1 *Let $\mathbf{M}$ be a finite algebra in $\mathcal{D} := \mathbb{ISP}\,\mathbf{D}$, let $\mathbf{D}$ be a subalgebra of $\mathbf{M}$ and assume that $\mathbf{D}$ is a retract of $\mathbf{M}$.*

(a) *If $\mathbf{D}$ is injective in $\mathcal{D}$ and is (n-almost) endodualisable, then $\mathbf{M}$ is (n-almost) endodualisable.*

(b) *If $\underset{\sim}{\mathbf{D}} = \langle D; G, R, \mathcal{T} \rangle$ dualises $\mathbf{D}$ for some subset G of $\operatorname{End} \mathbf{D}$ and the image of every map $g \in G$ lies in some injective subalgebra, $\mathbf{D}_g$, of $\mathbf{D}$, then there is a subset G' of $\operatorname{End} \mathbf{M}$, satisfying $\operatorname{im}(g) \subseteq D$ for all $g \in G'$, such that $\underset{\sim}{\mathbf{M}} = \langle M; G', R_D, \mathcal{T} \rangle$ dualises $\mathbf{M}$.*

Proof. Assume that $\underset{\sim}{\mathbf{D}} := \langle D; G, R, \mathcal{T} \rangle$ dualises $\mathbf{D}$ where $G \subseteq \operatorname{End} \mathbf{D}$. Assume that $\mathbf{D}$ is a subalgebra of a finite algebra $\mathbf{M} \in \mathcal{D}$ and that $\alpha : \mathbf{M} \to \mathbf{D}$ is a retraction. Thus by Theorem 3.1, $\underset{\sim}{\mathbf{M}} := \langle M; \Gamma, G_D, R_D, \mathcal{T} \rangle$ dualises $\mathbf{M}$ for some subset Γ of $\mathcal{D}(\mathbf{M}, \mathbf{D})$ with $\alpha \in \Gamma$. If the image of every map $g \in G$ lies in some injective subalgebra of $\mathbf{D}$, then each map $g_D \in G_D$ extends to an endomorphism g' of $\underline{\mathbf{M}}$ satisfying $\text{im}(g') \subseteq D$. Since $\alpha \in \Gamma$ and since D is the fixpoint set of α, it follows (as in the proof of Theorem 3.3) that we can replace the partial map g_D by the total map g' without destroying the duality. Thus $\underset{\sim}{\mathbf{M}}' := \langle M; G', R_D, \mathcal{T} \rangle$ dualises $\underline{\mathbf{M}}$, where $G' := \Gamma \cup \{\, g' \mid g \in G_D \,\}$. This proves (b) and (a) is an easy consequence—take $R = \varnothing$ in the endodualisable case and $R = \{s\}$ in the n-almost endodualisable case. $\quad\square$

We intend to apply Theorem 3.1 and so we seek conditions under which the endomorphisms of $\underline{\mathbf{M}}$ entail the relation s_D in the following sense. Let G, H and R be families of algebraic operations, partial operations and relations on $\underline{\mathbf{M}}$ and let s be an n-ary algebraic relation on $\underline{\mathbf{M}}$. Then $G \cup H \cup R$ **entails** s **on** $D(\mathbf{A})$ if every continuous map $u : D(\mathbf{A}) \to M$ which preserves G, H and R also preserves s. If $G \cup H \cup R$ entails s on $D(\mathbf{A})$ for each $\mathbf{A} \in \mathcal{D}$ then we write $G \cup H \cup R \vdash s$ and say that $G \cup H \cup R$ **entails** s. Without having formally introduced the concept, we have already used entailment in the proofs of Theorems 3.3 and 4.1. The following fundamental lemma is proved in Davey, Haviar and Priestley [7]. (See also [11].) We denote the relation s *qua* algebra by $\underline{\mathbf{s}}$ and we denote the n coordinate projections by $\rho_1, \ldots, \rho_n$. Note that $\rho_1, \ldots, \rho_n \in D(\underline{\mathbf{s}}) = \mathcal{D}(\underline{\mathbf{s}}, \underline{\mathbf{M}})$. A primitive positive formula is an existential conjunct of atomic formulæ.

Lemma 4.2 *Let $\underline{\mathbf{M}}$ be a finite algebra, let G, H and R be families of algebraic operations, partial operations and relations on $\underline{\mathbf{M}}$ and let s be an algebraic relation on $\underline{\mathbf{M}}$. Then the following are equivalent:*

(a) *$G \cup H \cup R$ entails s;*

(b) *$G \cup H \cup R$ entails s on $D(\underline{\mathbf{s}})$;*

(c) *$D(\underline{\mathbf{s}}) \models \Phi(\rho_1, \ldots, \rho_n)$ and $s = \{\, (c_1, \ldots, c_n) \mid M \models \Phi(c_1, \ldots, c_n)\,\}$ for some primitive positive formula $\Phi(x_1, \ldots, x_n)$ in the language $\mathcal{L}$ of $G \cup H \cup R$.*

The following lemma can be proved directly but is most easily obtained as an application of the previous lemma.

Lemma 4.3 *Let $\underline{\mathbf{s}} \leq \underline{\mathbf{M}}^n$. The following are related by (a) $\Leftrightarrow$ (b) $\Rightarrow$ (c).*

(a) *$\underline{\mathbf{s}}$ is a retract of $\underline{\mathbf{M}}$;*

(b) *there exist endomorphisms $e_1, \ldots, e_n$ of $\underline{\mathbf{M}}$ such that*

$$\mathcal{D}(\underline{\mathbf{s}}, \underline{\mathbf{M}}) \models (\exists y) \,\&_{i=1}^{n}\, e_i(y) = \rho_i$$

and

$$s = \{\, (c_1, \ldots, c_n) \in M^n \mid M \models (\exists a) \,\&_{i=1}^{n}\, e_i(a) = c_i \,\}.$$

(c) *there exist $e_1, \ldots, e_n \in \operatorname{End} \underline{\mathbf{M}}$ such that $\{e_1, \ldots, e_n\} \vdash s$.*

Moreover, if $\underline{\mathbf{D}} \leq \underline{\mathbf{M}}$ and $\underline{\mathbf{s}} \leq \underline{\mathbf{D}}^n \leq \underline{\mathbf{M}}^n$, then the endomorphisms in (b) and (c) can be chosen so that $\operatorname{im}(e_i) \subseteq D$ for all i.

Proof. Let $y : \underline{\mathbf{s}} \to \underline{\mathbf{M}}$ and $z : \underline{\mathbf{M}} \to \underline{\mathbf{s}}$ be homomorphisms satisfying $z \circ y = \mathrm{id}_{\underline{\mathbf{s}}}$ and define $e_i := \rho_i \circ z$. Clearly, $D(\underline{\mathbf{s}}) \models \Phi(\rho_1, \ldots, \rho_n)$, where $\Phi(x_1, \ldots, x_n) := (\exists y) \ \&_{i=1}^n e_i(y) = x_i$ and moreover, since $z \circ y = \mathrm{id}_{\underline{\mathbf{s}}}$,

$$
\begin{aligned}
&\{(c_1, \ldots, c_n) \mid M \models \Phi(c_1, \ldots, c_n)\} \\
&= \{(c_1, \ldots, c_n) \mid (\exists a \in M) \ \&_{i=1}^n e_i(a) = c_i\} \\
&= \{(c_1, \ldots, c_n) \mid (\exists a \in M) \ \&_{i=1}^n \rho_i(z(a)) = c_i\} \\
&= \{(c_1, \ldots, c_n) \mid (\exists c \in s) \ \&_{i=1}^n \rho_i(c) = c_i\} \\
&= s.
\end{aligned}
$$

Thus (a) implies (b). Conversely, assume that (b) holds. Let $y \in \mathcal{D}(\underline{\mathbf{s}}, \underline{\mathbf{M}})$ and $e_i \in \operatorname{End} \underline{\mathbf{M}}$ be the homomorphisms given by (b). The second part of (b) says that the obvious product map $z := e_1 \sqcap \cdots \sqcap e_n : \underline{\mathbf{M}} \to \underline{\mathbf{s}}$ is well defined while the first part says that $z \circ y = \mathrm{id}_{\underline{\mathbf{s}}}$, whence $\underline{\mathbf{s}}$ is a retract of $\underline{\mathbf{M}}$. Thus (b) implies (a). If (b) holds then, by Lemma 4.2, $\{e_1, \ldots, e_n\} \vdash s$, whence (c) holds. Finally, if $\underline{\mathbf{D}} \leq \underline{\mathbf{M}}$ and $\underline{\mathbf{s}} \leq \underline{\mathbf{D}}^n$, then $\operatorname{im}(e_i) = \operatorname{im}(\rho_i) \subseteq D$. $\square$

Theorem 4.1 and Lemma 4.3 combine to produce a plethora of endodualisable algebras.

Theorem 4.4 (a) *If $\underline{\mathbf{M}}$ is almost endodualisable via $\underset{\sim}{\mathbf{M}} := \langle M; G, s, \mathcal{T} \rangle$, with $G \subseteq \operatorname{End} \underline{\mathbf{D}}$, and $\underline{\mathbf{s}}$ is a retract of $\underline{\mathbf{M}}$, then $\underline{\mathbf{M}}$ is endodualisable.*

(b) *If $\underset{\sim}{\mathbf{D}} = \langle D; G, s, \mathcal{T} \rangle$ dualises $\underline{\mathbf{D}}$, where G is a set of constants, and $\underline{\mathbf{M}} \in \mathbb{ISP}\,\underline{\mathbf{D}}$ has both $\underline{\mathbf{D}}$ and $\underline{\mathbf{s}}$ as retracts, then $\underline{\mathbf{M}}$ is endodualisable.*

(c) *If $\underset{\sim}{\mathbf{D}} = \langle D; G, s, \mathcal{T} \rangle$ dualises $\underline{\mathbf{D}}$, where G is a set of constants, and $\underline{\mathbf{D}}$ is a retract of $\underline{\mathbf{s}}$, then $\underline{\mathbf{s}}$, and more generally every algebra, $\underline{\mathbf{M}}$, in $\mathbb{ISP}\,\underline{\mathbf{D}}$ which has $\underline{\mathbf{s}}$ as a retract, is endodualisable.*

(d) *Assume that $\underset{\sim}{\mathbf{D}} = \langle D; R, \mathcal{T} \rangle$ dualises $\underline{\mathbf{D}}$ for some finite set R of algebraic relations on $\underline{\mathbf{D}}$ and let s be the direct product of the relations in R. If $\underline{\mathbf{D}}$ is a retract of $\underline{\mathbf{s}}$ and, for each $r \in R$, the natural projection $\pi_r : \underline{\mathbf{s}} \to \underline{\mathbf{r}}$ is a retraction, then $\underline{\mathbf{s}}$, and more generally every algebra in $\mathcal{D}$ which has $\underline{\mathbf{s}}$ as a retract, is endodualisable.*

Proof. If $\underline{M}$ is almost endodualisable via $\underset{\sim}{M} := \langle M; G, s, \mathcal{T} \rangle$, with $G \subseteq$ End $\underline{D}$, and $\underline{s}$ is a retract of $\underline{M}$, then by Lemma 4.3, End $\underline{M}$ entails s and hence $\underline{M}$ is endodualisable. Thus (a) holds.

Assume that $\underset{\sim}{D} = \langle D; G, s, \mathcal{T} \rangle$ dualises $\underline{D}$, where G is a set of constants, and that $\underline{M} \in \mathbb{ISP} \underline{D}$ has both $\underline{D}$ and $\underline{s}$ as retracts. As we remarked earlier, the constants may be replaced by constant endomorphisms and consequently $\underline{D}$ is almost endodualisable. Since one-element algebras are injective, it follows from Theorem 4.1(b) that $\underset{\sim}{M} = \langle M; G, s_D, \mathcal{T} \rangle$ dualises $\underline{M}$. Since $\underline{s}_D$ is isomorphic to $\underline{s}$ and $\underline{s}$ is a retract of $\underline{M}$, it follows from (a) that $\underline{M}$ is endodualisable. This proves (b), and (c) follows at once.

Finally, assume that $\underset{\sim}{D} = \langle D; R, \mathcal{T} \rangle$ dualises $\underline{D}$ for some finite set R of algebraic relations on $\underline{D}$ and let s be the direct product of the relations in R. If $\pi_r : \underline{s} \to \underline{r}$ is a retraction for all $r \in R$, then, in the terminology of [7], r may be obtained from s via a retractive projection and hence $s \vdash r$ for all $r \in R$. Consequently, $\underset{\sim}{D}' := \langle D; s, \mathcal{T} \rangle$ dualises $\underline{D}$. Thus (d) is an immediate consequence of (c) in the case that $G = \varnothing$. □

When $\underline{D}$ is injective, we can allow non-constant endomorphisms.

Theorem 4.5 *Assume that $\underline{D}$ is almost endodualisable via $\underset{\sim}{D} = \langle D; G, s, \mathcal{T} \rangle$, with $G \subseteq$ End $\underline{D}$, and assume that $\underline{D}$ is injective in $\mathcal{D} := \mathbb{ISP} \underline{D}$. If the algebra $\underline{s}$ has a subalgebra isomorphic to $\underline{D}$, then $\underline{s}$, and more generally every algebra, $\underline{M}$, in $\mathcal{D}$ which has $\underline{s}$ as a retract, is endodualisable. Moreover, if $\underline{D}'$ is any subalgebra of $\underline{M}$ isomorphic to $\underline{D}$, then the endomorphisms of $\underline{M}$ which yield the duality may be chosen so that they all map into the subalgebra $\underline{D}'$.*

Proof. Assume that $\underset{\sim}{D} = \langle D; G, s, \mathcal{T} \rangle$ dualises $\underline{D}$, with $G \subseteq$ End $\underline{D}$ and assume that $\underline{D}$ is injective in $\mathcal{D}$. Let $\underline{M}$ be an algebra which has $\underline{s}$ as a retract. Since $\underline{s}$ has a subalgebra isomorphic to $\underline{D}$, so does $\underline{M}$. Let $\underline{M}'$ be an algebra in $\mathcal{D}$ which has $\underline{D}$ as a subalgebra and is isomorphic to $\underline{M}$. Since $\underline{D}$ is injective in $\mathcal{D}$, there is a retraction of $\underline{M}'$ onto $\underline{D}$. By Theorem 4.1, there is a subset G' of End $\underline{M}'$, satisfying $\mathrm{im}(g) \subseteq D$ for all $g \in G'$, such that $\underset{\sim}{M}'_1 := \langle M'; G', s_D, \mathcal{T} \rangle$ dualises $\underline{M}'$. By Lemma 4.3, since $\underline{s}$ is a retract of $\underline{M}'$ there is a subset E of End $\underline{M}'$, satisfying $\mathrm{im}(e) \subseteq D$ for all $e \in E$, such that $E \vdash s$. Hence $\underset{\sim}{M}' := \langle M'; G'', \mathcal{T} \rangle$ dualises $\underline{M}'$, where $G'' := G' \cup E$, and moreover $\mathrm{im}(g) \subseteq D$ for all $g \in G''$. Thus $\underline{M}'$ is endodualisable. Since $\underline{M}$ is isomorphic to $\underline{M}'$, the result follows. □

5   Examples and applications

We shall see that the results of the previous section produce an abundance of endodualisable and therefore endoprimal algebras. For example, if $\underline{D}$ is a finite lattice-based algebra which is injective in $\mathcal{D}$ and $\underline{\mathbf{D}}$ embeds into every algebra in $\mathcal{D}$, then $\mathcal{D}$ contains non-trivial endodualisable algebras. To see this, we argue as follows. As $\underline{\mathbf{D}}$ is lattice based, the NU Duality Theorem of [12], pp.139–142, guarantees that $\underset{\sim}{\mathbf{D}} := \langle D; \mathbb{S}(\underline{\mathbf{D}}^2), \mathcal{T} \rangle$ dualises $\underline{\mathbf{D}}$. Theorem 4.4(d) now shows that $\mathcal{D}$ contains a non-trivial endodualisable algebra, namely the product $\underline{\mathbf{s}}$ of all subalgebras of $\underline{\mathbf{D}}^2$: as $\underline{\mathbf{D}}$ embeds into $\underline{\mathbf{r}}$, for all $r \in \mathbb{S}(\underline{\mathbf{D}}^2)$, the homsets $\mathcal{D}(\underline{\mathbf{r}}_1, \underline{\mathbf{r}}_2)$ are nonempty, for $r_1, r_2 \in \mathbb{S}(\underline{\mathbf{D}}^2)$, which is easily seen to imply the existence of coretractions $\sigma_r : \underline{\mathbf{r}} \to \underline{\mathbf{s}}$ satisfying $\pi_r \circ \sigma_r = \mathrm{id}_{\underline{\mathbf{r}}}$, for each $r \in \mathbb{S}(\underline{\mathbf{D}}^2)$. By reducing $\mathbb{S}(\underline{\mathbf{D}}^2)$ to a minimal subset R which still dualises $\underline{\mathbf{D}}$ (that is, by reducing to an optimal duality in the sense of [11]), we can reduce the size of the induced endodualisable algebra $\underline{\mathbf{s}}$.

5.1 Abelian groups. Every finite abelian group $\underline{\mathbf{A}}$ is dualisable. By the representation theorem for finite abelian groups, there is a cyclic group $\underline{\mathbf{C}}_m$ such that $\underline{\mathbf{A}} \in \mathcal{A}_m := \mathbb{ISP}\,\underline{\mathbf{C}}_m$ and such that $\underline{\mathbf{C}}_m$ is a direct factor and therefore a retract of $\underline{\mathbf{A}}$. Since $\underset{\sim}{\mathbf{C}}_m := \langle C_m; \cdot, {}^{-1}, 1, \mathcal{T} \rangle$ yields a duality on $\mathcal{A}_m$ (see [12], pp.151–154), it follows immediately from Theorem 3.3 that $\underset{\sim}{\mathbf{A}} := \langle A; \mathrm{End}\,\underline{\mathbf{A}}, \cdot, {}^{-1}, 1, \mathcal{T} \rangle$ dualises $\underline{\mathbf{A}}$.

The operation ${}^{-1}$ is an endomorphism of $\underline{\mathbf{A}}$ and the constant 1 may be replaced by the corresponding constant endomorphism. Thus, since the graph of $\cdot$ is ternary, every finite abelian group is 3-almost endodualisable. As a group, the graph of the operation $\cdot$ is isomorphic to its domain, $\underline{\mathbf{C}}_m^2$. Since $\underline{\mathbf{C}}_m$ is injective in $\mathcal{A}_4$ so is $\underline{\mathbf{C}}_m^2$. Thus, by Theorem 4.5, if $\underline{\mathbf{A}} \in \mathcal{A}_m$ and $\underline{\mathbf{A}}$ has a subgroup isomorphic to $\underline{\mathbf{C}}_m^2$ then $\underline{\mathbf{A}}$ is endodualisable.

5.2 Vector spaces over a finite field. The variety $\mathcal{V}$ of vector spaces over a finite field $\mathbb{F}$ is $\mathbb{ISP}\,\underline{\mathbf{F}}$ where $\underline{\mathbf{F}} := \langle F; +, \{\lambda_a\}_{a \in F} \rangle$ is the one-dimensional vector space over $\mathbb{F}$. (Of course, λ_a denotes left multiplication by a.) By [12], pp.154–155, $\underline{\mathbf{F}}$ is dualised by the one-dimensional topological vector space, $\underset{\sim}{\mathbf{F}} := \langle F; +, \{\lambda_a\}_{a \in F}, \mathcal{T} \rangle$ over $\mathbb{F}$. Thus $\underline{\mathbf{F}}$ is 3-almost endodualisable since λ_a is an endomorphism of $\underline{\mathbf{F}}$ for all $a \in F$. Again, the binary operation $+$ as an algebra is isomorphic to its domain, $\underline{\mathbf{F}}^2$. Thus, by Theorem 4.5, if the dimension of $\underline{\mathbf{V}} \in \mathcal{V}$ is not 1 then $\underline{\mathbf{V}}$ is endodualisable. With very little extra effort we can prove that the following are equivalent for a finite vector space $\underline{\mathbf{V}}$ in $\mathcal{V}$:

(a) $\underline{\mathbf{V}}$ *is endodualisable;*

(b) $\underline{\mathbf{V}}$ *is endoprimal;*

(c) $\underline{\mathbf{V}}$ *is 2-endoprimal;*

(d) dim $\underline{\mathbf{V}} \neq 1$.

It remains to prove that $\underline{\mathbf{F}}$ is not 2-endoprimal. (Of course, $\underline{\mathbf{F}}$ is 1-endoprimal since if $e : F \to F$ preserves the left translations, λ_a, then it is also a left translation; indeed, $e = \lambda_{e(1)}$.) Define $f : F^2 \to F$ by $f(0, y) := y$ and $f(x, y) = x$ for $x \neq 0$. It is easily seen that f preserves the left translations (which are the endomorphisms of $\underline{\mathbf{F}}$). But f does not preserve $+$, since $f((0, 1) + (1, 0)) \neq f(0, 1) + f(1, 0)$, and hence f is not a term function of $\underline{\mathbf{F}}$.

5.3 Semilattices. $\underline{\mathbf{D}} = \langle \{0, 1\}; \vee, 0 \rangle$, the two-element lower-bounded semilattice, is dualised by $\underset{\sim}{\mathbf{D}} = \langle \{0, 1\}; \vee, 0, \mathcal{T} \rangle$, the two-element lower-bounded topological semilattice (see [12], p.157). Thus $\underset{\sim}{\mathbf{S}} := \langle S; \mathrm{End}\,\underline{\mathbf{S}}, \vee, 0, \mathcal{T} \rangle$ dualises any finite lower-bounded semilattice, $\underline{\mathbf{S}}$, by Theorem 3.3. Consequently, every finite lower-bounded semilattice is 3-almost endodualisable. For any finite lower-bounded semilattice, $\underline{\mathbf{S}}$, the following are equivalent:

(a) $\underline{\mathbf{S}}$ *is endodualisable;*

(b) $\underline{\mathbf{S}}$ *is endoprimal;*

(c) $\underline{\mathbf{S}}$ *is 2-endoprimal;*

(d) $\underline{\mathbf{S}}$ *is not a chain.*

Since $\underline{\mathbf{D}}$ is injective in $\mathcal{D}$, both Theorem 4.4(c) and Theorem 4.5 imply that if $\underline{\mathbf{S}}$ has a subalgebra isomorphic to $\underline{\mathbf{D}}^2$, then $\underline{\mathbf{S}}$ is endodualisable. Since $\underline{\mathbf{S}}$ has a subalgebra isomorphic to $\underline{\mathbf{D}}^2$ if and only if $\underline{\mathbf{S}}$ is not a chain, it remains to show that if $\underline{\mathbf{S}}$ is a chain then $\underline{\mathbf{S}}$ is not 2-endoprimal. (It is easily seen that a chain is 1-endoprimal.) Define $f : S^2 \to S$ by $f(x, y) := x \wedge y$. Since every semilattice endomorphism of a chain is a lattice endomorphism, it follows that f preserves the endomorphisms of $\underline{\mathbf{S}}$. As f is not a term function of $\underline{\mathbf{S}}$ (it does not preserve $\vee$), we conclude that $\underline{\mathbf{S}}$ is not 2-endoprimal, as required.

The two-element semilattice $\underline{\mathbf{D}} = \langle \{0, 1\}; \vee \rangle$, is dualised by the two-element bounded topological semilattice $\underset{\sim}{\mathbf{D}} = \langle \{0, 1\}; \vee, 0, 1, \mathcal{T} \rangle$ (see [12], p.158), and consequently $\underline{\mathbf{D}}$ is 3-almost endodualisable. It is easy to modify the argument above to characterise finite endodualisable = endoprimal semilattices: Condition (c) becomes '$\underline{\mathbf{S}}$ *is 3-endoprimal*' and Condition (d) becomes '$\underline{\mathbf{S}}$ *is not a tree*'. ($\underline{\mathbf{S}}$ is a tree if $\uparrow x$ is a chain for all $x \in S$.) The ternary map required for the contrapositive of (c) $\Rightarrow$ (d) is given by $f(x, y, z) = (x \vee y) \wedge (x \vee z)$, which is well defined since $\uparrow x$ is a chain.

5.4 Distributive lattices. As was observed above, the NU Duality Theorem implies that $\underset{\sim}{\mathbf{L}} := \langle L; \mathbb{S}(\mathbf{L}^2), \mathcal{T} \rangle$ dualises $\mathbf{L}$ for every finite distributive lattice $\mathbf{L}$. What Theorems 3.1 and 3.3 offer us is a very natural subset of $\mathbb{S}(\mathbf{L}^2)$ which still dualises $\mathbf{L}$. Let Γ be the set of all retractions of $\mathbf{L}$ onto the two-element sublattice $\mathbf{D}$ consisting of the bounds of $\mathbf{L}$. By Theorem 3.1, $\underset{\sim}{\mathbf{L}} := \langle L; \Gamma, 0, 1, \leq_D, \mathcal{T} \rangle$ dualises $\mathbf{L}$ and by Theorem 3.3, $\underset{\sim}{\mathbf{L}}_1 := \langle L; \Gamma, 0, 1, \leq_L, \mathcal{T} \rangle$ also dualises $\mathbf{L}$.

Modulo our general theory, and quite independently of the observations in the paragraph above, we may now give a very short proof the theorem of Davey, Haviar and Priestley [6] which states that the following conditions are equivalent for a nontrivial finite distributive lattice $\mathbf{L}$:

  (a) $\mathbf{L}$ *is endodualisable;*

  (b) $\mathbf{L}$ *is endoprimal;*

  (c) $\mathbf{L}$ *is 3-endoprimal;*

  (d) $\mathbf{L}$ *is nonboolean.*

Since $\underset{\sim}{\mathbf{D}} = \langle \{0, 1\}; 0, 1, \leq_D, \mathcal{T} \rangle$ dualises $\mathbf{D} = \langle \{0, 1\}; \vee, \wedge \rangle$ and since $\leq_D$ is a 3-element chain, Theorem 4.4(c) implies that any finite distributive lattice having a retract onto a 3-element chain is endodualisable. It is a very easy fact of lattice theory that a nontrivial distributive lattice has a retract onto a 3-element chain if and only if it is nonboolean. Thus it remains to prove that a nontrivial boolean lattice $\mathbf{L}$ is not 3-endoprimal. This is an easy observation due to Márki and Pöschel [15]: define $f : L^3 \to L$ by declaring $f(a, b, c)$ to be the relative complement of a in the interval $[a \wedge b \wedge c, a \vee b \vee c]$; then f preserves the action of the endomorphisms but is not order-preserving and so is not a term function on $\mathbf{L}$.

Of course, by using the appropriate form of Priestley duality (see [12], pp.169–170), corresponding results may be obtained for bounded distributive lattices (in which case Condition (c) becomes '$\mathbf{L}$ *is 1-endoprimal*') and for distributive lattices with a single bound (in which case Condition (c) becomes '$\mathbf{L}$ *is 2-endoprimal*').

It should be said that this paper and Davey, Haviar and Priestley [6] have a symbiotic relationship. The original proof of the equivalence of (a)–(d) in [6] used an early version of our Theorem 3.1, while the present version of Theorem 3.1 grew out of a careful study of the special case needed in [6]. Moreover, without the analysis of endodualisable distributive lattices carried out in [6] the general theory developed in Section 4 would never have seen the light of day.

5.5 Boolean algebras. Let $\underline{\mathbf{D}} = \langle\{0,1\}; \vee, \wedge, ', 0, 1\rangle$ be the two-element boolean algebra. The algebraic portion of Stone duality says precisely that $\underset{\sim}{\mathbf{D}} := \langle\{0,1\}; \mathcal{T}\rangle$ yields a duality on the class $\mathbb{ISP}\,\underline{\mathbf{D}}$ of all Boolean algebras. Thus, by Corollary 3.2, $\underset{\sim}{\mathbf{B}} := \langle B; \mathrm{End}\,\underline{\mathbf{B}}, \mathcal{T}\rangle$ dualises any finite boolean algebra $\underline{\mathbf{B}}$, whence every finite boolean algebra $\underline{\mathbf{B}}$ is endodualisable and therefore endoprimal.

5.6 Heyting algebras. The only Heyting algebras which were previously known to be endodualisable were the finite chains and $\mathbf{2}^2 \oplus \mathbf{1}$ (see [2], [12], pp.184–186, [5], [10]), and by [10], amongst subdirectly irreducible Heyting algebras, these are precisely the endodualisables. Of these, only $\mathbf{2}^2 \oplus \mathbf{1}$ and the two- and three-element chains are injective (see [2]). Consider the three-element chain $\underline{\mathbf{D}}_3 = \langle\{0, d, 1\}; \vee, \wedge, \rightarrow, 0, 1\rangle$ regarded as a Heyting algebra. By Theorem 4.1, every algebra in $\mathcal{L}_3 := \mathbb{ISP}\,\underline{\mathbf{D}}_3$ which has a subalgebra isomorphic to $\underline{\mathbf{D}}_3$ is endodualisable. But if $\underline{\mathbf{M}} \in \mathcal{L}_3$ has no subalgebra isomorphic to $\underline{\mathbf{D}}_3$, then $\underline{\mathbf{M}}$ is term equivalent to a boolean algebra and so is endodualisable by 5.5. Hence every finite algebra in $\mathcal{L}_3$ is endodualisable and therefore endoprimal. A similar analysis now shows shows that every finite algebra in the (quasi)variety generated by $\mathbf{2}^2 \oplus \mathbf{1}$ is endodualisable and consequently endoprimal.

While we know of no other endodualisable Heyting algebras we are able to apply Theorem 3.1 to produce a large class of finite Heyting algebras which are dualised by their endomorphisms and partial endomorphisms. For example, every finite relative Stone algebra $\underline{\mathbf{M}}$ is dualised by its endomorphisms and partial endomorphisms. To see this we argue by induction. The finite subdirectly irreducible relative Stone algebras are precisely the finite chains. Let $\mathcal{L}_n$ be the (quasi)variety generated by the n-element chain, $\underline{\mathbf{D}}_n$. If $\underline{\mathbf{M}} \in \mathcal{L}_3$, then $\underline{\mathbf{M}}$ is endodualisable by the previous paragraph. Now assume that $\underline{\mathbf{M}} \in \mathcal{L}_n \backslash \mathcal{L}_{n-1}$ for some $n > 3$. Then $\underline{\mathbf{M}}$ has $\underline{\mathbf{D}}_n$ as a homomorphic image. It is straightforward to prove that $\underline{\mathbf{D}}_n$ is projective in $\mathcal{L}_n$ and consequently $\underline{\mathbf{M}}$ has $\underline{\mathbf{D}}_n$ as a retract. Thus, by Theorem 3.1, since $\underline{\mathbf{D}}_n$ is endodualisable, $\underline{\mathbf{M}}$ is dualised by its endomorphisms and partial endomorphisms. In [13], pp.19–22, a general sufficient condition was given for a finite subdirectly irreducible Heyting algebra $\underline{\mathbf{D}}$ to be dualised by its endomorphisms and partial endomorphisms. Theorem 3.1 guarantees that any finite algebra in $\mathbb{ISP}\,\underline{\mathbf{D}}$ which has $\underline{\mathbf{D}}$ as a retract will also be dualised by its endomorphisms and partial endomorphisms.

5.7 Stone algebras. The (quasi)variety $\mathcal{S}$ of Stone algebras is generated by the three-element algebra $\underline{\mathbf{D}} := \langle\{0, d, 1\}; \vee, \wedge, {}^*, 0, 1\rangle$, where * is pseudocomplementation given by $0^* = 1$ and $d^* = 1^* = 0$, which is dualised by $\underset{\sim}{\mathbf{D}} := \langle\{0, d, 1\}; \preccurlyeq, e, \mathcal{T}\rangle$, where $\preccurlyeq$ is the order relation $\{(0,0), (d, d), (d, 1), (1, 1)\}$ and $e : \underline{\mathbf{D}} \rightarrow \underline{\mathbf{D}}$ is the endomorphism given by $e(x) = x^{**}$. (See [3], pp.90–95, [12], pp.180–182, [4].) Thus **D** is 2-almost endodualisable. Since **D** is injective in 𝒮, we may apply Theorem 4.5 to show that the following are equivalent for a finite Stone algebra **S**:

(a) **S** *is endodualisable;*

(b) **S** *is endoprimal;*

(c) **S** *is 2-endoprimal;*

(d) *either **S** is boolean or the dense set of **S** is nonboolean;*

(e) *either **S** is boolean or the 4-element chain is a retract (in 𝒮) of **S**;*

(f) *the length of the ordered set of prime filters of **S** is not 1.*

We shall first prove the equivalence of (d), (e) and (f). The easiest way to do this is (of course) via the duality. The natural dual $D(\underline{S}) := \mathcal{S}(\underline{S}, \underline{D})$ of a Stone algebra **S** is in one-to-one correspondence with the set $\mathcal{F}(\underline{S})$ of prime filters of **S**: the order $\preccurlyeq$ corresponds to set inclusion and the map e corresponds to the map which sends a prime filter to the unique maximal filter which contains it. The ordered set of prime filters of the dense set of **S** is isomorphic to the ordered set of non-maximal prime filters of **S**. It follows immediately that (d) and (e) are equivalent. To prove the equivalence of (e) and (f) we must prove that **3** is a $\{\preccurlyeq, e\}$-retract of $\mathcal{F}(\underline{S})$ if and only if the length of $\mathcal{F}(\underline{S})$ is greater than 1. The forward implication is trivial, so assume that the length of $\mathcal{F}(\underline{S})$ is greater than 1. Then $\mathcal{F}(\underline{S})$ contains a chain $u < v < w$ with w maximal in $\mathcal{F}(\underline{S})$, in which case the map $\varphi : \mathcal{F}(\underline{S}) \to \mathcal{F}(\underline{S})$, defined by

$$\varphi(x) = \begin{cases} w & \text{if } x \nsubseteq v, \\ v & \text{if } x \subseteq v \text{ and } x \nsubseteq u, \\ u & \text{if } x \subseteq u, \end{cases}$$

is a $\{\preccurlyeq, e\}$-retraction of $\mathcal{F}(\underline{S})$ onto $\{u, v, w\}$. As usual, (a) $\Rightarrow$ (b) $\Rightarrow$ (c) is trivial. Since the order $\preccurlyeq$ as an algebra is isomorphic to **4**, (e) implies (a) by Theorem 4.5 and the fact that every boolean algebra is endodualisable. It remains to prove that if **S** is nonboolean but the dense set of **S** is boolean then **S** is not 2-endoprimal. Denote the dense set of **S** by $d(\underline{S})$. Define $f_1 : A^2 \to d(\underline{S})^2$ by $f_1(a, b) := (a \vee a^*, b \vee b^*)$ and $f_2 : d(\underline{S})^2 \to d(\underline{S})$ by declaring $f(a, b)$ to be the complement of a in the interval $[a \wedge b, 1]$. If e is an endomorphism of **S** then e maps $d(\underline{S})$ into $d(\underline{S})$ and consequently $f := f_2 \circ f_1$ preserves the action of the endomorphisms of **S**. A simple induction shows that any term function of **S** is order-preserving when its domain is restricted to $d(\underline{S})$. Let $a \in d(\underline{S})$ with $a < 1$, then $f(a, a) = 1$ while $f(1, a) = a$. Thus f is not a term function and consequently **S** is not 2-endoprimal.

5.8 Problem. We close the paper with a problem which arises naturally from the examples presented here: *give an example of an endoprimal algebra which is dualisable but not endodualisable.*

References

[1] D.M. Clark and B.A. Davey, *Natural Dualities for the Working Algebraist*, Cambridge University Press, (in preparation).

[2] B.A. Davey, *Dualities for equational classes of Brouwerian and Heyting algebras*, Trans. Amer. Math. Soc. **221** (1976), 119–146.

[3] ———, *Topological duality for prevarieties of universal algebras*, Studies in Foundations and Combinatorics, Advances in Mathematics: Supplementary Studies 1, Academic Press, New York–London, 1978, pp. 61–99.

[4] ———, *Dualities for Stone algebras, double Stone algebras, and relative Stone algebras*, Colloq. Math. **46** (1982), 1–14.

[5] B.A. Davey, *Duality theory on ten dollars a day*, Algebras and Orders (I.G. Rosenberg and G. Sabidussi, eds.), Advanced Study Institute Series, Series C, Vol. 389, Kluwer Academic Publishers, 1993, pp. 71–111.

[6] B.A. Davey, M. Haviar, and H.A. Priestley, *Endoprimal distributive lattices are endodualisable*, to appear in *Algebra Universalis*.

[7] ———, *The syntax and semantics of entailment in duality theory*, to appear in *J. Symb. Logic*.

[8] B.A. Davey, L. Heindorf, and R. McKenzie, *Near unanimity: an obstacle to general duality theory*, Algebra Universalis **33** (1995), 428–439.

[9] B.A. Davey and H.A. Priestley, *Generalized piggyback dualities and applications to Ockham algebras*, Houston J. Math. **13** (1987), 151–197.

[10] ———, *Optimal dualities for varieties of Heyting algebras*, to appear in *Studia Logica*.

[11] ———, *Optimal natural dualities II: general theory*, to appear in *Trans. Amer. Math. Soc.*

[12] B.A. Davey and H. Werner, *Dualities and equivalences for varieties of algebras*, Contributions to lattice theory (Szeged, 1980) (A.P. Huhn and E.T. Schmidt, eds.), Colloq. Math. Soc. János Bolyai, Vol. 33, North–Holland, Amsterdam, 1983, pp. 101–275.

[13] B.A. Davey and H. Werner, *Piggyback-dualitäten*, Bull. Austral. Math. Soc. **32** (1985), 1–32.

[14] B.A. Davey and H. Werner, *Piggyback dualities*, (L. Szabó and A. Szendrei, eds.), Colloq. Math. Soc. János Bolyai, Vol. 43, North–Holland, Amsterdam, 1986, pp. 61–83.

[15] L. Márki and R. Pöschel, *Endoprimal distributive lattices*, Algebra Universalis **30** (1993), 272–274.

LA TROBE UNIVERSITY, BUNDOORA, VICTORIA, AUSTRALIA 3083
e-mail address: B.Davey@latrobe.edu.au

Hyperordinals and Nonstandard α-Models

Mauro Di Nasso

Abstract

In this paper we extend the traditional set-theoretic notion of standard models and nonstandard models going up to α levels in the cumulative hierarchy, α any given limit ordinal. A proof of the representation theorem is given and the structure of nonstandard models is studied where the "transfer principle" holds for every (not necessarily bounded) formula. These models preserve a stratified structure which is investigated by means of "pseudo-rank" functions taking linearly ordered values (hyperordinals). In particular, such functions show a "rigidity" property of the internal sets, in that each external set has a pseudo-rank which is greater than the pseudo-rank of any internal set.

1 Nonstandard α-Models

We shall work in ZF^-C, i.e. in Zermelo-Fraenkel axiomatic set theory ZF with choice but without the axiom Fnd of foundation (regularity). Thus the axioms of our set theory are precisely:

> Ext (Extensionality); $Pair$ (Pairing); Sep (Separation Schema); Un (Union); Pow (Power Set); Rep (Replacement Schema); Inf (Infinity) and AC (Choice).

For unexplained set-theoretic and model-theoretic notions and notations we refer to [6] and [2] respectively.

The traditional set-theoretic approach to nonstandard models is the one streamlined by A. Robinson and E. Zakon in [11] and [14]. It is limited to the first ω levels of the cumulative hierarchy. Here we extend that notion by replacing ω with any limit ordinal α. Namely, a *standard α-model* is an infinite $\in$-model $\langle V_\alpha(X), \in \rangle$ where $V_\alpha(X)$ is the α-superstructure over the set X [1].

---

[1] Remind the cumulative hierarchy defined by transfinite induction: $V_o(X) = X$; $V_{\beta+1}(X) = V_\beta(X) \cup \mathcal{P}(V_\beta(X))$; $V_\alpha(X) = \bigcup_{\beta < \alpha} V_\beta(X)$ if α is a limit ordinal.

For convenience, X is usually assumed to consist of urelements. However, as pointed out by J. Schmid and J. Schmidt in [12], in order to apply nonstandard methods to various structures one should then claim that the number of urelements available in the universe exceeds each fixed cardinality. Since we prefer not to postulate the existence of a proper class of urelements (indeed the existence of any) for the sole purpose of nonstandard methods, we only require X to be the set of the atoms relative to the standard model, i.e. $\emptyset \notin X$ and $\forall t \in X\ t \cap V_\alpha(X) = \emptyset$. This definition of standard model seems to be a natural generalization which allows the use of infinite ordinals. Notice that infinite sets are available even in the case of superstructures built up on the sole empty set, provided that $\alpha > \omega$. This will please the reductionists.

Let $\mathcal{M} = \langle M, \mathrm{E} \rangle$ be any model of the language of set theory with $\mathrm{E} = \in_{\mathcal{M}}$ the $\mathcal{M}$-interpretation of the membership relation symbol, and let ϑ be a mapping $\vartheta : \langle V_\alpha(X), \in \rangle \longrightarrow \mathcal{M}$. We shall say that ϑ satisfies the *transfer* principle for the formula φ if the equivalence

$$V_\alpha(X) \models \varphi(a_1, \ldots, a_n) \quad \Longleftrightarrow \quad \mathcal{M} \models \varphi(\vartheta(a_1), \ldots, \vartheta(a_n))$$

holds for every assignment $a_1, \ldots, a_n \in V_\alpha(X)$. ϑ is a *nonstandard embedding (NSE)* when the *transfer* principle holds for every *bounded* formula [2]. We shall call $\mathcal{M}$ the *nonstandard α-model (α-NSM)* relative to ϑ. When $\mathcal{M}$ is an $\in$-model, the additional property $\vartheta(\emptyset) = \emptyset$ is also required. We remark that only $\in$-models are considered as NSM's in the literature.

In order to simplify the notation, the subset $\subseteq$, the union $\cup$, the intersection $\cap$, the set-minus $\setminus$, the ordered pair $\langle\ ,\ \rangle$ and the cartesian product $\times$ symbols will be used. All of these abbreviations are bounded formulae. E.g., we shall write $x = \{a, b\}$ instead of $\forall t \in x\,(\,t = a \vee t = b\,) \wedge a \in x \wedge b \in x$; $x = \langle a, b \rangle$ instead of $x = \{\{a\}, \{a, b\}\}$ and so on. Other basic notions in elementary set theory can be expressed by bounded formulae too. For instance *"f is a function"*, *"A is the domain (or the range) of f"*, *"f is one-one (or onto)"* etc.[3]

For every $m \in \mathcal{M}$ denote $m_{\mathrm{E}} = \{m' \in \mathcal{M} : m' \mathrm{E} m\}$ the set of its E-elements. Every NSE ϑ preserves the basic operations on sets. Namely, if $A, B \in V_\alpha(X)$, then $A \subseteq B$ iff $\vartheta(A)_{\mathrm{E}} \subseteq \vartheta(B)_{\mathrm{E}}$; $\vartheta(A \cup B)_{\mathrm{E}} = \vartheta(A)_{\mathrm{E}} \cup \vartheta(B)_{\mathrm{E}}$; $\vartheta(A \cap B)_{\mathrm{E}} = \vartheta(A)_{\mathrm{E}} \cap \vartheta(B)_{\mathrm{E}}$; $\vartheta(A \setminus B)_{\mathrm{E}} = \vartheta(A)_{\mathrm{E}} \setminus \vartheta(B)_{\mathrm{E}}$ and $\vartheta(A \times B)_{\mathrm{E}} = \vartheta(A)_{\mathrm{E}} \times \vartheta(B)_{\mathrm{E}}$. Also, if R is an n-ary relation on A then $\vartheta(R)_{\mathrm{E}}$ is an n-ary relation on $\vartheta(A)_{\mathrm{E}}$ and if $f : A^n \to B$ is an n-ary function then $\vartheta(f)_{\mathrm{E}} : \vartheta(A)_{\mathrm{E}}^n \to \vartheta(B)_{\mathrm{E}}$ is an n-ary function where $\langle \vartheta(a_1), \ldots, \vartheta(a_n) \rangle \mapsto \vartheta(f(a_1, \ldots, a_n))$ for every $a_1, \ldots, a_n \in V_\alpha(X)$. Moreover, $\vartheta(f)_{\mathrm{E}}$ is one-one (resp. onto) iff f is.

---

[2] A formula φ in the language of set theory is a *bounded formula* if every quantifier in φ occurs in the form $\forall x\,(\,x \in y \to \ldots)$ or $\exists x\,(\,x \in y \wedge \ldots)$.

[3] See [4], Ch. 3 ,§2.

Membership minimality is also preserved, i.e. A is $\in$-minimal in $\langle V_\alpha(X), \in\rangle$ iff $\vartheta(A)_{\rm E} = \emptyset$. All these facts are straightforwardly proved by applying the *transfer* principle to suitable bounded formulae (see [14] §1.).

Call $b \in \mathcal{M}$ *standard* if $b = \vartheta(a)$ for some $a \in V_\alpha(X)$; *internal* if $b{\rm E}\vartheta(a)$ for some $a \in V_\alpha(X)$ and *external* if it is not internal. Denote $\mathcal{S}$ the collection of all standard sets and $\mathcal{I}$ the collection of all internal sets. Then $\mathcal{S} \subseteq \mathcal{I} = \bigcup_{\beta<\alpha} \vartheta(V_\beta(X))_{\rm E}$ and $\mathcal{I}$ is E-*transitive*, i.e. $b{\rm E}a$ and $a \in \mathcal{I}$ implies $b \in \mathcal{I}$. Moreover the *transfer* principle also holds between the nonstandard model and the internal submodel $\langle \mathcal{I}, {\rm E}\rangle \subseteq \mathcal{M}$. Thus restriction $\vartheta : \langle V_\alpha(X), \in\rangle \longrightarrow \langle \mathcal{I}, {\rm E}\rangle$ is a NSE too.

Keisler's limit ultrapower construction yields an algebraic characterization of the internal models. First, let's fix our notation. Let $V_\alpha(X)^I{}_D$ be any ultrapower of $V_\alpha(X)$. For every filter F over the set $I \times I$, the corresponding limit ultrapower $V_\alpha(X)^I{}_D|F \subseteq V_\alpha(X)^I{}_D$ is the submodel made up by those elements f^D such that $\{\langle i,j\rangle \in I \times I : f(i) = f(j)\} \in F$[4]. Notice that every ultrapower is a limit ultrapower (take $F = I \times I$ the trivial filter). An element $f^D \in V_\alpha(X)^I{}_D|F$ is *bounded* if $f^D \in V_\beta(X)^I{}_D|F$ for some $\beta < \alpha$. Call *bounded limit ultrapower* the submodel $^b V_\alpha(X)^I{}_D|F \subseteq V_\alpha(X)^I{}_D|F$ whose universe is made up by the bounded elements. It is easily seen that the diagonal immersion $d : \langle V_\alpha(X), \in\rangle \to {}^b V_\alpha(X)^I{}_D|F$ and the inclusion $\imath : {}^b V_\alpha(X)^I{}_D|F \hookrightarrow V_\alpha(X)^I{}_D|F$ are NSE's.

Theorem 1.1 *(REPRESENTATION THEOREM)*
The mapping $\vartheta : \langle V_\alpha(X), \in\rangle \longrightarrow \mathcal{M}$ is a NSE if and only if there exist

- *an ultrafilter D over the set I*

- *a filter F over the set $I \times I$*

- *an isomorphism $\pi : \langle \mathcal{I}, {\rm E}\rangle \cong {}^b V_\alpha(X)^I{}_D|F$*

such that the following diagram commutes:

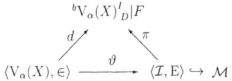

Moreover, the restrictions $\pi_A : \langle \vartheta(A)_{\rm E}, {\rm E}\rangle \longrightarrow A^I{}_D|F$ are isomorphisms for every $A \in V_\alpha(X)$.

---

[4]See [2], §6.4 and [7] as references on limit ultrapowers.

The above theorem generalizes a result proved by H.J. Keisler for $\alpha = \omega$ (see [8] § 1.E* and [2] §6.4). A detailed proof for the general case is given in the Appendix.

$\vartheta(\{a_1, \ldots, a_n\})_{\mathrm{E}} = \{\vartheta(a_1), \ldots, \vartheta(a_n)\}$ for every finite collection of elements in $V_\alpha(X)$. More generally, if $A \in V_\alpha(X)$ then $\vartheta(A)_{\mathrm{E}}$ includes the image set $\vartheta[A] = \{\vartheta(a) : a \in A\}$. With regard to this, the following result holds:

Proposition 1.2 *The following properties are equivalent:*
- *(i)* *For every infinite $A \in V_\alpha(X) \setminus X$, $\vartheta[A] \neq \vartheta(A)_{\mathrm{E}}$*
- *(ii)* *For every denumerable $A \in V_\alpha(X) \setminus X$, $\vartheta[A] \neq \vartheta(A)_{\mathrm{E}}$*
- *(iii)* *There exists a denumerable $A \in V_\alpha(X) \setminus X$ with $\vartheta[A] \neq \vartheta(A)_{\mathrm{E}}$*
- *(iv)* *There exists $A \in V_\alpha(X) \setminus X$ with $|A|$ less than the first uncountable measurable cardinal and $\vartheta[A] \neq \vartheta(A)_{\mathrm{E}}$.*

Moreover, if $|X|$ and $|\alpha|$ are less than the first uncountable measurable cardinal, also the following property is equivalent:
- *(v)* *There exist internal elements which are not standard.*

PROOF. Implications $(i) \Rightarrow (ii) \Rightarrow (iii) \Rightarrow (iv) \Rightarrow (v)$ are trivial. $(iii) \Rightarrow (i)$ Take A as in the hypothesis. For every infinite $B \in V_\alpha(X)$ take a function $f : B \to A$ onto A and consider $\vartheta(f)_{\mathrm{E}} : \vartheta(B)_{\mathrm{E}} \to \vartheta(A)_{\mathrm{E}}$. Suppose by absurd $\vartheta(B)_{\mathrm{E}} = \vartheta[B]$. Then for every $a \mathrm{E} \vartheta(A)$ there exists $b \in B$ such that $\vartheta(f)_{\mathrm{E}}(\vartheta(b)) = \vartheta(f(b)) = a$. We conclude that $\vartheta[A] = \vartheta(A)_{\mathrm{E}}$, a contradiction. $(iv) \Rightarrow (iii)$ Thanks to the Representation Theorem, up to isomorphisms, we can identify $\vartheta : V_\alpha(X) \to \mathcal{I}$ with a diagonal immersion $d : V_\alpha(X) \to {}^bV_\alpha(X)^I_D|F$. Now take A as in the hypothesis. $\vartheta[A] \neq \vartheta(A)_{\mathrm{E}}$ iff $d : A \to A^I_D|F$ is not onto and thus there exists a non-trivial ultrapower A^J_E and an elementary embedding $\tau : A^J_E \to A^I_D|F$ such that the following diagram commutes: [5]

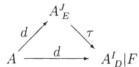

E is countably incomplete in that A^J_E is non-trivial and $|A|$ is less than the first uncountable measurable cardinal [6]. Thus for every infinite $B \subseteq A$, $d : B \to B^J_E \subseteq B^I_D|F$ is not onto, i.e. $\vartheta[B] \neq \vartheta(B)_{\mathrm{E}}$. $(v) \Rightarrow (iv)$ Take $A \in V_\alpha(X)$ with $\vartheta[A] \neq \vartheta(A)_{\mathrm{E}}$. From the hypotheses it follows that $|A|$ is less than the first uncountable measurable cardinal, in that every measurable cardinal is inaccessible [7].   ⊣

---

[5] This property is a known fact about limit ultrapowers: see [7] Theorem 3.6.
[6] See [2] Corollary 4.2.8.
[7] See [6] Lemma 27.2.

From now on, we shall only consider non-trivial NSE's, i.e. NSE's which satisfy the above properties.

Other properties of the internal model are in the following:

Proposition 1.3 *Let Q be the theory whose axioms are W.Ext, Pair, Δ_o-Sep, Un, Pow, AC, Fnd* [8]. *Then*
 (i) $\langle \mathcal{I}, \mathrm{E} \rangle \models Q$
 (ii) $\langle \mathcal{I}, \mathrm{E} \rangle \models Inf \iff \alpha > \omega$
 (iii) Σ_2^Q *formulae are preserved under* $\vartheta : \langle V_\alpha(X), \in \rangle \longrightarrow \langle \mathcal{I}, \mathrm{E} \rangle$, *i.e.*

$$\langle V_\alpha(X), \in \rangle \models \varphi(a_1, \ldots, a_n) \implies \langle \mathcal{I}, \mathrm{E} \rangle \models \varphi(\vartheta(a_1), \ldots, \vartheta(a_n))$$

for every $\varphi \in \Sigma_2^Q$ and for every assignment $a_1, \ldots, a_n \in V_\alpha(X)$.

Remind the *Levi's hierarchy*. The collections of formulae Σ_n, Π_n are defined by induction as follows. $\Delta_o = \Sigma_o = \Pi_o$ is the collection of the bounded formulae; a formula φ is a Σ_{n+1} (resp. Π_{n+1}) formula if $\varphi \equiv \exists x \psi$ (resp. $\varphi \equiv \forall x \psi$) where ψ is a Π_n (resp. Σ_n) formula. If T is any given theory, φ is a Σ_n^T (resp. Π_n^T) formula if there exists a Σ_n (resp. Π_n) formula such that $T \vdash \varphi \leftrightarrow \psi$. For a survey of basic results on Levy's hierarchy, see [4] Ch.3, §2.

PROOF OF THE PROPOSITION. *(i)* is proved by straightforward applications of the *transfer* principle. E.g. the *Pair* axiom is proved by applying the *transfer* principle to the following formulae

$$V_\alpha(X) \models \forall x, y \in V_\beta(X)\, \exists z \in V_{\beta+1}(X)\; z = \{x, y\}$$

for every $\beta < \alpha$. *(iii)* Since $\langle V_\alpha(X), \in \rangle \models Q$ and $\langle \mathcal{I}, \mathrm{E} \rangle \models Q$, one can suppose without loss of generality that $\varphi \in \Sigma_2 \subseteq \Sigma_2^Q$, namely $\varphi \equiv \exists x \forall y\, \psi$ where ψ is bounded. For every given assignment $a_1, \ldots, a_n \in V_\alpha(X)$, notice that $V_\alpha(X) \models \exists x \forall y\, \psi(x, y, a_1, \ldots, a_n)$ iff there exists $\hat{a} \in V_\alpha(X)$ such that

$$\langle V_\alpha(X), \in \rangle \models \forall y \in V_\beta(X)\, \psi(\hat{a}, y, a_1, \ldots, a_n)$$

for every $\beta < \alpha$. Then apply the *transfer* principle β times and get

$$\langle \mathcal{I}, \mathrm{E} \rangle \models \forall y\, \psi(\vartheta(\hat{a}), y, \vartheta(a_1), \ldots, \vartheta(a_n))$$

which yields the thesis. *(ii) Inf* is a Σ_1-formula:

---

[8] *W.Ext* is the following weak version of the axiom of extensionality:

$$\forall x \forall y\, [\exists t \in x \wedge \forall s(s \in x \leftrightarrow s \in y)] \rightarrow x = y$$

Contrarily to the axiom of extensionality, *W.Ext* allows a (possibly empty) collection of atoms. Δ_o-*Sep* denotes the axiom schema of separation restricted to bounded formulae.

$$\exists x [\emptyset \in x \wedge (\forall y \in x \quad y \cup \{y\} \in x)]$$

and $\neg Inf$ is a $\Pi_1^Q \subseteq \Sigma_2^Q$ formula. Thus one can apply *(iii)*.   ⊣

Property *(iii)* cannot be extended to Π_2^Q-formulae. The formula used in the proof of Prop. 3.1 will give a counter-example.

2   Hyperordinals

In this section we shall extend the notion of Von Neumann ordinal to NSM's. Let $Ord(x, \emptyset)$ be the following bounded formula:

$$\text{``}x \text{ is transitive''} \wedge \text{``}x \text{ is trichotomic''} \wedge$$
$$x \neq \emptyset \rightarrow \emptyset \in x$$

where *"x is trichotomic"* is the formula [9]

$$\text{``}\forall a, b \in x \ (a = b \ \underline{aut} \ a \in b \ \underline{aut} \ b \in a)\text{''}$$

Then

$$\langle V_\alpha(X), \in \rangle \models Ord(\beta, \emptyset) \quad \Longleftrightarrow \quad \beta < \alpha \text{ is a Von Neumann ordinal}$$

We call *hyperordinal* any element $m \in \mathcal{M}$ satisfying $\mathcal{M} \models Ord(m, \vartheta(\emptyset))$. Denote $Ord_\mathcal{M}$ the collection of all the hyperordinals and denote $Ord_\mathcal{I}$ the collection of the internal hyperordinals.

The next proposition itemizes some basic properties of $Ord_\mathcal{I}$. For the sake of simplicity, if Λ is any collection of elements in $\mathcal{M}$, we shall write $\Lambda \in \mathcal{M}$ ($\Lambda \in \mathcal{I}$ resp.) whenever $\Lambda = m_{\mathrm{E}}$ for some $m \in \mathcal{M}$ ($m \in \mathcal{I}$ resp.).

Proposition 2.1
  (i) $Ord_\mathcal{I} = \bigcup_{\beta < \alpha} \vartheta(\beta)_{\mathrm{E}}$
  (ii) $\langle Ord_\mathcal{I}, \mathrm{E} \rangle$ *is a linearly ordered set*
  (iii)   *Cofinality* $\langle Ord_\mathcal{I}, \mathrm{E} \rangle$ = *Cofinality* α
  (iv) $\xi' \mathrm{E} \xi \in Ord_\mathcal{I} \Rightarrow \xi' \in Ord_\mathcal{I}$
  (v)   *Every* $\xi \in Ord_\mathcal{I}$ *has immediate successor* $\xi + 1 \in Ord_\mathcal{I}$ *where*
      $(\xi + 1)_{\mathrm{E}} = \xi_{\mathrm{E}} \cup \{\xi\}_{\mathrm{E}}$
  (vi)   *Every non-empty internal collection* $\Lambda \in \mathcal{I}$ *of hyperordinals*
      admits minimum and supremum in $\langle Ord_\mathcal{I}, \mathrm{E} \rangle$
  (vii) $Ord_\mathcal{I} \notin \mathcal{I}$

---

[9] $\underline{aut}$ is the *"exclusive or"* connective, i.e. "$\varphi \ \underline{aut} \ \psi$" stands for "$(\varphi \vee \psi) \wedge (\neg \varphi \vee \neg \psi)$".

PROOF. $(i),(ii),(iv),(v)$ and (vi) are all proved by straightforward applications of the *transfer* principle to suitable formulae. As examples, let's see (i) and (vi) in detail. (i) For every $\beta < \alpha$

$$V_\alpha(X) \models \forall x \ (x \in V_\beta(X) \ \wedge \ Ord(x, \emptyset)) \ \leftrightarrow \ x \in \beta$$

Thus $Ord_\mathcal{M} \cap \vartheta(V_\beta(X))_E = \vartheta(\beta)_E$ and the thesis follows from the equalities $Ord_\mathcal{I} = \mathcal{I} \cap Ord_\mathcal{M} = \bigcup_{\beta<\alpha} \vartheta(V_\beta(X))_E \cap Ord_\mathcal{M} = \bigcup_{\beta<\alpha} \vartheta(\beta)_E.$ (vi) Suppose $\Lambda E \vartheta(V_\beta(X))$. Then apply the *transfer* principle to the formula

$$\begin{aligned}
V_\alpha(X) \models \forall x \in V_\beta(X) \ [\ x \neq \emptyset \wedge \forall t \in x \ Ord(t, \emptyset)] \ &\rightarrow \\
\{ \ [\ \exists \xi \in x \ \forall \xi' \in x \ (\xi = \xi' \vee \xi \in \xi')] \ &\wedge \\
[\ \exists \eta \in \beta \ (\ \forall \xi \in x \ (\xi = \eta \vee \xi \in \eta)) \ &\wedge \\
\forall \eta' \in \eta \ (\exists \xi \in x \ \eta' \in \xi)] \}
\end{aligned}$$

(iii) trivially follows from (i). (vii) Suppose $\xi_E = Ord_\mathcal{I}$ for some $\xi \in \mathcal{I}$. A glance at the definition of hyperordinal shows that $\xi \in Ord_\mathcal{M}$ and thus $\xi E \xi$, which contradicts $\langle \mathcal{I}, E \rangle \models Fnd.$ ⊣

We remark that (in general) E is not a total ordering on $Ord_\mathcal{M}$. E.g., take $\mathcal{N} = \langle {}^b V_\alpha(X)^I_D, \in_\mathcal{N} \rangle$ any non-trivial bounded ultrapower and consider the structure $\mathcal{M} = \langle M, \in_\mathcal{M} \rangle$ defined as follows:

- The universe M is the union ${}^b V_\alpha(X)^I_D \cup \{\Omega\}$.

- $m \in_\mathcal{M} m' \overset{def}{\Longleftrightarrow} (m, m' \in \mathcal{N}$ and $m \in_\mathcal{N} m')$ or $(m' = \Omega)$ and $m = d(n)$ for some $n < \omega)$.

Notice that $|\Omega| = \aleph_o$, hence Ω is an external set (see [2], Prop. 4.4.20). It is straightforwardly proved that $\mathcal{M} \models Ord(\Omega, \vartheta(\emptyset))$ while $\Omega \neq \xi$ and $\Omega \notin_\mathcal{M} \xi$ and $\xi \notin_\mathcal{M} \Omega$, for each $\xi \omega^I_D \setminus \{d(n) : n < \omega\}$. We conclude that hyperordinals in $\mathcal{M}$ exists which are not comparable.

Similarly as the internal model, the internal hyperordinals admit an algebraic characterization.

Theorem 2.2 *Let* $\vartheta : V_\alpha(X) \rightarrow \mathcal{M}$ *be a NSE. Take the ultrafilter D over I, the filter F over $I \times I$ and isomorphism $\pi : \mathcal{I} \rightarrow {}^b V_\alpha(X)^I_D | F$ as in the Representation Theorem. Then the restriction $\tau = \pi \rceil Ord_\mathcal{I} : Ord_\mathcal{I} \rightarrow {}^b \alpha^I_D | F$ is an isomorphism and the following diagram commutes:*

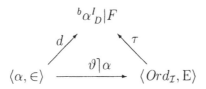

PROOF. Notice that for every $b \in \mathcal{I}$ the following properties are equivalent:

$$\langle \mathcal{I}, \mathrm{E} \rangle \models Ord(b, \vartheta(\emptyset))$$

$$^b\mathrm{V}_\alpha(X)^I{}_D | F \models Ord(\pi(b), d(\emptyset))$$

$$\{i \in I : \mathrm{V}_\alpha(X) \models Ord(\pi(b)(i), \emptyset)\} \in D$$

$$\{i \in I : \pi(b)(i) < \alpha \ \text{ is a V.N. ordinal}\} \in D$$

$$\pi(b) \in \alpha^I{}_D | F \cap {}^b\mathrm{V}_\alpha(X)^I{}_D | F = {}^b\alpha^I{}_D | F \qquad \dashv$$

A subset C of a linearly ordered set $\langle Y, < \rangle$ is a *Dedekind cut* if C is an initial segment (i.e. $y < c \in C \Rightarrow y \in C$) and C lacks maximum. A *gap* is a Dedekind cut C such that $Y \setminus C$ lacks minimum. The whole ordered set Y is a *final gap* if it is a gap of itself (i.e. if Y lacks maximum). An *initial gap* C is an empty gap. A gap C is called *proper* if it is neither final nor initial (i.e. if $C \neq \emptyset$ and $Y \setminus C \neq \emptyset$). It is a well-known fact in Robinsonian Analysis that the standard naturals are a gap of the hypernaturals. Similarly, the following holds:

Proposition 2.3 $\vartheta[\omega]$ *is a gap of* $\langle \vartheta(\omega)_\mathrm{E}, \mathrm{E} \rangle$.

In particular, there exist E-descending chains $\xi_1 \mathrm{E} \xi$; $\xi_2 \mathrm{E} \xi_1$; $\xi_3 \mathrm{E} \xi_2 \ldots$ for every $\xi \mathrm{E} \vartheta(\omega)$ which is not standard. Notice that if $\alpha = \omega$ then every internal hyperordinal is standard: apply the *transfer* principle to

$$V_\omega(X) \models \forall x \in n+1 \to (x = 0 \lor \cdots \lor x = n) \qquad n = 0, 1, 2, \ldots$$

Finally the internal model is well-founded if $\alpha = \omega$. In fact

$$x \mathrm{E} y \mathrm{E} \vartheta(V_n(X)) \ \Rightarrow \ x \mathrm{E} \vartheta(V_{n-1}(X))$$

This completes the proof of the

Proposition 2.4 *The following properties are equivalent:*
(i) $\alpha > \omega$.
(ii) $\langle \mathcal{I}, \mathrm{E} \rangle$ *is non-well-founded.*
(iii) *There exist internal hyperordinals which are not standard.*

3 Strong NSMs

We now turn to consider the case when the NSM $\mathcal{M}$ is an elementary extension of the standard model, i.e. when the *transfer* principle holds for every (not necessarily bounded) formula. Call such an $\mathcal{M}$ a *strong nonstandard α-model*.

We remark that this notion is not considered in the traditional approach to nonstandard models, because the existence of strong ∈-NSM's contradicts the Foundation axiom even for $\alpha = \omega$. In fact the following holds:

Proposition 3.1 *Every strong NSM $\mathcal{M}$ is non-well-founded. In particular, no strong NSE's $\vartheta : \langle V_\alpha(X), \in \rangle \longrightarrow \langle V_\alpha(Y), \in \rangle$ between superstructures exist.*

PROOF. Consider $\langle N, < \rangle \in V_\alpha(X)$ an order-isomorphic copy of the natural numbers. By taking A_n the Von Neumann ordinal corresponding to the natural number n, one proves that

$$V_\alpha(X) \models \forall n \in N \, \exists A_n \, \exists f : [0,n]_N \to A_n \, \forall a, b \in A_n \, (a < b \leftrightarrow f(a) \in f(b))$$

$f : [0,n]_N \to A_n$ denotes the formula "f *is a function* $\wedge$ *Range(f)*= A_n $\wedge$ *Domain(f)*$\subseteq N$ $\wedge$ (x ∈*Domain(f)* $\leftrightarrow x \leq n$)". The *transfer* principle yields a linearly ordered set $\langle \vartheta(N)_E, <_\vartheta \rangle$. Moreover, for every given $\xi E \vartheta(N)$ one gets an element $B \in \mathcal{M}$ and a function $F : [0,\xi]_{\vartheta(N)} \to B_E$ with the property "$x <_\vartheta y \leftrightarrow F(x) \in F(y)$". $<_\vartheta$ denotes the total ordering on $\vartheta(N)_E$ given by $\vartheta(<_N)_E \subseteq \vartheta(N)_E \times \vartheta(N)_E$ and $[0,\xi]_{\vartheta(N)}$ the initial segment $\{x E \vartheta(N) : x \leq_\vartheta \xi\}$. It is a well-known fact in nonstandard analysis that there are infinite descending chains $\xi >_\vartheta \xi - 1 >_\vartheta \xi - 2 >_\vartheta \ldots$ for every $\xi E \vartheta(N)$ which is not standard. Hence $\langle \mathcal{M}, E \rangle$ is non-well-founded. ⊣

The above result and Proposition 2.2 in the previous section, show that only non-strong ω-NSE's can yield well-founded NSM's.

Remark that the formula $\varphi(N)$ we used in this proof is a Π_2^Q-formula which is not preserved by the map $\vartheta : \langle V_\alpha(X), \in \rangle \to \langle \mathcal{I}, E \rangle$ (*cfr.* Prop. 1.3). In fact, $\langle \mathcal{I}, E \rangle \models \varphi(\vartheta(N))$ implies that $\langle \mathcal{I}, E \rangle$ is non-well-founded, while the internal model is well-founded for $\alpha = \omega$.

When $\mathcal{M}$ is a strong NSM, $\langle Ord_\mathcal{M}, E \rangle$ turns out to be a linearly ordered set satisfying similar properties as the internal hyperordinals in Proposition 2.1.

Proposition 3.2 *Suppose $\mathcal{M}$ is a strong NSM. Then*
  (i) $\langle Ord_\mathcal{M}, E \rangle$ *is a linearly ordered set*
  (ii) *Every $\xi \in Ord_\mathcal{M}$ has immediate successor $\xi + 1$ where*
        $(\xi + 1)_E = \xi_E \cup \{\xi\}$
  (iii) *Every non-empty collection $\Lambda \in \mathcal{M}$ of hyperordinals admits supremum $Sup\Lambda \in Ord_\mathcal{M}$ and infimum $Inf\Lambda \in Ord_\mathcal{M}$*
  (iv) $Ord_\mathcal{M} \notin \mathcal{M}$

PROOF. The *transfer* principle holds for every formula, thus one can proceed similarly as in the proof of proposition 2.1.   ⊣

In analogy to Von Neumann ordinals, when dealing with hyperordinals we shall often write $\xi' <_E \xi$ instead of $\xi' E \xi$. Putting together results in propositions 2.1 and 3.2, one gets the

Proposition 3.3 *Suppose $\mathcal{M}$ is a strong NSM. Then $Ord_\mathcal{I}$ is a Dedekind cut of $\langle Ord_\mathcal{M}, <_E \rangle$.*

The question whether the Dedekind cut $Ord_\mathcal{I}$ is a gap of $Ord_\mathcal{M}$ naturally arises. In §5 we show that this problem is closely related to a "rigidity" property of $\langle \mathcal{I}, E \rangle$ which is in turn related to the existence of non-regular uniform ultrafilters.

4   The Pseudorank Function

Every standard model is provided with a *rank function*, i.e. a function $\rho_X : V_\alpha(X) \to \alpha$ verifying:

$$\rho_X(A) = \begin{cases} 0 & \text{if } A \in X \text{ or } A = \emptyset ; \\ Sup\{\rho_X(a) + 1 : a \in A\} & \text{otherwise.} \end{cases}$$

Notice that ρ_X is well defined because X is the set of atoms relative to $\langle V_\alpha(X), \in \rangle$. The following properties straightforwardly follow from the definition of rank.

Proposition 4.1 *Let $\rho_X : V_\alpha(X) \to \alpha$ be the rank function. Then*
(i) $\rho_X(\beta) = \beta$ for every ordinal $\beta < \alpha$
(ii) $\rho_X(A) < \beta \Rightarrow A \in V_\beta(X) \Rightarrow \rho_X(A) < \beta + 1$
(iii) $a \in A \Rightarrow \rho_X(a) < \rho_X(A)$

We raise the question whether such a function can exist for NSMs. In general, the answer is negative because of non-well-foundedness.

Following an idea by V.M. Tortorelli [13], our aim is to relax the concept of rank so as to provide every strong NSM with a pseudorank function. Precisely, we shall consider functions whose range is the linearly ordered set of the hyperordinals.

Although the whole set ρ_X is not a member of $V_\alpha(X)$, suitable restrictions of ρ_X are elements of the standard model. Call $\rho_a \in V_\alpha(X)$ a *partial rank function (PRF)* for the set $a \in V_\alpha(X) \setminus X$ if the following conditions are verified:

(i) ρ_a is a function

(ii) $X \cup \{\emptyset, a\} \subseteq Domain(\rho_a)$, i.e. the domain of ρ_a contains a and the $\in$-minimals

(iii) $Domain(\rho_a)$ is transitive up to elements in X, i.e.
$$t \in x \in Domain(\rho_a) \wedge x \notin X \Rightarrow t \in Domain(\rho_a)$$

(iv) $\rho_a(t) = \begin{cases} 0 & \text{if } t \in X \text{ or } t = \emptyset ; \\ Sup\{\rho_a(s) + 1 : s \in t\} & \text{otherwise.} \end{cases}$

(v) $Range(\rho_a)$ is a Von Neumann ordinal.

Observe that condition *(iv)* implies condition *(v)*. The PRF is a definable notion. In fact

$$V_\alpha(X) \models \psi(x, f, t) \Leftrightarrow x = Sup\{f(s) + 1 : s \in t\}$$

where $\psi(x, f, t)$ denotes the following (bounded) formula

$$(\forall s \in t \ \ f(s) + 1 = x \vee f(s) + 1 \in x) \wedge$$
$$(\forall x' \in x \ \exists s \in t \ \ x' \in f(s) + 1)$$

Thus

$$V_\alpha(X) \models Prf(a, f, \emptyset) \Leftrightarrow f \text{ is a PRF for } a$$

where $Prf(a, f, \emptyset)$ is the following Π_1^Q formula:

$$\forall A \ \forall x \ \forall t \ \text{"}f \text{ is a function"} \wedge$$
$$\text{"}A \text{ is the range of } f\text{"} \rightarrow Ord(A, \emptyset) \wedge$$
$$\text{"}A \text{ is the domain of } f\text{"} \rightarrow$$
$$\{\emptyset \in A \wedge a \in A \wedge \text{"}A \text{ is transitive"} \wedge$$
$$[(\forall s \in t \ s \neq s) \rightarrow (t \in A \wedge f(t) = \emptyset)] \wedge$$
$$[(\exists s \in t \wedge t \in A \wedge f(t) = x) \rightarrow \psi(x, f, t)]\}$$

For every $\beta < \alpha$, $\rho_\beta : V_\beta(X) \rightarrow \beta + 1$ will denote the restriction $\rho_X \!\restriction\! V_\beta(X)$. If $\rho_X(a) = \beta$ then $\rho_{\beta+1}$ is a PRF for a. Vice versa if f is a PRF for a, then one can proceed by $\in$-induction and show that $f(t) = \rho_X(t)$ for every $t \in Domain(f)$. Thus the following is proved:

Proposition 4.2

(i) $V_\alpha(X) \models \forall a \ \exists f \ Prf(a, f, \emptyset)$

(ii) $V_\alpha(X) \models \forall a, b \ \forall f, g \ \forall t \ [Prf(a, f, \emptyset) \wedge Prf(b, g, \emptyset) \wedge$
$$[(t \in Domain(f) \cap Domain(g)) \rightarrow f(t) = g(t)]$$

We are ready to extend the notion of rank to strong NSMs.

Theorem 4.3 *There exists a pseudorank function* $\psi_{\mathcal{M}} : \mathcal{M} \to Ord_{\mathcal{M}}$ *such that*

(i) $\quad \psi_{\mathcal{M}}(m) = \begin{cases} \vartheta(0) & \text{if } m_{\mathrm{E}} = \emptyset \ ; \\ \mathrm{Sup}\{\psi_{\mathcal{M}}(m') + 1 : m'\mathrm{E}m\} & \text{otherwise}. \end{cases}$

$\quad\quad$ *where* $\{\psi_{\mathcal{M}}(m') + 1 : m'\mathrm{E}m\} \in \mathcal{M}$ *for every* $m \in \mathcal{M}$.

(ii) $\quad \psi_{\mathcal{M}}(\xi) = \xi$ *for every hyperordinal* ξ.

(iii) $\quad \psi_{\mathcal{M}}(m) <_{\mathrm{E}} \vartheta(\beta) \ \Rightarrow \ m \, \mathrm{E} \, \vartheta(V_{\beta}(X)) \ \Rightarrow \ \psi_{\mathcal{M}}(m) <_{\mathrm{E}} \vartheta(\beta + 1)$.

(iv) $\quad m'\mathrm{E}m \ \Rightarrow \ \psi_{\mathcal{M}}(m') <_{\mathrm{E}} \psi_{\mathcal{M}}(m)$.

(v) $\quad m \in \mathcal{I} \ \Leftrightarrow \ \psi_{\mathcal{M}}(m) \in Ord_{\mathcal{I}}$.

PROOF. Take the formulae in the previous proposition and apply the *transfer* principle. Then

- For every $m \in \mathcal{M}$, there exists a hyperordinal valued function ψ_m whose domain includes m_{E} and the set of the E-minimal elements, and satisfying

$$\psi_m(t) = \begin{cases} \vartheta(0) & \text{if } t_{\mathrm{E}} = \emptyset \ ; \\ \mathrm{Sup}\{\psi_m(s) + 1 : s\mathrm{E}t\} & \text{otherwise}. \end{cases}$$

  It is easily proved by the *transfer* principle that $\{\psi_m(s)+1 : s \in t\} \in \mathcal{M}$.

- Any two such functions coincide over the intersection set of the domains.

Thus, putting together the various ψ_m's, one gets a function $\psi_{\mathcal{M}} : \mathcal{M} \to Ord_{\mathcal{M}}$ satisfying *(i)*. Besides, *(ii)*, *(iii)* and *(iv)* follow by *transfer* from the properties *(i)*, *(ii)* and *(iii)* of Prop. 4.1 respectively. Finally, *(iii)* implies *(v)*. ⊣

If $\alpha = \omega$, the submodel of the internal sets is well-founded and thus it can be isomorphically collapsed into a transitive submodel of an ω-superstructure [10]. A similar result can be obtained also in our more general context $\alpha > \omega$. Precisely, the internal model is provided with a pseudorank function even for non-strong NSE's. Moreover, assuming a suitable Anti-Foundation Axiom, every internal model can be isomorphically collapsed into a transitive submodel of a *pseudosuperstructure*, which is a full and transitive $\in$-structure [11] provided with a pseudorank function (see [9]). See also the recent paper [1] by D. Ballard and K. Hrbacek, where a different but related anti-foundational approach is presented.

---

[10]This fact follows from a modification of arguments in [2] §4.4. A detailed proof can be found in [10]. Here, transitivity has to be intended up to atoms.

[11]An $\in$-structure S is *full* if "$\forall s \in S \ \mathcal{P}(s) \subseteq S$".

5 Regular Nonstandard Models

In this last section, we investigate the structure of strong NSM's by means of hyperordinals and show how the internal sets make up a "core" which is in some sense isolated from the rest of the model. A first result is the following

Proposition 5.1 *Suppose $\mathcal{M}$ is a strong nonstandard model. If $\Lambda \in \mathcal{M}$ is a bounded collection of internal sets, i.e. $\Lambda_E \subseteq \vartheta(V_\beta(X))_E$ for some $\beta < \alpha$, then Λ is internal.*

PROOF. Apply *transfer* to $V_\alpha(X) \models \forall x \; x \subseteq V_\beta(X) \to x \in V_{\beta+1}(X)$. ⊣

Thus every bounded collection of internal sets which cannot be in itself internal, is not a member of the nonstandard model. This is the case of many collections which play a central role in the practice of Robinsonian analysis, e.g. the standard reals, the infinite reals, the infinitesimals, the monads, the galaxies, etc.

We now raise the question whether the above property holds for every (not necessarily bounded) collection of internal sets. The next result shows that this "stability" property for the internal sets is closely related to the order structure of the hyperordinals.

Theorem 5.2 *Suppose $\mathcal{M}$ is a strong nonstandard model. Then the following are equivalent:*
  (i)   *For every $m \in \mathcal{M}$, $m_E \subseteq \mathcal{I} \Rightarrow m \in \mathcal{I}$*
  (ii) *Every external element contains external E-elements.*
  (iii) *$Ord_\mathcal{I}$ is a gap of the ordered set $\langle Ord_\mathcal{M}, <_E \rangle$*
  (iv) *$Ord_\mathcal{I} \notin \mathcal{M}$*
  (v)   *$\mathcal{I} \notin \mathcal{M}$*

PROOF. Equivalence $(i) \leftrightarrow (ii)$ is trivial. $(i) \Rightarrow (v)$ Otherwise, let $m \in \mathcal{M}$ be such that $m_E = \mathcal{I}$. Now $m_E \subseteq \mathcal{I}$ implies $m \in \mathcal{I}$, hence mEm which contradicts $\langle \mathcal{I}, E \rangle \models Fnd$. $(v) \Rightarrow (iv)$ Suppose $Ord_\mathcal{I} \in \mathcal{M}$ and apply *transfer* to

$$V_\alpha(X) \models \forall \Lambda \; [\forall \lambda \in \Lambda \; Ord(\lambda, \emptyset)] \to$$
$$\exists V \; \forall t \; [t \in V \; \leftrightarrow \; \exists \lambda \in \Lambda \; \forall f \; (\; Prf(t, f, \emptyset) \to f(t) \in \lambda \;)]$$

One gets an element $\Upsilon \in \mathcal{M}$ such that

$$\Upsilon_E = \{m \in \mathcal{M} : \exists \xi \in Ord_\mathcal{I} \; \psi_\mathcal{M}(m) <_E \xi \} = \mathcal{I}$$

$(iv) \Rightarrow (iii)$ Otherwise let $\xi = Min \, (Ord_\mathcal{M} \setminus Ord_\mathcal{I})$. Then $\xi \in \mathcal{M}$ and $\xi_E = Ord_\mathcal{I}$, which contradicts (iv). $(iii) \Rightarrow (i)$ Suppose $m \in \mathcal{M}$ be such that $m_E \subseteq \mathcal{I}$ while $m \notin \mathcal{I}$. Take $\xi = \psi_\mathcal{M}(m) = Sup\{\psi_\mathcal{M}(m') + 1 : m'Em\}$. ξ is an external supremum of a set of internal hyperordinals, hence $\xi = \text{Sup}\,Ord_\mathcal{I} = \text{Min}\,(Ord_\mathcal{M} \setminus Ord_\mathcal{I})$ and $Ord_\mathcal{I}$ is not a gap. $\dashv$

We shall call *regular* any strong nonstandard model which satisfies the above properties. The use of the adjective "regular" originates from theorem 5.6. For $\alpha = \omega$, the regularity question is settled by the

Theorem 5.3 *Every strong ω-NSM is regular.*

PROOF. Since every ordinal $0 < n < \omega$ is a successor, $\rho_x(A) = \text{Sup}\{\rho_x(a)+1 : a \in A\}$ is a maximum for every $A \in V_\omega(X)$ which is not $\in$-minimal. Thus

$$V_\omega(X) \models \forall A\ \forall f\ [\,\exists x \in A \wedge Prf(A, f, \emptyset)\,] \ \to\ (\,\exists a \in A\ f(A) = f(a) + 1\,)$$

By *transfer* it follows that for every given $m \in \mathcal{M}$ $\psi_\mathcal{M}(m) = \psi_\mathcal{M}(m') + 1$ for some $m'\mathrm{E}m$. In particular, if $m_\mathrm{E} \subseteq \mathcal{I}$ then $m \in \mathcal{I}$. $\dashv$

The above theorem yields a characterization for the internal sets of a strong ω-NSM. This result was first proved in [13] by other means.

Theorem 5.4 *Suppose $\mathcal{M}$ is a strong ω-NSM. Then the collection $\mathcal{I} \subseteq \mathcal{M}$ of the internal sets is characterized by the following properties:*
  (i)   *$\mathcal{I}$ contains the standard elements.*
  (ii) *$\mathcal{I}$ is E-transitive.*
  (iii) *$\langle \mathcal{I}, \mathrm{E} \rangle$ is well-founded.*

PROOF. It is known that $\mathcal{I}$ satisfies the above properties. Since $\mathcal{I}$ is the E-transitive closure of the standards, any other collection $\Lambda \subseteq \mathcal{M}$ satisfying (i) and (ii) contains the internal sets. Moreover no external set can be a member of Λ. Otherwise let $m \in \Lambda$ be external. The regularity of $\mathcal{M}$ yields the existence of an external element $m_1\mathrm{E}m$. By iteration, one gets an E-descending chain $m_1\mathrm{E}m$, $m_2\mathrm{E}m_1$, $m_3\mathrm{E}m_2$, ... which contradicts (iii). $\dashv$

Let's now concentrate on the regularity problem when $\alpha > \omega$. This problem happens to be in some way connected to the existence of non-regular uniform ultrafilters.

Remind that a filter D over the set I is *regular* [12] if it satisfies the following equivalent conditions: (i) There exists a family $\{D_i : i \in I\} \subseteq D$ of members of D such that $\bigcap\{D_i : i \in I_o\} = \emptyset$ for every infinite $I_o \subseteq I$; (ii) there exists a family $\{\mathcal{J}_i : i \in I\}$ of finite subsets of I such that $\{i \in I : j \in \mathcal{J}_i\} \in D$ for every $j \in I$. The filter D is called *uniform* if $|X| = |I|$ for every $X \in D$. Regular uniform ultrafilters exist over any given infinite set.

The notion of regularity of ultrafilters was isolated as a consequence of the following result:

---
[12]For a collection of basic results on regular ultrafilters, see [2] §4.3.

- Suppose $\mathcal{M}$ is any infinite structure of cardinality μ and D is a uniform regular ultrafilter on an index set I of cardinality κ. Then $\mathcal{M}'_D$ has the largest cardinality possible, i.e. μ^κ.

The next proposition is a generalization to limit ordinals of a known result [13].

Proposition 5.5 *Suppose α is a limit ordinal and D a regular ultrafilter over a set I with $|I| \geq |\alpha|$. If $f^D \in \alpha'_D$ is "unbounded", i.e. $f^D > d(\gamma)$ for every $\gamma < \alpha$, then $f^D > g^D$ for some unbounded g^D.*

PROOF. Let $\{\mathcal{J}_i : i \in I\}$ be a family of finite subsets of I such that $\{i \in I : j \in \mathcal{J}_i\} \in D$ for every $j \in I$. Fix $\sigma : \kappa \to \alpha$ a surjective map and defin g as follows

$$g(i) \ = \ Max\{\sigma(j) < f(i) : j \in \mathcal{J}_i\}$$

The $\mathcal{J}_i$'s are finite sets, thus g is well-defined as an element of α^I/D because

$$\Lambda = \{i \in I : \exists j \in \mathcal{J}_i \ \sigma(j) < f(i)\} \in D.$$

In fact, $\Lambda \supseteq \{i \in I : \sigma(j_0) < f(i)\} \cap \{i \in I : j_0 \in \mathcal{J}_i\} \in D$ for each $j_0 \in I$. $g^D < f^D$ follows straight from the definition of g. It remains to show that g is unbounded.

For each $\beta < \alpha$, let $\sigma(j_0) = \beta$. $\{i \in I : j_0 \in \mathcal{J}_i\} \in D$ implies

$$\Xi_1 = \{i \in I : \exists j \in \mathcal{J}_i \ \sigma(j) = \beta\} \in D$$

and f^D unbounded implies

$$\Xi_2 = \{i \in I : \beta < f(i)\} \in D.$$

The it follows that $g(i) \leq \beta$ for every $i \in \Lambda \cap \Xi_1 \cap \Xi_2 \in D$.   ⊣

Regular ultrafilters yield regular NSM's (and this originates the name).

Theorem 5.6 *Take D any regular ultrafilter over a set I with $|I| \geq |\alpha|$. Then the diagonal immersion $d : V_\alpha(X) \to V_\alpha(X)'_D$ yields a regular α-NSM.*

PROOF. Take $d : V_\alpha(X) \to V_\alpha(X)'_D$ the NSE given by the diagonal immersion. Notice that $\xi^D \in Ord_\mathcal{M} \Leftrightarrow \{i \in I : \xi(i)$ is a V.N. ordinal$\} \in D \Leftrightarrow \xi^D \in \alpha'_D$. Moreover $\xi^D \notin Ord_\mathcal{I} \Leftrightarrow \xi^D$ is unbounded. Thus one can apply the previous proposition and get the thesis.   ⊣

Using the core model, H.-D. Donder [3] proved that if ZFC is consistent, then so is $ZFC+$ *"every uniform ultrafilter over an infinite set is regular"*. Thus the following holds:

---

[13]See [6] §38.

Corollary 5.7 *There is a model of ZFC where every diagonal immersion* $d : V_\alpha(X) \to V_\alpha(X)^I_D$ *of the standard model into any of its ultrapowers with* $|I| \geq |\alpha|$ *yields a regular NSM.*

An extensive literature indicates that the existence problem for uniform ultrafilters which are not regular is closely connected with the existence of large cardinals [14]. We shall use the following: [15]

Proposition 5.8 *Suppose D is a uniform non-regular ultrafilter over ω_1. Then the D-equivalence class of the diagonal function $\Delta : \gamma \mapsto \gamma$ $\gamma < \alpha$, is the least unbounded element in $\omega_1{}^{\omega_1}_D$.*

A recent and deep result by M. Foreman, M. Magidor and S. Shelah [5], states that if a huge cardinal exists then there is a model provided with an ultrafilter D such that $|\omega_D^{\omega_1}| = \omega_1$. In particular such a D yields a non-regular uniform ultrafilter over ω_1. If one proceeds as in the proof of Theorem 5.6, this fact together with the previous proposition yields the

Theorem 5.9 *The existence of non-regular ω_1-NSM's is consistent relative to* ZFC+ *"there exists a huge cardinal".*

We conclude by raising two interesting problems:

- Is there a model of ZFC where every strong NSM is regular ?

- Which relative consistency relations hold between the existence of non-regular strong NSM's and various large cardinal axioms ?

Appendix.
PROOF OF THE REPRESENTATION THEOREM. Denote $\mathcal{L}_*$ the language of set theory whose set of variables is $\{x_b : b \in \mathcal{I}\}$ instead of $\{x_n : n < \omega\}$. For every model $\mathcal{N} = \langle N, \in_\mathcal{N} \rangle$, for every formula φ of $\mathcal{L}_*$ and for every assignment $\chi : \mathcal{I} \to N$, the satisfaction relation $\models_*$ is defined as follows:

$$\mathcal{N} \models_* \varphi(x_{b_1}, \dots, x_{b_n})\, [\chi] \overset{\text{def}}{\Longleftrightarrow} \mathcal{N} \models \varphi(\chi(b_1), \dots, \chi(b_n))$$

Now take $\mathcal{B}_* = \{\psi_i : i \in I\}$ the set of all the bounded formulae of $\mathcal{L}_*$ such that $\langle \mathcal{I}, E \rangle \models_* \psi_i\, [1]$ ($1 : \mathcal{I} \to \mathcal{I}$ denotes the identity map). Fix $\psi_i \in \mathcal{B}_*$ a formula all of whose free variables are among $x_{b_1}, \dots, x_{b_n}$. Without loss of generality, suppose that $b_s = \vartheta(a_s)$ is standard if and only if $s = 1, \dots, k$ and denote $\beta_s = \text{Min}\{\beta : b_s E \vartheta(V_\beta(X))\}$ $s = k+1, \dots, n$. Then apply *transfer* to $\langle \mathcal{I}, E \rangle \models_* \psi_i(x_{b_1}, \dots, x_{b_n})\, [1]$ and get that $\langle V_\alpha(X), \in \rangle \models \psi_i(a_1, \dots, a_k, a_{k+1}, \dots, a_n)$ for some $a_s \in V_{\beta_s+1}(X)$ $s = k+1, \dots, n$. This fact yields the following property:

---

[14] E.g. see [5].
[15] See [6] §38.

$\forall i \in I$ there exists an assignment $\chi_i : \mathcal{I} \to V_\alpha(X)$ such that

(i) $\quad \langle V_\alpha(X), \in \rangle \models_* \psi_i [\chi_i]$

(ii) $\quad \forall \beta < \alpha \ bE\vartheta(V_\beta(X)) \Rightarrow \chi_i(b) \in V_\beta(X)$

(iii) $\quad \forall a \in V_\alpha(X) \ \chi_i(\vartheta(a)) = a$

The family $\mathcal{D} = \{D_i : i \in I\}$ where $D_i = \{j \in I : \langle V_\alpha(X), \in \rangle \models_* \psi_i [\chi_j]\}$ has the *finite intersection property (FIP)* (notice that the formula $\psi_{i_1} \wedge \ldots \wedge \psi_{i_n}$ belongs to $D_{i_1} \cap \cdots \cap D_{i_n}$). Thus we can take an ultrafilter $D \supseteq \mathcal{D}$ on I and define $\pi : \mathcal{I} \to V_\alpha(X)^I_D$ as follows. For every $b \in \mathcal{I}$, let $\pi(b) = f_b^D$ be the D-equivalence class of the function $f_b : i \mapsto \chi_i(b)$. From *(ii)* it follows that $Range(\pi) \subseteq {}^bV_\alpha(X)^I_D$. Besides, *(iii)* yields that $\pi \circ \vartheta = d : V_\alpha(X) \hookrightarrow {}^bV_\alpha(X)^I_D|F$ is the diagonal immersion. Now take φ any bounded formula, $b_1, \ldots, b_n \in \mathcal{I}$ and suppose $\langle \mathcal{I}, E \rangle \models \varphi(b_1, \ldots, b_n)$. Then

$$ {}^bV_\alpha(X)^I_D \models \varphi(\pi(b_1), \ldots, \pi(b_n)) \iff $$

$$ \{j \in I : V_\alpha(X) \models \varphi(\pi(b_1)(j), \ldots \pi(b_n)(j)\} = $$

$$ \{j \in I : V_\alpha(X) \models \varphi(\chi_j(b_1), \ldots, \chi_j(b_n)\} = $$

$$ \{j \in I : V_\alpha(X) \models_* \varphi(x_{b_1}, \ldots, x_{b_n}) [\chi_j]\} = D_i \in D $$

and this last property is true by the definition of D. Our next step is to single out a suitable submodel of ${}^bV_\alpha(X)^I_D$ in such a way that the *transfer* principle still holds and π is onto (hence an isomorphism). The family $\mathcal{F} = \{\text{eq}(f) : f^D \in Range(\pi)\}$ where $\text{eq}(f) = \{\langle i, j \rangle \in I \times I : f(i) = f(j)\}$ has the FIP. Take F the filter over $I \times I$ generated by $\mathcal{F}$. We claim that $\pi : \mathcal{I} \to {}^bV_\alpha(X)^I_D|F$ is onto. Fix $f^D \in {}^bV_\alpha(X)^I_D$ with $\text{eq}(f) \in F$. By the definition, there exist $g_s^D = \pi(b_s) \ s = 1, \ldots, n$ such that

$$ \text{eq}(g_1) \cap \ldots \cap \text{eq}(g_n) \subseteq \text{eq}(f). $$

Take $\beta < \alpha$ such that $b_1, \ldots, b_n \in \vartheta(V_\beta(X))$ (hence $g_1^D, \ldots, g_n^D \in V_\beta(X)^I_D$) and also $f^D \in V_\beta(X)^I_D$. The above inclusion assures the existence of an n-ary function $\sigma : V_\beta(X)^n \to V_\beta(X)$, with $\sigma : \langle g_1(i), \ldots, g_n(i) \rangle \mapsto f(i) \ \forall i \in I$. Notice that $\sigma \in V_{\beta+2n+2} \subseteq V_\alpha(X)$. Thus

$$ {}^bV_\alpha(X)^I_D|F \models d(\sigma) : \langle g_1^D, \ldots, g_n^D \rangle \mapsto f^D $$

Take $\hat{b}E\vartheta(V_\beta(X))$ with $\vartheta(\sigma)_E : \langle b_1, \ldots, b_n \rangle \mapsto \hat{b}$ and apply the *transfer* to the mapping π so to get

$${}^{b}\mathrm{V}_{\alpha}(X){}^{I}_{D}|F \models \pi(\vartheta(\sigma)) : \langle \pi(b_1), \dots, \pi(b_n) \rangle \mapsto \pi(\widehat{b})$$

Recall that $\pi(\vartheta(\sigma)) = d(\sigma)$, $\pi(b_s) = g^{D}_{s}$ $s = 1, \dots, n$ and compare the two above formulae which hold in ${}^{b}\mathrm{V}_{\alpha}(X){}^{I}_{D}|F$. Then it necessarily follows that $f^D = \pi(\widehat{b}) \in \mathrm{Range}(\pi)$.

The last statement of the theorem is proved by noticing that for every $a \in \mathcal{I}$, $a\mathrm{E}\vartheta(A) \Leftrightarrow \pi(a)\mathrm{E}\pi(\vartheta(A)) = d(A) \Leftrightarrow \pi(a) \in A^{I}_{D}|F$.   ⊣

Acknowledgments. The author would like to thank Marco Forti for many helpful suggestions on preliminary drafts of this paper and Piero Ghilardi for his encouragement.

References

[1] D. Ballard and K. Hrbacek, *Standard Foundations for Nonstandard Analysis*, J. Symbolic Logic **57** (1992), 741–748.

[2] C.C. Chang and H.J. Keisler, *Model Theory*, 3rd ed., North-Holland, Amsterdam, 1990.

[3] H.-D. Donder, *Regularity of Ultrafilters and the Core model*, Israel J. Math. **63** (1988), 289–322.

[4] F.R. Drake, *Set Theory*, North-Holland, Amsterdam, 1974.

[5] M. Foreman, M. Magidor, and S. Shelah, *Martin's Maximum, Saturated Ideals and non-Regular Ultrafilters. Part II*, Ann. of Math. (2) **127** (1988), 521–545.

[6] T. Jech, *Set Theory*, Academic Press, New York, 1978.

[7] H.J. Keisler, *Limit Ultrapowers*, Trans. Amer. Math. Soc. **107** (1963), 382–408.

[8] ———, *Foundations of Infinitesimal Calculus*, Prindle, Weber & Schmidt, Boston, 1976.

[9] M. Di Nasso, *Pseudo-superstructures as Nonstandard Universes*, submitted.

[10] ———, *Universi nonstandard stratificati*, Ph.D. thesis, Università di Siena, 1995, (Italian).

[11] A. Robinson and E. Zakon, *A Set-Theoretical Characterization of Enlargements*, Applications of Model Theory to Algebra, Analysis and Probability (W.A.J. Luxemburg, ed.), Holt, Rinehart and Winston, New York, 1969, pp. 109-122.

[12] J. Schmid and J. Schmidt, *Enlargements without Urelements*, Colloq. Math. **52** (1987), 1–22.

[13] V.M. Tortorelli, *A characterization of Internal Sets*, Rend. Sem. Mat. Univ. Padova **81** (1989), 193–200.

[14] E. Zakon, *A New Variant of Nonstandard Analysis*, Victoria Symposium on Nonstandard Analysis (A.E. Hurd and P.A. Loeb, eds.), Lecture Notes in Mathematics **369**, Springer-Verlag, Berlin, 1974, pp. 313–339.

DIPARTIMENTO DI MATEMATICA, VIA DEL CAPITANO 15, 53100 SIENA, ITALY
e-mail address: dinasso@unisi.it

Some notes on subword quantification and induction thereof

Fernando Ferreira*

Abstract

The first section of this paper consists of a defense of the binary string notation for the formulation of weak theories of arithmetic which have computational significance. We defend that a string language is the most natural framework and that the usual arithmetic setting suffers from some troubles when dealing with very low complexity classes. Having introduced in the first section the theory $Th - FO$ - associated with a rather robust uniform version of the class of problems that can be decided by constant depth, polynomial size circuit families (the so-called AC^0-class) - we prove in the second section that the deletion of a crucial axiom from $Th - FO$ results in a theory which is unsuitable from the computational point of view.

Keywords: bounded arithmetic; computational complexity; AC^0; provably total functions.

1 The apology of a notation

In his Ph.D. Dissertation [2], Samuel Buss studies systems of arithmetic related to conspicuous classes of computational complexity. The main results of Buss show that the provability of certain sentences of the type "$\forall x \exists y A(x, y)$" in suitable sub-theories of Peano Arithmetic imply the existence of a function f such that $A(n, f(n))$, for all $n \in \omega$, and such that f has a certain computational complexity. Buss is interested in computational complexity classes consisting only of *feasible computable* functions and this requirement excludes, *a fortiori*, the exponential function. On the proof-theoretic side, this requirement compels the theories of arithmetic to be *weak theories*, i.e., theories that do not prove the totality of the exponential function.

---

*This work was partially supported by project 6E91 of CMAF (Portugal)

A good first example is the theory $I\Delta_0 + \Omega_1$. This theory consists of $I\Delta_0$, the main feature of which is the restriction of the induction scheme to bounded formulæof the language of $I\Delta_0$, plus a Π_2^0-axiom (the Ω_1-principle) saying that the function $\lambda x.x^{\lceil log_2(x+1)\rceil}$ is total ($\lceil w \rceil$ represents the smallest integer greater than or equal to w). The language of the theory $I\Delta_0 + \Omega_1$ is the usual language of arithmetic: a constant 0, a unary function symbol S, two binary function symbols $+$ and $\cdot$ and a binary relation symbol $\leq$. This language presents some technical difficulties for the study of the provable total functions of $I\Delta_0 + \Omega_1$. A first reason is that $I\Delta_0 + \Omega_1$ does not have a Π_1^0-axiomatization. Hence it is not suitable for the formulation of a Parikh type theorem (we recommend chapter V of [6] as a reference for this section).

Buss gave a reformulation of the theory $I\Delta_0 + \Omega_1$. His reformulation adds to the usual language of arithmetic two unary function symbols $\lfloor \frac{1}{2}x \rfloor$, $|x|$ (for $\lceil log_2(x+1)\rceil$, the length of the binary representation of x) and a binary function symbol $x\#y$ (for $2^{|x|\cdot|y|}$, the "smash" function)[1]. In this language Buss presents a Π_1^0-axiomatization consisting of thirty two basic open axioms, plus the usual scheme of induction for bounded formulae, resulting in the so-called theory T_2 (the induction scheme, whose instances are "$A(0) \wedge \forall x(A(x) \rightarrow A(x + 1)) \rightarrow \forall x A(x)$", where A is a bounded formula, does not consist of Π_1^0-sentences but can be easily reformulated as such by "$\forall x(A(0) \wedge \neg A(x) \rightarrow \exists y \leq x(A(y) \wedge \neg A(y + 1)))$"). Parikh's theorem applies:

Theorem. *If $T_2 \vdash \forall x \exists y A(x, y)$, where A is a bounded formula, then there is a term $t(x)$ of the language of T_2 such that $T_2 \vdash \forall x \exists y \leq t(x)A(x, y)$.*

From the truth of "$\forall x \exists y \leq t(x)A(x, y)$" alone, one easily concludes that the witness function $f(n) = \mu m A(n, m)$ is computable in polynomial space - indeed, even computable in polynomial time with an oracle in PH (the polynomial time hierarchy). More specifically, if the relation $A(x, y)$ is in Σ_n^p (and this is always the case for a certain n) then f is in $\square_{n+1}^p$. This analysis takes little advantage of the provability of "$\forall x \exists y A(x, y)$" in T_2, relying solely upon the very general fact of Parikh. As we will see, Buss is able to provide a deeper analysis.

The class of *sharply bounded formulae* is the smallest class of formulae of the language of T_2 that contains the atomic formulae and that is closed under Boolean operations and *sharply bounded* quantifications (these are quantifications of the form $\forall x \leq |t|(...)$ or $\exists x \leq |t|(...)$, where t is a term in which the variable x does not occur). The Σ_n^b-*formulae*, $n \geq 1$, are the bounded formulae

---

[1]The growth rate of the smash function entails that lengths of numbers are closed under multiplication. This enables the formulation of many standard constructions, in particular of polynomial time computations.

of the language of T_2 with the following form:

$$(BD) \qquad \exists x_1 \leq t_1 \forall x_2 \leq t_2 \exists x_3 \leq t_3 \ldots Q x_n \leq t_n \, A$$

where A is a sharply bounded formula, $t_1, \ldots, t_n$ are terms of the language and the quantifier Q is a $\forall$ or a $\exists$ depending on whether n is even or odd (respectively). The Σ_n^b-formulae define, in the standard model, the Σ_n^p-relations of PH. Hence, there is a matching between bounded formulae of the language of T_2 and the levels of the polynomial time hierarchy. Moreover, this is a *faithful* matching in the sense that the complexity of formula-construction goes hand-in-hand with the complexity levels of the polynomial hierarchy. In other words, a sub-formula of a Σ_n^b-formula can only define Σ_n^p-relations [2].

Let Ψ be a set of formulae. The $\Psi - PIND$ axioms are:

$$(PIND) \qquad A(0) \wedge \forall x (A(\lfloor \tfrac{1}{2} x \rfloor) \to A(x)) \to \forall x A(x)$$

where $A \in \Psi$, possibly with parameters. The theory S_2^n $(n \geq 1)$ consists of the thirty two basic open axioms plus the $\Sigma_n^b - PIND$ axioms (it is well known that $T_2 = \cup_{n \in \omega} S_2^n$). Buss' main result of his thesis is the following:

Theorem. *If $S_2^n \vdash \forall x \exists y A(x, y)$, where A is a Σ_n^b-formula, then there is $f \in \square_n^p$ such that $A(k, f(k))$, for all $k \in \omega$* [3].

In the above, $f(k)$ is not necessarily the least m such that $A(k, m)$. The function f is constructed by means of a careful analysis of the proof of "$\forall x \exists y A(x, y)$" in S_2^n, and the importance of the faithfulness of the matching between Σ_n^b-formulae and Σ_n^p-relations cannot be over-emphasized in this respect. It is this faithfulness that permits a successful application of Gentzen's Hauptsatz (*vulgo* cut-elimination) to proving the above theorem. In short, Buss' notation for $I\Delta_0 + \Omega_1$ is not only superior to the usual notation on the account of providing Π_1^0-axiomatizations but, more important, on the account of giving faithful representations of the polynomial hierarchy.

Nonetheless, a critique can be made of Buss' notation. The most important is that it is not faithful with respect to certain complexity classes below the class P of polynomial time decidable predicates (we will discuss this later). It also parts from the notation of choice of most computer scientists working on feasibility.

---

[2] The definition of Σ_n^b-formula presented here is what Buss calls a *strict* Σ_n^b-formula. Buss defines Σ_n^b-formulae in a more general way. For this latter definition we have to modify slightly the concept of faithfulness: it means that every positive (*resp.*, negative) occurrence of a sub-formula of a Σ_n^b-formula is a Σ_n^p-relation (*resp.*, a Π_n^p-relation).

[3] Buss also remarked that the conclusion of the theorem is provable in S_2^n (pace a suitable formalization).

In our Ph.D. Dissertation [3] we work with a language based on the operation of concatenation, following Quine [9] and Smullyan [11]. Our intended (standard) model is the tree $\{0,1\}^*$ of binary strings. The first-order stringlanguage of the binary tree consists of three constant symbols ϵ, 0 and 1, two binary function symbols $\frown$ (for *concatenation*, sometimes omitted) and $\times$, and a binary relation symbol $\subseteq$ (for *initial subwordness*). There are fourteen basic open axioms:

$$x \frown \epsilon = x \qquad x \times \epsilon = \epsilon$$
$$x \frown (y \frown 0) = (x \frown y) \frown 0 \qquad x \times (y \frown 0) = (x \times y) \frown x$$
$$x \frown (y \frown 1) = (x \frown y) \frown 1 \qquad x \times (y \frown 1) = (x \times y) \frown x$$
$$x \frown 0 = y \frown 0 \rightarrow x = y \qquad x \frown 1 = y \frown 1 \rightarrow x = y$$
$$x \subseteq \epsilon \;\leftrightarrow\; x = \epsilon$$
$$x \subseteq y \frown 0 \;\leftrightarrow\; x \subseteq y \vee x = y \frown 0$$
$$x \subseteq y \frown 1 \;\leftrightarrow\; x \subseteq y \vee x = y \frown 1$$
$$x \frown 0 \;\neq\; y \frown 1$$
$$x \frown 0 \;\neq\; \epsilon$$
$$x \frown 1 \;\neq\; \epsilon$$

In the standard model, $x \times y$ is the word x concatenated with itself length of y times (the growth rate of $\times$ corresponds exactly to the growth rate of Buss' smash function $\#$). *Subwordness* of x with respect to y, denoted by $x \subseteq^* y$, is defined by $\exists z \subseteq y \, (z \frown x \subseteq y)$. The class of *sw.q.-formulae* ("subword quantification formulae") is the smallest class of formulae containing the atomic formulae and closed under Boolean operations and subword quantification, i.e., quantification of the form $\forall x \subseteq^* t(...)$ or $\exists x \subseteq^* t(...)$, where t is a term in which the variable x does not occur. These formulae define in the standard model the so-called class FO of *first-order expressible* properties: this notion was introduced by Neil Immerman in [8] and was originally defined in terms of first-order definability in suitable finite structures with domain $\{0, 1, \cdots, n-1\}$. The FO-class is included in AC^0, the class of sets that can be decided by constant depth, polynomial size circuit families (one should view the class of FO-relations as a rather robust uniform version of AC^0). In the same paper, Immerman also discusses what we call FO-functions. In parallel with the set case, these functions can also be computed by constant depth, polynomial size circuit families (the so-called AC^0-functions). A more detailed rendering of these notions can be found in [5].

The relation of x being of length less than or equal to the length of y, denoted by $x \leq y$, is defined by $1 \times x \subseteq 1 \times y^4$; $x \equiv y$ abbreviates $x \leq y \wedge y \leq x$. The class of *bounded formulae*, also named the class of Σ_∞^b-formulae, is the smallest class of formulae containing the sw.q.-formulae and closed under Boolean operations and bounded quantification, i.e., quantification of the form $\forall x \leq t(...)$ or $\exists x \leq t(...)$, where t is a term in which the variable x does not occur. In the standard model these formulae define exactly the sets of the polynomial hierarchy. Mimicking Buss' terminology, we define the Σ_n^b-formulae $(n \geq 1)$ as the bounded formulae with the form (BD) above, where A is sw.q.-formula and $t_1, \ldots, t_n$ are terms of the new language. Not surprisingly, these Σ_n^b-formulae define exactly the Σ_n^p-relations.

The theory $\Sigma_n^b - NIA$ (for *Notation Induction Axioms*) consists of the basic axioms plus the induction scheme,

$$(NIA) \qquad A(\varepsilon) \wedge \forall x \, (A(x) \rightarrow A(x0) \wedge A(x1)) \rightarrow \forall x \, A(x)$$

where A is a Σ_n^b-formula, possibly with parameters. This theory is equivalent, in a sense that could be made precise, to Buss' theory S_2^n, and it is partly a matter of taste and habits the preference for working within the string setting.

In the sequel, we will be interested in the theory $Th - FO$. The axioms of this theory consist of the fourteen basic open axioms. the scheme of induction on notation (NIA) for sw.q.-formulae, and a string-building principle:

$$(SB) \qquad \forall u(tally(u) \rightarrow \exists x \equiv u \forall v \subset u(bit(x, v) = 1 \leftrightarrow A(v)))$$

where "$v \subset u$" abbreviates "$v \subseteq u \wedge v \neq u$", "$bit(x, v) = 1$" abbreviates "$\exists w \subseteq x(w \equiv v \wedge w1 \subseteq x)$", and A is a sw.q.-formula (possibly with parameters). This is a rather weak theory, but still an interesting one [5]. The theory $Th - FO$ is not an arithmetical theory in the usual sense. However, it is possible to do some arithmetic in $Th - FO$. More specifically, it is possible to "smoothly" introduce the successor and the addition functions in $Th - FO$. We are using the term "smoothly" in the following sense: it is possible to expand the language of $Th - FO$ with new function symbols S and $+$ such that the usual recursive defining

---

[4] This appropriation of the traditional symbol for the usual order of the natural numbers has, sometimes, been criticized. We maintain that it has virtualities from the point of view of computational complexity because the number of natural numbers less than or equal to n has the same order of growth as the number of elements of $\{0, 1\}^*$ with length less than or equal to the length of a word σ of length $|n|$.

[5] A reason for its interest, apart from being closely related to the FO-class, is its bearing on $I\Delta_0$ (see [5] for an explanation of this). The theory $Th - FO$ was also independently defined by Zambella [16]. In his setting, it is called $\Sigma_0^p - comp$. Recently P. Clofe and G. Takeuti gave yet another independent definition in *First Order Bounded Arithmetic and Small Boolean Circuit Complexity* (Feasible Mathematics II, P. Clofe and J.B. Remmel eds., Birkhäuser, 1995); in their setting the theory is called TAC^0 and the authors also mention a previous version T^0AC^0 (unpublished) by P. Clofe (1990).

relations are provable and such that these new symbols count as primitive when making an analysis of whether a particular formula of the extended language falls within a pertinent class of formulae (e.g., whether a particular formulae is a sw.q.-formula). Let us briefly see how arithmetic notions can be introduced in the theory $Th - FO$.

The models of $Th - FO$ have a canonical linear ordering $<_\ell$, which is formally defined by a sw.q.-formula:

$$x <_\ell y := (x \leq y \wedge x \not\equiv y) \vee (x \equiv y \wedge \exists z \subseteq x(z0 \subseteq x \wedge z1 \subseteq y))$$

In the standard model this yields a ω-like ordering:

$$
\begin{array}{cccccccccccc}
0 & 1 & 2 & 3 & 4 & 5 & 6 & 7 & 8 & 9 & 10 & 11 & \cdots \\
\updownarrow & \updownarrow & \updownarrow & \updownarrow & \updownarrow & \updownarrow & \updownarrow & \updownarrow & \updownarrow & \updownarrow & \updownarrow & \updownarrow & \\
\epsilon & 0 & 1 & 00 & 01 & 10 & 11 & 000 & 001 & 010 & 011 & 100 & \cdots
\end{array}
$$

Having this correspondence in mind, we introduce arithmetic notions in models of $Th - FO$. The graph of the successor function is easily defined by the following sw.q.-formula $\theta_1(x, y)$:

$$(x = 1 \times x \wedge y = 0 \times x1) \vee \exists w \subseteq x \exists z \subseteq x(x = w0 \frown (1 \times z) \wedge y = w1 \frown (0 \times z)).$$

Moreover,

$(I\theta_1)$ $\qquad\qquad\qquad Th - FO \vdash \forall x \exists^1 y\, \theta_1(x, y)$

$(II\theta_1)$ $\qquad\qquad\qquad\qquad Th - FO \vdash \theta_1(\epsilon, 0)$

$(III\theta_1)$ $\qquad Th - FO \vdash \forall x \forall y(\theta_1(x0, x1) \wedge (\theta_1(x, y) \rightarrow \theta(x1, y0)))$

The proofs of these three facts are elementary and can be obtained by means of judicious uses of the axioms. Some basic properties recur in these proofs (and in proofs of similar nature), e.g.: $\epsilon \subseteq x$; $x \subseteq x$; $x \subseteq y \rightarrow x \subseteq^* y$; $x \subseteq^* y \wedge y \subseteq^* z \rightarrow x \subseteq^* z$; $x \subseteq^* y \frown z \rightarrow \exists x_1 \subseteq^* y \exists x_2 \subseteq z\, (x = x_1 \frown x_2)$; and $x \subseteq^* y \times z \rightarrow \exists x_1 \subseteq^* y \exists x_3 \subseteq z \exists x_3 \subseteq y\, (x = x_1 \frown (y \times x_2) \frown x_3)$. Actually, all these properties are provable without the use of the string-building principle. (Proofs of the previous statements were worked out in detail in [3].) The properties $(I\theta_1)$, $(II\theta_1)$ and $(III\theta_1)$ permit to "smoothly" introduce in the theory $Th - FO$ a new unary function symbol S (for the *successor* function) satisfying the following recursive specifications: $S(\epsilon) = 0$; $S(x0) = x1$; and $S(x1) = S(x) \frown 0$. Almost dually, it is also possible to introduce the corresponding predecessor function: $pred(\epsilon) = pred(0) = \epsilon$; $pred(x0) = pred(x) \frown 1$, for $x \neq \epsilon$; and $pred(x1) = x0$.

The introduction of "addition" is not so straightforward. Perhaps the easiest way to introduce it consists in reducing this operation of addition to the usual addition in binary number notation. The configuration of a number in binary notation is a non-empty string of zeroes and ones which does not begin by a zero, with the sole exception of the string 0 itself. The relation β that is the graph of the order-preserving, no-gap, bijection between $\{0,1\}^*$ and the set of binary number configurations is sw.q.-definable. In fact:

$$\beta(x,y) \leftrightarrow (x = \epsilon \wedge y = 0) \vee (x \neq \epsilon \wedge y = 1 \frown pred(x))$$

Here is a picture:

ϵ	0	1	00	01	10	11	000	001	010	011	100	$\cdots$
$\updownarrow$	$\updownarrow$	$\updownarrow$	$\updownarrow$	$\updownarrow$	$\updownarrow$	$\updownarrow$	$\updownarrow$	$\updownarrow$	$\updownarrow$	$\updownarrow$	$\updownarrow$	
0	1	10	11	100	101	110	111	1000	1001	1010	1011	$\cdots$

It is not difficult to see that,

$$Th - FO \vdash \forall x \exists^1 y \, \beta(x,y)$$

$$Th - FO \vdash \forall y (bin(y) \rightarrow \exists^1 x \, \beta(x,y))$$

where $bin(y)$ abbreviates $y = 0 \vee \exists z \subseteq^* y(y = 1z)$. This permits to "smoothly" introduce two unary function symbols β_f and β_b in the theory $Th - FO$ such that the sentences $\forall x(\beta(x, \beta_f(x)))$, $\forall y(bin(y) \rightarrow \beta(\beta_b(y), y))$, $\forall x(x = \beta_b(\beta_f(x)))$, and $\forall y(bin(y) \rightarrow y = \beta_f(\beta_b(y)))$ are provable in $Th - FO$. The definition of binary addition can be obtained by a sw.q.-formula with the help of the ternary *carry* predicate (the following is a standard construction; see [12]):

$$carry(x,y,u) := tally(u) \wedge \exists v \subset u(bit(x,v) = 1 \wedge bit(y,v) = 1 \wedge$$

$$\wedge \forall w \subset u(v \subseteq w \rightarrow bit(x,w) = 1 \vee bit(y,w) = 1))$$

Now, if we let $sum(x,y,z)$ be

$$bin(x) \wedge bin(y) \wedge bin(z) \wedge (\forall u \subset 1 \times z(bit(z,u) = 1 \leftrightarrow (bit(x,u) = 1 \dot\vee bit(y,u)$$
$$= 1 \dot\vee carry(x,y,u))))$$

where $\dot\vee$ stands for the exclusive *or*, we get the graph of binary addition. Define $\theta_2(x,y,z)$ by $\beta_b(sum(\beta_f(x), \beta_f(y), \beta_f(z)))$. It is a matter of careful attention to details and familiarity with the axioms to show that,

$(I\theta_2)$ $\qquad\qquad Th - FO \vdash \forall x \forall y \exists^1 z \, \theta_2(x,y,z)$

$(II\theta_2)$ $Th - FO \vdash \forall x \theta_2(x, \epsilon, x)$

$(III\theta_2)$ $Th - FO \vdash \forall x \forall y \forall z (\theta_2(x, y, z) \to \theta_2(x, S(y), S(z)))$

(The string-building principle is used to proving the existence part of $(I\theta_2)$.)
The previous three properties permit to "smoothly" introduce in the theory
$Th - FO$ a new binary function symbol $+$ such that $x + \epsilon = x$ and $x + S(y) = S(x + y)$.

The next natural step towards defining arithmetic notions in $Th - FO$
consists in introducing the multiplication function. Well, at this point we are
faced with a stumbling block, since there is no Σ_1^b-formula θ_3 such that,

$(I\theta_3)$ $Th - FO \vdash \forall x \forall y \exists^1 z \, \theta_3(x, y, z)$

$(II\theta_3)$ $Th - FO \vdash \forall x \theta_3(x, \epsilon, \epsilon)$

$(III\theta_3)$ $Th - FO \vdash \forall x \forall y \forall z (\theta_3(x, y, z) \to \theta_3(x, S(y), z + x))$

The reason for the inexistence of such a formula rests on deep work in circuit
complexity theory. If such a formula existed, then the unique binary function f
such that $\theta_3(\sigma, \tau, f(\sigma, \tau))$, for all σ, τ in $\{0, 1\}^*$, would be a FO-function (this
is a consequence of a theorem in [5] which characterizes the provably total
functions, with Σ_1^b-graphs, of the theory $Th - FO$). Hence, *a fortiori*, f is
an AC^0-function, and this easily entails that the usual multiplication function
on the setting of binary number configurations is also an AC^0-function. Well,
this latter fact contradicts work of M. Ajtai and, independently, of M. Furst,
J. Saxe and M. Sipser [6].

The previous discussion explains why Buss' arithmetic language is inadequate for the formulation of the theory $Th - FO$, since the representation of
sw.q.-relations in Buss' language uses the multiplication function pervasively.
In short, Buss' notation is not faithful with respect to sw.q.-relations [7]. This is
the main point of the present section.

---

[6] In groundbreaking papers [1] and [10], these authors proved that "parity" is not an
AC^0-predicate. ("Parity" is the set of elements of $\{0, 1\}^*$ with an even number of 1's.) The
fact that "parity" is AC^0-reducible to "multiplication" is explained in [14].

[7] Buss' class of sharply bounded formulae is, in certain aspects, a bizarre class: in effect,
all the initial functions and relations of this class, as well as all its closure operations, are
of AC^0-character, with the sole exception of the multiplication function. (For an in-depth
study of sharply bounded predicates, consult [7].) A similar remark applies to the class SR
of *strictly rudimentary* formulae, as introduced by Wilkie and Paris in [15].

2   The unsuitability of a theory

The theory S_2^0 consists of the thirty two basic open axioms of Buss plus the *PIND* axioms for sharply bounded formulae. Gaisi Takeuti proved in [13] that the sentence "$\forall x \exists y (x = 0 \lor x = y + 1)$" cannot be deduced in S_2^0. In our opinion, this result shows that the theory S_2^0 is uninteresting and artificial. Namely, it shows that S_2^0 is too sensible to the basic open axioms and to the exact language chosen. Let $sw.q. - NIA$ be the theory in the stringlanguage formed by the fourteen open axioms listed in the previous section and the scheme of induction on notation (NIA) for sw.q.-formulae (the reader should compare this theory with $Th - FO$). Is $sw.q. - NIA$ an interesting theory? Similarly, the answer is no. We show that the function $\lambda \sigma . \bar{\sigma}$, where the string $\bar{\sigma}$ is obtained from σ by changing zeroes into ones and ones into zeroes, is not provably total in the theory $sw.q. - NIA$. More specifically, let "$y = \bar{x}$" abbreviate the sw.q.-formula,

$$y \equiv x \land \forall x' \subseteq x \exists y' \subseteq y \, (y' \equiv x' \land (x'0 \subseteq x \to y'1 \subseteq y) \land (x'1 \subseteq x \to y'0 \subseteq y))$$

We prove the following result:

Theorem. *The theory $sw.q. - NIA$ does not prove $\forall x \exists y$ "$y = \bar{x}$".*

We need to prepare the ground for the proof of this theorem. In what follows we will use at ease the provability of some simple facts in the theory $sw.q. - NIA$, e.g., those listed in the previous section during the discussion of the properties $(I\theta_1)$, $(II\theta_1)$ and $(III\theta_1)$.

Proposition 1. *Let $A(x)$ be a sw.q.-formula, possibly with parameters. Then the theory $sw.q. - NIA$ proves the following sentence,*

$$A(\epsilon) \land \neg A(y) \to \exists x \subseteq y \, (A(x) \land ((x0 \subseteq y \land \neg A(x0)) \lor (x1 \subseteq y \land \neg A(x1)))).$$

Proof : Consider the formula $B(x) := x \subseteq y \to A(x)$ and assume that $A(\epsilon)$ and $\neg A(y)$. Then $B(\epsilon)$ and $\neg B(y)$. By the scheme (NIA) for sw.q.-formulae we may conclude that there is x such that either $B(x) \land \neg B(x0)$ or $B(x) \land \neg B(x1)$. Such an x does the job.   □

Given two structures M and N for the stringlanguage, we say that N is a *weak end-extension* of M, and write $M \subseteq_w N$, if M is a substructure of N and the following implication holds: $a \in M$, $b \in N$, $b \subseteq^* a \Rightarrow b \in M$.

Proposition 2. *Let $M \subseteq_w N$. Then the sw.q.-formulae are absolute between M and N, i.e., given any sw.q.-formula $A(\vec{x})$, with the free variables as shown, and given $\vec{a}$ in M, we have the equivalence $M \models A(\vec{a}) \Leftrightarrow N \models A(\vec{a})$.*

Proof : The proof is by a straightforward induction on the complexity of the formulae A. □

Corollary. *Let $M \subseteq_w N$, with N a model of $sw.q. - NIA$. Then M is also a model of $sw.q. - NIA$.*

Proof : It is clear that M satisfies the basic axioms, since these are open axioms. The validity of the induction scheme (NIA) for sw.q.-formulae is a consequence of the reformulation of this scheme given in proposition 1, and of the absoluteness of sw.q.-formulae. □

Let k be a positive integer. We define by induction on n $(n \in \omega)$ the following $(k+1)$-predicates:

$$cl_0(y, x_1, \ldots, x_k) := y = 0 \lor y = 1 \lor y \subseteq^* x_1 \lor \ldots \lor y \subseteq^* x_k$$

$$\begin{aligned} cl_{n+1}(y, x_1, \ldots, x_k) \ := \ &\exists z \exists w (cl_n(z, x_1, \ldots, x_k) \\ \land \ &cl_n(w, x_1, \ldots, x_k) \land (y = z \frown w \lor y = z \times w)). \end{aligned}$$

Lemma.

1. *For all $n \in \omega$, $sw.q. - NIA \vdash \forall \vec{x}\, cl_n(\epsilon, \vec{x})$.*

2. *If $n, m \in \omega$ and $n \leq m$ then $sw.q. - NIA \vdash \forall \vec{x} \forall y (cl_n(y, \vec{x}) \to cl_m(y, \vec{x}))$.*

3. *For all $n \in \omega$, $sw.q. - NIA \vdash \forall \vec{x} \forall y \forall z (cl_n(z, \vec{x}) \land y \subseteq^* z \to cl_{3n}(y, \vec{x}))$.*

Proof : The first part is clear. For the second part, it is enough to show that $sw.q. - NIA \vdash \forall \vec{x} \forall y (cl_n(y, \vec{x}) \to cl_{n+1}(y, \vec{x}))$. This follows immediately from part 1.

The third part is proved by induction on n. The base case is clear. Assume the conclusion for n. We reason inside $sw.q. - NIA$. Fix $\vec{x}$, y and z and suppose that $cl_{n+1}(z, \vec{x})$ and $y \subseteq^* z$. Then there are elements z_1 and z_2 such that $cl_n(z_1, \vec{x})$, $cl_n(z_2, \vec{x})$ and either $z = z_1 \frown z_2$ or $z = z_1 \times z_2$. First, let us consider the case when $z = z_1 \frown z_2$. Take y_1 and y_2 with $y_1 \subseteq^* z_1$, $y_2 \subseteq z_2$, and $y = y_1 \frown y_2$. By induction hypotheses, we have $cl_{3n}(y_1, \vec{x})$ and $cl_{3n}(y_2, \vec{x})$. Hence $cl_{3n+1}(y_1 \frown y_2, \vec{x})$. With more reason (see part 2), we have $cl_{3n+3}(y, \vec{x})$. Lastly, consider the case when $z = z_1 \times z_2$. Take y_1, y_2 and y_3 such that $y_1 \subseteq^* z_1$, $y_2 \subseteq z_2$, $y_3 \subseteq z_1$, and $y = y_1 \frown (z_1 \times y_2) \frown y_3$. By induction hypothesis, we have $cl_{3n}(y_1, \vec{x})$, $cl_{3n}(y_2, \vec{x})$, and $cl_{3n}(y_3, \vec{x})$. Hence, we successively get $cl_{3n+1}(z_1 \times y_2, \vec{x})$, $cl_{3n+2}(y_1 \frown (z_1 \times y_2), \vec{x})$ and, finally, $cl_{3n+3}(y_1 \frown (z_1 \times y_2) \frown y_3), \vec{x})$. □

Given a model M of $sw.q. - NIA$ and $\vec{a}$ a sequence of elements of M, we define $cl_n^M(\vec{a}) := \{b \in M : M \models cl_n(b, \vec{a})\}$ and $cl_\infty^M(\vec{a}) := \cup_{n \in \omega} cl_n^M(\vec{a})$. By the previous lemma, it is clear that $cl_\infty^M(\vec{a})$ is a weak substructure of M and, hence, a model of $sw.q. - NIA$.

Proposition 3. *Suppose that* $sw.q. - NIA \vdash \forall \vec{x} \exists y A(\vec{x}, y)$, *where A is a sw.q.-formula. Then there is* $n \in \omega$ *such that* $sw.q. - NIA \vdash \forall \vec{x} \exists y (cl_n(y, \vec{x}) \wedge A(\vec{x}, y))$.

Proof : Assume, to obtain a contradiction, that for all $n \in \omega$, $sw.q. - NIA \not\vdash \forall \vec{x} \exists y (cl_n(y, \vec{x}) \wedge A(\vec{x}, y))$. Add to the stringlanguage a new sequence of constant symbols $\vec{c}$ (one for each corresponding variable of $\vec{x}$). It is easy to see that the theory $sw.q. - NIA \cup \{\forall y (cl_n(y, \vec{c}) \rightarrow \neg A(\vec{c}, y)) : n \in \omega\}$ is finitely consistent. Hence, by compactness, this theory has a model M. Let us use the very same $\vec{c}$ for the interpretation of the constant symbols $\vec{c}$ in M. As we have remarked, $cl_\infty^M(\vec{c})$ is a model of $sw.q. - NIA$. So, by hypothesis, $cl_\infty^M(\vec{c}) \models \forall \vec{x} \exists y A(\vec{x}, y)$. Take $b \in cl_\infty^M(\vec{c})$ such that $cl_\infty^M(\vec{c}) \models A(\vec{c}, b)$. By absoluteness, $M \models A(\vec{c}, b)$. On the other hand, $b \in cl_n^M(\vec{c})$ for some $n \in \omega$, i.e., $M \models cl_n(b, \vec{c})$. This contradicts the definition of M. □

Let us introduce some easy combinatorial notions. For a positive integer i, denote by b_i the word $00\ldots01$ consisting of $(i+1)$ initial zeroes followed by 1. We say that a word σ has a *k-block*, $k \geq 1$, if there is j, $j \geq 1$, such that $b_j \frown b_{j+1} \frown \ldots \frown b_{j+k-1} \subseteq^* \sigma$.

Lemma. *If the word* $\sigma_1 \frown \sigma_2$ *has a 2k-block, then either σ_1 or σ_2 has a k-block.*

Proof : Let β, with $\beta \subseteq^* \sigma_1 \sigma_2$, be a 2k-block. If $\beta \subseteq^* \sigma_1$ or $\beta \subseteq^* \sigma_2$, there is nothing to prove. Otherwise, $\beta = \rho_1 \rho_2$ with $\sigma_1 = \sigma_1' \rho_1$ and $\sigma_2 = \rho_2 \sigma_2'$, for some σ_1', σ_2'. Consider β_1' the largest initial sub-block of β such that $\rho_1 = \beta_1' \alpha_1$, for some α_1, and consider β_2' the largest final sub-block of β such that $\rho_2 = \alpha_2 \beta_2'$, for some α_2. Clearly, $\alpha_1 \alpha_2$ is either ϵ or a 1-block. Hence, if β_1' is a i-block and β_2' is a j-block, we have $i + j \geq 2k - 1$. This implies that either $i \geq k$ or $j \geq k$. □

Lemma. *If the word* $\sigma_1 \times \sigma_2$ *has a 4k-block, then σ_1 has a k-block.*

Proof : Let β, with $\beta \subseteq^* \sigma_1 \times \sigma_2$, be a 4k-block. We claim that $\beta \subseteq^* \sigma_1 \sigma_1 \sigma_1$. Note that if this is the case, then the result follows from the previous lemma.

In order for $\sigma_1 \times \sigma_2$ to have a 4k-block, σ_1 must have at least two 1's. So, consider a fixed subword τ of σ_1 of the form $100\ldots01$. If $\beta \subseteq^* \overbrace{\sigma_1 \sigma_1 \ldots \sigma_1}^{j\text{-times}}$, with $j > 3$, and $\beta \not\subseteq^* \sigma_1 \sigma_1 \sigma_1$, then $\sigma_1 \sigma_1 \subseteq^* \beta$. This implies that τ occurs twice in β, contradicting the form of β. □

We are now ready to prove the theorem.

Proof of the theorem: In order to obtain a contradiction, assume that $sw.q. - NIA \vdash \forall x \exists y\,\text{"}y = \bar{x}\text{"}$. Then, by proposition 3, there is $n \in \omega$ such that $sw.q. - NIA \vdash \forall x \exists y (cl_n(y,x) \wedge \text{"}y = \bar{x}\text{"})$. In particular, this is true in the standard model $\{0,1\}^*$. For this n consider the word $\rho = \bar{b}_1 \frown \bar{b}_2 \frown \ldots \frown \bar{b}_{4^n}$. We claim that the 4^n-block $b_1 \frown b_2 \frown \ldots \frown b_{4^n}$ is not in $cl_n^{\{0,1\}^*}(\rho)$. This is, of course, a contradiction.

We actually prove that for any $k \in \omega$, if $\sigma \in cl_k^{\{0,1\}^*}(\rho)$ then σ does not contain 4^k-blocks. The claim follows from the case $k = n$. We show this by induction on k. If $k = 0$ then either $\sigma = 0$ or $\sigma = 1$ or $\sigma \subseteq^* \rho$: in all these cases $001 \not\subseteq^* \sigma$ and, hence, σ does not have 4^0-blocks. Assume the result for k and suppose that $\sigma \in cl_{k+1}^{\{0,1\}^*}(\rho)$. Then there are $\sigma_1, \sigma_2 \in cl_k^{\{0,1\}^*}(\rho)$ such that either $\sigma = \sigma_1 \frown \sigma_2$ or $\sigma = \sigma_1 \times \sigma_2$. Now, if σ has 4^{k+1}-blocks then, by the previous two lemmas, we can conclude that either σ_1 or σ_2 has 4^k-blocks, which contradicts the induction hypothesis. $\qquad\square$

The results of this second section appeared in our Ph.D. Dissertation [3]. An abstract reporting them was published in The Journal of Symbolic Logic (see [4]). A preliminary result towards proving the main result of this section was first obtained by Mantzivis. Mantzivis' result is concerned with a theory even weaker than $sw.q. - NIA$, namely the theory of the stringlanguage consisting of the fourteen basic open axioms together with the scheme of induction on notation (NIA) for the smallest class of formulae that contains the atomic formulae and it is closed under Boolean operations and *initial* subword quantification. Mantzivis showed that the function that associates to a non-empty word the one obtained by deleting its first bit, is not provably total in this theory.

References

[1] M. Ajtai, Σ_1^1-*formulae on finite structures*, Ann. Pure Appl. Logic **24** (1983), 1–24.

[2] S. Buss, *Bounded arithmetic*, Ph.D. thesis, Princeton University, June 1985, A revision of this thesis was published by Bibliopolis, Naples in 1986.

[3] F. Ferreira, *Polynomial Time Computable Arithmetic and Conservative Extensions*, Ph.D. thesis, Pennsylvania State University, December 1988.

[4] ———, *Subword quantification and the complement function*, J. Symbolic Logic **57** (1992), 295, abstract.

[5] ———, *On end-extensions of models of $\neg exp$*, Tech. Report 5, CMAF: Universidade de Lisboa, 1994.

[6] P. Hájek and P. Pudlák, *Metamathematics of First-Order Arithmetic*, Perspectives in Mathematical Logic, Springer-Verlag, 1993.

[7] S.-G. Mantzivis, *Circuits in bounded arithmetic (part I)*, Annals of Mathematics and Artificial Intelligence **6** (1992), 127–156.

[8] N.Immerman, *Descriptive and computational complexity*, Computational Complexity Theory (J.Hartmanis, ed.), Proceedings of Symposia in Applied Mathematics **38**, American Mathematical Society, 1989.

[9] W. V. Quine, *Concatenation as a basis for arithmetic*, J. Symbolic Logic **11** (1946), 105–114.

[10] J. Saxe, M. Furst, and M. Sipser, *Parity, circuits and the polynomial time hierarchy*, Mathematical Systems Theory **17** (1984), 13–27.

[11] R. Smullyan, *Theory of Formal Systems*, Princeton University Press, 1961.

[12] L. Stockmeyer and U. Vishkin, *Simulation of parallel random access machines by circuits*, SIAM J. Control Optim. **13** (1984), 409–422.

[13] G. Takeuti, *Sharply bounded arithmetic and the function $a\dot{-}1$*, Logic and Computation (W. Sieg, ed.), American Mathematical Society, 1990, pp. 281–288.

[14] I. Wegener, *The Complexity of Boolean Functions*, Wiley-Teubner, 1987.

[15] A. Wilkie and J. Paris, *On the scheme of induction for bounded arithmetic formulas*, Ann. Pure Appl. Logic **35** (1987), 261–302.

[16] D. Zambella, *Chapters on Bounded Arithmetic & Provability Logic*, Ph.D. thesis, Universiteit van Amsterdam, September 1994.

DEPARTEMENTO DE MATEMÁTICA, UNIVERSIDADE DE LISBOA, RUA ERNESTO DE VASCONCELOS, BLOCO C1, P-1700 LISBOA, PORTUGAL
e-mail address: `ferferr@lmc.fc.ul.pt`

Research in automated deduction as a basis for a probabilistic proof-theory

Paola Forcheri   Paolo Gentilini   Maria Teresa Molfino

Abstract

In this paper the following problem is explored: if **T** is an equational fragment of Arithmetic **PRA** and S is a theorem proved by an automated prover through rewrite rules and induction rule, i.e. through a proof **P** in **T** +**Induction**, to assign a *probability* $p(S)$ *for S to belong to* **T**. The notion of *probability* is produced via the new concepts of *entropy* or *quantity of information of a proof* and of *virtual proof*. The *entropy of a proof* is defined through an extension to a logical context of Shannon's Information Theory; the notion of *virtual proof* is based on the possibility to estimate the length and the set of proper **T**-axioms of a possible proof **Q** of S in **T**. To this end a proof-theory for equational logic is proposed, a normal form for proofs is defined and a normal form theorem is given; properties of normal proofs are investigated and criteria to estimate the length of proofs are given.

Introduction

The starting point of the research is the design of an automatic prover for theories **T** which can be immersed in Primitive Recursive Arithmetic extended to integers **PRA(Z)** (Section 1), by using rewrite rules (i.e. equational logic) and induction. The induction rule *Ind* increases proof capability and can produce shorter proofs, though usually the proof is not in **T**. As a consequence, given a theory **T** and a sequent S proven in **PRA(Z)** + **AxT**, **AxT** set of the axioms of **T**, through a proof **P** in which Induction rules occur, the prover should be able to decide whether S is a theorem of **T**. To this end, we wish to enable the prover to assign a *probability* $p(S)$ for S to belong to **T**. This probability is defined through the following process:

(1) by stating proof-theoretical criteria which estimate the length of an equational proof **Q** in **T** of a given sequent S and the possible structure and axioms of **Q** (Section 2);

(2) by introducing the new notion of quantity of information or entropy of a proof, through a parallel with Shannon's Information Theory, which allows us to formally define the concept of virtual proof **Q** of a given sequent S in a given system **F** (Section 3);

(3) by setting: $p(S) = H(\mathbf{Q})/H(\mathbf{P}^*)$, where $H(\mathbf{Q})$ and $H(\mathbf{P}^*)$ are the entropies of two kinds of virtual proof of S; the a priori virtual proof **Q** in **T**, and the consequent virtual proof $\mathbf{P}^*$ in **PRA(Z)** without Ind given the effective proof **P** (Section 3).

The two new concepts on which the work is based are those of entropy of a proof and of virtual proof. The concept of entropy we present is a way to measure the actual information contained in a syntactical structure. We define the entropy of a proof as the sum of the entropies of proper axioms occurring in it; the entropy of an axiom is defined as follows: Shannon's Information Theory states that the *entropy* or *quantity of information* of a finite set of events $A = \{A_1, \ldots, A_n\}$ with probabilities $p_1, \ldots, p_n$, $\sum_i p_i = 1$, is the number: $H(A) = -\sum_i (p_i \log p_i)$. Our extension of this notion to a logical context is: let Ax be a finite equational axiom system in a fixed language $L(Ax)$ in which m axioms occur; let $Te(Ax)$ be the set of terms which occur in Ax, which we call the *term-alphabet* of Ax ; if $B \in Ax$ then we define entropy of B the number

$$H(B) = -\sum_i (p_i \log_{|Te(Ax)|m} p_i)$$

where: $t_1, \ldots, t_n$ are the terms occurring in B and

$$p_i = p(t_i) = \frac{(\text{number of occurrences of } t_i \text{ in } B)}{(\text{number of occurrences of terms in } B)}.$$

(We note that such definition can be extended to each logical system i.e. to the whole predicate calculus and so on). *We regard this definition of entropy of an axiom as the best measure of the "syntactical variety" contained in it.* Entropy is in many cases an increasing function of the various classical measures of the complexity of a syntactical structure, but in a certain sense it principally expresses the quality of the structure and not the quantity of syntactical components occurring in it. Trivial structures are penalized, even if they have a big complexity in a classical sense: a formula of the form $f = f$ has the lowest entropy among the formulas with the same complexity, i.e. with the same number of occurrences of function letters, variables, constants, and may have a lower entropy than a formula with lower complexity. Moreover, such measure is naturally linked to the whole theoretical setting of Information Theory ([7], [12],[16]) and, therefore, allows us to stress the parallel between Proof Theory and Information Theory up to the formulation of the following hypothesis : *a proof-tree* **P** *in a theory* F *with the logic* L *can be seen as a machine which transmits to the root the specific information included in the proper F-axioms which are its leaves; the axioms can be seen as information inputs and the root as information output; the demonstration power of a rule* I *of* L *in* **P** *is given by the real information which* I *transmits, i.e. by the information included in the F-axioms above* I *in* **P** *which are really necessary to prove the conclusion of* I. Section 3 presents a formalization of this idea.

The second starting point of our research was the consideration that, in the past few years, investigations on the length of proofs, i.e. on the possibility of estimating the length of the proof of a given sequent S in a fixed system **T**, have become more and more important in logic. Beginning with the works of Parikh, many logicians (see for example [14],[15],[18]) have produced important results in the field; in addition, through Kreisel conjecture (see [15],[21]) the concept of length of a proof has acquired a substantial proof-theoretical meaning. As a result, we can state that today, in several significant cases, given a sequent S and a system **T**, we are able to produce a non trivial estimate of the length of the proof **Q** which S would have in **T** and of the set of the proper **T** - axioms occurring in **Q**; in particular, we present here in Section 2 a set of proof-theoretical results which give the estimate for equational logic, i.e. for all equational fragments of **PRA**; *such estimate can be produced even if a real proof of S in* **T** *does not exist*. Therefore, the premises exist which allow us to introduce the new concept of *virtual proof*: let S be a sequent in the language of **T**, let us suppose we have a criterion **CR** for estimating the set of the instances of proper axioms of **AxT** which would be necessary for proving S in **T**, and for estimating the length of a possible proof of S in **T**; then we call an *a priori virtual proof* **Q** of S in **T** the 4-ple $(Ax(\mathbf{Q}), H(\mathbf{Q}), l(\mathbf{Q}), v(\mathbf{Q}))$ where: $Ax(\mathbf{Q})$ is the *estimated set of the proper axiom instances* occurring in **Q**; $H(\mathbf{Q})$ is the *entropy* of **Q** i.e. $H(Ax(\mathbf{Q}))$; $l(\mathbf{Q})$ is the *estimated length* of **Q**; $v(\mathbf{Q}) = H(\mathbf{Q})/l(\mathbf{Q})$ is the *information transfer speed* of **Q**. It is the concept of virtual proof, in which entropy plays a central role, which produces the definition of *probability* $p(S)$ for S to belong to **T**. In fact , we translate the real proof **P** of S in **T** $+ Ind$, in which the induction rule occurs, into a suitable virtual proof of S in **T**, the *consequent virtual proof* **P***, and we can compare it with the a priori virtual proof **Q** of S in **T**: *this is possible since they are homogeneous mathematical objects*, and $p(S) = H(\mathbf{Q})/H(\mathbf{P}^*)$ is a good definition. The translation is based on the assumption that a link exists between the demonstration power of an Induction rule occurrence I in **P** and the set of proper axiom instances occurring above it in the proof, i.e. with the entropy of the segment of **P** above I. This assumption is also justified by the proof-theoretical results of Section 2.

This research is a starting point for the following general goals:

- to define a good notion of *probabilistic inference* which can be used both in an automatic theorem proving context and in a proof-theoretical context.

- to develop a *proof-theoretical information theory* which can explore classical problems with new tools (for example consistency problems).

The results of Section 2 are also independently interesting : a proof-theory for equational logic is proposed (2.1), a normal form for proofs is defined (2.3) and a normal form theorem is given (2.4); properties of normal proofs are investigated and some criteria for estimating the length of proofs are given (2.19,2.21).

For the implementation we use *rewrite rules* (see [1], [2], [3], [6],[11]) and a version of induction similar to that used by Boyer and Moore (see [4]).

1   Preliminary definitions

We consider theories **T** so that:

(1) **T** admits as model a reduction of the standard model **Z** of Primitive Recursive Arithmetic extended to integers, that is, **T** can be immersed in **PRA(Z)**, i.e.: let **T** be a theory with language L_T and with a model $A_T = (Z, F, R)$, where:

  (i)   the set **Z** of integers is the universe;
  (ii) for every n-ary operation $*$ of L_T, $F(*) = f$, where f is a primitive recursive n-ary function which is the interpretation of a function letter $\overline{f}$ of **PRA(Z)** in the standard model M_Z of **PRA(Z)**;
  (iii) for every m-ary predicate **R** of L_T, $R(\mathbf{R}) = P$, where P is an m-ary recursive relation which is the interpretation of a recursive predicate $\overline{P}$ of **PRA(Z)** in M_Z .

Then we say that A_T can be identified as an L_T-reduction of the standard model M_Z of **PRA(Z)** and that **T** can be immersed in **PRA(Z)**.

(2) **T** admits a sequent version so that:

  (i)   the axioms **AxT** are sequents of equations;
  (ii) the proofs admit a normal form in which the important part is the equational part, in which axioms and cut rules only occur; the Induction *Ind* of **PRA(Z)**, acting on atomic formulas, is added in this segment of proof.

Interesting mathematical theories, such as group theory, ring theory, lattice theory are included in the above class.

In the following we will identify ***T*** *with its immersion in* ***PRA(Z)***. Moreover, we include among the systems **T** which we consider all equational arithmetical fragments of **PRA** or **PRA(Z)**. For sequent calculus and proof-theory our reference is [21] with the following slight change (which is also employed in [8]): a sequent is an expression of the form $X \vdash Y$ in which X, Y are *sets* of formulas. For Primitive Recursive Arithmetic see [19], [20] .

Axiomatization of **PRA(Z)**:

Atomic formulas have the form $s = t$, since we replace each recursive predicate with its corresponding characteristic function; we add to **PRA** a function p, called predecessor, semantically defined as:

$$p : \begin{array}{c} \mathbf{Z} \longrightarrow \mathbf{Z} \\ m \longmapsto m-1 \end{array}$$

p has to behave according to S, the extended function successor:

$$S : \begin{array}{c} \mathbf{Z} \longrightarrow \mathbf{Z} \\ m \longmapsto m+1 \end{array}$$

We define formally the functional letters $\overline{S}$, $\overline{p}$ by the following axioms:

$$\vdash \overline{S}(\overline{p}(t)) = t$$

$$\vdash \overline{p}(\overline{S}(t)) = t$$

$$\overline{S}(t) = \overline{0} \vdash t = \overline{p}(\overline{0})$$

1.1 $\qquad \overline{p}(t) = \overline{0} \vdash t = \overline{S}(\overline{0}) \qquad\qquad t, t_1, t_2$ arbitrary terms

$$\overline{S}(t_1) = \overline{S}(t_2) \vdash t_1 = t_2$$

$$\overline{p}(t_1) = \overline{p}(t_2) \vdash t_1 = t_2$$

The functions $\overline{Z}$ (zero), the projections $\overline{P}_i^k$, an *extended recursion scheme* and the *the composition scheme*, can be added. The Induction rule for **PRA** is (see [21]):

$$\frac{F(a), X \vdash Y, F(\overline{S}(a))}{F(\overline{0}), X \vdash Y, F(t)} \qquad \mathbf{I}$$

where t is an arbitrary term and a is a free variable which does not occur in X, Y, t; we will always suppose that F is an atomic formula.

If we define the functions

$$cg_+ : \begin{array}{c} \mathbf{Z} \longrightarrow \mathbf{Z} \\ m \longmapsto \end{array} \begin{cases} 0 & \text{if } m=0 \\ m & \text{if } m>0 \\ -m & \text{if } m<0 \end{cases} \qquad\qquad cg_- : \begin{array}{c} \mathbf{Z} \longrightarrow \mathbf{Z} \\ m \longmapsto \end{array} \begin{cases} 0 & \text{if } m=0 \\ m & \text{if } m<0 \\ -m & \text{if } m>0 \end{cases}$$

then we can define the following one-premise induction rules for **PRA(Z)**

$$1.2 \quad \frac{F(a), X \vdash Y, F(\overline{S}(a))}{F(\overline{0}), X \vdash Y, F(\overline{cg}_+(t))} \; \mathbf{I}_1 \qquad 1.3 \quad \frac{F(a), X \vdash Y, F(\overline{p}(a))}{F(\overline{0}), X \vdash Y, F(\overline{cg}_-(t))} \; \mathbf{I}_2$$

where t is an arbitrary term, a is a free variable which does not occur in X, Y, t and F is an atomic formula.

But the induction rule which gives the best formalization of the induction employed in the automated proofs produced by the prover is the following two -premise rule:

$$1.4 \qquad \frac{F(a), X \vdash Y, F(\overline{p}(a)) \qquad F(a), X \vdash Y, F(\overline{S}(a))}{F(\overline{0}), X \vdash Y, F(t)} \textbf{ I}$$

where t is an arbitrary term, a is a free variable which does not occur in X, Y, t and F is an atomic formula.

2   Estimate of virtual proofs in equational systems: estimated length of an equational proof and estimated set of proper axioms of an equational proof

Let $\mathbf{T}$ be an arithmetical equational fragment as defined in Section 1; the equational logic $\mathbf{EQ}$ which we wish to apply is the following.

2.1. Definition of EQ. $X, Y, Z, W, U \ldots$ are sets of equations in the language of $\mathbf{PRA(Z)}$; $A, B, C, D \ldots$ are equations; we identify the equations $s = t$ and $t = s$.

EQ-Axioms :

    A1: $s_1 = t_1, \ldots, s_n = t_n \vdash f(s_1, \ldots, s_n) = f(t_1, \ldots, t_n)$,
        s_i, t_i arbitrary terms, f n-ary function letter;
    A2: $\vdash t = t$, t arbitrary term;
    A3 (transitivity axioms): $s_1 = t_1, s_2 = t_2, s_1 = s_2 \vdash t_1 = t_2$,
        s_i, t_i arbitrary terms;
    A4 (logical axioms): $B \vdash B$.

EQ-Rules:

R1 (Weakening rules): $\dfrac{X \vdash Y}{B, X \vdash Y}$ $\qquad$ $\dfrac{X \vdash Y}{X \vdash Y, B}$

R2 (Cut rule): $\dfrac{X \vdash Y, B \qquad B, U \vdash W}{X, U \vdash Y, W}$

R3 (Reduction rule, Red): $\dfrac{f(s_1, \ldots, s_n) = f(t_1, \ldots, t_n), X \vdash Y}{s_1 = t_1, \ldots, s_n = t_n, X \vdash Y}$

f n-ary function letter, s_i, t_i arbitrary terms;

R4 (Expansion rule, Exp): $\dfrac{X \vdash Y, t = s}{X \vdash Y, g(t) = g(s)}$

f 1-place function letter, s, t arbitrary terms.

2.2. Conventions.

2.2.1. We suppose to replace each occurrence of axioms **A1** in a **EQ**-proof with suitable instances of **R3**-rule:

$$\dfrac{f(s_1, \ldots, s_n) = f(t_1, \ldots, t_n) \vdash f(s_1, \ldots, s_n) = f(t_1, \ldots, t_n)}{s_1 = t_1, \ldots, s_n = t_n \vdash f(s_1, \ldots, s_n) = f(t_1, \ldots, t_n)}$$

2.2.2. We suppose that **AxT** is closed w.r.t. cuts and w.r.t. Red, Exp rules, once a bound k for the complexity of terms in a proof has been fixed. The *complexity of a term t* is the number of occurrences of function letters, variables and constants in t. So our equational system can be indexed by k: a proof in $(\mathbf{AxT}, \mathbf{EQ}_k)$ is an $(\mathbf{AxT}, \mathbf{EQ})$-proof in which the complexity of terms is $\leq k$. We note that :

(i) closure w.r.t. cuts is trivial for all theories **T** axiomatized through axioms of the form $\vdash s = t$, such as group theory, ring theory, lattice theory, equational part of **PA** and so on;

(ii) the assumption of a complexity bound k is substantial in each automated deduction context.

Moreover, we suppose that formulas of the form $a = a$ do not occur in **AxT**.

2.2.3. Identification of theorems: we accept a proof **P** of a theorem of the form $a = a, X \vdash Y$ as a proof of $X \vdash Y$. This means that cuts in which the axiom A2 is a premise are not included in the proofs we consider.

For the standard notions of proof-theory and sequent calculus our reference is Takeuti [21]; in particular, for the definitions of *ancestor* and *descendent* of a formula in a proof and of *explicit* and *implicit* formula in a proof, see [21] pp. 78–79. An *integral descendent* of a formula B is a descendent of B which is equal to B. An *auxiliary formula* of a Red or Exp rule is the formula in the premise on which the rule works; *principal formulas* are the formulas in the conclusion which the rule introduces. If the sequent S can be written $X, U \vdash V, Y$ we say that the sequent $U \vdash V$ is a *sub-sequent* of S and that S is an *over-sequent* of it.

We define a normal form for proofs in $(\mathbf{AxT}, \mathbf{EQ})$, called a *non redundant form*, as follows.

2.3. Definition. Let $\mathbf{Q}$ be a proof in $(\mathbf{AxT}, \mathbf{EQ})$ of the sequent S. $\mathbf{Q}$ is a *non redundant form* if the following conditions are valid:

(i) a cut in which both cut-formulas have an ancestor in a transitivity axiom **A3** occurs in $\mathbf{Q}$ above each occurrence of Red, Exp rules and of axioms of **AxT**;

(ii) Exp. Red rules occur in $\mathbf{Q}$ above each occurrence of axioms of **AxT**;

(iii) cuts in which one cut-formula has an ancestor which is principal formula of a Red or Exp rule and one cut-formula has not been introduced by an axiom of **AxT**, do not occur in $\mathbf{Q}$;

(iv) weakening rules do not occur in $\mathbf{Q}$;

(v) a proof-segment H with the sequent N as a leaf and the sequent N as the root cannot exist in $\mathbf{Q}$; this implies that logical axioms $A \vdash A$ can occur in $\mathbf{Q}$ only as premises of a Red or Exp rules;

(vi) proper over-sequents of an axiom cannot occur in $\mathbf{Q}$;

(vii) a formula A can be a cut-formula of at most one cut in $\mathbf{Q}$;

(viii) a formula of the form $t = t$ cannot be an auxiliary formula of a Red or Exp rule.

A proof in normal non redundant form can be seen as being subdivided in three kinds of segments; starting from the top we have: *Q1-segments* in which only occur cuts between transitivity axioms; *Q2-segments* in which only occur Red, Exp rules starting from the root of *Q1* or from an axiom which is not in **AxT**; *Q3-segments* in which can only appear formulas introduced by axioms of **AxT** cut formulas of *Q2*-roots or formulas introduced by transitivity axioms **A3** cut formulas introduced by axioms of **AxT**.

2.4. Theorem. (Normal Form Theorem) *Let $\boldsymbol{P}$ be a proof in $(\boldsymbol{AxT}, \boldsymbol{EQ}_k)$ of the sequent S, k bound for the complexity of terms. Then we can always construct in $(\boldsymbol{AxT}, \boldsymbol{EQ}_k)$ a proof $\boldsymbol{Q}$ in non redundant form of S or of a sub-sequent of S.*

Proof. Let us preliminarily suppose transforming **P** into a proof of **P'** of a subsequent of S which respects the points (iv), (v), (vi), (vii), (viii) of Definition 2.3. Such reductions are:

(iv): we delete the weakenings and the rules which work on the descendent of a weakening formula;

(v): if a segment H of **P** has the sequent N as a leaf and the sequent N as the root, then we delete H and replace it with N; (vi): if the sequent N in **P** is a proper over-sequent of an axiom A we delete the proof of N and replace it with A;

(vii): if we have:

$$\frac{X \vdash Y, B \qquad\qquad B, U \vdash W}{X, U \vdash Y, W}$$

$$\vdots$$

$$\vdots \qquad\qquad\qquad \vdots$$

$$\frac{V \vdash J, B \qquad\qquad B, Z \vdash T}{V, Z \vdash J, T}$$

then we produce:

$$V \vdash Y, B$$

$$\vdots \qquad\qquad\qquad \vdots$$

$$\frac{V^* \vdash J^*, B \qquad\qquad B, Z \vdash T}{V^*, Z \vdash J^*, T}$$

where $V \supseteq V^*, J \supseteq J^*$;

(viii): we delete the rules working on $t = t$.

We now produce the reductions which transform **P'** into a proof **Q** in non redundant form; we work by induction on the length $l(\mathbf{P'})$ of **P'**, $l(\mathbf{P'}) =$ *(greatest number of proof-lines in a branch in **P'**)*.

Basis: $l(\mathbf{P'}) = 1$. Through the properties of **P'** and the hypotheses on **AxT** the thesis is straightforward.

Induction step: $l(\mathbf{P'}) > 1$.

(1) The lowermost rule of $\mathbf{P'}$ is an Exp:

$$\vdots (R)$$

$$\frac{X \vdash Y, t = s}{X \vdash Y, g(t) = g(s)}$$

Through the induction hypothesis, the proof R is in normal form. Therefore, if the premise $X \vdash Y, t = s$ is the conclusion of a Red or Exp rule, **P'** is in normal form. Let us suppose that $X \vdash Y, t = s$ is the conclusion of a cut C:

$$\frac{\overset{\vdots\,(\mathrm{H1})}{V \vdash Z, B} \qquad \overset{\vdots\,(\mathrm{H2})}{B, U \vdash W}}{\dfrac{X \vdash Y, t = s}{X \vdash Y, g(t) = g(s)}} \; \mathrm{C}$$

The critical cases are the ones in which one cut-formula of C has been introduced by an axiom of **AxT**. In this cases we apply Exp rules in suitable parts of H1, H2 to the ancestors $t = s$ of the explicit formula $t = s$ and we have a proof **Q**:

$$\frac{\overset{\vdots}{V \vdash Z^*, B} \qquad \overset{\vdots}{B, U \vdash W^*}}{X \vdash Y, g(t) = g(s)}$$

where Z^*, W^* are Z, W with the replacement of $t = s$ with $g(t) = g(s)$.

(2) The lowermost rule of **P'** is a Red: analogous to (1).

(3) The lowermost rule of **P'** is a cut C:

$$\frac{\overset{\vdots\,(\mathrm{H1})}{X \vdash Y, a = b} \qquad \overset{\vdots\,(\mathrm{H2})}{a = b, U \vdash W}}{X, U \vdash Y, W} \; \mathrm{C}$$

Y,W where through induction hypothesis H1, H2 are non redundant proofs. We have the following cases.

(3.0) Both cut-formulas have been introduced by axioms of **AxT**, let them be A1 and A2. Then we replace the axioms introducing the cut-formulas in H1 with the conclusion A3 of a cut between A1 and A2, which is in **AxT** by hypotheses on **AxT**, and we obtain a proof in normal form of a subsequent M of $Z, X, U \vdash Y, W, V$ where Z, V contain possible formulas which occur in A2 only. We apply the corresponding cuts of the $Q3$ part of H2 and we obtain a proof in normal form of a sub-sequent of $X, U \vdash Y, W$.

(3.1) H1 and H2 consist of $Q1$-segments only: then **P'** is in normal form.

(3.2) One cut -formula has been introduced by an axiom of **AxT** and one cut-formula has not been introduced by an axiom of **AxT**: then **P'** is in normal form.

(3.3) Either in H1 or in H2 Red, Exp rules occur and no cut-formula of C has been introduced by an axiom of **AxT**. Then we consider the following subcases:

(3.3.1) No ancestor of the cut-formulas is the principal formula of an Exp or Red rule. Let $(T \vdash a = b)_k$ $k = 1, \ldots, r$ and $(a = b, V \vdash M)_i$ $i = 1, \ldots, w$ be the $Q1$-roots introducing the cut-formulas ancestors in H1 and in H2. Then we fix one sequent $a = b, V \vdash M$ in the second set and we replace in H1 each $Q1$-root $(T \vdash a = b)_k$ with the following proof:

$$\frac{\overset{\vdots}{(T \vdash a = b)_k} \qquad \overset{\vdots}{a = b, V \vdash M}}{(T, V \vdash M)_k}$$

To each conclusion $(T, V \vdash M)_k$ we apply in a suitable manner a sequence of proof segments in the most general case of the form :

$(Q2\text{-segment of H1})/(Q2\text{- segment of H2})/(Q3\text{-segment of H1})/(Q3\text{-segment of H2})$

by supposing that rules working on deleted formulas have been deleted; so we obtain a proof in normal form of a subsequent of $X, U \vdash Y, W$.

(3.3.2) The cut-formulas of C are descendants of the principal formula of an Exp rule on the left and of a Red rule on the right. Then $a = b$ is $G(c) = G(d)$, G suitable sequence of *unary* function letters. Then we delete all Exp rules in H1 which work on ancestors of $a = b$ and suitably apply new Red rules in H2 to the ancestors of the right cut formula obtaining :

$$\frac{\overset{\vdots}{X \vdash Y, c = d} \qquad \overset{\vdots}{c = d, U \vdash W}}{X, U \vdash Y, W}$$

where the left cut formula $c = d$ is introduced by an axiom; so we are in case (3.3.3) exposed below.

(3.3.3) The left cut -formula is the integral descendent of an axiom-formula in H1 and the right cut -formula is the descendent of a principal formula of a Red rule in H2. By induction on the cardinality of the set RED of the Red rules in H2 which introduce an ancestor of the right cut-formula. Card(RED) $=1$: Let $F(t_1, \ldots, t_n, a) = F(s_1, \ldots, s_n, b)$, $Z \vdash V$ be the $Q1$-segment root in H2 introducing an ancestor of the right cut-formula, F a suitable sequence of function letters; let $(J \vdash a = b)_k$ be the $Q1$-segment roots introducing in H1 an ancestor of the left cut-formula. We state the following Sub-Lemma: *for each k we can obtain a transformed* $Q1$-*root of the form* $(J^* \vdash F(t_1, \ldots, t_n, a)) = F(s_1, \ldots, s_n, b)_k$ *where if a formula occurs in* J^* *then it belongs to the set* : $\{$*auxiliary formulas of Red rules in which the principal formulas are either of the form* $t_1 = s_1, \ldots, t_n = s_n$ *or occur in* $J\}$.

Proof: By induction on the length of the considered $Q1$-segment : if $l(Q1) = 1$ we have :

$$\frac{e = a, v = d, e = v \vdash a = d \qquad\qquad a = d, m = b, d = m \vdash a = b}{e = a, v = d, e = v, m = b, d = m \vdash a = b} \text{ cut}$$

Then we replace the premises with suitable different transitivity axioms (here and to the end of the proof we will often write **t** for $t_1, \ldots, t_n$, **s** for $s_1, \ldots, s_n$ and so on)

$$\frac{\begin{array}{l} F(\mathbf{t}, a) = F(\mathbf{s}, e), F(\mathbf{t}, v) = F(\mathbf{s}, d), F(\mathbf{t}, v) = F(\mathbf{s}, e) \vdash \\ F(\mathbf{t}, a) = F(\mathbf{s}, d) \\ F(\mathbf{t}, a) = F(\mathbf{s}, d), F(\mathbf{t}, m) = F(\mathbf{s}, b), F(\mathbf{t}, d) = F(\mathbf{s}, m) \vdash \\ F(\mathbf{t}, a) = F(\mathbf{s}, b) \end{array}}{\begin{array}{l} F(\mathbf{t}, a) = F(\mathbf{s}, e), F(\mathbf{t}, v) = F(\mathbf{s}, d), F(\mathbf{t}, v) = F(\mathbf{s}, e), \\ F(\mathbf{t}, m) = F(\mathbf{s}, b), F(\mathbf{t}, d) = F(\mathbf{s}, m) \vdash F(\mathbf{t}, a) = F(\mathbf{s}, b) \end{array}} \text{ cut}$$

and we have the thesis. Induction step $l(Q1) > 1$:

$$\frac{\overset{\vdots}{\Omega \vdash c = d} \qquad\qquad \overset{\vdots}{c = d, \Gamma \vdash a = b}}{\Omega, \Gamma \vdash a = b}$$

Through induction hypothesis we can produce:

$$\frac{\overset{\vdots}{\Omega^* \vdash F(\mathbf{t}, c) = F(\mathbf{s}, d)} \qquad\qquad \overset{\vdots}{F(\mathbf{t}, c) = F(\mathbf{s}, d), \Gamma^* \vdash F(\mathbf{t}, a) = F(\mathbf{s}, b)}}{\Omega^*, \Gamma^* \vdash F(\mathbf{t}, a) = F(\mathbf{s}, b)} \text{ cut}$$

and we have the thesis of the Sub-Lemma.

Let us now consider the $Q1$-root $F(t_1, \ldots, t_n, a) = F(s_1, \ldots, s_n, b), Z \vdash V$ above the Red rule introducing the right cut-formula in H2; we replace in H1 the $Q1$-roots $(J \vdash a = b)_k$ with the following proofs:

$$\frac{\overset{\vdots}{(J^* \vdash F(\mathbf{t}, a) = F(\mathbf{s}, b))_k} \qquad\qquad \overset{\vdots}{F(\mathbf{t}, a) = F(\mathbf{s}, b), Z \vdash V}}{(J^*, Z \vdash V)_k} \text{ cut}$$

where the proof of the left cut-premise, which is the transformed root of the k-th $Q1$-segment introducing the left cut-formula in H1, is given by the Sub-Lemma proved above. To each conclusion $(J^*, Z \vdash V)_k$ we apply in a suitable manner a sequence of proof segments of the form:

($Q2$-segment of H1)/($Q2$- segment of H2)/($Q3$-segment of H1)/($Q3$-segment of H2)

by supposing that rules working on deleted formulas have been deleted and by adding the suitable Red rules which must reduce the formulas of $(J^*)_k$; so we obtain a proof in normal form of a sub-sequent of $X, U \vdash Y, W$.
Card(RED)=$m > 1$: we choose in the $Q1$-root in H2 one ancestor

$$G(t_1, \ldots, t_l, a) = G(s_1, \ldots, s_l, b)$$

of a rule of RED and through the same transformations of the basis case we can obtain a proof H3 in normal form of a sub-sequent of $a = b, X, U \vdash Y, W$ and then a proof **P''** of the form:

$$
\begin{array}{cc}
\vdots\,(\text{H1}) & \vdots\,(\text{H3}) \\
X \vdash Y, a = b \qquad & a = b, X, U \vdash Y, W \\
\hline
\multicolumn{2}{c}{X, U \vdash Y, W}
\end{array}
$$

where H1, H3 are normal proofs and Card(RED)=$m - 1$; so we can apply induction hypothesis to get the conclusion.

(3.3.4) The left cut-formula $a = b \equiv G(c) = G(d)$, G unary function letter, is the descendent of the principal formula of an Exp rule in H1 and the right cut-formula is the integral descendent of an axiom-formula in H2

$$
\begin{array}{cc}
\vdots & \\
\Gamma \vdash \Delta, c = d & \\
\hline
\Gamma \vdash \Delta, G(c) = G(d) & \qquad G(c) = G(d), \Omega \vdash \Lambda \\
\vdots & \qquad \vdots \\
X \vdash Y, G(c) = G(d) & \qquad G(c) = G(d), U \vdash W \\
\hline
\multicolumn{2}{c}{X, U \vdash Y, W \qquad \underline{C}}
\end{array}
$$

then we delete the Exp in H1, we apply a new unary Red in H2 to $G(c) = G(d)$ and we obtain a end-cut C' as in the previous case (3.3.3).

We can give criteria to estimate the length of a non redundant equational proof **Q** in $(\mathbf{AxT}, \mathbf{EQ}_k)$ given the root S. We define *length* $l(\mathbf{Q})$ of **Q** the greatest number of proof-lines in a branch.

2.5. Definition. In the formula $s = t$ we call s and t the *maximal terms* of the formula; we call s the *maximal symmetric term* of t and vice versa. We call *minimal terms* variables and constants of the language.

2.6. Lemma. *Let Q be a non redundant proof of $X \vdash a = b$ in EQ in which only cuts between transitivity axioms occur. Then both a and b occur in X as maximal terms.*
Proof. By induction on the length of Q. Basis, $l(Q) = 1$:

$$\frac{e = a, v = d, e = v \vdash a = d \qquad a = d, m = b, d = m \vdash a = b}{e = a, v = d, e = v, m = b, d = m \vdash a = b} \; cut$$

and we can see that the thesis holds. The other cut-cases of the basis are similar.

Induction step: let us consider the end-rule of Q

$$\frac{\begin{matrix}\vdots\\U \vdash c = d\end{matrix} \qquad \begin{matrix}\vdots\\c = d, W \vdash a = b\end{matrix}}{U, W \vdash a = b}$$

If a is not in $\{c, d\}$ then, by induction hypothesis, a occurs in W as maximal. If a is equal to c or d then a occurs in U as maximal, by induction hypothesis; analogously for b.

2.7. Corollary. (Propagation property of terms in a *Q1*-segment) *Let Q be a non redundant proof in EQ in which only the Q1-segment occurs: then each maximal term occurring in Q is preserved, i.e. it occurs as maximal in the antecedent of the root; moreover, each term occurring in Q occurs in the antecedent of the root of Q.*

2.8. Corollary *Let Q be a non redundant proof in EQ. Then each variable or constant occurring in Q also occurs in the antecedent of the root of Q.*

The estimated length of a *Q1*-segment given its root is a polynomial function of the number of formulas in the antecedent of the root.

2.9. Proposition. *Let Q be a non redundant proof in EQ consisting of a Q1-segment. Then, if m formulas in the antecedent of the root S of Q occur, we have $l(Q) \leq m(m-1)/2$.*
Proof. Through properties of Q no cuts occur with the same cut-formula. By Corollary 2.7 the maximal terms of each cut-formula in Q must occur as maximal in the antecedent of the Q-root. Then the number of cut-rules occurring in Q is $\leq \text{Card}\{f = g : f, g$ maximal terms occurring in the antecedent of $S, g \neq f\} = m(m-1)/2$.

2.10. Lemma. *Let C be a cut in a Q1-segment of a non-redundant proof Q so that:*

(i) *in at least one premise of C a formula of the form $f = f$ occurs in the antecedent;*

(ii) *the cut-formula is of the form $f = d$.*

Then the term f occurs as maximal in the antecedent of the conclusion of C in at least two different formulas.

Proof. By a main induction on the length $l(Q1)$ of the considered $Q1$-segment and a subordinate induction on the rank $r(C)$ of C. The rank is so defined: $r(C) = r_L + r_R$, where the left rank r_L is the greatest number of consecutive sequents in the proof of the left premise M of C, starting from M, with $f = f$ in the antecedent, and the right rank r_R is the greatest number of consecutive sequents in the proof of the right premise N of C, starting from N, with $f = f$ in the antecedent. Basis $l(Q1) = 1$:

$$\frac{f = f, c = d, f = c \vdash f = d \qquad\qquad f = d, t = u, d = u \vdash f = t}{f = f, c = d, f = c, t = u, d = u \vdash f = t}$$

where $c \neq f$, for the non redundant form, and we have the thesis. The other cases of the basis (i.e. $r_L = 0, r_R = 1$ and $r_L = 1, r_R = 1$) are similar.

Basis of the subordinate induction: $l(Q1) > 1$, $r(C) = 1$. We observe that if $r_L = 1$ then the left premise is an axiom, since formulas of the form $f = f$ are never cut-formulas in a normal proof; analogously for r_R. We have the following cases:

(1) $l(Q1) > 1, r_L = 1, r_R = 0$:

$$\frac{f = f, c = d, f = c \vdash f = d \qquad\qquad \overset{\vdots}{f = d, X \vdash l = m}}{f = f, c = d, f = c, X \vdash l = m}$$

where $c \neq f$, for the non redundant form, and we have the thesis.

(2) $l(Q1) > 1, r_L = 0, r_R = 1$:

$$\frac{\overset{\vdots}{Y \vdash f = d} \qquad\qquad f = f, f = d, d = q \vdash f = q}{Y, f = f, d = q \vdash d = t}$$

We have the thesis since $r_L = 0$ implies $f = f \notin Y$ and through Lemma 2.6 f must occur in Y as maximal.

Induction step : $l(Q1) > 1, r(C) > 1$.

$$
\begin{array}{cccc}
\vdots\,(\mathrm{H1}) & \vdots\,(\mathrm{H2}) & \vdots\,(\mathrm{H3}) & \vdots\,(\mathrm{H4})
\end{array}
$$

$$
\dfrac{\dfrac{V \vdash k = v \qquad k = v, \Gamma \vdash f = d}{X \vdash f = d} \qquad \dfrac{J \vdash h = s \qquad h = s, \Delta \vdash l = m}{f = d, U \vdash l = m}\ \mathrm{C}}{f = f, X^{*}, U^{*} \vdash l = m}
$$

where $f = f \in X \cup U$, X^{*} and U^{*} are X and U with the elimination of one possible occurrence of $f = f$: H1, H2, H3, H4 are normal $Q1$-segments. We have the following cases:

(1) $r_L = 0$, $r_R > 1$: through Lemma 2.6 f must occur in X as maximal and if $f = f \notin X$ we have the thesis.

(2) $r_L > 0$, $r_R \geq 0$: if $r_L = 1$ the left premise of C is an axiom $f = f, c = d, f = c \vdash f = d$ and we have the thesis as in the basis cases. Let us suppose that $r_L > 1$; we have the following sub-cases:

(2.1) $f = f \notin \Gamma$: this forces $f = f \in V$; moreover, through Lemma 2.6, f must occur as maximal in $\{k = v\} \cup \Gamma$ in a formula different from $f = f$, since in a non redundant proof $k = v \not\equiv f = f$. If $f \notin \{k, v\}$ we have the thesis. Let us suppose that $f \in \{k, v\}$; then the left cut above C has the form

$$
\begin{array}{cc}
\vdots\,(\mathrm{H1}) & \vdots\,(\mathrm{H2})
\end{array}
$$

$$
\dfrac{V \vdash f = v \qquad f = v, \Gamma \vdash f = d}{X \vdash f = d}
$$

and through the main induction on the length we have the thesis.

(2.2) $f = f \notin \Gamma$: if $f \in \{k, v\}$ we are in case (2.1). Let us suppose that $f \notin \{k, v\}$; then we consider the following proof R:

$$
\begin{array}{cc}
 & \begin{array}{cc} \vdots\,(\mathrm{H3}) & \vdots\,(\mathrm{H4}) \end{array}
\end{array}
$$

$$
\dfrac{\begin{array}{c}\vdots\,(\mathrm{H2})\\[2pt] k = v, \Gamma \vdash f = d\end{array} \qquad \dfrac{J \vdash h = s \qquad h = s, \Delta \vdash l = m}{f = d, U \vdash l = m}}{k = v, \Gamma, U \vdash l = m}\ \mathrm{C^{\wedge}}
$$

where $l(R) \leq l(Q1)$, $r(C) \geq r(C^\wedge) + 1$ and then $r(C) > r(C^\wedge)$; then, through the induction on the rank, f must occur as maximal in $\{k = v\} \cup \Gamma \cup U$ in a formula different from $f = f$; therefore, since we assume $f \notin \{k, v\}$, f must occur as maximal in $\Gamma \cup U$ in a formula different from $f = f$.

2.11. Lemma. *Let $X \vdash s = t$ be the root of a Q1-segment of a non redundant proof Q in EQ. Let f be a maximal term occurring in the formula $f = v \in X$; then either f occurs in $X \vdash s = t$ as maximal in at least one formula different from $f = v$ or its symmetric maximal term v occurs as maximal in X in at least two different formulas.*

Proof. By induction on the length of the considered $Q1$-segment. Basis: $l(Q1) = 1$:

$$\frac{a = b, c = d, a = c \vdash b = d \qquad b = d, e = f, b = e \vdash d = f}{a = b, c = d, a = c, e = f, b = e \vdash d = f} \text{ cut}$$

The thesis is straightforward for f and d; in order to verify the thesis for a, b, c, e we must examine the possible contractions between formulas in the antecedent of the conclusion:

(1) if $a = b$ is equal to $a = c$, then $b \equiv c$ and the left premise is $a = c, c = d, a = c \vdash c = d$ which is absurd in **Q**.
(2) $c = d$ is equal to $a = c$: the case is similar to (1).
(3) if $e = f$ is equal to $b = e$ then the conclusion is $a = b, c = d, a = c, e = b \vdash d = b$ and we have the thesis.
(4) if $a = b$ is equal to $b = e$ then the conclusion is $c = d, e = c, e = f, b = e \vdash d = f$ and we have the thesis, since if $c \equiv f \equiv b$ then the right premise is $b = d, e = b, e = b \vdash d = b$ which is absurd in **Q**.

Induction step:

$$\frac{X \vdash g = h \qquad g = h, U \vdash l = m}{X, U \vdash l = m}$$

Let us suppose that $f = v \in X \cup U$. We subdivide the proof in two main cases.

(1) $f = v \in X$; then, through induction hypothesis, either f occurs in $X \vdash g = h$ as maximal in at least two different formula or v occurs in X as maximal in at least two different formulas. In the second case we have the thesis; in the first case, if $f \notin \{g, h\}$ we have the thesis. Let us suppose that f occurs in $X \vdash g = h$ in two different formulas only , that $f \in \{g, h\}$, and that v does not occur in X as maximal in at least two different formulas. Therefore, if we apply to v the induction hypothesis, we have $v \in \{g, h\}$, and then $f, v \in \{g, h\}$; moreover, by properties of **Q**, $g \neq h$ and $l \neq m$. Then we have $v \equiv f \equiv g$ and $f = v$ is of the form $f = f$; then we have the thesis through Lemma 2.10.

(2) Let us suppose that $f = v \in U$ and then that $(f = v) \not\equiv (g = h)$; if $f \in \{l, m\}$ we have the thesis. Let us suppose that $f \notin \{l, m\}$: then, by induction hypothesis: either f occurs in $\{g = h\} \cup U$ in a formula $f = q$, $q \neq v$, or v occurs in $\{g = h\} \cup U$ in at least two different formulas, let them be $k = v$, $p = v$. The critical cases are the following:

  (i) $(f = q) \equiv (g = h)$: this implies $q \equiv h$, $f \equiv g$, but then, by Lemma 2.6 f, q must occur in X as maximal and we are in the previous case;

  (ii) $(f = v) \equiv (k = v)$ and $(p = v) \equiv (g = h)$: then, by Lemma 2.6, v must occur in X as maximal in a formula $e = v$, and so the formulas $e = v$ and $f = v$ occur in $X \cup U$; if $f \equiv e$, then $f = v$ occurs in X and we are in the previous case; if $f \not\equiv e$ we have the thesis.

2.12 Lemma. *Let Q2 be a Q2-segment in a non redundant proof* **Q** *in the system* **EQ**. *Then we can always transform Q2 into a Q2-segment in which no Red rules have the same auxiliary formula.*

Proof. If we have

$$\frac{F(a_1, \ldots, a_n) = F(b_1, \ldots, b_n), X \vdash Y}{a_1 = b_1, \ldots, a_n = b_n, X \vdash Y} \quad \text{Red1}$$

$$\vdots(\text{H})$$

$$\frac{F(a_1, \ldots, a_n) = F(b_1, \ldots, b_n), Z \vdash W}{a_1 = b_1, \ldots, a_n = b_n, Z \vdash W} \quad \text{Red2}$$

we delete Red1 and we produce:

$$\frac{F(a_1, \ldots, a_n) = F(b_1, \ldots, b_n), X \vdash Y}{}$$

$$\vdots (\text{H1})$$

$$\frac{F(a_1, \ldots, a_n) = F(b_1, \ldots, b_n), Z^* \vdash W}{a_1 = b_1, \ldots, a_n = b_n, Z^* \vdash W} \text{ Red2}$$

$$\vdots (\text{H2})$$

$$a_1 = b_1, \ldots, a_n = b_n, Z \vdash W$$

where H1 is the segment of H which does not work on descendants of the auxiliary formula of Red1 and H2 is the segment of H which works on descendants of the principal formulas of Red1.

2.13. Note. Different Red rules in $Q2$ may have equal principal formulas. This makes the estimate of $l(Q2)$ not easy.

2.14. Theorem. (Conservation Property for **EQ** proofs) *Let Q be a non redundant proof in $\boldsymbol{EQ}$. Then each function letter and each minimal term occurring in Q occurs in the root S of Q.*
Proof. Through Corollary 2.7 it is sufficient to prove that if a function letter occurs in the $Q1$-root, then it also occurs in S, since the conservation of minimal terms through a $Q2$-segment is obvious. Let us suppose ad absurdum that the function letter f, occurring in the antecedent of the $Q1$-root in a formula $G(\ldots f \ldots) = G(\ldots d \ldots)$, does not occur in S, since it has been deleted by Red rules. Through Lemma 2.11, one between $G(\ldots f \ldots)$ and $G(\ldots d \ldots)$, e.g. $G(\ldots f \ldots)$, occurs in the $Q1$-root antecedent in a different formula. If we suppose that each occurrence of the maximal term $G(\ldots f \ldots)$ in different formulas of the $Q1$-root antecedent has been deleted, through properties of Red rule, such formulas must be of the form $G(\ldots f \ldots) = G(\ldots h \ldots)$; since the number of formulas in the $Q1$-root antecedent is finite, through point (viii) of Definition 2.3 and Lemma 2.11, the iteration of the process forces an occurrence of a maximal term of the form $G(\ldots q \ldots)$ in the succedent of the $Q1$-root; but no term can be deleted in the succedent through a $Q2$-segment, and we have an absurd.

2.15. Definition. We say that the formula $G(\ldots f \ldots) = G(\ldots p \ldots)$ is *reducible* to the formula $f = q$ if $f = q$ can be a principal formula of a Red and has $G(\ldots f \ldots) = G(\ldots p \ldots)$ as ancestor. We say that $f = q$ is a *sub-formula* of $t = s$ if f is a sub-term of t and q is a sub-term of s.

2.16. Proposition. *Let $Q2$ be a $Q2$-segment in a non redundant proof $\boldsymbol{Q}$ in the system $\boldsymbol{EQ}_k$, k bound for the complexity of terms. Then if in the antecedent of the root S of $Q2$ r formulas occur, we have $l(Q2) \leq k + u$ where $u = Card\{formulas\ reducible\ to\ formulas\ in\ the\ antecedent\ of\ S,with\ term\ complexity \leq k,\ function\ letters\ occurring\ in\ S,\ and\ all\ non-reducible\ sub-formulas occurring\ in\ the\ antecedent\ of\ S\}$, $u = u(r,v,k)$ with $v = \{number\ of\ different symbols\ in\ the\ Q2\text{-}root\}$; moreover, in the antecedent of the uppermost sequent L of $Q2$ at most $r + u$ formulas occur.*

Proof. Rules of $Q2$ change the complexity of formulas. The uppermost sequent L of $Q2$ is either an $\mathbf{EQ}_k$ axiom or the root of a $Q1$-segment; in both cases in the succedent of L, and then of each sequent in $Q2$, a single formula occurs. So in $Q2$ can occur at most $k - 1$ Exp rules; moreover, if $r = 0$, $l(Q2) \leq k$ and in the uppermost sequent of $Q2$ the antecedent is empty. Let us suppose that $r > 0$. We must estimate the number of Red rules in $Q2$, which, through Lemma 2.12, is equal to the number of possible different auxiliary formulas . Each auxiliary formula must be of the form $G(\ldots f \ldots) = G(\ldots p \ldots)$; so,through 2.14, Card$\{$Red rules in $Q2$ $\} \leq u = $ Card$\{$ formulas reducible to formulas in the antecedent of S,with complexity $\leq k$, function letters occurring in S, and all non-reducible sub-formulas occurring as sub-formulas in the antecedent of $S\}$. Therefore, $l(Q2) \leq k + u$. Moreover, u is the greatest possible number of Red-auxiliary formula ancestors in the antecedent L of the $Q1$-root; so in the antecedent of L at most $r + u$ formulas occur.

Corollary 2.17.

(i) *In the same hypotheses of 2.16, if each formula in the antecedent of S have a complexity $= k$ and the succedent formula has a complexity $h \leq k$ then $l(Q2) \leq k - h$ and in the antecedent of the uppermost sequent of $Q2$ r formulas occur;*

(ii) *in a non-redundant proof in $\boldsymbol{EQ}_k$ a formula B in the $Q1$-root antecedent can be an auxiliary formula in the $Q2$-segment only if all its non-reducible sub-formulas occur in the antecedent of S as formulas or as sub-formulas of reducible formulas different from B and of complexity $< k$.*

In several concrete cases the above condition 2.17.(ii) gives a strong bound for the parameter u. Moreover, formulas of the form $f = f$ are excluded from the set of formulas from which we calculate u, since they cannot be auxiliary formulas of Red rules.

In addition to the assumption on the bound on the complexity of terms we will assume that variables occurring in the proofs vary in a finite given set $\{x_1, \ldots, x_n\}$. *So we have indexed systems* $(AxT, EQ_{k,n})$, *and in* **AxT** *only variables of the set* $\{x_1, \ldots, x_n\}$ *can occur.*

2.18. Proposition. *Let Q3 be a Q3-segment of a non redundant proof* **Q** *in the system* $(AxT, EQ_{k,n})$. *Let* $g(k, n)$ *be number of different formulas in* **AxT**, *under our standard hypotheses on* **AxT**. *Then* $l(Q3) \leq g(k, n)$.

Proof. Through properties of **Q** two cuts with the same cut-formulas cannot exist. Since in *Q3* cuts in which one cut-formula is introduced by an axiom of **AxT** only occur, the greatest number of possible cuts in *Q3* is $g(k, n)$.

2.19. Theorem. (Weak estimate of the length of a non redundant proof **Q**) *Let* **Q** *be a non redundant proof of a sequent* $S = Z \vdash W$ *with m formulas in the antecedent in the system* $(AxT, EQ_{k,n})$. *Then if* $g = g(k, n) = $ *number of different formulas in* **AxT**, *we have :*

$$l(\boldsymbol{Q}) \leq g + (u + k) + (m + g + u)[(m + g + u) - 1]/2$$

where $u = u(m + g, v, k)$ *is the function u of Proposition 2.16 calculated from a Q2-root of the form:* $\{formulas\ of\ \boldsymbol{AxT}\}, Z \vdash W$.

Proof. Starting from the **Q**-root S, through Proposition 2.18, the upper bound of the length of the *Q3*-segment is g; moreover, the highest number of formulas in the antecedent of a *Q2*-segment root is $m + g$, since g formulas at the most can be eliminated in *Q3*, and by supposing the extreme case in which all formulas in the S antecedent were in the *Q2*- roots antecedents. Through Proposition 2.16, for each *Q2*-segment, $l(Q2) \leq k + u$ where u is calculated from a possible *Q2*-root in **Q** with the largest antecedent and containing all symbols which are not in **AxT** and which occur in S: we see that the sequent $\{formulas\ of\ \mathbf{AxT}\}, Z \vdash W$ satisfies these conditions. Moreover, through Proposition 2.16 in the antecedent of an uppermost sequent of a *Q2*-segment occur $m + g + u$ formulas at the most; therefore, through Proposition 2.9, for each *Q1*-segment in **Q**, $l(Q1) \leq (m + g + u)[(m + g + u) - 1]/2$ and we have the thesis.

Our next goal is to select the hypotheses which allow us to employ the following Conservation Principle **CP**.

2.20. *Let* **Q** *be a non redundant proof of a sequent* S *in* $(AxT, EQ_{k,n})$; *then we can transform* **Q** *into a non redundant proof* **Q**' *of* S *such that:*

2.20.(i) (CP)   *if in Q' an axiom A of $\mathbf{AxT}$ occurs, then all function letters and minimal terms of A occur in the root S of Q'.*

We will employ also a weaker version **CP1**:

2.20.(ii) (CP1)   *if in Q' an axiom A of $\mathbf{AxT}$ occurs, then all function letters of A occur in the root S of Q'.*

CP for equational logic is similar to the subformula property for predicative logic, i.e. it is close to a sub-term property for equational proofs in normal form. If **CP** holds, we can give the following strong estimate of $l(\mathbf{Q})$.

2.21. Theorem.   (Strong estimate of the length of a not redundant proof) *Let Q be a not redundant proof of the sequent $S = Z \vdash W$ with m formulas in the antecedent in the system $(AxT, EQ_{k,n})$: let*

$$SYM(S) = \{f_1, \ldots, f_p, x_1, \ldots, x_n, c_1, \ldots, c_w\}$$

be the set of different function letters, variables and constants occurring in S; let $g^* = g^*(k, p, n, w)$ be the number of different formulas in the subset $\mathbf{AxT}(SYM(S))$ of $\mathbf{AxT}$ in which symbols of $SYM(S)$ only occur; then, if we assume that $\mathbf{CP}$ holds we have

$$l(\mathbf{Q}) \leq g^* + (u^* + k) + (m + g^* + u^*)[(m + g^* + u^*) - 1]/2$$

where $u^* = u^*(m + g^*, v^*, k)$ is the function u of Proposition 2.16 and Theorem 2.19 calculated from a Q2-root of the form $\{formulas\ of\ \mathbf{AxT}(SYM(S))\}, Z \vdash W$.

Proof. The thesis follows from Theorem 2.19.

We observe that in the estimate of $l(\mathbf{Q})$ given in 2.21 both the parameters $g^*(k, p, n, w)$ and $u^*(g^* + m, v^*, k)$ are strictly dependent on the given $\mathbf{Q}$-root S, since we also have $v^* = $ (number of different symbols in S) $= \mathrm{Card}(SYM(S))$.

If we use the **CP1** form of the Conservation Principle then 2.21 holds with the replacement of $SYM(S)$ with

$$SYM'(S) = \{f_1, \ldots, f_p\} = SYM(S) - \{x_1, \ldots, x_n, c_1, \ldots, c_w\}.$$

We note that $SYM'(S) \leq SYM(S)$ but $\mathbf{AxT}(SYM'(S)) \geq \mathbf{AxT}(SYM(S))$.

Definition 2.22. Let $S = X \vdash l = m$ be the root of a $Q1$-segment and let $f = v \in X$; then we call *linked maximal terms* the occurrences of f, v as maximal terms in X which are forced in S by Lemma 2.11. We call *linked term sequence* a sequence of maximal terms of a $Q1$-root in which two consecutive elements are maximal linked terms.

We can prove the following version of the Conservation Principle **CP**:

Theorem 2.23. *Let **Q** be a non redundant proof in $(\textbf{AxT}, \textbf{EQ}_{k,n})$; if*

(i) *axioms of **AxT** are of the form $\vdash a = b$ with {function letters and minimal terms of a} = {function letters and minimal terms of b} i.e. $SYM(a) = SYM(b)$, and*

(ii) *no proper sub-formula of an axiom of **AxT** occurs as formula in an axiom of **AxT**,*

*then **CP** holds for **Q**.*

Proof. Let C be the lowermost cut in **Q**:

$$\vdots (H)$$

$$\frac{\vdash s = t \qquad s = t, V \vdash c = d}{V \vdash c = d} \; C$$

where the left premise is an axiom of **AxT** and the succedent $c = d$ has the $Q1$-root succedent as sub-formula. Let

$l(Q_3)$ = greatest number of proof lines in a branch in a Q_3-segment.

$l(Q3) = 1$: the proof H includes only $Q1$ and $Q2$-segments at most. Let us suppose that the function letter g occurring in the term s does not occur in the conclusion of C. Through hypotheses on **AxT**, g occurs both in s and in t; the right cut-formula has an ancestor in the $Q1$-root in H which is in the most general case of the form $F(\ldots s \ldots) = F(\ldots t \ldots)$, where F may be empty. Through Lemma 2.11 at least one of the two maximal terms of the considered ancestor, e.g. $F(\ldots s \ldots)$ must occur in the $Q1$-root in a different formula; our assumption forces such formula to be in the antecedent. If we suppose that the letter g has been deleted by Red rules in each maximal linked term of the linked term-sequence starting from $F(\ldots s \ldots)$, each element of the sequence must be reducible and with formulas of the form $g(e) = g(p)$ as sub-formulas. Since the number of formulas in the $Q1$-root is finite, and since formulas of the form $d = d$ cannot be either reduced or cut, we have that g must occur in the succedent and this is absurd.

$l(Q3) > 1$: if we suppose that letter g has been completely deleted either by cuts or by Red rules, through Lemma 2.11 and the same considerations of the basis case we must conclude that g occurs in the succedent, and we have an absurd.

Corollary 2.24. *Let $\boldsymbol{Q}$ be a non redundant proof in $(\boldsymbol{AxT}, \boldsymbol{EQ}_{k,n})$; if*

(i) axioms of $\boldsymbol{AxT}$ are of the form $\vdash a = b$,
(ii) no principal formula of a Red rule in $\boldsymbol{Q}$ is a cut-formula in $\boldsymbol{Q}$,

then $\boldsymbol{CP}$ holds in $\boldsymbol{Q}$ for the set of symbols $\bigcap_i SYM(t_i)$, $\{t_i\}$ set of maximal terms in $\boldsymbol{AxT}$.

The following theorem establishes a link between the demonstrative power of an Induction rule occurrence $\mathbf{I}$ in a proof $\mathbf{P}$ and the set of proper axioms occurring in $\mathbf{P}$ above $\mathbf{I}$.

2.25.Theorem. *Let the sequent $L = F(0), X \vdash Y, F(t)$ be a theorem of the equational system $(\boldsymbol{AxT}, \boldsymbol{EQ}_{k,n})$, k bound for the complexity of terms and $\{x_1, \ldots, x_n\}$ finite fixed set of variables. Let us suppose that L is the conclusion of an uppermost occurrence $\mathbf{I}$ of Induction rule with premise $N = F(x), X \vdash Y, F(S(x))$ in a proof $\mathbf{P}$ in $(\boldsymbol{EQ}_{k,n} + Ind, \boldsymbol{AxPRA} + \boldsymbol{AxT})$; let us suppose that the conservation principle $\boldsymbol{CP}$ holds for normal proofs in $(\boldsymbol{AxT}, \boldsymbol{EQ}_{k,n})$ and for the sub-proof H of the premise N of $\mathbf{I}$ in $\mathbf{P}$. Let $\boldsymbol{AxH}$ be the set of proper axiom instances in H, $\boldsymbol{AxH}$ sub-set of $\boldsymbol{AxT} + \boldsymbol{AxPRA}$. Then if symbols occurring in the term t of L occur also in H, a normal proof $\boldsymbol{Q}$ of the conclusion L of $\mathbf{I}$ in $(\boldsymbol{AxT}, \boldsymbol{EQ}_{k,n})$ exists, so that $l(\boldsymbol{Q}) \leq M(h^*)$, M increasing function of the cardinality h^* of the set $SYM(AxH)$ of symbols occurring in axioms of $\boldsymbol{AxH}$.*

Proof. We have

$$\vdots\, (H)$$

$$\frac{F(x), X \vdash Y, F(S(x))}{F(0), X \vdash Y, F(t)} \; \mathbf{I}$$

If $\mathbf{CP}$ holds , through Theorem 2.21 we have :

$$l(\mathbf{Q}) \leq g^* + (u^* + k) + (m + g^* + u^*)[(m + g^* + u^*) - 1]/2 = E(g^*),$$

increasing function of $g^* = \mathrm{Card}\{\text{formulas of } \mathbf{AxT}(SYM(L))\}$, $\mathbf{AxT}(SYM(L))$ sub-set of $\mathbf{AxT}$ in which occur symbols occurring in L only. But, through the hypotheses, we also have:

$$AxT(SYM(L)) \leq AxT(SYM(N)) \leq (AxT + AxPRA)(SYM(N)),$$

where in each set of axioms the complexity bound k is assumed; but, through **CP**, $(AxT + AxPRA)(SYM(N)) \geq (AxT + AxPRA)(SYM(AxH))$; therefore, through **CP**, the cardinality of the set $(AxT + AxPRA)(SYM(N))$ is an increasing function $D(h^*)$ of $h^* = \text{Card}(SYM(AxH))$; since $g^* \leq D(h^*)$ we have: $l(\mathbf{Q}) \leq E(g^*) \leq E(D(h^*)) = M(h^*)$, increasing function of h^*.

Theorem 2.25 shows that the bound for the length of the equational proof of the conclusion of **I** is an increasing function of the cardinality of the set of symbols of the proper axiom instances above **I**: that is, of the *quantity of information* included in the axiom instances , and this is a motivation of the concepts introduced in Section 3.

3   Entropy of an axiom system, virtual proofs, probability

In this Section we discuss the process which enables the prover to decide the "distance" between a theorem S and the system $\mathbf{T}$, when S is proven in $\mathbf{PRA(Z)+AxT}$ with a proof $\mathbf{P}$ in which the induction rule I occurs; our aim is to reach a good definition of the probability for S to belong to $\mathbf{T}$. Our first approach was:

a) to estimate the length $l(\mathbf{Q})$ of a proof of S in $\mathbf{T}$;
b) to define a translation $l(I)$ of each occurrence of the Induction rule in the given proof $\mathbf{P}$ into a suitable number of equational proof-lines, and so to obtain a translated length $lt(\mathbf{P})$ of $\mathbf{P}$, as a number of equational proof-lines;
c) to define the probability for S to belong to $\mathbf{T}$ as $\mathbf{P}(S) = l(\mathbf{Q})/lt(\mathbf{P})$.

This approach, even if apparently very linear, left many problems unsolved. problems arising also from the application of the process to many concrete cases:

(1) Since a real proof of S in $\mathbf{T}$ in general does not exist, and since a real translation of I into an equational proof-segment in general is not possible, $l(\mathbf{Q})$ and $l(I)$ had a virtual character which was not sufficiently expressed in a formal manner: it was therefore necessary to define mathematical objects which expressed the concept of virtual proof in a formal manner;

(2) $l(\mathbf{Q})$ and $lt(\mathbf{P})$ however remained non homogeneous and substantially not comparable: in fact the estimated $l(\mathbf{Q})$ was a virtual length, $lt(\mathbf{P})$ was a sum between the effective length $l(\mathbf{P})$ of the real proof $\mathbf{P}$ and the virtual estimated lengths $l(I)$;

(3) the proof-theoretical results on the Induction in equational proofs, stated in Theorem 2.25, suggested the following criterion for the calculation of $l(I)$:

$$l(I) \;=\; k + (\textit{number of proper axioms of } \boldsymbol{AxT} \textit{ and } \boldsymbol{PRA(Z)} \textit{ occurring}$$
$$\textit{in } \boldsymbol{P} \textit{ above the conclusion of I})$$

where k is the bound for the complexity of terms in $\mathbf{P}$; this criterion however, though motivated, produced unsatisfactory results. The intuition provided by concrete experiments was actually the following: there certainly exists a connection between the demonstration power of I in $\mathbf{P}$ and the set of proper axioms occurring above I in $\mathbf{P}$; this connection is however too coarsely expressed if we measure directly the equational virtual length of $\mathbf{P}$ with the number of proper axioms occurring above I in $\mathbf{P}$.

Finally, after examining many examples of real automated proofs, and encouraged by various proof-theoretical results obtained on the structure of equational proofs, we formulated the following hypothesis:

a proof-tree $\boldsymbol{P}$ in a theory F with the logic L can be seen as a machine which transmits to the root the specific information included in the proper F-axioms which are its leaves; the axioms can be seen as information inputs and the root as information output; the demonstration power of a rule I of L in $\boldsymbol{P}$ is given by the real information which I transmits, i.e. by the information included in the F-axioms above I in $\boldsymbol{P}$ which are really necessary for proving the conclusion of I.

However to express the above hypothesis in a formal version, a very important question must be answered: what is the "quantity of information" in an axiom or in a proof? It is clear that the core of the theory we are proposing is a good mathematical definition of such a concept. A natural answer was: to adapt to a proof-theoretical context the fundamental concepts and methods of Shannon's Information Theory. So we give the definitions 3.1., 3.4., of entropy H of an axiom system A and of a proof $\mathbf{Q}$. *It is the definition of entropy which allows an acceptable formal definition of probability.*

In fact, by thinking a proof as essentially characterized by its entropy, we can define virtual proofs as mathematical objects and introduce for the given S two kinds of virtual proof, which are comparable: the a priori virtual proof $\mathbf{Q}$ and the consequent virtual proof $\mathbf{P}^*$ given $\mathbf{P}$, in which the lengths are defined in an homogeneous manner; so we can assume $p(S) = l(\mathbf{Q})/l(\mathbf{P}^*)$, but with a new and strong motivation, since through the definition of virtual proof we also have $p(S) = H(\mathbf{Q})/H(\mathbf{P}^*)$, i.e. *the probability is the ratio between the two entropies.*

Such definition of probability is also providing acceptable results in the automated deduction experiments we are performing. An example of the process is given in Section 4. In the following we suppose that $\mathbf{T}$ is a fragment of $\mathbf{PRA(Z)}$ as defined in Section 1. Most of the following considerations are valid for the general case in which $\mathbf{T}=(\mathbf{EQ},\mathbf{AxT})$ is a sub-system of $(\mathbf{EQ+I}, \mathbf{AxT+B})$, $\mathbf{I}$ a generic rule extending $\mathbf{EQ}$, $\mathbf{B}$ a set of new equational axioms extending $\mathbf{AxT}$.

We recall in a very synthetic manner the basic concepts of Shannon's Information Theory (for a synthetic presentation we refer to [12]; for more complete expositions we refer to [7], [16]).

- *Entropy or quantity of Information* of a finite set of events $A = \{A_1, \ldots, A_n\}$ with probabilities $p_1, \ldots, p_n$, $\sum p_i = 1$, is the number:

$$H(A) = -\sum(p_i \log p_i)$$

where the log-base is fixed in a suitable manner and $p_k \log p_k$ is 0 if $p_k = 0$; moreover, $p_i \leq 1$ implies $H \geq 0$.
- *Source of Information*: it is a pair $[A, m]$ where A is a finite alphabet and m a probability measure on the set A^N of sequences of symbols of A.
- *Channel*: it is a triple $[A, n_x, B]$ where A is the input alphabet, B the output alphabet and n_x a probability measure such that $n_x(S)$ is the probability that the output sequence $y \in B^N$ belongs to $S \subset B^N$ if the input sequence $x \in A^N$ is transmitted.
- The entropy H of a source $[A, m]$ and the capacity of a channel $[A, n_x, B]$ (which can be seen as a compound source) are formally defined. The *entropy H of a source* $[A, m]$ represent the mean quantity of information in a transmitted symbol; if $S_{t_n} \subset A^N$, $S_{t_n} = \{x \in A^N : \text{symbols } a_{t_1}, \ldots, a_{t_n} \text{ of } A \text{ are in } x\}$, then we set $H_n = -\sum_{S_{t_n}} m(S_{t_n}) \log(m(S_{t_n}))$ and we define $H = \lim_n(H_n/n)$. The *capacity of a channel* is the maximum of information transmission speed R through the channel. R is so defined: if the source $[A, m]$ feeds the channel $[A, n_x, B]$ we can define a compound source $[C, w]$ which represents their link, and an output source $[B, e]$. If $H(X)$, $H(X, Y)$, $H(Y)$ are the respective entropies, then the *information transmission speed* is: $R(X, Y) = H(X) + H(Y) - H(X, Y)$.

We propose a parallel between Axioms and Proofs in proof-theory and Sources and Channels in information-theory. The crucial starting point is the definition of a notion of quantity of information in a formal logical context.

3.1. Definition. (Entropy of an Axiom System Ax) Let Ax be a finite equational axiom system in a fixed language $L(Ax)$ in which m axioms occur; let $Te(Ax)$ be the set of terms which occur in Ax, which we call the *term-alphabet* of Ax; if $B \in Ax$ then we define *entropy of B* the number:

$$H(B) = - \sum (p_i \log_{|Te(Ax)|m} p_i)$$

where: $t_1, \ldots, t_n$ are the terms occurring in B; $p_i = p(t_i) = $ *(number of occurrences of t_i in B) / (number of occurrences of terms in B)*. The *entropy of the system Ax* is:

$$H(Ax) = \sum (H(B_j)), \qquad B_j \in Ax.$$

The log-base $|Te(Ax)|m$ expresses the choice of penalizing the redundancy in an axiom system.

3.2. Example. Let
   $Ax = \{+(0, x) = x, +(x, -(x)) = 0\} \qquad L(Ax) = \{0, x, +(,), -()\}$
   $Te(Ax) = \{0, x, +(0, x), -(x), +(x, -(x))\}$
   $m = 2 \quad |Te(Ax)| = 5 \qquad \text{log-base} = m|Te(Ax)| = 10$
   $H(B1) : ocTe(B1) = \{x, x, 0, +(0, x)\} \qquad p[x] = 2/4 = 1/2 p[0] = 1/4$
   $p[+(0, x)] = 1/4 \qquad H(B1) = - \sum (p_i \log p_i) = 0.4515$
   $H(B2) : ocTe(B2) = \{x, x, 0, -(x), +(x, -(x))\} \qquad p[x] = 2/5$
   $p[0] = p[-(x)] = p[+(x, -(x))] = 1/5 \qquad H(B2) = - \sum (p_i \log p_i) = 0.5785$
   $H(Ax) = H(B1) + H(B2) = 1.301$
   We note that $H(+(0, x) = x) < H(+(x, -(x)) = 0)$.

3.3. Note. We note that $H(x = x) = 0$ for each term alphabet and for each log-base.

We wish to define the entropy of a proof $\mathbf{Q}$ of a sequent S in a system $\mathbf{T}$.

3.4. Definition. If $\mathbf{Q}$ is a proof-tree of the sequent S in a system $\mathbf{T}$ then the entropy $H(\mathbf{Q})$ of $\mathbf{Q}$ is:

$$H(\mathbf{Q}) = H(Ax(\mathbf{Q}))$$

where $Ax(\mathbf{Q})$ is the set of the instances of proper axioms of $\mathbf{T}$ which occur in $\mathbf{Q}$.

We shall have to compare the entropies of two different sets of axioms. To this purpose we must introduce a minimal homogeneity principle, which avoids situations such as the following:

Let us consider the sets $Ax_1 = \{x + 0 = x, y + 0 = y, z + 0 = z\}$ and $Ax_2 = \{x + 0 = x\}$; it is clear that we cannot state that Ax_1 contains more information than Ax_2, nevertheless if we compare the entropies without any preventive condition we have $H(Ax_1) > H(Ax_2)$. In order to avoid these phenomena, we introduce the Comparison Principle for Axiom Systems.

3.4.1. Let Ax_1 and Ax_2 be two different sets of axioms in the language L; we call *minimal terms* the constants and variables of L. Let $Te(Ax_1)$ and $Te(Ax_2)$ be the respective term alphabets; if a term t occurs in $Te(Ax_1)$, t not a minimal term, so that a term t' in $Te(Ax_2)$ exists, which is different from t only since variables of t are replaced by other minimal terms of L, then we complete $Te(Ax_1)$ by adding t', the new possible variables occurring in t which are not in $Te(Ax_1)$ and the possible new compound terms which we can obtain through replacement of t with t' in elements of $Te(Ax_1)$; moreover, we complete Ax_1 with the possible new axiom instances given by the replacement of t with t' in axiom of Ax_1; we call *completion $Ax_1\hat{}$ of Ax_1 for the comparison with Ax_2* the set so obtained. Analogously we obtain $Ax_2\hat{}$. We say that the entropies $H(Ax_1\hat{})$ and $H(Ax_2\hat{})$ are *comparable*.

3.4.2 Example. If $Ax_1 = \{x + 0 = x, y + 0 = y\}$ and $Ax_2 = \{x + 0 = x\}$ then $Ax_1\hat{} = \{x + 0 = x, y + 0 = y\}$ and $Ax_2\hat{} = \{x + 0 = x, y + 0 = y\}$; if $Ax_3 = \{0 + 0 = 0\}$ and $Ax_4 = \{z + 0 = z\}$ then $Ax_3\hat{} = \{0 + 0 = 0\}$ and $Ax_4\hat{} = \{z + 0 = 0, 0 + 0 = 0\}$.

In the last case we note that the Comparison Principle makes explicit the obvious fact that the information in Ax_4 is not only greater than the information in Ax_3, but it also strictly includes the information of Ax_3.

In the following we will always suppose that, each time the entropies of two axiom systems are compared or involved in the same mathematical expression, the completion procedure has been performed. The entropy of a proof enables us to characterize the concept of virtual proof which will be the basis of our definition of probability.

3.5 Definition. Let S be a sequent in the language of $\mathbf{T}$, let us suppose we have a criterion $\mathbf{CR}$ for estimating the set of the instances of proper axioms of $\mathbf{AxT}$ which would be necessary for proving S in $\mathbf{T}$, and for estimating the length of a possible proof of S in $\mathbf{T}$. Then we call *a priori virtual proof $\mathbf{Q}$* of S in $\mathbf{T}$ the 4-ple $(Ax(\mathbf{Q}), H(\mathbf{Q}), l(\mathbf{Q}), \nu(\mathbf{Q}))$ where

$Ax(\mathbf{Q})$ is the *estimated set of proper axiom instances* occurring in $\mathbf{Q}$;
$H(\mathbf{Q})$ is the *entropy* of $\mathbf{Q}$ i.e. $H(Ax(\mathbf{Q}))$;
$l(\mathbf{Q})$ is the *estimated length* of $\mathbf{Q}$;
$\nu(\mathbf{Q}) = H(\mathbf{Q})/l(\mathbf{Q})$ is the *information transfer speed* of $\mathbf{Q}$.
We say that two virtual proofs of S are *homogeneous* if they possess the same transfer speed ν.

3.6 Note. We emphasize that in the significant cases the virtual proof $\mathbf{Q}$ cannot exist as a real proof; therefore, the exact computation of $Ax(\mathbf{Q})$ and $l(\mathbf{Q})$ cannot exist. Moreover the a priori virtual proof $\mathbf{Q}$ of S depends on the criterion $\mathbf{CR}$ which we employ, and so it is not unique.

We now wish to define a second kind of virtual proof, the consequent virtual proof. Let $\mathbf{P}$ be an effective given proof of a sequent S; now we want to consider $\mathbf{P}$ as an information channel and to define the capacity of information transfer of an occurrence I^* of a rule I in $\mathbf{P}$. The most general definition is the following:

3.7 Definition. Let $\mathbf{P}$ be an effective give proof of a sequent S in a system F and let us suppose having a criterion $\mathbf{CR}$, which is independent of $\mathbf{P}$, which allows us to estimate the a priori virtual proofs $\mathbf{Q}_i$ of a given sequent L_i in a sub-system $\mathbf{T}_i$ of F. Let I_1^* be an uppermost occurrence of the rule I_1 in $\mathbf{P}$ with premise N and conclusion L; then we define *capacity* of I_1^* the number:

$$C(I_1^*) = H(\mathbf{P}_0) - H(\mathbf{Q}_1)$$

where
$\mathbf{P}_0$ is the segment of $\mathbf{P}$ with N as root and
$\mathbf{Q}_1$ is the a priori virtual proof of L in the system F without I_1, $F - I_1$;
moreover, we call *consequent virtual proof* $\mathbf{P}_1^*$ of L in the system $F - I_1$ given the proof $\mathbf{P}$ the one characterized by:
entropy $H(\mathbf{P}_1^*) = C(I_1^*) + H(\mathbf{Q}_1)$;
transfer speed $\nu(\mathbf{P}_1^*) = \nu(\mathbf{Q}_1)$;
length $l(\mathbf{P}_1^*) = H(\mathbf{P}_1^*)/\nu(\mathbf{P}_1^*)$;
the set of proper axiom instances $Ax(\mathbf{P}_1^*)$ is the set $Ax(I_1^*)$ of proper F-axiom instances occurring in $\mathbf{P}_0$ above I_1^*.
Let I_j^* be an occurrence of a rule I_j in $\mathbf{P}$, with premise N_j and conclusion L_j and let $\{I_1^*, \ldots, I_{j-1}^*\}$ be the set of occurrences of rules above I_j^* in $\mathbf{P}$; then the *capacity* of I_j^* is:

$$C(I_j^*) = H(\mathbf{P}_{j-1}^*) + H(Ax(I_j^*)) - H(Qj)$$

where: $\mathbf{P}_{j-1}^*$ is the consequent virtual proof of the premise N_j in the system $F - I_1 - I_2 - \ldots - I_{j-1}$; $\mathbf{Q}_j$ is the a priori virtual proof of the conclusion

L_j in the system $F - I_1 - I_2 - \ldots - I_{j-1} - I_j$; $Ax(I_j^*)$ is the set of proper F-axiom instances occurring in $\mathbf{P}$ above I_j^* and which are not in $Ax(\mathbf{P}_{j-1}^*)$. Moreover, the *consequent virtual proof $\mathbf{P}_j^*$ of the conclusion L_j in the system $F - I_1 - I_2 - \ldots - I_{j-1} - I_j$ given $\mathbf{P}$* is defined by:

> *entropy* $H(\mathbf{P}_j^*) = C(I_j^*) + H(\mathbf{Q}_j)$;
> *transfer speed* $\nu(\mathbf{P}_j^*) = \nu(\mathbf{Q}_j)$;
> *length* $l(\mathbf{P}_j^*) = H(\mathbf{P}_j^*)/\nu(\mathbf{P}_j^*)$;
> *the set of proper axiom instances* $Ax(\mathbf{P}_j^*)$ *is the set of proper axiom instances occurring in* $\mathbf{P}$ *above* I_j^*.

The *capacity* of the given proof $\mathbf{P}$ is the capacity of its end rule.

We want however to apply here the above mentioned notions in a concrete context in which F is $\mathbf{PRA(Z)}$ or $\mathbf{PRA}$, $\mathbf{P}$ is a proof in $\mathbf{EQ+Ind}$ and $\mathbf{T}$ is a fixed arithmetical fragment with proper axiom set $\mathbf{AxT}$; the set of proper axiom instances occurring in $\mathbf{P}$ is in general a set of arithmetical axioms which properly extends $\mathbf{AxT}$; the only rule I in which we are interested to define capacity is the induction rule.

3.8 Definition. Let $\mathbf{T}$ be an equational fragment of $\mathbf{PRA(Z)}$ or $\mathbf{PRA}$; let $\mathbf{P}$ be a real given proof in $\mathbf{PRA(Z)}$ or $\mathbf{PRA}$ of a sequent S in the language of $\mathbf{T}$ in which induction rule occurs; let I^* be an occurrence of induction rule in $\mathbf{P}$ with N as premise and L as conclusion; let us suppose we have a criterion $\mathbf{CR}$ for estimating the a priori virtual proof $\mathbf{Q} = (Ax(\mathbf{Q}), H(\mathbf{Q}), l(\mathbf{Q}), \nu(\mathbf{Q}))$ in $\mathbf{T}$ of a sequent in the language of $\mathbf{T}$. Then the *capacity of I^** is: $C(I^*) = H(Ax(I^*)) - H(\mathbf{Q})$, where $Ax(I^*)$ is the set of proper axiom instances, both arithmetical and of $\mathbf{AxT}$, above I^* in $\mathbf{P}$, and $\mathbf{Q}$ is the a priori virtual proof of L in $\mathbf{T}$. The *consequent virtual proof $\mathbf{R}^*$ of L in $\mathbf{PRA(Z)}$ without I given $\mathbf{P}$* is defined by:

> *entropy* $H(\mathbf{R}^*) = H(\mathbf{Q}) + C(I^*) = H(Ax(I^*))$;
> *set of proper axiom instances* $Ax(I^*)$;
> *length* $l(\mathbf{R}^*) = H(\mathbf{R}^*)/v(\mathbf{R}^*)$.

The *consequent virtual proof $\mathbf{P}^*$ of S in $\mathbf{PRA(Z)}$ without Ind* is defined in the same fashion with:

> $Ax(\mathbf{P}^*)$= set of proper axiom instances in $\mathbf{P}$,
>   both arithmetical and of $\mathbf{AxT}$;
> $H(\mathbf{P}^*) = H(Ax(\mathbf{P}^*))$;
> $\nu(\mathbf{P}^*) = \nu(\mathbf{Q})$, $\mathbf{Q}$ a priori virtual proof of S in $\mathbf{T}$;
> $l(\mathbf{P}^*) = H(\mathbf{P}^*)/v(\mathbf{P}^*)$.

3.9 Note. Also the consequent virtual proof $\mathbf{P}^*$ of S in general cannot exist as a real proof, since in general a proof of S without Induction rule does not exist .

Our definition of capacity of information of an occurrence of an induction rule in a given **P** expresses the following assumptions:

(i) the quantity of information included in the rule I is an increasing function of the cardinality of the set of the proper axiom instances occurring in **P** above I; nevertheless we wish to penalize it with the addend $-H(\mathbf{Q})$ which measures the possibility that the conclusion of the induction I can be proved without induction:

(ii) the consequent virtual proof **P*** of S is defined in order to obtain an abstract translation of the given **P** into an equational proof, and to be able to compare it with the estimated equational proof **Q**. The property $H(\mathbf{P}^*) = H(\text{proper axioms in } \mathbf{P}) = H(\mathbf{P})$ expresses the fact that **P*** is an abstract translation of **P**.

Having produced the estimated equational proof **Q** of S in **T** and the translated real proof of S in **PRA(Z)** as two homogeneous objects, we can give our definition of probability.

3.10 Definition.   Let **T** be an equational fragment of **PRA(Z)** or **PRA**; let **P** be a real given proof in **PRA(Z)** or **PRA** of a sequent S in the language of **T** in which induction rule I occurs; let us suppose we have a criterion **CR** to estimate the a priori virtual proof $\mathbf{Q} = (Ax(\mathbf{Q}), H(\mathbf{Q}), l(\mathbf{Q}), \nu(\mathbf{Q}))$ in **T** of a sequent in the language of **T**; then we define *probability of S being provable in* **T** the number:

$p(S) = length(a\ priori\ virtual\ proof\ \boldsymbol{Q}\ of\ S\ in\ \boldsymbol{T})/length(consequent\ virtual\ proof\ \boldsymbol{P}^*\ of\ S\ in\ \boldsymbol{PRA(Z)}\ without\ I\ given\ \boldsymbol{P})$

3.11 Note   We have: $l(\mathbf{Q})/l(\mathbf{P}^*) = l(\mathbf{Q})\nu(\mathbf{Q})/H(\mathbf{P}^*)$ i.e. the probability is the ratio between the two entropies. So we have two heuristic justifications to our definition:

(i) $l(\mathbf{P}^*)$ is in some sense the translation of the induction rules occurring in **P** into a suitable number of equational proof lines: then the induction rules in **P** are really necessary to prove S only if their resulting length is greater than the length $l(\mathbf{Q})$ of the possible equational proof;

(ii) if the induction rules give a real increase of information for proving S, then their contribution must produce for **P** an entropy greater than the entropy of the possible equational proof **Q** of S.

However, our definition of capacity of an induction rule occurring in **P**, and thus the link between the proof-theoretical power of an induction rule and the entropy of the set of proper axiom instances occurring above it in the proof, has also a strictly proof-theoretical justification, which is expressed through Theorem 2.25.

4   Example of implementation

Let **T** be a sub-theory of group theory:

$$\mathbf{AxT} = \{+(0, x) = x, +(x, 0) = x, +(-(x), x) = 0, +(x, -(x)) = 0\}$$

$L_{\mathbf{T}} = \{0, +(\ ,\ ), -(\ )\}$.

Let S be the sequent: $\vdash +(x, +(y, z)) = +(+(x, y), z)$. The effective proof **P** given by the prover in **PRA(Z)** is the following:

```
The theorem is:
add(add(A9,B0),B1) = add(A9,add(B0,B1))
Begin induction on A9
Induction on A9 case 0
  (L) add(add(0,B0),B1)               << subst A9 with 0 <<
  (R) add(0,add(B0,B1))              << subst A9 with 0 <<

  (R) add(B0,B1)              << reduct by ax_1 <<
  (L) add(B0,B1)              << reduct by ax_1 <<
*** equality obtained ***
Induction on A9 case succ(B2)
  (L) add(add(succ(B2),B0),B1)   << subst A9 with succ(B2) <<
  (R) add(succ(B2),add(B0,B1))   << subst A9 with succ(B2) <<
  (R) succ(add(B2,add(B0,B1)))   << reduct by ax_2 <<
  (L) add(succ(add(B2,B0)),B1)   << reduct by ax_2 <<
  (L) succ(add(add(B2,B0),B1))   << reduct by ax_2 <<
  (L) succ(add(B2,add(B0,B1)))   << ind. hyp. on A9 for B2 <<
*** equality obtained ***
Induction on A9 case pred(B3)
  (L) add(add(pred(B3),B0),B1)   << subst A9 with pred(B3) <<
  (R) add(pred(B3),add(B0,B1))   << subst A9 with pred(B3) <<
  (R) pred(add(B3,add(B0,B1)))   << reduct by ax_3 <<
  (L) add(pred(add(B3,B0)),B1)   << reduct by ax_3 <<
  (L) pred(add(add(B3,B0),B1))   << reduct by ax_3 <<
  (L) pred(add(B3,add(B0,B1)))   << ind. hyp. on A9 for B3 <<
*** equality obtained ***
End induction on A9
QED
```

Computation of the a priori virtual proof **Q** *of S in* **T** *(Def. 3.5) by assuming as* **CR** *the criteria stated in Section 2 with the Conservation Principle* **CP1** *(2.20.(ii)). (We recall that* **Q** *is not unique, since depends on* **CR**).

We set: complexity bound for terms: $k = 5$; set of variables: $\{x, y, z\}$; so we assume as system $\mathbf{AxT+EQ}_{5,3}$. The resulting set $Ax(\mathbf{Q})$ of axiom instances for $\mathbf{Q}$ is the following:

```
axiom(ax1,add(0,x),x).                    axiom(ax2,add(x,0),x).
axiom(ax3,add(0,y),y).                    axiom(ax4,add(y,0),y).
axiom(ax5,add(0,z),z).                    axiom(ax6,add(z,0),z).
axiom(ax7,add(0,0),0).                    axiom(ax8, add(add(x,x),0),add(x,x)).
axiom(ax9,add(add(y,y),0),add(y,y)).      axiom(ax10,add(add(z,z),0),add(z,z)).
axiom(ax11,add(add(0,0),0),add(0,0)).     axiom(ax12,add(add(x,y),0),add(x,y)).
axiom(ax13,add(add(x,z),0),add(x,z)).     axiom(ax14,add(add(y,z),0),add(y,z)).
axiom(ax15,add(add(y,x),0),add(y,x)).     axiom(ax16,add(add(z,x),0),add(z,x)).
axiom(ax17,add(add(z,y),0),add(z,y)).     axiom(ax18,add(add(x,0),0),add(x,0)).
axiom(ax19,add(add(y,0),0),add(y,0)).     axiom(ax20add(add(z,0),0),add(z,0)).
axiom(ax21,add(0,add(x,0)),add(x,0)).     axiom(ax22,add(0,add(y,0)),add(y,0)).
axiom(ax23,add(0,add(z,0)),add(z,0)).     axiom(ax24,add(0,add(y,x)),add(y,x)).
axiom(ax25,add(0,add(z,x)),add(z,x)).     axiom(ax26,add(0,add(z,y)),add(z,y)).
axiom(ax27,add(0,add(y,y)),add(y,y)).     axiom(ax28,add(0,add(z,z)),add(z,z)).
axiom(ax29,add(0,add(x,x)),add(x,x)).     axiom(ax30,add(0,add(0,0)),add(0,0)).
axiom(ax31,add(0,add(x,y)),add(x,y)).     axiom(ax32,add(0,add(x,z)),add(x,z)).
axiom(ax33,add(0,add(y,z)),add(y,z)).     axiom(ax34,add(add(0,x),0),add(x,0)).


axiom(ax35,add(add(0,y),0),add(y,0)).     axiom(ax36,add(add(0,z),0),add(z,0)).
axiom(ax37,add(0,add(0,x)),add(0,x)).     axiom(ax38,add(0,add(0,y)),add(0,y)).
axiom(ax39,add(0,add(0,z)),add(0,z)).
```

$|Ax(\mathbf{Q})| = 39$; the term alphabet $Te(Ax(\mathbf{Q}))$ is:

$$\{x, y, z, 0, +(u_i, u_j) + (+(u_i, u_j), 0), +(0, +(u_i, u_j))\}$$

where $u_i, u_j \in \{0, x, y, z\}$ and $|Te(Ax(\mathbf{Q}))| = 52$. The log-base is $52 * 39 = 2028$; the computation of the entropy $H(\mathbf{Q}) = H(Ax(\mathbf{Q}))$ is the following:

```
ax_en(ax1,0.136539).     ax_en(ax2,0.136539).     ax_en(ax3,0.136539).
ax_en(ax4,0.136539).     ax_en(ax5,0.136539).     ax_en(ax6,0.136539).
ax_en(ax7,0.0738476).    ax_en(ax8,0.159296).     ax_en(ax9,0.159296).
ax_en(ax10,0.159296).    ax_en(ax11,0.118224).    ax_en(ax12,0.204809).
ax_en(ax13,0.204809).    ax_en(ax14,0.204809).    ax_en(ax15,0.204809).
ax_en(ax16,0.204809).    ax_en(ax17,0.204809).    ax_en(ax18,0.173463).
ax_en(ax19,0.173463).    ax_en(ax20,0.173463).    ax_en(ax34,0.19622).
ax_en(ax35,0.19622).     ax_en(36,0.19622).       ax_en(ax21,0.173463).
ax_en(ax22,0.173463).    ax_en(ax23,0.173463).    ax_en(ax24,0.204809).
ax_en(ax25,0.204809).    ax_en(ax26,0.204809).    ax_en(ax27,0.159296).
ax_en(ax28,0.159296).    ax_en(ax29,0.159296).    ax_en(ax30,0.118224).
ax_en(ax31,0.204809).    ax_en(ax32,0.204809).    ax_en(ax33,0.204809).
ax_en(ax37,0.173463).    ax_en(ax38,0.173463).    ax_en(ax39,0.173463).
total_entropy(6.6928).
```

The estimate of the length $l(\mathbf{Q})$ is, through theorem 2.21:

$$l(\mathbf{Q}) \leq g^* + (u^* + k) + (m + g^* + u^*)[(m + g^* + u^*) - 1]/2$$

where $g = 39$, $k = 5$, $m = 0$, $u = 0$ (see Corollary 2.17) and then $l(\mathbf{Q}) \leq 785$. The transfer speed is: $\nu(\mathbf{Q}) = H(\mathbf{Q})/l(\mathbf{Q}) = 6.6928/785 = 0.0085$. Therefore, the a priori virtual proof $\mathbf{Q}$ of S in $\mathbf{T}$ is:

$$\mathbf{Q} = (\mathbf{Ax}(\mathbf{Q}), 6.6928, 785, 0.0085).$$

Computation of the consequent virtual proof $\boldsymbol{P}^*$ *of S in* $\boldsymbol{PRA(Z)}$ *without Ind given the effective* $\boldsymbol{P}$. The axioms $Ax(\mathbf{P}^*)$, after the completion procedure for the comparison with $A(\mathbf{Q})$ (see 3.4.1) are the following:

```
schema(ax1,add(p(u_k),add(u_i,u_j, p(add(u_k,add(u_i,u_j)))))).
schema(ax2,add(p(0),add(u_i,u_j)),p(add(0,add(u_i,u_j)))).
schema(ax3,add(p(u_i),b_j),p(add(u_i,u_j))).
schema(ax4,add(p(add(u_i,u_j)),y),p(add(add(u_i,u_j),y))).
schema(ax5,add(p(add(u_i,u_j)),0),p(add(add(u_i,u_j),0))).
schema(ax6,add(0,add(u_i,u_j)),add(u_i,u_j)).
schema(ax7,add(0,u_i),u_i).
schema(ax8,add(s(z),add(u_i,u_j)),s(add(z,add(u_i,u_j)))).
schema(ax9,add(s(0),add(u_i,u_j)),s(add(0,add(u_i,u_j)))).
schema(ax10,add(s(u_i),u_j),s(add(u_i,u_j))).
schema(ax11,add(s(add(u_i,u_j)),y),s(add(u_i,u_j))).
schema(ax12,add(s(add(u_i,u_j),0),s(add(add(u_i,u_j),0)))).
```

$u_i, u_j \in \{0, x, y, z\}$.

We have $|Ax(\mathbf{P}^*)| = 180$, $|Te(Ax(\mathbf{P}^*))| = 412$; the resulting entropy is: $H(\mathbf{P}^*) = H(Ax(\mathbf{P}^*)) = 28.1849$; the transfer speed is $\nu(\mathbf{P}^*) = \nu(\mathbf{Q}) = 0.0085$; the resulting length is: $l(\mathbf{P}^*) = 3315$. Therefore, we have:

$$\mathbf{P}^* = (Ax(\mathbf{P}^*), 28.1849, 3315, 0.0085).$$

The probability for S to belong to $\boldsymbol{T}$, *with complexity bound $k = 5$ and var $= \{x, y, z\}$, is:*

$$p(s) = l(\mathbf{Q})/l(\mathbf{P}^*) = 0.2368.$$

5 Open problems

The probability $p(S) = l(\mathbf{Q})/l(\mathbf{P}^*)$ depends on the criterion **CR** which we use for the computation of the a priori virtual proof $\mathbf{Q}$ of S in $\mathbf{T}$. A step for the next research could consist of the following problem: which hypotheses on **CR** allow us to conclude that a suitable $p^* \in (0, 1)$ exists so that $p(S) < p^*$ implies S is not **T**-provable and $p(S) > p^*$ implies that S is **T**-provable?

A second open direction is the following: entropy is a syntactical notion but we cannot exclude that it can also acquire a syntactical meaning. We can start from the following very simple result: *let A be a set of equations in* **PRA** *involving only minimal terms . Then A is consistent if and only if the entropy H(A)=0.*

References

[1] S. Antoy, *Design strategies for rewrite rules*, Proceedings of 2th Int. Workshop on Conditional and Typed Rewriting Systems, 1990.

[2] S. Antoy, P. Forcheri, and M.T. Molfino, *Specification-based Code generation* Proceedings of Hawaii Int. Conf., Systems Sci. (1990), 165–170.

[3] J. Avenhaus and K. Madlener, *Term Rewriting and Equational Reasoning*, Formal Techniques in Artificial Intelligence, North Holland, 1990, pp. 1–43.

[4] R.S. Boyer and J.S. Moore, *Proving theorems about LISP functions*, J. Assoc. Comput. Mach. **22**, 129–144.

[5] C. Chang and J. Keisler, *Model Theory*, North Holland, 1990.

[6] H. Ehrig and B. Mahr, *Fundamentals of Algebraic Specification 1*, Springer Verlag, 1985.

[7] R. Gallager, *Information Theory and Reliable Communication*, Wiley, New York, 1968.

[8] P. Gentilini, *Provability Logic in the Gentzen Formulation of Arithmetic*, Z. Math. Logik Grundlag. Math. **38** (1992), 536–550.

[9] J.Y. Girard, *Proof-Theory and logical complexity*, Bibliopolis, 1987.

[10] R. Gurevic, *Equational Theory of Positive Numbers with Exponentiation is not finitely axiomatizable*, Ann. Pure Appl. Logic **49** (1990), 1–30.

[11] G. Huet and D.C. Oppen, *Equations and rewrite rules, a survey*, Formal Language Theory: perspectives and open problems (R. Book, ed.), Academic Press, 1980.

[12] A.I. Khinchin, *Mathematical Foundations of Information Theory*, Dover Pub., New York, 1957.

[13] D.E. Knuth and P.B. Bendix, *Simple word problems in universal algebras*, Computational Problems in Abstract Algebras (J. Leech, ed.), Pergamon Press, Oxford, 1970, pp. 263–297.

[14] J. Krajicek, *On the Number of Steps in Proofs*, Ann. Pure Appl. Logic **41** (1989), 153–178.

[15] J. Krajicek and P. Pudlak, *The Number of Proof Lines and the Size of Proofs in First Order Logic*, Arch. Math. Logik Grundlag. (1988), 69–84.

[16] G. Longo, *Teoria dell'Informazione*, Boringhieri, Torino, 1980.

[17] D. Monk, *Mathematical Logic*, Springer Verlag, 1976.

[18] R.J. Parikh, *Some results on the length of proofs*, Trans. Amer. Math. Soc. **177** (1973), 29–36.

[19] C. Smorynski, *The Incompleteness Theorems*, Handbook of Mathematical Logic, North Holland, 1977.

[20] ———, *Self-reference and modal logic*, Springer Verlag, 1985.

[21] G. Takeuti, *Proof Theory*, North Holland, 1987.

ISTITUTO PER LA MATEMATICA APPLICATA DEL C.N.R., VIA DE MARINI 6, TORRE DI FRANCIA, GENOVA, ITALY

Idempotent Simple Algebras [*]

Keith A. Kearnes [†]

Abstract

We show that every idempotent simple algebra which has a skew congruence in a power either has an absorbing element or is a subreduct of an affine module. We refine this result for idempotent algebras which have no nontrivial proper subalgebras. One corollary we obtain is a new proof of Á. Szendrei's classification theorem for minimal locally finite idempotent varieties. We partially extend this classification to non-locally finite varieties. Another corollary is a complete classification of all minimal varieties of modes.

1 Introduction

An operation $f(x_1, \ldots, x_n)$ on a set U is said to be **idempotent** if $f(u, u, \ldots, u) = u$ is true of any $u \in U$. An algebra is idempotent if each of its fundamental operations is idempotent. Equivalently, **A** is idempotent if every constant function $c : A \to A$ is an endomorphism. As is usual, we say that **A** is **simple** if it has exactly two congruences. In our paper we prove the following theorem concerning idempotent simple algebras.

THEOREM 1.1 *If* **A** *is an idempotent simple algebra, then exactly one of the following conditions is true.*

(a) **A** *has a unique absorbing element.*

(b) **A** *is a subalgebra of a simple reduct of a module.*

(c) *Every finite power of* **A** *is skew-free.*

[*]1991 *Mathematical Subject Classification* Primary 08A05; Secondary 08A30, 06F25.
[†]Research supported by a fellowship from the Alexander von Humboldt Stiftung.

We call $0 \in A$ an **absorbing element** for $\mathbf{A}$ if whenever $t(x, \bar{y})$ is an $(n+1)$-ary term operation of $\mathbf{A}$ such that $t^{\mathbf{A}}$ depends on x and $\bar{a} \in A^n$, then $t^{\mathbf{A}}(0, \bar{a}) = 0$. We say that $\mathbf{A}^n$ is **skew-free** if the only congruences on $\mathbf{A}^n$ are the product congruences. Since $\mathbf{A}$ is simple in the above theorem, the product congruences on $\mathbf{A}^n$ are just the canonical factor congruences. Thus, the statement that "every finite power of $\mathbf{A}$ is skew-free" is exactly the claim that $\mathbf{Con\,A}^n \cong \mathbf{2}^n$ for each $n < \omega$.

The class of simple algebras of type (c) includes the class of all idempotent functionally complete algebras and this class is interdefinable with the class of all idempotent algebras. Hence the simple algebras of type (c) resist classification. A similar statement may be made about the simple algebras of type (a). Therefore, the theorem is properly viewed as a criterion for an idempotent simple algebra to be a subreduct of a module. For, if an idempotent simple algebra has a skew congruence in a power but has no absorbing element, then it is a subreduct of a module.

It is plausible that our theorem will lead to a complete classification of all minimal varieties of idempotent algebras. By Magari's Theorem, any variety contains a simple algebra (see [8]); hence a minimal idempotent variety is generated by an algebra of type (a), (b) or (c). It is not hard to prove that any minimal variety containing a simple algebra of type (a) is equivalent to the variety of semilattices. Any minimal variety containing a simple algebra of type (b) is equivalent to the variety of sets or to a variety of affine modules over a simple ring. (In this paper, all rings have a unit element, an **affine R-module** is the idempotent reduct of an **R**-module and an **affine vector space** is the idempotent reduct of a vector space.) The minimal varieties containing only simple algebras of type (c) are difficult to analyze. The only examples that we know are congruence distributive. If such a variety is not congruence distributive, it cannot be locally finite.

The **entropic law** for an algebra $\mathbf{A}$ is the statement that for $f \in \mathrm{Clo}_m \mathbf{A}$ and $g \in \mathrm{Clo}_n \mathbf{A}$ and an $m \times n$ array of elements of A,

$$
\begin{bmatrix}
u_1^1 & u_1^2 & \cdots & u_1^n \\
u_2^1 & u_2^2 & \cdots & \\
\vdots & \vdots & \ddots & \\
u_m^1 & u_m^2 & & u_m^n
\end{bmatrix},
$$

we have $f(\overline{g(\bar{u}_i)}) = g(\overline{f(\bar{u}^j)})$. Idempotent entropic algebras are called **modes** (see [11]). We prove that there is no simple mode of type (c) and that the only one of type (a), up to equivalence, is the 2-element semilattice. This leads to a good description of simple modes and a classification of all minimal varieties of modes. Up to equivalence, the minimal varieties are: the variety of semilattices, the variety of sets and any variety of affine vector spaces.

Our notation for congruences in powers shall roughly follow [2]. The projection homomorphism from $\mathbf{A}^\kappa$ onto a sequence of coordinates σ will be denoted π_σ. For example, in $\mathbf{A}^3$ we have $\pi_{01}((a, b, c)) = (a, b)$ while $\pi_{10}((a, b, c)) = (b, a)$. We will write η_σ for the kernel of π_σ and write α_σ for $\pi_\sigma^{-1}(\alpha)$ where α is a congruence on $\mathbf{A}^\sigma$. We write η_σ' to denote the canonical factor congruence which complements η_σ. We prefer to write 0 in place of $\eta_{01\cdots(n-1)}$ and 1 in place of $\eta_\emptyset$. A congruence on $\mathbf{A}^n$ of the form

$$\alpha_0 \wedge \beta_1 \wedge \cdots \wedge \nu_{n-1}$$

will be called a **product congruence**. A congruence which is not a product congruence is a **skew congruence**. Hence a power of $\mathbf{A}$ is skew-free iff it has no skew congruences.

2 Skew Congruences in Powers

In this section we show that if $\mathbf{A}$ is a simple idempotent algebra which has a skew congruence in a finite power, then $\mathbf{A}$ has a unique absorbing element or else $\mathbf{A}$ is abelian. We shall find it convenient to restrict our arguments to algebras that have at least three elements. The reason for this is that the 2-element set with no operations is an algebra which is abelian, but has an absorbing element. In fact, each of the two elements of the 2-element set is an absorbing element, so this algebra does not contradict Theorem 1.1; it belongs in class (b). But the fact that it has an absorbing element and is at the same time abelian makes it difficult to separate classes (a) and (b) of Theorem 1.1 when $|A| = 2$. So, we now observe simply that Theorem 1.1 can be established by hand in the 2-element case and we leave it to the reader to do this. Alternately, one can refer to Post's description of 2-element algebras in [10] or to J. Berman's simplification of Post's argument in [1]. From these sources one finds that a 2-element idempotent algebra must be, up to term equivalence, one of the following:

(i) a semilattice,

(ii) a set,

(iii) an affine vector space or

(iv) a member of a congruence distributive variety.

A semilattice belongs only to class (a) of Theorem 1.1, a set or an affine vector space belongs only to class (b) of Theorem 1.1 and a member of a congruence distributive variety belongs only to class (c) of Theorem 1.1.

Here is how we shall use our hypothesis that $|A| > 2$. We shall need to know that any idempotent simple algebra of cardinality greater than two (i) has at most one absorbing element and (ii) is not essentially unary. We prove this now.

LEMMA 2.1 *If* **A** *is an idempotent simple algebra, then the following are equivalent.*

(i) **A** *has more than one absorbing element.*

(ii) **A** *is essentially unary.*

(iii) **A** *is term equivalent to the 2-element set.*

Proof: We shall prove that $(i) \Rightarrow (ii) \Rightarrow (iii) \Rightarrow (i)$. Assume (i) and deny (ii). Then **A** has distinct absorbing elements 0 and 1 and a term $t(x, y, \bar{z})$ which depends on at least x and y in **A**. Since 0 and 1 are absorbing we have

$$0 = t^{\mathbf{A}}(0, 1, 1, \ldots, 1) = 1,$$

a contradiction.

Now assume (ii). If $s(x, \bar{y})$ is a term that depends on x in **A**, then since **A** is essentially unary and idempotent we get

$$s^{\mathbf{A}}(x, y_0, \ldots, y_k) = s^{\mathbf{A}}(x, x, \ldots, x) = x.$$

Hence any term depending on a variable is projection onto that variable. It is impossible for **A** to have a term depending on no variables, since **A** is idempotent. Hence every term of **A** is projection onto some variable. (iii) follows from this and the simplicity of **A**.

If (iii) holds, then each of the elements of **A** is an absorbing element. Hence (i) holds. This completes the lemma. $\square$

Henceforth we assume that $|A| > 2$.

Definition 2.2 If α and β are congruences on **A**, then α **centralizes** β if whenever $p(x, \bar{y}) \in \mathrm{Pol}_{n+1}\mathbf{A}$ and $(u, v), \in \alpha$, $(a_i, b_i) \in \beta$ the implication

$$p(u, \bar{a}) = p(u, \bar{b}) \Rightarrow p(v, \bar{a}) = p(v, \bar{b})$$

holds. We denote the fact that α centralizes β by writing $C(\alpha, \beta)$. An algebra **A** is said to be **abelian** if $C(1_{\mathbf{A}}, 1_{\mathbf{A}})$.

We remark that the implication described in this definition is called the α, β-**term condition**. When $\alpha = \beta = 1$, then it is simply called the **term condition**. "**A** satisfies the term condition" is synonymous with "**A** is abelian." For additional properties of abelian congruences and algebras, see Chapter 3 of [4]. Here are a few facts we shall need.

(*i*) An algebra **A** is abelian if and only if the diagonal of $\mathbf{A}^2$ is a congruence class.

(*ii*) There is always a largest congruence α which centralizes a given β. This congruence is called the **centralizer** of β and it is denoted $(0 : \beta)$.

(*iii*) It is the case that $\alpha \wedge \beta = 0 \Rightarrow C(\alpha, \beta) \Leftrightarrow \alpha \leq (0 : \beta)$.

(*iv*) If **B** is a subalgebra of **A** and α and β are congruences on **A**, then $C(\alpha, \beta) \Rightarrow C(\alpha|_B, \beta|_B)$.

All of the properties of $C(x, y)$ mentioned in this paragraph have easy proofs.

We begin with a first approximation to the main theorem.

THEOREM 2.3 *If* **A** *is an idempotent simple algebra, then exactly one of the following is true.*

(*a*) **A** *has a unique absorbing element.*

(*b*) **A** *is abelian.*

(*c*) *Every finite power of* **A** *is skew-free.*

Proof: The theorem is true when $|A| = 2$, so we focus only on the case where $|A| > 2$.

First we argue that **A** belongs to at least one of the classes $(a), (b)$, or (c) described in the theorem. Later we explain why **A** cannot belong to two different classes. Assume that **A** is not in class (c). Then for some $m > 1$ there is a skew-congruence on $\mathbf{A}^m$. We assume that m is chosen minimally for this. Let θ be a skew congruence of $\mathbf{A}^m$. We may choose i so that $\eta_i \not\geq \theta$ since $\theta \neq 0$. By permuting coordinates if necessary, we may assume that $i = 0$. If $\theta \geq \eta_0'$, then θ/η_0' is a skew congruence on $\mathbf{A}^m/\eta_0' \cong \mathbf{A}^{m-1}$. This contradicts the minimality of m. Hence $\theta \not\geq \eta_0'$. We shall split our argument into two cases. Case 1: $\mathbf{A}^m$ has a congruence δ such that $0 < \delta < \eta_0'$. Case 2: $\mathbf{A}^m$ has no congruence δ such that $0 < \delta < \eta_0'$.

PROOF FOR CASE 1: Assume that $\mathbf{A}^m$ has a congruence δ such that $0 < \delta < \eta_0'$. We shall argue that **A** has an absorbing element. Each η_0'-class of $\mathbf{A}^m$ is a subalgebra of $\mathbf{A}^m$ since $\mathbf{A}$ is idempotent. Furthermore, each of these subalgebras is isomorphic to $\mathbf{A}$ via the restriction of the first coordinate projection. If $\mathbf{B} = \mathbf{A} \times \{u\}$, $u \in A^{m-1}$, is such a subalgebra, then $\delta|_{\mathbf{B}}$ is a congruence on $\mathbf{B}$. But, as $\mathbf{B} \cong \mathbf{A}$, $\mathbf{B}$ is simple. Hence $\delta|_{\mathbf{B}} = 0_{\mathbf{B}}$ or $1_{\mathbf{B}}$. Define

$$U = \{u \in A^{m-1} \mid \delta|_{A \times \{u\}} = 1_{A \times \{u\}}\}.$$

We cannot have $U = \emptyset$, for this is equivalent to $\delta = 0$. We cannot have $U = A^{m-1}$, for this is equivalent to $\delta = \eta_i'$. Hence there is a $u \in U$ and a $v \in A^{m-1} - U$. If $u = (b_1, \ldots, b_{m-1})$ and $v = (c_1, \ldots, c_{m-1})$, then by sequentially changing each b_j to the value c_j in these $(m-1)$-tuples we find that for some i we must have

$$(c_1, \ldots, c_{i-1}, b_i, b_{i+1}, \ldots, b_{m-1}) \in U \quad \text{and} \quad (c_1, \ldots, c_{i-1}, c_i, b_{i+1}, \ldots, b_{m-1}) \notin U.$$

Consider the subalgebra of $\mathbf{C} \leq \mathbf{A}^m$ which has universe

$$A \times \{c_1\} \times \cdots \times \{c_{i-1}\} \times A \times \{b_{i+1}\} \times \cdots \times \{b_{m-1}\}.$$

Note that $\mathbf{C} \cong \mathbf{A}^2$ via projection onto coordinates 0 and i. Furthermore, $\delta|_{\mathbf{C}}$ is a congruence which has

$$A \times \{c_1\} \times \cdots \times \{c_{i-1}\} \times \{b_i\} \times \{b_{i+1}\} \times \cdots \times \{b_{m-1}\}$$

as a congruence class, but not

$$A \times \{c_1\} \times \cdots \times \{c_{i-1}\} \times \{c_i\} \times \{b_{i+1}\} \times \cdots \times \{b_{m-1}\}.$$

It follows that

$$0_{\mathbf{C}} < \delta|_{\mathbf{C}} < \eta_0'|_{\mathbf{C}}.$$

From the isomorphism $\mathbf{C} \cong \mathbf{A}^2$ we deduce that $\mathbf{A}^2$ has a skew congruence which lies below a projection kernel. By the minimality assumption on m, it must be that $m = 2$. We proceed with the knowledge that $m = 2$ retaining the definitions of δ and U from above. Note that $U \subseteq A^{m-1} = A$ now.

Claim. If $t(x, \bar{y})$ is an $(n+1)$-ary term which depends on x in $\mathbf{A}$ and $\bar{s} \in A^n$, then $t^{\mathbf{A}}(U, \bar{s}) \subseteq U$.

Proof of Claim: Assume otherwise that $t(x, \bar{y})$ depends on x in $\mathbf{A}$ and for some $u \in U$ we have $t^{\mathbf{A}}(u, \bar{s}) = v \notin U$. Since $t^{\mathbf{A}}(x, \bar{y})$ depends on x, there is some $\bar{r} \in A^n$ such that $t^{\mathbf{A}}(x, \bar{r})$ is non-constant. Recall that $u \in U$ means precisely that $A \times \{u\}$ is a nontrivial δ-class. Applying the polynomial $t^{\mathbf{A} \times \mathbf{A}}(x, \overline{(r_i, s_i)})$ to this class, we obtain that the set

$$t^{\mathbf{A} \times \mathbf{A}}(A \times \{u\}, \overline{(r_i, s_i)}) = t^{\mathbf{A}}(A, \bar{r}) \times \{v\}$$

is a nontrivial subset contained in $A \times \{v\}$. Since this subset is a polynomial image of a δ-class, δ restricts nontrivially to $A \times \{v\}$. As argued above, this implies that $A \times \{v\}$ is a δ-class. But this forces $v \in U$ which is contrary to our assumption. This proves the Claim.

The Claim just proven implies that U is a class of a congruence on $\mathbf{A}$. Since we have shown in the first paragraph of the argument for Case 1 that $U \neq A^{m-1} = A$. we cannot have U a class of the universal congruence, 1. Since $\mathbf{A}$ is simple, U is a 0-class. Thus $U = \{0\}$ for some element $0 \in A$. By the Claim, if $t(x, \bar{y})$ depends on x in $\mathbf{A}$ and $\bar{s} \in A^n$, then $t^{\mathbf{A}}(0, \bar{s}) = 0$. This is precisely what it means for 0 to be an absorbing element for $\mathbf{A}$. By Lemma 2.1 and our assumption that $|A| > 2$, 0 is the unique absorbing element for $\mathbf{A}$. Hence, any algebra in Case 1 has a unique absorbing element and therefore belongs to class of algebras of type (a) of the theorem.

PROOF FOR CASE 2: Now assume that $\mathbf{A}^m$ has no congruence δ such that $0 < \delta < \eta_0'$. Since $\theta \not\geq \eta_0'$. we have $\theta \wedge \eta_0' < \eta_0'$. Because we are in Case 2, this means that $\theta \wedge \eta_0' = 0$. Hence $C(\theta, \eta_0')$ and so $\theta \leq (0 : \eta_0')$. As η_0 and η_0' are complements, we also have $\eta_0 \wedge \eta_0' = 0$ and so $C(\eta_0, \eta_0')$ also holds. Thus $\eta_0 \leq (0 : \eta_0')$. Altogether this implies that $\theta \vee \eta_0 \leq (0 : \eta_0')$. But $\eta_0 \not\geq \theta$, by choice, and $\eta_0 \prec 1$ since $\mathbf{A}$ is simple. Hence $\theta \vee \eta_0 = 1$ and so $(0 : \eta_0') = 1$ and it follows that $C(1, \eta_0')$ holds. If $\mathbf{B} = \mathbf{A} \times \{u\}$, $u \in A^{m-1}$, then $\mathbf{B}$ is a subalgebra of $\mathbf{A}^m$ isomorphic to $\mathbf{A}$ as we explained in the argument for Case 1. Now, restricting congruences to $\mathbf{B}$ we get that $C(1_{\mathbf{A}}|_{\mathbf{B}}, \eta_0'|_{\mathbf{B}})$ holds. But

$$(1_{\mathbf{A}})|_{\mathbf{B}} = 1_{\mathbf{B}} = \eta_0'|_{\mathbf{B}},$$

so $C(1_{\mathbf{B}}, 1_{\mathbf{B}})$ holds and $\mathbf{B}$ is abelian. Since $\mathbf{A}$ is isomorphic to $\mathbf{B}$, $\mathbf{A}$ is abelian, too.

Our arguments have shown that a skew congruence below some η_0' implies the existence of an absorbing element for $\mathbf{A}$. If a finite power of $\mathbf{A}$ has a skew congruence, but has no skew congruence below any η_0', then $\mathbf{A}$ is abelian. Hence any idempotent simple algebra $\mathbf{A}$ must belong to at least one of the classes described in the theorem. Now we show that no idempotent simple algebra belongs to two classes.

If $\mathbf{A}$ has an absorbing element 0. then $\mathbf{A}^2$ has an **ideal congruence** which is skew. That is, for $I = A \times \{0\}$ it is the case that $\theta = \mathrm{Cg}(I \times I)$ has I as its only nontrivial congruence class. This congruence satisfies $0 < \theta < \eta_1$ and is therefore skew. If $\mathbf{A}$ is abelian, then $\mathbf{A}^2$ has a congruence which has the diagonal $D = \{(x, x)|x \in A\}$ as a congruence class. This congruence is clearly distinct from the product congruences $0, \eta_0, \eta_1$ and 1, so it is skew. Hence any algebra in classes (a) or (b) cannot belong to class (c). To show that classes (a) and (b) are disjoint, assume that $\mathbf{A}$ is abelian and has an absorbing element

$0 \in A$. Since we have assumed that $|A| > 2$ we can also assume, by Lemma 2.1, that $\mathbf{A}$ is not essentially unary. Let $t(x, y, \bar{z})$ be a term depending on x and y in $\mathbf{A}$ and choose $a \in A - \{0\}$. 0 is an absorbing element, so

$$t^{\mathbf{A}}(\underline{0}, 0, a, \ldots, a) = 0 = t^{\mathbf{A}}(\underline{0}, a, a, \ldots, a).$$

Changing the underlined occurrences of 0 to a we obtain

$$t^{\mathbf{A}}(a, 0, a, \ldots, a) = 0 \neq a = t^{\mathbf{A}}(a, a, a, \ldots, a).$$

This is a failure of the term condition. It contradicts our assumption that $\mathbf{A}$ is abelian. We conclude that classes (a) and (b) of the theorem are disjoint. This completes the proof of the theorem. $\square$

COROLLARY 2.4 *If* $\mathbf{A}$ *is an idempotent simple algebra and* $\mathbf{A}^2$ *is skew-free, then every finite power of* $\mathbf{A}$ *is skew-free.*

Proof: Theorem 2.3 shows that if some finite power of $\mathbf{A}$ has a skew congruence, then $\mathbf{A}$ has an absorbing element or is abelian. In the former case, $\mathbf{A}^2$ has skew congruences associated with ideal congruences while in the latter case $\mathbf{A}^2$ has a skew congruence associated with the "diagonal congruence". Hence if some finite power of $\mathbf{A}$ has a skew congruence, then $\mathbf{A}^2$ has a skew congruence. $\square$

3   An Affine Module Embedding

In [3], C. Herrmann proves that every abelian algebra in a congruence modular variety is affine. This implies that an idempotent abelian algebra which generates a congruence modular variety is an affine module. Starting with an abelian algebra, Herrmann constructs a 1-1 function into an affine algebra in the first half of his paper. In the second half, he proves that this function is an isotopy. For idempotent algebras, his function is an isomorphism. Herrmann's proof depends quite heavily on congruence modularity, but we shall see in this section that it is possible to filter out much of this dependency in the first half of his proof. In the few remaining places where it seems impossible to avoid at least some modularity requirement, we shall see that simplicity and idempotence can substitute for modularity. In this way we will have modified the first half of his proof to produce an embedding into an affine module. Hence we prove that any idempotent simple algebra which is abelian is a subreduct of an affine module.

For this section, **A** denotes a simple, idempotent, abelian algebra of more than two elements. We single out a part of the proof of Theorem 2.3 that we shall need later.

LEMMA 3.1 *If η_i, $i = 0, 1$ are the kernels of the coordinate projections of* $\mathbf{A}^2$, *then* $0 \prec \eta_i$.

Proof: In the proof of Case 1 of Theorem 2.3 we show that when $|A| > 2$ and there is a congruence $0 < \delta < \eta_i$, then **A** has a unique absorbing element. We show later in the proof of Theorem 2.3 that such an algebra cannot be abelian. $\square$

LEMMA 3.2 *There is a largest proper congruence Δ on $\mathbf{A} \times \mathbf{A}$ which contains $D = \{(x, x) \mid x \in A\}$ in a Δ-class. This Δ has D as a congruence class and satisfies $\Delta\eta_0 = \Delta\eta_1 = 0$. Furthermore, $\mathbf{B} = (\mathbf{A} \times \mathbf{A})/\Delta$ is simple and an extension of* **A**.

Proof: Saying that D is equal to a class of some congruence is just another way of saying that **A** is abelian. Hence, there *is* a proper congruence which contains D in a congruence class. Now suppose that there is some congruence θ which properly contains D in a θ-class. Since the containment is proper, we can find $a \neq b$ in A such that $(a, b) \, \theta \, (a, a)$ and so $\theta\eta_0 > 0$. By Lemma 3.1 and the fact that **A** is simple, each η_i is a coatom and also an atom in $\mathbf{Con}\mathbf{A}^2$. It follows $\theta = \eta_0$ or else $\theta = 1$. Since $D \times D \subseteq \theta$ and $D \times D \nsubseteq \eta_0$, we have $\theta = 1$. Hence for any proper congruence γ on $\mathbf{A} \times \mathbf{A}$ we have that γ contains D in a γ-class iff D is a γ-class.

Let Σ be the set of all congruences on $\mathbf{A} \times \mathbf{A}$ which have D as a congruence class. Define Δ to equal the join of all members of Σ. Clearly Δ has D as a congruence class, so Δ is the largest element in Σ. Since Δ is the largest congruence which has D as a congruence class and no proper congruence contains D in a congruence class we have $\Delta \prec 1$ in $\mathbf{Con}\, \mathbf{A} \times \mathbf{A}$. Hence $\mathbf{B} = (\mathbf{A} \times \mathbf{A})/\Delta$ is simple. Since $\eta_i \nleq \Delta$, clearly, and each η_i is an atom, we must have

$$\Delta\eta_0 = \Delta\eta_1 = 0.$$

In this paragraph we have justified all the claims of the lemma except that we haven't yet shown that **B** is an extension of **A**.

To show that **A** embeds in **B**, choose an element $0 \in A$ and consider the function $\mathbf{A} \to \mathbf{B}$ defined by $x \mapsto (x, 0)/\Delta$. This function is a homomorphism since **A** is idempotent and it is an embedding since $\Delta\eta_0 = 0$. $\square$

We shall think of the fourth power of $\mathbf{A}$ as 2×2 matrices of elements of $\mathbf{A}$ and write $\mathbf{A}^{2 \times 2}$ to denote this. We order the entries as follows:

$$\begin{bmatrix} 0 & 1 \\ 2 & 3 \end{bmatrix}$$

According to our conventions this means that

$$\pi_{01}\left(\begin{bmatrix} a & b \\ c & d \end{bmatrix}\right) = (a, b) \text{ while } \pi_{10}\left(\begin{bmatrix} a & b \\ c & d \end{bmatrix}\right) = (b, a).$$

Thus Δ_{01}, for example, is the congruence on $\mathbf{A}^{2 \times 2}$ which contains all pairs of 2×2 matrices of the form

$$\left\langle \begin{bmatrix} a & b \\ c & d \end{bmatrix}, \begin{bmatrix} e & f \\ g & h \end{bmatrix} \right\rangle, \quad \langle (a, b), (e, f) \rangle \in \Delta.$$

LEMMA 3.3 *For Δ as described in Lemma 3.2, the following properties hold.*

(i) $(x, y) \, \Delta \, (u, v) \Leftrightarrow (y, x) \, \Delta \, (v, u)$.

(ii) $(x, y) \, \Delta \, (u, v) \Leftrightarrow (x, u) \, \Delta \, (y, v)$.

Proof: Let † denote the canonical involutory automorphism of $\mathbf{A}^2$. If one applies † coordinatewise to $\mathbf{A}^{2 \times 2} = (\mathbf{A}^2)^2$, then one obtains an automorphism of $\mathbf{A}^{2 \times 2}$ which permutes the subalgebras of $\mathbf{A}^{2 \times 2}$. This automorphism permutes the congruences of $\mathbf{A}^2$ considered as subalgebras. We shall use † to denote both the described automorphism of $\mathbf{A}^2$ and the induced permutation of congruences of $\mathbf{A}^2$.

Clearly $\Delta^\dagger$ is a congruence on $\mathbf{A}^2$ which contains the diagonal as a congruence class. It follows that $\Delta^\dagger \subseteq \Delta$ by Lemma 3.2. But now applying † to both sides of this inclusion we get $\Delta^{\dagger\dagger} \subseteq \Delta^\dagger$ and therefore $\Delta = \Delta^{\dagger\dagger} \subseteq \Delta^\dagger \subseteq \Delta$. Hence $\Delta = \Delta^\dagger$ which means precisely that (i) holds.

Now let ‡ denote the automorphism of $\mathbf{A}^{2 \times 2}$ which may be described as "transposing matrices" or equivalently as "switching coordinates 1 and 2". Viewing Δ as a subset of $\mathbf{A}^{2 \times 2}$, part (ii) of the lemma is the statement that $\Delta^\ddagger = \Delta$. By the same argument we used in the last paragraph, to prove that $\Delta^\ddagger = \Delta$ it will suffice for us to prove that $\Delta^\ddagger \subseteq \Delta$.

$\Delta^\ddagger$ is a subalgebra of $\mathbf{A}^{2 \times 2}$ since Δ is. Since Δ satisfies part (i) of the lemma, $\Delta^\ddagger$ is a symmetric relation. Since $D \times D \subseteq \Delta$, we get that $\Delta^\ddagger$ is reflexive and since Δ is reflexive, we get $D \times D \subseteq \Delta^\ddagger$. Hence $\Delta^\ddagger$ is a reflexive, symmetric, compatible relation on $\mathbf{A}^2$ which contains $D \times D$. Let θ denote the transitive closure of $\Delta^{\ddagger}$. θ is a congruence on $\mathbf{A}^2$ which contains $D \times D$ and also contains $\Delta^{\ddagger}$. Since $\Delta\eta_0 = 0$, we get

$$(a, b) \,\Delta\, (a, c) \Rightarrow b = c.$$

Hence

$$(a, a) \,\Delta^{\ddagger}\, (b, c) \Rightarrow b = c.$$

From this it follows that

$$(a, a) \,\theta\, (b, c) \Rightarrow b = c.$$

Thus, θ is a proper congruence on $\mathbf{A}^2$ and D is a θ-class. By the maximality of Δ, we get that $\Delta^{\ddagger} \subseteq \theta \subseteq \Delta$ which implies that $\Delta^{\ddagger} = \Delta$. Thus (ii) holds. $\square$

LEMMA 3.4 *The following hold.*

I. $\mathbf{A}^2/\Delta$ *is abelian.*

II. If χ is a coatom of $\mathbf{Con}\ \mathbf{A}^{2 \times 2}$, then the following conditions are equivalent.

(i) $\Delta_{01}\Delta_{23} \leq \chi$ *and χ contains some pair of the form*

$$\langle M, N \rangle = \left\langle \begin{bmatrix} x & y \\ x & y \end{bmatrix}, \begin{bmatrix} u & z \\ u & z \end{bmatrix} \right\rangle$$

where $\langle (x, y), (u, z) \rangle \notin \Delta$.

(ii) $\Delta_{02}\Delta_{13} \leq \chi$ *and χ contains some pair of the form*

$$\langle P, Q \rangle = \left\langle \begin{bmatrix} x & x \\ y & y \end{bmatrix}, \begin{bmatrix} u & u \\ z & z \end{bmatrix} \right\rangle$$

where $\langle (x, y), (u, z) \rangle \notin \Delta$.

(iii) $\Delta_{01}\Delta_{23} + \Delta_{02}\Delta_{13} \leq \chi$.

III. There is a unique congruence χ on $\mathbf{A}^{2 \times 2}$ which is a coatom of $\mathbf{Con}\ \mathbf{A}^{2 \times 2}$ and satisfies the equivalent conditions in part II.

Proof: If $\mathbf{B} = \mathbf{A}^2/\Delta$, then $\mathbf{B}$ is simple by Lemma 3.2 and $\mathbf{B} \times \mathbf{B}$ is naturally isomorphic to $\mathbf{A}^{2 \times 2}/(\Delta_{01}\Delta_{23})$. Throughout this proof we shall identify $\mathbf{B} \times \mathbf{B}$ and $\mathbf{A}^{2 \times 2}/(\Delta_{01}\Delta_{23})$. Choose any pair $\langle M, N \rangle$ as described in items $II(ii)$ above. Define

$$\psi = [(\Delta_{01}\Delta_{23}) + \mathrm{Cg}(M, N)]/(\Delta_{01}\Delta_{23})$$

and let $\theta = (\Delta_{01}\Delta_{23} + \Delta_{02}\Delta_{13})/(\Delta_{01}\Delta_{23})$. Since $\langle M, N \rangle \in \Delta_{02}\Delta_{13}$ we have $\psi \leq \theta$. We shall prove that $\psi \leq \theta < 1$ in **Con B^2** and that there is a largest proper congruence in the interval $[\psi, 1]$ of **Con B^2**. Together, these two facts imply that the largest proper congruence above ψ is also the largest proper congruence above θ. This will prove that in **Con $A^{2\times 2}$**, a coatom satisfies $II(i)$ if and only if it satisfies $II(iii)$. By symmetry, we will get that a coatom satisfies $II(ii)$ if and only if it satisfies $II(iii)$. This will prove II. Along the way we will establish parts I and III of the lemma.

If $E = \{(x, x) | x \in B\}$ is the diagonal of B^2, then the congruence ψ on B^2 contains $E \times E$. The reason for this is that ψ contains the pair

$$\langle M/\Delta_{01}\Delta_{23}, N/\Delta_{01}\Delta_{23} \rangle = \left\langle \left(\begin{array}{c} (x\ y)/\Delta \\ (x\ y)/\Delta \end{array} \right), \left(\begin{array}{c} (u\ v)/\Delta \\ (u\ v)/\Delta \end{array} \right) \right\rangle$$

which is a pair of elements of E. Our assumption that $\langle (x, y), (u, z) \rangle \notin \Delta$ means that this pair is a pair of distinct elements of E. The fact that the diagonal subalgebra $E \leq B^2$ is simple (since it is isomorphic to B) together with $\psi|_E \neq 0$ implies that $E \times E \subseteq \psi \leq \theta$.

Next we prove that in **Con B^2** we have $\theta < 1$. This is equivalent to the claim that

$$\Delta_{01}\Delta_{23} + \Delta_{02}\Delta_{13} < 1$$

in **Con $A^{2\times 2}$**. We prove this by exhibiting a proper subset of $A^{2\times 2}$ which is a union of $(\Delta_{01}\Delta_{23} + \Delta_{02}\Delta_{13})$-classes. That subset is $\Delta \subseteq A^{2\times 2}$. Δ is a proper subset of $A^{2\times 2}$ since $\Delta < 1$ in **Con A^2**. We first argue that Δ, as a subset of $A^{2\times 2}$, is a union of $\Delta_{01}\Delta_{23}$-classes. Applying the automorphism $\ddagger$ to this statement then yields that $\Delta^{\ddagger}$ is a union of $\Delta_{02}\Delta_{13}$-classes. But $\Delta = \Delta^{\ddagger}$ by Lemma 3.3 (ii). Hence it will follow that Δ is a union of $(\Delta_{01}\Delta_{23} + \Delta_{02}\Delta_{13})$-classes.

To see that Δ is a union of $\Delta_{01}\Delta_{23}$-classes, take matrices

$$R = \left[\begin{array}{cc} a & b \\ c & d \end{array} \right], \ S = \left[\begin{array}{cc} e & f \\ g & h \end{array} \right]$$

such that $R \in \Delta$ and $\langle R, S \rangle \in \Delta_{01}\Delta_{23}$. The fact that $R \in \Delta$ means precisely that $(a, b) \equiv_{\Delta} (c, d)$. The fact that $\langle R, S \rangle \in \Delta_{01}\Delta_{23}$ means precisely that $(a, b) \equiv_{\Delta} (e, f)$ and $(c, d) \equiv_{\Delta} (g, h)$. By transitivity, we get $(e, f) \equiv_{\Delta} (g, h)$, so $S \in \Delta$. As explained in the last paragraph, this proves that $\Delta_{01}\Delta_{23} + \Delta_{02}\Delta_{13} < 1$ in **Con $A^{2\times 2}$** and so $\theta < 1$ in **Con B^2**.

We now prove that in **Con B^2** we have $\theta\eta_0 = \theta\eta_1 = 0$. Suppose instead that for some $b \neq c$ we have $\langle (a, b), (a, c) \rangle \in \theta\eta_0$. Then θ restricts nontrivially to the simple subalgebra $F = \{a\} \times B$. Hence $F \times F \subseteq \theta$. But now for every $x, y \in B - \{a\}$ we have $(a, x) \equiv_{\theta} (a, a) \equiv_{\theta} (x, x)$ and $(a, y) \equiv_{\theta} (a, a) \equiv_{\theta} (y, y)$, so the same type of argument shows that $G \times G, H \times H \subseteq \theta$ when $\mathbf{G} = \mathbf{B} \times \{x\}$ and $\mathbf{H} = \mathbf{B} \times \{y\}$ if $x \neq a \neq y$. Using the fact that $|B| \geq |A| > 2$, we choose $x, y \in B$ so that $|\{a, x, y\}| = 3$ and we get that θ restricts nontrivially to every subalgebra of the form $\{u\} \times \mathbf{B}$ since θ contains $\langle (u, x), (u, y) \rangle$. This implies that $\theta = 1$ which we know to be false. Our conclusion is that $\theta \eta_0 = 0$. Similarly $\theta \eta_1 = 0$.

We have shown that $\mathbf{B}^2$ has congruences ψ and θ where $E \times E \subseteq \psi \leq \theta < 1$ for E equal to the diagonal of $\mathbf{B}^2$. We have also shown that $\theta \eta_0 = \theta \eta_1 = 0$. These conditions imply that E is a θ-class (and also a ψ-class), so $\mathbf{B}$ is abelian. This establishes part I of the lemma. Applying Lemma 3.2 to $\mathbf{B}$ we see that there is a largest proper congruence $\Delta^{\mathbf{B}}$ on $\mathbf{B}^2$ which contains E in a congruence class. Lemma 3.2 proves that $\Delta^{\mathbf{B}} \prec 1$. We have $\psi \leq \theta \leq \Delta^{\mathbf{B}} \prec 1$ since θ is a proper congruence which contains E in a congruence class. $\Delta^{\mathbf{B}}$ is the largest proper congruence in $\mathbf{Con}\ \mathbf{B}^2$ which has E as a congruence class, so $\Delta^{\mathbf{B}}$ is the largest proper congruence above ψ and is also the largest proper congruence above θ. Hence a coatom in $\mathbf{Con}\ \mathbf{B}^2$ is above ψ if and only if it is above θ. This holds for the coimages of ψ and θ in $\mathbf{Con}\ \mathbf{A}^{2 \times 2}$, so a coatom satisfies condition $II(i)$ if and only if it satisfies condition $II(iii)$. By symmetry we get that II holds.

For III, our only choice is to take χ to be the coimage of $\Delta^{\mathbf{B}}$. From the equivalence of the conditions in II, it doesn't matter if we take this coimage with respect to the natural map

$$\mathbf{A}^{2 \times 2} \to \mathbf{A}^{2 \times 2} / (\Delta_{01} \Delta_{23}) \cong \mathbf{B}^2$$

or with respect to the natural map

$$\mathbf{A}^{2 \times 2} \to \mathbf{A}^{2 \times 2} / (\Delta_{02} \Delta_{13}) \cong \mathbf{B}^2.$$

In either case we get the largest proper congruence of $\mathbf{A}^{2 \times 2}$ above $\Delta_{01} \Delta_{23} + \Delta_{02} \Delta_{13}$. $\square$

Now we define a sequence of algebras and embeddings between them.

(*i*) Set $\mathbf{A}_0 = \mathbf{A}$ and choose $0_0 \in A_0$.

(*ii*) Let $\mathbf{A}_{n+1} = (\mathbf{A}_n \times \mathbf{A}_n) / \Delta^{\mathbf{A}_n}$, $0_{n+1} = (0_n, 0_n) / \Delta^{\mathbf{A}_n}$.

(*iii*) Let $\delta_n : \mathbf{A}_n \to \mathbf{A}_n^2 : x \mapsto (x, 0_n)$.

(*iv*) Let $t_n : \mathbf{A}_n^2 \to \mathbf{A}_{n+1} : (x, y) \mapsto (x, y) / \Delta^{\mathbf{A}_n}$. (The canonical map.)

(*v*) Let $\epsilon_n = t_n \circ \delta_n : \mathbf{A}_n \to \mathbf{A}_{n+1}$. Let $\epsilon_{ij} = \epsilon_{j-1} \circ \cdots \circ \epsilon_{i+1} \circ \epsilon_i$ if $j > i$ and let $\epsilon_{ii} = \mathrm{id}_{A_i}$.

LEMMA 3.5 *If* $\mathcal{V} = \mathcal{V}(\mathbf{A})$, *then the following hold.*

(i) $\mathbf{A}_n \in \mathcal{V}$ *for all* n.

(ii) $\epsilon_n : \mathbf{A}_n \to \mathbf{A}_{n+1}$ *is an embedding for all* n.

(iii) *Each* $\mathbf{A}_n$ *is simple and abelian.*

(iv) *For all* $a, b, c, d \in A_n$,

$$t_{n+1}(t_n(a,b), t_n(c,d)) = t_{n+1}(t_n(a,c), t_n(b,d)).$$

Proof: Item (i) is immediate from the fact that $\mathcal{V} = \mathbf{HSP}(\mathcal{V})$. Item ($ii$) follows from induction based on our proof in Lemma 3.2 that the function

$$\mathbf{A} \to (\mathbf{A} \times \mathbf{A})/\Delta : x \mapsto (x,0)/\Delta$$

is an embedding. Item (iii) follows by induction from our proofs that $(\mathbf{A} \times \mathbf{A})/\Delta$ is simple (Lemma 3.2) and abelian (Lemma 3.4).

Item (iv) is the statement that the following diagram commutes:

$$
\begin{array}{ccc}
\mathbf{A}_n^{2\times2} & \xrightarrow{\;t_n\;} & \mathbf{A}_{n+1}^2 \\
t_n \downarrow & & t_{n+1} \downarrow \\
\mathbf{A}_{n+1}^2 & \xrightarrow{\;t_{n+1}\;} & \mathbf{A}_{n+2}
\end{array}
$$

The map across the top is t_n applied to the rows of a matrix in $\mathbf{A}_n^{2\times2}$. The map on the left is t_n applied to the columns of a matrix in $\mathbf{A}_n^{2\times2}$. Since the composite maps in both directions are the canonical maps, it suffices to prove that they have the same kernel. But the kernels of these two composite maps are each coatoms of **Con** $\mathbf{A}^{2\times2}$ since $\mathbf{A}_{n+2}$ is simple. The map across the top and down the right side has a kernel which satisfies condition $II(i)$ of Lemma 3.4 while the map down the left side and across the bottom has a kernel which satisfies $II(ii)$ of Lemma 3.4. The result of Lemma 3.4 proves that these kernels are the same. □

We now define the direct limit $\hat{\mathbf{A}}$ in $\mathcal{V}$ of the system $\{\epsilon_n : \mathbf{A}_n \to \mathbf{A}_{n+1}\}$ in a concrete way. Let X denote the set of partial functions from the natural numbers into $\bigcup A_n$ which satisfy the following conditions.

1. If $a \in X$ and $n \in \mathrm{dom}\, a$, then $a(n) \in A_n$.

2. If $a \in X$ and $n \in \mathrm{dom}\, a$, then $n + 1 \in \mathrm{dom}\, a$ and $a(n+1) = \epsilon_n(a(n))$.

Let Θ be the equivalence relation on X defined by $a \equiv_\Theta b$ if $a(n) = b(n)$ for all $n \in (\text{dom } a \cap \text{dom } b)$ (equivalently, for some $n \in (\text{dom } a \cap \text{dom } b)$). The underlying set of $\hat{\mathbf{A}}$ will be taken to be $\hat{A} = X/\Theta$.

If f is an m-ary $\mathcal{V}$-operation, then there is an induced operation $f^{\hat{\mathbf{A}}}$ on $\hat{A}$ which we define as follows. For $a_0, \ldots, a_{m-1} \in X$, let $f^{\hat{\mathbf{A}}}(a_0/\Theta, \ldots, a_{m-1}/\Theta)$ be the Θ-class of the partial function whose domain is $\cap \text{dom } a_i$ and which is defined by

$$f^{\hat{\mathbf{A}}}(a_0, \ldots, a_{m-1})(n) \overset{\text{def}}{=} f^{\mathbf{A}_n}(a_0(n), \ldots, a_{m-1}(n))$$

for $n \in \cap \text{dom } a_i$. Now we make some further definitions.

(i) Let $i_m : \mathbf{A}_m \to \hat{\mathbf{A}}$ be the function which assigns to each $a \in A_m$ the Θ-class of the partial function $i_m(a)$ defined by $i_m(a)(n) = \epsilon_{mn}(a)$ whenever $n \geq m$.

(ii) Let $0 \in \hat{A}$ to be the Θ-class of the everywhere-defined function $0(n) = 0_n$.

(iii) Define $t : \hat{\mathbf{A}}^2 \to \hat{\mathbf{A}}$ as follows. For $R, S \in \hat{A}$ choose $r \in R$ and $s \in S$. Let t be the function that assigns to the pair (R, S) the Θ-class of the partial function $t(R, S)$ defined by $t(R, S)(n + 1) = t_n(r(n), s(n))$ whenever $n \in (\text{dom } r \cap \text{dom } s)$.

LEMMA 3.6 *If $\mathcal{V} = \mathcal{V}(\mathbf{A})$, then the following hold.*

(i) *If f is a $\mathcal{V}$-operation, then $f^{\hat{\mathbf{A}}}$ is a well-defined operation on $\hat{A}$.*

(ii) *$i_n : \mathbf{A}_n \to \hat{\mathbf{A}}$ is an embedding for all n and $\hat{A} = \bigcup i_n(A_n)$.*

(iii) *$\hat{\mathbf{A}} \in \mathcal{V}$.*

(iv) *$\hat{\mathbf{A}}$ is simple and abelian.*

(v) *0 is a well-defined element of $\hat{A}$ and t is a well-defined binary operation on $\hat{\mathbf{A}}$ and the following equations hold:*

(a) $t(x, 0) = x$.

(b) $t(x, x) = 0$.

(c) $t(t(x, y), t(z, u)) = t(t(x, z), t(y, u))$.

(vi) *$t : \hat{\mathbf{A}}^2 \to \hat{\mathbf{A}}$ is a homomorphism.*

Proof: To show that $f^{\hat{\mathbf{A}}}$ is a well-defined function, we need to show that if $a_i \in X$, $i < m$, then the partial function defined by

$$f^{\hat{\mathbf{A}}}(a_0, \ldots, a_{m-1})(n) = f^{\mathbf{A}_n}(a_0(n), \ldots, a_{m-1}(n))$$

is also in X and that if $a_i \equiv_\Theta a'_i$, $i < m$, then $f^{\hat{\mathbf{A}}}(\bar{a}) \equiv_\Theta f^{\hat{\mathbf{A}}}(\bar{a}')$. The first part requires showing that

$$\begin{aligned}
\epsilon_n(f^{\mathbf{A}_n}(a_0(n), \ldots, a_{m-1}(n))) &= f^{\mathbf{A}_{n+1}}(a_0(n+1), \ldots, a_{m-1}(n+1)) \\
&= f^{\mathbf{A}_{n+1}}(\epsilon_n(a_0(n)), \ldots, \epsilon_n(a_{m-1}(n))).
\end{aligned}$$

This is just the statement that ϵ_n is a homomorphism which is immediate from its definition. The second part follows immediately, since $a_i \equiv_\Theta a'_i$ means $a_i(m) = a'_i(m)$ wherever they are both defined, so $f^{\hat{\mathbf{A}}}(\bar{a})(m) = f^{\hat{\mathbf{A}}}(\bar{a}')(m)$ wherever they are both defined.

To show (ii) we first need to show that for $a \in A_m$ the partial function $i_m(a)$ is in X. This means that we must show that $\epsilon_n(i_m(a)(n)) = i_m(a)(n+1)$ for $n \geq m$. This just requires observing that $\epsilon_n \circ \epsilon_{mn}(a) = \epsilon_{mn+1}(a)$. Now we must show that each i_m is a 1-1 homomorphism. To see that i_m is 1-1, it suffices to note that if $a \neq b$ in $\mathbf{A}_m$, then $i_m(a)(m) = a \neq b = i_m(b)(m)$, so $i_m(a)$ and $i_m(b)$ differ on a natural number that belongs to their common domain. Thus, $i_m(a) \not\equiv_\Theta i_m(b)$. Checking that i_m is a homomorphism reduces to checking that

$$i_m(f^{\mathbf{A}_m}(a_0, \ldots, a_k))(m) = f^{\mathbf{A}_m}(i_m(a_0)(m), \ldots, i_m(a_k)(m)).$$

Clearly both sides equal $f^{\mathbf{A}_m}(a_0, \ldots, a_k)$. To finish (ii) we must show that $\hat{A} = \bigcup i_n(A_n)$. For this it is enough to show that for each $a \in X$ there is an $n < \omega$ and an element $b \in A_n$ such that $i_n(b)(n) = a(n)$. (Then we will have $i_n(b) = a$.) We may simply choose $b = a(n)$.

To show that $\hat{\mathbf{A}} \in \mathcal{V}$ it suffices to observe that a variety fails to contain $\hat{\mathbf{A}}$ if and only if it fails to contain some finitely generated subalgebra of $\hat{\mathbf{A}}$. By (ii), $\hat{A} = \bigcup A_n$, so any such variety will also fail to contain some $\mathbf{A}_n$. $\mathcal{V}$ contains all $\mathbf{A}_n$, so $\mathcal{V}$ contains $\hat{\mathbf{A}}$.

Statement (iv) follows from (ii), since any algebra which is the nested union of simple or abelian subalgebras has the same properties. (The union $\hat{A} = \bigcup A_n$ is nested since $i_n = i_{n+1} \circ \epsilon_n$. Hence the image of i_n is contained in the image of i_{n+1}.)

0 is a well-defined element of $\hat{A}$, since $\epsilon_n(0_n) = t_n(0_n, 0_n) = 0_{n+1}$. To see that t is well-defined, we must first show that for $R, S \in \hat{A}$ and $r \in R$ and $s \in S$ we have $t(R, S) \in X$ where $t(R, S)(n+1) = t_n(r(n), s(n))$ whenever $n \in (\text{dom } r \cap \text{dom } s)$. Then we need to verify that the Θ-class of $t(R, S)$

doesn't depend on our choice of r and s. The first part follows from Lemma 3.5 (iv) as we show here.

$$
\begin{aligned}
\epsilon_{n+1}(t(R,S)(n+1)) &= \epsilon_{n+1}(t_n(r(n),s(n))) \\
&= t_{n+1} \circ \delta_{n+1} \circ t_n(r(n),s(n)) \\
&= t_{n+1}(t_n(r(n),s(n)),0_{n+1}) \\
&= t_{n+1}(t_n(r(n),s(n)),t_n(0_n,0_n)) \\
&= t_{n+1}(t_n(r(n),0_n),t_n(s(n),0_n)) \\
&= t_{n+1}(t_n(\delta_n(r(n))),t_n(\delta_n(s(n)))) \\
&= t_{n+1}(\epsilon_n(r(n)),\epsilon_n(s(n))) \\
&= t_{n+1}(r(n+1),s(n+1)) \\
&= t(R,S)(n+2).
\end{aligned}
$$

For the second part of this verification, it is clear that if r and r' agree on their common domain and s and s' agree on their common domain, then $t_n(r(n),s(n)) = t_n(r'(n),s'(n))$ whenever n belongs to the domain of r,r',s and s'.

To show that $t(x,0) = x$, choose $R \in \hat{A}$ and $r \in R$. We must show that $t_n(r(n),0_n) = r(n+1)$ for some (equivalently all) $n \in$ dom r. This is accomplished with

$$
\begin{aligned}
t_n(r(n),0_n) &= t_n(\delta_n(r(n))) \\
&= \epsilon_n(r(n)) \\
&= r(n+1).
\end{aligned}
$$

To show that $t(x,x) = 0$, choose $R \in \hat{A}$ and $r \in R$. We must show that $t_n(r(n),r(n)) = 0_{n+1}$ for some $n \in$ dom r. We show this with: $t_n(r(n),r(n)) = t_n(0_n,0_n) = 0_{n+1}$. Here we used that fact that $(r(n),r(n)) \equiv_{\Delta \mathbf{A}_n} (0_n,0_n)$.

To establish the equation $t(t(x,y),t(z,u)) = t(t(x,z),t(y,u))$ choose $P,Q,R,S \in \hat{A}$ and $p \in P$, $q \in Q$, $r \in R$ and $s \in S$. Then we must show that

$$
t_{n+1}(t_n(p(n),q(n)),t_n(r(n),s(n))) = t_{n+1}(t_n(p(n),r(n)),t_n(q(n),s(n)))
$$

for some n where p,q,r and s are all defined. This equation follows from Lemma 3.5 (iv).

To show that $t : \hat{\mathbf{A}}^2 \to \hat{\mathbf{A}}$ is a homomorphism, we must show that if f is an m-ary fundamental operation, then t and f commute on any $2 \times m$ matrix of elements of $\hat{\mathbf{A}}$. This verification reduces to showing that each $t_n : \mathbf{A}_n^2 \to \mathbf{A}_{n+1}$ is a homomorphism. But by definition t_n is the canonical homomorphism between these two algebras. $\square$

LEMMA 3.7 If $p(x,y,z) = t(x,t(y,z))$, then

(*i*) $p(x, y, y) = x$ and $p(x, x, y) = y$.

(*ii*) *p commutes with itself and with all the operations of* $\hat{\mathbf{A}}$.

(*iii*) $\langle \hat{\mathbf{A}}; p(x, y, z) \rangle$ *is an idempotent, abelian, simple algebra.*

Proof: For (*i*), note that $p(x, y, y) = t(x, t(y, y)) = t(x, 0) = x$ by Lemma 3.6 (*v*)(*a*) and (*v*)(*b*). Using Lemma 3.6 (*v*)(*c*) as well, we get

$$
\begin{aligned}
p(x, x, y) &= t(x, t(x, y)) \\
&= t(t(x, 0), t(x, y)) \\
&= t(t(x, x), t(0, y)) \\
&= t(0, t(0, y)) \\
&= t(t(y, y), t(0, y)) \\
&= t(t(y, 0), t(y, y)) \\
&= t(y, 0) = y.
\end{aligned}
$$

Property (*ii*) follows from the fact that p is in the clone on $\hat{A}$ generated by t and t has the properties mentioned in (*ii*) by Lemma 3.6.

Since $\hat{\mathbf{A}}$ is idempotent and simple and $p(x, y, z)$ is idempotent, the expansion $\langle \hat{\mathbf{A}}; p(x, y, z) \rangle$ is an idempotent simple algebra. This expansion generates a congruence permutable variety by part (*i*) of the lemma. Part (*ii*) proves that $p(x, y, z)$ commutes with all the operations of $\langle \hat{\mathbf{A}}; p(x, y, z) \rangle$, so by Proposition 5.7 of [2] we get that $\langle \hat{\mathbf{A}}; p(x, y, z) \rangle$ is abelian. □

THEOREM 3.8 *If* **A** *is a simple, idempotent, abelian algebra, then there is an embedding* $i : \mathbf{A} \to \hat{\mathbf{A}}$ *where* $\hat{\mathbf{A}} \in \mathcal{V}(\mathbf{A})$ *is a simple reduct of an affine module. Furthermore, the subgroup of this affine module generated by* $i(A)$ *is* $\hat{A}$.

Proof: Let $\langle \hat{\mathbf{A}}; p(x, y, z) \rangle$ be the expansion described in Lemma 3.7 of the simple algebra $\hat{\mathbf{A}}$ of Lemma 3.6. $\langle \hat{\mathbf{A}}; p(x, y, z) \rangle$ is an idempotent, simple, abelian algebra by Lemma 3.7. By Theorem 9.16 of [2] and the fact that $\langle \hat{\mathbf{A}}; p(x, y, z) \rangle$ is idempotent and simple, we get that $\langle \hat{\mathbf{A}}; p(x, y, z) \rangle$ is a simple affine module. It follows that $\hat{\mathbf{A}}$ is a reduct of a simple affine module. Since **A** $= \mathbf{A}_0$ and $i_0 : \mathbf{A}_0 \to \hat{\mathbf{A}}$ is an embedding (by Lemma 3.6 (*ii*)) we have proven the first claim of this theorem (taking $i = i_0$).

We may assume that the ring associated with the module structure on $\langle \hat{\mathbf{A}}; p(x, y, z) \rangle$ acts faithfully. If we take 0 to be the zero element of this module, then the equations $t(x, x) = 0$ and $t(x, 0) = x$ force $t(x, y) = x - y$ in this module. If one now looks back to the definition of t and i_n, one will find that $i_{n+1}(A_{n+1}) = t(i_n(A_n), i_n(A_n))$. Since $\hat{A} = \bigcup i_n(A_n)$, we get that $\hat{A}$ is generated under $t(x, y) = x - y$ by the set $i_0(A_0) = i(A)$. This proves the second claim of the theorem. $\square$

Theorem 1.1 is a consequence of Theorems 2.3 and 3.8.

4   Restrictions on Subalgebras

In this section we refine Theorem 1.1 in the case that **A** has no nontrivial proper subalgebras.

THEOREM 4.1 *If* **A** *is an idempotent algebra with no nontrivial proper subalgebras, then exactly one of the following conditions is true.*

(*i*) **A** *is term equivalent to the 2-element set,*

(*ii*) **A** *is term equivalent to the 2-element semilattice,*

(*iii*) **A** *is a reduct of an affine module and* **A** *is not essentially unary or*

(*iv*) *Every member of* $\mathsf{HSP}_{fin}(\mathbf{A})$ *is congruence distributive.*

   Proof: Our proof shall be quite similar to the proof of Theorem 2.3 except that we shall apply our assumption that **A** has no nontrivial proper subalgebras in appropriate spots in order to strengthen our conclusion. Note that since **A** is idempotent and has no proper nontrivial subalgebras, **A** is simple; for any nontrivial proper congruence class is a subuniverse.

   We omit the argument for the case $|A| = 2$; this case can be handled as we described at the beginning of Section 2. Also, we will only explain why **A** must satisfy "at least one" of the conditions $(i) - (iv)$. The argument that shows that **A** satisfies no more than one of these conditions can be found in the proof of Theorem 2.3.

   We will see that the condition in (iv) can be strengthened to $(iv)'$: If **B** is isomorphic to a subalgebra of $\mathbf{A}^m$ but not isomorphic to a subalgebra of $\mathbf{A}^{m-1}$, then $\mathbf{ConB} \cong \mathbf{2}^m$. From this it follows that every member of $\mathsf{SP}_{fin}(\mathbf{A})$ (and thus every member of $\mathsf{HSP}_{fin}(\mathbf{A})$) is congruence distributive. To prove the theorem it will suffice to prove that if **A** does not satisfy $(iv)'$ and $|A| > 2$, then **A** is a reduct of an affine module. Assume that **A** does not satisfy condition $(iv)'$. Then for some $m > 1$ there is some subalgebra $\mathbf{B} \leq \mathbf{A}^m$ where **B** is not isomorphic to a subalgebra of $\mathbf{A}^{m-1}$, but where $\mathbf{ConB}$ is not isomorphic to $\mathbf{2}^m$. We assume that m is chosen minimally for this. In particular, the representation $\mathbf{B} \leq \mathbf{A}^m$ is irredundant and subdirect. If η_i is the kernel of the coordinate i^{th} projection of $\mathbf{A}^m$, then we use the same symbol,

η_i, to denote the restriction of this congruence to $\mathbf{B}$. The irredundance of the representation $\mathbf{B} \leq \mathbf{A}^m$ implies that $0 < \eta_i'$ for each i and this implies that all product congruences are distinct. Furthermore, the (meet semilattice) of product congruences on $\mathbf{B}$ is order-isomorphic to $\mathbf{2}^m$. It follows that $\mathbf{B}$ has a congruence θ which is not a product congruence. We may choose i so that $\eta_i \not\geq \theta$ since $\theta \neq 0$. Arguing as in the proof of Theorem 2.3, we may assume that $i = 0$ and $\theta \not\geq \eta_0'$. Again we split our argument into two cases. Case 1: $\mathbf{B}$ has a congruence δ such that $0 < \delta < \eta_0'$. Case 2: $\mathbf{B}$ has no congruence δ such that $0 < \delta < \eta_0'$.

PROOF FOR CASE 1: Write $\mathbf{A}^m$ as $\mathbf{A} \times \mathbf{A}^{m-1}$. Each η_0'-class of $\mathbf{B}$ is a subuniverse of the form $(A \times \{c\}) \cap B$ where $c \in A^{m-1}$. Since $A \times \{c\}$ is a subuniverse of $\mathbf{A} \times \mathbf{C}$ which generates an algebra isomorphic to $\mathbf{A}$ and $\mathbf{A}$ has no proper nontrivial subalgebras, it follows that either $A \times \{c\} \subseteq B$ or else $|(A \times \{c\}) \cap B| = 1$. Define

$$C = \{c \in A^{m-1} \mid A \times \{c\} \subseteq B\}.$$

We claim that C is a subuniverse of $\mathbf{A}^{m-1}$. To see this, choose a fundamental operation f, say of arity ℓ, and elements $c_0, \ldots, c_{\ell-1} \in C$. Choose distinct $a, b \in A$. Then from the definition of C we have $(a, c_i), (b, c_i) \in B$ for all i, so

$$p := f((a, c_0), \ldots, (a, c_{\ell-1})) = (a, f(c_0, \ldots, c_{\ell-1})) \in B$$

and similarly $q := (b, f(c_0, \ldots, c_{\ell-1})) \in B$. But p and q are distinct elements of B contained in the subuniverse $A \times \{f(c_0, \ldots, c_{\ell-1})\}$. Since the subalgebra generated by this subuniverse is isomorphic to $\mathbf{A}$ and $\mathbf{A}$ has no proper nontrivial subalgebras, it must be that

$$A \times \{f(c_0, \ldots, c_{\ell-1})\} \subseteq B$$

which means that $f(c_0, \ldots, c_{\ell-1}) \in C$. This proves that C is a subuniverse. Define $U \subseteq C$ by

$$U = \{u \in C \mid \delta|_{A \times \{u\}} = 1_{A \times \{u\}}\}.$$

Since $0 < \delta < \eta_0'$ it follows that $\mathbf{B}$ has a nontrivial δ-class (which equals an η_0'-class) and also $\mathbf{B}$ has a nontrivial η_0'-class which is not a δ-class. The nontrivial η_0'-classes of $\mathbf{B}$ are precisely the sets of the form $A \times \{c\}$ where $c \in C$. Hence $0 < \delta < \eta_0'$ implies that U is a nonempty, proper subset of C.

Claim. If $t(x, \bar{y})$ is an $(n+1)$-ary term which depends on x in $\mathbf{A}$ and $\bar{s} \in C^n$, then $t^{\mathbf{C}}(U, \bar{s}) \subseteq U$.

Proof of Claim: Assume otherwise that $t(x, \bar{y})$ depends on x in $\mathbf{A}$ and for some $u \in U$ we have $t^{\mathbf{C}}(u, \bar{s}) = v \notin U$. Since $t^{\mathbf{A}}(x, \bar{y})$ depends on x, there is some $\bar{r} \in A^n$ such that $t^{\mathbf{A}}(x, \bar{r})$ is non-constant. Recall that $u \in U$ means precisely that $A \times \{u\}$ is a nontrivial δ-class. Applying the polynomial $t^{\mathbf{A} \times \mathbf{C}}(x, \overline{(r_i, s_i)})$ to this class, we obtain that the set

$$t^{\mathbf{A} \times \mathbf{C}}(A \times \{u\}, \overline{(r_i, s_i)}) = t^{\mathbf{A}}(A, \bar{r}) \times \{v\}$$

is a nontrivial subset contained in $A \times \{v\}$. Since this subset is a polynomial image of a δ-class, δ restricts nontrivially to $A \times \{v\}$. This implies that $A \times \{v\}$ is a δ-class. But this forces $v \in U$ which is contrary to our assumption. This proves the Claim.

The Claim just proven implies that U is a class of a congruence on $\mathbf{C}$. Let ψ be a congruence on $\mathbf{C}$ which is maximal for the property that U is a ψ-class. Since $U \neq C$ we get that $\psi < 1$. Now $\mathbf{C}$ is a subalgebra of $\mathbf{A}^{m-1}$, so there exists a $k < m$ such that $\mathbf{C}$ is isomorphic to a subalgebra of $\mathbf{A}^k$ but not isomorphic to a subalgebra of $\mathbf{A}^{k-1}$. By the minimality of m we have that $\mathbf{Con}\,\mathbf{C} \cong \mathbf{2}^k$ for this value of k. In $\mathbf{2}^k$ there is a unique way to represent the bottom element as a meet of meet-irreducible elements. Since $\mathbf{C}$ is a subdirect power of $\mathbf{A}$, it follows that the maximal congruences of $\mathbf{C}$ are exactly the kernels of homomorphisms of $\mathbf{C}$ onto $\mathbf{A}$. Choose a maximal proper congruence $\beta \in \mathbf{Con}\,\mathbf{C}$ such that $\psi \leq \beta$. Let α be a complement of β in $\mathbf{Con}\,\mathbf{C}$ ($\cong \mathbf{2}^k$). We have $\alpha \wedge \psi = 0$, so $\psi < \alpha \vee \psi$. If U was a union of α-classes, then U would be a union of $\alpha \vee \psi$-classes. But this would violate the maximality of ψ. We conclude that U is not a union of α-classes. There must be some α-class V such that $V \not\subseteq U$, but $V \cap U \neq \emptyset$. V is a subuniverse since it is a congruence class. The facts that α complements β and β is the kernel of a homomorphism onto $\mathbf{A}$ implies that the subalgebra $\mathbf{V} \leq \mathbf{C}$ which is generated by V is isomorphic to a nontrivial subalgebra of $\mathbf{A}$: i.e., $\mathbf{V} \cong \mathbf{A}$. Now $U \cap V$ is a proper subuniverse of V, so this set must contain only one element. Say $U \cap V = \{0\}$. Let's interpret what the previous Claim means when $\bar{s}$ is chosen from V^n (we use the fact that $\mathbf{V} \cong \mathbf{A}$): If $t(x, \bar{y})$ is an $(n+1)$-ary term which depends on x in $\mathbf{V}$ and $\bar{s} \in V^n$, then $t^{\mathbf{V}}(0, \bar{s}) = 0$. This means precisely that 0 is an absorbing element for $\mathbf{V}$. Choose $v \in V - \{0\}$. From the definition of an absorbing element, it follows that $\{0, v\}$ is a subuniverse of V. This is impossible, since $\mathbf{V} \cong \mathbf{A}$ and $\mathbf{A}$ is an algebra with more than two elements which has no proper nontrivial subalgebras. This shows that Case 1 can never occur when $|A| > 2$.

PROOF FOR CASE 2: Now assume that $\mathbf{B}$ has no congruence δ such that $0 < \delta < \eta_0'$. Since $\theta \not\geq \eta_0'$, we have $\theta \wedge \eta_0' < \eta_0'$. Because we are in Case 2, this means that $\theta \wedge \eta_0' = 0$. Hence $C(\theta, \eta_0')$ and so $\theta \leq (0 : \eta_0')$. As η_0 and η_0' are complements, we also have $\eta_0 \wedge \eta_0' = 0$ and so $C(\eta_0, \eta_0')$ also holds. Thus $\eta_0 \leq (0 : \eta_0')$. Altogether this implies that $\theta \vee \eta_0 \leq (0 : \eta_0')$. But $\eta_0 \not\geq \theta$, by choice, and $\eta_0 \prec 1$ since $\mathbf{A}$ is simple. Hence $\theta \vee \eta_0 = 1$ and so $(0 : \eta_0') = 1$ and it follows that $C(1, \eta_0')$ holds. If V is a nontrivial η_0'-class, then the subalgebra of $\mathbf{B}$ generated by V is isomorphic to a subalgebra of $\mathbf{B}/\eta_0 \cong \mathbf{A}$, so $\mathbf{V} \cong \mathbf{A}$. But V is a class of an abelian congruence. Hence $\mathbf{V}$, and therefore $\mathbf{A}$, is an abelian algebra. Theorem 1.1 proves that $\mathbf{A}$ is a subreduct of an affine module. We shall prove that $\mathbf{A}$ is a reduct of an affine module.

As proved in the last section, there is a largest proper congruence Δ on $\mathbf{A}^2$ which has the diagonal of $\mathbf{A}^2$ contained in a single Δ-class. Choose $0 \in A$ arbitrarily and let $\mathbf{A}' = \mathbf{A}^2/\Delta$ and define

(i) $\delta : \mathbf{A} \to \mathbf{A}^2 : a \mapsto (a, 0)$,

(ii) $t : \mathbf{A}^2 \to \mathbf{A}' : (a, b) \mapsto (a, b)/\Delta$ and

(iii) $\epsilon = t \circ \delta$.

We proved (in Lemma 3.2 and we reiterated in Lemma 3.5) that ϵ is a 1-1 homomorphism. We claim that ϵ is onto as well. This amounts to proving that $t(A, A) = t(A, 0)$. To prove this, choose any $b \in A$ with $b \neq 0$. Let $U = A \times \{0\}$ and let $V = \{b\} \times A$. U and V are subuniverses of $\mathbf{A}^2$ which generate subalgebras isomorphic to $\mathbf{A}$. Since $\delta(A) = U$ we get that $\epsilon(A) = t \circ \delta(A) = t(U)$. The homomorphism ϵ is non-constant, so it must be that t is non-constant on U. $\mathbf{U} \cong \mathbf{A}$ is simple, so we get that t is 1-1 on U. Now $(b, 0) \in U \cap V$ and

$$t(b, 0) \neq t(0, 0) = t(b, b)$$

where the non-equality follows from the fact that $b \neq 0$ and t is 1-1 on U. This proves that t is non-constant on V since $t(b, 0) \neq t(b, b)$. Since $\mathbf{V} \cong \mathbf{A}$ we conclude that t is 1-1 on V.

$t(\mathbf{U})$ and $t(\mathbf{V})$ are subalgebras of $\mathbf{A}'$ which are isomorphic to $\mathbf{A}$. Furthermore, $t(U) \cap t(V)$ contains the distinct elements $t(b, 0)$ and $t(b, b)$. In this case we must have $t(U) = t(V)$ since otherwise one of the algebras $t(\mathbf{U})$ or $t(\mathbf{V})$, both of which are isomorphic to $\mathbf{A}$, has a nontrivial proper subuniverse: $t(U) \cap t(V)$. We have proven the following statement: For any distinct elements $0, b \in A$ it is the case that $t(A, 0) = t(b, A)$. This together with our assumption that $|A| > 2$ implies that $t(A, 0) = t(A, A)$.

Now that we know ϵ is onto, we get that each ϵ_n in Lemma 3.5 is an isomorphism. It follows that $\mathbf{A} \cong \hat{\mathbf{A}}$ in Lemma 3.6. This proves that $\mathbf{A}$ is a reduct of a module in Case 2 and concludes the proof of the theorem. $\square$

One may wonder if in Theorem 4.1 (iii) it must be that $\mathbf{A}$ is itself an affine module. The answer is "yes" if $\mathbf{A}$ is finite. This can be deduced from theorems in the next section. But in general the answer is "no". Another question that arises is whether **A** must generate a congruence distributive variety in Theorem 4.1 (iv). The answer is again "yes" when **A** is finite, but "no" in general. Here are examples which support these claims.

Example. Let $\mathbf{F}_2$ denote the 2-element field and let V be the $\mathbf{F}_2$-space with basis $\mathcal{B} = \{e_n \mid n < \omega\}$. Let $\mathbf{R} = \mathbf{F}_2\langle \alpha, \beta \rangle$ be the free $\mathbf{F}_2$-algebra in the noncommuting variables α and β. We define an **R**-module structure on V by defining actions of α and β on $\mathcal{B}$ and extending these definitions to V by linearity. We define $\alpha(e_n) = e_{n-1}$ if $n > 0$ and $\alpha(e_0) = 0$; $\beta(e_n) = e_{n^2+1}$. V is a faithful simple **R**-module as is proved on pages 196–197 of [7]. We define **A** to be the algebra which is the reduct of $_\mathbf{R}V$ to the operations $rx + (1 - r)y$, $r \in R$. Clearly, **A** is idempotent.

The translations $x \mapsto x + a$ form a transitive group of automorphisms of **A**, so to see that **A** has no proper nontrivial subalgebras it suffices to prove that if $b \neq 0$, then the subuniverse generated by $\{0, b\} \subseteq V$ is all of V. Certainly the subuniverse generated by $\{0, b\}$ contains all elements of the form $rb + (1 - r)0 = rb$. Hence this subuniverse contains all elements of V which belong to the submodule of $_\mathbf{R}V$ generated by b. But $Rb = V$ since $_\mathbf{R}V$ is simple, so we are done.

We have shown that **A** is an algebra of the sort described in Theorem 4.1 (iii). We now show that **A** is not affine. Let I be the ideal of **R** generated by $\{\alpha, \beta\}$. Since $\mathbf{R}/I \cong \mathbf{F}_2$ it follows that for all $r \in R$ we have $r \in I$ or $(1 - r) \in I$, but not both. Hence the set of all idempotent operations of the form

$$r_1 x_1 + \cdots + r_n x_n \quad \text{where } r_i \in I \text{ for all but one } i$$

is a clone on V which contains each $rx + (1 - r)y$, $r \in R$, but does not contain $x - y + z$. Since $_\mathbf{R}V$ has $x - y + z$ as its unique Mal'cev operation and this operation is not in the clone of **A**, it follows that **A** is not affine. $\square$

The next example was supplied by Ágnes Szendrei.

Example. Let A be an infinite set and consider all operations f on A which have the following properties:

(i) $f(x, \ldots, x) = x$ for all $x \in A$.

(ii) Assume that the variables on which f depends are $x_{i_0}, \ldots, x_{i_{n-1}}$. There is a finite set $S \subseteq A$ such that $f(\bar{a}) \in S$ whenever $a_{i_j} \neq a_{i_k}$ for some $j, k < n$.

When f depends on all of its variables, these properties say that f is idempotent, but has finite range "off the diagonal". We let **A** be the algebra with universe A and basic operations chosen so that each operation on A satisfying (i) and (ii) is the interpretation of a basic operation. The following facts are easily verified.

- All operations in the clone of **A** are the interpretation of basic operations. (That is, the operations on A satisfying (i) and (ii) are closed under composition and contain the projections.)

- **A** is idempotent.

- **A** has no proper nontrivial subuniverse.

- **A** is not abelian.

- **A** has no ternary operation $m(x, y, z)$ which depends on its middle variable and satisfies $m(x, y, x) = x$ on A.

These items imply that **A** is an algebra which falls under case (iv) of Theorem 4.1. If $\mathcal{V}(\mathbf{A})$ is a congruence distributive variety, then for some q this variety would have terms $m_i(x, y, z)$, $i < q$, such that the following list of equations hold in $\mathcal{V}(\mathbf{A})$:

(i) $m_0(x, y, z) = x$, $m_{q-1}(x, y, z) = z$,

(ii) $m_i(x, y, x) = x$ for all i,

(iii) $m_i(x, x, z) = m_{i+1}(x, x, z)$ for even i, and

(iv) $m_i(x, z, z) = m_{i+1}(x, z, z)$ for odd i.

The second equation implies that each m_i is independent of its middle variable. This can be coupled with equations (iii) and (iv) to deduce that $m_i(x, y, z) = m_j(x, y, z)$ for all i and j. Finally, applying (i) we get that

$$x = m_0(x, y, z) = m_{q-1}(x, y, z) = z$$

is an equation of $\mathcal{V}(\mathbf{A})$. But this is false, since **A** is nontrivial. We conclude that $\mathcal{V}(\mathbf{A})$ is not congruence distributive. $\square$

Of course, if **A** is finite and falls under case (iv) of Theorem 4.1, then the fact that $\mathcal{V}(\mathbf{A})$ is locally finite and $\mathsf{HSP}_{fin}(\mathbf{A})$ is congruence distributive is enough to imply that $\mathcal{V}(\mathbf{A})$ is congruence distributive.

The last two pathological examples indicate the limits of Theorem 4.1. However, we can give a more complete description of idempotent algebras without proper subalgebras satisfying additional hypotheses. The next theorem is an example of such a result.

THEOREM 4.2 *Let* **A** *be a non-unary algebra with a minimal clone which has no nontrivial proper subalgebras. Then exactly one of the following conditions is true.*

(*i*) **A** *is term equivalent to the 2-element semilattice,*

(*ii*) **A** *is term equivalent to an affine vector space over a prime field, or*

(*iii*) **A** *is term equivalent to a 2-element algebra with a majority operation.*

Proof: Since **A** is non-unary and has a minimal clone, it is idempotent. According to Lemma 2.1 and the proof of Lemma 3.3 in [14], if **A** is idempotent, has no nontrivial proper subalgebras and $|A| > 2$, then **A**

(a) has a local discriminator operation,

(b) is locally affine, or

(c) has an element 0 and a local operation $x * y$ such that $0 * x = x * 0 = 0$ and $x * y = x$ when $x \neq 0 \neq y$.

We assume that $|A| > 2$ and analyze each of these three cases.

Assume that **A** has a local discriminator operation. Choose distinct $a, b \in A$ and an operation $d(x, y, z)$ in $\mathrm{Clo}_3(\mathbf{A})$ whose restriction to $\{a, b\}$ is the discriminator on that set. It cannot be that $d(x, y, z)$ is a projection, so d generates $\mathrm{Clo}(\mathbf{A})$. The fact that $\{a, b\}$ is closed under d implies that $\{a, b\}$ is closed under all members of $\mathrm{Clo}(\mathbf{A})$. Hence $\{a, b\}$ is a subuniverse. **A** has no nontrivial proper subuniverses, so $|A| = 2$ which contradicts the assumption in the last paragraph.

The contradiction in Case (*c*) is obtained in exactly the same way: choose $a \in A - \{0\}$. Then choose $b(x, y) \in \mathrm{Clo}_2(\mathbf{A})$ so that b interpolates the local term operation $x * y$ on $\{0, a\}$. The operation $b(x, y)$ is not a projection, so it generates the clone of **A**. But $\{0, a\}$ is closed under b, hence it is a subuniverse. This implies that $|A| = 2$.

It must be that **A** is locally affine. Fix an affine representation of **A** over the appropriate ring, denoted **R**, and choose an operation $r_1 x_1 + \cdots + r_n x_n \in \mathrm{Clo}(\mathbf{A})$ which is not a projection. We may assume that $r_1 \notin \{0, 1\}$, so by setting x_1 equal to x and all other variables equal to y we get that $r_1 x + (1 - r_1) y$ is an operation in the clone which is not a projection. This operation generates the clone. It is a consequence of this that **R** is generated as a ring by the element r_1. The expansion of **A** to an **R**–module is still simple. There are no infinite simple modules over 1–generated rings. (1–generated rings are commutative, simple modules over commutative rings are 1–dimensional vector spaces, 1–dimensional vector spaces over 1–generated rings are finite.) But a finite, locally affine algebra is affine. Hence, **A** is a 1-dimensional affine vector space over a finite field. The minimality of the clone implies that field of scalars is a prime field.

What remains to check is the case where $|A| = 2$. In this case, the classification of clones on a 2-element set given in [10] proves that **A** must be affine over the 2-element field, equivalent to the 2-element semilattice or to the 2-element majority algebra. □

5   Minimal Varieties

As we described in the Introduction, the study of simple algebras is closely related to the study of minimal varieties. A full classification of the simple algebras in a given variety leads immediately to a full classification of the minimal subvarieties, but in fact the determination of minimal subvarieties can usually be accomplished with only a partial knowledge of the simple algebras. For example, one does not need a description of all simple groups to see that the minimal varieties of groups are precisely the varieties of elementary abelian p-groups. It suffices to know that a 1-generated simple group has prime cardinality. It turns out that the partial description we have of idempotent simple algebras is detailed enough to serve as a basis for a classification of all *locally finite*, idempotent, minimal varieties. However, our only achievement would be to reproduce the main result of [15]: Call a minimal idempotent variety **exceptional** if it is not

  (*i*) term equivalent to the variety of sets,

  (*ii*) term equivalent to the variety of semilattices,

  (*iii*) a variety of affine modules, or

  (*iv*) a congruence distributive variety.

THEOREM 5.1 (Á. Szendrei) *There is no exceptional, locally finite, idempotent, minimal variety.*

This result can be obtained as a corollary to Theorem 4.1. Any locally finite minimal variety is generated by a finite simple algebra which has no proper nontrivial subalgebras. Such an algebra must be term equivalent to a 2-element set, a 2-element semilattice, a reduct of an affine module or it must be that

$\mathrm{HSP}_{fin}(\mathbf{A}) = \mathcal{V}(\mathbf{A})$ is congruence distributive. The only detail we must attend to is a proof that when $\mathbf{A}$ is a finite simple reduct of an affine module and $\mathbf{A}$ has no proper nontrivial subalgebras, then $\mathbf{A}$ is an affine module or is term equivalent to the 2-element set. This fact follows immediately from the results of this section, but it is only a byproduct of our efforts. Our real goal is not to duplicate the characterization of locally finite, idempotent, minimal varieties described above. We are interested only in whether there exist exceptional, idempotent, minimal varieties and this goal necessarily leads us into the realm of non-locally finite varieties. The principal result of this section is that no exceptional variety contains a subreduct of an affine module. This result can be used to complete the proof of Theorem 5.1 that we have just sketched.

Theorem 5.1 indicates that an exceptional, idempotent, minimal variety is not locally finite. Referring to Theorem 1.1, it is clear that an exceptional variety will not contain a simple algebra of type (a). For if $\mathbf{A}$ is such a simple algebra and $0 \in A$ is absorbing, then for any $a \in A - \{0\}$, $\{0, a\}$ is a subuniverse. Hence, there is no non-locally finite minimal variety which contains a simple algebra of type (a). This means that an exceptional variety must be generated by an infinite simple algebra of type (b) or (c). This gives us the following result about non-locally finite minimal varieties. (In this proposition we say that $\mathbf{A}$ is an $(x - y + z)$-**reduct** of the affine module $\mathbf{M}$ if $\mathbf{A}$ is a reduct of $\mathbf{M}$, but the subclone of $\mathbf{M}$ generated by $x - y + z$ and the operations of $\mathbf{A}$ is the full clone of $\mathbf{M}$. $\mathbf{A}$ is a **proper** $(x - y + z)$-reduct of $\mathbf{M}$ if $x - y + z$ is not in the clone of $\mathbf{A}$.)

PROPOSITION 5.2 *Let $\mathcal{V}$ be an exceptional, idempotent, minimal variety. Then $\mathcal{V}$ is generated by an infinite simple algebra $\mathbf{A}$ where*

(i) *$\mathbf{A}$ is a simple, proper $(x - y + z)$-reduct of an affine module or*

(ii) *Every finite power of $\mathbf{A}$ is skew-free.* $\square$

The argument for this has essentially already been given. If $\mathbf{A}$ is of type (c) from Theorem 1.1, then we are in case (ii) of this proposition. If $\mathbf{A}$ is of type (b), then we can replace $\mathbf{A}$ with the simple algebra $\hat{\mathbf{A}} \in \mathcal{V}$ which we described in Theorem 3.8. Since $\mathcal{V}$ is exceptional, $\hat{\mathbf{A}}$ is a proper $(x - y + z)$-reduct of an affine module. Now we are in case (i) of this proposition.

In this section we show that there is no exceptional, idempotent, minimal variety generated by an algebra of the type described in Proposition 5.2 (i).

LEMMA 5.3 *If a variety contains a nontrivial subreduct of an affine algebra, then it contains one which is a reduct of a simple module over a simple ring.*

Proof: Assume that the variety $\mathcal{V}$ contains $\mathbf{A}$, which is a subreduct of the affine algebra $\mathbf{M}$. Fix an affine representation for the operations of $\mathbf{A}$. We may assume that the clone of $\mathbf{M}$ is generated by the operations of $\mathbf{A}$ and the operation $p(x, y, z) = x - y + z$. Furthermore, we may assume that $\mathbf{M}$ is generated under these operation by the set A. If we take the reduct of $\mathbf{M}$ to the operations of $\mathbf{A}$, and call this reduct $\mathbf{M}'$, then we may assume that A is maximal under inclusion as a subuniverse of $\mathbf{M}'$ which generates an algebra belonging to $\mathcal{V}$. Under all these (permissible) assumptions, $\mathbf{A}$ is a reduct of $\mathbf{M}$ (not just a subreduct). Here is why. The operation $p(x, y, z)$ is a homomorphism $p : \mathbf{M}^3 \to \mathbf{M}$ which is a left inverse to the diagonal homomorphism $\delta : \mathbf{M} \to \mathbf{M}^3 : x \mapsto (x, x, x)$. Both p and δ remain homomorphisms when we replace $\mathbf{M}$ with its reduct $\mathbf{M}'$. Hence we have a homomorphism $p \circ \delta|_{\mathbf{A}} : \mathbf{A} \to \mathbf{M}'$ which is the identity on $\mathbf{A}$, since $p \circ \delta = \mathrm{id}_{\mathbf{M}}$. It follows that

$$A = p \circ \delta|_{\mathbf{A}}(A) \subseteq p(A, A, A).$$

Since $\mathbf{A}^3 \in \mathcal{V}$ and $(\ker p)|_{\mathbf{A}^3}$ is a congruence on $\mathbf{A}^3$, we get that

$$p(\mathbf{A}, \mathbf{A}, \mathbf{A}) \cong \mathbf{A}^3 / (\ker p)|_{\mathbf{A}^3} \in \mathcal{V}.$$

But $p(\mathbf{A}, \mathbf{A}, \mathbf{A})$ is the subalgebra of $\mathbf{M}'$ with universe $p(A, A, A) \supseteq A$. The maximality condition on A implies that $A = p(A, A, A)$ and therefore that A is closed under $x - y + z$. Since this operation and the operations of $\mathbf{A}$ generate the clone of $\mathbf{M}$, A is a subuniverse of $\mathbf{M}$. We chose $\mathbf{M}$ so that A generates $\mathbf{M}$, so $A = M$.

Let $\mathbf{N}$ be a simple algebra generating a minimal subvariety of $\mathsf{HSP}(\mathbf{M})$. Replacing $\mathbf{N}$ by its linearization if necessary, we may assume that $\mathbf{N}$ is a reduct of a module. There exists a cardinal κ, a subalgebra $\mathbf{P} \leq \mathbf{M}^\kappa$ and a congruence θ such that $\mathbf{P}/\theta \cong \mathbf{N}$. Let $\mathbf{P}'$ be the reduct of $\mathbf{P}$ to the operations of $\mathbf{A}$. Then $\mathbf{P}'$ is a subalgebra of $\mathbf{A}^\kappa$ and the equivalence relation θ is a congruence on $\mathbf{P}'$. $\mathbf{N}' := \mathbf{P}'/\theta$ is the reduct of $\mathbf{N}$ to the operations of $\mathbf{A}$. This makes $\mathbf{N}'$ a reduct of the simple affine algebra $\mathbf{N}$ which is itself a reduct of a module. Since $\mathbf{N}$ generates a minimal variety, the corresponding ring is simple. Furthermore, $\mathbf{N}' \in \mathsf{HSP}(\mathbf{A}) \subseteq \mathcal{V}$. Hence, $\mathbf{N}'$ fulfills the claims of the lemma. $\square$

In Lemma 5.3, the algebra $\mathbf{N}'$ constructed is an $(x - y + z)$-reduct of a simple, linear, affine algebra over a simple ring. Expanding $\mathbf{N}'$ by $x - y + z$ and a constant 0 interpreted as a trivial subalgebra, one obtains a simple module over a simple ring.

Observe that if we attempt to apply Lemma 5.3 to an abelian idempotent simple algebra, then we gain some information and we lose some information. Starting with a simple subreduct of an affine module, we end up with a reduct of a simple affine module over a simple ring. What we gain is the simplicity of the coefficient ring. What we lose is the simplicity of our abelian algebra. We shall find that the simplicity of the coefficient ring is more important for us than the simplicity of the abelian algebra.

For the rest of this section, we assume that $\mathbf{A}$ is a reduct of a simple affine $\mathbf{R}$-module $\mathbf{M}$ where $\mathbf{R}$ is a simple ring. We assume also that $\mathbf{A}$ generates an exceptional minimal variety, so $\mathbf{A}$ is a proper $(x - y + z)$-reduct of $\mathbf{M}$. We do not assume that $\mathbf{A}$ is simple. Each term operation of $\mathbf{A}$ may be expressed uniquely as

$$t^{\mathbf{A}}(x_0, \ldots, x_{n-1}) = r_0 x_0 + \cdots + r_{n-1} x_{n-1}$$

where each $r_i \in R$ and $\Sigma_{i<n} r_i = 1$ (since $t(x, \ldots, x) = x$). If some $r \in R$ appears as the ith coefficient of some term t, then it occurs as the coefficient of x in the term $t(y, \ldots, y, x, y, \ldots, y)$ where x is in the ith position. Hence $r \in R$ is a coefficient in some term iff $rx + (1-r)y$ represents a term operation of $\mathbf{A}$. The set of distinct binary term operations of $\mathbf{A}$ may be identified with a subset $U \subseteq R$ under the correspondence $rx + (1 - r)y \leftrightarrow r$. This set U of coefficients of terms must satisfy some very special properties.

Definition 5.4 If $\mathbf{R}$ is a ring, then a set $U \subseteq R$ will be called a **unit interval** if the following conditions are met.

(*i*) U is a proper subset of R.

(*ii*) $0, 1 \in U$.

(*iii*) $U \neq \{0, 1\}$.

(*iv*) If $a, b, c \in U$, then $ab + (1 - a)c \in U$.

(*v*) $\mathbf{R}$ is generated by U as an abelian group.

Note that condition (*iv*) implies that U is closed under multiplication (take $c = 0$) and also under the function $x \mapsto (1 - x)$ (take $b = 0, c = 1$).

THEOREM 5.5 *Assume that $\mathbf{A}$ is not essentially unary. If $\mathbf{A}$ is a proper $(x - y + z)$-reduct of an affine module over a simple ring, then the set of coefficients of term operations of $\mathbf{A}$ is a unit interval.*

Proof: The set U of coefficients of binary term operations $ax + (1 - a)y$ satisfies (*ii*) since the projection operations $p_0(x, y) = x = 1x + 0y$ and

$p_1(x, y) = y = 0x + 1y$ are term operations. (iv) is satisfied, since if $a, b, c \in U$, then $ax + (1 - a)y$, $bx + (1 - b)y$ and $cx + (1 - c)y$ represent binary terms, so

$$a(bx + (1-b)y) + (1-a)(cx + (1-c)y) = (ab + (1-a)c)x + (1 - (ab + (1-a)c))y$$

also represents a binary term. This proves that $ab + (1 - a)c \in U$. Using the fact that $x - y + z$ commutes with all term operations of $\mathbf{A}$, we see that the collection of binary terms of $\mathbf{A}$ generates all binary terms of $\langle \mathbf{A}; x - y + z \rangle$ under the operation $x - y + z$. Thus the additive subgroup of $\mathbf{R}$ generated by U contains all elements of R which occur as the coefficient in x in some binary term of $\langle \mathbf{A}; x - y + z \rangle$. But the binary terms of $\langle \mathbf{A}; x - y + z \rangle$ up to equivalence are the set of all operations $rx + (1 - r)y$, $r \in R$. Hence U generates $\mathbf{R}$ under $x - y + z$ and (v) holds.

Assume that $U = \{0, 1\}$. Since $\mathbf{A}$ is not essentially unary, there is a term which depends on all $n > 1$ of its variables: $r_0 x_0 + \cdots + r_{n-1} x_{n-1}$. Since all coefficients come from $U - \{0\} = \{1\}$, this term is simply $x_0 + \cdots + x_{n-1}$. But now if we set the first two variables equal to x_1, we get a new term $2x_1 + \cdots + x_{n-1}$. Since all coefficients belong to $U = \{0, 1\}$ we must have $2 = 0$ or $2 = 1$ in $\mathbf{R}$. But $2 = 1$ forces $\mathbf{R}$, and therefore $\mathbf{A}$ to be trivial. We must have $2 = 0$. Now property (v) implies that $\mathbf{R}$ is the 2-element field. $\mathbf{A}$ can only be a reduct of a 1-dimensional affine vector space over $\mathbf{R}$. But the clone of a 1-dimensional affine vector space over the 2-element field is a minimal clone. Hence, the only reducts of a 1-dimensional affine vector space over the 2-element field are the full clone of the affine vector space and the clone of projection operations. Our assumptions tell us that $\mathbf{A}$ is not essentially unary and $\mathbf{A}$ is not affine. This is a contradiction to $U = \{0, 1\}$.

To finish we must prove (i). Suppose that $U = R$. Then the clone of $\mathbf{A}$ contains every operation of $\mathbf{M}$ of the form $rx + (1 - r)y$, $r \in U = R$, but it does not contain $x - y + z$. Corollary 2 of [12] proves that $x - y + z$ is in the clone generated by $\{rx + (1 - r)y \mid r \in R\}$ if and only if $\mathbf{R}$ has no homomorphism onto a 2-element field. Hence $U = R$ implies that our simple ring $\mathbf{R}$ has a homomorphism onto a 2-element field. This means that $\mathbf{R}$ is a 2-element field. Now we have $U = R = \{0, 1\}$ even though we have already shown that $U = \{0, 1\}$ is false. This contradiction proves that $U \subset R$. $\square$

A simple ring may have more than one unit interval. For example, if U is a unit interval of $\mathbf{R}$ and α is an automorphism of $\mathbf{R}$, then $\alpha(U)$ is another unit interval. If $\mathbf{R}$ is the field obtained by adjoining to the field of rationals the number $\sqrt{2}$ and α is the nontrivial automorphism of this field, then $U := \mathbf{Q}[\sqrt{2}] \cap [0, 1]$, $U' := \alpha(U)$ and in this case $U'' := U \cap U'$ are all unit intervals of $\mathbf{Q}[\sqrt{2}]$. So, there may be many unit intervals, some properly containing others. (The observations of this paragraph are due to G.M. Bergman.)

Which simple rings have unit intervals? We focus on this question now. First, we explain how to construct some examples.

Definition 5.6 Let **R** be a ring. A **positive cone** of an ordering of **R** is a subset $P \subseteq R$ satisfying

(*i*) $P + P \subseteq P$,

(*ii*) $P \cdot P \subseteq P$ and

(*iii*) $P \cap -P \subseteq \{0\}$.

That is, P is closed under addition and multiplication and does not contain both r and $-r$ if $r \neq 0$. The **ordering of R** associated with P is

$$a \leq b \Leftrightarrow b - a \in P.$$

We call $\langle \mathbf{R}; \leq \rangle$ a **partially ordered ring**.

The words "positive", "negative" and the symbol $<$ have their usual meanings, as does the interval notation $[x, y] := \{z \in R \mid x \leq z \leq y\}$. "Ordering" will always mean "partial ordering". If n is an integer, then the element $n \cdot 1 \in R$, defined to be a sum of n copies of 1, will be denoted simply by n.

Definition 5.7 Let $\langle \mathbf{R}; \leq \rangle$ be a partially ordered ring. **R** is **archimedean** if $na \leq b$ for all positive integers n implies $a \leq 0$ whenever $a, b \in R$. An element $e \in R$ is a **strong order unit** if whenever $a \in R$ there is a positive integer n such that $a \leq ne$. We say that $\langle \mathbf{R}; \leq \rangle$ is **strongly archimedean** if it is archimedean and 1 is a positive strong order unit.

If **R** is a subfield of the real numbers and $\leq$ is the restriction to **R** of the usual ordering, then $\langle \mathbf{R}; \leq \rangle$ is an example of a strongly archimedean partially ordered, simple ring. The next theorem describes one half of the connection between unit intervals and strongly archimedean partial orderings.

THEOREM 5.8 *A simple ring with a strongly archimedean partial ordering has a unit interval.*

Proof: Let $\langle \mathbf{R}; \leq \rangle$ be a simple ring with a strongly archimedean partial ordering. We denote the center of $\mathbf{R}$, which is a field, by $Z(\mathbf{R})$. Define

$$P' = \left\{ \frac{p}{f} \in R \mid p \in P, \ f \in (P \cap Z(\mathbf{R})) - \{0\} \right\}.$$

P' is the closure of P under division by positive elements of $\mathbf{F}$. An easy exercise shows that P' is the positive cone of a partial ordering which extends the one given by P. Any extension of a strongly archimedean partial ordering is another strongly archimedean partial ordering, so P' determines such an ordering. P' is closed under division by positive central elements (even by division by the central elements that are P'-positive but not P-positive). Replacing P with P' we may assume that our ordering has the property that the positive cone is closed under division by positive central elements.

Let $U = \{r \in R \mid 0 \leq r \leq 1\}$. We claim that U is a unit interval in $\mathbf{R}$. We must show that

(i) U is a proper subset of R.

(ii) $0, 1 \in U$.

(iii) $U \neq \{0, 1\}$.

(iv) If $a, b, c \in U$, then $ab + (1 - a)c \in U$.

(v) $\mathbf{R}$ is generated by U as an abelian group.

Since $U \subseteq P \subset R$, (i) is satisfied. (ii) clearly holds. (iii) holds since all rationals between 0 and 1 belong to U. For (iv), assume that $a, b, c \in U$. Then $0 \leq a, b, c \leq 1$. We get that $0 \leq 1 - a \leq 1$, so each of $a, b, c, (1 - a)$ is between 0 and 1. From this we deduce that

$$
\begin{aligned}
0 \ & \leq ab + (1 - a)c \\
& \leq a \cdot 1 + (1 - a) \cdot 1 = 1
\end{aligned}
$$

Hence $ab + (1 - a)c \in U$. Finally, assume that $r \in R$. Since our ordering is a strongly archimedean partial ordering, we can find positive integers $m, n \in F$ such that $-r < m$ and $m + r < n$. These inequalities imply that $0 < m + r < n$. Hence $(m + r)/n \in U$. The abelian group generated by U contains all multiples of $(m + r)/n$, so it contains $m + r$. But this group also contains all multiples of $1 \in U$, so this group contains m. Since it contains m and $m + r$ it also contains r. Our choice of $r \in R$ was arbitrary, so the abelian group generated by U is R. This proves that U is a unit interval in $\mathbf{R}$. $\square$

Strongly archimedean simple rings give us some examples of simple rings with unit intervals. Our next goal it to prove the converse of Theorem 5.8, which is that all simple rings with unit intervals are strongly archimedean partially ordered rings. The converse implication is really the only one that is useful to us; we will find the partial ordering easier to work with than unit intervals. (Theorem 5.8 is included only to assure us that we lose no generality in considering orderings rather than unit intervals.)

Let $\mathbf{R}$ be a simple ring with a unit interval U. To continue the analysis, it will be convenient to associate to $\mathbf{R}$ an auxiliary algebraic structure $\widehat{\mathbf{R}}$. The universe of $\widehat{\mathbf{R}}$ is R and the basic operations of $\widehat{\mathbf{R}}$ are just the binary operations $[u](x,y) := ux + (1-u)y$, $u \in U$. We will refer to $\widehat{\mathbf{R}} = \langle R; [u], u \in U \rangle$ as simply 'the auxiliary structure'. Note that $\widehat{\mathbf{R}}$ is a reduct of the affine $\mathbf{R}$-module structure of $_\mathbf{R}\mathbf{R}$. It is worth pointing out that property (iv) of a unit interval implies that the binary term operations of $\widehat{\mathbf{R}}$ are exactly those of the form $[u](x,y)$, $u \in U$. This implies, that no term operation of $\widehat{\mathbf{R}}$ agrees with $x - y + z$, for then

$$[u](x,y) - [v](x,y) + [w](x,y) = [u-v+w](x,y)$$

would force U to be closed under $x - y + z$. Since $0 \in U \neq R$ and U generates R as an abelian group, U is not closed under $x - y + z$. Thus, the auxiliary structure $\widehat{\mathbf{R}}$ is a proper $(x - y + z)$-reduct of $_\mathbf{R}\mathbf{R}$.

LEMMA 5.9 *If $a_0 x_0 + \cdots + a_{n-1} x_{n-1}$ represents a term operation of $\widehat{\mathbf{R}}$ and $a_i = 1$ for some i, then $a_j = 0$ for all $j \neq i$.*

Proof: This proof is based on an idea from [13]. Assume that, say, $a_0 = 1$. Then

$$a_0 x + a_1 z + a_2 y + \cdots + a_{n-1} y = x - a_1 y + a_1 z$$

represents a term operation of $\widehat{\mathbf{R}}$. But, we shall argue, the set $I = \{r \in R | \ x - ry + rz \in \mathrm{Clo}_3 \widehat{\mathbf{R}}\}$ is a proper ideal of $\mathbf{R}$. To see that I is an ideal, note that $a, b \in I$ implies that

$$(x - bz + by) - ay + az = x - (a-b)y + (a-b)z,$$

so $a - b \in I$. Since I is closed under subtraction it is an abelian group. I is an additive subgroup of $\mathbf{R}$ and U generates $\mathbf{R}$ as an additive subgroup, so to show that I is an ideal it suffices to show that if $a \in I$ and $u \in U$ we have $au, ua \in I$. This is proved by the lines

$$x - a(uy + (1-u)z) + az = x - auy + auz$$

and

$$u(x - ay + az) + (1 - u)x = x - uay + uaz.$$

The ideal I must be proper, for if $1 \in I$ then the definition of I would force $x - y + z$ into the clone of $\widehat{\mathbf{R}}$ and we know this to be false. By the simplicity of $\mathbf{R}$, $I = \{0\}$. Hence, $a_1 = 0$. The same type of argument proves that $a_i = 0$ for all $i > 0$. $\square$

THEOREM 5.10 *A simple ring with a unit interval has a strongly archimedean partial ordering.*

Proof: Let U be a unit interval of the simple ring $\mathbf{R}$. Let P be the set of elements of R that can be expressed as sums of elements of U. Our goal will be to show that P is a positive cone for a partial ordering which is extendible to a strongly archimedean partial ordering.

Clearly P is closed under sums. Furthermore, the fact that U is closed under products and P is the set of sums of elements of U implies that P is closed under products. The bulk of our argument will be to prove that $P \cap -P \subseteq \{0\}$. If this is not the case, then there is an $r \in P - \{0\}$ such that $r \in -P$. From the definition of P this means that there are elements $u_1, \dots, u_m \in U$ and $v_1, \dots, v_n \in U$ such that

$$u_1 + \cdots + u_m = -(v_1 + \cdots + v_n) = r \neq 0.$$

It follows that the u_i, v_i are not all zero even though $(\Sigma u_i) + (\Sigma v_i) = 0$. We set

$$V = \{u \in U \mid \exists w_1, \dots, w_k \in U(u + w_1 + \cdots + w_k = 0)\}.$$

V contains $\{u_1, \dots, u_m, v_1, \dots, v_m\}$ and so V properly contains $\{0\}$. Furthermore, since U is closed under multiplication, it follows that $U \cdot V \subseteq V$ and $V \cdot U \subseteq V$. If I is the abelian group generated by V, then our conclusions mean that I is a nonzero ideal of $\mathbf{R}$. To repeat and amplify, $P \cap -P \not\subseteq \{0\}$ implies that $I = R$.

Claim. $I \neq R$.

Proof of Claim: We begin with a little detour to uncover a strange property of the elements of V. We will show that if $u \in V$, then there exists a positive integer n such that $r(1 - r)^n u = 0$ for any $r \in U$. (This fact was discovered by Ágnes Szendrei.) To show this, fix some $u \in V$. Since $u \in V$ there exist $w_1, \dots, w_k \in U$ such that $u + w_1 + \cdots + w_k = 0$. We claim that $r(1 - r)^n u = 0$ for $n = k$ and for any $r \in U$.

Fix specific choices of r, u and w_i as described in the previous paragraph. Since $r, u, w_i \in U$, we get that $rx + (1 - r)y$, $ux + (1 - u)y$ and $w_i x + (1 - w_i)y$

are operations of the auxiliary structure. $\widehat{\mathbf{R}}$. We can compose the operation $rx+(1-r)y$ with itself repeatedly in order to produce operations of larger arity. We will change variables to those of the form x_σ where σ is a sequence of 0's and 1's in order to make it clearer how the composition should be done. Begin with $rx_0 + (1-r)x_1$. If we substitute $rx_{\epsilon 0} + (1-r)x_{\epsilon 1}$ in for the occurrence of x_ϵ in $rx_0 + (1-r)x_1$, we obtain a 4-ary operation:

$$r(rx_{00} + (1-r)x_{01}) + (1-r)(rx_{10} + (1-r)x_{11})$$
$$= r^2 x_{00} + r(1-r)x_{01} + r(1-r)x_{10} + (1-r)^2 x_{11}.$$

If $t(x_{00}, x_{01}, x_{10}, x_{11})$ denotes this 4-ary operation, we can build an 8-ary operation by substituting $t(x_{\epsilon 00}, x_{\epsilon 01}, x_{\epsilon 10}, x_{\epsilon 11})$ in for x_ϵ in $rx_0 + (1-r)x_1$. Continuing, for any n we can build a 2^n-ary operation of the variables x_σ, $\sigma \in \{0,1\}^n$, where the coefficient of x_σ is $r^m(1-r)^{n-m}$ precisely when m is the number of 0's in the string σ. We let $s(\bar{x})$ be such an operation for $n = k+1$, Define τ_i, $i = 0, 1, \ldots, k$, to be the element of $\{0,1\}^{k+1}$ which has a 0 in the ith position and 1's elsewhere. For each i, the coefficient of x_{τ_i} is $r(1-r)^k$. Substitute into the variable x_{τ_i} of $s(\bar{x})$ the operation $w_i z_i + (1-w_i)x$ if $i > 0$ and substitute $uz_0 + (1-u)x$ into x_{τ_0}. If σ is a sequence which does not have exactly one occurrence of 0, then set x_σ equal to x. With these substitutions into $s(\bar{x})$ we end up with a new $(k+2)$-ary operation in the clone generated by the operations $rx + (1-r)y$. $ux + (1-u)y$ and $w_i x + (1-w_i)y$:

$$r(1-r)^k u z_0 + r(1-r)^k w_1 z_1 + \cdots + r(1-r)^k w_k z_k + Cx$$

for some $C \in R$. This operation is in the clone of $\widehat{\mathbf{R}}$. If you set all variables equal to x and use the fact that $u + w_1 + \cdots + w_k = 0$, you get

$$r(1-r)^k ux + r(1-r)^k w_1 x + \cdots + r(1-r)^k w_k x + Cx = Cx.$$

But this operation is idempotent. so $Cx = x$. Thus,

$$r(1-r)^k u z_0 + r(1-r)^k w_1 z_1 + \cdots + r(1-r)^k w_k z_k + x$$

is in the clone of $\widehat{\mathbf{R}}$. Now we apply Lemma 5.9 to this operation: the coefficient of x equals 1, so $r(1-r)^k u = r(1-r)^k w_i = 0$ for all i. (All that we care about is that $r(1-r)^k u = 0$, though.)

Now we return to our proof that $I \neq R$. Assume instead that $I = R$. Then

$$1 = \pm u_1 \pm \cdots \pm u_\ell$$

for some choice of elements $u_i \in V$. By the results of the last paragraph, there exist m_i such that $r(1-r)^{m_i} u_i = 0$ for any $r \in U$ and for each i. Since U is closed under the operation $x \mapsto (1 - x)$, what we have said about r also holds for $(1 - r)$: $(1 - r)r^{m_i}u_i = 0$ for all i. Choose any N greater than all m_i. Then for any $r \in U$ we have

$$r(1 - r)^N u_i = 0 = (1 - r)r^N u_i$$

for all i. Hence we have

$$r(1 - r)^N = r(1 - r)^N(u_1 + \cdots + u_\ell) = 0 = (1 - r)r^N$$

for all $r \in U$.

We claim that there are polynomials with integer coefficients, $p(x), q(x) \in \mathbf{Z}[x]$, such that

$$x^{N-1} \cdot p(x) + (1 - x)^{N-1} \cdot q(x) = 1.$$

Otherwise, the ideal $J = (x^{N-1}, (1 - x)^{N-1})$ is proper in $\mathbf{Z}[x]$, so $\mathbf{Z}[x]$ has a maximal ideal M containing J. $\mathbf{Z}[x]/M$ is a field which has an element $\bar{x} := x/M$ such that $\bar{x}^{N-1} = 0 = (1 - \bar{x})^{N-1}$. This is impossible because it implies that $\bar{x} = 0 = (1 - \bar{x})$ or $1 = 0$. Now, using p and q we see that

$$
\begin{aligned}
r(1 - r) &= r(1 - r)[r^{N-1} \cdot p(r) + (1 - r)^{N-1} \cdot q(r)] \\
&= r(1 - r)^N \cdot p(r) + (1 - r)r^N \cdot q(r) \\
&= 0
\end{aligned}
$$

whenever $r \in U$. This proves that for any $r \in U$ we have $r^2 = r$. We write $U \models x^2 = x$ for this.

Choose $a, b \in U$. Recall that U is closed under multiplication and the function $x \mapsto (1 - x)$. Hence

$$1 - ((1 - a)(1 - b)) = a + b - ab \in U.$$

Furthermore, taking $c = 1 - b$ in property (iv) of a unit interval we get $ab + (1 - a)(1 - b) \in U$, so

$$1 - (ab + (1 - a)(1 - b)) = a + b - 2ab \in U.$$

For $x = a + b - ab$ and for $x = a + b - 2ab$ we must have $x^2 = x$. We calculate (using $a^2 = a$, $b^2 = b$, $(ab)^2 = ab$) that

$$
\begin{aligned}
a + b - ab &= (a + b - ab)^2 \\
&= a^2 + ab - a^2b + ba + b^2 - bab - aba - ab^2 + abab \\
&= a + ab - ab + ba + b - bab - aba - ab + ab \\
&= a + ba + b - bab - aba.
\end{aligned}
$$

Hence $aba + bab = ab + ba$. A further calculation shows that

$$
\begin{aligned}
a + b - 2ab &= (a + b - 2ab)^2 \\
&= a^2 + ab - 2a^2b + ba + b^2 - 2bab - 2aba - 2ab^2 + 4abab \\
&= a + ab - 2ab + ba + b - 2bab - 2aba - 2ab + 4ab \\
&= a + ab + ba + b - 2bab - 2aba.
\end{aligned}
$$

This yields $2aba + 2bab = 3ab + ba$. Our two conclusions together give us that

$$
3ab + ba = 2(aba + bab) = 2(ab + ba) = 2ab + 2ba.
$$

This simplifies to $ab = ba$. Since $a, b \in U$ were chosen arbitrarily, $U \models xy = yx$. As U generates $\mathbf{R}$ as an abelian group we must have $\mathbf{R} \models xy = yx$. $\mathbf{R}$ is a commutative simple ring, so $\mathbf{R}$ is a field. Since $U \models x^2 = x$, U is contained in the set of idempotents of the field $\mathbf{R}$. Thus $U \subseteq \{0, 1\}$. This contradicts the fact that U is a unit interval: U must properly contain $\{0, 1\}$. This contradiction finishes the proof of the Claim.

The impact of the Claim is that P is the positive cone of a partial ordering of $\mathbf{R}$. In this ordering, if $u \in U$, then $u, 1 - u \in U \subseteq P$ so $0 \leq u \leq 1$ as one might hope. We argue that 1 is a positive strong order unit. Positivity of 1 follows from $1 \in U \subseteq P$. Now we must show that every element of R is majorized by a positive integer. Fix $r \in R$. Since $\mathbf{R}$ is generated as an abelian group by U, we can find $u_1, \ldots, u_i \in U$ and $v_1, \ldots, v_j \in U$ such that

$$
-r = u_1 + \cdots + u_i - v_1 - \cdots - v_j.
$$

For each $v_k \in U$ we have $(1 - v_k) \in U$, so we can re-express this as

$$
-r = u_1 + \cdots + u_i + (1 - v_1) + \cdots + (1 - v_j) - j.
$$

Since $u_1 + \cdots + u_i + (1 - v_1) + \cdots + (1 - v_j) \in P$, this expresses $-r$ as $p - j$ where $p \in P$ and j is a nonnegative integer. We conclude that $r \leq j$. Since r is majorized by a nonnegative integer and 1 is positive, r is majorized by a positive integer. This proves that 1 is a positive strong order unit under the ordering determined by P.

It can be shown by example that the P-ordering does not have to be archimedean. So replace P with a maximal extension. This does not affect the fact the 1 is a positive strong order unit, but it implies extra properties of P. For example, by maximality, P must be closed under division by positive central elements. Define P' to be the set of all $x \in R$ such that $-\frac{1}{n} \leq x$ holds for all positive integers n. Then $P \subseteq P'$, $P' + P' \subseteq P'$ and (since 1 is a positive strong order unit) $P' \cdot P' \subseteq P'$. Furthermore, using the fact that 1 is a positive strong order unit, it is easy to see that

$$
K := P' \cap -P' = \bigcap_{n < \omega} \left[-\frac{1}{n}, \frac{1}{n} \right]
$$

is an ideal of **R**. Since $1 \notin K$, we have $K = \{0\}$. P' is the positive cone of an ordering which extends the P-ordering, so $P = P'$.

Since 1 is a strong order unit and P is closed under division by positive integers, $na \leq b$ holds for all positive n iff $a \leq \frac{1}{n}$ holds for all positive n. The latter condition is equivalent to $a \in -P' = -P$ which may be expressed as $a \leq 0$. This proves that P is the positive cone of an archimedean partial ordering, so **R** is a strongly archimedean partially ordered ring. $\square$

For us, the most important fact about strongly archimedean partial orderings of simple rings is contained in the next theorem. The proof evolved out of email correspondence with K.H. Leung.

THEOREM 5.11 *A simple ring with a strongly archimedean partial ordering has no nontrivial zero divisors.*

Proof: Replacing P by an extension if necessary, we may assume that P is closed under division by positive central elements. Assume that $r^2 = 0$ in **R**. Using the fact that 1 is a strong order unit, select and fix a positive integer q such that $-r \leq q$. Since $0 \leq r + q$, we get that, for any positive integer m,

$$0 \leq (r + q)^m = mrq^{m-1} + q^m = mq^{m-1}(r + \frac{q}{m}).$$

Since P is closed under division by positive central elements, the integer mq^{m-1} is positive and central, and $0 \leq mq^{m-1}(r + \frac{q}{m})$, we get that $0 \leq r + \frac{q}{m}$. Thus, $-\frac{q}{m} \leq r$ for fixed q and increasing m. It follows that $-\frac{1}{n} \leq r$ for all n, so $r \in P$ by the archimedean property. But if $r^2 = 0$, then we also have $(-r)^2 = 0$. We get $-r \in P$ as well. Hence, $r \in P \cap -P \subseteq \{0\}$ which means that $r = 0$. This proves that **R** has no nonzero nilpotent elements.

Now assume that for certain $a, b \in R$ we have $ab = 0$. Then each element of bRa is nilpotent. From our earlier conclusion, $bRa = \{0\}$. This implies that the product of the ideals generated by b and a is

$$(b)(a) = (RbR)(RaR) = R(bRa)R = \{0\}.$$

But the only ideals of **R** are $\{0\}$ and R. Since $RR = R \neq \{0\}$ we conclude that $(a) = \{0\}$ or $(b) = \{0\}$. Hence $a = 0$ or $b = 0$. $\square$

We are prepared to prove the main result of this section.

THEOREM 5.12 *Let $\mathcal{V}$ be a minimal idempotent variety and assume that $\mathcal{V}$ contains a nontrivial subreduct of an affine module. Then $\mathcal{V}$ is equivalent to the variety of sets or else is affine. In particular, $\mathcal{V}$ is not exceptional.*

Proof: Assume that $\mathcal{V} = \mathcal{V}(\mathbf{A})$ where $\mathbf{A}$ is a nontrivial subreduct of an affine module. By Lemma 5.3, we may assume that $\mathbf{A}$ is a reduct of an affine module $\mathbf{N}$ over a simple ring, $\mathbf{R}$. Fix such a representation for $\mathbf{A}$. If $\mathbf{A}$ is affine or essentially unary, then we are done, so assume that $\mathbf{A}$ is neither. Theorem 5.5 proves that $\mathbf{R}$ has a unit interval U, and that the binary term operations of $\mathbf{A}$ are precisely the operations $ux + (1 - u)y$, $u \in U$. Using the same argument as in the proof of Lemma 5.3 we may adjust our choice of $\mathbf{A}$ so that it is a reduct of any affine $\mathbf{R}$-module in $\mathsf{HSP}(\mathbf{N})$. Let $\mathbf{B}$ be the reduct of $_{\mathbf{R}}\mathbf{R}$ to the operations of $\mathbf{A}$. Since the binary term operations of $\mathbf{B}$ are the operations $ux + (1 - u)y$, $u \in U$. it is clear that the universe of the subalgebra $\mathbf{C} \le \mathbf{B}$ generated by $0, 1 \in B = R$ is

$$C = \{u \cdot 1 + (1 - u) \cdot 0 = u \mid u \in U\} = U \subseteq R.$$

We define $I \subseteq C$ to be an **ideal** in $\mathbf{C}$ if whenever $t(x, \bar{y})$ is an $(n+1)$-ary term that depends on x in $\mathbf{C}$ and $u \in I$ we have $t^{\mathbf{A}}(u, \bar{c}) \in I$ for any $\bar{c} \in C^n$. (So, an absorbing element is nothing more than a 1-element ideal.)

Claim. If $I = C - \{0\}$, then I is an ideal of $\mathbf{C}$.

Proof of Claim: We need to show that if $p(\bar{x}) = c_0 x_0 + \cdots + c_{m-1} x_{m-1}$ is the interpretation of a term operation of $\mathbf{C}$ which depends on all of its variables and $\bar{d} \in C^m$ with at least one $d_i \in I = C - \{0\} = U - \{0\}$. then $p(\bar{d}) \in I$. This is trivial if $m = 1$, so we assume that $m > 1$. Since p depends on all variables, $c_i \in U - \{0\}$ for each i. Now $c_i, d_i \in U = C$ for all i, so $c_i d_i \in U$ for all i. This implies that $0 \le c_i d_i$ for all i. Furthermore, $c_i \ne 0$ for all i and $d_j \ne 0$ for at least one j. Since $\mathbf{R}$ has no nontrivial zero divisors, $0 < c_j d_j$ for some j. We conclude that $0 < \Sigma_{i < m} c_i d_i = p(\bar{d})$, so $p(\bar{d}) \in C - \{0\} = I$. Since $p(\bar{x})$ and $\bar{d} \in C^m$ were chosen arbitrarily, the claim is proved.

The relation $\theta = (I \times I) \cup \{(0, 0)\}$ is a congruence of $\mathbf{C}$. since $I = C - \{0\}$ is an ideal, and the quotient $\mathbf{C}/\theta$ has two elements, $0/\theta$ and I/θ. The latter element is a 1-element ideal (otherwise known as an absorbing element). Thus $\mathbf{C}/\theta$ is a 2-element idempotent simple algebra in $\mathcal{V}$ with at least one absorbing element. It must be equivalent to either a 2-element semilattice or a 2-element set. We have assumed that $\mathcal{V}$ is minimal, so $\mathcal{V}$ must be equivalent to the variety of sets or the variety of semilattices. Since $\mathbf{A}$ is not essentially unary it must be that $\mathcal{V}$ is not equivalent to the variety of sets. But $\mathbf{A} \in \mathcal{V}$ is a nontrivial abelian algebra: $\mathcal{V}$ cannot be equivalent to the variety of semilattices either. This contradiction finishes the proof. $\square$

Let $\mathcal{V}$ be an exceptional idempotent minimal variety. If $\mathcal{V}$ does not satisfy the commutator equation $[\alpha, \beta] = \alpha \wedge \beta$, equivalent to the implication $[\alpha, \alpha] = 0 \Rightarrow \alpha = 0$, then some member of $\mathcal{V}$ has a nontrivial abelian congruence.

A nontrivial congruence class of an abelian congruence generates an abelian subalgebra, by idempotence, so a failure of $[\alpha, \beta] = \alpha \wedge \beta$ implies the existence of a nontrivial abelian algebra in the variety. It is our conjecture that any nontrivial idempotent abelian algebra has a nontrivial homomorphic image which is a subreduct of a module. If this conjecture is true, then Theorem 5.12 can be applied to obtain a contradiction to $\mathcal{V}$ being exceptional. To summarize: if every nontrivial idempotent abelian algebra has a nontrivial homomorphic image which is a subreduct of a module, then any exceptional idempotent minimal variety satisfies the commutator equation $[\alpha, \beta] = \alpha \wedge \beta$. This equation implies congruence meet semidistributivity. A proof of the conjecture stated in this paragraph would be a good first step toward proving that there is no exceptional idempotent minimal variety.

6   Simple Modes

As mentioned in Section 1, a mode is an idempotent, entropic algebra. Of the numerous articles have been written about modes, most develop specific examples and restrict attention to only those examples. Hard evidence that the current list of examples is comprehensive is lacking. In response to this we initiated in [6] a classification all locally finite varieties of modes. This project was not completed in [6]. Here is what we accomplished. We showed that when $\mathcal{V}$ is a locally finite variety of modes, then

$$\mathcal{V} = (\mathcal{V}_1 \times \mathcal{V}_2) \circ \mathcal{V}_5$$

where $\mathcal{V}_1$ is locally strongly solvable, $\mathcal{V}_2$ is affine and $\mathcal{V}_5$ has a semilattice term. $(\mathcal{V}_1 \times \mathcal{V}_2)$ denotes the varietal product of $\mathcal{V}_1$ and $\mathcal{V}_2$ while the previous displayed line means that $\mathcal{V}$ is a Mal'cev product of $(\mathcal{V}_1 \times \mathcal{V}_2)$ and $\mathcal{V}_5$. The structure of $\mathcal{V}_2$ may be said to be well-understood. $\mathcal{V}_2$ is term equivalent to a variety of affine modules over a finite commutative ring. We investigated $\mathcal{V}_5$ and its non-locally finite analogues in [5] and now these varieties are fairly well-understood as well. (We know all such varieties up to term equivalence. A mode with a semilattice term turns out to be a subreduct of a semimodule over a commutative semiring satisfying $1 + r = 1$.) Varieties of the form $\mathcal{V}_1$ are not yet well-understood. The classification of locally finite modes may be completed by (i) classifying mode varieties of the form $\mathcal{V} = \mathcal{V}_1$ and (ii) fully describing the nature of the product of the subvarieties $\mathcal{V}_1 \times \mathcal{V}_2$ and $\mathcal{V}_5$. This product appears to be a generalization of the regularization of a variety – indeed it is exactly the generalization in the case of groupoids – and so the algebras in $(\mathcal{V}_1 \times \mathcal{V}_2) \circ \mathcal{V}_5$ may turn out to be 'generalized Płonka sums of locally solvable algebras'.

Let us give a more explicit description of what the decomposition theorem of [6] means. Let $\mathbf{A}$ be an idempotent algebra and let $\alpha \prec \beta$ be a covering pair of congruences on **A**. If B is a β-class, then B is a subuniverse. Let **B** be the subalgebra generated by B. We define a **relative quotient of A** to be any algebra of the form $\mathbf{B}/(\alpha|_B)$. The algebras in an idempotent variety $\mathcal{V}$ which are relative quotients are precisely those algebras whose universe is a minimal congruence class of an algebra in $\mathcal{V}$. In particular, any simple algebra in $\mathcal{V}$ is a relative quotient. It happens that the relative quotients of a finite mode come in three types. A relative quotient of a finite mode is term equivalent to

(**1**) a set;

(**2**) a 1-dimensional affine vector space over a finite field; or

(**5**) a 2-element semilattice.

The three-fold classification of relative quotients of finite modes are numbered according to the tame congruence theory labeling scheme delineated in [4]. Types **3** and **4** do not appear in modes. Now we are in a position to give a more informative description of the decomposition $\mathcal{V} = (\mathcal{V}_1 \times \mathcal{V}_2) \circ \mathcal{V}_5$. Say that a finite algebra **has type i**, for $i = \mathbf{1}, \mathbf{2}$ or **5** if all its relative quotients are of type i. Say that a locally finite algebra **has type i** if all its finitely generated subalgebras have type i. The main result of [6] is that if $\mathcal{V}$ is a locally finite variety of modes, then the collection of all algebras in $\mathcal{V}$ of type i comprises a subvariety, $\mathcal{V}_i$. Lastly, these subvarieties combine so that $\mathcal{V} = (\mathcal{V}_1 \times \mathcal{V}_2) \circ \mathcal{V}_5$.

A classification of arbitrary mode varieties, including those that are not locally finite, might begin with a full description of those modes which arise as relative quotients. A smaller project would be to classify the simple modes. It was this project that led to the results in this paper. Beginning with the complete characterization of the finite simple modes (which is an immediate consequence of the results in [6]) it is natural to call a simple mode **standard** if it is term equivalent to a 2-element set, a 2-element semilattice or a 1-dimensional affine vector space. Otherwise, we shall call it **non-standard**. We shall show that any non-standard simple mode is a subreduct of a 1-dimensional vector space. This result, together with Theorem 5.12 implies that the minimal varieties of modes are precisely those varieties generated by a standard simple mode.

According to remarks made at the beginning of this paper, we only need to consider simple modes with more than two elements. Clearly every non-standard simple mode has more than two elements (in fact, has infinitely many). Hence, we must show that every simple mode of more than two elements is a subreduct of a 1-dimensional vector space.

THEOREM 6.1 *If* **A** *is a simple mode with more than two elements, then* **A** *is a subalgebra of a simple reduct of a 1-dimensional affine vector space.*

Proof: A simple mode is an idempotent simple algebra and therefore must belong to one of the three classes described in Theorem 1.1. We shall prove that there is no simple mode of more than two elements in classes (a) and no simple mode in class (c). This will prove at least that a simple mode with more than two elements is a subalgebra of a simple reduct of a module. A little more work will establish that this module is a vector space.

It follows from Lemma 2.1 and our assumption that $|A| > 2$ that $\mathbf{A}$ is not essentially unary. Let $t(\bar{x})$ be a term which depends on all $n > 1$ of its variables. Since $\mathbf{A}$ is entropic, $t^{\mathbf{A}} : \mathbf{A}^n \to \mathbf{A}$ is a homomorphism. If we restrict $t^{\mathbf{A}}$ to the diagonal of $\mathbf{A}^n$, then $t^{\mathbf{A}}$ is a bijection since $t^{\mathbf{A}}$ is idempotent. Hence $t^{\mathbf{A}} : \mathbf{A}^n \to \mathbf{A}$ is onto, but not 1-1. It follows that ker $t^{\mathbf{A}}$ is a proper, nonzero congruence of $\mathbf{A}^n$. If ker $t^{\mathbf{A}} = \eta_\sigma$ where σ is a proper, non-empty subsequence of n, then $t^{\mathbf{A}}$ depends only the variables whose subscripts appear in σ. This is impossible, since t was chosen so that $t^{\mathbf{A}}$ depends on all variables. Hence ker $t^{\mathbf{A}} \neq \eta_\sigma$ for any σ which means that ker $t^{\mathbf{A}}$ is a skew congruence on $\mathbf{A}^n$. This proves that there is no simple mode in class (c) of Theorem 1.1.

Now we must show that there is no simple mode with more than two elements in class (a) of Theorem 1.1. Assume that $\mathbf{B}$ is such a mode and that 0 is the unique absorbing element of $\mathbf{B}$. Choose $a \neq b$ such that $a, b \in B - \{0\}$. As $\mathbf{B}$ is simple we have $(0, a) \in \mathrm{Cg}(a, b)$, so by Mal'cev's congruence generation lemma there exists $p(x) \in \mathrm{Pol}_1 \mathbf{B}$ such that $p(a) = 0 \neq p(b)$ or $p(b) = 0 \neq p(a)$. Without loss of generality we may assume that $p(a) = 0 \neq p(b)$. $\mathbf{B}$ is a mode, so $p : \mathbf{B} \to \mathbf{B}$ is a homomorphism. Since $p(a) \neq p(b)$, p is not a constant homomorphism; the simplicity of $\mathbf{B}$ forces p to be a 1-1 homomorphism. Thus, p is an isomorphism from $\mathbf{B}$ to $p(\mathbf{B})$. But $0 \in p(B)$ is clearly an absorbing element for $p(\mathbf{B})$ since it is an absorbing element for $\mathbf{B}$. Since $a \mapsto 0$ under the isomorphism $p : \mathbf{B} \to p(\mathbf{B})$, it follows that a is an absorbing element for $\mathbf{B}$. Hence 0 and a are distinct absorbing elements for $\mathbf{B}$. This is impossible for algebras in class (a) of Theorem 1.1.

To finish the proof of the theorem, we must show that if $\mathbf{C}$ is a simple mode which is a subalgebra of a simple reduct of a module, then $\mathbf{C}$ is in fact a subalgebra of a simple reduct of a vector space. From Theorem 3.8, $\mathbf{C}$ is a subalgebra of a simple algebra $\widehat{\mathbf{C}} \in \mathcal{V}(\mathbf{C})$ and $\langle \widehat{\mathbf{C}} \; ; \; p(x, y, z) \rangle$ is a simple affine module. Thus $\widehat{\mathbf{C}}$ is a simple $(x - y + z)$-reduct of a 1-dimensional vector space. The fundamental operations of $\widehat{\mathbf{C}}$ are idempotent and commute with themselves, since $\mathcal{V}(\mathbf{C})$ is a variety of modes. The operation $p(x, y, z)$ is idempotent, commutes with itself and all the fundamental operations of $\mathbf{C}$ by Lemma 3.7. Hence $\langle \widehat{\mathbf{C}} \; ; \; p(x, y, z) \rangle$ is an affine module which is at the same time a mode. If we assume that $\langle \widehat{\mathbf{C}} ; p(x, y, z) \rangle$ is a faithful $\mathbf{R}$-module (which we may assume), then $\mathbf{R}$ is a commutative ring. (For when $\mathbf{R}$ acts faithfully on $\langle \widehat{\mathbf{C}} \; ; \; p(x, y, z) \rangle$ and $r, s \in R$, then the affine module operations $r(x) + (1 - r)(y)$

and $s(x) + (1 - s)(y)$ commute iff $rs = sr$.) It is well-known that any simple (affine) module over a commutative ring is a 1-dimensional vector space, so this completes our proof. $\square$

The simple modes of the form $\hat{\mathbf{A}}$ are either 1-dimensional affine vector spaces or they are simple, proper $(x - y + z)$-reducts of 1-dimensional affine vector spaces. By Theorem 5.5, the corresponding field must have a unit interval. By Theorem 5.10, the field must have a strongly archimedean partial order. We obtain the following corollary to these theorems.

COROLLARY 6.2 *If* $\mathbf{A}$ *is a non-standard simple mode, then* $\mathbf{A}$ *is a sub-algebra of a simple mode* $\hat{\mathbf{A}} \in \mathcal{V}(\mathbf{A})$ *where* $\hat{\mathbf{A}}$ *is a proper* $(x - y + z)$-reduct *of a 1-dimensional affine vector space over a strongly archimedean ordered field* $\langle \mathbf{F}; \leq \rangle$. *If* $t(\bar{x}) = \Sigma a_i x_i$ *is the affine representation of a term of* $\hat{\mathbf{A}}$, *then* $0 \leq a_i \leq 1$ *for each* i *and* $\Sigma a_i = 1$. $\square$

The important open question that concerns us most in this section is:

Question: Which fields have a strongly archimedean partial ordering?

This question must be answered in order to classify the non-standard simple modes. It may be that any strongly archimedean partial ordering of a field can be extended to a total ordering. If this is so, then the answer to the previous question is "the subfields of the real numbers" since, as was shown by Hilbert, any totally ordered archimedean ring is isomorphic to a subring of the real numbers. One can show fairly easily that to prove the statement "If a field $\mathbf{F}$ has a strongly archimedean partial ordering, then $\mathbf{F}$ is isomorphic to a subfield of the real numbers", it suffices to prove the statement when $\mathbf{F}$ is a subfield of the complex numbers. If every field with a strongly archimedean partial ordering is isomorphic to a subfield of the real numbers, then every non-standard simple mode is a subreduct of a 1-dimensional affine vector space over the real numbers.

Theorem 5.12 implies that a non-standard simple mode does not generate a minimal variety. This has the following consequence.

COROLLARY 6.3 *If* $\mathcal{V}$ *is a minimal variety of modes, then* $\mathcal{V}$ *is term equivalent to the variety of sets, the variety of semilattices or a variety of affine vector spaces.* $\square$

References

[1] J. Berman, *A proof of Lyndon's finite basis theorems*, Discrete Math. **29** (1980), 229–233.

[2] R. Freese and R. McKenzie, *Commutator Theory for Congruence Modular Varieties*, LMS Lecture Notes, no. 125. Cambridge University Press, 1987.

[3] C. Herrmann, *Affine algebras in congruence modular varieties*, Acta Sci. Math. (Szeged) **41** (1979), 119–125.

[4] D. Hobby and R. McKenzie, *The Structure of Finite Algebras*, Contemporary Mathematics, American Mathematics Society, Providence, Rhode Island, 1988.

[5] K. Kearnes, *Semilattice modes i & ii*, to appear in Algebra Universalis.

[6] ———, *The structure of finite modes*, in preparation.

[7] T.Y. Lam, *A First Course in Noncommutative Rings*, Graduate Texts in Mathematics, no. 131, Springer Verlag, 1991.

[8] R. Magari, *Una dimostrazione del fatto che ogni varietà ammette algebre semplici*, Ann. Univ. Ferrara Sez. VII (N.S.) **14** (1969), 1–4.

[9] R. McKenzie, G. McNulty, and W. Taylor, *Algebras, Lattices and Varieties*, vol. *1*, Wadsworth & Brooks/Cole, 1987.

[10] E.L. Post, *The Two-Valued Iterative Systems of Mathematical Logic*, Annals of Math. Studies, no. 5, Princeton Univ. Press, 1941, Reprinted by Kraus Reprint Corp., 1965.

[11] A. Romanowska and J.D.H. Smith, *Modal Theory - an Algebraic Approach to Order, Geometry and Convexity*, Heldermann Verlag, Berlin, 1985.

[12] Á. Szendrei, *On the arity of affine modules*, Colloq. Math. **38** (1977), 1–4.

[13] ———, *On the idempotent reducts of modules, I & II*, Coll. Math. Soc. J. Bolyai **29** (1977), 753–780.

[14] ———, *Idempotent algebras with restrictions on subalgebras*, Acta Sci. Math. (Szeged) **51** (1987), 251–268.

[15] ———, *Every idempotent plain algebra generates a minimal variety*, Algebra Universalis **25** (1988), 36–39.

DEPARTMENT OF MATHEMATICAL SCIENCES, UNIVERSITY OF ARKANSAS, FAYET-TEVILLE, AR 72701, USA

   e-mail address: `kearnes@comp.uark.edu`

A revision of the mathematical part of Magari's paper on "Introduction to metamorality"

Roberto Magari   Giulia Simi

Abstract

In [2], Roberto Magari studies a mathematical problem concerning moralities. Therein, mathematical considerations are very synthetic, and there are some minor errors. Our aim here is to present a detailed exposition of Magari's results, and to frame his paper in the context of pure decision theory.

1 Introduction

As stated in the summary, the aim of this note is to give a detailed proof of the main mathematical result of [2] (proofs were very sketchy there). Even though the mathematical content of the paper had been completed before, this introduction has been entirely written by the second author after Magari's death. For this reason, the philosophical motivations of this work are not included here; for these, the reader is warmly invited to read [1] or [2]. Instead in this section we informally introduce the main concepts occurring in the paper, and we give a rather informal statement of the main problem of this note. A formal treatment is given in Section 2.

Our first formulation of the main problem is very close to the original one [2]. A simplified version will be introduced in Section 2. Consider a finite set $S = \{s_1, \ldots, s_n\}$ of mutually incompatible situations, a set A of possible actions and an ordered field K (we can assume K to be the field of real numbers if we wish to, but this is not necessary). Let T denote the set of instants. We assume that to each $a \in A$ and to each $t \in T$, an n-tuple $(p_1(a, t), \ldots, p_n(a, t))$ is associated such that for each i, $p_i(a, t) \in K$, $p_i(a, t) \geq 0$, and $\sum_{i=1}^{n} p_i(a, t) = 1$. Indeed $p_i(a, t)$ is supposed to represent the probability that situation s_i occurs at instant $t + 1$ if we perform action a at instant t.

Roughly speaking, a morality is a way of choosing, among a set X of possible actions, the subset Y of X consisting of those actions which are considered as "the best ones among all actions in X": of course, the intuition is that the choice of the best possible actions should take the probability distribution (i.e., the functions $p_i(a,t), i = 1,\ldots,n$) into account. Note that the domain of a morality is a *family* of subsets of A: we suggest the following intuitive interpretation of this fact: an action can be possible a priori, but not possible in a particular context; a subset of A can be thought as the set of actions which are possible in a given context (e.g., in a chess game, a move, say moving the queen along a diagonal, can be possible a priori, but whether it is possible or not during the game depends on the position of the chess pieces). Thus, a morality is a way of choosing not among all possible actions, but among all actions which are possible in a given context. A morality can fail to be defined on all subsets of possible actions: this corresponds to the case where the person is not able to make a choice: moreover it is possible that many actions from a given set X are selected: indeed, we do not exclude the possibility that different actions are regarded as "equally good", or simply incomparable.

Now, assume that our person is equipped with a function ν from S into K, assigning to each s_i the value $\nu_i = \nu(s_i)$ that the person attributes to situation s_i. Then, a possible morality is illustrated by the following example.

Example 1.1 (The roulette player) In this example, the situation at instant t can be described by the money the player owns, the past behavior of the roulette wheel, and so on. The player has a choice between various actions. Suppose he predicts that, if he carries out the action a, then situation s_i occurs with probability p_i ($i = 1,\ldots,n$). This information is collected in the function $p(s_i,t,a)$. Now, suppose that the player can calculate the value ν_i of situations s_i ($i = 1,\ldots,n$) on the basis of the money he might lose or win in each case. How should the gambler choose how to act and justify it to himself? A possible "morality" for him is the following: $a(t)$, the action which the gambler carries out at instant t, should make maximal the expression $\sum_{i=1}^{n} p(s_i,t,a)\nu(s_i)$.

In the sequel, this kind of morality (i.e., a morality which selects, among all elements of a set X of actions in its domain, the subset of those actions a of X which make the expression $\sum_{i=1}^{n} p(s_i,t,a)\nu(s_i)$ maximal) will be called a "Pascalian morality". One of the arguments which can be invoked against Pascalian moralities is that such moralities do not allow for "inestimable values" or absolute imperatives (cf. also the intuitive remark at the end of this paper). This sort of objection is made invalid if we allow our function ν to take values not necessarily in K, but possibly in an unbounded extension field L of K. This of course allows for values which are infinite with respect to K, i.e. not bounded by any element of K. The aim of this paper consists of a characterization of Pascalian moralities.

2   Moralities

In order to simplify our treatment, we shall reformulate the original problem in a slightly different way. First of all, since dependence on time is irrelevant for our purposes, we shall simply ignore it (we can assume time to be included e.g. in a situation or in a context); thus, our functions p_i will depend only on actions and not on time. Moreover, from an abstract mathematical point of view, we can identify two actions a and b if $(p_1(a), \ldots, p_n(a)) = (p_1(b), \ldots, p_n(b))$, i.e. if the probability distribution of $s_1, \ldots, s_n$ after performing actions a, b are the same; thus, we think of an action as a n-tuple $(p_1, \ldots, p_n)$ such that for $i = 1, \ldots, n$, $p_i \geq 0$ and $\sum_{i=1}^n p_i = 1$.

Notation The set of all actions, i.e. the set

$$\{(p_1, \ldots, p_n) : \text{for } i = 1, \ldots, n, \ p_i \geq 0, \text{ and } \sum_{i=1}^n p_i = 1\}$$

is denoted by P^n. In the sequel, if $a \in P^n$ and $i \leq n$, a_i denotes the i-th component of a.

Definition 2.1 Let S be a set of cardinality n. A **morality on** S (or n-**morality**) is a partial function f from $\mathcal{P}(P^n)$, the power set of P^n, into $\mathcal{P}(P^n)$ such that, letting $D_f = Dom(f)$, the following condition holds:

> if $X \in D_f$, then $X \neq \emptyset$, $f(X) \neq \emptyset$, and $f(X)$ is a subset of X.

A morality f is said to be **complete** if every finite non empty subset of P^n belongs to D_f.

Example 2.2 (The chess player). Consider a chess game; let S be the set of situations which may occur during the game. If we imagine that a player's move, allowing for his adversary will also move, does not lead him to a certain situation previously determinable, but, in his opinion, leads to situation s_i with probability p_i $(i = 1, \ldots, n)$ we can abstractly identify the move with the n-tuple $(p_1, \ldots, p_n)$. Of course, in a chess game not all moves are possible. This might suggest to isolate a subset of P^n (the set of possible actions); however, we do not need this: e.g., if only actions from a subset A of P^n are possible, we can assume each set in the domain of the morality f to be a subset of A, or simply we can consider irrelevant the values of f on subsets of P^n which are not subsets of A. In the example under consideration, the chess player in each situation has a choice among actions (moves) from a certain subset X of the set A of possible actions, and his morality (style of playing) is determined if, for each subset X of A, we have a corresponding $f(X)$ representing the singleton of the move the player would choose among the moves in X.

As we said before, given a morality f and a set $X \in D_f$, $f(X)$ does not need in general to be a singleton; we call **regular** a morality f such that, for all $X \in D_f$, $f(X)$ is a singleton.

Definition 2.3 An n-**evaluation** (or, simply, an **evaluation**) is a map ν from a set S of cardinality n into an extension field L of K. A **Pascalian pair** is a pair $\langle f, \nu \rangle$ such that:

(i) f is a morality, and ν is an n-evaluation.

(ii)   Let, for $a = (a_1, \ldots, a_n) \in P^n$, $\nu^*(a) = \sum_{i=1}^n a_i \nu_i$. Then;

   (iia)   If $X \in D_f$, $a, b \in X$, and $a, b \in f(X)$, then $\nu^*(a) = \nu^*(b)$.
   (iib)   If $X \in D_f$, $a, b \in X$, and $b \in f(X)$ and $a \notin f(X)$, then $\nu^*(a) < \nu^*(b)$.

A morality f is said **Pascalian** if it is the first component of some Pascalian pair $\langle f, \nu \rangle$.

Remark 2.4 We shall see (cf. Lemma 2.18) that, given a Pascalian morality, there are infinitely many evaluations such that $\langle f, \nu \rangle$ is a Pascalian pair. Any such ν will be called an **evaluation** for f.

Definition 2.5 A Pascalian morality is called **standard** if there is some evaluation ν for f such that for $s \in S$, $\nu(s) \in K$.

We shall prove (Example 4.3) that there are nonstandard Pascalian moralities. As one might expect, nonstandard Pascalian moralities correspond to the case where "absolute imperatives" are present.

Definition 2.6 Let f, g be n-moralities. We say that g is a **completion** of f if g is a complete morality which extends f (i.e., g is complete and $g(X) = f(X)$ for all $X \in D_f$).

Lemma 2.7 *A morality f is Pascalian iff it has a Pascalian completion.*

Proof. The right to left implication is obvious. Conversely, let f be a Pascalian morality, and let ν be an evaluation for f. Define $g(X) = f(X)$ if $X \in D_f$; $g(X) = \{a \in X : \forall x \in X \; \nu^*(x) \le \nu(a)\}$ if X is finite and $X \notin D_f$; $g(X)$ undefined otherwise. Clearly, g is complete and extends f. Moreover, it is easily seen that $\langle g, \nu \rangle$ is a Pascalian pair, therefore g is Pascalian.

As we said before, our aim is a characterization of Pascalian moralities. By Lemma 2.7, our problem reduces to that of characterizing those complete moralities which are Pascalian. In the sequel, even without explicit mention, *we always refer to complete moralities.*

Definition 2.8 Let f be a complete morality; define, for $a, b \in P^n$, $a <_f b$ (to be read as "b is f-**preferable**, or, simply, **preferable**, to a") iff there is a subset X of P^n such that $a, b \in X$, $a \notin f(X)$, $b \in f(X)$; $<_f$ will be called the **preferability relation associated to** f; the subscript "f" will be often omitted when there is not danger of confusion.

Note that there are moralities f such that, for some $a, b \in P^n$, $a <_f b$ and $b <_f a$; these correspond to the behaviour of a person who prefers a to b in some contexts and b to a in other contexts.

Definition 2.9 A morality f is called **stable** if, whenever $a, b \in X \in D_f$, and $a <_f b$, one has $a \notin f(X)$ (hence, in particular, if $a <_f b$, then $b \not<_f a$).

Clearly, any morality f determines $<_f$ uniquely; conversely, if f is a stable complete morality, its behaviour on finite subsets of P^n is completely determined by $<_f$: given a finite subset X of P^n,

$$f(X) = \{z \in X : \forall u \in X \neg (z <_f u)\}.$$

Since a Pascalian morality is evidently stable, by the observation above, our problem reduces to characterizing those preference relations which are associated to a complete Pascalian morality.

Lemma 2.10 *The pair $\langle f, \nu \rangle$ (f a complete morality) is Pascalian iff the following condition holds:*

$$(*) \qquad \text{for } a, b \in P^n, \; a < b \text{ iff } \nu^*(a) < \nu^*(b).$$

Thus, a (complete) morality f is Pascalian iff there is a linear function ν^* from K^n to an extension field L of K (where both K^n and L are regarded as vector spaces over the field K) such that the preference relation $<_f$ associated with f satisfies condition $(*)$.

Proof. That condition (∗) implies that $\langle f, \nu \rangle$ is Pascalian, is an easy consequence of Definition 2.3. Conversely, assume that $\langle f, \nu \rangle$ is Pascalian. If $a < b$, then, by condition (iib) in Definition 2.3, one has $\nu^*(a) < \nu^*(b)$. Viceversa, if $\nu^*(a) < \nu^*(b)$, consider $f(\{a, b\})$ (which is defined, as f is complete, and is non-empty, by the definition of morality). By condition (iib) in Definition 2.3, $f(\{a, b\})$ can not be $\{a\}$; by condition (iia) of Definition 2.3, $f(\{a, b\})$ cannot be $\{a, b\}$. Thus, $f(\{a, b\}) = \{b\}$, and, again by condition (iib), $\nu^*(a) < \nu^*(b)$.

Definition 2.11 . A binary relation $<$ on P^n is said to be a **Pascalian** n-**preference relation** if there is an n-evaluation ν such that, for $a, b \in P^n$, we have $a < b$ iff $\nu^*(a) < \nu^*(b)$.

By Lemma 2.10, our problem reduces to the characterization of Pascalian preference relations.

Definition 2.12 . Let $<$ be a binary relation on P^n; we define, for $a, b \in P^n$, $a \sim_< b$ iff $a \not< b$ and $b \not< a$. The subscript "$<$" will be often omitted when there is no danger of confusion. In the sequel, we write $a \leq b$ as an abbreviation for $(a < b) \vee (a \sim_< b)$.

Proposition 2.13 *If $<$ is a Pascalian n-preference relation, then:*

(i) *$<$ is transitive and antireflexive;*

(ii) *$\sim$ is an equivalence relation;*

(iii) *for $a, b \in P^n$, one has $a \sim b$ iff for every $c \in P^n$, $c < a$ iff $c < b$, and $a < c$ iff $b < c$;*

(iv) *for $a, b \in P^n$, exactly one of the following holds: $a < b$, $b < a$, $a \sim b$.*

Proof. Obvious.

In the sequel, a binary relation $<$ satisfying conditions (i), (iii) and (iv) in Proposition 2.13 will be called a n-**near order**.

Notation. Let $n > 0$ be given. In the sequel, for every $i \leq n$, ϵ_i denotes the element of P^n whose j-th component $(j \leq n)$ is 1 if $j = i$ and 0 otherwise. Given an n-evaluation ν, we write ν_i as an abbreviation for $\nu(s_i)$ (where we assume $S = D_\nu = \{s_1, \ldots, s_n\}$). Note that $\nu_i = \nu^*(\epsilon_i)$.

The requirement that $<$ is a n-near order for some n does not exhaust the properties that common sense would expect of any morality, as the next example suggests.

Example 2.14 Let $n = 3$, and assume $(0, 1, 0) < (1, 0, 0)$. One would probably expect that the first component is, in a sense, more relevant than the second one. Thus, common sense would probably consider "incoherent" that $(1/3, 1/3, 1/3) < (1/6, 1/2, 1/3)$.

Definition 2.15 An n-preference relation $<$ is said to be **coherent** if, whenever $\epsilon_i < \epsilon_h$ and a, b are elements of P^n which differ only on components i and h, then $a < b$ iff $a_h < b_h$ (and, therefore, iff $b_i < a_i$, since the sums of all components of a and b respectively is 1).

Lemma 2.16 *Any Pascalian preference relation is coherent.*

Proof. Left to the reader. The reader who is not willing to do this proof can observe that Lemma 2.16 follows immediately from Lemmas 2.21 and 2.24 below.

Definition 2.17 Let ν be an n-evaluation. We define, for $a, b \in P^n$, $a <_\nu b$ iff $\nu^*(a) < \nu^*(b)$. Note that $<_\nu$ is a Pascalian preference relation.

Lemma 2.18 *Let ν be an n-evaluation, $u, k \in L$, $u > 0$; let, for $s \in S$, $\nu'(s) = \nu(s) + u$; $\nu''(s) = k\nu(s)$. Then, the relations $<_\nu$, $<_{\nu'}$, and $<_{\nu''}$ coincide.*

Proof. Obvious.

One might expect that every coherent n-near order is a Pascalian preference relation. This conjecture is refuted by the following example.

Example 2.19 Let us identify each $(p, q, r) \in P^3$ with the point of the plane of coordinates (p, r) (note that q is uniquely determined from (p, r) by the condition $p + q + r = 1$). According to this convention, P^3 can be represented as the set of elements of the triangle T of vertices $B(0, 0)$, $A(1, 0)$ and $C(0, 1)$ (cf. Figure 1). Let $O(a, c)$ be such that $a > 1$, $b > 0$. Define a near order on T as follows: given $P, Q \in T$, let P', Q' be the intersections of the line BC with OP and OQ respectively. We define $P < Q$ iff P' precedes Q' in the order from B to C (cf. Figure 1).

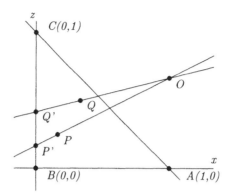

Figure 1.

It is easy to verify that $<$ is a near order. Note that $A < B < C$, therefore, $\epsilon_1 < \epsilon_2 < \epsilon_3$. Moreover, it is easily seen that, given $P(p_1, p_2, p_3)$ and $Q(q_1, q_2, q_3)$, one has: if $p_1 = q_1$, or $p_2 = q_2$, then $P < Q$ iff $p_3 < q_3$; if $p_3 = q_3$, then $P < Q$ iff $p_2 < q_2$. Thus, $<$ is coherent. We prove that $<$ is not Pascalian. Suppose it is; let ν be a 3-evaluation such that $<=<_\nu$. By Lemma 2.18, we may assume $\nu_1 = \nu(\epsilon_1) = 0$; $\nu_2 = \nu(\epsilon_2) = 1$; $\nu_3 = \nu(\epsilon_3) > 1$. Let k be such that $0 \le k \le 1$, and let $a = (0, 1 - k, k)$; if $(x, y, z) \sim a$, then we must have $\nu^*(x, y, z) = \nu^*(a)$, i.e. $y + \nu_3 z = 1 - k + \nu_3 k$. Since $y = 1 - x - z$, the previous condition is equivalent to:

$$(**) \qquad x + (1 - \nu_3)z = k(1 - \nu_3).$$

Thus, all the points $Q(x, z)$ corresponding to elements $a \in P^3$ such that $a \sim (0, 1 - k, k)$ lie on the line of equation $(**)$. Note that this line contains the point $P(0, k)$. Now, since $P \not< Q$ and $Q \not< P$, the intersections P' and Q' of the line BC with the lines OP and OQ respectively must coincide; it follows that O, P and Q lie on the same line, which is necessarily the line of equation $(**)$. Since this is true independently of the choice of $k \in [0, 1]$, we conclude that O lies in all the lines of equation $(**)$ as k varies in $[0, 1]$. This is clearly impossible if O is proper, as these lines are parallel. Note that if we assume O to be not proper, we can obtain in this way a Pascalian preference relation.

Notation. For $a, b \in P^n$, $r \in K$, we let $a + b = (a_1 + b_1, \ldots, a_n + b_n)$, $ra = (ra_1, \ldots, ra_n)$. Note that, if $a, b \in P^n$, and $r \in [0, 1]$, then $ra + (1 - r)b \in P^n$.

In order to characterize Pascalian preference relations, we start from the following intuitive consideration: let $a, b, c, d \in P^n$ be such that $a < b$ and

$c \leq d$. The action $h = (1/3)a + (2/3)c$ can be regarded as one yielding the same result as a with probability $1/3$ and the same result as c with probability $2/3$. Similarly, the action $k = (1/3)b + (2/3)d$ can be regarded as one yielding the same result as b with probability $1/3$ and the same result as d with probability $2/3$. Most uses of the notion of preference in morality suggest that it should be $h < k$. This observation suggests the following:

Definition 2.20 (Condition of linearity) A n-near order $<$ is said to be **linear** iff, whenever $a, b, c, d \in P^n$, $r \in [0, 1]$, the following hold:

(i) if $a \leq b$, $c \leq d$, then $ra + (1 - r)c \leq rb + (1 - r)d$;

(ii) if $a < b$, $c \leq d$, and $r > 0$, then $ra + (1 - r)c < rb + (1 - r)d$.

Lemma 2.21 *Any Pascalian preference relation is a linear near order.*

Proof. Let ν^* be a linear function from K^n to an extension field L of K such that, for $a, b \in P^n$, one has: $a < b$ iff $\nu^*(a) < \nu^*(b)$. We verify condition (i) in Definition 2.20 (verification of (ii) is similar). If $r \in [0, 1]$, $a \leq b$ and $c \leq d$, then $\nu^*(a) \leq \nu^*(b)$, $\nu^*(c) \leq \nu^*(d)$, and, by the linearity of ν^*, $\nu^*(ra + (1 - r)c) = r\nu^*(a) + (1 - r)\nu^*(c) \leq r\nu^*(b) + (1 - r)\nu^*(d) = \nu^*(rb + (1 - r)d)$.

Lemma 2.22 *Let $<$ be a n-linear near order, $b^i, c^i \in P^n$ $(i = 1, \ldots, k)$. Let $r_i \in [0, 1]$ $(i = 1, \ldots, k)$ be such that $\sum_{i=1}^{k} r_i = 1$. Then:*

(i) *if, for $i = 1, \ldots, k$, $b^i \leq c^i$, then $\sum_{i=1}^{k} r_i b^i \leq \sum_{i=1}^{k} r_i c^i$.*

(ii) *if, for $i = 1, \ldots, k$, $b^i \leq c^i$, and there is an $i \leq k$ such that $r_i > 0$, and $b^i < c^i$, then $\sum_{i=1}^{k} r_i b^i < \sum_{i=1}^{k} r_i c^i$.*

Proof. (i) Induction on k. For $k = 2$, the claim follows from the definition of linear near order. Let $k > 2$, and assume that (i) holds for $m < k$. Let $r = \sum_{i=1}^{k-1} r_i$. If $r = 0$, then $r_k = 1$, and the claim is obvious. Thus assume $r > 0$. By the Induction Hypothesis, $\sum_{i=1}^{k-1}(r_i/r)b^i \leq \sum_{i=1}^{k-1}(r_i/r)c^i$. By linearity,

$$
\begin{aligned}
\sum_{i=1}^{k} r_i b^i &= r \sum_{i=1}^{k-1}(r_i/r)b^i + r_k b^k \\
&= r \sum_{i=1}^{k-1}(r_i/r)b^i + (1 - r)b^k \leq r \sum_{i=1}^{k}(r_i/r)c^i + (1 - r)c^k \\
&= \sum_{i=1}^{k} r_i c^i.
\end{aligned}
$$

(ii) Induction on k. Again, the claim is trivial if $k = 2$. Let $k > 2$, and assume that (ii) holds for $m < k$. After renaming indexes, we can assume w.l.o.g. $r_k > 0$, and $b^k < c^k$. Let r be as in the proof of (i), and assume w.l.o.g. $r > 0$. Again, we get $\sum_{i=1}^{k-1}(r_i/r)b^i \leq \sum_{i=1}^{k-1}(r_i/r)c^i$. By linearity,

$$
\begin{aligned}
\sum_{i=1}^{k} r_i b^i &= r \sum_{i=1}^{k-1}(r_i/r)b^i + (1-r)b^k \\
&< r \sum_{i=1}^{k-1}(r_i/r)c^i + (1-r)c^k = \sum_{i=1}^{k} r_i c^i.
\end{aligned}
$$

Corollary 2.23 *Let $<$ be a n-linear near order, $b^i, c^i \in P^n$ be such that, for $i = 1, \ldots, k$, $b^i \sim c^i$. Let r_i $(i = 1, \ldots, k)$ be as in Lemma 2.22. Then $\sum_{i=1}^{k} r_i b^i \sim \sum_{i=1}^{k} r_i c^i$.*

Lemma 2.24 *Every linear near order is coherent.*

Proof. Let $\epsilon_i < \epsilon_h$, and let $a, b \in P^n$ differ only on components i and h. Assume $a_h < b_h$. Note that $a_h + a_i = b_h + b_i$, therefore, $a_i - b_i = b_h - a_h$. By Lemma 2.22, we obtain:

$$
\begin{aligned}
a &= \sum_{r \neq i,h} a_r \epsilon_r + b_i \epsilon_i + (a_i - b_i)\epsilon_i + a_h \epsilon_h < \sum_{r \neq i,h} a_r \epsilon_r + b_i \epsilon_i + (a_i - b_i)\epsilon_h + a_h \epsilon_h \\
&= \sum_{r \neq i,h} b_r \epsilon_r + b_i \epsilon_i + (b_h - a_h)\epsilon_h + a_h \epsilon_h = b.
\end{aligned}
$$

3 Main result

This section is entirely devoted to the proof of the main result of this paper.

Theorem 3.1 *A (complete) morality f is Pascalian iff its associated relation $<$ is a linear near order.*

Proof. One direction is given by Lemma 2.21. Thus, assume that $<$ is a linear near order on P^n, and let us prove that f is Pascalian. By Lemma 2.10, it suffices to prove that $<$ is a Pascalian preference relation. We reason by induction on n. If $n = 1$ or $n = 2$, the claim is trivial. Let $n > 2$, and assume that the claim holds for $m < n$. After renaming the $\epsilon_i s$, we can assume $\epsilon_1 \leq \epsilon_2 \leq \ldots \leq \epsilon_n$. For $a = (a_1, \ldots, a_{n-1}) \in P^{n-1}$, let $a^* = (a_1, \ldots, a_{n-1}, 0)$, and let, for $a, b \in P^{n-1}$, $a <' b$ iff $a^* < b^*$. Clearly, $<'$ is a linear near order. By the Induction Hypothesis, there are an extension field L of K and elements $\nu_1, \nu_2, \ldots, \nu_{n-1}$ of L such that letting, for $(a_1, \ldots, a_{n-1}) \in P^{n-1}$, $\nu'(a) = \sum_{i=1}^{n-1} a_i \nu_i$, one has $a <' b$ iff $\nu'(a) < \nu'(b)$. By Lemma 2.18, we can also assume w.l.o.g. $\nu_{n-1} = 1$. Of course, it suffices to find a field extension L^* of L and a $\nu_n \in L^*$ such that, letting, for $a = (a_1, \ldots, a_n) \in P^n$, $\nu^*(a) = \sum_{i=1}^{n} a_i \nu_i$, one has:

$$(+) \qquad a < b \text{ iff } \nu^*(a) < \nu^*(b).$$

Clearly, condition $(+)$ is satisfied (independently of the choice of L^* and ν_n), if a, b have the form c^*, d^* respectively for some $c, d \in P^{n-1}$ (in other words is satisfied by those pairs $c, d \in P^n$ whose last component is 0). If $\epsilon_n \sim \epsilon_{n-1}$, we put $L^* = L$, $\nu_n = \nu_{n-1}$, and using linearity, we can easily prove that condition $(+)$ is satisfied. Thus, suppose $\epsilon_{n-1} < \epsilon_n$. First of all, we prove:

Claim 1. If $a_n = b_n$, then $a < b$ iff $\sum_{i=1}^{n-1} a_i \nu_i < \sum_{i=1}^{n-1} b_i \nu_i$.

Proof of Claim 1. The claim is obvious if $a = \epsilon_n$ or $b = \epsilon_n$. Thus suppose $a, b \neq \epsilon_n$. Let, for $a \in P^n - \{\epsilon_n\}$, $a^+ = (1/(1 - a_n))(a_1, \ldots, a_{n-1}, 0)$. Clearly, for $a \in P^n - \{\epsilon_n\}$, one has $a = (1 - a_n)a^+ + a_n \epsilon_n$. Applying linearity to the pairs a^+, b^+ and ϵ_n, ϵ_n, and using the assumption $a_n = b_n$, we obtain: $a < b$ iff $a^+ < b^+$ iff $\nu^*(a^+) < \nu^*(b^+)$ iff $\sum_{i=1}^{n-1} a_i \nu_i < \sum_{i=1}^{n-1} b_i \nu_i$. This proves Claim 1.

Now, define, for $a \in P^n$, $k \in L$, $w(a, k) = \sum_{i=1}^{n-1} a_i \nu_i + k a_n$. Also, define:

$$W^- = \{k \in L : \exists a, b \in P^n \ (a < b, a_n < b_n, \text{ and } w(a, k) \geq w(b, k))\}$$
$$W = \{k \in L : \exists a, b \in P^n \ (a \sim b, a_n \neq b_n, \text{ and } w(a, k) = w(b, k))\}$$
$$W^+ = \{k \in L : \exists a, b \in P^n \ (a < b, a_n > b_n, \text{ and } w(a, k) \geq w(b, k))\}.$$

Note that, if there are an extension field L^* of L and a $\nu_n \in L^*$ such that condition $(+)$ is satisfied, then, for $h \in W^-$, $k \in W$, $m \in W^+$, we must have $h < \nu_n = k < m$. The plan of the proof is to show that, if W is non empty, then it has a unique element, say k, and $\nu_n = k$ satisfies our requirements; otherwise, there are an extension field L^* of L and a $\nu_n \in L^*$ such that ν_n is an upper bound of W^- and a lower bound of W^+; any such ν_n will meet our requirements.

It is easily seen that W^- is non empty, because $\epsilon_{n-1} < \epsilon_n$, and $1 = \nu_{n-1} = w(\epsilon_{n-1}, 1) = w(\epsilon_n, 1)$. Thus, $1 \in W^-$. On the other hand, it follows from Example 4.3 below that in some cases W and W^+ might be empty.

It is readily seen that, if $h \in W^-$ and $k < h$, then $k \in W^-$, and that, if $k \in W^+$ and $k < h$, then $h \in W^+$. The following claim is crucial to our proof:

Claim 2.

(i)   If $h \in W^-$ and $k \in W$, then $h < k$;

(ii) if $h \in W \cup W^-$ and $k \in W^+$, then $h < k$;

(iii) if $h, k \in W$, then $h = k$.

Proof of Claim 2. We prove only (i); proofs of (ii) and (iii) are similar. Let $h \in W^-$, $k \in W$. Let $a, b, c, d \in P^n$ be such that $a < b$, $a_n < b_n$, $c \sim d$, $c_n \neq d_n$, (say, $c_n < d_n$), $w(a, h) \geq w(b, h)$, and $w(c, k) = w(d, k)$. Let $t = b_n - a_n + d_n - c_n$, $r = (b_n - a_n)/t$, $s = 1 - r = (d_n - c_n)/t$. By linearity, $rd + sa < rc + sb$. Now, $rd + sa$ and $rc + sb$ have the same n-th component, namely $(b_n d_n - c_n a_n)/t$. By Claim 1,

$$\sum_{i=1}^{n-1}(rd + sa)_i \nu_i \quad < \quad \sum_{i=1}^{n-1}(rc + sb)_i \nu_i$$

$$r\sum_{i=1}^{n-1}(d_i - c_i)\nu_i \quad < \quad s\sum_{i=1}^{n-1}(b_i - a_i)\nu_i.$$

By the definition of r and s, it follows

$$(1/(d_n - c_n))\sum_{i=1}^{n-1}(d_i - c_i)\nu_i < (1/(b_n - a_n))\sum_{i=1}^{n-1}(b_i - a_i)\nu_i,$$

therefore

$$(1/(b_n - a_n))\sum_{i=1}^{n-1}(a_i - b_i)\nu_i < (1/(d_n - c_n))\sum_{i=1}^{n-1}(c_i - d_i)\nu_i$$

From $w(a, h) \geq w(b, h)$, and $w(c, k) = w(d, k)$ it follows:

$$h \leq (1/(b_n - a_n))\sum_{i=1}^{n-1}(a_i - b_i)\nu_i \qquad k = (1/(d_n - c_n))\sum_{i=1}^{n-1}(c_i - d_i)\nu_i.$$

Thus, $h < k$. The proofs of (ii) and (iii) are similar.

Now, suppose first that W is non empty; by Claim 2(iii), W has exactly one element, k say. We put $\nu_n = k$, and we prove (+) according to this choice of ν_n. It follows immediately from Claim 1 if $a_n = b_n$, hence suppose $a_n < b_n$. If $a < b$, then $\nu^*(a) = w(a, \nu_n) < w(b, \nu_n)$, otherwise $\nu_n \in W^-$, and by Claim 2(i), we would conclude $\nu_n < \nu_n$. Thus, $\nu^*(a) < \nu^*(b)$. Conversely, if $\nu^*(a) < \nu^*(b)$, by what we have just shown, it is not the case that $b < a$ (otherwise, $\nu^*(b) < \nu^*(a)$); we can also exclude that $a \sim b$. Indeed, the linear equation $w(a, t) = w(b, t)$ has a solution t_0 (since $a_n \neq b_n$, the coefficient of t is $\neq 0$). Since $w(a, \nu_n) < w(b, \nu_n)$, $t_0 \neq \nu_n$. If $a \sim b$, then $t_0 \in W$, $\nu_n \in W$, and $t_0 \neq \nu_n$, against Claim 2(iii). The argument for the case $a_n > b_n$ is quite similar.

Now, assume $W \neq \emptyset$. We can find an extension field L^* of L and an element $k^* \in L^*$ which is an upper bound of W^- and a lower bound of W^+ (the second condition is vacuously satisfied if W^+ is empty). A possible way to do this is the following: let D be the elementary diagram of K, let k be a constant symbol not in D, and consider the theory

$$T = D \cup \{h < k : h \text{ a symbol for an element of } W^-\}$$
$$\cup \{k < m : m \text{ a symbol for an element of } W^+\}.$$

By the Compactness Theorem, we can easily prove that T has a model, call it L^*. Let k^* be the interpretation of k in L^*. It is easily seen that L^* and k^* meet our requirements. We claim that, letting $\nu_n = k^*$, condition $(+)$ holds. First, note that, if $a_n = b_n$, then, by Claim 1, we have $a < b$ iff $\sum_{i=1}^{n-1} a_i \nu_i < \sum_{i=1}^{n-1} b_i \nu_i$ iff for all t, $w(a, t) < w(b, t)$. Thus, independently of our choice of ν_n, condition $(+)$ is satisfied by those a, b such that $a_n = b_n$; moreover, if $a \sim b$, then $a_n = b_n$, otherwise, letting t_0 be the unique solution of the linear equation $w(a, t) = w(b, t)$, t_0 would belong to W. Therefore, independently of our choice of ν_n, condition $(+)$ is satisfied by those a, b such that $a \sim b$. Thus, we have only to worry about those a, b such that $a < b$ and $a_n \neq b_n$. Let $F(t) = w(a, t) - w(b, t)$, and consider the linear equation $F(t) = 0$. Let t_0 be the unique solution of that equation. Note that $t_0 \in L$. If $a_n < b_n$, then $t_0 \in W^-$, therefore $t_0 < \nu_n$. Moreover $F(t)$ is increasing, therefore $F(\nu_n) > 0$ and $\nu^*(a) < \nu^*(b)$; if $a_n > b_n$, then $t_0 \in W^+$, therefore $t_0 > \nu_n$. Moreover, $F(t)$ is decreasing, therefore $F(\nu_n) > 0$ and $\nu^*(a) < \nu^*(b)$.

4   Standard Pascalian moralities

Definition 4.1 A preference relation $<$ is said to have the **intermediate value property** (in short: i.v.p.) iff, whenever $a \leq b \leq c$, there is a $t \in [0, 1]$ such that $ta + (1 - t)c \sim b$.

Theorem 4.2 *A morality f is a standard Pascalian morality iff the preference relation $<$ associated to it is linear and has the i.v.p.*

Proof. Let f be a standard Pascalian morality, ν be a standard evaluation for f defined on $S = \{s_1, \ldots, s_n\}$; let $\nu_i = \nu(s_i)$ (thus, $\nu_i \in K$). Let $a \leq b \leq c$. Consider the linear function $F(t) = \nu^*(ta + (1 - t)c)$, $t \in [0,1]$. $F(0) = c \geq \nu^*(b)$, and $F(1) = a \leq \nu^*(b)$. It follows that there is $t_0 \in [0,1]$ such that $F(t_0) = \nu^*(b)$. Thus, $t_0 a + (1 - t_0)c \sim b$, and $<$ has the i.v.p. Conversely, assume that $<$ is linear and has the i.v.p. We prove the $<$ is standard Pascalian by induction on n. For $n = 2$, the claim is easy. (If, say, $\epsilon_1 < \epsilon_2$, it suffices to choose $\nu_1, \nu_2 \in K$ such that $\nu_1 < \nu_2$). Let $n > 2$, and assume that the claim holds for $m < n$. After renaming $\epsilon_i s$, we can assume $\epsilon_1 \leq \epsilon_n \leq \epsilon_2$.

Let, for $a = (a_1, \ldots, a_{n-1}) \in P^{n-1}$, a^* be defined as in the proof of Theorem 3.1 by $a^* = (a_1, \ldots, a_{n-1}, 0)$; again, let for $a, b \in P^{n-1}$, $a <' b$ iff $a^* < b^*$. $<'$ is linear and has the i.v.p. By the Induction Hypothesis, there are $\nu_1, \nu_2, \ldots, \nu_{n-1} \in K$ such that, letting, for $(a_1, \ldots, a_{n-1}) \in P^{n-1}$, $\nu'(a) = \sum_{i=1}^{n-1} a_i \nu_i$, one has $a <' b$ iff $\nu'(a) < \nu'(b)$. Now, let t_0 be such that $t_0 \epsilon_1 + (1 - t_0)\epsilon_2 \sim \epsilon_n$, and let $e = t_0 \epsilon_1 + (1 - t_0)\epsilon_2$, $\nu_n = \nu^*(e)$. As in the proof of Theorem 3.1, we put, for $a \in P^n - \{\epsilon_n\}$, $a^+ = (1/1 - a_n)(a_1, \ldots, a_{n-1}, 0)$. Recall that, for $a \in P^n - \{\epsilon_n\}$, $a = (1 - a_n)a^+ + a_n \epsilon_n$; thus, letting

$$a' = (1 - a_n)a^+ + a_n e, \text{ if } a \neq \epsilon_n,$$
$$a' = e, \text{ if } a = \epsilon_n,$$

we get $a \sim a'$. Moreover, since $\nu^*(\epsilon_n) = \nu(e) = \nu_n$, we can easily deduce $\nu^*(a) = \nu^*(a')$. Finally, the last component of a' is 0, therefore, for $a, b \in P^n$, one has $a' < b'$ iff $\nu^*(a') < \nu^*(b')$. Thus, $a < b$ iff $a' < b'$ iff $\nu^*(a') < \nu^*(b')$ iff $\nu^*(a) < \nu^*(b)$. This completes the proof.

We conclude with an example of a nonstandard Pascalian preference relation.

Example 4.3 Define, for $a, b \in P^3$, $a < b$ iff either $a_3 < b_3$, or $a_3 = b_3$ and $a_2 < b_2$. Clearly, $\epsilon_1 < \epsilon_2 < \epsilon_3$, but there is no $t \in [0,1]$ such that $\epsilon_2 \sim t\epsilon_1 + (1 - t)\epsilon_3$. Indeed, for $t < 1$, $t\epsilon_1 + (1 - t)\epsilon_3 > \epsilon_2$, and for $t = 1$, $t\epsilon_1 + (1 - t)\epsilon_3 < \epsilon_2$. Thus, $<$ has not the i.v.p. and therefore it is not standard. Note that $<$ is a Pascalian preference relation: let L be a nonstandard extension of K, and let $\nu_1 = 0$, $\nu_2 = 1$, $\nu_3 \in L$ be such that $\nu_3 > k$, for all $k \in K$. It is easily seen that, for $a, b \in P^3$, one has: $a < b$ iff $\nu^*(a) < \nu^*(b)$. Finally, note that $<$ is a linear order, therefore the morality f associated to $<$ is regular ($f(X)$ is the singleton of the maximum of X w.r.t. $<$, if such a maximum exists, and is undefined otherwise).

The following intuitive observation shows that standard Pascalian moralities are not completely satisfactory: let $S = \{s_1, s_2, s_3\}$, where s_1 is an indifferent situation, s_2 is a situation in which one gets a toothache, and s_3 is a situation in which one saves a human life; (assume s_1, s_2, s_3 to be mutually exclusive); let $p \in (0,1)$ be any (arbitrarily small) nonzero element of K, and consider two actions a_1 and a_2 such that a_1 yields situation s_1 with probability 1, and a_2 yields situation s_3 with probability p and situation s_2 with probability $1 - p$. In our opinion, independently on p, one should choose the second action.

References

[1] R. Magari, *Morale e matamorale*, CLUEB, Bologna, 1986.

[2] _____ , *Introduction to Metamorality*, Knowledge, belief and strategic interaction (C. Bicchieri and M.L. Dalla Chiara, eds.), Cambridge U. Press, 1992.

DIPARTIMENTO DI MATEMATICA, VIA DEL CAPITANO 15, I-53100, SIENA, ITALY

Some aspects of the categorical semantics for the polymorphic λ-calculus

Maria Emilia Maietti

Abstract

After introducing the syntax for a version of second order typed lambda calculus (Girard's system **F** plus product types and a terminal type), we give a categorical presentation of the coherence spaces model of polymorphism as a PLF-category, a generalization of Seely's PL-categories, whose class provides a complete semantics for second order lambda calculus.

Introduction

The polymorphic lambda calculus (or second order typed lambda calculus) is an extension of simply typed lambda calculus. Its main novelty is the introduction of "*type variables*", which allow to parameterize types and abstract them.

It was discovered, and given the name of system **F**, by Girard in 1971 from proof-theoretic motivations and, independently, by Reynolds in 1974 in order to extend conventional typed programming languages. Here, in order to obtain immediately a completeness theorem, we will present a version of system **F** plus product types and a terminal type, via a sequential calculus approach, where contexts are ordered lists of type variables and ordinary variables.

We know that second order lambda calculus is very expressive, as it can represent all the functions provably total in second order Peano Arithmetic, nevertheless a strong normalization theorem holds for it.

The difficulty in providing models of polymorphism is its impredicative nature: in the universal abstraction type $\Pi X.T$ the type variable X is intended to range over all types including $\Pi X.T$ itself. In fact there are many different models of polymorphism. We know there are particularly simple models of second order lambda calculus generated by sufficiently complete small categories in certain toposes ([13], [5], [15], [7]). These models solve the problem of impredicativeness by means of an uniformity principle [18].

Besides the models mentioned above, there are many models based on algebraic structures, like Scott domains [2] or coherence spaces [4], which solve impredicativeness by means of algebraic representability.

In 1987, Seely showed that PL-categories are a complete class of models for higher order lambda calculus [16].

It is possible to show that the elimination of generic orders and operators from higher order lambda calculus to obtain second order lambda calculus corresponds semantically to passing from PL-categories to categories indexed over a category with only finite products, in which the fibers are just cartesian closed categories and there are the indexed adjunctions that one needs in order to interpret universal abstraction and application. Hence these last categories, which we call PLF-categories and are a special kind of Lawvere's hyperdoctrines, provide a complete semantics for second order lambda calculus.

One of the main problems in the semantics for polymorphic lambda calculus is today to unify the algebraic models with Seely's categorical semantics and eventually with the models based on complete small categories in the effective topos. As a first step in this direction we will show that the model of coherence spaces give rise to a PLF-category.

In order to relate the various kinds of categorical models we hope that this example of PLF-category is useful to determine whether it is possible to prove an embedding theorem for any PLF-category into an appropriate small complete category of locally cartesian closed categories.

Second order lambda calculus syntax

The system λ_2 presented here is a version of Girard's system **F** enriched with product types and a terminal type. We show the rules to form types (denoted by the capital letters $T, U, V \ldots$), terms (denoted by the small letters $t, u, v \ldots$) and the equality rules. The rules state when a judgment can be derived from other judgments. We consider three kinds of judgments:

$$(\text{type judgment}) \quad \Gamma \vdash U \; type$$

whose intended meaning is "U is a type under the context Γ"

$$(\text{term judgment}) \quad \Gamma \vdash t : T$$

whose intended meaning is "t is a term of type T under the context Γ"

$$(\text{conversion judgment}) \quad \Gamma \vdash s = u : U$$

whose intended meaning is "the term s converts to the term u of type U under the context Γ".

In all these judgments the context is an ordered list of variables which contains all the type variables and term variables which occur free in the type or in the term(s) on the right. More formally, contexts are defined recursively, together with the rules, starting from *type variable declarations*: $X typeVar$, $Y typeVar$, $Z typeVar$, ... whose intended meaning is "X is a type variable, Y is a type variable, Z is a type variable, ...". In the following we suppose that there are countably many type variables. We can now introduce the first two rules of context construction:

$$(1C) \quad \emptyset \quad cont$$

whose intended meaning is "the empty list is a context."

$$(2C) \quad \frac{\Gamma \quad cont \quad X \ typeVar}{\Gamma, X \quad cont} \ (X \notin \Gamma)$$

whose intended meaning is "If X is a type variable which does not appear in Γ and Γ is a context then Γ, X is a context".

Rules for type judgments

Now we present the inference rules to construct type judgments:

$$(\text{type variable}) \quad \Gamma \vdash X \ type \quad (\ X \in \Gamma)$$

$$(\text{product}) \quad \frac{\Gamma \vdash U \ type \quad \Gamma \vdash V \ type}{\Gamma \vdash U \times V \ type}$$

$$(\text{terminal}) \quad \Gamma \vdash \top \ type$$

$$(\text{implication}) \quad \frac{\Gamma \vdash U \ type \quad \Gamma \vdash V \ type}{\Gamma \vdash U \to V \ type}$$

$$(\text{abstraction}) \quad \frac{\Gamma, X \vdash V \ type}{\Gamma \vdash \Pi X.V \ type}$$

We can note that the form of the type in the type judgment establishes the last rule used in its derivation.

Rules for term judgments

In order to construct term judgments we need the following *declarations of term variables*:

$x : U$ *Var*, $y : U$ *Var*, $z : U$ *Var*, ... provided that $\Gamma \vdash U$ *type*, whose intended meaning is "x is a variable of type U, y is a variable of type U, z is a variable of type U, ...".

With the above declarations a new rule of context construction can be added:

$$3C) \quad \frac{\Gamma \vdash U \text{ type} \quad x : U \text{ Var}}{\Gamma, x : U \text{ cont}} \quad (x : U \notin \Gamma)$$

whose intended meaning is "If U is a type under Γ and x is a variable of type U, which doesn't appear in Γ, then $\Gamma, x : U$ is a context".

This rule completes the definition of what a context is; hence contexts are ordered lists of type variables and term variables without repetitions.

Now we can introduce the inference rules to construct term judgments:

$$\text{(variable)} \quad \Gamma \vdash x : U \quad \text{if } x : U \in \Gamma$$

$$\text{(terminal)} \quad \Gamma \vdash \star : \top$$

$$\text{(abstraction)} \quad \frac{\Gamma, x : U \vdash v : V}{\Gamma \vdash \lambda x : U.v : U \to V}$$

$$\text{(application)} \quad \frac{\Gamma \vdash t : U \to V \quad \Gamma \vdash u : U}{\Gamma \vdash t(u) : V}$$

$$\text{(pair)} \quad \frac{\Gamma \vdash u : U \quad \Gamma \vdash v : V}{\Gamma \vdash \langle u, v \rangle : U \times V}$$

$$\text{(projections)} \quad \frac{\Gamma \vdash t : U \times V}{\Gamma \vdash \pi_1(t) : U} \quad \frac{\Gamma \vdash t : U \times V}{\Gamma \vdash \pi_2(t) : V}$$

$$\text{(universal abstraction)} \quad \frac{\Gamma, X \vdash w : W}{\Gamma \vdash \bigwedge X.w : \Pi X.W}$$

$$\text{(universal application)} \quad \frac{\Gamma \vdash t : \Pi X.W \quad \Gamma \vdash V \text{ type}}{\Gamma \vdash t(V) : W[X := V]}$$

Note that in the universal abstraction rule the abstracted variable X is the right most one in the context, which is possible only if no type of a term variable in Γ depends on it.

One can prove that if $\Gamma \vdash U$ or $\Gamma \vdash u : U$ is derived by the previous rules then the context Γ contains all the free variables in U or u.

As before for the case of type judgments, also here the form of the term determines the last rule applied in the derivation of a term judgment.

Rules for conversion judgments

Finally we consider the conversion judgments; they express an equivalence relation among the typed terms which represents the computational contents of the calculus. Hence, besides the rules which make conversion an equivalence relation compatible with the construction of terms, we have:

$$(\beta\text{-conversion}) \quad \frac{\Gamma, x : U \vdash t : V \quad \Gamma \vdash u : U}{\Gamma \vdash (\lambda x : U.t)(u) = t[x := u] : V}$$

$$(\eta\text{-conversion}) \quad \frac{\Gamma \vdash t : U \to V}{\Gamma \vdash \lambda x : U.t(x) = t : U \to V} \quad (x : U \notin \Gamma)$$

$$(\text{first projection}) \quad \frac{\Gamma \vdash u : U \quad \Gamma \vdash v : V}{\Gamma \vdash \pi_1(\langle u, v \rangle) = u : U}$$

$$(\text{second projection}) \quad \frac{\Gamma \vdash u : U \quad \Gamma \vdash v : V}{\Gamma \vdash \pi_2(\langle u, v \rangle) = v : V}$$

$$(\text{surjective pairing}) \quad \frac{\Gamma \vdash t : U \times V}{\Gamma \vdash \langle \pi_1(t), \pi_2(t) \rangle = t : U \times V}$$

$$(\text{terminal}) \quad \frac{\Gamma \vdash t : \top}{\Gamma \vdash t = \star : \top}$$

$$(\text{universal } \beta\text{-conversion}) \quad \frac{\Gamma, X \vdash t : T \quad \Gamma \vdash U \, type}{\Gamma \vdash (\bigwedge X.t)(U) = t[X := U] : T[X := U]}$$

$$(\text{universal } \eta\text{-conversion}) \quad \frac{\Gamma \vdash t : \Pi X.W}{\Gamma \vdash \bigwedge X.t(X) = t : \Pi X.W} \quad (X \notin \Gamma)$$

We recall that for this calculus a strong normalization theorem holds, according to which each term reduces to one of the following normal forms:
$x, \quad \lambda x : U.t, \quad \langle u, v \rangle, \quad \bigwedge X.t, \quad \star$.
Moreover a completeness theorem with respect to a suitable class of categories, which we call PLF-categories, can be proved.

Categorical semantics for λ_2

In this section we describe PLF-categories, a generalization of Seely's PL-categories, that are the correct structures to give a complete semantics of λ_2. A PL-category is a two-dimensional universe of cartesian closed categories indexed over a cartesian closed category and Seely showed that they provide a complete semantics for higher order lambda calculus. A PLF-category generalizes a PL-category, since we require only that the cartesian closed categories are indexed by a cartesian category instead of a cartesian closed category. From a syntactical point of view passing from PL-categories to PLF-categories corresponds to the elimination of generic orders and operators from higher order lambda calculus in order to obtain λ_2.

In the sequel $CCCat$ denotes the category of small cartesian closed categories and cartesian closed functors, and $Ob : CCCat \rightarrow Set$ is the forgetful functor, i.e. for every small category Ob considers the set of its objects.

Definition: PLF-Category

A PLF-category is a pair (S, G) where S is a small category with finite products and $G : S^{OP} \rightarrow CCCat$ is a functor, i.e. (S, G) is a strictly indexed category, and these satisfy the following two conditions:

(i) the composite functor $Ob \cdot G : S^{OP} \rightarrow Set$ is representable, in other words in S there is an object Ω such that $Ob \cdot G$ is naturally isomorphic to $S(-, \Omega)$.

(ii) (S, G) is *weakly complete*, i.e. for every C in S the S-indexed functor $K_C : G \rightarrow G \cdot (- \times C)$ defined at each A as $G(\pi_A) : G(A) \rightarrow G(A \times C)$ has an S-indexed right adjoint $\Pi_C : G \cdot (- \times C) \rightarrow G$.

In the sequel G^C denotes $G \cdot (- \times C)$.

Hence a PLF-category is a family of cartesian closed categories where each one provides an interpretation for first order lambda calculus with type variables. Moreover the condition of weak completeness allows to interpret second order abstraction.

It is possible to show that PLF-categories, as a special kind of Lawvere's hyperdoctrines, provide a complete semantics for λ_2 [10], [13]. In fact, if (S, G) is a PLF-category, an interpretation $I : \lambda_2 \rightarrow (S, G)$ can be given as follows. The interpretation of a type judgment $\Gamma \vdash T\,type$ is a morphism of S, whose codomain is Ω, such that if the type variables which appear in Γ are $X_1, \ldots, X_n$ then

$$I(\Gamma \vdash T\,type) : \Omega^n \rightarrow \Omega.$$

The first order type judgments with a context containing n type variables are interpreted in the corresponding objects of the cartesian closed category $G(\Omega^n)$, while abstraction type judgment with a context containing n type variables is interpreted via the functor $\Pi_\Omega(\Omega^n)$.

Substitution of type variable within a type judgment is interpreted by composition of morphisms in S.

A term judgment $\Gamma \vdash t : T$, whose context contains n type variables, is interpreted as a morphism of $G(\Omega^n)$. In fact, since every $G(\Omega^n)$ is a cartesian closed category, it is a model of first order typed lambda calculus for terms derived with contexts of n fixed type variables. If the term variables of the context Γ are $x_1 : U_1, \ldots, x_m : U_m$ where Γ has n type variables then

$$I(\Gamma \vdash t : T) : I(\Gamma \vdash U_1 \, type) \times \ldots \times I(\Gamma \vdash U_m \, type) \to I(\Gamma \vdash T \, type)$$

is a morphism of $G(\Omega^n)$.

In particular universal abstraction in a term judgment with a context of n type variables is interpreted by $\Pi_\Omega(\Omega^n)$ and universal application in the same condition is interpreted by the counit of the adjunction between $K_\Omega(\Omega^n)$ and $\Pi_\Omega(\Omega^n)$.

Substitution of type variables in a term judgment is interpreted via the functor G applied to the morphism of S which interprets the type judgments that must be substituted in the type variables of the term in question. Substitution of term variables in a term judgment with a context of n type variables is interpreted by composition of morphisms in the appropriate fiber $G(\Omega^n)$.

Besides $K_\Omega(\Omega^n)$ interprets the weakening rule by adding a type variable in the context of n type variables between type judgments when it is applied to the objects of $G(\Omega^n)$ and between term judgments when it is applied to the morphisms of $G(\Omega^n)$.

As usual conversion judgments are interpreted as equality between morphisms of $G(\Omega^n)$.

So the syntax that we adopt for this version of second order lambda calculus allows to prove immediately the following theorem:

Theorem: Completeness. $\Gamma \vdash t = s : T$ is a conversion judgment provable in λ_2 if and only if for every PLF-category (S, G) and for every interpretation $I : \lambda_2 \to (S, G)$ we have $I(\Gamma \vdash t : T) = I(\Gamma \vdash s : T)$.

The PLF-category of coherence spaces

Now we present a concrete example of a PLF-category based on coherence spaces, that is the categories *Gem* and *Stab* [4], and stable functors. This construction is the categorical version of the model for λ_2 of coherence spaces made by Girard and Taylor .

We recall that *Stab* is the category of coherence spaces and stable functions and it is a cartesian closed category. A coherence space X is a downward closed set of subsets, i.e. $a \in X$ and $a' \subset a$ imply $a' \in X$, which is closed under unions of pairwise compatible families, i.e. for each $M \subset X$, if $a_1 \cup a_2 \in X$ for every $a_1, a_2 \in M$, then $\bigcup M \in X$.

Given two coherence spaces A and B, a function $f : A \to B$ is called *stable* if it preserves inclusions, directed joins and pullbacks. Moreover we say that a stable function $f : A \to B$ is a *rigid embedding* if there is a stable function $p : B \to A$ such that

$$p \cdot f = id_A \quad \text{and} \quad f \cdot p \leq_B id_B$$

(the function p is necessarily unique, denoted usually f^{-1}, and is called a *rigid projection*).

Gem is the subcategory of *Stab*, whose objects are coherence spaces and whose morphisms are rigid embeddings. Note that *Gem* has pullbacks and directed colimits. We say that a functor $H : Gem^m \to Gem^n$ is *stable* if it preserves pullbacks and directed colimits.

To build the PLF- category (S, G) of coherence spaces, let S be the small category built on finite powers of the category *Gem*, whose arrows are the stable functors. Obviously S has finite products, as pullbacks and directed colimits are computed pointwise in each power of *Gem*.

The next step in the construction is the definition of the functor $G : S^{OP} \to CCCat$:

for every $n \in IN$ the objects of the category $G(Gem^n)$ are the stable functors $F : Gem^m \to Gem$ and a morphism α between F_1 and F_2 in $G(Gem^n)$ is $\alpha : \langle \Pi_1, \ldots, \Pi_n \rangle \to [F_1 \to F_2]$ defined as a family of stable functions

$$\alpha(\underline{A}) \in [F_1(\underline{A}) \to F_2(\underline{A})]$$

as $\underline{A}$ varies among the objects of Gem^n (where Π_i is the stable functor i-th projection and $[F_1(\underline{A}) \to F_2(\underline{A})]$ is the exponent object in *Stab*), that satisfies the *uniformity equation*:

for every morphism $\underline{f} : \underline{A} \to \underline{B}$ in Gem^n

$$F_2(\underline{f})^{-1} \cdot \alpha(\underline{B}) = \alpha(\underline{A}) \cdot F_1(\underline{f})^{-1}.$$

The composition between the morphisms is defined as follows:
for every $\alpha \in G(Gem^n)(F, H)$, $\beta \in G(Gem^n)(H, K)$, for every $\underline{A} \in ObGem^n$

$$(\beta \cdot \alpha)(\underline{A}) = \beta(\underline{A}) \cdot \alpha(\underline{A})$$

and the unit is:
for $F \in Mor S$, for $\underline{A} \in ObGem^n$

$$id_F(\underline{A}) = id_{F(\underline{A})}.$$

One can now prove that the fibers of G are cartesian closed categories and hence it is possible to interpret the generic first order lambda calculus with at most n type variables into $G(Gem^n)$.

Proposition
For every $n \in IN$, $G(Gem^n)$ is a cartesian closed category.

Proof The properties of cartesian closed category are inherited from *Stab*. Here we give only the definitions of the various adjoints.
Terminal object.
The terminal object is the stable functor $T_n : Gem^n \to Gem^0$ with constant value $\{\emptyset\}$.
Cartesian products.
For every $F_1, F_2 : Gem^n \to Gem$ in $G(Gem^n)$ the cartesian product $F_1 \times F_2$ is obtained by taking the pointwise product which takes an object $\underline{A}$ in $ObGem^n$ to $F_1(\underline{A}) \times F_2(\underline{A})$, the product in *Stab*, and for every morphism $\underline{f}$ in Gem^n $(F_1 \times F_2)(\underline{f}) = F_1(\underline{f}) \times F_2(\underline{f})$.
Besides the projections $\pi_F : F \times H \to F$ and $\pi_H : F \times H \to H$ and the couple $\langle \alpha, \beta \rangle : K \to F \times H$, for every $\alpha \in G(Gem^n)(K, F)$ and $\beta \in G(Gem^n)(K, H)$, are defined pointwise through the cartesian structure of *Stab* and they satisfy the uniformity equation as $(F_1(\underline{f}) \times F_2(\underline{f}))^{\dashv} = F_1(\underline{f})^{\dashv} \times F_2(\underline{f})^{\dashv}$.
Closure.
For every F_1, F_2 in $ObG(Gem^n)$ the exponent $F_1 \to F_2$ is the stable functor obtained by composing the product of F_1 and F_2 with the bifunctor " implication" " $\to$ " : $Gem \times Gem \to Gem$, which takes every object (A, B) in $ObGem^2$ to $[A \to B]$, and every morphism (f, g) in Gem^2 to the rigid embedding $f \to g$: $[A \to B] \to [A' \to B']$ defined as follows:
for every trace t in $[A \to B]$ $(f \to g)(t) = Tr(g \cdot t^{\vee} \cdot f^{\dashv})$, where $t^{\vee}$ is the stable map whose trace is t and $f^{\dashv}$ is the rigid projection of f.
Moreover for every H, K, F in $ObG(Gem^n)$ and $\alpha \in G(Gem^n)(K \times H, F)$, the evaluation $Ap_H : (H \to F) \times H \to F$ and the abstraction $\alpha^{\wedge} : K \to [H \to F]$ are defined pointwise through the closure of *Stab*. $\qquad \square$

The indexing functor $G : S^{OP} \to CCCat$ is given by composition on the left (=semantical substitution), that is for every $F : Gem^n \to Gem^m$ in S let $G(F) : G(Gem^m) \to G(Gem^n)$ be the following functor:

for every $H \in ObG(Gem^m)$ and for every $\alpha \in G(Gem^m)(H_1, H_2)$

$$G(F)(H) = H \cdot F \text{ and } G(F)(\alpha) = \alpha_F$$

where for every $\underline{A} \in ObGem^n$

$$\alpha_F(\underline{A}) = \alpha(F(\underline{A})) \in [H_1 \cdot F(\underline{A}) \rightarrow H_2 \cdot F(\underline{A})].$$

One can easily prove that G is a functor and that for every $m, n \in IN$ and for every $F \in S(Gem^n, Gem^m)$ $G(F)$ preserves the cartesian closed structure.

According to the abstract definition of PLF-category we must consider the S-indexed functor K_{Gem} such that for every $n \in IN$,

$$K_{Gem}(Gem^n) = G(\langle \Pi_1, \dots \Pi_n \rangle) : G(Gem^n) \rightarrow G(Gem^{n+1}),$$

whose right S-indexed adjoint is necessary to interpret universal abstraction. So we can prove the following proposition:

Proposition
For every $n \in IN$ the S-indexed functor K_{Gem^m} has an S-indexed right adjoint

$$\Pi_{Gem^m} : G^{Gem^m} \rightarrow G.$$

Proof Let $\Pi_{Gem}(Gem^m) : G^{Gem}(Gem^m) \rightarrow G(Gem^m)$ be the following map on objects:
for every $F : Gem^m \times Gem \rightarrow Gem$

$$\Pi_{Gem}(Gem^m)(F) = TrF : Gem^m \rightarrow Gem.$$

where TrF is the trace of the functor F (see [4]).
For every $F : Gem^m \rightarrow Gem$, $H : Gem^m \times Gem \rightarrow Gem$ we define the bijection of the adjunction of $K_{Gem}(Gem^m)$

$$\phi_{Gem^m} : G^{Gem}(Gem^m)(K_{Gem}(Gem^m)(F), H) \rightarrow G(Gem^m)(F, \Pi_{Gem}(Gem^m)(H))$$

by letting: $\phi_{Gem^m}(\alpha) = Tr(\alpha)$, where for every $\underline{A} \in ObGem^n$

$$(Tr(\alpha))(\underline{A}) = Tr(\alpha(\underline{A}, .))$$

since $\alpha(\underline{A}, B) \in [F \cdot \langle \Pi_1, \dots \Pi_n \rangle(\underline{A}, B) \rightarrow H(\underline{A}, B)]$ (see [4]).
For every $\beta \in G(Gem^m)(F, \Pi_{Gem}(Gem^m)(H))$ one can show that the inverse map is

$$\phi^{-1}_{Gem^m}(\beta) = \tilde{\beta} \in G^{Gem}(Gem^m)(K_{Gem}(Gem^m)(F), H)$$

where for every $B \in ObGem$ $\tilde{\beta}(\underline{A}, B) = \beta(\underline{A})(B)$ (see [4]).

Moreover ϕ_{Gem^n} is natural in the first variable as naturality corresponds to the well-definedness of substitution of variables in terms obtained by universal abstraction. Now we have just proved that for every $n \in IN$ there is a functor $\Pi_{Gem}(Gem^n)$, defined on objects as above, which is the right adjoint of $K_{Gem}(Gem^n)$ through ϕ_{Gem^n} (see cor.2 p.83 [11]).

In order to complete the proof that there is an S-indexed adjunction between K_{Gem} and Π_{Gem}, we must show that the family of adjunctions between K_{Gem} and Π_{Gem} is natural over the morphisms of S. This last naturality on indexes corresponds to the well-definedness of substitution of type variables in term judgments with abstraction and application of second order . As stable functors between coherence spaces are a model of λ_2, this kind of substitution must be well defined, that is the naturality on indexes holds in our indexed category of coherence spaces. Note that $K_{Gem} \dashv_S \Pi_{Gem}$ is sufficient to yield weak completeness as for every $n \in IN$

$$K_{Gem^n} \equiv \underbrace{K_{Gem} \cdot \ldots \cdot K_{Gem}}_{n-times}$$

Hence

$$\Pi_{Gem^n} \equiv \underbrace{\Pi_{Gem} \cdot \ldots \cdot \Pi_{Gem}}_{n-times}$$

is the right S- indexed adjoint of K_{Gem^n}. □

In the construction of the PLF-category of coherence spaces most definitions are forced by the very definition of PLF-category. The only important choice was the morphisms of the fiber $G(\Omega^n)$. We started from the idea that the morphisms $G(\Omega^n)(F, H)$ should be similar to natural transformations between $\langle \Pi_1, \ldots \Pi_n \rangle$ and $[F \rightarrow H]$. As we said, our solution was to choose the family of stable functions which satisfy the uniformity equation

$$G(\Omega^n)(F, H) =$$

$$= \{\alpha : \langle \Pi_1, \ldots \Pi_n \rangle \rightarrow [F \rightarrow H] \mid \alpha(\underline{A}) \in [F(\underline{A}) \rightarrow H(\underline{A})] \text{ for } \underline{A} \in ObGem^n$$

$$\text{s.t. for} \underline{f} \, inGem^n(\underline{A}, \underline{B}) \,\, H(\underline{f})^{\dashv} \cdot \alpha(\underline{B}) = \alpha(\underline{A}) \cdot F(\underline{f})^{\dashv}\}.$$

It expresses semantically the concept of a parametric polymorphic function, somehow a function whose behaviour is the same over all types. More precisely, the choice of one stable morphism for each $\underline{A} \in ObGem^n$ connects with the idea of when one considers a λ_2 term without universal abstraction and application: for every finite number of types which one substitutes for the free type variables, one obtains a proof codified by a first order term. Similarly for every $\underline{A} \in ObGem^n$ we obtain uniformly a morphism of the cartesian closed category $G(\Omega^n)$. Uniformity is a way to express the dependencies of polymorphic types since universal abstraction and application operate on families of first order lambda terms in a parametric way.

For this PLF-category we have taken the model of coherence spaces into the group of categorical models à la Seely and this work can suggest what to do with other algebraic models.

As we have already noticed in the introduction, there is another group of categorical models built in the toposes, as small complete categories, [15], [5], [7]. By studying the PLF-category of coherence spaces, we would like to prove an embedding theorem for PLF-categories as appropriate small complete categories of locally cartesian closed categories in order to see if it is possible to correlate the two groups of models.

I would like to thank Pino Rosolini and Silvio Valentini for their precious help.

References

[1] J. Benabou, *Fibered categories and the foundations of naive category theory*, J. Symbolic Logic **50** (1985), 10–37.

[2] T. Coquand, C. Gunter, and G. Winskel, *Domain theoretic models of polymorphism*, Information and Computation **81** (1989), 123–167.

[3] J. Y. Girard, *The system F of variable types, fifteen years later*, Theoret. Comput. Sci. **45** (1989), 159–192.

[4] J. Y. Girard, Y. Lafont, and P. Taylor, *Proofs and types*, Cambridge tracts in Theoretical Computer Science, vol. 7, Cambridge University Press, 1989.

[5] M. Hyland, *A small complete category*, Ann. Pure Appl. Logic **90** (1989), 135–165.

[6] M. Hyland and A. Pitts, *The theory of constructions: categorical semantics and topos theoretic models*, Categories in computer science and logic (Gray and Scedrov, eds.), AMS, 1989, pp. 137–199.

[7] M. Hyland, E. Robinson, and G. Rosolini, *The discrete object in the effective topos*, Proceedings of the London Mathematical Society, 60, 1990, pp. 1–36.

[8] B. Jacobs, *Categorical type theory*, Ph.D. thesis, Katholieke Universiteit Nijmegen, 1991.

[9] J. Lambek and P. J. Scott, *An introduction to higher order categorical logic*, Studies in Advanced Mathematics, vol. 7, Cambridge University Press, 1986.

[10] F. W. Lawvere, *Equality in hyperdoctrines and comprehension schema as an adjoint functor*, New York Symp. on Applications of Categorical Algebra (A. Heller, ed.), 1970, pp. 1–14.

[11] S. Mac Lane, *Categories for the working mathematician*, Graduate text in Mathematics, vol. 5, Springer, 1971.

[12] S. MacLane and R. Paré, *Coherence for bicategories and indexed categories*, J. Pure Appl. Algebra **37** (1985), 59–80.

[13] A. Pitts, *Polymorphism is set-theoretic, constructively*, Conference on category theory and computer science, LNCS no. 283 (Pitts, Poigné, and Rydeheard, eds.), Springer, 1987, pp. 12–39.

[14] J.C. Reynolds, *Polymorphism is not set-theoretic*, Semantics of data types, LNCS 173 (Kahn et al., ed.), Springer, 1984, pp. 145–156.

[15] G. Rosolini, *About modest sets*, International Journal of Foundations of Computer Science **1** (1990), 341–353.

[16] R. Seely, *Categorical semantics for higher order polymorphic lambda calculus*, J. Symbolic Logic **52** (1987), 969.

[17] P. Taylor, *Recursive domains, indexed category theory and polymorphism*, Ph.D. thesis, University of Cambridge, 1987.

[18] A. Troelstra, *Intuitionistic formal systems*, Metamathematical investigations of intuitionistic aritmetic and analysis, LNM 344, Springer, 1973, pp. 275–323.

DIPARTIMENTO DI MATEMATICA PURA E APPLICATA, UNIVERSITA' DI PADOVA, VIA BELZONI 7, 35131 PADOVA, ITALY

e-mail address: maietti@pdmat1.math.unipd.it

Reflection using the derivability conditions

Sean Matthews   Alex K. Simpson

Abstract

We extend arithmetic with a new predicate, Pr, giving axioms for Pr based on first-order versions of Löb's derivability conditions. We hoped that the addition of a reflection schema mentioning Pr would then give a non-conservative extension of the original arithmetic theory. The paper investigates this possibility. It is shown that, under special conditions, the extension is indeed non-conservative. However, in general such extensions turn out to be conservative.

1 Introduction

In any recursively axiomatized theory of arithmetic, T, one can follow Gödel's construction to obtain a 'provability predicate', a Σ_1-formula $Bew_T(x)$ such that $Bew_T(\ulcorner A \urcorner)$ is true if and only if $T \vdash A$, where $\ulcorner A \urcorner$ is the Gödel number of the formula A. Moreover, if T is sufficiently strong then Bew_T satisfies the following predicate (or 'uniform') versions of Löb's derivability conditions [7]:

$(D1)$ $\qquad\qquad$ $if \ \ T \vdash \forall x A \ \ then \ \ T \vdash \forall x Bew_T(\ulcorner A\langle x\rangle \urcorner),$

$(D2)$ $\quad$ $T \vdash \forall x (Bew_T(\ulcorner (A \to B)\langle x\rangle \urcorner) \to (Bew_T(\ulcorner A\langle x\rangle \urcorner) \to Bew_T(\ulcorner B\langle x\rangle \urcorner))),$

$(D3)$ $\qquad$ $T \vdash \forall x (Bew_T(\ulcorner A\langle x\rangle \urcorner) \to Bew_T(\ulcorner Bew_T(\ulcorner A\langle x\rangle \urcorner)\langle x\rangle \urcorner)),$

where we write $\ulcorner A\langle x\rangle \urcorner$ for a term with a free variable x 'disquoting' any occurrence of x in A (see Section 2). Solovay, [9], showed that the original propositional versions of the derivability conditions identify all the valid 'modal' schematic properties of Bew_T (the other modal axiom, the formalization of Löb's theorem, is derivable from $(D1)$–$(D3)$ using the diagonalization lemma). Although the first-order derivability conditions above do not capture all the valid first-order schematic properties of Bew_T (see [1]), they do isolate a natural class of 'modal' properties satisfied by Bew_T.

All the aforementioned work treats the derivability conditions as *descriptive* in that their purpose is to describe properties of the Bew predicate. In this paper we consider them in an alternative *prescriptive* rôle. We define a language, $\mathcal{L}'$, by adding a new unary predicate symbol, Pr, to the original language $\mathcal{L}$. Then we define an $\mathcal{L}'$-theory, T', as the least theory containing T that is closed under the following analogues of $(D1)$–$(D3)$:

$(C1)$ $\qquad\qquad\qquad$ *if* $\ T' \vdash \forall x A \quad$ *then* $\quad T' \vdash \forall x Pr(\ulcorner A\langle x\rangle\urcorner)$,

$(C2)$ $\qquad T' \vdash \forall x(Pr(\ulcorner(A \to B)\langle x\rangle\urcorner) \to Pr(\ulcorner A\langle x\rangle\urcorner) \to Pr(\ulcorner B\langle x\rangle\urcorner))$,

$(C3)$ $\qquad\quad T' \vdash \forall x(Pr(\ulcorner A\langle x\rangle\urcorner) \to Pr(\ulcorner Pr(\ulcorner A\langle x\rangle\urcorner)\langle x\rangle\urcorner))$,

where we assume that Gödel numbering has been extended to $\mathcal{L}'$. It is natural to ask how much of the behaviour of $Bew_{T'}$ is forced upon Pr by the satisfaction of $(C1)$–$(C3)$.

As remarked in Boolos and Jeffrey [3, p. 185], there are many 'predicates' other than Bew_T that satisfy $(D1)$–$(D3)$; for example, the predicate expressing the property of being (the Gödel number of) a well-formed formula. Therefore it does not hold that $T' \vdash \forall x(Pr(x) \to Bew_{T'}(x))$. We shall see below that the converse implication fails too.

However, it occurred to us to consider the effect of adjoining the following analogue of the uniform reflection schema to T':

(R) $\qquad\qquad\qquad\qquad\qquad \forall x(Pr(\ulcorner A\langle x\rangle\urcorner) \to A)$.

The question we were interested in was whether $T' + R$ is a non-conservative extension of the original theory T.

The possibility that $T' + R$ might not be conservative over T was initially plausible for the following reason. There is an evident 'intended' interpretation of T' in T under which Pr is (modulo some mapping of Gödel numbers) translated as Bew_T. Although this interpretation can be used to prove that T' is a conservative extension of T, it cannot be used to show that $T' + R$ is. Furthermore, no other translation of Pr can be used for this purpose either (Theorem 2.3).

On the other hand, the same interpretation can be used to establish that any $\mathcal{L}$-formula entailed by $T' + R$ is a theorem in the theory obtained by extending T with its uniform reflection schema:

(Rfn) $\qquad\qquad\qquad\qquad\qquad \forall x(Bew_T(\ulcorner A\langle x\rangle\urcorner) \to A)$.

By Gödel's second incompleteness theorem, $T' + Rfn$ is a non-conservative extension of T. Our initial hope was that $T' + R$ might be a (necessarily conservative) extension of $T + Rfn$.

This possibility would be of practical interest. If $T' + R$ were an extension of $T + Rfn$, then the definition of $T' + R$ would provide a feasible way of extending the reasoning powers of T without having to go through the laborious construction of Gödel's Bew_T predicate (although admittedly the definition of $T' + R$ does still require a Gödel numbering of formulae). Unfortunately, it turns out that $T' + R$ is always conservative over T (Theorem 3.1). (This shows that, as claimed above, $T' \not\vdash \forall x (Bew_{T'}(x) \rightarrow Pr(x))$.) Thus our construction of $T' + R$ does not give the *general* method of achieving a non-conservative extension of T that we hoped for.

Nevertheless, a slight and natural modification of the construction of $T' + R$ does lead to a non-conservative extension in one notable case. Since $Pr(t)$ is intended to mimic $Bew_T(t)$ it ought to be treated as a Σ_1-formula. So if T supports induction over Σ_1-formulae then it is reasonable to include induction over atomic formulae of the form $Pr(t)$ in T'. In this case $T' + R$ provides full induction over formulae of $\mathcal{L}'$, and thus contains Peano Arithmetic (Theorem 4.1). So for any T containing Σ_1-induction but not full induction, a non-conservative extension can be obtained by our method.

Unfortunately, the non-conservative effect does not extend beyond Peano Arithmetic. Since Peano Arithmetic supports induction over arbitrary formulae of $\mathcal{L}$ it is natural to allow induction over arbitrary formulae of $\mathcal{L}'$ in T'. However, even allowing such induction, if T is Peano Arithmetic then $T' + R$ is conservative over T (Theorem 4.2).

The paper is structured as follows. In Section 2 we give the technical background to our work. In Section 3 we give a semantic proof that, in general, $T' + R$ is conservative over T. In Section 4 we consider extending induction to the new language, proving the non-conservativity result for arithmetic with Σ_1-induction and the conservativity result for Peano Arithmetic. Finally, Section 5 contains some concluding remarks.

2 Preliminaries

Throughout the paper we work, for convenience, with the language, $\mathcal{L}$, of Primitive Recursive Arithmetic (PRA) [4]. Thus when we refer to Peano Arithmetic (PA) we mean a definitional extension in $\mathcal{L}$ of the usual Peano Arithmetic (which is in the language of elementary arithmetic). As in Section 1, $\mathcal{L}'$ is the language obtained by adding a new unary predicate symbol, Pr, to $\mathcal{L}$.

A Gödel-numbering of $\mathcal{L}'$ is an injective mapping from $\mathcal{L}'$ into the natural numbers. We assume some such mapping. We denote the number standing for a formula A of $\mathcal{L}'$ by $\ulcorner A \urcorner$, and similarly for terms, etc. We assume that all the relevant operations and predicates on formulae/terms are primitive recursive. In particular there is a primitive recursive function $sub(\cdot, \cdot, \cdot)$, such that for any formula A (or term t), and number n:

$$sub(\ulcorner A \urcorner, \ulcorner x \urcorner, n) \;=\; \ulcorner A[\bar{n}/x] \urcorner$$

where $\bar{n}$ is the numeral $s^n(0)$. We write $\ulcorner A\langle t\rangle \urcorner$ as an abbreviation for $sub(\ulcorner A \urcorner, \ulcorner x \urcorner, t)$. The restriction of $\ulcorner \cdot \urcorner$ to $\mathcal{L}$ gives us also a Gödel-numbering of $\mathcal{L}$.

Let T be any consistent, recursively axiomatized theory in $\mathcal{L}$ extending PRA (thus T supports quantifier-free induction). Let $Bew_T(x)$ be Gödel's provability predicate for T. As T extends PRA, the formula Bew_T does indeed satisfy the properties $(D1)$–$(D3)$ of Section 1. Define the $\mathcal{L}'$-theory T' as in Section 1.

Proposition 2.1 *T' is a conservative extension of T.*

Proof. We define a translation $(\cdot)^*$ from formulae of $\mathcal{L}'$ to formulae of $\mathcal{L}$. By the second recursion theorem, there is a number r such that (writing $\{r\}$ for the r-th partial recursive function):

$$\{r\}(\ulcorner A \urcorner) \;=\; \ulcorner A^* \urcorner$$

where $(\cdot)^*$ commutes with connectives and quantifiers and is defined on atomic formulae by:

$$P(t_1, \ldots, t_n)^* \;=\; P(t_1, \ldots, t_n) \quad \text{(where } P \neq Pr\text{)}$$
$$Pr(t)^* \;=\; \exists y(\, T(\bar{r}, t, y) \wedge Bew_T(U(y)))$$

(here T and U are Kleene's primitive-recursive T predicate and result-extraction function). By definition $\{r\}$ is primitive recursive, so there is a function symbol, *star*, such that, by the formalized recursion theorem and quantifier-free induction:

(1) $\qquad\qquad T \vdash \forall x \exists y(\, T(\bar{r}, x, y) \wedge U(y) = star(x)),$

(2) $\qquad\qquad T \vdash \forall x(star(\ulcorner A\langle x\rangle \urcorner) = \ulcorner A^*\langle x\rangle \urcorner).$

We now show that for all $\mathcal{L}'$-formulae A, if $T' \vdash A$ then $T \vdash A^*$; which, since $(\cdot)^*$ is the identity on $\mathcal{L}$-formulae, establishes the desired conservativity result. The proof is a straightforward induction on the closure conditions of T':

(C1) Assume that $T' \vdash \forall x A$. By the induction hypothesis we have that $T \vdash (\forall x A)^*$, and therefore that $T \vdash \forall x A^*$. We need to show that $T \vdash (\forall x Pr(\ulcorner A\langle x\rangle \urcorner))^*$; i.e., that

$$T \vdash \forall x \exists y(\, T(\bar{r}, \ulcorner A\langle x\rangle \urcorner, y) \wedge Bew_T(U(y))).$$

However, $T \vdash \forall xyzw\,(\, T(x, y, z) \wedge T(x, y, w)) \rightarrow z = w$. Therefore, by (1) and (2), the above formula is equivalent to $T \vdash \forall x Bew_T(\ulcorner A^*\langle x\rangle \urcorner)$. And this, in turn, follows from $(D1)$ and the fact that $T \vdash \forall x A^*$.

(C2) We have to show that

$$T \vdash (\forall x\ Pr(\ulcorner (A \to B)\langle x\rangle \urcorner) \to (Pr(\ulcorner A\langle x\rangle \urcorner) \to Pr(\ulcorner B\langle x\rangle \urcorner)))^*$$

which, in the same way as $(C1)$ above, reduces to

$$T \vdash \forall x(Bew_T(\ulcorner (A \to B)^*\langle x\rangle \urcorner) \to Bew_T(\ulcorner A^*\langle x\rangle \urcorner) \to Bew_T(\ulcorner B^*\langle x\rangle \urcorner)),$$

an instance of $(D2)$.

(C3) Similar to $(C2)$ only making use of $(D3)$ instead. □

Proposition 2.2 *For any $\mathcal{L}$-formula A, if $T' + R \vdash A$ then $T + Rfn \vdash A$.*

Proof. Let $(\cdot)^*$ be the translation from $\mathcal{L}'$ to $\mathcal{L}$ defined in the last proof. We already know that if $T' \vdash A$ then $T \vdash A^*$ and hence $T + Rfn \vdash A^*$. So we need only show that $T + Rfn \vdash R^*$. However, as in the proof above, this translates to showing that:

$$T + Rfn \vdash \forall x(Bew_T(\ulcorner A^*\langle x\rangle \urcorner) \to A^*),$$

which is an instance of Rfn. □

The above translation cannot be used to prove the conservativity of $T' + R$ over T, because it is not in general the case that $T \vdash \forall x(Bew_T(\ulcorner A^*\langle x\rangle \urcorner) \to A^*)$. One might wonder whether there is a cleverer translation that works instead. We now give a quite general proof that in fact there is none.

We consider a general notion of translation useful for proving conservativity. A *retraction* of $\mathcal{L}'$ onto $\mathcal{L}$ is a function, $(\cdot)^\dagger$, from $\mathcal{L}'$-formulae to $\mathcal{L}$-formulae that: commutes with connectives and quantifiers; maps atomic formulae in $\mathcal{L}$ to themselves; and maps $Pr(t)$ to $H(t)$, where $H(x)$ is some fixed $\mathcal{L}$-formula. (It is a retraction in the appropriate category of languages and translations.) Clearly $(\cdot)^\dagger$ is determined by the choice of $H(x)$. Note that the translation, $(\cdot)^*$, used in the above proofs is the retraction determined by the formula $\exists y(T(\bar{r}, x, y) \wedge Bew_T(U(y)))$.

Let S be any $\mathcal{L}$-theory and S' be any $\mathcal{L}'$-theory extending S. A *retraction* of S' onto S is a retraction, $(\cdot)^\dagger$, from $\mathcal{L}'$ to $\mathcal{L}$ such that, for any $\mathcal{L}'$-formula A, it holds that $S' \vdash A$ implies $S \vdash A^\dagger$. (It is a retraction in the appropriate category of theories and interpretations.) It is clear that the existence of a retraction from S' to S implies that S' is a conservative extension of S. Indeed the proof of Proposition 2.1 worked by establishing that $(\cdot)^*$ is a retraction of T' onto T. The impossibility of obtaining a similar translational proof of the conservativity of $T' + R$ over T is given by:

Theorem 2.3 *There is no retraction of $T' + R$ onto T.*

Proof. Suppose, for contradiction, that $(\cdot)^\dagger$ is a retraction of $T' + R$ onto T in which Pr is translated to $H(x)$. By the diagonalization lemma, there is an $\mathcal{L}$-sentence A such that:

$$(3) \qquad\qquad\qquad T \vdash A \leftrightarrow \neg H(\ulcorner A \urcorner).$$

However, we claim that:

$$(4) \qquad\qquad\qquad if \ \ T \vdash A \ \ \ then \ \ \ T \vdash H(\ulcorner A \urcorner),$$
$$(5) \qquad\qquad\qquad T \vdash H(\ulcorner A \urcorner) \rightarrow A.$$

To see that (4) holds, suppose that $T \vdash A$. Then $T' \vdash A$. So, by $(C1)$, it follows that $T' \vdash Pr(\ulcorner A \urcorner)$. Therefore $T \vdash (Pr(\ulcorner A \urcorner))^\dagger$. So $T \vdash H(\ulcorner A \urcorner)$ as required. For (5), we have that $T' + R \vdash Pr(\ulcorner A \urcorner) \rightarrow A$. So $T \vdash (Pr(\ulcorner A \urcorner) \rightarrow A)^\dagger$. Thus indeed $T \vdash H(\ulcorner A \urcorner) \rightarrow A$.

But from (3)–(5) it is easy to derive that T is inconsistent — a contradiction. $\qquad\qquad\qquad\qquad\qquad\qquad\qquad\qquad\qquad\qquad\qquad\qquad\qquad\quad\square$

This proof is similar to Montague's proof of the inconsistency of giving syntactic interpretations to certain modal logics [8].

3   The general conservativity proof

Theorem 2.3 gives hope that $T' + R$ might be non-conservative over T. Unfortunately, this turns out not to be the case. The main theorem of this section is:

Theorem 3.1 $T' + R$ *is a conservative extension of* T.

The proof of the theorem involves some analysis of properties of Gödel-numbering when formalized in T. Recall that all the relevant operations and predicates on Gödel-numbers have been assumed to be primitive recursive. More specifically, we require primitive recursive 'constructors' for all function symbols, predicate symbols, connectives and quantifiers, which can be used to assemble terms and formulas. As T supports quantifier-free induction, each constructor is provably injective. Furthermore it is provable in T that the Gödel-number of a compound term/formula has a unique decomposition into the components out of which it is built. We also require a primitive recursive function $free\text{-}in(\cdot, \cdot)$, such that $free\text{-}in(\ulcorner A \urcorner, \ulcorner x \urcorner)$ if and only if x is free in A (and similarly for terms). Again, quantifier-free induction suffices to ensure that:

$$(S1) \qquad T \ \vdash \ \ulcorner A[s(x)/x]\langle y \rangle \urcorner = \ulcorner A\langle s(y) \rangle \urcorner,$$

$(S2)$ $T \vdash \neg\textit{free-in}(\ulcorner A \urcorner, \ulcorner x \urcorner) \rightarrow \ulcorner A\langle y \rangle \urcorner = \ulcorner A\langle z \rangle \urcorner,$

$(S3)$ $T \vdash (\textit{free-in}(\ulcorner A \urcorner, \ulcorner x \urcorner) \wedge \ulcorner A\langle y \rangle \urcorner = \ulcorner A\langle z \rangle \urcorner) \rightarrow y = z,$

$(S4)$ $T \vdash (\textit{free-in}(\ulcorner t \urcorner, \ulcorner x \urcorner) \wedge \ulcorner t \urcorner \neq \ulcorner x \urcorner \wedge y \leq z) \rightarrow \ulcorner x\langle y \rangle \urcorner \neq \ulcorner t\langle z \rangle \urcorner.$

The meanings of $(S1)$–$(S3)$ are clear (and, of course, analogous properties hold for substitution in terms). The more cumbersome $(S4)$ reflects the fact that if t is different from, but contains, x and $m \leq n$, then $\overline{m}$ is different from $t[\overline{n}/x]$ (since the former is a strict subterm of the latter).

Theorem 3.1 will be proved semantically. Let $\mathfrak{M} = (D, \leq, 0, s, \dots)$ be an arbitrary model of T. We expand $\mathfrak{M}$ to a $\mathcal{L}'$-structure, $\mathfrak{M}'$, by defining, for $d \in D$:

> $Pr(d)$ if   there exists an $\mathcal{L}'$-formula A and an element $d' \in D$ such that
> $$d = \ulcorner A\langle d' \rangle \urcorner \text{ and } T' \vdash \forall x A.$$

We shall prove a sequence of results aiming to show that $\mathfrak{M}'$ is a model of $T' + R$.

First we make some useful observations. If x occurs free in A then, by $(S3)$, the function $x \mapsto \ulcorner A\langle x \rangle \urcorner$ tends to infinity. Moreover, for any n, there exists m such that $T \vdash \forall x \geq m \ \ulcorner A\langle x \rangle \urcorner \geq n$ (by quantifier free induction). Thus if d is any non-standard element in D then the element $\ulcorner A\langle d \rangle \urcorner$ is also non-standard. On the other hand, if x does not occur free in A then, by (S2), $\ulcorner A\langle d \rangle \urcorner$ is standard and equal to $\ulcorner A \urcorner$.

Lemma 3.2 *If $d \in D$ is non-standard, $d \leq d' \in D$ and $\ulcorner A\langle d \rangle \urcorner = \ulcorner B\langle d' \rangle \urcorner$ then there exists n such that A is syntactically identical to $B[s^n(x)/x]$ (notation $A \equiv B[s^n(x)/x]$).*

Proof. The proof is by induction on the structure of B. Suppose that $d \in D$ is non-standard and $d \leq d' \in D$.

We first show, by induction on the structure of terms t, that if $\ulcorner t'\langle d \rangle \urcorner = \ulcorner t\langle d' \rangle \urcorner$ then:

1. if x does not occur free in t, then $t' \equiv t$;

2. if x occurs free in t then there exists n such that $d' = s^n(d)$ and $t' \equiv t[s^n(x)/x]$.

Suppose the term is a variable, y, different from x, and $\ulcorner t'\langle d \rangle \urcorner = \ulcorner y\langle d' \rangle \urcorner$. Now x is not free in y so, by (S2), $\ulcorner y\langle d' \rangle \urcorner = \ulcorner y \urcorner$ and is standard. Thus $\ulcorner t'\langle d \rangle \urcorner$ is standard, which implies that x does not occur free in t'. So $\ulcorner t'\langle d \rangle \urcorner = \ulcorner t' \urcorner$. Therefore, by the injectivity of Gödel numbering, $t' \equiv y$ as required.

Suppose the term is x and $\ulcorner t'\langle d\rangle\urcorner = \ulcorner x\langle d'\rangle\urcorner$. We prove, by induction on the structure of t', that there exists n such that $d' = s^n(d)$ and $t' \equiv s^n(x)$. First, t' cannot be a variable y different from x because then $\ulcorner t'\langle d\rangle\urcorner$ would be standard whereas $\ulcorner x\langle d'\rangle\urcorner$ is non-standard. If t' is x then we are done with $n = 0$, as $d = d'$ by (S3). Lastly, suppose that t' is of the form $f'(t_1', \ldots, t_h')$ (with h possibly zero). Now $d' \in D$ is non-standard so it has a predecessor $d'' \in D$. Thus $\ulcorner x\langle d'\rangle\urcorner = \ulcorner x\langle s(d'')\rangle\urcorner = \ulcorner s(x)\langle d''\rangle\urcorner$, the last equality by (S1). But then $\ulcorner t'\langle d\rangle\urcorner = \ulcorner s(x)\langle d''\rangle\urcorner$. So, by the formalized injectivity of Gödel numbering, t' is of the form $s(t'')$ for some t'' such that $\ulcorner t''\langle d\rangle\urcorner = \ulcorner x\langle d''\rangle\urcorner$. Then, by the induction hypothesis, there exists n such that $d'' = s^n(d)$ and $t'' \equiv s^n(x)$. Thus $n + 1$ is the number required as $d' = s^{n+1}(d)$ and $t' \equiv s^{n+1}(x)$.

Suppose that the term is $f(t_1, \ldots, t_k)$ (where k is possibly zero) and $\ulcorner t'\langle d\rangle\urcorner = \ulcorner f(t_1, \ldots, t_k)\langle d'\rangle\urcorner$. Then t' cannot be a variable y different from x. If t' is x then x must occur free in some t_i (otherwise $\ulcorner f(t_1, \ldots, t_k)\langle d'\rangle\urcorner$ would be standard). However, $d \leq d'$ so, by (S4), $\ulcorner x\langle d\rangle\urcorner \neq \ulcorner f(t_1, \ldots, t_k)\langle d'\rangle\urcorner$, a contradiction. So t' must be of the form $f'(t_1', \ldots, t_h')$. But then, by formalized injectivity, we have that $f \equiv f'$. So $h = k$ and, for all i $(1 \leq i \leq k)$ $\ulcorner t_i'\langle d\rangle\urcorner = \ulcorner t_i\langle d'\rangle\urcorner$. If x does not occur free in any t_i then, by the induction hypothesis, $t_i' \equiv t_i$ for all i and thus $t' \equiv f(t_1, \ldots, t_k)$ as required. If x does occur free in some t_i then, by the induction hypothesis, there exists n such that $d' = s^n(d)$ and, for all i, $t_i' \equiv t_i[s^n(x)/x]$. So indeed $t' \equiv f(t_1, \ldots, t_k)[s^n(x)/x]$.

It remains only to extend the induction to formulae. One proves, by induction on the structure of B, that $\ulcorner A\langle d\rangle\urcorner = \ulcorner B\langle d'\rangle\urcorner$ implies that if x does not occur free in B then $A \equiv B$ and if x does occur free in B then there exists n such that $d' = s^n(d)$ and $A \equiv B[s^n(x)/x]$. The straightforward argument, similar to the case for $f'(t_1', \ldots, t_h')$ and $f(t_1, \ldots, t_k)$ above, is omitted. The result follows. □.

Lemma 3.3

1. If $d \in D$ is standard then $\mathfrak{M}' \models Pr(\ulcorner A\langle d\rangle\urcorner)$ if and only if $T' \vdash A[\bar{d}/x]$.

2. If $d \in D$ is non-standard then $\mathfrak{M}' \models Pr(\ulcorner A\langle d\rangle\urcorner)$ if and only if there exists n such that $T' \vdash \forall x(A[s^n(x)/x])$.

Proof.

1. Suppose $d \in D$ is standard and $\mathfrak{M}' \models Pr(\ulcorner A\langle d\rangle\urcorner)$. Then $\ulcorner A[\bar{d}/x]\urcorner = \ulcorner A\langle d\rangle\urcorner = \ulcorner B\langle d'\rangle\urcorner$ for some $d' \in D$ and B such that $T' \vdash \forall xB$ (by the definition of the extension of Pr in $\mathfrak{M}'$). Now if d' is standard then $T' \vdash B[\bar{d'}/x]$ and $\ulcorner A[\bar{d}/x]\urcorner = \ulcorner B[\bar{d'}/x]\urcorner$ so $A[\bar{d}/x] \equiv B[\bar{d'}/x]$. Thus indeed $T' \vdash A[\bar{d}/x]$. If, however, d' is non-standard then x cannot occur

free in B. Therefore $T' \vdash B$ and $\ulcorner A[\overline{d}/x] \urcorner = \ulcorner B \urcorner$ so $A[\overline{d}/x] \equiv B$. Thus again $T' \vdash A[\overline{d}/x]$ as required.

Conversely, suppose that $T' \vdash A[\overline{d}/x]$. Then trivially $T' \vdash \forall x A[\overline{d}/x]$. It follows that $\mathfrak{M}' \models Pr(\ulcorner A[\overline{d}/x] \urcorner)$. Thus indeed $\mathfrak{M}' \models Pr(\ulcorner A\langle d \rangle \urcorner)$.

2. Suppose that $d \in D$ is non-standard, and that $\mathfrak{M}' \models Pr(\ulcorner A\langle d \rangle \urcorner)$. Then $\ulcorner A\langle d \rangle \urcorner = \ulcorner B\langle d' \rangle \urcorner$ for some $d' \in D$ and B such that $T' \vdash \forall x B$. If $d \leq d'$ then, by Lemma 3.2, $A \equiv B[s^m(x)/x]$ for some m. So clearly $T' \vdash \forall x A$, and the n we are required to find is zero. If $d' < d$ and d' is non-standard then, by Lemma 3.2, $A[s^n(x)/x] \equiv B$ for some n. But then we have found an n such that $T' \vdash \forall x A[s^n(x)/x]$. Lastly, if d' is standard then $\ulcorner B\langle d' \rangle \urcorner$ is standard, so x cannot occur free in A. Thus $A \equiv B[\overline{d'}/x]$ and $T' \vdash B[\overline{d'}/x]$. Therefore $T' \vdash \forall x A$ and again n is zero.

Conversely, suppose there exists n such that $T' \vdash \forall x A[s^n(x)/x]$. As d is non-standard, there exists $d' \in D$ such that $d = s^n(d')$. By the definition of the extension of Pr, $\mathfrak{M}' \models Pr(\ulcorner A[s^n(x)/x]\langle d' \rangle \urcorner)$. But, by (S1), $\ulcorner A\langle d \rangle \urcorner = \ulcorner A[s^n(x)/x]\langle d' \rangle \urcorner$. So indeed $\mathfrak{M}' \models Pr(\ulcorner A\langle d \rangle \urcorner)$. $\qquad \square$

Proposition 3.4 $\mathfrak{M}'$ *is a model of* T'.

Proof. We must show that $\mathfrak{M}'$ validates (C1)–(C3).

(C1) Suppose $T' \vdash \forall x A$ and $d \in D$. Then it is immediate from the definition of the extension of Pr in $\mathfrak{M}'$ that $\mathfrak{M}' \models Pr(\ulcorner A\langle d \rangle \urcorner)$ as required.

(C2) Suppose $d \in D$, $\mathfrak{M}' \models Pr(\ulcorner (A \to B)\langle d \rangle \urcorner)$ and $\mathfrak{M}' \models Pr(\ulcorner A\langle d \rangle \urcorner)$. If d is standard then, by Lemma 3.3(1), $T' \vdash (A \to B)[\overline{d}/x]$ and $T' \vdash A[\overline{d}/x]$. So $T' \vdash B[\overline{d}/x]$ whence, by Lemma 3.3(1), $\mathfrak{M}' \models Pr(\ulcorner B\langle d \rangle \urcorner)$ as required. If d is non-standard then, by Lemma 3.3(2), there exists m such that $T' \vdash \forall x (A \to B)[s^m(x)/x]$ and there exists m' such that $T' \vdash \forall x A[s^{m'}(x)/x]$. Therefore $T' \vdash \forall x B[s^n(x)/x]$ where n is the maximum of m and m'. So, by Lemma 3.3(2), $\mathfrak{M}' \models Pr(\ulcorner B\langle d \rangle \urcorner)$ as required.

(C3) Suppose that $d \in D$ and $\mathfrak{M}' \models Pr(\ulcorner A\langle d \rangle \urcorner)$. We omit the easy argument if d is standard. If d is non-standard then, by Lemma 3.3(2), there exists n such that $T' \vdash \forall x A[s^n(x)/x]$. Whence, by (C1),

$$T' \vdash \forall x Pr(\ulcorner A[s^n(x)/x]\langle x \rangle \urcorner).$$

Now, by n applications of (S1), $T' \vdash \forall x (Pr(\ulcorner A\langle s^n(x) \rangle \urcorner))$. So, by Lemma 3.3(2), it follows that $\mathfrak{M}' \models Pr(\ulcorner Pr(\ulcorner A\langle x \rangle \urcorner)\langle d \rangle \urcorner)$ as required. $\qquad \square$

We now have a second proof of Proposition 2.1. We have shown that any model $\mathfrak{M}$ of T expands to a model $\mathfrak{M}'$ of T'. It follows that T' is a conservative extension of T.

Proposition 3.5 $\mathfrak{M}'$ *is a model of* $T' + R$.

Proof. We need only verify R. Suppose then that $d \in D$ and $\mathfrak{M}' \models Pr(\ulcorner A\langle d \rangle \urcorner)$. If d is standard then, by Lemma 3.3(1), $T' \vdash A[\overline{d}/x]$. Thus, by Proposition 3.4, $\mathfrak{M}' \models A[d/x]$ as required. If, however, d is non-standard then, by Lemma 3.3(2), there exists n such that $T' \vdash \forall x(A[s^n(x)/x])$. By Proposition 3.4, $\mathfrak{M}' \models \forall x(A[s^n(x)/x])$. But d is non-standard, so there exists $d' \in D$ such that $d = s^n(d')$. Therefore $\mathfrak{M}' \models A[d/x]$ as required. $\qquad\square$

We have shown that any model of T expands to a model of $T' + R$. This completes the proof of Theorem 3.1.

4 Extending induction to $\mathcal{L}'$

The conservativity result of the last section is very general, as the proof works for an arbitrary T extending PRA. Nevertheless, one important possibility has been overlooked: that of extending induction to the language $\mathcal{L}'$. However, the rules of how one ought to do this are not immediately clear. For example, if T is PRA then it only has induction over quantifier-free formulae. Given that we are thinking of Pr as a Σ_1-formula in disguise, it does not seem reasonable to give T' any instances of induction not already available in PRA. Thus although uniform reflection together with PRA gives PA, there is no analogous result using T' and R.

The situation becomes a good deal more interesting if we consider PRA together with Σ_1-induction as the initial theory. We shall refer to this theory as $I\Sigma_1$.

With $I\Sigma_1$ as the base theory it seems reasonable to give the extended theory induction over some appropriate analogue of Σ_1 in $\mathcal{L}'$. To this end, we extend the arithmetical hierarchy to $\mathcal{L}'$. We define sets Σ'_n, Π'_n ($1 \le n$) as the least sets closed under:

1. $\Sigma'_n \subseteq \Sigma'_{n+1}, \Pi'_{n+1}$ and $\Pi'_n \subseteq \Sigma'_{n+1}, \Pi'_{n+1}$

2. If P is not Pr then $P(t_1, \ldots, t_n) \in \Sigma'_1, \Pi'_1$.

3. $Pr(t) \in \Sigma'_1$.

4. If $A, B \in \Sigma'_n$ then $A \wedge B, \exists x A \in \Sigma'_n$ and $\neg A \in \Pi'_n$.

5. If $A, B \in \Pi'_n$ then $A \wedge B, \forall x A \in \Pi'_n$ and $\neg A \in \Sigma'_n$.

The motivation is that Pr is supposed to be emulating a Σ_1 (but not Π_1) formula.

We now give the extended theory, $I\Sigma'_1$, the evident definition: $I\Sigma'_1$ is the smallest $\mathcal{L}'$-theory containing $I\Sigma_1$ and Σ'_1-induction and closed under $(C1)$–$(C3)$. Again we consider adding the analogue of uniform reflection, R. This time we do get the desired non-conservativity.

Theorem 4.1 $I\Sigma'_1 + R$ *contains PA.*

Proof. Suppose that $I\Sigma'_1 \vdash A[0/x]$ and $I\Sigma'_1 \vdash \forall x\, (A \to A[s(x)/x])$. Applying $(C1)$ we get that $I\Sigma'_1 \vdash Pr(\ulcorner A[0/x]\urcorner)$ and $I\Sigma'_1 \vdash \forall x Pr(\ulcorner (A \to A[s(x)/x])[x]\urcorner)$. The former gives immediately:

$$I\Sigma'_1 \vdash Pr(\ulcorner A\langle 0 \rangle \urcorner).$$

The latter gives, by $(C2)$, $I\Sigma'_1 \vdash \forall x (Pr(\ulcorner A\langle x\rangle\urcorner) \to Pr(\ulcorner A[s(x)/x]\langle x\rangle\urcorner))$ whence, by $(S1)$:

$$I\Sigma'_1 \vdash \forall x (Pr(\ulcorner A\langle x\rangle\urcorner) \to Pr(\ulcorner A\langle s(x)\rangle\urcorner)).$$

We can now apply Σ'_1-induction to derive $I\Sigma'_1 \vdash \forall x Pr(\ulcorner A\langle x\rangle\urcorner)$. Therefore, by one application of R, we have that $I\Sigma'_1 + R \vdash \forall x A$.

It is now easy to see that $I\Sigma'_1 + R$ derives induction for any $\mathcal{L}'$-formula, B. Just apply the above argument to the formula:

$$A \;\equiv\; (B[0/x] \wedge \forall y(B[y/x] \to B[s(y)/x])) \to B.$$

The result follows. $\qquad\qquad\qquad\qquad\qquad\qquad\qquad\qquad\qquad\qquad\square$

It is a special case of Lemma 4.4 below that $I\Sigma'_1 + R$ is actually conservative over PA.

The above argument can be translated back to give an elegant proof, using only $(D1)$, $(D2)$ and $(S1)$, that $I\Sigma_1 + Rfn$ is a theory as strong as PA. Note that condition $(C3)$ was not needed in the proof. Also, R was used only as a rule. We conjecture that if any of $(C1)$, $(C2)$ and R are weakened to their propositional versions then the resulting extension of $I\Sigma_1$ is conservative.

We conclude by showing that the trick used to prove Theorem 4.1 cannot be generalized to derive stronger principles than full induction. Define PA' to be the least $\mathcal{L}'$-theory containing PA and induction over every $\mathcal{L}'$-formula and closed under $(C1)$–$(C3)$.

Theorem 4.2 $PA' + R$ *is a conservative extension of PA.*

We write $I\Sigma_n$ for the $\mathcal{L}$-theory obtained by extending PRA with Σ_n-induction. Following the definition of $I\Sigma'_1$ above, define $I\Sigma'_n$ to be the least

$\mathcal{L}'$-theory containing PRA and Σ'_n-induction and closed under $(C1)$–$(C3)$. The proof of Theorem 4.2 uses the observation that:

$$(6) \qquad\qquad PA' = \bigcup_n I\Sigma'_n.$$

The inclusion $\bigcup_n I\Sigma'_n \subseteq PA'$ is obvious. For the converse, it is easy to show that $\bigcup_n I\Sigma'_n$ contains PRA, contains induction for arbitrary $\mathcal{L}'$-formulae and is closed under $(C1)$–$(C3)$. Thus $\bigcup_n I\Sigma'_n$ satisfies the closure conditions of PA'. Therefore it contains PA'.

Lemma 4.3 *For all n, the theory $I\Sigma'_n$ is a conservative extension of $I\Sigma_n$.*

Proof. Taking T to be $I\Sigma_n$, consider the translation $(\cdot)^*$ from $\mathcal{L}'$-formulae to $\mathcal{L}$-formulae defined in the proof of Proposition 2.1. We claim that for all $\mathcal{L}'$-formulae A, if $I\Sigma'_n \vdash A$ then $I\Sigma_n \vdash A^*$. The claim is shown by a straightforward modification of the proof of Proposition 2.1. The only additional case is to show that if A is an instance of Σ'_n-induction then $I\Sigma_n \vdash A^*$. But this holds because $(\cdot)^*$ maps Σ'_n-formulae to Σ_n-formulae, so A^* is an instance of Σ_n-induction. □

Lemma 4.4 *For all n, the theory $I\Sigma'_n + R$ is a conservative extension of PA.*

Proof. By Theorem 4.1, $I\Sigma'_n + R$ contains PA. Let $(\cdot)^*$ be the translation used in the last proof. We claim that $I\Sigma'_n + R \vdash A$ implies $PA \vdash A^*$. We already know that if $I\Sigma'_n \vdash A$ then $I\Sigma_n \vdash A^*$ and hence $PA \vdash A^*$. So we need only show that $PA \vdash R^*$. However, as in the proof of Proposition 2.2, this follows from the following fact about PA [6]:

$$\text{for all } n, \quad PA \vdash \forall x (Bew_{I\Sigma_n}(\ulcorner A\langle x\rangle \urcorner) \to A). \qquad\qquad □$$

By (6), it is clear that $PA' + R = \bigcup_n (I\Sigma'_n + R)$. It follows from Lemma 4.4 that $PA' + R$ is indeed conservative over PA. This completes the proof of Theorem 4.2.

5 Conclusions

In this paper we have investigated the potential of using the derivability conditions to induce properties of a provability predicate without having to go to the effort of following Gödel's construction. In particular we have focused on the possibility of obtaining non-conservative extensions using an extra axiom mimicking the uniform reflection schema.

Unfortunately, our results have been mainly negative. Although we have obtained a non-conservative extension in one notable case, the resulting theory, *PA*, can be obtained much more easily just by giving the full induction schema. Nevertheless, we believe that our results (both of non-conservativity and of conservativity) are interesting.

One natural question is whether a more general method of obtaining non-conservative extensions could be obtained by using more powerful axioms than $(C1)$–$(C3)$. It is clear that the proof of Theorem 4.2 is general enough to apply to any T' generated by a collection of axioms based on arithmetically valid formulae of predicate provability logic (see [1]). Nevertheless, the possibility remains that a more general method could be obtained by going beyond the modal paradigm of provability logic (for example, by replacing $(C2)$ and $(C3)$ with single axioms quantifying over the Gödel numbers of formulae). We believe it to be an interesting programme to investigate such generalizations.

There are other ways of adding a new predicate to the language to obtain non-conservative extensions. For example, one can axiomatize the property of being a *satisfaction class* as in the work of Robinson, Kotlarski and others (see [5, Ch. 15]). Also, Feferman has obtained non-conservative extensions by axiomatizing a partial truth predicate [2]. It is unclear how such semantic approaches relate to the provability based approach of this paper.

Acknowledgements

The first author was supported in part by a BFT grant (no. ITR ITS 9103). The second author was supported by an SERC studentship (no. 90311820) and by an EPSRC postdoctoral fellowship. Both authors thank Alan Smaill for encouraging this work.

References

[1] G. Boolos, *The logic of provability*, Cambridge University Press, 1994.

[2] S. Feferman, *Reflecting on incompleteness*, J. Symbolic Logic **56** (1991), 1–49.

[3] G.Boolos and R.Jeffrey, *Computability and Logic*, 3rd ed., Cambridge University Press, 1989.

[4] J.-Y. Girard, *Proof Theory and Logical Complexity, Volume 1*, Studies in Proof Theory, Bibliopolis, Naples, 1987.

[5] R. Kaye, *Models of Peano Arithmetic*, Clarendon Press, Oxford, 1991.

[6] G. Kreisel and A. Lévy, *Reflection principles and their use for establishing the complexity of axiomatic systems*, Z. Math. Logik Grundlag. Math. **14** (1968), 97–142.

[7] M. Löb, *Solution of a problem of Leon Henkin*, J. Symbolic Logic **20** (1955), 115–118.

[8] R. Montague, *Syntactical treatment of modality, with corollaries on reflexion principles and finite axiomatizability*, Acta Phil. Finnica **16** (1963), 153–167 (Reprinted in: Formal Philosophy, selected papers of Richard Montague, (R.Thomason, ed.), Yale University Press, 1974).

[9] R. Solovay, *Provability interpretations of modal logic*, Israel J. Math. **25** (1976), 287–304.

MAX-PLANCK-INSTITUT FÜR INFORMATIK, SAARBRÜCKEN, GERMANY
    e-mail address: SEAN@MPI-SB.MPG.DE
DEPARTMENT OF COMPUTER SCIENCE, UNIVERSITY OF EDINBURGH, SCOTLAND
    e-mail address: ALEX.SIMPSON@DCS.ED.AC.UK

Stone bases, alias the constructive content of Stone representation

Sara Negri

Abstract

This paper provides a constructive proof of Stone representation theorem for distributive lattices, in the framework of G. Sambin's formal topologies. In order to formalize the result wholly inside Martin- Löf's Intuitionistic Type Theory, the notion of *Stone base* is introduced, and it is proved to be equivalent to that of formal topology in which any cover of a basic open admits a finite subcover. The main theorem states that the category of distributive lattices with apartness is equivalent to the category of Stone bases. Finally the results are related to those of Johnstone and to the classical point-set representation.

1 Introduction

The purpose of this paper is to analyze the constructive content of Stone representation theorem for distributive lattices (cf. [13]), asserting that any distributive lattice is isomorphic to the compact opens of a suitable topological space, called *coherent space*.

This is done in the setting of formal topologies, which have been introduced by G. Sambin in [7] with the purpose of developing pointfree topology inside the constructive foundational framework of Martin-Löf's Intuitionistic Type Theory, abbreviated ITT. The demand of expressibility inside ITT often brings to very "elementary" definitions and proofs. This is particularly evident in dealing with Stone representation theory for distributive lattices.

It is well known (cf. [4]) that distributive lattices form a category which is equivalent to the category of coherent frames. It will be shown here that the category of coherent frames, enriched with an intuitionistic predicate of apartness from 0 corresponding to the positivity predicate of formal topologies, is equivalent to the category of *Stone formal topologies*. Stone formal topologies are formal topologies in which every basic open is compact. Indeed we will prove that any Stone formal topology is uniquely determined by its behaviour on the finite subsets of the base. We are thus led to the notion of *Stone bases*, which represent the finitary content of Stone formal topologies, with the advantage of being wholly expressible inside ITT. In fact a cover, which is an infinitary relation, i.e. a relation between elements and subsets, is replaced by a *finitary cover*, which is a relation between elements of types, with no reference to arbitrary subsets. So the representation theorem for distributive lattices here takes the form of an equivalence with the category of Stone bases.

This ground result allows us to obtain the usual pointfree representation via coherent frames (cf. [4]) simply by "set-theoretic sugaring".

Finally, in order to obtain the usual point-set Stone representation by means of coherent spaces, we need to use the Prime Filter Theorem, which is seen to be equivalent to extensionality of Stone formal topologies.

Coherent spaces arise as spaces of (formal) points on a Stone formal topology, or equivalently on a Stone base; so we might say that (cf. [12]) Stone bases are to coherent spaces what Information bases are to Scott domains.

Even if we do not use the ITT notation explicitly throughout the paper, we have been careful to distinguish which notions are expressible inside ITT and which are not. The various equivalences, proved outside ITT, tell us that our notion of Stone base is a good replacement, inside ITT, of the notion of coherent frame. We can thus forget about coherent frames, whose treatment in the constructive context of ITT would require an awkward amount of specifications, and retain the simpler notion of Stone base.

We conclude with observing that this work represents a continuation of some ideas already in [7] and [9].

2   Stone formal topologies and the category of Stone bases

We start with some preliminary definitions concerning our approach to pointfree topology. For references, further details and notation the reader is referred to [7] (where sometimes a different notation is used), [9], [12].

Definition 2.1 *A* formal topology *is a structure* $\mathcal{A} = (S, \cdot, 1, \lhd_{\mathcal{A}}, Pos_{\mathcal{A}})$ *where*[1]

- $(S, \cdot, 1)$ *is a* (formal) base, *i.e. a commutative monoid with unit;*

- $\lhd_{\mathcal{A}}$ *is a relation between elements and subsets of* S, *called* (formal) cover, *satisfying:*

---

[1] From now on we will omit subscripts when clear from the context.

$$\frac{a \in U}{a \triangleleft_A U} \ reflexivity,$$

$$\frac{a \triangleleft_A U \quad U \triangleleft_A V}{a \triangleleft_A V} \ transitivity,$$

$$\frac{a \triangleleft_A U \quad a \triangleleft_A V}{a \triangleleft_A U \cdot V} \ \text{-}right,$$

$$\frac{a \triangleleft_A U}{a \cdot b \triangleleft_A U} \ \text{-}left,$$

where $U \triangleleft_A V$ denotes a derivation $x \triangleleft_A V [x \in U]$ of $x \triangleleft_A V$ from the assumption $x \in U$ and $U \cdot V \equiv \{a \cdot b : a \in U, b \in V\}$;

- *Pos_A is a predicate, called* positivity predicate *satisfying:*

$$\frac{Pos_A(a) \quad a \triangleleft_A U}{Pos_A(U)} \ monotonicity,$$

$$\frac{Pos_A(a) \to a \triangleleft_A U}{a \triangleleft_A U} \ positivity,$$

where $Pos_A(U) \equiv (\exists b \in U) Pos_A(b)$.

Formal opens are subsets of S, up to the equivalence $U =_A V \equiv U \triangleleft_A V \& V \triangleleft_A U$. Morphisms between formal topologies are the formal counterpart of the inverse of continuous maps, and thus they are defined as follows:

Definition 2.2 *A morphism between formal topologies $A \equiv (S, \cdot, 1, \triangleleft_A, Pos_A)$ and $B \equiv (T, \cdot, 1, \triangleleft_B, Pos_B)$ is an application $f : S \to \mathcal{P}T$, where $\mathcal{P}T$ denotes the power set of T, such that:*

1. *$f(1) =_B 1$;*

2. *$f(a \cdot b) =_B f(a) \cdot f(b)$;*

3. *$a \triangleleft_A U \to f(a) \triangleleft_B f(U)$, where $f(U) \equiv \cup_{b \in U} f(b)$;*

4. *$Pos_B(f(a)) \to Pos_A(a)$.*

Equality of morphisms $f, g : A \to B$ is defined by $f = g \equiv (\forall a \in S)(f(a) =_B g(a))$.

It is easy to prove that formal topologies and morphisms of formal topologies, with composition given by $(f \circ g)(a) \equiv f(g(a))$ and identity defined by $1(a) \equiv \{a\}$, form a category, called the category of formal topologies and denoted with **FTop**.

In the sequel we will be mostly concerned with a particular full subcategory **SFTop** of **FTop** whose objects are Stone formal topologies, defined below:

Definition 2.3 *A formal topology $\mathcal{A}$ is said to be* Stone *if $\lhd$ is a Stone cover, i.e. whenever $a \lhd U$ there exists a finite subset K of U such that $a \lhd K$.*

The notation $\mathcal{P}_\omega S$ will be used to denote the collection of finite subsets of S as defined in [11]. If S is a set (or a type) in the ITT framework, then so is $\mathcal{P}_\omega S$.

It is well known (cf. e.g. [7] for details) that a cover $\lhd_\mathcal{A}$ on S is associated with a closure operator $\mathcal{A}$ on $\mathcal{P}S$, defined by $\mathcal{A}U \equiv \{a \in S : a \lhd_\mathcal{A} U\}$. Here, we say that a subset U is $\mathcal{A}$-saturated if $\mathcal{A}U = U$. A bijective correspondence holds between covers on S and closure operators on $\mathcal{P}S$ satisfying the equality $\mathcal{A}(U \cdot V) = \mathcal{A}U \cap \mathcal{A}V$. Thus Stone formal topologies can be characterized in terms of closure operators by the following:

Lemma 2.4 *For any formal topology $\mathcal{A}$, $\mathcal{A}$ is Stone iff $\mathcal{A}U = \cup\{\mathcal{A}K : K \subseteq_\omega U\}$.*

Proof: $\mathcal{A}$ Stone means by definition that $a \lhd U \to (\exists K \subseteq_\omega U)(a \lhd K)$, which is equivalent to $a \lhd U \leftrightarrow (\exists K \subseteq_\omega U)(a \lhd K)$, which is just another way of expressing the equality $\mathcal{A}U = \cup\{\mathcal{A}K : K \subseteq_\omega U\}$. $\square$

The interest of the above characterization of Stone topologies is that closure operators satisfying the corresponding condition are a well known object of study: they are just *algebraic closure operators*. By a well known result, which can be provided with a constructive proof, a closure operator $\mathcal{C}$ is algebraic if and only if it is *inductive* i.e. the union of directed families of $\mathcal{C}$- saturated subsets is $\mathcal{C}$-saturated.

One of the aims of this paper is to develop a strictly finitary treatment of Stone formal topologies. For any cover $\lhd$ on S, we define the *finite trace* $\prec_\lhd$ of $\lhd$ as the restriction of $\lhd$ to elements and finite subsets of S:

$$a \prec_\lhd K \equiv a \lhd K \ (a \in S, K \in \mathcal{P}_\omega S).$$

As we shall see, any Stone formal topology is uniquely determined by its behaviour on finite subsets, that is by its finite trace.

Our basic definition is obtained by abstraction on finite traces:

Definition 2.5 *A* Stone base *is a quintuple $\mathcal{S} \equiv (S, \cdot, 1_\mathcal{S}, \prec_\mathcal{S}, Pos_\mathcal{S})$ where:*

- $(S, \cdot, 1_\mathcal{S})$ *is a (formal) base;*

- $\prec_\mathcal{S}$ *is a finitary cover, i.e. a relation between elements and finite subsets of S formally satisfying the same conditions of covers, i.e. for any $a, b \in S$ and $K, L \in \mathcal{P}_\omega S$:*

$$\frac{a \in K}{a \prec K} \ reflexivity,$$

$$\frac{a \prec K \quad K \prec L}{a \prec L} \ transitivity,$$

$$\frac{a \prec K}{a \cdot b \prec K} \ \cdot\text{-}left,$$

$$\frac{a \prec K \quad a \prec L}{a \prec K \cdot L} \ \cdot\text{-}right:$$

- Pos_S *is a predicate on* S *satisfying monotonicity and positivity w.r.t.* $\prec$*, i.e.:*

$$\frac{Pos_S(a) \quad a \prec K}{(\exists b \in K)Pos_S(b)} \ monotonicity,$$

$$\frac{Pos_S(a) \to a \prec K}{a \prec K} \ positivity.$$

For any formal topology $\mathcal{A}$, the finite trace of $\vartriangleleft_A$ determines a Stone base, by the following:

Proposition 2.6 *Let* $\mathcal{A} \equiv (S, \cdot, 1, \vartriangleleft, Pos)$ *be a formal topology and let* $\prec_\vartriangleleft$ *be the finite trace of* $\vartriangleleft$*. Then the structure* $\mathbf{S}(\mathcal{A}) \equiv (S, \cdot, 1, \prec_\vartriangleleft, Pos)$ *is a Stone base.*

Proof: Straightforward from definitions. As for $\cdot R$, observe that $U \cdot V$ is finite whenever U and V are finite. $\square$

Conversely, we have the following:

Proposition 2.7 *Given a Stone base* $\mathcal{S} \equiv (S, \cdot, 1, \prec, Pos)$*, let* $\vartriangleleft_\prec$ *be the minimal cover with trace* $\prec$*, that is the cover inductively generated by taking* $a \vartriangleleft_\prec K$ *as axioms whenever* $a \prec K$ *and by closing under the rules for cover, i.e. reflexivity, transitivity,* $\cdot$*-left,* $\cdot$*-right. Then for any* $a \in S$ *and* $U \subseteq S$*:*

$$a \vartriangleleft_\prec U \ \textit{iff} \ \textit{there exists} \ K \subseteq_\omega U \ \textit{such that} \ a \prec K. \tag{1}$$

Therefore $\mathbf{A}(\mathcal{S}) \equiv (S, \cdot, 1, \vartriangleleft_\prec, Pos)$ *is a Stone formal topology.*

Proof: Trivially, $\vartriangleleft_\prec$ is a cover.

Suppose there exists $K \subseteq_\omega U$ such that $a \prec K$. Then $a \vartriangleleft_\prec K$ by the axioms; from $K \subseteq_\omega U$ it follows by reflexivity $K \vartriangleleft_\prec U$ and hence $a \vartriangleleft_\prec U$ by transitivity.

The converse is also simple, but it illustrates the method of induction on covers, in this case on the generation of $\vartriangleleft_\prec$. For axioms and reflexivity the claim is trivial. Assume $a \vartriangleleft_\prec V$ is obtained from $a \vartriangleleft_\prec U$ and $U \vartriangleleft_\prec V$ by transitivity. By inductive hypothesis, $a \prec K$ for some $K \subseteq_\omega U$ and for all $x \in K$, $x \prec L_x$ for some $L_x \subseteq_\omega V$. By reflexivity, $L_x \prec \cup_{x \in K} L_x$, hence by transitivity of $\prec$, $x \prec \cup_{x \in K} L_x$ for any $x \in K$, hence $K \prec \cup_{x \in K} L_x$ and finally $a \prec \cup_{x \in K} L_x$ by transitivity. For $\cdot L$ and $\cdot R$ the proof is straightforward.

Since by axioms $a \prec K$ gives $a \vartriangleleft_\prec K$. $\vartriangleleft_\prec$ is a Stone cover by (1).

To conclude, it remains only to be shown that *Pos* satisfies monotonicity and positivity w.r.t. the cover $\vartriangleleft_\prec$. To prove monotonicity, assume $Pos(a)$ and $a \vartriangleleft_\prec U$. Then there exists $K \subseteq_\omega U$ such that $a \prec K$. By monotonicity of *Pos* w.r.t. $\prec$, $Pos(K)$ holds, thus a fortiori $Pos(U)$ holds too. Before proving positivity, define, for all $U \subseteq S$, $U^+ \equiv \{a \in U : Pos(a)\}$ and for all $a \in S$ let $a^+ \equiv \{a\}^+$. Then the premise of positivity, i.e. $Pos(a) \to a \vartriangleleft_\prec U$ is equivalent to $a^+ \vartriangleleft_\prec U$. Moreover, from $a^+ \prec a^+$, we have $Pos(a) \to a \prec a^+$, which, by positivity of *Pos* w.r.t. $\prec$, gives $a \prec a^+$ and therefore $a \vartriangleleft_\prec a^+$. This, together with $a^+ \vartriangleleft_\prec U$, yields by transitivity $a \vartriangleleft_\prec U$. $\square$

Let $\vartriangleleft$ and $\vartriangleleft'$ be covers on the same base S. We say that $\vartriangleleft'$ is a *quotient* of $\vartriangleleft$ if for all $a \in S$, $U \subseteq S$,

$$a \vartriangleleft U \to a \vartriangleleft' U.$$

This relation corresponds to set-theoretic inclusion between the extensions of $\vartriangleleft$ and $\vartriangleleft'$, thus we also say that $\vartriangleleft'$ is *greater* than $\vartriangleleft$, and write $\vartriangleleft \leq \vartriangleleft'$. Given a cover $\vartriangleleft$, let $\prec_\vartriangleleft$ be its finite trace. By proposition 2.6 and 2.7, $\prec_\vartriangleleft$ generates a cover $\vartriangleleft_{\prec_\vartriangleleft}$ which is Stone and therefore does not coincide in general with $\vartriangleleft$. Indeed $\vartriangleleft$ is a quotient of $\vartriangleleft_{\prec_\vartriangleleft}$. In fact, if $a \vartriangleleft_{\prec_\vartriangleleft} U$ then there exists $K \subseteq_\omega U$ such that $a \prec_\vartriangleleft K$ and therefore, since $\prec_\vartriangleleft$ is a restriction of $\vartriangleleft$, $a \vartriangleleft K$, thus $a \vartriangleleft U$. Moreover, $\vartriangleleft_{\prec_\vartriangleleft}$ is the greatest Stone cover of which $\vartriangleleft$ is a quotient. In order to prove this, consider a Stone cover $\vartriangleleft'$ of which $\vartriangleleft$ is a quotient. Then $a \vartriangleleft' U$ implies $a \vartriangleleft' K$ for some $K \subseteq_\omega U$, and therefore also $a \vartriangleleft K$. Thus, by definition of trace of a cover, $a \prec_\vartriangleleft K$ and by definition of cover generated by a trace $a \vartriangleleft_{\prec_\vartriangleleft} U$.

The cover $\vartriangleleft$ coincides with $\vartriangleleft_{\prec_\vartriangleleft}$ iff $\vartriangleleft$ is a Stone cover. In fact, suppose that $\vartriangleleft = \vartriangleleft_{\prec_\vartriangleleft}$. Then $\vartriangleleft$ is Stone since $\vartriangleleft_{\prec_\vartriangleleft}$ is Stone by the above remarks. Conversely, $\vartriangleleft_{\prec_\vartriangleleft} \leq \vartriangleleft$ since $\vartriangleleft$ is a quotient of $\vartriangleleft_{\prec_\vartriangleleft}$. Since $\vartriangleleft_{\prec_\vartriangleleft}$ is the greatest Stone cover of which $\vartriangleleft$ is a quotient, then also the opposite inequality holds and the conclusion follows.

We denote $\vartriangleleft_{\prec_\vartriangleleft}$ with $\vartriangleleft_\omega$ and call it the *Stone compactification* [2] *of* $\vartriangleleft$. Then the above discussion proves the following:

Proposition 2.8 *Any cover $\vartriangleleft$ admits a Stone compactification $\vartriangleleft_\omega$, which is the cover induced by its finite trace, $\vartriangleleft_{\prec_\vartriangleleft}$. This is the greatest Stone cover of which $\vartriangleleft$ is a quotient and it coincides with $\vartriangleleft$ iff $\vartriangleleft$ is a Stone cover.*

We also have:

---

[2] Not to be confused with the familiar Stone-Čech compactification!

Corollary 2.9 *A cover* $\lhd$ *is a Stone cover iff it is finitarily axiomatizable, i.e. there exists a relation* R *between elements and finite subsets of* S *such that* $\lhd$ *coincides with the cover* $\lhd_R$ *generated by* R, *which is obtained from the axioms* $a \lhd_R U$ *whenever* $R(a, U)$ *holds, by closing under the rules for covers.*

For any formal topology $\mathcal{A}$, let $\mathcal{A}_\omega \equiv \mathbf{AS}(\mathcal{A})$, that is, by definitions of $\mathbf{A}$, $\mathbf{S}$ and $\lhd_\omega$, $\mathcal{A}_\omega \equiv (S. \cdot, 1, \lhd_\omega, Pos)$. Then, by propositions 2.6 and 2.7, $\mathcal{A}_\omega$ is a Stone formal topology, called the *Stone compactification of* $\mathcal{A}$.

Given two formal topologies with the same base monoid and the same positivity predicate, $\mathcal{A} \equiv (S. \cdot, 1, \lhd, Pos)$ and $\mathcal{A}' \equiv (S, \cdot, 1, \lhd', Pos)$, we say that $\mathcal{A}$ is a quotient of $\mathcal{A}'$ if the cover $\lhd'$ is quotient of the cover $\lhd$.

Summing up we have proved:

Theorem 2.10 *Any formal topology* $\mathcal{A} \equiv (S, \cdot, 1, \lhd, Pos)$ *admits a Stone compactification* $\mathcal{A}_\omega \equiv (S, \cdot, 1, \lhd_\omega, Pos)$. *This is the greatest Stone formal topology of which* $\mathcal{A}$ *is a quotient. Moreover* $\mathcal{A} = \mathcal{A}_\omega$ *iff* $\mathcal{A}$ *is a Stone formal topology.*

An important corollary of the above results is the fact that any Stone formal topology is determined by its finite trace, which is a relation between elements and finite subsets of S and is therefore completely expressible inside the foundational framework of ITT.

The definition of morphism between Stone bases is obtained in the natural way, similarly to the definition of morphism between formal topologies, except for the fact that an element is always mapped into a finite subset of the base. For any Stone base $\mathcal{S}$, let $=_S$ denote the equivalence relation on $\mathcal{P}_\omega S$ defined by $K =_S L \equiv K \prec_S L \,\&\, L \prec_S K$. Then we can define:

Definition 2.11 *A morphism between two Stone bases* $\mathcal{S} \equiv (S, \cdot, 1_S, \prec_S, Pos_S)$ *and* $\mathcal{T} \equiv (T, \cdot, 1_T, \prec_T, Pos_T)$ *is an application* $f : S \to \mathcal{P}_\omega T$ *formally satisfying the same conditions of morphisms between formal topologies, i.e.*

1. $f(1_S) =_T 1_T$;

2. $f(a \cdot b) =_T f(a) \cdot f(b)$;

3. $a \prec_S K \to f(a) \prec_T f(K)$;

4. $Pos_T(f(a)) \to Pos_S(a)$.

Likewise, equality of morphisms $f, g : \mathcal{S} \to \mathcal{T}$ *is defined by* $f = g \equiv (\forall a \in S)(f(a) =_T g(a))$.

As for formal topologies, composition of morphisms $f : \mathcal{S} \to \mathcal{T}$ and $g : \mathcal{T} \to \mathcal{V}$ is defined by $(g \circ f)(a) \equiv g(f(a))$ and the identity $1 : \mathcal{S} \to \mathcal{S}$ is the morphism mapping each element into its singleton, that is $1(a) \equiv \{a\}$. Thus Stone bases form a category which we denote with **SB**.

Let **SFTop*** be the category of Stone formal topologies with *coherent morphisms* i.e. morphisms from $\mathcal{A}$ to $\mathcal{B}$ induced by applications mapping elements of the base of $\mathcal{A}$ into finite subsets of the base of $\mathcal{B}$. Then we can prove that the bijective correspondence determined by **A** and **S** between Stone bases and Stone formal topologies extends to an equivalence between **SFTop*** and **SB**.

Observe that any coherent morphism f between Stone formal topologies $\mathcal{A}$ and $\mathcal{B}$ *is* a morphism between the Stone bases $\mathbf{S}(\mathcal{A})$ and $\mathbf{S}(\mathcal{B})$. In fact, condition *1* and *2* hold since on finite subsets $=_\mathcal{B}$ and $=_{\mathbf{S}(\mathcal{B})}$ coincide. As for condition *3*, suppose that $a \prec_{\lhd_\mathcal{A}} K$. Then by definition $a \lhd_\mathcal{A} K$. Since f is a morphism in **SFTop***, $f(a) \lhd_\mathcal{B} f(K)$, and since $f(K)$ is finite, $f(a) \lhd \prec_{\lhd_\mathcal{B}} f(K)$. Condition *4* holds trivially since the positivity predicate is the same.

Conversely, any morphism g between Stone bases $\mathcal{S}$ and $\mathcal{T}$ *is* a morphism between the Stone formal topologies $\mathbf{A}(\mathcal{S})$ and $\mathbf{A}(\mathcal{T})$. The proof is trivial, except for the third condition. In fact, let $a \lhd_{\prec_\mathcal{S}} U$. Then $a \prec_\mathcal{S} K$ for some $K \subseteq_\omega U$ and therefore, since g is a morphism of Stone bases, $g(a) \lhd_{\prec_\mathcal{T}} g(K)$. Since $g(K) \subseteq_\omega g(U)$, then $g(a) \lhd_{\prec_\mathcal{T}} g(U)$.

Thus **A** and **S** extend to functors which are the identity on morphisms between **SB** and **SFTop***. Then the functor **A** is a bijection on morphisms, i.e. is full and faithful. Moreover the functor is dense, since for all Stone formal topologies $\mathcal{A}$, $\mathcal{A} \cong \mathbf{A}(\mathbf{S}(\mathcal{A}))$.

We have therefore proved:

Theorem 2.12 *The category* **SFTop*** *of Stone formal topologies with coherent morphisms is equivalent to the category* **SB** *of Stone bases.*

Observe that the proof of this equivalence, as well as the proofs of the other equivalences throughout the paper, is constructive since we actually define the inverse functor, in this case **S**. Moreover the constructive character of the category **SB** consists in the fact that objects are types and morphisms are functions between types in ITT. Anyway, it would be easy to give a more general notion of morphism between Stone bases, with $f : S \to \mathcal{P}T$, in such a way that a category equivalent to the whole of **SFTop** is obtained.

3 Pointfree representation of coherent frames and distributive lattices

In this section we will show how Stone bases give a constructive representation of distributive lattices and relate this with the usual pointfree version of Stone representation, based on coherent frames (see [4], II.3.4).

Since Stone bases are endowed with an intuitionistic predicate of positivity, we will deal with a richer structure than that of distributive lattices, namely distributive lattices with a predicate which reflects the properties of *Pos*. We say that $\cdot \#0$ is a predicate of apartness from 0 in a distributive lattice L if the following properties hold for $a \in L$ and $K \subseteq_\omega L$:

$$\frac{a\#0 \quad a \leq \vee K}{(\exists b \in K)(b\#0)} \ (m), \qquad \frac{a\#0 \to a \leq \vee K}{a \leq \vee K} \ (p).$$

The name "apartness" is justified by the fact that the predicate $\cdot \#0$ is easily seen to satisfy the following

$$\neg(a\#0) \leftrightarrow a = 0, \tag{2}$$

which is an instance of one of the defining properties of apartness (cf. [14]). Thus a lattice with apartness from 0 has a *stable* equality to 0, i.e. $\neg\neg(a = 0)$ holds iff $a = 0$ holds. On the other hand, we can prove *classically* that the predicate defined by $a\#0 \equiv \neg(a = 0)$ is a predicate of apartness from 0. This is the coarsest predicate of apartness from 0 since, for every such predicate $\cdot \#0$ and for all $a \in L$, it follows from (2) that $a\#0 \to \neg(a = 0)$.

A morphism between distributive lattices with apartness $f : (L_1, \#_1) \to (L_2, \#_2)$ is a lattice morphism such that for all $a \in L_1$, $f(a)\#_2 0 \to a\#_1 0$. The category of distributive lattices with apartness will be denoted with $\mathbf{DLat}_\#$.

Given a distributive lattice with apartness from 0, $(L, \#)$, let $\mathcal{S}_L$ be the *Stone base associated to L*, defined by $\mathcal{S}_L \equiv (L, \wedge, 1, \prec_L, Pos_L)$, with

$$a \prec_L K \equiv a \leq \vee K$$
$$Pos_L(a) \equiv a\#0.$$

Conversely, given a Stone base $\mathcal{S} \equiv (S, \cdot, 1, \prec, Pos)$, we obtain a distributive lattice with apartness from 0 defining $Sat(\mathcal{S}) \equiv \{\mathcal{S}K : K \subseteq_\omega S\}$, where $\mathcal{S}K \equiv \{a \in S : a \prec K\}$. Then $Sat(\mathcal{S})$, with operations given by

$$\mathcal{S}K \vee \mathcal{S}L \equiv \mathcal{S}(K \cup L)$$
$$\mathcal{S}K \wedge \mathcal{S}L \equiv \mathcal{S}K \cap \mathcal{S}L$$

top and bottom $\mathcal{S}\{1\}$ and $\mathcal{S}\emptyset$ respectively and apartness $\mathcal{S}K\#0 \equiv Pos(K)$, is a distributive lattice with apartness from 0. $Sat(\mathcal{S})$ is called the lattice of *saturated subsets* of the Stone base $\mathcal{S}$.

Then we have:

Proposition 3.1 *Any distributive lattice L is isomorphic to the lattice of saturated subsets of the Stone base $\mathcal{S}_L$.*

Proof: First observe that any saturated subset of $\mathcal{S}_L$ is the saturation of a singleton, since for all $K \subseteq_\omega L$, $\mathcal{S}K = \mathcal{S}\{\vee K\}$. Therefore the map

$$\phi: \quad L \quad \to \quad Sat(\mathcal{S}_L)$$
$$a \quad \longmapsto \quad \mathcal{S}\{a\}$$

is surjective. Since $\mathcal{S}\{a\} = \{b \in L : b \leq a\}$, which will be denoted with $\downarrow a$ for short, ϕ is clearly injective. This map is a morphism of lattices because $\phi(a \vee b) \equiv \mathcal{S}\{a \vee b\} = \mathcal{S}\{a,b\} \equiv \mathcal{S}\{a\} \vee \mathcal{S}\{b\}$; then ϕ preserves meet since $\downarrow (a \wedge b) = (\downarrow a) \cap (\downarrow b)$. The top is preserved by definition and so does the bottom since $\mathcal{S}\{0\} = \mathcal{S}\emptyset$. Finally ϕ is an isomorphism of lattices with apartness since for all $a \in L$, $\mathcal{S}\{a\}\#0$ iff $Pos_L(a)$ iff $a\#0$. $\square$

Let $f : \mathcal{S} \to \mathcal{T}$ be a morphism of Stone bases. Then it is easy to prove that the map

$$Sat(f): \quad Sat(\mathcal{S}) \quad \to \quad Sat(\mathcal{T})$$
$$\mathcal{S}K \quad \mapsto \quad \mathcal{T}(f(K))$$

is a morphism of lattices with apartness, and functoriality is straightforward. By the above proposition, Sat is a dense functor. Moreover we have:

Lemma 3.2 *The functor Sat establishes a bijective correspondence between $\mathbf{SB}(\mathcal{S}, \mathcal{T})$ and $\mathbf{DLat}_\#(Sat(\mathcal{S}), Sat(\mathcal{T}))$.*

Proof: Let $f, g \in \mathbf{SB}(\mathcal{S}, \mathcal{T})$ and suppose $Sat(f) = Sat(g)$. Then for all $a \in S$, $Sat(f)(\mathcal{S}\{a\}) = Sat(g)(\mathcal{S}\{a\})$, that is $\mathcal{T}f(a) = \mathcal{T}g(a)$, i.e. $f(a) =_\mathcal{T} g(a)$. By definition of equality between coherent morphisms, $f = g$, thus the correspondence is 1-1. As for surjectivity, let $h \in \mathbf{DLat}_\#(Sat(\mathcal{S}), Sat(\mathcal{T}))$. Since for all $a \in S$, $h(\mathcal{S}\{a\}) \in Sat(\mathcal{T})$, then we know [3] that $h(\mathcal{S}\{a\}) = \mathcal{T}(K_a)$, for a suitable finite subset K_a of T. Define $f : S \to \mathcal{P}_\omega T$ by $f(a) \equiv K_a$. It is then routine to verify that f is a morphism of Stone bases. $\square$

The above lemma and theorem 2.12 prove the following:

Theorem 3.3 *The category $\mathbf{DLat}_\#$ of distributive lattices with apartness is equivalent to the category $\mathbf{SB}$ of Stone bases with coherent morphisms and to the category $\mathbf{SFTop}^*$ of Stone formal topologies with coherent morphisms.*

---

[3] A careful formalization in ITT would substitute $Sat(\mathcal{S})$, which is not a set but a set-indexed family, with $(\mathcal{P}_\omega(S), =_\mathcal{S})$, which is a set together with an equality relation and which is isomorphic to $Sat(\mathcal{S})$. In particular, this step of the proof would be more explicit.

Before going on with the representation of distributive lattices by means of coherent frames, we introduce the category of frames with apartness from 0. Due to the presence of arbitrary suprema, the notion of frame cannot be directly formalized in ITT. This is the reason why we base our approach to pointfree topology on the equivalent notion of formal topology, which is expressible in the ITT framework. In fact, we will provide an extension of the representation theorem for frames (cf. [1]) by showing that the category of frames with apartness from 0 is equivalent to the category of formal topologies.

Apartness from 0 in a frame is defined like apartness from 0 in a lattice, except that arbitrary suprema must be considered, i.e. we require:

$$\frac{a\#0 \quad a \leq \vee U}{(\exists b \in U)(b\#0)} \ (m), \qquad \frac{a\#0 \rightarrow a \leq \vee U}{a \leq \vee U} \ (p).$$

Given a predicate $\cdot \#0$ of apartness from 0 on a base S of a frame F, we can extend it to the whole frame F by defining, for $a \in F$,

$$a\#_F 0 \equiv (\exists b \in S)(b \leq a \ \& \ b\#0).$$

In fact it is easy to prove that for all $a \in S$, $a\#_F 0$ iff $a\#0$, i.e. $\cdot \#_F 0$ extends $\cdot \#0$, and that $\cdot \#_F 0$ is a predicate of apartness from 0 as well.

We say that a morphisms of frames is a morphism between two given frames with apartness from 0, $(F_1, \#_1)$ and $(F_2, \#_2)$, if for all $a \in S_1$ (base of F_1), $f(a)\#_2 0 \rightarrow a\#_1 0$. We will denote with $\mathbf{Frm}_\#$ the category thus obtained and call it the category of frames with apartness. Let Sat be the functor which associates to a formal topology $\mathcal{A} \equiv (S, \cdot, 1, \triangleleft, Pos)$ the frame of its saturated subsets, $Sat(\mathcal{A}) \equiv \{\mathcal{A}U : U \subseteq S\}$ with operations of finite meet and arbitrary join given by

$$\mathcal{A}U \wedge \mathcal{A}V \equiv \mathcal{A}U \cap \mathcal{A}V,$$
$$\vee_{i \in I} \mathcal{A}U_i \equiv \mathcal{A}(\cup_{i \in I} U_i).$$

A base of $Sat(\mathcal{A})$ is given by the saturations of singletons from S, i.e. $\{\mathcal{A}\{a\} : a \in S\}$. On such a base a predicate of apartness from 0 defined by:

$$\mathcal{A}\{a\}\#_{Sat} 0 \equiv Pos(a).$$

If $f : \mathcal{A} \rightarrow \mathcal{B}$ is a morphism of formal topologies, then $Sat(f) : Sat(\mathcal{A}) \rightarrow Sat(\mathcal{B})$ is defined by

$$Sat(f)(\mathcal{A}U) \equiv \mathcal{B}(f(U))$$

and it is seen to be a morphism between frames with apartness.

The functor Sat is dense since, for all $(F, \#) \in \mathbf{Frm}_\#$, we can construct a formal topology $\mathcal{A}$ such that $(F, \#) \cong (Sat(\mathcal{A}), \#_{Sat})$. Let S be a base of F and define, for all $a \in S, U \subseteq S$:

$$a \triangleleft U \equiv a \leq \vee U,$$
$$Pos(a) \equiv a\#0.$$

Then $\mathcal{A} \equiv (S, \cdot, 1, \triangleleft, Pos)$ is a formal topology and the map

$$(Sat(\mathcal{A}), \#_{Sat}) \rightarrow (F, \#)$$
$$\mathcal{A}U \mapsto \vee U$$

establishes an isomorphism of frames with apartness. Moreover the assignment

$$Sat : \mathbf{FTop}(\mathcal{A}, \mathcal{B}) \rightarrow \mathbf{Frm}_{\#}(Sat(\mathcal{A}), Sat(\mathcal{B}))$$
$$f \mapsto Sat(f)$$

is a bijection, i.e. *Sat* is full and faithful. We may thus conclude:

Theorem 3.4 *The category of formal topologies* **FTop** *is equivalent to the category of frames with apartness from 0* **Frm**$_{\#}$.

We will obtain the equivalence between **SFTop*** and the category of coherent frames as a special case of theorem 3.4. We first recall from the literature some definitions and basic properties concerning coherent frames (cf. [4]).

Definition 3.5 *Let F be a frame. We say that an element a of F is* compact *(finite) if for every $U \subseteq F$ such that $a \leq \vee U$ there exists $K \subseteq_\omega U$ such that $a \leq \vee K$.*

If F is the frame of open sets of a topological space, we obtain the usual definition of compact open. Compact elements of a frame form a join semilattice, in fact we have:

Lemma 3.6 *Let F be a frame. Then:*

1. *0 is compact;*

2. *If a and b are compact, then $a \vee b$ is compact.*

In general, compact elements of a frame are not closed under meet. This fact justifies the following definition:

Definition 3.7 *A frame F is* coherent *if:*

1. *The compact elements $K(F)$ of F form a sublattice of F;*

2. *Every element of F is a join of compact elements, i.e. $K(F)$ generates F.*

Coherent frames can be made into a category **CohFrm** whose objects are coherent frames and whose morphism $f : F_1 \to F_2$ are coherent morphisms of frames, i.e. maps preserving finite meets and arbitrary joins and mapping $K(F_1)$ into $K(F_2)$. By endowing coherent frames with a predicate of apartness from 0, we obtain the category **CohFrm$_\#$** of coherent frames with apartness.

We are now going to prove that the equivalence stated in 3.4 restricts to an equivalence between **SFTop*** and **CohFrm$_\#$**. This will permit us to use the notion of Stone formal topology (or equivalently of Stone base) as the version expressible inside ITT of the notion of coherent frame with apartness from 0.

In order to prove that for any Stone formal topology $\mathcal{A}$, $Sat(\mathcal{A})$ is a coherent frame, we start with a preliminary lemma characterizing compact elements of $Sat(\mathcal{A})$ in terms of covers.

Lemma 3.8 *Let $\mathcal{A}$ be a formal topology and let $U \in Sat(\mathcal{A})$. Then U is compact in $Sat(\mathcal{A})$ iff for all $V \in Sat(\mathcal{A})$,*

$$U \lhd V \to U \lhd L$$

for a suitable finite subset L of V.

Proof: Suppose that U is compact in $Sat(\mathcal{A})$ and assume $U \lhd V$. Since $\mathcal{A}V = \vee_{a \in V} \mathcal{A}\{a\}$, this hypothesis can be rephrased by the inequality $U \leq \vee_{a \in V} \mathcal{A}\{a\}$ holding in the frame $Sat(\mathcal{A})$. By compactness of U in $Sat(\mathcal{A})$, there exists $L \subseteq_\omega V$ such that $U \leq \vee_{a \in L} \mathcal{A}(a)$, which is equivalent to $U \lhd L$. Conversely, let $U \leq \vee_{i \in I} V_i$, where $V_i \in Sat(\mathcal{A})$. Then $U \lhd \cup_{i \in I} V_i$, hence by hypothesis there exists $L \subseteq_\omega \cup_{i \in I} V_i$ with $U \lhd L$. Let I_0 be the finite subset of I such that $L \subseteq \cup_{i \in I_0} V_i$. Then $U \lhd \cup_{i \in I_0} V_i$, that is $U \leq \vee_{i \in I_0} V_i$, hence U is compact in the frame $Sat(\mathcal{A})$. $\square$

As a corollary, let U be compact in $Sat(\mathcal{A})$. Then there exists $K \subseteq_\omega U$ such that $U = \mathcal{A}(K)$. In fact, by reflexivity, $U \lhd U$, and thus, by the above lemma, there exists $K \subseteq_\omega U$ such that $U \lhd K$. Since the opposite cover relation holds in general, we have $U =_\mathcal{A} K$, that is, as U is $\mathcal{A}$-saturated, $U = \mathcal{A}(K)$.

Observe however that the converse need not hold, i.e. U may be the saturation of a finite subset of S without being compact.

Anyway, if the formal topology $\mathcal{A}$ is Stone, the saturations of the singletons are compact, by lemma 3.8 and definition of Stone cover. Thus, by lemma 3.6, the saturation of any finite subset of S is compact. We may thus conclude:

Proposition 3.9 *Let $\mathcal{A}$ be a Stone formal topology. Then U is compact in $Sat(\mathcal{A})$ iff there exists $K \subseteq_\omega U$ such that $U = \mathcal{A}(K)$.*

We define $Sat_\omega(\mathcal{A}) \equiv \{\mathcal{A}K : K \subseteq_\omega S\}$. Then the following result is easily obtained:

Proposition 3.10 *Let $\mathcal{A}$ be a formal topology. Then $\mathcal{A}$ is a Stone formal topology iff the compact elements of $Sat(\mathcal{A})$ coincide with $Sat_\omega(\mathcal{A})$.*

We are now in the position to prove:

Theorem 3.11 *For any Stone formal topology $\mathcal{A}$, $Sat(\mathcal{A})$ is a coherent frame.*

Proof: In order to prove that $Sat_\omega(\mathcal{A})$ is a sublattice of $Sat(\mathcal{A})$, we have to check closure under finite meet. Of course the top element of $Sat(\mathcal{A})$, $\mathcal{A}\{1\}$, is compact. Then suppose $U, V \in Sat_\omega(\mathcal{A})$. By proposition 3.9, there exist $K \subseteq_\omega U$ and $L \subseteq_\omega V$ such that $U = \mathcal{A}K$ and $V = \mathcal{A}L$. Recall that in any formal topology $\mathcal{A}$ the equality $\mathcal{A}U \cap \mathcal{A}V = \mathcal{A}(U \cdot V)$ holds for all $U, V \subseteq S$. Then $U \wedge V \equiv U \cap V = \mathcal{A}(U \cdot V)$, and also $U \wedge V = \mathcal{A}(K \cdot L)$, where $K \cdot L \subseteq_\omega U \cdot V$. Therefore, by proposition 3.9 again, $U \wedge V$ is compact. Finally, by lemma 2.4, $Sat_\omega(\mathcal{A})$ generates $Sat(\mathcal{A})$. $\square$

Observe that $\mathcal{S}$-saturation and $\mathbf{A}(\mathcal{S})$- saturation coincide on finite subsets of S, that is for all $K \subseteq_\omega S$. $\mathcal{S}K = (\mathbf{A}(\mathcal{S}))K$. Thus the compact elements of the coherent frame $Sat(\mathbf{A}(\mathcal{S}))$ are precisely the elements of $Sat(\mathcal{S})$. Since $Sat(\mathcal{S})$ generates (by closure under directed unions) the coherent frame with apartness $Sat(\mathbf{A}(\mathcal{S}))$. proposition 3.1 together with theorem 3.11 gives:

Corollary 3.12 Pointfree Stone's representation theorem for distributive lattices. *Any distributive lattice (with apartness) is isomorphic to the compact elements of a coherent frame (with apartness).*

The reader should notice that the proof of the above pointfree Stone representation theorem is completely constructive, and no prime ideal theorem or equivalents are involved.

We are now in the position to prove the converse of 3.11, i.e. that any coherent frame with apartness is isomorphic to $Sat(\mathcal{A})$ for a suitable Stone formal topology $\mathcal{A}$. Let F be a coherent frame with apartness $\cdot \#0$, and consider the Stone base $\mathcal{S}_L$ associated to the sublattice L of compact elements of F. By the results in section 1. a Stone formal topology $\mathcal{A}$ is obtained by defining $\mathcal{A} \equiv \mathbf{A}(\mathcal{S}_L)$. Observe that for all $a \in L$. $U \subseteq L$, $a \lhd_\mathcal{A} U$ holds iff there exists $K \subseteq_\omega U$ such that $a \leq \vee K$ that is iff $a \in \mathcal{I}(U)$, where $\mathcal{I}(U)$ denotes the ideal generated by U in L. Thus $Sat(\mathcal{A})$ coincides with the frame of ideals on L, denoted $Idl(L)$. Consider now the map

$$\phi: \quad F \quad \to \quad Sat(\mathcal{A})$$
$$a \quad \mapsto \quad \downarrow_L a \equiv \{b \in L : b \leq a\}.$$

Since L generates F, ϕ is injective. As for surjectivity, observe that for all $\mathcal{A}$-saturated subsets $\mathcal{A}U$. $\mathcal{A}U = \downarrow \vee U$ since for any $b \in L$, $b \in \mathcal{A}U$ iff there exists $K \subseteq_\omega U$ such that $b \leq \vee K$ iff (by compactness of b) $b \leq \vee U$.

The bijection ϕ is a frame morphism since it is an order preserving map with order preserving inverse, given by $\psi(I) \equiv \vee I$ for all $I \in Idl(L)$. Indeed ϕ is a morphism of frames with apartness since for all $a \in K(F)$, $\phi(a)\#_{Sat}0$ iff $\downarrow_L a\#_{Sat}0$ iff $a\#_F 0$. We can summarize the above result with the following:

Proposition 3.13 *For any coherent frame with apartness F there exists a Stone formal topology $\mathcal{A}$ such that $F \cong Sat(\mathcal{A})$. Such formal topology $\mathcal{A}$ is given by the topology of ideals on the distributive lattice of compact elements of F.*

If F is the coherent frame of saturated subsets of a Stone formal topology, by using the characterization of compact elements of $Sat(\mathcal{A})$ provided in 3.10, we obtain:

Corollary 3.14 *If $\mathcal{A}$ is a Stone formal topology, the map*

$$\phi : Sat(\mathcal{A}) \rightarrow Idl(Sat_\omega(\mathcal{A}))$$
$$U \mapsto \downarrow U \equiv \{\mathcal{A}K : K \subseteq_\omega U\}$$

is a frame isomorphism with inverse $\psi(I) \equiv \vee I$.

Proposition 3.13 together with theorem 3.11 gives (by observing that coherence is preserved under frame isomorphism):

Theorem 3.15 *Let F be a frame with apartness from 0. Then F is coherent iff there exists a Stone formal topology $\mathcal{A}$ such that $F \cong Sat(\mathcal{A})$.*

By proposition 3.13, $Sat : \mathbf{SFTop}^* \rightarrow \mathbf{CohFrm}_\#$ is a dense functor; moreover it is easy to see that the bijection on morphisms given by theorem 3.4 restricts to a bijection between $\mathbf{SFTop}^*(\mathcal{A}, \mathcal{B})$ and $\mathbf{CohFrm}_\#(Sat(\mathcal{A}), Sat(\mathcal{B}))$. Therefore we have:

Theorem 3.16 *The category $\mathbf{CohFrm}_\#$ of coherent frames with apartness is equivalent to the category $\mathbf{Sftop}^*$ of Stone formal topologies with coherent morphisms.*

The above result gives, together with theorem 2.12 and theorem 3.3:

Corollary 3.17 *The category $\mathbf{CohFrm}_\#$ of coherent frames with apartness is equivalent to the category $\mathbf{SB}$ of Stone bases and to the category $\mathbf{DLat}_\#$ of distributive lattices with apartness from 0.*

Finally, we observe that in all the categories introduced so far, morphisms between two given objects can be partially ordered pointwise and moreover the functors yielding the equivalences preserve this order. This means that all the equivalences are indeed equivalences of **2**-categories.

4   Stone formal spaces and topological representation

In order to connect our pointfree approach to representation theory with the traditional one, the notion of point has to be recovered. Since we reverse the usual conceptual order between points and opens, and take the opens as *primitive*, points will be defined as particular, well behaved, collections of opens. We recall here the definition of (formal) point on a formal topology, and specialize it to the Stone case.

Definition 4.1 *Let $\mathcal{A} \equiv (S, \cdot, 1, \lhd, Pos)$ be a formal topology. A subset α of S is said to be a* formal point *if for all $a, b \in S$, $U \subseteq S$ the following conditions hold:*

1. $1 \in \alpha$;

2. $\dfrac{a \in \alpha \quad b \in \alpha}{a \cdot b \in \alpha}$;

3. $\dfrac{a \in \alpha \quad a \lhd U}{(\exists b \in U)(b \in \alpha)}$;

4. $\dfrac{a \in \alpha}{Pos(a)}$.

Observe that this definition connects to the usual intuition on points by reading "$a \in \alpha$" as "α is a point of a", 1 as the whole space, product as intersection, "$a \lhd U$" as "a is included in the union of U", and "$Pos(a)$" as "a is inhabited", that is a positive way of saying that a is not empty.[4]

Now let $\mathcal{A}$ be a Stone formal topology. Then, by definition, $a \lhd U$ holds iff there exists a finite subset K of U such that $a \lhd K$. Then condition *3* above can be weakened to the following:

3'. $\dfrac{a \in \alpha \quad a \lhd K}{(\exists b \in K)(b \in \alpha)}$ for all $a \in S$, $K \subseteq_\omega S$.

and thus points can be defined directly in terms of the Stone base $\mathbf{S}(\mathcal{A})$.

The definition of formal point of a Stone formal topology can be rewritten in such a way that it results in the more familiar notion of prime filter on a suitable lattice.

The notion of filter in a lattice is defined as usual. In a lattice L endowed with a predicate of apartness from 0, the property that a filter F is proper is expressed in a stronger form by

$$a \in F \rightarrow a \# 0.$$

---

[4]For a more exhaustive treatment of this topic cf. [9].

As usual, a filter F is said to be *prime* if for all $a, b_i \in L$

$$a \in F, a \le b_1 \vee \ldots \vee b_n \to (\exists i \le n)(b_i \in F).$$

As announced we have:

Proposition 4.2 *There exists an order preserving bijection between formal points on a Stone formal topology $\mathcal{A}$ and prime filters on $Sat_\omega(\mathcal{A})$.*

Proof: Let α be a formal point on $\mathcal{A}$ and define $\hat{\alpha} \equiv \{\mathcal{A}\{a\} : a \in \alpha\}$. Then $\hat{\alpha}$ is a prime filter on $Sat_\omega(\mathcal{A})$. In fact $\mathcal{A}\{1\} \in \hat{\alpha}$ since $1 \in \alpha$. It is closed under meet since $\mathcal{A}\{a\} \wedge \mathcal{A}\{b\} = \mathcal{A}\{a \cdot b\}$ and α is closed under product. Then suppose $\mathcal{A}\{a\} \in \hat{\alpha}$ and $\mathcal{A}\{a\} \le \mathcal{A}\{b_1\} \vee \ldots \vee \mathcal{A}\{b_n\}$, that is $a \in \alpha$ and $a \triangleleft \{b_1, \ldots, b_n\}$. Since α is a formal point, there exists $i \le n$ such that $b_i \in \alpha$, that is such that $\mathcal{A}\{b_i\} \in \hat{\alpha}$. Finally, if $\mathcal{A}\{a\} \in \hat{\alpha}$, then $a \in \alpha$, hence $Pos(\{a\})$ holds and therefore $\mathcal{A}\{a\} \#_{Sat} 0$.

Similarly one proves that if F is a prime filter on $Sat_\omega(\mathcal{A})$, then $F^\circ \equiv \{a \in S : \mathcal{A}\{a\} \in F\}$ is a formal point on $\mathcal{A}$. The correspondence is clearly order preserving and it is bijective since, for all formal points α on $\mathcal{A}$ and for all filters F on $Sat_\omega(\mathcal{A})$, the equalities $(\hat{\alpha})^\circ = \alpha$ and $\hat{F^\circ} = F$ hold by definitions. $\square$

For any formal topology $\mathcal{A}$, the formal space $Pt(\mathcal{A})$ of formal points on $\mathcal{A}$ can be endowed with a topology, called the *extensional topology*. Define, for $a \in S$

$$ext(a) \equiv \{\alpha \in Pt(\mathcal{A}) : a \in \alpha\}.$$

The family $\{ext(a)\}_{a \in S}$ is a base for a topology on $Pt(\mathcal{A})$. In fact, from definition of formal point we have $ext(1) = Pt(\mathcal{A})$, thus the whole space is a basic open, and $ext(a) \cap ext(b) = ext(a \cdot b)$, thus the family is closed under intersection. If we denote $\cup_{b \in U} ext(b)$ with $ext(U)$, then the generic open is of the form $ext(U)$ for $U \subseteq S$.

Let $\Omega Pt(\mathcal{A})$ be the topology so obtained. Then the map

$$
\begin{aligned}
\phi : \quad Sat(\mathcal{A}) \quad &\to \quad \Omega Pt(\mathcal{A}) \\
U \quad &\longmapsto \quad ext(U)
\end{aligned}
$$

is clearly a surjective frame homomorphism, therefore it is a frame isomorphism iff it is injective. Injectivity of ϕ amounts to the condition

$$ext(U) = ext(V) \to U = V$$

which is equivalent to $ext(U) \subseteq ext(V) \to U \subseteq V$, which in turn is equivalent to

$$ext(a) \subseteq ext(V) \to a \in V. \tag{3}$$

When condition (3) holds, we say that the formal topology is *extensional* or *has enough points*, since it is classically equivalent to

$$a \notin V \rightarrow (\exists \alpha \in Pt(\mathcal{A}))(a \in \alpha \ \& \ V \cap \alpha = \emptyset)$$

which intuitively says that formal points are enough to "separate" different opens.

If $\mathcal{A}$ is a Stone formal topology extensionality can be proved by admitting classical logic and the prime filter theorem. Indeed, by corollary 3.14 and by proposition 4.2, respectively characterizing saturated subsets of $\mathcal{A}$ as ideals on $Sat_\omega(\mathcal{A})$ and points on $\mathcal{A}$ as prime filters on $Sat_\omega(\mathcal{A})$, the above condition holds iff for all ideals I of $Sat_\omega(\mathcal{A})$ and for all $a \in Sat_\omega(\mathcal{A})$, $a \notin I$, there exists a prime filter F such that $a \in F$ and $F \cap I = \emptyset$. The reader may here recognize a well known non constructive principle (cf. e.g. [3]):

Theorem 4.3 The Prime Filter Theorem. *Let I be an ideal of a distributive lattice L and let a be an element of L such that $a \notin I$. Then there exists a prime filter F of L such that $I \subseteq L$ and $F \cap I = \emptyset$*

With the above discussion we have proved:

Theorem 4.4 *Any Stone formal topology $\mathcal{A}$ is extensional iff the Prime Filter Theorem holds.*

We say that a topological space $(X, \Omega X)$ is *sober* if it is homeomorphic to the space of formal points on its frame of opens, that is if $X \cong Pt(\Omega X)$ where the right hand side is provided with the extensional topology. A sober space is said to be *coherent* (cf. [4]) if the collection $K(\Omega X)$ of compact opens of X is closed under intersection and forms a base for ΩX. Thus, simply by "adding points" to our pointfree results of section 3, we obtain:

Theorem 4.5 Stone representation theorem for distributive lattices. *Any distributive lattice with apartness is isomorphic to the lattice of compact opens of a coherent topological space.*

Proof: By corollary 3.12 any distributive lattice with apartness L is isomorphic to the compact elements of a coherent frame with apartness, thus, by proposition 3.10, to the compact saturated subsets of a Stone formal topology $\mathcal{A}$, that is

$$L \cong K(Sat(\mathcal{A})).$$

By theorem 4.4 such an $\mathcal{A}$ is extensional and therefore there exists a frame isomorphism between $Sat(\mathcal{A})$ and $\Omega Pt(\mathcal{A})$. Since compactness is preserved under frame isomorphisms (and compact opens of a space X are just the compact elements of the frame ΩX) the claim is proved. $\square$

The above result can be formulated in a categorical language. Given two coherent spaces $(X, \Omega X)$ and $(Y, \Omega Y)$, a map $f : X \to Y$ is called a *coherent map* if f^{-1} is a coherent frame morphism between ΩY and ΩX. Coherent spaces with coherent maps form a category which we denote with **CohSpa**. By a basic result in pointfree topology, the category of sober spaces is dual to the category of *spatial* frames, that is frames arising from extensional formal topologies. Thus, by restriction of the equivalence, the category of coherent spaces with coherent maps is dual to the category of coherent frames with coherent morphism, which is spatial by 4.4. Then, by composing this duality with the equivalence given by theorem 3.17, we have:

Theorem 4.6 *The category* **DLat**$_\#$ *of distributive lattices with apartness is dual to the category* **CohSpa** *of coherent spaces with coherent maps.*

5   Acknowledgements

I wish to thank Giovanni Sambin for encouraging me to write this paper and Pino Rosolini for reading a preliminary version.

References

[1] G. Battilotti and G. Sambin, *Pretopologies and a uniform presentation of sup-lattices, quantales and frames*, to appear.

[2] M. P. Fourman and R.J. Grayson, *Formal Spaces*, "The L. E. J. Brouwer Centenary Symposium" (A. S. Troelstra and D. van Dalen, eds.), North Holland, Amsterdam, 1982, pp. 107–122.

[3] G. Grätzer, *Lattice theory*, Birkhauser, Basel, 1978.

[4] P. T. Johnstone, *Stone Spaces*, Cambridge University Press, 1982.

[5] S. Mac Lane, *Categories for the Working Mathematician*, Graduate Texts in Mathematics, Springer Verlag, 1971.

[6] P. Martin-Löf, *Intuitionistic type theory*, Bibliopolis, Napoli, 1984, notes by Giovanni Sambin of a series of lectures given in Padua, Jun. 1980.

[7] G. Sambin, *Intuitionistic formal spaces — a first communication*, Mathematical logic and its applications (D. Skordev, ed.), Plenum, 1987, pp. 187–204.

[8] _____, *Intuitionistic formal spaces and their neighbourhood*, Logic Colloquium 88 (R. Ferro, C. Bonotto, S. Valentini, and G. Zanardo, eds.), North Holland, Amsterdam, 1989, pp. 261–285.

[9] _____, *Notes for a course in Intuitionistic Pointfree Topology*, Università di Padova, 1990.

[10] _____, *The semantics of pretopologies*, Substructural Logics (P. Schroeder-Heister and K. Došen, eds.), Clarendon Press, 1993, pp. 293–307.

[11] G. Sambin and S. Valentini, *Building up a toolbox for Martin–Löf intuitionistic type theory*, to appear.

[12] G. Sambin, S. Valentini, and P. Virgili, *Constructive Domain Theory as a branch of Intuitionistic Pointfree Topology*, to appear.

[13] M.H. Stone, *Topological representation of distributive lattices and Brouwerian logics*, Časopis Pěst. Mat. **67** (1937), 1–25.

[14] A.S. Troelstra and D. van Dalen, *Constructivism in mathematics, an introduction, vol. II*, Studies in logic and the foundations of mathematics, Springer Verlag, 1988.

DIPARTIMENTO DI MATEMATICA PURA ED APPLICATA, VIA BELZONI 7, 35131 PADOVA, ITALY
    e-mail address: negri@pdmat1.math.unipd.it

On k-permutability for categories of T-algebras[*]

M. Cristina Pedicchio

Abstract

Quasi-monadic categories over *Set* are characterized by means of the existence of a regular projective generator. This result is used to describe k- permutability of equivalence relations.

Introduction

In this paper I study monadic categories over *Set*, i.e. categories Set^T of T-algebras for a triple T over *Set* [7], and their (regular-epi) reflective subcategories (that I will call quasi-monadic). My aim is to show that the well known characterization for monadic categories over *Set* in terms of exact categories with a regular projective generator P with copowers (see [3], [12]) can be easily generalized to the quasi-monadic situation simply by dropping the effectiveness of equivalence relations. So, I will prove that a quasi-monadic category over *Set* can be completely described as a regular category with coequalizers of equivalence relations and a regular, projective generator with copowers.

This approach in terms of exactness or regularity plus the existence of a suitable "good" object P, will allow in Section 2 to describe k-permutability of equivalence relations by means of an appropriate coalgebra structure on P.

I wish to thank W. Tholen for interesting conversations on the subject; I am also indebted to York University for the support during my visit.

1 Regular and monadic categories

I recall the following basic categorical properties of regularity and exactness; for more details see [1] or [3].

[*]Research supported by the Italian Topology grant 40% and by the NATO grant CRG941330.

Definition 1.1 A category $\mathcal{E}$ is regular if and only if it verifies the following properties:

    (1) $\mathcal{E}$ has finite limits;
    (2) $\mathcal{E}$ has coequalizers of kernel pairs;
    (3) regular epimorphisms are stable under pull-backs.

It is well known that in any regular category it is possible to define in a canonical way a (regular-epi)-mono factorization of morphisms: such a factorization is stable under pull-backs and this will allow to define an associative composition of relations.

Definition 1.2 A category $\mathcal{E}$ is exact if and only if it is regular and equivalence relations are effective (i.e. every equivalence relation is the kernel pair of some morphism).

Now, given a triple $T = (T, \eta, \mu)$ on a category $\mathcal{S}$, I recall that $\mathcal{E}$ is called *monadic over $\mathcal{S}$* if it is equivalent to the category $\mathcal{S}^T$ of T-algebras; I denote by $F^T : \mathcal{S} \to \mathcal{S}^T$ and $U^T : \mathcal{S}^T \to \mathcal{S}$ the corresponding free and forgetful functors:

$$(U^T, F^T) : \mathcal{E} \simeq \mathcal{S}^T \rightleftarrows \mathcal{S}$$

where $F^T \vdash U^T$ [7].

    If $\mathcal{S}$ is complete, cocomplete, exact and each regular epimorphism in $\mathcal{S}$ has a section, then, for any triple T on $\mathcal{S}$, $\mathcal{S}^T$ is still complete, cocomplete and exact [3]. In particular, all these properties apply when $\mathcal{S} = Set$ (assuming the axiom of choice). As a consequence, since varieties are precisely those monadic categories over Set where the triple T has finite rank, we get that any variety is a complete, cocomplete, exact category.

    Consider now a (regular-epi) reflective subcategory $\mathcal{B}$ of $\mathcal{E}$; denote by $\Phi : \mathcal{B} \to \mathcal{E}$ the inclusion, by $\Psi : \mathcal{E} \to \mathcal{B}$ the reflection functor and by $\nu_X : X \to \Phi\Psi(X)$ the unit of the adjoint pair, for any $X \in \mathcal{E}$.

Lemma 1.3 *Any (regular-epi) reflective subcategory $\mathcal{B}$ of a regular category $\mathcal{E}$, is regular.*

    Proof. Any regular epimorphism $f : A \to B$ in $\mathcal{B}$ can be described as $\Phi(f) \simeq \nu_Q \cdot q$, where $q : \Phi(A) \twoheadrightarrow Q$ is the coequalizer in $\mathcal{E}$ of the kernel pair of f, and $\nu_Q : Q \twoheadrightarrow \Phi\Psi(Q) \simeq B$; since the composite of regular epimorphisms is a regular epimorphism we get that Φ preserves regular epimorphisms.

    Then the result follows since regular epimorphisms are stable under pull-backs in $\mathcal{E}$ and they are reflected by Φ. ∎

By 1.3 we obtain that any epireflective subcategory of a monadic one over *Set*, is regular, complete and cocomplete; in particular the result applies to any quasi-variety.

Observe that the exactness condition is not in general preserved by reflections. For example, the category of torsion free abelian groups is a (regular-epi) reflective, regular, but not exact subcategory of that one of abelian groups.

Definition 1.4 A category equivalent to a (regular-epi) reflective subcategory of a monadic category over S, is called quasi-monadic over S.

One of the basic facts on monadic categories is that they can be characterized by Beck's theorem in terms of properties of the associated forgetful functor U^T. In the case where the base category is *Set*, they also admit an intrinsic description in terms of the exactness axiom, without referring to U^T. We will prove this result and a similar weaker version corresponding to the quasi-monadic case.

First, I recall that an object $P \in \mathcal{E}$ is called *regular projective* if and only if, for any regular epimorphism $f : A \twoheadrightarrow B$, and for any $h : P \to B$, there exists a morphism $g : P \to A$ such that $f \cdot g = h$. The following Lemma is a trivial consequence of Definition 1.4:

Lemma 1.5 *An object P is a regular projective if and only if the corresponding functor $Hom(P, -) : \mathcal{E} \to Set$ preserves regular epimorphisms.*

Consider $G \in \mathcal{E}$, suppose $\mathcal{E}$ admits all copowers $\sqcup_Y G$ of $\mathcal{E}$, for any set Y, and write $e_y : G \to \sqcup_Y G$, $y \in Y$, for the canonical inclusions. Then we can give the following:

Definition 1.6 G is a regular generator in $\mathcal{E}$, if and only if the morphism

$$\varepsilon_X : \sqcup_{Hom(G,X)} G \twoheadrightarrow X$$

defined by $\varepsilon_X \cdot e_f = f$, for any $f \in Hom(G, X)$, $X \in \mathcal{E}$, is a regular epimorphism.

I recall that G is called a *generator* if ε_X is an epimorphism. Now, we can give the two main results:

Theorem 1.7 *[3] $\mathcal{E}$ is a monadic category over Set, if and only if it verifies the following conditions:*

(1) $\mathcal{E}$ is exact;

(2) $\mathcal{E}$ has a regular projective object P;

(3) the copowers $\sqcup_Y P$ of P exist, for all $Y \in Set$;

(4) P is a regular generator.

Proof. Let $\mathcal{E}$ be equivalent to Set^T for a triple $T = (T, \eta, \mu)$ on Set, then it suffices to choose $P = F^T(1)$, where 1 denotes a one object set, to get all the desired properties. Conversely, let $\mathcal{E}$ verify conditions (1) to (4); define $U : \mathcal{E} \to Set$ by $U = Hom(P, -)$ and $F : Set \to \mathcal{E}$ by $F(Y) = \sqcup_Y P$: F is then easily defined on arrows by the universal property of copowers. It is straightforward to verify that $F \dashv U$ and that the morphism $\varepsilon_X : \sqcup_{Hom(P,X)} P \to X$ defined by $\varepsilon_X \cdot e_f = f$, for any $X \in \mathcal{E}$, $f : P \to X$, is the counit of the adjunction. Now, let $T = UF$ be the induced triple on Set, $(U^T, F^T) : Set^T \rightleftarrows Set$ the corresponding adjoint pair and $\Phi : \mathcal{E} \to Set^T$ the comparison morphism defined by $\Phi(X) = (U(X), U(\varepsilon_X))$ and $\Phi(f) = U(f)$. By construction, U preserves regular epimorphisms since P is projective and it also preserves limits being a right adjoint; then, since $U^T \Phi = U$ and U^T reflects limits and regular epimorphisms, we get that Φ preserves limits and regular epimorphisms.

Now, we will show that Φ is an equivalence. Clearly Φ is faithful since U is faithful being P a generator. To show that Φ is full (see also [2]), first notice that since $U^T \Phi = U$ and $\Phi F = F^T$, by adjointness we get:

$$Hom_{\mathcal{E}}(F(Y), X) \simeq Hom_{Set^T}(\Phi F(Y), \Phi(X))$$

for $Y \in Set$, $X \in \mathcal{E}$.

To verify

$$Hom_{\mathcal{E}}(E, X) \simeq Hom_{Set^T}(\Phi(E), \Phi(X))$$

for any $X, E \in \mathcal{E}$, consider a morphism $h : \Phi(E) \to \Phi(X)$. By the previous remark, there exists $k : FU(E) \to X$ such that $\Phi(k) = h \cdot \Phi(\varepsilon_E)$. Denote by $(u, v) : N \to FU(E)$ the kernel pair of ε_E in $\mathcal{E}$, hence ε_E is the coequalizer of (u, v) being a regular epimorphism.

Since $\Phi(k) \cdot \Phi(u) = \Phi(k) \cdot \Phi(v)$ and Φ is faithful $k \cdot u = k \cdot v$, so there exists $l : E \to X$ with $l \cdot \varepsilon_E = k$. $\Phi(l) = h$ follows since Φ preserves regular epimorphisms. Now, we can suppose up to equivalences that $\mathcal{E}$ is a full subcategory of Set^T.

It remains to verify that Φ is dense, i.e. that for any $Z \in Set^T$, there exists $X \in \mathcal{E}$ with $\Phi(X) \simeq Z$. Consider the following presentation for Z:

$$F^T U^T(M) \overset{\varepsilon_M^T}{\to} M \overset{(s,t)}{\rightrightarrows} F^T U^T(Z) \overset{\varepsilon_Z^T}{\twoheadrightarrow} Z$$

where ε^T is the counit of the adjoint pair $F^T \dashv U^T$, and $(s, t) : M \rightrightarrows F^T U^T(Z)$ is the kernel pair of ε_Z^T in Set^T.

Since $\Phi F = F^T$ and Φ is full, we get two morphisms

(1.7.1.) $(s', t') : FU^T(M) \rightrightarrows FU^T(Z)$

in $\mathcal{E}$, such that $\Phi(s') = s \cdot \varepsilon_M^T$ and $\Phi(t') = t \cdot \varepsilon_M^T$.

Let I be the image in the (regular epi)-mono factorization in $\mathcal{E}$ of (s', t'). Since Φ preserves limits and regular epimorphisms, it also preserves factorizations, hence $\Phi(I) \simeq M$.

Now, let $q : FU^T(Z) \twoheadrightarrow Q$ be the coequalizer of (s', t') in $\mathcal{E}$; it exists since it coincides with the coequalizer of I and I is an equivalence relation, being M an equivalence relation.

But, $\mathcal{E}$ is exact, hence I is the kernel pair of q so. since Φ preserves limits and regular epimorphisms $\Phi(q)$ is the coequalizer of $\Phi(I)$. Hence we can conclude with $\Phi(q) \simeq \varepsilon_Z^T$ and $\Phi(Q) \simeq Z$. ∎

Theorem 1.8 *A category $\mathcal{B}$ is quasi-monadic over Set, if and only if it verifies the following conditions:*

  (1) $\mathcal{B}$ is regular;
  (2) $\mathcal{B}$ has coequalizers for equivalence relations;
  (3) $\mathcal{B}$ has a regular projective object P;
  (4) the copowers $\sqcup_Y P$ of P exist, for all $Y \in Set$;
  (5) P is a regular generator.

Proof. Let $\mathcal{B}$ be quasi-monadic, then it is complete, cocomplete and regular by Lemma 1.3. P is defined as the reflection of the regular projective generator of the corresponding monadic category.

For the converse. apply the same proof as in 1.7 to show that there exists an adjoint pair $(U, F) : \mathcal{B} \rightleftarrows Set$ such that the comparison functor $\Phi : \mathcal{B} \to Set^T$ where $T = UF$. is full. faithful and preserves limits and regular epimorphisms.

To show that $\mathcal{B}$ is epireflective in Set^T define $\Psi : Set^T \to \mathcal{B}$ by $\Psi(Z) = Q$, for $Z \in Set^T$, where (Q, q) is the coequalizer of 1.7.1 in $\mathcal{B}$.

Observe that the coequalizer exists by assumption (2). since I is an equivalence relation.

Moreover, by the universal property of coequalizers, we can easily define Ψ on morphisms.

To verify $\Psi \dashv \Phi$ define the unit $\nu_Z : Z \to \Phi\Psi(Z)$, by the coequalizer ε_Z^T, as the unique morphism such that $\nu_Z \cdot \varepsilon_Z^T = \Phi(q)$.

So $\mathcal{B}$ is a reflective subcategory of Set^T; furthermore, since Φ preserves regular epimorphisms, ν_Z is a regular epimorphism, for any $Z \in Set^T$. hence $\mathcal{B}$ is epireflective. ∎

Observe that the only difference between the proofs of 1.7 and 1.8 consists in the fact that, since equivalence relations are not effective in the second case, we cannot conclude that I is the kernel pair of q. I recall that quasi-monadic categories over Set have been also studied by Tholen in [11], by using different kinds of axioms.

Remark 1.9 We conclude by noticing that in Theorem 1.8 it suffices to assume as hypothesis conditions (2) to (5) plus the following:

(1′) $\mathcal{E}$ has finite limits

In fact, the stability of regular epimorphisms under pull-backs, hence the regularity of $\mathcal{E}$ can be obtained by the assumptions since $U = Hom(P, -)$ preserves limits, preserves and reflects regular epimorphisms and *Set* is regular. The proof that U reflects regular epimorphisms is given in 2.2.

Also, it is still equivalent to take as assumptions conditions (3) to (5) plus the following:

(2′) $\mathcal{E}$ has coequalizers (see [11]).

2   Internal k-permutable algebras

I recall that an exact category $\mathcal{E}$ is called a *Mal'cev category* if and only if, for any $X \in \mathcal{E}$ and R, S equivalence relations on X, $R \cdot S = S \cdot R$ (see [5]).

Similarly, if we denote by $R \cdot^{(k-1)} S$, for $k \geq 2$, the product $R \cdot S \cdot R \dots$ iterated $(k-1)$ times, then, a *k-permutable category* $\mathcal{E}$ is an exact category such that $R \cdot^{(k-1)} S = S \cdot^{(k-1)} R$, for any R, S equivalence relations on X. The Mal'cev case clearly corresponds to 2-permutability; for more details see [4].

Our aim is to show that for monadic and quasi-monadic categories over *Set*, the property of being k-permutable is completely characterized by the behaviour of the projective generator $P = F(1)$.

An object $X \in \mathcal{E}$ is an *internal Mal'cev algebra* if and only if there exists a morphism $p : X^3 \to X$ such that the following diagram is commutative:

$$
\begin{array}{ccccc}
X^2 & \xrightarrow{\Delta \times 1} & X^3 & \xleftarrow{1 \times \Delta} & X^2 \\
\downarrow{\scriptstyle \pi_2} & & \downarrow{\scriptstyle p} & & \downarrow{\scriptstyle \pi_1} \\
X & \xrightarrow{Id} & X & \xleftarrow{Id} & X
\end{array}
$$

where π_i denotes the canonical projection, and Δ the diagonal morphism.

Similarly, we say that X is a *internal k-permutable algebra* if and only if there exist $(k-1)$ morphisms $p_i : X^3 \to X$, $i = 1 \dots k - 1$, such that, if we define $p_0 = \pi_1$, $p_k = \pi_3$, then the following diagram commutes for any i, $0 \leq i \leq k - 1$:

$$
\begin{array}{ccc}
X^2 & \xrightarrow{\Delta \times 1} & X^3 \\
\downarrow{\scriptstyle 1 \times \Delta} & & \downarrow{\scriptstyle p_i} \\
X^3 & \xrightarrow{p_{i+1}} & X
\end{array}
$$

The corresponding notions of internal coalgebras are then obtained by duality; so $A \in \mathcal{E}$ is an *internal k-permutable coalgebra* if and only if there exist $(k-1)$ morphisms $p_i : A \to A \sqcup A \sqcup A$ such that, if we define $p_0 = e_1$, $p_k = e_3$ where e_i denotes the canonical inclusion, the following diagram is commutative for any i, $0 \le i \le k - 1$:

$$
\begin{array}{ccc}
A & \xrightarrow{\;\;p_i\;\;} & A \sqcup A \sqcup A \\
{\scriptstyle p_{i+1}}\Big\downarrow & & \Big\downarrow {\scriptstyle \nabla \sqcup 1} \\
A \sqcup A \sqcup A & \xrightarrow{\;\;1 \sqcup \nabla\;\;} & A \sqcup A
\end{array}
$$

∇ obviously denotes the codiagonal morphism. Then, we get the following results:

Theorem 2.1 $\mathcal{E}$ *is a k-permutable monadic category over* Set, *if and only if it verifies the following conditions:*

  (1) $\mathcal{E}$ *is exact;*
  (2) $\mathcal{E}$ *has a regular projective object* P;
  (3) *the copowers* $\sqcup_Y P$ *of* P *exist, for all* $Y \in$ Set;
  (4) P *is a regular generator;*
  (5) P *is an internal k-permutable coalgebra.*

Proof. By 1.7 conditions (1) to (4) are equivalent to $\mathcal{E} \simeq Set^T$, for a suitable triple T on *Set*. Now, suppose $P = F(1)$ is an internal k-permutable coalgebra in $\mathcal{E}$, then since $U \simeq Hom(P, -) : \mathcal{E} \to Set$, by Yoneda Lemma we get a family of ternary natural transformations

$$\bar{p}_i : U \times U \times U \to U, \qquad 0 \le i \le k$$

verifying the following axioms:

$$(\bar{p}_0)_X(x, y, z) = x$$

$$(\bar{p}_i)_X(x, x, y) = (\bar{p}_{i+1})_X(x, y, y) \quad 0 \le i \le k - 1$$

$$(\bar{p}_k)_X(x, y, z) = z$$

for any $X \in \mathcal{E}$, and $x, y, z \in U(X)$.

Now, given R, S equivalence relations on X in $\mathcal{E}$, to show $R \cdot^{(k-1)} S = S \cdot^{(k-1)} R$, consider $U(R)$ and $U(S)$. Since U preserves limits and regular epimorphisms, it also preserves the composition of relations. Moreover, we get the following:

Lemma 2.2 *If P is a regular generator in $\mathcal{E}$, and $U = Hom(P, -) : \mathcal{E} \to Set$, then U reflects isomorphisms.*

Proof. U being faithful reflects monomorphisms. Moreover U reflects regular epimorphisms; in fact, given $f : A \to B$ in $\mathcal{E}$, such that $Hom(P, f) : Hom(P, A) \to Hom(P, B)$ is a regular epi, consider for any $m : P \to B$ the following diagram:

where $n_m : P \to A$ with $f \cdot n_m = m$, exists by hypothesis. Then we get a morphism $\xi : \sqcup_{Hom(P,B)} P \to A$ such that $\xi \cdot e_m = n_m$, for any $m : P \to B$.

This implies $f \cdot \xi \cdot e_m = f \cdot n_m = m = \varepsilon_B \cdot e_m$, hence $f \cdot \xi = \varepsilon_B$ so f is a regular epimorphism. ∎

Now, by 2.2 and the previous remarks it suffices to show that $U(R) \,^{.(k-1)} \, U(S) = U(S) \,^{.(k-1)} U(R)$, and this last result follows by the usual set-theoretical arguments: in fact, if

$$a_0 U(R) a_1 U(S) a_2 U(R) \ldots a_k$$

with $a_i \in U(X)$ define $b_i = (\bar{p}_i)_X (a_{i-1}, a_i, a_{i+1})$ for $1 \leq i \leq k - 1$ to get

$$a_0 U(S) b_1 U(R) b_2 U(S) \ldots a_k.$$

Conversely, if $\mathcal{E} \simeq Set^T$ is a k-permutable category, consider $F(k + 1)$, the free object on $(k + 1)$ generators. Since F preserves colimits, $F(k + 1) \simeq \sqcup_{(k+1)} F(1)$.

Define a morphism $\lambda : F(k+1) \to F(k+1)$, by $\lambda_0 = \lambda_1 = e_0$, $\lambda_2 = \lambda_3 = e_2$ and so on, where $\lambda_i : F(1) \to F(k+1)$, $0 \leq i \leq k$ denote the $(k+1)$ components of λ, and e_i are the inclusions in the copower.

Similarly, define $l : F(k+1) \to F(k+1)$, by $l_0 = e_0, l_1 = l_2 = e_1, l_3 = l_4 = e_3$ and so on. Let R and S be the kernel pairs of λ and l respectively. Since $F(k + 1)$ has k-permuting equivalence relations and U preserves composition of relations, we get $U(R) \,^{.(k-1)} \, U(S) = U(S) \,^{.(k-1)} U(R)$. Now, let $k + 1 = \{x_0, x_1, \ldots x_k\}$ and $\eta_{k+1} : (k + 1) \to UF(k + 1)$ be the unit of the adjunction.

$\eta_{k+1}(x_i) \in UF(k + 1)$, corresponds by adjunction to the inclusion $e_i : F(1) \to F(k + 1)$, so by definition of R and S we get:

$$(\eta(x_0)) U(R) (\eta(x_1)) U(S) (\eta(x_2)) \ldots$$

Hence, there exist $q_i \in UF(k + 1)$, $1 \leq i \leq k - 1$, such that:

(2.1.1) $(\eta(x_0)) U(S) q_1 U(R) q_2 \ldots$

Each q_i determines by adjunction a morphism $\bar{q}_i : F(1) \to F(k + 1)$; then (2.1.1) implies the validity of the following axioms:

$$l \cdot e_0 = l \cdot \bar{q}_1$$
$$\lambda \cdot \bar{q}_1 = \lambda \cdot \bar{q}_2$$
$$l \cdot \bar{q}_2 = l \cdot \bar{q}_3$$

and so on.

Now, if we define $p_i : F(1) \to F(3)$ as the composition $\xi_i \cdot \bar{q}_i$, $1 \leq i \leq k-1$, where ξ_i is given by:

$$(\xi_i)_j = \begin{cases} e_0 & \text{if } j < 1 \\ e_1 & \text{if } j = 1 \\ e_2 & \text{if } j > 1 \end{cases}$$

with $0 \leq j \leq k$, routine verifications show that the equations (2.1.2) imply $(F(1), p_i)$ $1 \leq i \leq k-1$, is an internal k-permutable coalgebra. ■

Theorem 2.3 $\mathcal{B}$ *is a k-permutable quasi-monadic category over Set, if and only if it verifies the following conditions:*

 (1) $\mathcal{B}$ *is regular;*
 (2) $\mathcal{B}$ *has coequalizers for equivalence relations;*
 (3) $\mathcal{B}$ *has a regular projective object P;*
 (4) *the copowers $\sqcup_Y P$ of P exist, for all $Y \in Set$;*
 (5) P *is a regular generator;*
 (6) P *is an internal k-permutable coalgebra.*

Proof. By 1.8, conditions (1) to (5) are equivalent to $\mathcal{B}$ quasi-monadic over *Set*. Then we can apply the same arguments as in 2.1 to show that (6) is equivalent to k-permutability, in fact we never used effectiveness of equivalence relations. ■

Corollary 2.4 *A monadic (quasi-monadic) category $\mathcal{E}$ over Set is k-permutable if and only if the free algebra $F(k+1)$ has k-permutable equivalence relations.*

Observe that same arguments as in 2.1 and 2.3 could be also used to describe monadic and quasi-monadic categories verifying other kinds of assumptions on equivalence relations like modularity, distributivity and so on.

The Mal'cev case for monadic categories has also been considered in [10] and [11].

References

[1] M. Barr, P.A. Grillet, and D.H. van Osdal, *Exact categories and categories of sheaves*, LNM **236**, Springer-Verlag, Berlin, 1971.

[2] M. Barr and C. Wells, *Toposes, Triples and Theories*, Grundlehren der math. Wissenschaften **278**, Springer-Verlag, Berlin, 1985.

[3] F. Borceux, *A handbook in categorical algebra*, Cambridge Press, to appear.

[4] A. Carboni, G.M. Kelly, and M.C. Pedicchio, *Some remarks on Maltsev and Goursat categories*, Applied Categorical Structures **1** (1993), 385–421.

[5] A. Carboni, J. Lambek, and M.C. Pedicchio, *Diagram chasing in Mal'cev categories*, J. Pure Appl. Algebra **69** (1991), 271–284.

[6] J. Hagemann and A. Mitschke, *On n-permutable congruences*, Algebra Universalis **3** (1973), 8–12.

[7] S. Mac Lane, *Categories for the Working Mathematicians*, Graduate texts in Mathematics 5, Springer-Verlag, Berlin, 1972.

[8] R. McKenzie, G. McNulty, and W. Taylor, *Algebras, Lattices, Varieties, Volume I*, Wadsworth and Brooks Cole, Monterey, CA, 1987.

[9] M.C. Pedicchio, *Maltsev categories and Maltsev operations*, J. Pure Appl. Algebra (1994), to appear.

[10] G. Richter, *Maltsev conditions for categories*, Proc. Conference Toledo 1983, Heldermann-Verlag, 1984, pp. 453–469.

[11] W. Tholen, *Relative Bildzerlegungen und algebraische Kategorien*, Ph.D. thesis, Universität Munster, 1974.

[12] E. Vitale, *On the characterization of monadic categories over Set*, 1994, to appear.

DIPARTIMENTO DI MATEMATICA, UNIVERSITÀ DI TRIESTE, P.LE EUROPA 1, I-34100 TRIESTE, ITALY.

e-mail address: PEDICCHI@UNIV.TRIESTE.IT

Weak vs. Strong Boethius' Thesis: a Problem in the Analysis of Consequential Implication

Claudio Pizzi

Abstract

Consequential implication is a refinement of Angell-McCall's connexive implication lacking the law of factor and having as a characteristic axiom the weak Boethius' Thesis $(A \to B) \supset \neg(A \to \neg B)$. The problem discussed is whether the strong Boethius' Thesis $(A \to B) \to \neg(A \to \neg B)$, which has been originally proposed as a typical connexive truth, is provable in the basic system CI.O*Eq of consequential implication. The answer could be given by passing through the definitional equivalence of the fragment CI.O with the modal system KT and then applying the tableaux method for KT. The author suggests, however, that a quick solution can be given by employing the so-called double cancellation property of KT, by virtue of which it turns out that $A \to B$ is a CI.O-theorem iff $A \equiv B$ is a KT-theorem. An application of this result leads to the consequence that Strong Boethius' Thesis cannot be CI.O*Eq-theorem since this would imply $\Box A \equiv \Diamond A$ being a KT-theorem, which is impossible. It is also proved that we cannot have as CI.O*Eq-theorem any variant of strong Beothius' Thesis having in place of $\to$ the operators $\Rightarrow$ and $>$ defined by $A \Rightarrow B =_{Df} \Box(A \supset B) \wedge (\Diamond A \supset \Diamond B)$ and by $A > B =_{Df} {}^{*}A \Rightarrow B$. where * is a properly axiomatized circumstantial operator.

1. A system of *consequential implication* (see Pizzi [6],[7],[5],[8]) is defined to be any system axiomatizing a notion of implication **CI** with the following properties:

(1) It must hold $\neg(A \textbf{ CI } \neg A)$ (Aristotle's Thesis) and equivalently $(A \textbf{ CI } B) \supset \neg(A \textbf{ CI } \neg B)$ (Weak Boethius' Thesis[1]).

---

[1] This wff has received several names in the literature. It is called *law of conditional contrariety* in Angell [2] and it is called *Strawson's principle* in Routley et al. [9].

(2) The law of conjunctive simplification $(A \wedge B)$ **CI** A does not hold or holds in weakened form.

(3) The law of factor $(A \textbf{ CI } B) \supset ((A \wedge R) \textbf{ CI } (B \wedge R))$ does not hold or holds in weakened form.

(4) The language and the axioms of the system allow to define an *analytic* and a *synthetic* variant of the same connective **CI**, **CI'** and **CI''**, having the following properties: if $\Box A$ is definable in the reference system and has the minimal properties of standard logical necessity then

a) $(A \textbf{ CI' } B)$ implies $\Box(A \supset B)$;

b) $(A \textbf{ CI'' } B)$ does not imply $\Box(A \supset B)$;

c) $(A \textbf{ CI' } B)$ implies $(A \textbf{ CI'' } B)$ but not viceversa.

The problem we want to focus in the analysis of consequential implication is the following. Given that consequential implication aims to be a refined version of so-called connexive implication (see [1],[2] and [4]), we ask which is the position in systems of consequential implication of the law which is the cornerstone of connexive implication - namely the so called (strong) Boethius' Thesis

(BTO) $$(A \textbf{ CI } B) \textbf{ CI } \neg(A \textbf{ CI } \neg B).$$

In order to discuss the problem we take into account two basic axiomatizations of consequential implication, CI.O and CI.O⇒ (see Pizzi [7] and [5]).

The language of CI.O is the language of the standard propositional calculus PC extended with the dyadic connective "→" and endowed with standard formation rules.

Definitions: $\top =_{Df} p \supset p$; $\bot =_{Df} \neg\top$; $\Box A =_{Df} \top \to A$; $\Diamond A =_{Df} \neg\Box\neg A$.

Axioms: Standard axioms for the classical propositional calculus PC.

a) $((p \to q) \wedge (q \to r)) \supset (p \to r)$

b) $\Diamond(p \wedge r) \supset ((p \to q) \supset ((p \wedge r) \to (q \wedge r)))$

c) $(\Box(p \supset q) \wedge \Diamond p \wedge \neg\Box q) \supset (p \to q)$

d) $(\neg p \to \neg q) \supset (q \to p)$

e) $(p \to \bot) \supset (\bot \to p)$

f) $(\bot \to p) \supset (p \to \bot)$

g) $(p \to q) \supset \neg(p \to \neg q)$

h) $(p \to q) \supset (p \supset q)$

i) $p \to p$.

Rules: Uniform Substitution (US), Modus Ponens for $\supset$ (MP), Replacement of Proved Material Equivalents (Eq).

The language of CI.O⇒ is the same as in CI.O with ⇒ in place of →.

Definitions: $\top, \bot$ as in CI.O; $\square^{\circ}A =_{Df} \top \Rightarrow A$; $\diamond^{\circ}A =_{Df} \neg\square^{\circ}\neg A$.

Axioms: Standard axioms for the classical propositional calculus PC.

a') $((p \Rightarrow q) \wedge (q \Rightarrow r)) \supset (p \Rightarrow r)$
b') $\diamond^{\circ}(p \wedge r) \supset ((p \Rightarrow q) \supset ((p \wedge r) \Rightarrow (q \wedge r)))$
c') $(\square^{\circ}(p \supset q) \wedge \diamond^{\circ}p) \supset (p \Rightarrow q)$
d') $(p \Rightarrow \bot) \supset (\bot \Rightarrow p)$
e') $(\bot \Rightarrow p) \supset (p \Rightarrow \bot)$
f') $(p \Rightarrow q) \supset \neg(p \Rightarrow \neg q)$
g') $(p \Rightarrow q) \supset (p \supset q)$
h') $p \Rightarrow p$.

Rules: as in CI.O.

CI.O and CI.O$\Rightarrow$ axiomatize two connectives of analytical consequential implication, $\rightarrow$ and $\Rightarrow$, which may be shown to have the properties required for the analytic consequential implication **CI'**; $\rightarrow$ and $\Rightarrow$ are different with respect to contraposition, (see axiom d), which is lacking for $\Rightarrow$. Thanks to the definition $\square A =_{Df} \top \rightarrow A$ in CI.O we are able to show that this system is definitionally equivalent to the well known modal system KT extended with the definition

$$A \rightarrow B =_{Df} \square(A \supset B) \wedge (\diamond B \supset \diamond A) \wedge (\square B \supset \square A)$$

while adding the definition $\diamond A =_{Df} \top \Rightarrow A$ to CI.O$\Rightarrow$ yields a system which is definitionally equivalent to KT extended with the definition

$$A \Rightarrow B =_{Df} \square^{\circ}(A \supset B) \wedge (\diamond^{\circ}B \supset \diamond^{\circ}A).$$

By identifying $\square A$ and $\square^{\circ}A$ we may then easily show that CI.O and CI.O$\Rightarrow$ are definitionally equivalent systems.

In order to take care of the synthetic variants of consequential implication we have to introduce two axioms for a monadic *"ceteris paribus"* operator $*$ which are as follows:

C*1 $*p \supset p$
C*2 $\diamond p \supset \diamond * p$

and the rule (*Eq) $\vdash A \equiv B \longrightarrow \vdash *A \equiv *B$.

Let us call CI.O*Eq the system CI.O + C*1 + C*2+ *Eq. Then we may introduce several formal counterparts of the synthetic variant of **CI**, the most plausible of which is

$$A > B =_{Df} \square(*A \supset B) \wedge (\diamond B \supset \diamond * A).$$

If we extend CI.O$^*Eq$ by Def$\Rightarrow$ and identify $\square$ and $\square^\circ$, an alternative definition of ">" could be given by

$$A > B =_{Df} *A \Rightarrow B.$$

While the decision procedure for CI.O and CI.O$\Rightarrow$ reduces to the one of KT, the decision procedure for CI.O$^*Eq$ is a more complicated tableaux procedure devised in Pizzi [6], [7].

Now our problem is to check the validity, in CI.O$^*Eq$, of Boethius' Thesis $(A \; \mathbf{CI} \; B) \; \mathbf{CI} \; \neg(A \; \mathbf{CI} \; \neg B)$ since we know that CI.O$^*Eq$ grants Weak Boethius' Thesis $(A \; \mathbf{CI} \; B) \supset \neg(A \; \mathbf{CI} \; \neg B)$ in at least three variants according to taking $\rightarrow, \Rightarrow, >$ as formal counterparts of $\mathbf{CI}$:

$$(A \rightarrow B) \supset \neg(A \rightarrow \neg B)$$
$$(A \Rightarrow B) \supset \neg(A \Rightarrow \neg B)$$
$$(A > B) \supset \neg(A > \neg B)$$

are in fact theorems of CI.O$^*Eq$. Furthermore, by applying the rule of Necessitation, we obtain the same theorems with the strict implication connective "-3" in place of "$\supset$".

Since we have at least three connectives satisfying the conditions required for $\mathbf{CI}$, we have to take into account several variants of (strong) Boethius' Thesis, and we could employ the tableaux procedure worked out in Pizzi [6], [7] to solve the problem. For wffs shorn of *-operators the procedure consists in reducing every subformula of form $A \rightarrow B$ into $\square(A \supset B) \wedge (\diamondsuit B \supset \diamondsuit A) \wedge (\square B \supset \square A)$, and then in testing the resulting wff by the tableaux method for KT. This procedure, however, turns out to be lengthy and unpractical when the nested arrows of the wff under test exceeds a minimal threshold, as it happens just in the case of such a second degree wff as strong Boethius' Thesis. The procedure is even more complicated for second degree wffs containing the circumstantial operator. Luckily, we may find a shortcut thanks to a property of KT which has recently received attention.

We follow the terminology of Lemmon-Scott [3] and Williamson [10], [11] in defining a modal system S as having the *single cancellation property* just in case $A \equiv B$ is a thesis of S whenever $\square A \equiv \square B$ is, and having the *double cancellation property* just in case $A \equiv B$ is a thesis whenever both $\square A \equiv \square B$ and $\diamondsuit A \equiv \diamondsuit B$ are. If a system has the single cancellation property it has also the double, but the converse is not true. It has been proved that KT has the double cancellation property but not the single one, while K and KD have the single cancellation property (see [11]).

It has to be noticed that the cancellation property holds for KT, which is not equivalent to CI.O$^*Eq$ but to its analytic fragment CI.O. However, CI.O contains all and only the CI.O*Eq-theses shorn of the star-operators, as it follows from the fact that the tableaux for wffs shorn of star-operators are coincident with the ones for KT (see Pizzi[6]). It follows then that a wff shorn of star-operator (hence shorn of >) is a non-theorem of CI.O if and only if it is a non-theorem of CI.O*Eq. This allows us to restrict the relevant part of our investigation to CI.O.

The shortcut procedure rests then on the following two metatheorems.

Theorem 1 *Any wff of form $A \to B$ is a theorem of CI.O iff $A \equiv B$ is such.*

Proof. From right to left the implication is obvious since CI.O contains a rule of replacement for proved material equivalents. From left to right we must reason as follows. Suppose a wff of form $A \to B$ is a CI.O-theorem. Then by the definitional equivalence between CI.O and KT, $\Box(A \supset B) \wedge (\Diamond B \supset \Diamond A) \wedge (\Box B \supset \Box A)$ is a schema of theorem of KT. But by the normality of KT the preceding wff equals $\Box(A \supset B) \wedge (\Diamond B \equiv \Diamond A) \wedge (\Box B \equiv \Box A)$. Thus if a wff of form $A \to B$ is a theorem of KT+Def$\to$ then $\Diamond B \equiv \Diamond A$ and $\Box B \equiv \Box A$ are such. Hence by the double cancellation property of KT a wff of form $A \to B$ is a theorem only if $A \equiv B$ is a KT+Def-theorem [2].

Let us now call BTO the primary variant of strong Boethius' Thesis, namely $(p \to q) \to \neg(p \to \neg q)$. Then we may prove this result.

Theorem 2 *BTO is not a CI.O*Eq-theorem.*

Proof. Suppose it is. Being shorn of star-operators, BTO should also be a CI.O-theorem. Then by Theorem 1 $(p \to q) \equiv \neg(p \to \neg q)$ should also be such. But by US we should also have $\top \to q \equiv \neg(\top \to \neg q)$ and then, by Def.$\Box$, $\Box q \equiv \Diamond q$, which is a non-theorem of KT.

2. We may generalize our result to any variant of strong Boethius' Thesis, where "$\Rightarrow$" and ">" replace one or more occurrences of "$\to$". To begin with, we prove the following lemma.

Lemma 1 *$\Box q \Rightarrow \Diamond q$ is not a theorem of CI.O*Eq $+$ Def$\Rightarrow$.*

Proof. Suppose by Reductio $\Box q \Rightarrow \Diamond q$ is a CI.O*Eq-theorem. Then by Def$\Rightarrow$ we should have also that $\Diamond\Diamond q \equiv \Diamond\Box q$ is such. From this we would have that $\neg\Diamond\Diamond q \equiv \neg\Diamond\Box q$ is such, so by US also $\neg\Diamond\Diamond\neg q \equiv \neg\Diamond\Box\neg q$ and, by standard

---

[2] This theorem leads to the equivalence between $\vdash A \to B$ and $\vdash B \to A$. This stresses the difference between "$\to$" when it is the main operator and when it occurs in other positions.

interchanges, $\Box\Box q \equiv \Box\Diamond q$. Hence by Def$\to$, $\Box q \to \Diamond q$ should also be a CI.O*Eq -theorem, so that we should have again $\Box q \equiv \Diamond q$ as a KT-theorem, which is impossible.

We are able then to prove this general result.

Theorem 3 *Let $\to^x$, $\to^y$, $\to^z$ be arbitrary connectives in the set $\{\to, \Rightarrow, >\}$. Then for any choice of $\to^x$, $\to^y$, $\to^z$. $(p \to^x q) \to^y \neg(p \to^z \neg q)$ is not a CI.O*Eq-theorem.*

Proof. Let A^- be the wff in the language of CI.O*Eq obtained by dropping from A every occurrence of the operator $*$. Then by a standard inductive argument we may prove that, if A is provable in CI.O*Eq, then A^- is a KT-theorem. It follows then, by the definitions of $\to, \Rightarrow, >$ and the substitution of $\top$ to p, that $\Box q \Rightarrow \Diamond q$ should be in any case a theorem of CI.O*Eq, which is impossible by Lemma 1.

3. The result which has been reached in the preceding section is then that no variant of strong Boethius' Thesis can be a theorem of the basic system of consequential implication CI.O*Eq. However, BTO and every variant of it are consistent with CI.O*Eq. In order to grasp this point we simply have to introduce a map f from wffs of CI.O*Eq into PC-wffs defined by letting $f(*A) = f(A)$; $f(\neg A) = \neg f(A)$; $f(A \wedge B) = f(A) \wedge f(B)$; $f(A \supset B) = f(A) \supset f(B)$; $f(A \to B) = f(A) \equiv f(B)$ and $f(A) = A$ otherwise. Thus by a standard argument all the f-images of theorems of CI.O*Eq turn out to be PC-valid. Since BTO is mapped into the PC-thesis $(A \equiv B) \equiv \neg(A \equiv \neg B)$, we conclude that BTO can be consistently added to CI.O*Eq. We have to notice that

$$
\begin{aligned}
f(A \Rightarrow B) &= f(\Box(A \supset B) \wedge (\Diamond A \equiv \Diamond B)) \\
&= f((\top \to (A \supset B)) \wedge (\neg(\top \to \neg B) \supset \neg(\top \to \neg A)) \\
&= f(A \supset B) \wedge f(B \supset A) = A \equiv B.
\end{aligned}
$$

Thus the argument works for every variant of BTO.

An open question which can be sketchily discussed here is to know which modification of the axiomatic basis of CI.O*Eq can yield BTO. A very simple but unfortunately trivial answer is the following. Let KD! be a system which is sound and complete with respect of a class of models in which the accessibility relation is *functional*, namely such that for every world m_i there is one and only one m_j such that $m_i \, R \, m_j$. This system is axiomatized by K + $\Box p \equiv \Diamond p$ and is seen to be closed with respect of Eq. Notice that $\Box p \equiv p$ cannot be derived since $\Box p \supset p$ is not a theorem. If we drop from CI.O the axiom (h) $(p \rightarrow q) \supset (p \supset q)$ and add as an axiom the converse of axiom (d), namely $\neg(p \rightarrow \neg q) \supset (p \rightarrow q)$, we obtain a system - let us call it CI.O! - which is definitionally equivalent to KD! + Def$\rightarrow$. Since we have $p \rightarrow q \equiv \neg(p \rightarrow \neg q)$ then BTO can be derived in a straightforward manner by applying replacement to $(p \rightarrow q) \rightarrow (p \rightarrow q)$. The price to pay for this result is, of course, that we lose the asymmetrical implication between $(p \rightarrow q)$ and $\neg(p \rightarrow \neg q)$, so that the genuine flavor of consequential implication rests simply on the Aristotle's Thesis $\neg(p \rightarrow \neg p)$ and its variants.

The most interesting result about CI.O! is perhaps the fact that CI.O! is not, strictly speaking, a system of consequential implication. In fact we may prove, thanks to the decision procedure for KD!, that in this system the so-called law of factor $(p \rightarrow q) \supset ((p \wedge r) \rightarrow (q \wedge r))$ turns out to be a theorem. It is so violated the condition 3) required for the definition of **CI**. Since this law belongs to the so-called connexive logics, we have thus found a system which is not properly a system of consequential implication but is an intermediate one between systems of consequential implication and systems of connexive implication.

CI.O! may then be considered an useful reference point in the investigations about the relations between connexive implication and consequential implication[3].

References

[1] R.B. Angell, *A Propositional Logic with Subjunctive Conditionals*, J. Symbolic Logic **27**, 327–343.

[2] _____, in *Tre logiche dei condizionali congiuntivi, Leggi di Natura, Modalità, Ipotesi*, (C. Pizzi, ed.), Feltrinelli, Milano, 1978, pp. 156–180.

[3] J. Lemmon and D.S. Scott, *The "Lemmon Notes": An Introduction to Modal Logic*, (K. Segerberg, ed.), Oxford Blackwell, 1977.

[4] S. McCall, *Connexive Implication and the Syllogism*, Mind **76** (1967), 346–356.

[5] C. Pizzi, *Modal Operators in Logics of Consequential Implication*, Proceedings of the IX SLAMLS (Bahia Blanca 1992), forthcoming.

[6] _____, *Decision Procedures for Logics of Consequential Implication*, Notre Dame J. Formal Logic **32** (1991), 618–636.

[7] _____, *Consequential Implication: a correction*, Notre Dame J. Formal Logic **34** (1993), 621–624.

---

[3]The author wishes to thank Timothy Williamson for drawing his attention to the relevance of his results on cancellation rules for the analysis of consequential implication, and the anonymous referee for the suggested simplifications.

[8] _____, *L'implicazione crisippea e i suoi critici contemporanei*, in Semantica aristotelica e semantica stoica (W. Cavini, ed.), L. Olskii, Firenze, 1995, forthcoming.

[9] R. Routley, G. Plumwood, R.K. Meyer, and R. Brady, *Relevant Logic and its Rivals*, Ridgeview, Atascadero CA, 1982.

[10] T. Williamson, *Assertion, Denial and some Cancellation Rules in Modal Logic*, J. Philos. Logic **17** (1988), 299–318.

[11] _____, *Verification, Falsification and Cancellation in KT*, Notre Dame J. Formal Logic **31** (1990), 286–290.

DIPARTIMENTO DI FILOSOFIA E SCIENZE SOCIALI, UNIVERSITÀ DI SIENA, VIA ROMA 47, 53100 SIENA, ITALY

A new and elementary method to represent every complete Boolean algebra

Giovanni Sambin

Abstract

For any semilattice $(S, \wedge, 1)$ and any $X, Z \subseteq S$, define X^Z as $\{a \in S : (\forall b \in X)(a \wedge b \in Z)\}$ and say that X is Z-stable if $X = X^{ZZ}$. We prove that for any S and any downward closed $Z \subseteq S$, Z-stable subsets form a complete boolean algebra, and that all cBa's can isomorphically be represented in this way. The proof is obtained by specializing a similar representation theorem for boolean quantales, already known in the literature. Contrary to Stone's representation theorems, the proof is carried over in a fully constructive set theory (that is, no prime filter principle or similar is necessary, and no argument by reductio ad absurdum or proof by cases is used).

Foreword

The aim of this paper is to present in all details[1] a new method to represent every complete boolean algebra. Such a method is substantially different from the classic representation by Stone: contrary to Stone's, where the prime filter principle is necessary, it does not need any set-theoretical principle.

This is not pursued just as a debatable curiosity; more positively, all definitions and results can be carried over in a specific highly constructive set theory, namely Martin-Löf's intuitionistic type theory (see e.g. [4] or [7] chap. 11). However, little attention is here payed to details of formalization, with the aim of showing, by an example, that such theory can be put into practice also by the working mathematician.

---

[1] A first proof was obtained as a by-product of the development of a uniform method to prove completeness of several different logics w.r.t. the many valued semantics given by the notion of pretopology (cf. [6], thm. 14); I here give a simpler proof, and pruned of any reference to logic and pretopologies. I am deeply grateful to Silvio Valentini who has played an important role in the writing of this paper, through conversation and encouragement.

For the purpose of the proof of the theorem, it is convenient to look at cBa's as quantales satisfying some additional conditions. A unital commutative quantale (cf. e.g. [5]), here simply a *quantale*, is a structure $\mathcal{Q} = (Q, \cdot, 1, \vee)$ where $(Q, \cdot, 1)$ is a commutative monoid, $(Q, \vee)$ is a complete lattice and infinite distributivity $(\vee_{i \in I} a_i) \cdot b = \vee_{i \in I}(a_i \cdot b)$ holds for any family of elements $(b_i)_{i \in I}$ and any element a of Q. We denote the top and bottom element of $\mathcal{Q}$ by $\top$ and 0 respectively. For example, given any monoid $(M, \cdot, 1)$ (here and in the whole paper commutativity is understood), the structure $(\mathcal{P}M, \cdot, \{1\}, \cup)$ is a quantale, where $\mathcal{P}M$ is the powerset of M, $\cdot$ is defined on subsets by $A \cdot B \equiv \{a \cdot b : a \in A, b \in B\}$ and $\cup$ is set-theoretic union. A boolean quantale is a quantale equipped with an operation $-$ such that $a = --a$ for any a. In section 1 we review basic properties of quantales and of boolean quantales up to the point where it can be proved that cBa's are the same thing as boolean quantales satisfying two simple additional conditions.

A phase space $\mathcal{P}M_Z$ (cf. [5], p. 142) is a structure formed by those subsets of M which are stable under double Z-complementation (defined below). In section 2 we repeat a theorem in [5] showing that a quantale is boolean iff it is isomorphic to a phase space. Putting such representation theorem together with the characterization of cBa's, in section 3 the main theorem is easily obtained: phase spaces of the form $\mathcal{P}S_Z$, where S is a semilattice and Z is any downward closed subset of S, exhaust, up to isomorphism, all complete boolean algebras.

Contrary to Stone's, the method presented here unfortunately does not seem to be applicable to complete distributive lattices without modifications.[2] Also, the question remains to obtain Stone's representation of a cBa $\mathcal{B}$ directly from ours, thus detecting precisely the role of the prime filter principle. To make the paper self contained and readable also by the non-specialist, we review all properties of quantales which are needed (even if most of them are in [5]) and assume only a little acquaintance with lattices.

1   Preliminaries on quantales and boolean quantales

The next four lemmas contain the standard facts we need about quantales.

Lemma 1.1 *In any quantale $\mathcal{Q}$, the following hold for any a, b, c, $d \in Q$:*

1. localization: if $a \le b$ then $a \cdot c \le b \cdot c$

---

[2]However, I thank John L. Bell for pointing out a connection with the method of polarities, cf. [1], p. 122.

2. *stability: if $a \leq b$ and $c \leq d$ then $a \cdot c \leq b \cdot d$*

Proof. Localization: $a \leq b$ means that $a \vee b = b$, hence also $(a \vee b) \cdot c = b \cdot c$; but by distributivity $(a \vee b) \cdot c = a \cdot c \vee b \cdot c$, hence $a \cdot c \vee b \cdot c = b \cdot c$, that is $a \cdot c \leq b \cdot c$. Stability: by localization, from $a \leq b$ we have $a \cdot c \leq b \cdot c$ and similarly $b \cdot c \leq b \cdot d$ from $c \leq d$, so by transitivity $a \cdot c \leq b \cdot d$.

In any quantale, the operation $\rightarrow$ is defined putting, for any $a, b \in Q$,

$$a \rightarrow b \equiv \vee\{c : a \cdot c \leq b\}$$

Lemma 1.2 *(Basic Properties of $\rightarrow$) In any quantale, the operation $\rightarrow$ satisfies the following properties:*

1. $a \cdot (a \rightarrow b) \leq b$

2. $c \cdot a \leq b$ *iff* $c \leq a \rightarrow b$

3. $1 \rightarrow a = a$

4. $a \rightarrow (b \rightarrow c) = a \cdot b \rightarrow c$

Proof. (1) By definition, $a \cdot (a \rightarrow b) \equiv a \cdot \vee\{c : a \cdot c \leq b\}$; by distributivity, $a \cdot \vee\{c : a \cdot c \leq b\} \leq \vee\{a \cdot c : a \cdot c \leq b\}$, which by definition of sup is less than b. (2) Assume $a \cdot c \leq b$; then, since $c \in \{c : a \cdot c \leq b\}$, it follows that $c \leq \vee\{c : a \cdot c \leq b\} \equiv a \rightarrow b$. Conversely, if $c \leq a \rightarrow b$ then by localization $a \cdot c \leq a \cdot (a \rightarrow b)$, so that $a \cdot c \leq b$ by (1). (3) $a = a \cdot 1 \leq a$ hence $a \leq 1 \rightarrow a$; conversely, $1 \rightarrow a = 1 \cdot (1 \rightarrow a) \leq a$ (4) By (1), $(a \cdot b \rightarrow c) \cdot a \cdot b \leq c$, hence by (2) $(a \cdot b \rightarrow c) \cdot a \leq b \rightarrow c$ and $a \cdot b \rightarrow c \leq a \rightarrow (b \rightarrow c)$; similarly, $(a \rightarrow (b \rightarrow c)) \cdot a \leq b \rightarrow c$ gives $(a \rightarrow (b \rightarrow c)) \cdot a \cdot b \leq c$ hence $a \rightarrow (b \rightarrow c) \leq a \cdot b \rightarrow c$.

In any quantale, for any element d the operation of *d-complementation* is defined putting

$$-_d a \equiv a \rightarrow d.$$

If d is such that $-_d -_d a = a$ for any a, then d is called a *dualizer* for Q. A quantale is called *boolean*[3] if it has a dualizer d; in any case the d-complement of a is abbreviated by $-a$.

Lemma 1.3 *In any quantale, for any d, d-complementation satisfies:*

1. $a \leq b$ *implies* $-b \leq -a$

2. $-1 = d$

---

[3]Following a suggestion in [5], where however it is called 'Girard'.

3. $-0 = \top$

4. $a \leq - - a$

5. $-a = - - -a$

6. $a \cdot - - b \leq - - (a \cdot b)$

7. $- - (a \cdot - - b) = - - (a \cdot b)$

Proof. (1) holds because $a \leq b$ implies $a \cdot (b \to d) \leq b \cdot (b \to d) \leq d$ and hence $b \to d \leq a \to d$. (2) since $-1 \equiv 1 \to d$ and $1 \to d = d$, $-1 = d$ holds. (3) To show $-0 = \top$ it is enough to see that $\top \leq -0$: from $0 \leq -\top$ one has $0 \cdot \top \leq d$ hence $\top \leq -0$. (4) follows immediately from $a \cdot -a \leq d$. (5) holds since the inequality $-a \leq - - -a$ is an application of (5), while $- - -a \leq -a$ comes from $a \leq - -a$ by (1). (6) By (1), $(a \to (b \to d)) \cdot a \leq b \to d$, that is, by (4) of lemma 2, $-(a \cdot b) \cdot a \leq -b$; by localization, $-(a \cdot b) \cdot a \cdot - - b \leq -b \cdot - - b \leq d$, hence by (2) of lemma 2 the claim $a \cdot - - b \leq - - (a \cdot b)$. (7) is an easy consequence of (4) and (6).

Lemma 1.4 *In any boolean quantale, d-complementation satisfies:*

1. $a \leq b$ *iff* $-b \leq -a$

2. $-1 = d, \ -d = 1$

3. $-\top = 0, \ -0 = \top$

4. $-(a \wedge b) = -a \vee -b$ *(de Morgan law)*

Proof. (1) By properties of $\to$, from $a \leq b$ we have $-b \leq -a$; conversely, by the same reason, $-b \leq -a$ gives $- - a \leq - - b$, i.e. $a \leq b$. (3) From $-0 = \top$, which holds in any quantale, one has $0 = - - 0 = -\top$. (2) Similarly, from $-1 = d$ one has $1 = - - 1 = -d$ (4) $a \wedge b \leq a$ gives $-a \leq -(a \wedge b)$ and similarly $-b \leq -(a \wedge b)$, so $-a \vee -b \leq -(a \wedge b)$. Conversely, $-a \leq -a \vee -b$ gives $-(-a \vee -b) \leq a$ and similarly for b, hence $-(-a \vee -b) \leq a \wedge b$, from which $-(a \wedge b) \leq -a \vee -b$.[4]

Lemma 1.5 *In any boolean quantale, the following are equivalent:*

1. $a \cdot b \leq a$ *for any* a, b

---

[4] Note that also the second de Morgan law $-(a \vee b) = -a \wedge -b$ holds, but we don't need it in this paper.

   2. $a \cdot b \leq a \wedge b$ *for any* a, b

   3. $a \leq 1$ *for any* a

   4. $\top \leq 1$

   5. $d \leq 0$

Proof. From (1) it follows that $a \cdot b \leq a$ and $a \cdot b \leq b$, hence $a \cdot b \leq a \wedge b$, i.e. (2); the converse is trivial. By taking $a = 1$ in (1) one has $b = b \cdot 1 \leq 1$ for any b, i.e. (3); the converse holds by localization. Finally, (3) iff (4) because $a \leq \top$ for any a and (4) iff (5) because by lemma 4 $d \leq 0$ iff $-0 \leq -d$ iff $\top \leq 1$. Note that the equivalences (1)-(4) hold in any quantale.

Lemma 1.6 *In any quantale, the following are equivalent:*

   1. $a \wedge b \leq a \cdot b$ *for any* a, b

   2. $a \leq a \cdot a$ *for any* a

Proof. From (1) by taking $b = a$ it follows that $a = a \wedge a \leq a \cdot a$. Conversely, $a \wedge b \leq a$ and $a \wedge b \leq b$ by stability give $(a \wedge b) \cdot (a \wedge b) \leq a \cdot b$, from which (1) since $a \wedge b \leq (a \wedge b) \cdot (a \wedge b)$.

Corollary 1.7 *In a boolean quantale, if 0 is a dualizer, then it is the unique one.*

Proof. By lemma 5. if $d = 0$ then $\top = 1$ and hence $-d' = 1 = \top$ for any dualizer d', so that $d' = - - d' = -\top = 0$.

We are now ready to characterize complete boolean algebras as a special class of boolean quantales. *Boolean algebras* are here defined as usual (apart from notation), namely as distributive lattices with bottom 0 and top $\top$ and with an operation of complementation ν s.t. $a \wedge \nu a = 0$ and $a \vee \nu a = 1$. We say that a quantale $\mathcal{Q}$ is *idempotent* if all its elements are idempotent, that is $a \cdot a = a$ holds for any $a \in \mathcal{Q}$. Then we have:

Theorem 1.8 *Complete boolean algebras can be characterized as idempotent boolean quantales in which 0 is a dualizer.*

Proof. It is an exercise on boolean algebras to show that any cBa $\mathcal{B} = (B, \wedge, \vee, 0, \top, \nu)$ is a quantale with the desired properties; however, here is a hint. Trivially, $(B, \wedge, \top)$ is a semilattice, that is a monoid in which $a = a \wedge a$ holds, and $(B, \vee, 0, \top)$ is a complete lattice. Infinite distributivity is easily derivable from the fact that $\rightarrow$ is definable, putting

$$a \rightarrow b \equiv \nu a \vee b$$

It is easy to check that $\rightarrow$ so defined satisfies $a \wedge c \leq b$ iff $c \leq a \rightarrow b$; in fact, if $a \wedge c \leq b$ then $c \leq (\nu a \vee a) \wedge (\nu a \vee c) = \nu a \vee (a \wedge c) \leq \nu a \vee b$, and conversely if $c \leq \nu a \vee b$ then $c \wedge a \leq a \wedge (\nu a \vee b) = (a \wedge \nu a) \vee (a \wedge b) \leq b$. Therefore $(B, \wedge, \top, \vee)$ is an idempotent quantale. Finally, since $\nu a = \nu a \vee 0 = a \rightarrow 0$, the equation $\nu \nu a = a$, which holds in any boolean algebra, means that $a = (a \rightarrow 0) \rightarrow 0$, i.e. 0 is a dualizer.

Conversely, assume $\mathcal{Q} = (Q, \cdot, \vee, 1)$ is a boolean quantale. The assumptions $a \cdot a = a$ and $0 = d$ by lemmas 1.5 and 1.6 mean that $a \cdot b = a \wedge b$; moreover $1 = \top$ by lemma 1.5, so that $(Q, \cdot, \vee, \top)$ is a complete distributive lattice. Finally $a \wedge -a = 0$ because $a \wedge -a = a \cdot -a \leq d = 0$ and $a \vee -a = \top$ because $a \vee -a = --a \vee -a = (\text{by de Morgan law}) - (-a \wedge a) = -0 = \top$, and so $-a$ is the (boolean) complement of a.

2 Phase spaces and the representation of Boolean quantales

As the last result of the preceding section suggests, the representation of cBa's is obtained by specializing the representation of boolean quantales (see [5], Theorem 6.1.1); we here repeat it shortly. We need a lemma which shows how any quantale can be transformed into a boolean quantale:

Lemma 2.1 *For any quantale $\mathcal{Q}$ and any $d \in Q$, define d-complementation $-a \equiv a \rightarrow d$ as usual, and say that an element $a \in Q$ is d-stable if $a = --a$. Then the set of d-stable elements of Q forms a boolean quantale $\mathcal{Q}_d$ with dualizer d and with operations defined by:*

$$\bigvee_{i \in I}^d a_i \equiv -- \left(\bigvee_{i \in I} a_i \right) \qquad a \cdot^d b \equiv --(a \cdot b) \qquad 1^d \equiv --1$$

Proof. The proof is obtained by standard applications of the properties of d-complementation (cf. lemma 1.3). In fact, $\bigvee_{i \in I}^d a_i$ is the supremum, since $a_i \leq \bigvee_{i \in I} a_i \leq --\left(\bigvee_{i \in I} a_i \right)$ and if $a_i \leq b$ for all $i \in I$, then $\bigvee_{i \in I} a_i \leq b$ from which $--\left(\bigvee_{i \in I} a_i \right) \leq --b = b$. The operation $\cdot^d$ is associative, because

$(a \cdot^d b) \cdot^d c = a \cdot^d (b \cdot^d c)$ is by definition $--(--(a \cdot b) \cdot c) = --(a \cdot --(b \cdot c))$, which is obtained from $(a \cdot b) \cdot c = a \cdot (b \cdot c)$ by repeated use of (7) of lemma 1.3. The element $--1$ is the unit element for $\cdot^d$. since $--1 \cdot^d a = --(--1 \cdot a) = --(1 \cdot a) = --a = a$. Distributivity follows from distributivity of $\mathcal{Q}$, with an argument quite similar to that for associativity above.

Finally, d is d-stable, because $d \cdot (d \to d) \leq d$ gives $d \leq (d \to d) \to d$ and $1 \cdot d \leq d$ gives $1 \leq d \to d$, hence $(d \to d) \to d \leq 1 \to d = d$; trivially $--a = a$ for any d-stable a. So d is a dualizer in $\mathcal{Q}_d$ and $\mathcal{Q}_d$ is a boolean quantale.

For any monoid M and any $Z \subseteq M$, the quantale $\mathcal{P} M_Z$, obtained from the quantale $\mathcal{P} M$ as in the preceding lemma. is called the *phase space* on M with dualizer Z. We denote by $F, G, \ldots$, the Z-stable subsets, sometimes called facts; the explicit definition of operations is:

$$\vee_{i \in I}^Z F_i \equiv --(\cup_{i \in I} F_i) \qquad F \cdot^Z G \equiv --(F \cdot G) \qquad 1^Z \equiv --1$$

Note that by the definition of $\to$ in a quantale. it is $A \to B \equiv \cup \{C : C \cdot A \subseteq B\}$ and it is easily seen that $A \to B = \{a \in M : a \cdot A \subseteq B\}$. So in particular Z-complementation is

$$-_Z A \equiv A^Z = \{a \in M : a \cdot A \subseteq Z\}$$

We say that a submonoid M of a quantale $\mathcal{Q}$ is a *base* for $\mathcal{Q}$ if for any $q \in \mathcal{Q}$, $q = \vee \{a \in M : a \leq q\}$. Note that any quantale has a base, at worst $\mathcal{Q}$ itself[5].

Theorem 2.2 *Any quantale is boolean iff it is isomorphic to a phase space. More specifically, any phase space $\mathcal{P} M_Z$ is a boolean quantale, and for any boolean quantale $\mathcal{Q}$, for any submonoid M of $\mathcal{Q}$ which is a base for $\mathcal{Q}$, there exists $Z \subseteq M$ s.t. $\mathcal{Q} \cong \mathcal{P} M_Z$.*

Proof. $\mathcal{P} M_Z$ is a boolean quantale by lemma 2.1. Conversely, let $\mathcal{Q}$ be a boolean quantale and let M be a base for $\mathcal{Q}$. For any $A \subseteq M$, we put $jA \equiv \{a \in M : a \leq \vee A\}$. We now see that $jA = -_Z -_Z A$, where we have put $Z \equiv \{a \in M : a \leq d\}$ for a dualizer d of $\mathcal{Q}$. In fact, for any $a \in M$, by definition $a \in --A$ iff $a \cdot (A \to Z) \subseteq Z$ iff $(\forall b)(b \in A \to Z \Rightarrow a \cdot b \in Z)$. Now $a \cdot b \in Z$ iff $a \cdot b \leq d$ iff $a \leq -b$ and similarly $b \in A \to Z$ iff $b \cdot A \subseteq Z$ iff $(\forall c \in A)(b \cdot c \in Z)$ iff $(\forall c \in A)(c \leq -d)$ iff $\vee A \leq -d$. So $a \in --A$ iff $(\forall b)(\vee A \leq -b \Rightarrow a \leq -b)$, which is the same as $a \leq \vee A$, i.e. $a \in jA$.

---

[5]Actually all results below could be stated without mentioning bases. Here some care is taken towards keeping the highest possible degree of constructivity, and this is why bases are essential: in the terminology of [4], a base M for $\mathcal{Q}$ could be a set, even if $\mathcal{Q}$ is a proper category.

So the Z-stable subsets of $\mathcal{P}M_Z$ are all of the form jA for some $A \subseteq M$. Note that for any $A \subseteq M$, it is $\vee(--A) = \vee A$, since $--A = jA$ and $\vee(jA) = \vee A$.

Then the isomorphism $k : \mathcal{P}M_Z \to \mathcal{Q}$ is given by $k : F \mapsto \vee F$. In fact, since obviously $F \subseteq G$ iff $\vee F \leq \vee G$, k is one-one and preserves inclusion. Since M is a base, any $q \in Q$ is equal to $\vee A$ where $A \equiv \{a \in M : a \leq q\}$ and hence $k(jA) = \vee A = q$, that is k is onto. A few equalities show that k preserves suprema:

$$k(\vee_{i \in I} F_i) \equiv k(--(\cup_{i \in I} F_i)) = \vee(\cup_{i \in I} F_i) = \vee_{i \in I}(\vee F_i) \equiv \vee_{i \in I} k F_i$$

and that k preserves the monoid operation:

$$k(F \cdot_Z G) \equiv k(--(F \cdot G)) = \vee(F \cdot G) = \vee F \cdot \vee G \equiv kF \cdot kG$$

3   The representation of complete Boolean algebras

We begin with a corollary to theorem 2.2 above:

Corollary 3.1 *$\mathcal{Q}$ is a boolean quantale in which 0 is a dualizer iff it is isomorphic to $\mathcal{P}M_Z$ for some monoid M and some $Z \subseteq M$ s.t. $Z = --\emptyset$.*

Proof. By the theorem, it is enough to show that 0 is a dualizer of $\mathcal{P}M_Z$ iff $Z = --\emptyset$. Recall that $--\emptyset$ is the bottom of $\mathcal{P}M_Z$ and that Z is always a dualizer of $\mathcal{P}M_Z$. Then the claim follows from corollary 1.7.

The condition $Z = --\emptyset$ does not give a direct description of how Z should be. However, since obviously $-\emptyset = M$, we can bring $Z = --\emptyset$ to the form

$$Z = M \to Z$$

which is a sort of fixed point equation to be solved. The solutions are all of the form $M \to A$ for any $A \subseteq M$. In fact, if $Z = M \to Z$ then Z is of such form, and conversely for any A it is $M \to (M \to A) = M \cdot M \to A = M \to A$. It also follows that $Z = \emptyset^{ZZ}$ iff $Z = \emptyset^{AA}$ for some A. Note also that, since the inclusion $M \to Z \subseteq Z$ holds for any Z (because $a \in M \to Z \equiv a \cdot M \subseteq Z$ implies $a = a \cdot 1 \in Z$), the equation $Z = M \to A$ is equivalent to $Z \subseteq M \to Z$, and hence to $Z \cdot M \subseteq Z$; that is, Z is an ideal in the monoid M. So, in the case M is in fact a semilattice $(S, \wedge, 1)$, the condition becomes $S \wedge Z \subseteq Z$, which is equivalent to $\downarrow Z \subseteq Z$, where $\downarrow Z \equiv \{a \in S : (\exists z \in Z)(a \leq z)\}$ is the downward closure of Z. Summing up, we have:

Lemma 3.2 *For any monoid M, the following are equivalent for any $Z \subseteq M$:*

1. *Z is the bottom in $\mathcal{P}M_Z$, i.e. $Z = --\emptyset$*

2. *$Z = \emptyset^{AA}$ for some $A \subseteq M$*

3. *$Z = M \to Z$*

4. *$Z = M \to A$ for some $A \subseteq M$*

5. *Z is an ideal, i.e. $M \cdot Z \subseteq Z$*

In a semilattice S, any of the above is equivalent to:

6. *Z is downward closed, i.e. $\downarrow Z \subseteq Z$*

We are finally ready to prove the main theorem:

Theorem 3.3 *A structure $\mathcal{B}$ is a complete boolean algebra iff it is isomorphic to a phase space on a semilattice with a downward closed dualizer. More specifically, for any semilattice S and any downward closed $Z \subseteq S$, $\mathcal{P}S_Z$ is a cBa. Conversely, for any cBa $\mathcal{B}$, there exists a semilattice S and a downward closed $Z \subseteq S$ s.t. $\mathcal{B} \cong \mathcal{P}S_Z$.*

Proof. First we prove that, for any semilattice S and any downward closed $Z \subseteq S$, $\mathcal{P}S_Z$ is a complete boolean algebra. By theorem 2.2, $\mathcal{P}S_Z$ is a boolean quantale. Since Z is downward closed, 0 is a dualizer by lemmas 1 and 2. So, by theorem 1.8, it only remains to prove that $\mathcal{P}S_Z$ is idempotent. Since 0 is a dualizer, by lemma 1.5 $U \cdot^Z U \subseteq U$ holds for any Z-stable subset $U \subseteq S$; for the other inclusion $U \subseteq U \cdot^Z U$ it is enough to show that $U \subseteq U \cdot U$, which is immediate since for any $a \in S$, $a \in U$ implies $a = a \cdot a \in U \cdot U$.

Now let $\mathcal{B}$ be a cBa. By theorem 1.8 $\mathcal{B}$ is a boolean quantale, and hence by theorem 2.2 it is $\mathcal{B} \cong \mathcal{P}S_Z$ for some base S and $Z \subseteq S$; note that S, like any base for $\mathcal{B}$, is a semilattice. By the isomorphism $\mathcal{B} \cong \mathcal{P}S_Z$, since 0 is a dualizer in $\mathcal{B}$, it follows that Z is equal to bottom in $\mathcal{P}S_Z$, that is $Z = --\emptyset$, which is equivalent to Z downward closed by lemma 2.

It may be worthwhile to notice that, given a cBa $\mathcal{B}$, it may well happen that $Z \equiv \{a \in M : a \leq 0\}$ is empty, which happens if $0 \notin S$. In this case, it is easy to check that $\mathcal{P}S_Z$ is (isomorphic to) the two-element boolean algebra $\{0, \top\}$; in fact, for any $A \subseteq S$, it is $A^\emptyset = \emptyset$ if $A \neq \emptyset$, and $A^\emptyset = S$ if $A = \emptyset$, so that $A^{\emptyset\emptyset}$ is either $\emptyset$ or S. This does not affect the theorem, however, nor its uniformity; in fact, one finds out (as suggested by Silvio Valentini) that when $\mathcal{B}$ is different from $\{0, \top\}$, then any base S must contain 0:

Proposition 3.4 *If a cBa $\mathcal{B}$ contains some element different from bottom and top, then any base S for $\mathcal{B}$ must contain 0.*

Proof. If $\mathcal{B}$ is different from $\{0, \top\}$, then any base S must be different from the singleton $\{\top\}$; in fact, $\{\top\}$ can generate only $\top$ and 0. So let $b \in S$ with $b \neq \top$; then $0 = b \wedge -b = b \wedge \vee\{c \in S : c \leq -b\}$, and since $b \neq \top$, it must be $-b \neq 0$, hence $\{c \in S : c \leq -b\} \neq \emptyset$. Let c be an element of S s.t. $c \leq -b$; then $0 = b \wedge -b \geq b \wedge c$, hence $0 \in S$ because $b, c \in S$.

Concluding general remark

As a concluding remark, note that the fact that the proof of the representation theorem does not need any set-theoretic principle is not in contrast with the well known situation for representation theorems in the style of Stone; the constructivity of the present proof is a trade off of the fact that our presentation is not a field of sets, that is we have given up with the idea that $\emptyset$ and $\bigcup$ should be the zero and sup, respectively[6]. So, while it is true that $\mathcal{P}S_Z$ is a family of subsets of S, it is *not* true that it is a field of sets, i.e. a subalgebra of the powerset $\mathcal{P}S$ with usual set-theoretic operations.

This is perfectly in line with the spirit of pointfree (or pointless) topology, where for example Tychonoff's theorem on products of compact topological spaces does not need the axiom of choice, at the price of giving up points (for a discussion on why a mathematician should be interested in pointless topology, see [3]). Actually, also our example can be interpreted in this way: a constructive representation is obtained by giving up the requirement that an element of the boolean algebra is to be represented as a set of prime filters, i.e. points. The intriguing problem remains to explain why such a representation seems to work only for the "classical" case of boolean algebras, rather than for an arbitrary lattice of opens.

References

[1] G. Birkhoff, *Lattice theory*, 3rd ed., Amer. Math. Soc. Colloquium Publications XXV, 1967.

[2] J-Y. Girard, *Linear logic*, Theoret. Comput. Sci. **50** (1987), 1–102.

---

[6]For the logician: this should be confronted with the fact that, to obtain a constructive proof of completeness of Kripke models for intuitionistic logic, one has to give up the idea that absurdity is forced in the empty set of points.

[3] P. Johnstone, *The point of pointless topology*, Bull. Amer. Math. Soc. (N.S.) **8** (1983), 41–53.

[4] P. Martin-Löf, *Intuitionistic type theory, Notes by G. Sambin of a series of lectures given in Padua, June 1980*, Bibliopolis, 1984.

[5] K. Rosenthal, *Quantales and their applications*, Longman, 1990.

[6] G. Sambin, *Pretopologies and completeness proofs*, J. Symbolic Logic, to appear.

[7] A. S. Troelstra and D. Van Dalen, *Constructivism in mathematics. an introduction*, North Holland, 1988.

DIPARTIMENTO DI MATEMATICA PURA ED APPLICATA, VIA BELZONI 7, 35131 PADOVA, ITALY

e-mail address: SAMBIN@PDMAT1.MATH.UNIPD.IT

On finite intersections
of intermediate predicate logics

Dmitrij Skvortsov

Abstract

S. Miura [4] proved that the intersection of any two finitely axiomatizable intermediate propositional logics is finitely axiomatizable. Here
we establish that this result cannot be transferred to the predicate case.
We use S. Ghilardi's functor semantics [1] generalizing standard Kripke
semantics for predicate logics. Weaker generalizations of Kripke semantics [6] are not sufficient here and so we obtain some new examples
of predicate logics which are incomplete in Kripke semantics and its
generalizations.

1 Basic syntactical notions

1.1 We fix countable sets Vr (of individual variables) and Pr^n (of n-place
predicate letters), $n \geq 0$ (letters from Pr^0 are also called propositional ones).
Atomic formulas have the form $P(x_1, \ldots, x_n)$ with $P \in \mathrm{Pr}^n$, $x_1, \ldots, x_n \in$ Vr.
Predicate formulas are built from atomic ones using $\bot, \rightarrow, \wedge, \vee, \forall, \exists$. *Propositional formulas* are predicate ones without occurrences of individual variables
(i.e. they have no occurrences of quantifiers and n-place predicate letters with
$n > 0$). *Free* and *bound* occurrences of variables in a formula are defined as
usual. The *universal closure* of a formula A with free variables $x_1, \ldots, x_n$ is :
$\overline{\forall} A = \forall x_1 \ldots \forall x_n A$. As usual $\neg A$ denotes $(A \rightarrow \bot)$.

Let A be a formula with n_i-place predicate letters P_i, $i = 1, \ldots, k$, B_i
be predicate formulas and $\mathbf{x}^i = (x_1, \ldots, x_{n_i})$ be lists of distinct variables.
Then a *substitution instance* $A[P_i(\mathbf{x}^i)/B_i(\mathbf{x}^i) : i = 1, \ldots, k]$ of a formula A is
obtained by replacing in A each occurrence of atomic subformula $P_i(\mathbf{y})$ with
$B_i[\mathbf{x}^i/\mathbf{y}]$ where $\mathbf{y} = (y_1, \ldots, y_{n_i})$ and $B[\mathbf{x}^i/\mathbf{y}]$ denotes the result of simultaneous
replacement of free occurrences of each x_j in B_i by y_j (with possible renaming
of bound variables, whenever clashes might appear).

Let $A^m = A[P_i(\mathbf{x}^i)/Q_i(\mathbf{x}^i, \mathbf{z}) : i = 1, \ldots, k]$ for $m \geq 0$, $\mathbf{z} = (z_1, \ldots, z_m)$ being the list of distinct variables not occurring in A, Q_i being distinct $(n_i + m)$-place predicate letters. Let also $A^{[m]} = \overline{\forall} A^m$.

1.2 *Superintuitionistic predicate logic* (or merely, a *logic*) is a set of predicate formulas L closed under the rules of detachment $(A, A \to B \,/\, B)$, generalization $(A \,/\, \forall x \, A)$, substitution and containing all axioms (and therefore all theorems) of Heyting's predicate calculus H. Analogously, a *propositional logic* is a set of propositional formulas closed under detachment and substitution of propositional letters by propositional formulas and containing all axioms of Heyting's propositional calculus PH. The *propositional fragment* PL of a predicate logic L is the set of all propositional formulas from L.

Let Γ be a set of predicate formulas and A be a formula. Then $(H + \Gamma)$ denotes the least superintuitionistic predicate logic containing Γ and $(H + A)$ denotes $(H + \{A\})$. Logics of the form $(H + A)$ are called *finitely axiomatizable*. Note that $(H + A) = (H + \overline{\forall} A)$. Similar notations are used for propositional logics: $(PH + \Gamma)$ and $(PH + A)$. Also we write $L \vdash A$ iff $A \in L$.

We will consider the following propositional and predicate formulas:

$J : \neg Q \vee \neg\neg Q$

$J^- : \neg\neg Q \vee (\neg\neg Q \to Q)$

$P_n : Q_n \vee (Q_n \to Q_{n-1} \vee (Q_{n-1} \to \ldots \to Q_2 \vee (Q_2 \to Q_1 \vee \neg Q_1) \ldots))$

$R_2 : P \vee (P \to Q) \vee \neg Q$

$Z : (P \to Q) \vee (Q \to P)$

$D : \forall x \, (P(x) \vee Q) \to \forall x \, P(x) \vee Q$

$K : \neg\neg\forall x \, (P(x) \vee \neg P(x))$.

$(PH + P_1)$ and $(H + P_1)$ are classical propositional and predicate logic respectively. A propositional logic L is called *intermediate* iff $L \subseteq (PH + P_1)$. An intermediate propositional logic is *properly intermediate* iff $L \neq PH$ and $L \neq (PH + P_1)$. It is known that $(PH + R_2) = (PH + (P_2 \wedge Z))$ is the greatest properly intermediate propositional logic.

1.3 The following lemma (the Deduction Theorem for predicate logics) is well-known (cf. [5], Lemma 5.2):

Lemma 1.1 $(H + A) \vdash C$ *iff* $H \vdash \bigwedge_{i=1}^{k}(\overline{\forall} A_i) \to C$ *for some substitution instances* $A_1, \ldots, A_k$ *of a formula* A.

Proposition 1.2 *Let A and B be closed predicate formulas having no common predicate letters.*

(1) $(\mathsf{H} + A) \cap (\mathsf{H} + B) = (\mathsf{H} + \{(A^{[m]} \vee B^{[m']}) : m, m' \in \omega\}) = (\mathsf{H} + \{(A^{[m]} \vee B^{[m]}) : m \in \omega\})$.

(2) *(cf. [5], Theorem 5.5)* If $(\mathsf{H} + A) \vdash D$ and $(\mathsf{H} + B) \vdash D$ then $(\mathsf{H} + A) \cap (\mathsf{H} + B) = (\mathsf{H} + (A \vee B) \wedge D)$

Proof. (1) Let $(\mathsf{H} + A) \cap (\mathsf{H} + B) \vdash C$, i.e. $\mathsf{H} \vdash (\bigwedge_{i=1}^{k} (\overline{\forall} A_i) \to C)$ and $\mathsf{H} \vdash (\bigwedge_{j=1}^{k} (\overline{\forall} B_j) \to C)$ for some substitution instances A_i, B_j of A and B respectively. Then $\mathsf{H} \vdash (\bigwedge_{i,j} ((\overline{\forall} A_i) \vee (\overline{\forall} B_j)) \to C)$ and $(\overline{\forall} A_i) \vee (\overline{\forall} B_j)$ are substitution instances of $(A^{[m]} \vee B^{[m']})$ (here A_i and B_j are supposed to contain m and m' free variables respectively).

And now $(\mathsf{H} + (A^{[n]} \vee B^{[n]})) \vdash (A^{[m]} \vee B^{[m']})$ where $n = \max(m, m')$ since in fact $(\mathsf{H} + (A^{[n]} \vee B^{[n']})) \vdash (A^{[m]} \vee B^{[m']})$ if $m \leq n$, $m' \leq n'$ (by substitution and omitting vacuous quantifiers).

(2) Let

$$A^{[m]} = \forall u_1 \ldots \forall u_m \, A^m$$
$$B^{[m]} = \forall v_1 \ldots \forall v_m \, B^m$$

and $C = \forall u_1 \forall v_1 \ldots \forall u_m \forall v_m (A^m \vee B^m)$. Then $(\mathsf{H} + (A \vee B)) \vdash C$ and $(\mathsf{H} + D) \vdash (C \to (A^{[m]} \vee B^{[m]}))$. $\boxtimes$

Analogously it can proved for propositional logics that

$$(*) \qquad (\mathsf{PH} + A) \cap (\mathsf{PH} + B) = (\mathsf{PH} + (A \vee B))$$

for any propositional formulas A and B without common letters (see [4]). Thus the intersection of two finitely axiomatizable propositional logics is always finitely axiomatizable. In the predicate case this holds for extensions of $(\mathsf{H}+D)$. H. Ono ([5], Theorem 5.6) gave an example of predicate logics for which an analogue of $(*)$ fails:

$$(\mathsf{H} + R_2) \cap (\mathsf{H} + D) \neq (\mathsf{H} + (R_2 \vee D)).$$

But H. Ono did not know whether this intersection is finitely axiomatizable. Here we prove that the predicate logic $(\mathsf{H} + R_2) \cap (\mathsf{H} + D)$ is not finitely axiomatizable and construct an infinite family of analogous examples.

Theorem 1 *Let* $(\mathsf{PH}+A)$ *be a properly intermediate propositional logic. Then the predicate logic* $(\mathsf{H} + A) \cap (\mathsf{H} + K)$ *is not finitely axiomatizable.*

Theorem 2 *Let* $(\mathsf{PH}+A)$ *be a properly intermediate propositional logic. Then the predicate logic* $(\mathsf{H} + A) \cap (\mathsf{H} + D)$ *is not finitely axiomatizable.*

Theorem 3 *Let* A *be a propositional formula such that the logics* $(\mathsf{PH} + A)$ *and* $(\mathsf{PH} + J)$ *are incomparable. Then the predicate logic* $(\mathsf{H} + A) \cap (\mathsf{H} + J)$ *is not finitely axiomatizable.*

We prove these theorems in Section 3. The proof of Theorem 1 is essentially simpler than the proofs of Theorems 2 and 3. Theorem 2 is interesting in connection with Proposition 1.2 (2) (and with H. Ono's example). Theorem 3 gives similar negative examples for predicate extensions of propositional logics (and definitely contrasts with Miura's result [4]).

2   Basic semantical notions

We will consider some generalizations of Kripke semantics for predicate logics ([1] and [6]). To uniformize the exposition, it will be convenient to use the notion of a metaframe from [7]. To begin with, we recall the main definitions concerning Kripke semantics for propositional logics.

2.1   A *propositional Kripke frame* (or merely, a *frame*) is a non-empty pre-ordered set F (with a reflexive and transitive relation $\leq$). A set $Z \subseteq F$ is *conic* iff $\forall u \in Z \forall v \geq u(v \in Z)$; a *cone* is $F^u = \{v \in F : u \leq v\}$ (for $u \in F$). The pre-order $\leq$ induces an equivalence relation $(u \approx v) \iff (u \leq v) \land (v \leq u)$ on F. By factorizing the frame F through $\approx$, we obtain the partially ordered set $S(F) = (F/\approx)$, which is called the *skeleton* of F.

A *p-morphism* of a frame $\langle F', \leq' \rangle$ onto a frame $\langle F, \leq \rangle$ is an onto map $h : F' \longrightarrow F$ such that

(1) $\forall u, v \in F' ((u \leq' v) \Rightarrow (h(u) \leq h(v)))$;
(2) $\forall u \in F' \forall w \in F ((h(u) \leq w) \Rightarrow \exists v \in F' ((u \leq' v) \land (h(v) = w)))$.

A *valuation* in a frame F is a function ξ assigning conic subsets of F to propositional letters. A valuation ξ gives rise to the *forcing relation* $u \vDash A$ between $u \in F$ and a propositional formula A; it is defined by the following conditions:

$$u \nvDash \bot;$$
$$u \vDash P \iff (u \in \xi(P));$$
$$u \vDash (B \land C) \iff (u \vDash B) \land (u \vDash C);$$
$$u \vDash (B \lor C) \iff (u \vDash B) \lor (u \vDash C);$$
$$u \vDash (B \to C) \iff \forall v \geq u ((v \vDash B) \to (v \vDash C)).$$

For a propositional formula A, the set $\{u \in F : u \vDash A\}$ is conic.

A formula A is *true* under the valuation ξ iff $\forall u \in F (u \vDash A)$. A formula A is *valid* if F if it is true under any valuation in F. The set of all propositional formulas valid in F is denoted by $\mathsf{PL}(F)$ and is called the *propositional logic* of the frame F. We may observe that $\mathsf{PL}(F) = \mathsf{PL}(S(F))$ and $\mathsf{PL}(F) = \bigcap \{\mathsf{PL}(F^u) : u \in F\}$.

Let us state some well-known properties of Kripke frames; they will be used in the sequel.

Let $\mathcal{F}_H$ be the class of all partially ordered frames with least element v_0. $\mathcal{F}_J$ be the class of all frames F from $\mathcal{F}_H$ such that $\forall u_1, u_2 \in F \exists v \in F((u_1 \leq v) \wedge (u_2 \leq v))$ and $\mathcal{F}_H^f$. $\mathcal{F}_J^f$ be the classes of all finite frames from $\mathcal{F}_H$ and $\mathcal{F}_J$ respectively (thus $\mathcal{F}_J^f$ is the class of all frames from $\mathcal{F}_H^f$ with greatest element v_1). Let $\mathcal{F}_Z$ be the class of linearly ordered frames from $\mathcal{F}_H$ and $\mathcal{F}_{P_n}$ be the class of all frames from $\mathcal{F}_H$ of height $\leq n$ (i.e. containing no $(n+1)$-element chains), $n > 0$. It is well-known that for $F \in \mathcal{F}_H$ holds that $J \in \mathsf{PL}(F) \Longleftrightarrow F \in \mathcal{F}_J$, and similarly for Z and P_n.

The following completeness results are also well-known.

Lemma 2.1 *(1)* $\mathsf{PH} = \bigcap(\mathsf{PL}(F) : F \in \mathcal{F}_H^f)$.
(2) $(\mathsf{PH} + J) = \bigcap(\mathsf{PL}(F) : F \in \mathcal{F}_J^f)$.
(3) $(\mathsf{PH} + R_2) = \mathsf{PL}(F_1)$ *(see Figure 1)*.

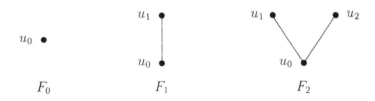

Figure 1.

Lemma 2.2 *For a propositional formula A*

(1) $(\mathsf{PH} + A) \vdash J$ *iff* $A \notin \mathsf{PL}(F_2)$ *(see Figure 1)*.
(2) $(\mathsf{PH} + A) \vdash J^-$ *iff* $A \notin \mathsf{PL}(F_3)$ *and* $A \notin \mathsf{PL}(F_4)$ *(see Figure 2)*.

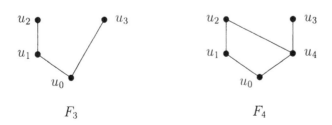

Figure 2.

The proof makes use of Jankov's characteristic formulas [3] for Heyting algebras corresponding to frames F from $\mathcal{F}_H^f$. The main property of Jankov's formula $X(F)$ is: $(\mathsf{PH} + A) \vdash X(F)$ iff $A \notin \mathsf{PL}(F)$ (for a formula A). We can observe that $(\mathsf{PH} + J) = (\mathsf{PH} + X(F_2))$ and $(\mathsf{PH} + J^-) = (\mathsf{PH} + X(F_3) \wedge X(F_4))$.

2.2 A *predicate frame* is a pair $\mathbb{F} = (F, \overline{D})$ in which F is a frame and $\overline{D} = (D_u : u \in F)$ is a family of nonempty disjoint sets. Let $D_n = \bigcup_{u \in F}(D_u)^n$ for $n > 0$ and let $D_0 = F$. A *metaframe* (a Cartesian $m^=$-metaframe, in the terminology from [7]) is a triple $\mathbb{F} = (F, \overline{D}, (\leq_n : n \geq 0))$ in which $(F, \overline{D})$ is a predicate frame and $\leq_n$ is a pre-ordering on D_n (we suppose that $\leq_0$ coincides with the pre-ordering on the frame F) such that for every function $\sigma : \{1, \ldots, m\} \longrightarrow \{1, \ldots, n\}$ the mapping $\pi_\sigma(a_1, \ldots, a_n) = (a_{\sigma(1)}, \ldots, a_{\sigma(m)})$ is a p-morphism of a frame $(D_n, \leq_n)$ onto $(\pi_\sigma(D_n), \leq_m)$ (here we suppose that $\pi_\sigma(D_n)$ is a conic subset of D_m; in fact $\pi_\sigma(D_n) = D_m$ for an injective σ; if $m = 0$, i.e. σ is the empty function, we set $\pi_\sigma(a_1, \ldots, a_n) = u$ for any $(a_1, \ldots, a_n) \in (D_u)^n$).

A *valuation* in a metaframe $\mathbb{F}$ is a function ξ sending every n-place predicate letter P to a $\leq_n$-conic subset $\xi(P)$ of D_n. A valuation ξ in $\mathbb{F}$ gives rise to the forcing relation $u \vDash A(\mathbf{a})$ between $u \in F$ and a D_u-formula $A(\mathbf{a})$, $\mathbf{a} = (a_1, \ldots, a_n) \in (D_u)^n$ (where a D_u-formula is a closed formula of the predicate language enriched by elements of D_u as individual constants); $\vDash$ is defined inductively:

$u \vDash P(\mathbf{a}) \Longleftrightarrow \mathbf{a} \in \xi(P)$, where $P \in Pr^n, u \nvDash \bot$;

$u \vDash (B \wedge C)(\mathbf{a}) \Longleftrightarrow (u \vDash B(\mathbf{a})) \wedge (u \vDash C(\mathbf{a}))$;

$u \vDash (B \vee C)(\mathbf{a}) \Longleftrightarrow (u \vDash B(\mathbf{a})) \vee (u \vDash C(\mathbf{a}))$;

$u \vDash (B \to C)(\mathbf{a}) \Longleftrightarrow \forall v \geq u \forall \mathbf{b} \in (D_v)^n ((\mathbf{a} \leq_n \mathbf{b}) \wedge (v \vDash B(\mathbf{b})) \Rightarrow (v \vDash C(\mathbf{b})))$;

$u \vDash \forall x\, B(\mathbf{a}, x) \Longleftrightarrow \forall v \geq u \forall \mathbf{b} \in (D_v)^n \forall c \in D_v ((\mathbf{a} \leq_n \mathbf{b}) \Rightarrow (v \vDash B(\mathbf{b}, c)))$;

$u \vDash \exists x\, B(\mathbf{a}, x) \Longleftrightarrow \exists c \in D_u (u \vDash B(\mathbf{a}, c))$.

A predicate formula A with free variables $\mathbf{x} = (x_1, \ldots, x_n)$ is *true* w.r.t. ξ iff $u \vDash A(\mathbf{a})$ for any $u \in F$, $\mathbf{a} \in (D_u)^n$. A formula A is *valid* in a metaframe $\mathbb{F}$ (notation: $\mathbb{F} \vDash A$) iff it is true w.r.t. any valuation in $\mathbb{F}$.

The *predicate logic of a metaframe* $\mathbb{F}$ is the set $\mathsf{L}(\mathbb{F}) = \{A : \forall m \geq 0, \mathbb{F} \vDash A^{[m]}\}$ (the formulas $A^{[m]}$ are involved here in order to make $\mathsf{L}(\mathbb{F})$ substitution closed, cf. [7]).

The following lemma is rather obvious (see [7], Lemma 2.5.0).

Lemma 2.3 *Let $\mathbb{F}$ be a metaframe. Then the propositional fragment of the logic $\mathsf{L}(\mathbb{F})$ is $\mathsf{PL}(\mathbb{F}) = \bigcap_m \mathsf{PL}(D_m, \leq_m)$ (more precisely $\mathbb{F} \vDash A^{[m]}$ iff $A \in \mathsf{PL}(D_m, \leq_m)$).*

Lemma 2.4 $D \in L(\mathbb{F})$ *if*

$$\forall n \geq 0 \, \forall \mathbf{b} \in D_n \, \forall (\mathbf{c}, c) \in D^{n+1} \, ((\mathbf{b} \leq_n \mathbf{c}) \Rightarrow \exists b((\mathbf{b}, b) \leq_{n+1} (\mathbf{c}, c))).$$

Proof. Let $u \models \forall x \, (P(\mathbf{b}, x) \vee Q(\mathbf{b}))$, $u \not\models \forall x \, P(\mathbf{b}, x) \vee Q(\mathbf{b})$, where $\mathbf{b} \in (D_u)^n$ (for some valuation ξ in $\mathbb{F}$). Then $u \not\models Q(\mathbf{b})$. $v \not\models P(\mathbf{c}, c)$ for some $v \geq u$, $(\mathbf{c}, c) \in (D_v)^{n+1}$ such that $\mathbf{b} \leq_n \mathbf{c}$. Let $(\mathbf{b}, b) \leq_{n+1} (\mathbf{c}, c)$. Then $u \not\models P(\mathbf{b}, b)$ and $u \not\models (P(\mathbf{b}, b) \vee Q(\mathbf{b}))$. This is a contradiction. $\boxtimes$

For a metaframe $\mathbb{F}$ over a frame F and for $u \in F$ the *cone* $\mathbb{F}^u$ is defined in an obvious way, as the restriction of $\mathbb{F}$ to F^u. It is clear that $L(\mathbb{F}) = \cap (L(\mathbb{F}^u) : u \in F)$.

2.3. Recall that a *predicate Kripke frame* (in standard sense) can be defined as $\mathbb{F} = (F, (D'_u : u \in F))$ where $\forall u \leq v (D_u' \subseteq D_v')$. By replacing each D'_u by $D_u = \{u\} \times D'_u$ we obtain a predicate frame in the sense of 2.2. The latter corresponds to a metaframe in which

$$(\langle u, a_1 \rangle, \ldots, \langle u, a_n \rangle) \leq_n (\langle v, b_1 \rangle, \ldots, \langle v, b_n \rangle) \iff ((u \leq v) \wedge \bigwedge_{i=1}^{n} (a_i = b_i)).$$

The predicate logic of this metaframe is $L(\mathbb{F}) = \{A : \mathbb{F} \models A\}$; the same logic can be obtained with the standard definition for predicate Kripke frames. Also note that $PL(\mathbb{F}) = PL(F)$.

A *Kripke bundle* [6] is a triple $\mathbb{F} = (F, \overline{D}, \leq_1)$ in which $(F, \overline{D})$ is a predicate frame (in the sense of 2.2) and $\leq_1$ is a pre-ordering of the set $D = \cup(D_u : u \in F)$ such that a mapping $\pi(a) = u$ for $a \in D_u$ is a p-morphism of the frame $(D, \leq_1)$ onto F. A bundle corresponds to a metaframe in which

$$(a_1, \ldots, a_n) \leq_n (b_1, \ldots, b_n) \quad \text{iff} \quad \bigwedge_{i=1}^{n} (a_i \leq_1 b_i) \wedge \bigwedge_{i<j} ((a_i = a_j) \Rightarrow (b_i = b_j))$$

for $n \geq 1$. It is clear that the definition for the predicate Kripke frames is a particular case of this one: it is sufficient to take $\langle u, a \rangle \leq_1 \langle v, b \rangle$ iff $(u \leq v) \wedge (a = b)$.

Now let us consider the functor semantics [1]. Let $\mathcal{C}$ be a category with the frame representation F. It means that $F = Ob\mathcal{C}$ is the set of objects of $\mathcal{C}$ and $u \leq v$ iff $\mathcal{C}(u, v) \neq \emptyset$ (i.e. there exist morphisms from u to v). A $\mathcal{C} - set$ (or a SET-valued functor over $\mathcal{C}$) is a triple $\mathbb{F} = (F, \overline{D}, E)$ consisting of a predicate frame $(F, \overline{D})$ (in the sense of 2.2) and of a family of functions $E = (E_\mu : \mu \in Mor(\mathcal{C}))$ with $E_\mu : D_u \longrightarrow D_v$ whenever $\mu \in \mathcal{C}(u, v)$ (i.e. μ is a morphism from u to v). As usual it is required that $E_{\mu \circ \mu'} = E_\mu \circ E_{\mu'}$

and that $E_{1_u} = 1_{D_u}$ (the identical function on D_u corresponds to the identical morphism $1_u \in \mathcal{C}(u, u)$, $u \in F$). A $\mathcal{C}$-set corresponds to a metaframe in which

$$(a_1, \ldots, a_n) \leq_n (b_1, \ldots, b_n) \text{ iff } \exists\, \mu \in \mathrm{Mor}(\mathcal{C}) \, (\bigwedge_{i=1}^{n} (E_\mu(a_i) = b_i)).$$

The bundle semantics is a particular case of the functor semantics: a bundle $\mathbb{F}$ corresponds to a category $\mathcal{C}$ and a $\mathcal{C}$-set $\mathbb{F}^0$ such that $\mathsf{L}(\mathbb{F}^0) = \mathsf{L}(\mathbb{F})$ (see [7], §1, 1.3). Note also that the functor semantics is equivalent to the metaframe semantics for the countable case: speaking more precisely, every metaframe $\mathbb{F}$ with at most countable domains (i.e. $\forall u \in F$, $\mathrm{Card}(D_u) \leq \aleph_0$) corresponds to some $\mathcal{C}$-set (see [7] Theorem 4.4.1).

2.4 Finally let us consider some properties of bundles which will be used in Section 4 (the following two lemmas may be skipped by the readers who are only interested in Section 3).

Lemma 2.5 *Let $\mathbb{F}$ be a bundle such that $D \in \mathsf{L}(\mathbb{F})$, $u \in F$, $b_1, b_2 \in D_u$, $b_1 \neq b_2$, $b_1 \leq_1 b_2$. Then for any $n > 0$ there exist $v_n \in F$ and n distinct elements $c_i^n \in D_{v_n}$ such that $u \approx v_n$ and $c_i^n \leq_1 b_1$ for $i \leq n$.*

Proof. By induction on n. For the given $v_n (\approx u)$ and $\mathbf{c} = (c_1^n, \ldots, c_n^n) \in (D_{v_n})^n$ take $\mathbf{d} = (d_1^n, \ldots, d_n^n) \in (D_u)^n$ such that $\mathbf{c} \leq_n \mathbf{d}$, $b_1 \notin \{d_1^n, \ldots, d_n^n\}$ (namely: $d_i^n = b_2$ whenever $c_i^n \leq_1 b_1$).

Let $D^n(\mathbf{c}) = \forall x (P(\mathbf{c}, x) \vee Q(\mathbf{c})) \to \forall x P(\mathbf{c}, x) \vee Q(\mathbf{c})$. Fix a valuation $\xi(Q) = \{\mathbf{a} \in D_n : \neg(\mathbf{a} \leq_n \mathbf{c})\}$. $\xi(P) = \{(\mathbf{a}, a) \in D_{n+1} : \neg((\mathbf{a}, a) \leq_{n+1} (\mathbf{d}, b_1))\}$. Then $v_n \nVdash \forall x \, (P(\mathbf{c}, x) \vee Q(\mathbf{c}))$ since $v_n \nVdash \forall x P(\mathbf{c}, x) \vee Q(\mathbf{c})$. Thus there exist $v_{n+1} \geq v_n$ and $(\mathbf{e}, e) \in (D_{v_{n+1}})^{n+1}$ such that $\mathbf{e} \leq_n \mathbf{c}$, $(\mathbf{e}, e) \leq_{n+1} (\mathbf{d}, b_1)$ and so $v_{n+1} \leq v_n$ and all $n + 1$ components of $(\mathbf{e}, e)$ are distinct elements from $D_{v_{n+1}}$. $\boxtimes$

A bundle $\mathbb{F}$ is called a *quasisheaf* [6] iff $\forall u \in F \, \forall b_1, b_2 \in D_u ((b_1 \leq_1 b_2) \Rightarrow (b_1 = b_2))$ (i.e. if the premise of Lemma 2.5 is never satisfied for $\mathbb{F}$). If $\mathbb{F}$ is a quasisheaf then $\forall n > 0 \, \forall u, v \in F \, \forall \mathbf{c} \in (D_u)^n \, \forall \mathbf{d} \in (D_v)^n \, ((u \approx v) \wedge (\mathbf{c} \leq_n \mathbf{d}) \Rightarrow (\mathbf{d} \leq_n \mathbf{c}))$.

Lemma 2.6 *Let $\mathbb{F}$ be a bundle such that $D \in \mathsf{L}(\mathbb{F})$ and $n > 0$.*

(1) If $J \notin \mathsf{L}(\mathbb{F})$ then $\mathbb{F} \nVdash J^{[1]}$.
(2) If $Z \notin \mathsf{L}(\mathbb{F})$ then $\mathbb{F} \nVdash Z^{[3]}$.
(3) If $P_n \notin \mathsf{L}(\mathbb{F})$ then $\mathbb{F} \nVdash (P_n)^{[n+1]}$.

Proof. (1) we have to show that the condition $\forall a \in D\,(S(D^a) \in \mathcal{F}_J)$ implies $\forall n > 0\,\forall \mathbf{a} \in D_n\,(S((D_n)^{\mathbf{a}}) \in \mathcal{F}_J)$. This follows from the claim:

if $(\mathbf{a}, \mathbf{a}') \leq_n (\mathbf{b}, \mathbf{b}')$ and $(\mathbf{a}, \mathbf{a}') \leq_n (\mathbf{b}, \mathbf{b}'')$

then $(\mathbf{b}, \mathbf{b}') \leq_n (\mathbf{c}, \mathbf{c}')$ and $(\mathbf{b}, \mathbf{b}'') \leq_n (\mathbf{c}, \mathbf{c}')$ for some $(\mathbf{c}, \mathbf{c}')$,

which is proved by induction on the length of $\mathbf{a}'$.

(2) If $\mathbb{F}$ is not a quasisheaf then (by Lemma 2.5) there exist $v \in F$ and distinct $c_0, c_1, c_2 \in D_v$ such that $c_1 \leq_1 c_0$, $c_2 \leq_1 c_0$. Let $\mathbf{c} = (c_0, c_1, c_2)$. Then $S((D_3)^{\mathbf{c}}) \notin \mathcal{F}_Z$, since $\mathbf{c} \leq_3 (c_0, c_1, c_0)$ and $\mathbf{c} \leq_3 (c_0, c_0, c_2)$. If $\mathbb{F}$ is a quasisheaf and $\exists a \in D_u\,\exists b_1, b_2 \in D_v\,(a \leq_1 b_1, a \leq_1 b_2, b_1 \neq b_2)$ for some $u \leq v$, then $S(D^a) \notin \mathcal{F}_Z$. Finally, if $\forall u \leq v\,\forall a \in D_u\,\exists! b \in D_v\,(a \leq_1 b)$ and $S((D_k)^{\mathbf{c}}) \notin \mathcal{F}_Z$ for some $\mathbf{c} \in (D_u)^k$, then $S(F^u) \notin \mathcal{F}_Z$.

(3) If $\mathbb{F}$ is not a quasisheaf then (by Lemma 2.5) there exists $v \in F$ and distinct $c_0, c_1, \ldots, c_n \in D_v$ such that $\bigwedge_{i=1}^n (c_i \leq_1 c_0)$. Let $\mathbf{c} = (c_0, c_1, \ldots, c_n)$. Then $S((D_{n+1})^{\mathbf{c}}) \notin \mathcal{F}_{P_n}$, since $\mathbf{c} <_{n+1} (c_0, c_0, c_2, \ldots, c_n) <_{n+1} \cdots <_{n+1} (c_0, c_0, \ldots, c_0, c_n) <_{n+1} (c_0, c_0, \ldots, c_0, c_0)$. On the other hand, if $\mathbb{F}$ is a quasisheaf and $S((D_k)^{\mathbf{c}}) \notin \mathcal{F}_{P_n}$ for some $\mathbf{c} \in (D_u)^k$, then $S(F^u) \notin \mathcal{F}_{P_n}$. $\boxtimes$

Remark. If we omit the condition $D \in \mathsf{L}(\mathbb{F})$ (for a bundle $\mathbb{F}$) then the following holds:

(1) $J \notin \mathsf{L}(\mathbb{F}) \Rightarrow \mathbb{F} \nvdash J^{[1]}$;

(2) $Z \notin \mathsf{L}(\mathbb{F}) \Rightarrow \mathbb{F} \nvdash Z^{[4]}$;

(3) $P_n \notin \mathsf{L}(\mathbb{F}) \Rightarrow \mathbb{F} \nvdash P_n^{[2n]}\,(n > 0)$.

The proof of (2) and (3) is rather laborious and is not considered here.

3   Proofs of the main theorems

Theorem 1 is a consequence of the following

Proposition 3.1 *Let A and A' be propositional formulas such that* $\mathsf{PH} \nvdash A'$ *and* $(\mathsf{PH}+R_2) \vdash A$. *Then* $(\mathsf{H}+(A^{[m']} \vee K^{[m]})) \nvdash (A' \vee K^{[m+1]})$ *for any $m, m' \geq 0$.*

In fact, if $(\mathsf{H} + A) \cap (\mathsf{H} + K) = (\mathsf{H} + B)$ for some propositional formula $A \in (\mathsf{PH}+R_2)-\mathsf{PH}$ and some predicate formula B, then $(\mathsf{H}+(A^{[m]} \vee K^{[m]})) \vdash B$ for some $m \geq 0$ and hence $(\mathsf{H}+(A^{[m]} \vee K^{[m]})) \vdash (A \vee K^{[m+1]})$ which contradicts Proposition 3.1 (with $A' = A$).

The previous proposition becomes false if we replace K by D, since one can observe that $(\mathsf{H} + (A \vee D)) \vdash (A \vee D^{[m]})$ for any predicate formula A (having no common symbols with D). So for the proof of Theorem 2 we apply

Proposition 3.2 *Let A be a propositional formula (without occurrences of Q) such that $\mathsf{P} \nvdash A$ and $(\mathsf{PH} + R_2) \vdash A$. Then $(\mathsf{H} + (A^{[m]} \vee D^{[m']})) \nvdash (A^{[m+1]} \vee D)$ for any $m, m' \geq 0$.*

Similarly Theorem 3 follows from

Proposition 3.3 *Let A be a propositional formula (without occurrences of Q) such that $(\mathsf{PH} + J) \nvdash A$ and $(\mathsf{PH} + A) \nvdash J$. Then $(\mathsf{H} + (A^{[m]} \vee J^{[m']})) \nvdash (A^{[m+1]} \vee J)$ for any $m, m' \geq 0$.*

The proof of Propositions 3.2 and 3.3 is more difficult than the proof of Proposition 3.1 (because we have to "separate" $A^{[m+1]}$ from $A^{[m]}$ for an arbitrary propositional formula A).

3.2   For the sake of clarity we will give the proof of Proposition 3.1 for a particular case $A' = J$ and then show how to modify it for an arbitrary A'.

Fix a countable set $D_0 = \{b_i : i \in \omega\}$ and let T_0 be the set of one-to-one functions $\mu : D_0 \longrightarrow D_0$ such that $\{b_0, b_1, \ldots, b_{m+1}\} \not\subseteq \mathrm{Val}(\mu)$. Define a category $\mathcal{C}$ whose frame representation is F_2 (see Figure 1) and $\mathcal{C}(u_2, u_2) = \{1_{u_2}\} \cup T_0$ (otherwise $\mathcal{C}(u, v)$ is one-element). Consider a $\mathcal{C}$-set $\mathbb{F} = (F_2, \overline{D}, E)$ such that $D_{u_i} = \{a_i\}$ for $i \leq 1$, $D_{u_2} = \{a_2\} \cup D_0$, $E_\mu(a_2) = a_2$, $E_\mu \upharpoonright_{D_0} = \mu$ for $\mu \in T_0$ and $E_\mu(a_0) = a_2$ for $\mu \in \mathcal{C}(u_0, u_2)$. It is clear that for any $\mathbf{b}, \mathbf{b}' \in (D_0)^{m+1} : (\mathbf{b} \leq_{m+1} \mathbf{b}') \Longleftrightarrow (\mathbf{b}' \leq_{m+1} \mathbf{b})$.

Let us show that $\mathbb{F} \nvDash (J \vee K^{[m+1]}) = \neg Q \vee \neg\neg Q \vee \forall \mathbf{y} \, \neg\neg\forall x \, (P(x, \mathbf{y}) \vee \neg P(x, \mathbf{y}))$ (here $\mathbf{y} = (y_1, \ldots, y_{m+1})$). Fix $\mathbf{b} = (b_1, \ldots, b_{m+1}) \in (D_{u_2})^{m+1}$ and consider a valuation $\xi(Q) = \{u_2\}$, $\xi(P) = (D_{u_2})^{m+2} - \{(b_0, b_1, \ldots, b_{m+1})\}$ (the latter is obviously $\leq_{m+2}$-conic). Then $u_0 \nvDash \neg Q \vee \neg\neg Q$ and $u_2 \vDash \neg\forall x \, (P(x, \mathbf{b}) \vee \neg P(x, \mathbf{b}))$ (since $\mathbf{b} \leq_{m+1} \mathbf{b}' \Rightarrow \mathbf{b}' \leq_{m+1} \mathbf{b}$ and $u_2 \nvDash P(b_0, \mathbf{b})$, $u_2 \nvDash \neg P(b_0, \mathbf{b})$), thus $u_0 \nvDash \forall \mathbf{y} \, \neg\neg\forall x \, (P(x, \mathbf{y}) \vee \neg P(x, \mathbf{y}))$.

Let us show that

$$\mathbb{F} \vDash (A^{[m']} \vee K^{[m]})^k = \forall \mathbf{z}' \, A^{[m'+k]}(\mathbf{z}, \mathbf{z}') \vee \forall \mathbf{y} \, \neg\neg\forall x \, (P(x, \mathbf{y}, \mathbf{z}) \vee \neg P(x, \mathbf{y}, \mathbf{z}))$$

(here $\mathbf{z} = (z_1, \ldots, z_k)$, $\mathbf{z}' = (z_1', \ldots, z_{m'}')$, $\mathbf{y} = (y_1, \ldots, y_m)$). First, observe that $\mathbb{F}^{u_2} \vDash A^{[m'+k]}$ since $(\mathsf{PH} + R_2) \vdash A$ and for any $\mathbf{c} = (c_1, \ldots, c_{m'+k}) \in (D_{u_2})^{m'+k}$ the skeleton of the cone $(D_{m'+k})^{\mathbf{c}}$ is isomorphic to F_1 if $\{b_0, \ldots, b_{m+1}\} \subseteq \{c_1, \ldots, c_{m'+k}\}$ and to F_0 otherwise (see Figure 1). Thus it is sufficient to show that $u_i \vDash \forall \mathbf{y} \, \neg\neg\forall x \, (P(x, \mathbf{y}, \mathbf{a}_i) \vee \neg P(x, \mathbf{y}, \mathbf{a}_i))$ for any valuation ξ in $\mathbb{F}$ and $i \leq 1$, $\mathbf{a}_i = (a_i, \ldots, a_i) \in (D_{u_i})^k$. The case $i = 1$ is obvious. Take $i = 0$ and suppose the contrary. Then $u_2 \vDash \neg\forall x (P(x, \mathbf{c}, \mathbf{a}_2) \vee \neg P(x, \mathbf{c}, \mathbf{a}_2))$ for some $\mathbf{c} \in (D_{u_2})^m$ and thus $u_2 \nvDash (P(c, \mathbf{c}, \mathbf{a}_2) \vee \neg P(c, \mathbf{c}, \mathbf{a}_2))$ for some $(c, \mathbf{c}) \in (D_{u_2})^{m+1}$, which is impossible, since

$$(c, \mathbf{c}, \mathbf{a}_2) \leq_{m+k+1} (c', \mathbf{c}', \mathbf{a}_2) \Rightarrow (c', \mathbf{c}', \mathbf{a}_2) \leq_{m+k+1} (c, \mathbf{c}, \mathbf{a}_2).$$

A modification for arbitrary A' is trivial. As $\mathsf{PH} \not\vdash A'$ we can take a frame F from $\mathcal{F}_H^f$ such that $A' \notin \mathsf{PL}(F)$. Then we repeat the previous reasoning, using F instead of F_2 and some fixed maximal element of F instead of u_2.

3.3 To prove Propositions 3.2 and 3.3. we need some additional constructions. For a set X and $k \in \omega$, let $P_k(X) = \{Y \subseteq X : \mathrm{card}(Y) = k\}$ (therefore $P_0(X) = \{\emptyset\}$ and $P_k(X) = \emptyset$ iff $\mathrm{card}(X) < k$). An *m-colouring* is a triple $\rho = (X, Z, \lambda)$ in which X and Z are sets and λ is a function from $P_{m+1}(X)$ to Z. For $k \leq \omega$ we define the *index set* $I_k = \{i \in \omega : i < k\}$ (that is. $I_\omega = \omega$, $I_0 = \emptyset$ and $I_k = \{0, 1, \ldots, k-1\}$ for $k \in \omega$, $k \neq 0$). Let I be an index set, $a_i \in X$ for $i \in I$ and $\forall i \neq j \, (a_i \neq a_j)$; then $\mathbf{a} = (a_i : i \in I)$ is called an I, X-*sequence*. For an I, X-sequence $\mathbf{a}$ and m-colouring $\rho = (X, Z, \lambda)$ we define a function $\lambda_{\mathbf{a}} : P_{m+1}(I) \longrightarrow Z$ by $\lambda_{\mathbf{a}}(\{i_0, \ldots, i_m\}) = \lambda(\{a_{i_0}, \ldots, a_{i_m}\})$. Let $I \subseteq I'$ be two index sets and $\lambda' : P_{m+1}(I') \longrightarrow Z$; then $(\lambda'|I)$ denotes the restriction of λ' to the set $P_{m+1}(I)$.

An m-colouring $\rho = (X, Z, \lambda)$ is called *correct* if the set X is countable and the following holds:

let I, I' be index sets such that $I \subseteq I'$ and let $\lambda' : P_{m+1}(I') \longrightarrow Z$; then every I, X-sequence $\mathbf{a} = (a_i : i \in I)$ such that $\lambda_{\mathbf{a}} = (\lambda'|I)$ can be prolonged to an I', X-sequence $\mathbf{a}' = (a_i : i \in I')$ such that $\lambda_{\mathbf{a}'} = \lambda'$.

This condition in fact can be reduced to the case when $\mathrm{card}(I' - I) = 1$ (i.e. $I = I_k$, $I' = I_{k+1}$, $k \in \omega$). This condition is vacuous for $k < m$ since $P_{m+1}(I_{k+1}) = \emptyset$ if $k < m$ and for $k = m$ it means that for every I_m, X-sequence $\mathbf{a}$ and every $v \in Z$ there exists an I_{m+1}, X-sequence $(\mathbf{a}, a)$ such that $\lambda(\mathbf{a}, a) = v$ (and thus λ is an onto map).

Lemma 3.4 *Let Z be a finite set, $m \in \omega$. Then a correct m-colouring $\rho = (X, Z, \lambda)$ exists.*

Proof. Let $X = \bigcup(X_n : n \in \omega)$ be such that $\mathrm{card}(X_0) = m$, $X_{n+1} = X_n \cup X_n'$, $X_n' = \{[k, \mathbf{a}, \lambda^*] : m \leq k \in \omega, \mathbf{a} \text{ is an } I_k, X_n\text{-sequence}, \lambda^* : P_m(I_k) \longrightarrow Z\}$ and let $\lambda(Y) = \lambda^*(\{i_1, \ldots, i_m\})$ if $Y = \{a, a_{i_1}, \ldots, a_{i_m}\}$, $a = [k, \mathbf{a}, \lambda^*] \in X_n'$, $\mathbf{a} = \{a_i : i < k\}$, $\{i_1, \ldots, i_m\} \in P_m(I_k)$ (that is $\{a_{i_1}, \ldots, a_{i_m}\} \in P_m(X_n)$), arbitrary otherwise.

This m-colouring is correct. In fact, for $k \geq m$, $\mathbf{a} = (a_i : i < k) \in (X_n)^k$, $\lambda' : P_{m+1}(I_{k+1}) \longrightarrow Z$, $(\lambda'|I_k) = \lambda_{\mathbf{a}}$ we can put: $\lambda^*(\{i_1, \ldots, i_m\}) = \lambda'(\{i_1, \ldots, i_m, k\})$ (for $\{i_1, \ldots, i_m\} \in P_m(I_k)$), $a_k = [k, \mathbf{a}, \lambda^*] \in X_n'$ and $\mathbf{a}' = (a_i : i \leq k)$. $\boxtimes$

Let $\rho = (X, Z, \lambda)$ be an m-colouring and Z be a frame with pre-ordering $\leq$. We define the set T_ρ of all one-to-one functions $\mu : X \longrightarrow X$ such that

$$\forall Y = \{a_i : 0 \leq i \leq m\} \in P_{m+1}(X)\,(\lambda(Y) \leq \lambda(\mu Y))$$

where $\mu Y = \{\mu(a_i) : 0 \leq i \leq m\} \in P_{m+1}(X)$. It is clear that T_ρ is closed under composition and $1_X \in T_\rho$. For $k > 0$ we define a relation $\leq_{k\rho}$ on X^k:

$$(a_1, \ldots, a_k) \leq_{k\rho} (b_1, \ldots, b_k) \Longleftrightarrow \exists \mu \in T_\rho(\bigwedge_{i=1}^{k} (\mu(a_i) = b_i)).$$

Lemma 3.5 *Let $\rho = (X, Z, \lambda)$ be a correct m-colouring. Then $(a_0, \ldots, a_{k-1}) \leq_{k\rho} (b_0, \ldots, b_{k-1})$ iff*

(i) $\forall i < j\,((a_i = a_j) \Leftrightarrow (b_i = b_j))$ *and*

(ii) *for every $\{i_0, \ldots, i_m\} \in P_{m+1}(I_k)$, if $\{a_{i_0}, \ldots, a_{i_m}\} \in P_{m+1}(X)$ then $\lambda(\{a_{i_0}, \ldots, a_{i_m}\}) \leq \lambda(\{b_{i_0}, \ldots, b_{i_m}\})$ in Z.*

In particular, for $k \leq m$

$$(a_0, \ldots, a_{k-1}) \leq_{k\rho} (b_0, \ldots, b_{k-1}) \Longleftrightarrow \forall i < j\,((a_i = a_j) \Leftrightarrow (b_i = b_j)),$$

and for $k = (m+1)$, $\{a_0, \ldots, a_m\} \in P_{m+1}(X)$, $\{b_0, \ldots, b_m\} \in P_{m+1}(X)$ we have:

$$(a_0, \ldots, a_m) \leq_{(m+1)\rho} (b_0, \ldots, b_m) \Longleftrightarrow (\lambda(\{a_0, \ldots, a_m\}) \leq \lambda(\{b_0, \ldots, b_m\})\ \text{in } Z).$$

Proof. Suppose that (i) and (ii) hold. We may assume that $\forall i < j\,((a_i \neq a_j) \wedge (b_i \neq b_j))$. Enumerate the set X as an I_ω-sequence $\mathbf{a} = \{a_i : i \in \omega\}$, beginning from our a_i, $i < k$. Consider a function $\lambda' : P_{m+1}(I_\omega) \longrightarrow Z$ such that $\lambda'(\{i_0, \ldots, i_m\}) = \lambda(\{b_{i_0}, \ldots, b_{i_m}\})$ if $\forall j \leq m\,(i_j < k)$ and $\lambda'(\{i_0, \ldots, i_m\}) = \lambda(\{a_{i_0}, \ldots, a_{i_m}\})$ otherwise. Since $(\lambda'|I_k) = \lambda_{\mathbf{b}'}$ (where $\mathbf{b}' = (b_0, \ldots, b_{k-1})$) and ρ is correct, there exists an I_ω, X-sequence $\mathbf{b} = (b_i : i \in \omega)$ (beginning from our b_i, $i < k$) and such that $\lambda(\{b_{i_0}, \ldots, b_{i_m}\}) = \lambda'(\{i_0, \ldots, i_m\}) \geq \lambda(\{a_{i_0}, \ldots, a_{i_m}\})$ for all $\{i_0, \ldots, i_m\} \in P_{m+1}(\omega)$. Thus the function $\mu(a_i) = b_i$ ($i \in \omega$) belongs to T_ρ.   ⊠

3.4. Proof of Proposition 3.2   Fix a frame $F \in \mathcal{F}_H^f$ with least element v_0 such that $A \notin \mathsf{PL}(F)$ and a correct m-colouring $\rho = (X, F, \lambda)$. Consider a category $\mathcal{C}$ whose frame representation is F_1 (see Figure 1), $\mathcal{C}(u_1, u_1) = T_\rho$ (and one-element $\mathcal{C}(u_0, u_i)$ for $i \leq 1$). Consider a $\mathcal{C}$-set $\mathbb{F} = (F, \overline{D}, E)$ such that $D_{u_0} = \{a_0\}$, $D_{u_1} = \{a_1\} \cup X$. $E_\mu(a_1) = a_1$, $E_\mu|X = \mu$ for $\mu \in T_\rho$ and $E_\mu(a_0) = a_i$ for $\mu \in \mathcal{C}(u_0, u_i)$ ($i \leq 1$).

Let us show that $\mathbb{F} \nvDash \forall \mathbf{x}\, A^{m+1}(\mathbf{x}) \vee D$, $\mathbf{x} = (x_0, x_1, \ldots, x_m)$. Take $\mathbf{b} = (b_0, b_1, \ldots, b_m) \in (D_{u_1})^{m+1}$ such that $\lambda(\{b_0, b_1, \ldots, b_m\}) = v_0$. Then $u_1 \nvDash A^{m+1}(\mathbf{b})$ (for some valuation in $\mathbb{F}$) by Lemma 2.3, since the skeleton of the cone $(D_{m+1})^{\mathbf{b}}$ is isomorphic to F (see Lemma 3.5). Also consider a valuation ξ such that $\xi(Q) = \{u_1\}$, $\xi(P) = \{a_0, a_1\}$, then $u_0 \nvDash D$ under this valuation (since $u_o \vDash \forall x\, (P(x) \vee Q)$, $u_0 \nvDash \forall x\, P(x)$, $u_0 \nvDash Q$).

On the other hand let us show that $\mathbb{F} \vDash \forall \mathbf{x}\, A^{m+k}(\mathbf{x}, \mathbf{z}) \vee \forall \mathbf{y}\, D^{m'+k}(\mathbf{y}, \mathbf{z})$ (here $\mathbf{x} = (x_1, \ldots, x_m)$, $\mathbf{y} = (y_1, \ldots, y_{m'})$, $\mathbf{z} = (z_1, \ldots, z_k)$).

First we prove that $D \in \mathsf{L}(\mathbb{F}^{u_1})$ using Lemma 2.4. Let $\mathbf{b} = (b_0, \ldots, b_{n-1}) \in (D_{u_1})^n$, $(\mathbf{c}, c_n) = (c_0, \ldots, c_{n-1}, c_n) \in (D_{u_1})^{n+1}$, $\mathbf{b} \leq_n \mathbf{c}$. Since $\forall i\, ((b_i = a_1) \Leftrightarrow (c_i = a_1))$ and $\forall i < j\, ((b_i = b_j) \Leftrightarrow (c_i = c_j))$, it is sufficient to assume that $\forall i\, ((b_i \in X) \wedge (c_i \in X))$ and $\forall i < j\, ((b_i \neq b_j) \wedge (c_i \neq c_j))$. Define a function $\lambda': P_{m+1}(I_{n+1}) \longrightarrow Z$ as follows:

$$\lambda'(\{i_0, \ldots, i_m\}) = \begin{cases} \lambda(\{b_{i_0}, \ldots, b_{i_m}\}) & \text{if } \forall j\, (i_j \neq n) \\ \lambda(\{c_{i_0}, \ldots, c_{i_m}\}) & \text{if } \exists j\, (i_j = n) \end{cases}$$

Since ρ is correct and $(\lambda'|I_n) = \lambda_{\mathbf{b}}$, there exists an I_{n+1}, X-sequence $\mathbf{b}' = (\mathbf{b}, b_n)$ such that $\lambda_{\mathbf{b}'} = \lambda'$. Then $(\mathbf{b}, b_n) \leq_{n+1} (\mathbf{c}, c_n)$ by Lemma 3.5.

Now it is sufficient to show that $u_0 \vDash \forall \mathbf{x}\, A^{m+k}(\mathbf{x}, \mathbf{a}_0)$ for an arbitrary valuation in $\mathbb{F}$ and $\mathbf{a}_0 = (a_0, \ldots, a_0) \in (D_{u_0})^k$. This follows from Lemma 2.3. In fact $u_i \vDash A^{m+k}(\mathbf{a}'_i)$ for $i \leq 1$ where $\mathbf{a}'_0 = (a_0, \ldots, a_0) \in (D_{u_0})^{m+k}$, $\mathbf{a}'_1 = (b_1, \ldots, b_m, a_1, \ldots, a_1) \in (D_{u_1})^{m+k}$ since the skeletons of the cones $(D_{m+k})^{\mathbf{a}'_i}$ are isomorphic to F_{1-i} (by Lemma 3.5) and $(\mathsf{PH} + R_2) \vdash A$. $\boxtimes$

3.5. Proof of Proposition 3.3

By Lemmas 2.2 and 2.1, $(\mathsf{PH} + A) \nvdash J$ implies that $A \in \mathsf{PL}(F_2)$ (Figure 1) and $(\mathsf{PH} + J) \nvdash A$ implies that $A \notin \mathsf{PL}(F)$ for some frame F from $\mathcal{F}_J^f$ with the greatest element v_1. Take a correct m-colouring $\rho = (X, F, \lambda)$ and consider a category $\mathcal{C}$ whose frame representation is F_2, such that $\mathcal{C}(u_2, u_2) = T_\rho$ and the remaining $\mathcal{C}(u, v)$ (for $u \leq v$) are one-element. Consider a $\mathcal{C}$-set $\mathbb{F} = (F_2, \overline{D}, E)$ such that $D_{u_i} = \{a_i\}$ for $i \leq 1$, $D_{u_2} = \{a_2\} \cup X$, $E_\mu|X = \mu$ for $\mu \in T_\rho$ and $E_\mu(a_i) = a_j$ for $\mu \in \mathcal{C}(u_i, u_j)$, $u_i \leq u_j$. Then we obtain that $\mathbb{F} \nvDash A^{[m+1]} \vee J$ since F is a skeleton of some cone in $(D_{u_2})^{m+1}$ (cf. the proof of Proposition 3.2).

Now let us show that $\mathbb{F} \vDash \forall \mathbf{x}\, A^{m+k}(\mathbf{x}, \mathbf{z}) \vee \forall \mathbf{y}\, J^{m'+k}(\mathbf{y}, \mathbf{z})$, $\mathbf{x} = (x_1, \ldots, x_m)$, $\mathbf{y} = (y_1, \ldots, y_{m'})$, $\mathbf{z} = (z_1, \ldots, z_k)$. Since $A \in \mathsf{PL}(F_2)$ then $u_i \vDash \forall \mathbf{x}\, A^{m+k}(\mathbf{x}, \mathbf{a}_i)$ for $i \leq 1$, $\mathbf{a}_i = (a_i, \ldots, a_i) \in (D_{u_i})^k$ and an arbitrary valuation ξ in $\mathbb{F}$ (cf. the proof of Proposition 3.2). It remains to prove that $J \in \mathsf{PL}(\mathbb{F}^{u_2})$. It is sufficient to show that every cone in $(D_{u_2})^n$ has a greatest element. Let $\mathbf{b} = (b_0, \ldots, b_{n-1}) \in (D_{u_2})^n$. Without loss of generality we may assume that $n > m$, $\forall i\, (b_i \in X)$ and $\forall i < j\, (b_i \neq b_j)$. Since ρ is correct, there exists $\mathbf{c} =$

$(c_0, \ldots, c_{n-1}) \in (D_{u_2})^n$ such that $\forall i < j \, (c_i \neq c_j)$ and $\lambda(\{c_{i_0}, \ldots, c_{i_m}\}) = v_1$ for any $\{i_0, \ldots, i_m\} \in P_{m+1}(I_n)$. Thus, by Lemma 3.5, $\forall \mathbf{d}((\mathbf{b} \leq_n \mathbf{d}) \Rightarrow (\mathbf{d} \leq_n \mathbf{c}))$. ⊠

3.6. Remark From Proposition 3.1 one can easily deduce the following stronger version of Theorem 1: if L is a predicate logic, $\mathsf{L} \subseteq (\mathsf{H} + R_2)$ and $\mathsf{PL} \neq \mathsf{PH}$, then the predicate logic $\mathsf{L} \cap (\mathsf{H} + K)$ is not finitely axiomatizable. And similarly one can prove an analogous modification for $\mathsf{L} \cap (\mathsf{H} + D)$.

4   Some incompleteness results

4.1.   The results of Section 3 allow us to obtain numerous examples of predicate logics which are incomplete in Kripke-type semantics. For the standard Kripke semantics incompleteness comes easily.

Lemma 4.1 *Let A and B be closed predicate formulas without common letters and $\mathbb{F}$ be a predicate Kripke frame (in the sense of 2.3). If $(A \vee B) \in \mathsf{L}(\mathbb{F})$ then $(\mathsf{H} + A) \cap (\mathsf{H} + B) \subseteq \mathsf{L}(\mathbb{F})$.*

Proof. If $\mathbb{F}^u \vDash A \vee B$ then $\mathbb{F}^u \vDash A$ or $\mathbb{F}^u \vDash B$.   ⊠

Therefore if $(\mathsf{H} + A) \cap (\mathsf{H} + B) \neq (\mathsf{H} + (A \vee B))$ then the predicate logic $\mathsf{H} + (A \vee B)$ is incomplete w.r.t. the traditional Kripke semantics.

Now let us discuss incompleteness in the semantics of bundles.

4.2   First consider the logic $(\mathsf{H} + J) \cap (\mathsf{H} + K)$ (cf. 3.2). The following lemma shows that this logic can be axiomatized by formulas $(J \vee K^{[m]})$, $m \geq 0$.

Lemma 4.2 $(\mathsf{H} + (J \vee K^{[m]})) \vdash (J^{[n]} \vee K^{[m]})$ *and therefore:* $(\mathsf{H} + (J^{[m']} \vee K^{[m]})) \vdash (J^{[n']} \vee K^{[n]})$ *iff* $n \leq m$ *(see Proposition 3.1).*

Proof. Since $\mathsf{H} \vdash (\neg \neg K^{[m]} \to K^{[m]})$ we have: $(\mathsf{H} + (J \vee K^{[m]})) \vdash (\neg K^{[m]} \vee K^{[m]})$. On the other hand, by substitution in $(J \vee K^{[m]})$ we obtain $\forall \mathbf{z} \, (J^n(\mathbf{z}) \vee K^{[m]})$, $\mathbf{z} = (z_1, \ldots, z_n)$, which implies $(\neg K^{[m]} \to J^{[n]})$. Hence $(\mathsf{H} + (J \vee K^{[m]})) \vdash (J^{[n]} \vee K^{[m]})$.   ⊠

Theorem 4 *Let $\mathbb{F}$ be a bundle. If $(J \vee K^{[1]}) \in \mathsf{L}(\mathbb{F})$ then $(J \vee K^{[m]}) \in \mathsf{L}(\mathbb{F})$ for all $m \in \omega$.*

Therefore the predicate logic $(\mathsf{H} + (J \vee K^{[1]}))$ (and $\mathsf{H} + (J \vee K^{[m]})$ for every $m \geq 1$) is incomplete w.r.t. the semantic of bundles. Note that $(J \vee K) \in \mathsf{L}(\mathbb{F})$ and $(J \vee K^{[1]}) \notin \mathsf{L}(\mathbb{F})$ for the following bundle:

$$\mathbb{F}_0 = (F_2, \overline{D}, \leq_1)$$
$$D_{u_2} = \{a_2, b_0, b_1\}$$
$$D_{u_i} = \{a_i\}$$
$$((a_i \leq_1 a_j) \Longleftrightarrow (u_i \leq u_j))$$
$$(b_i \leq_1 b_{i'}) \quad \text{for } i, i' \leq 1, j \leq 2 \quad \text{(Figure 3)}$$

We do not know whether the logic $(\mathsf{H} + (J \vee K))$ is complete w.r.t. the semantics of bundles.

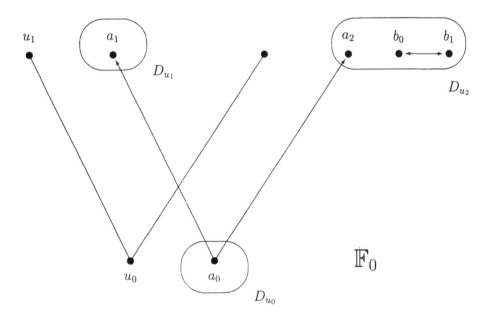

Figure 3.

Proof. We consider only the case $m = 2$ (the proof in the general case is similar). Assume that $u_0 \nvDash (J^k(\mathbf{c}^0) \vee \forall \mathbf{y} \, K^{k+2}(\mathbf{y}, \mathbf{c}^0))$ for some valuation ξ, $\mathbf{y} = (y_0, y_1)$, $\mathbf{c}^0 = (c_1^0, \ldots, c_k^0) \in (D_{u_0})^k$. Then there exists $u_1 \geq u_0$, $\mathbf{c}^1 = (c_1^1, \ldots, c_k^1) \in (D_{u_1})^k$, $\mathbf{d}^1 = (d_0^1, d_1^1) \in (D_{u_1})^2$ such that $\mathbf{c}^0 \leq_k \mathbf{c}^1$, $u_1 \vDash \neg \forall x \, (P(x, \mathbf{d}^1, \mathbf{c}^1) \vee \neg P(x, \mathbf{d}^1, \mathbf{c}^1))$. Now consider the two cases.
(I) Suppose that $J \notin \mathsf{L}(\mathbb{F}^{u_1})$ i.e. $u_2 \nvDash J^n(\mathbf{e})$ for some $n \in \omega$, $u_2 \geq u_1$, $\mathbf{e} \in (D_{u_2})^n$ and some valuation ξ'. Then $u_1 \nvDash (\forall \mathbf{z} \, J^{n+k+2}(\mathbf{z}, \mathbf{d}^1, \mathbf{c}^1) \vee K^{k+2}(\mathbf{d}^1, \mathbf{c}^1))$ for an appropriate valuation ξ'', that is $(J^{[n]} \vee K) \notin \mathsf{L}(\mathbb{F}^{u_1})$. Therefore $(J \vee K) \notin \mathsf{L}(\mathbb{F}^{u_1})$ by Lemma 4.2.

(II) Suppose that $J \in \mathsf{L}(\mathbb{F}^{u_1})$. Then all skeletons of cones in $(D_n, \leq_n)$ (in the bundle $\mathbb{F}^{u_1}$) belong to $\mathcal{F}_J$ (for all $n \in \omega$)). Also we may assume that

(α) $\quad \forall (d_0, d_1, \mathbf{c}) \in D_{k+2}\, [(\mathbf{d}^1, \mathbf{c}^1) \leq_{k+2} (d_0, d_1, \mathbf{c}) \Rightarrow (d_0 \neq d_1) \wedge \bigwedge_{i=0}^{1}(d_i \notin \mathbf{c})]$

(where $d \notin (c_1, \ldots, c_k)$ denotes $\bigwedge_{i=1}^{k}(d \neq c_i)$), since otherwise $(J \vee K^{[1]}) \notin \mathsf{L}(\mathbb{F}^{u_0})$ (in fact it is clear that $u_2 \vDash \neg \forall x\, (P(x, d_0, d_1, \mathbf{c}) \vee \neg P(x, d_0, d_1, \mathbf{c}))$ and $d_{1-i} \in (d_i, \mathbf{c})$ implies that $u_2 \vDash \neg \forall x\, (P'(x, d_i, \mathbf{c}) \vee \neg P'(x, d_i, \mathbf{c}))$ for some valuation ξ').

At first let us suppose that the following holds (for some $i \in \{0, 1\}$):

(β) $\quad \forall (d, \mathbf{c}) \in D_{k+1}\, [((d_i^1, \mathbf{c}^1) \leq_{k+1} (d, \mathbf{c})) \Rightarrow$

$\Rightarrow \exists (b, d', \mathbf{c}'), (d'', d'', \mathbf{c}'')\, [(b, d', \mathbf{c}') \leq_{k+2} (d'', d'', \mathbf{c}''), (d, \mathbf{c}) \leq_{k+1} (d', \mathbf{c}'), b \neq d']]$.

Then $(J \vee K^{[1]}) \notin \mathsf{L}(\mathbb{F}^{u_0})$ since $u_1 \vDash \neg \forall x\, (P'(x, d_i^1, \mathbf{c}) \vee \neg P'(x, d_i^1, \mathbf{c}))$ for the valuation $\xi'(P') = \{(d, d, \mathbf{c}) : (d_i^1 \mathbf{c}^1) \leq_{k+1} (d, \mathbf{c})\}$.

If (β) is not true for $i = 0$ and $i = 1$ it is sufficient to assume that

(γ) $\quad \neg \exists (b, d_0, d_1, \mathbf{c}), (b', d_0', d_1', \mathbf{c}') \in D^{k+3}\, [(\mathbf{d}^1, \mathbf{c}^1) \leq_{k+2} (d_0, d_1, \mathbf{c}),$

$\qquad\qquad (b, d_0, d_1, \mathbf{c}) \leq_{k+3} (b', d_0', d_1', \mathbf{c}'), b \notin \{d_0, d_1\}, b' \in \{d_0', d_1'\}]$

(since the set $\{(\ldots) : \ldots \vDash \neg \forall x\, (P(x, \ldots) \vee \neg P(x, \ldots))\}$ is conic, we can replace our u_1 by an appropriate higher point).

Let us prove that

(δ) $\qquad\qquad \forall \mathbf{c} \in D_k\, [(\mathbf{c}^1 \leq_k \mathbf{c}) \Rightarrow \exists! \mathbf{d} \in D_2((\mathbf{d}^1, \mathbf{c}^1) \leq_{k+2} (\mathbf{d}, \mathbf{c}))]$

Suppose the contrary, i.e. that $(\mathbf{d}^1, \mathbf{c}^1) \leq_{k+2} (\mathbf{d}^j, \mathbf{c})$, $\mathbf{d}^j = (d_0^j, d_1^j)$ for $j \in \{2, 3\}$ and $d_i^2 \neq d_i^3$ for some $i \in \{0, 1\}$, e.g. for $i = 1$. Since $J \in \mathsf{L}(\mathbb{F}^{u_1})$, there exists $(d_0', d_1', \mathbf{c}')$ such that

$$\bigwedge_{j=2}^{3} ((\mathbf{d}^j, \mathbf{c}) \leq_{k+2} (d_0', d_1', \mathbf{c}')).$$

Due to (α), $d_1^2 \notin (\mathbf{d}^3, \mathbf{c})$ (in fact, if we had $d_1^2 = d_0^3$ then $(\mathbf{d}^1, \mathbf{c}^1) \leq_{k+2} (d_2^3, d_2^3, \mathbf{c})$ since $(d_0^1, \mathbf{c}^1) \leq_{k+1} (d_0^3, \mathbf{c})$, $d_1^1 \leq_1 d_1^2 = d_0^3$, $d_1^1 \notin (d_0^1, \mathbf{c}^1)$). Thus $(d_1^2, \mathbf{d}^3, \mathbf{c}) \leq_{k+3} (d_1', d_0', d_1', \mathbf{c}')$ which contradicts (γ).

Now, by using (δ) we can define a valuation $\xi'(P') = \{(b, \mathbf{c}) : \mathbf{c}^1 \leq_k \mathbf{c}$ and $(b, \mathbf{d}, \mathbf{c}) \in \xi(P)$ for the unique $\mathbf{d}$ such that $(\mathbf{d}^1, \mathbf{c}^1) \leq_{k+2} (\mathbf{d}, \mathbf{c})\}$. Then $u_1 \vDash \neg \forall x\, (P'(x, \mathbf{c}^1) \vee \neg P'(x, \mathbf{c}^1))$ and thus $(J \vee K) \notin \mathsf{L}(\mathbb{F}^{u_0})$. $\boxtimes$

4.3 Lemma 4.2 can be extended to many other propositional formulas.

Proposition 4.3 *Let* $(\mathsf{PH} + A)$ *be a properly intermediate propositional logic and* $m, m', n, n' \in \omega$.

(1) If $(\mathsf{PH} + A) \vdash J^-$ *then*

$$(\mathsf{H} + (A^{[m']} \vee K^{[m]})) \vdash (A^{[n']} \vee K^{[n]}) \text{ iff } n \leq m.$$

(2) If $(\mathsf{PH} + A) \not\vdash J^-$ *then*

$$(\mathsf{H} + (A^{[m']} \vee K^{[m]})) \vdash (A^{[n']} \vee K^{[n]}) \text{ iff } n \leq m, \ n' \leq m'.$$

Due to Proposition 3.1 it is sufficient to prove the following two lemmas.

Lemma 4.4 *If* $(\mathsf{PH} + A) \vdash J^-$ *then* $(\mathsf{H} + (A \vee K^{[m]})) \vdash (A^{[n]} \vee K^{[m]})$.

Lemma 4.5 *If* $(\mathsf{PH} + A) \not\vdash J^-$ *(and* $\mathsf{PH} \not\vdash A$*), then*

$$(\mathsf{H} + (A^{[m]} \vee K^{[m']})) \not\vdash (A^{[m+1]} \vee K).$$

Proof of Lemma 4.4. It is sufficient to assume that $(\mathsf{PH} + P_1) \vdash A$, that is $\mathsf{PH} \vdash \neg\neg A$. Let $\mathsf{L} = (\mathsf{H} + (A^{[n]} \vee K^{[m]}))$. Let us show that $\mathsf{L} \vdash (A^{[n+1]} \vee K^{[m]})$.

Since $(\mathsf{H}+K) \vdash (\forall x \, \neg\neg Q(x) \to \neg\neg \forall x \, Q(x))$, we have $(\mathsf{H}+K) \vdash \neg\neg \forall \mathbf{x} \, A^{n+1}(y, \mathbf{x})$, $\mathbf{x} = (x_1, \ldots, x_n)$. By the Deduction Theorem (Lemma 1.1), it is easily proved that $\mathsf{L} \vdash (\neg \forall \mathbf{x} \, A^{n+1}(y, \mathbf{x}) \to \neg\neg \forall \mathbf{x} \, A^{n+1}(y, \mathbf{x}))$, and thus $\mathsf{L} \vdash \neg\neg \forall \mathbf{x} A^{n+1}(y, \mathbf{x})$. Since $(\mathsf{PH} + A) \vdash J^-$, we can conclude that $\mathsf{L} \vdash (J^- \vee K^{[m]})$ (again by the Deduction Theorem). Let $B = (A^{[n+1]} \wedge K^{[m]})$. We have $\mathsf{L} \vdash (\neg\neg B \to B) \vee \neg\neg B \vee K^{[m]}$. Also $\mathsf{H} \vdash (\neg\neg B \to K^{[m]})$ since $\mathsf{H} \vdash (\neg\neg K^{[m]} \to K^{[m]})$. Thus,

$$\mathsf{L} \vdash (\neg\neg B \to B) \vee K^{[m]}. \tag{1}$$

Let $C = \neg\neg \forall y \, (\forall \mathbf{x} \, A^{n+1}(y, \mathbf{x}) \vee \neg \forall \mathbf{x} \, A^{n+1}(y, \mathbf{x}))$. Then $\mathsf{L} \vdash \forall \mathbf{x} \, A^{n+1}(y, \mathbf{x}) \vee C$ since $\mathsf{L} \vdash \forall \mathbf{x} \, A^{n+1}(y, \mathbf{x}) \vee K$. Also we have that $(\mathsf{H} + \forall y \, \neg\neg \forall \mathbf{x} \, A^{n+1}(y, \mathbf{x})) \vdash (C \to \neg\neg A^{[n+1]})$ and thus $\mathsf{L} \vdash (C \to \neg\neg A^{[n+1]})$, $\mathsf{L} \vdash \forall \mathbf{x} \, A^{n+1}(y, \mathbf{x}) \vee (\neg\neg A^{[n+1]} \wedge K^{[m]})$ and $\mathsf{L} \vdash \forall y (\forall \mathbf{x} \, A^{n+1}(y, \mathbf{x}) \vee \neg\neg B)$. On the other hand, $\mathsf{H} \vdash (B \to \forall \mathbf{x} \, A^{n+1}(y, \mathbf{x}))$, and thus

$$\mathsf{L} \vdash (\neg\neg B \to B) \to A^{[n+1]} \tag{2}$$

Finally (1) and (2) imply $\mathsf{L} \vdash (A^{[n+1]} \vee K^{[m]})$. $\boxtimes$

Proof of Lemma 4.5. We use a construction similar to the one in the proof of Proposition 3.2 in Section 3, with some modifications (because of the formula K).

Since $(\mathsf{PH} + A) \not\vdash J^-$. we obtain that $A \in \mathsf{PL}(F_i)$ for some $i \in \{3,4\}$ (see Lemma 2.2). On the other hand. $\mathsf{PH} \not\vdash A$ implies that $A \notin \mathsf{PL}(F)$ for some frame F from $\mathcal{F}_H^f$. Let Z_2 be the set of all maximal elements of F and $Z_1 = F - Z_2$ (thus $F = Z_1 \cup Z_2$ and $Z_1 \neq \emptyset$). Take two correct m-colourings $\rho_j = (X_j, Z_j, \lambda_j)$, $j = 1, 2$. Consider a category $\mathcal{C}$ whose frame representation is F_i and such that $\mathcal{C}(u_3, u_3) = \{1_{u_3}, \mu^*\}$. $\mathcal{C}(u_j, u_j) = T_{\rho_j}$ for $j = 1, 2$ and $\mathcal{C}(u_1, u_2) = T_{\rho_{12}}$ is the set of one-to-one functions $\mu : X_1 \longrightarrow X_2$ such that $\forall Y \in P_{m+1}(X_1)\,(\lambda_1(Y) \leq \lambda_2(\mu Y))$ (otherwise $\mathcal{C}(u, v)$ is one-element). Consider a $\mathcal{C}$-set $\mathbb{F} = (F_i, \overline{D}, E)$ in which $D_{u_j} = (\{a_j\} \cup X_j)$ $(X_j = \{b', b''\}$ for $j = 3$ and $X_j = \emptyset$ for $j = 0, 4)$, $E_\mu | X_j = \mu$ for $\mu \in T_{\rho_j}$ $(j = 1, 2)$ or $\mu \in T_{\rho_{12}}$ $(j = 1)$, $E_{\mu^*}(b') = E_{\mu^*}(b'') = b''$ (and $E_\mu(a_j) = a_k$ for all $\mu \in \mathcal{C}(u_j, u_k)$, $u_j \leq u_k$). Then $\mathbb{F} \not\models (A^{[m+1]} \vee K)$ since $u_1 \not\models A^{[m+1]}$ and $u_3 \not\models K$ (for some valuation in $\mathbb{F}$), cf. Section 3.

On the other hand. $(A^{[m]} \vee K^{[m']}) \in \mathsf{L}(\mathbb{F})$. since we have

(1) $K \in \mathsf{L}(\mathbb{F}^{u_1})$ (note that the pre-ordering $\leq_k$ is symmetric on $(D_{u_2})^k$ for every $k \in \omega$. since Z_2 is an antichain in F);

(2) $A \in \mathsf{L}(\mathbb{F}^{u_j})$ for $j \geq 3$ (note that the skeletons of $\leq_k$-cones in $\mathbb{F}^{u_j}$ are isomorphic to F_n for $n \leq j - 2$):

(3) $u_0 \models \forall \mathbf{x}\, A^{m+k}(\mathbf{x}, \mathbf{a}_0)$ (similarly to the corresponding proof in Section 3. since all skeletons of $\leq_{m+k}$-cones are isomorphic to cones in F_i).

$$\boxtimes$$

4.4 Now let us consider examples of incomplete logics. related to Theorem 2.

Lemma 4.6 *Let A be a predicate formula (having no common symbols with D) and $n \geq 0$. Then $(\mathsf{H} + (A \vee D)) \vdash (A \vee D^{[n]})$.*

Proof. (induction on n). Let $B = \forall z\,(A \vee \forall \mathbf{x}\, D^{n+1}(\mathbf{x}, z)) \to A \vee \forall z \forall \mathbf{x}\, D^{n+1}(\mathbf{x}, z)$, $\mathbf{x} = (x_1, \ldots, x_n)$. Then $(\mathsf{H} + (A \vee D)) \vdash B$. Also $(\mathsf{H} + (A \vee D)) \vdash \forall z\,(A \vee \forall \mathbf{x}\, D^{n+1}(\mathbf{x}, z))$ (since $(\mathsf{H} + (A \vee D)) \vdash A \vee D^{[n]}$ by induction hypothesis). Thus $(\mathsf{H} + (A \vee D)) \vdash A \vee D^{[n+1]}$. $\boxtimes$

Therefore, By Proposition 3.2. for any properly intermediate propositional logic $(\mathsf{PH} + A)$ and $m, m', n, n' \in \omega$ we have

$$(\mathsf{H} + (A^{[m]} \vee D^{[m']})) \vdash (A^{[n]} \vee D^{[n']}) \qquad \text{iff} \quad n \leq m.$$

So we can consider only the formulas $(A^{[m]} \vee D)$.

Theorem 5 *Let $\mathbb{F}$ be a bundle.*

(1) If $(J^{[1]} \vee D) \in \mathsf{L}(\mathbb{F})$, then $\forall m \in \omega\,((J^{[m]} \vee D) \in \mathsf{L}(\mathbb{F}))$.

(2) If $(Z^{[3]} \vee D) \in \mathsf{L}(\mathbb{F})$, then $\forall m \in \omega \, ((Z^{[m]} \vee D) \in \mathsf{L}(\mathbb{F}))$.

(3) Let $n \geq 2$. If $((P_n)^{[n+1]} \vee D) \in \mathsf{L}(\mathbb{F})$, then $\forall m \in \omega \, (((P_n)^{[m]} \vee D) \in \mathsf{L}(\mathbb{F}))$.

Therefore predicate logics $(\mathsf{H} + (J^{[1]} \vee D))$, $(\mathsf{H} + (Z^{[3]} \vee D))$, $(\mathsf{H} + ((P_n)^{[n+1]} \vee D))$ for $n \geq 2$ are incomplete w.r.t. the semantics of bundles. Similarly the incompleteness can be proved for the logics $(\mathsf{H} + ((P_n \wedge J^{(-)})^{[n+1]} \vee D))$, $(\mathsf{H} + ((P_n \wedge Z^{[3]}) \vee D))$, $(\mathsf{H} + ((J^-)^{[3]} \vee D))$ and also for the logics $(\mathsf{H} + (A^{[3]} \vee D))$ for any logic $(\mathsf{PH} + A)$ between $(\mathsf{PH} + P_2)$ and $(\mathsf{PH} + R_2)$ (i.e. for any logic from the second slice, Hosoi, [2]).

The values of the indices $(1, 3, n + 1)$ in Theorem 5 cannot be reduced and corresponding counterexamples are in Figure 4. The bundle $\mathbb{F}_1$ separates $(J^{[1]} \vee D)$ from $(J \vee D)$ and the bundle $\mathbb{F}_2$ separates $(Z^{[3]} \vee D)$ from $(Z^{[2]} \vee D)$ and $((P_n)^{[n+1]} \vee D)$ from $((P_n)^{[n]} \vee D)$ for $n \geq 2$ (here $\forall i, j \in \omega \, (b_i \leq_1 b_j)$ in D_{u_1}). Note also that $(A^{[1]} \vee D)$ is separated from $(A \vee D)$ for any properly intermediate logic $(\mathsf{PH} + A)$. We do not know if there exists a propositional formula A such that all $(A^{[m]} \vee D)$ are distinguishable in the semantics of bundles.

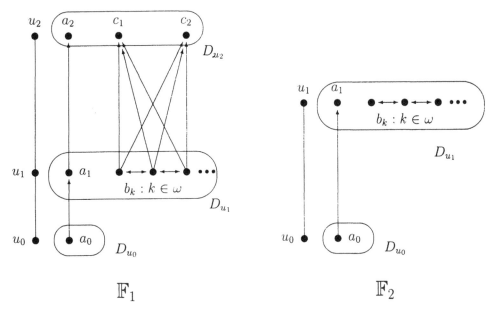

$$\mathbb{F}_1 \qquad\qquad \mathbb{F}_2$$

Figure 4.

Proof of Theorem 5. Let A be one of the formulas J, Z or P_n $(n \geq 2)$ and let $(A^{[m]} \vee D) \notin \mathsf{L}(\mathbb{F})$. Then there exists a valuation ξ such that $u_0 \not\vDash$

$(\forall \mathbf{x}\, A^{m+k}(\mathbf{x}, \mathbf{c}^0) \vee D^k(\mathbf{c}^0))$, $u_1 \nvdash A^{m+k}(\mathbf{d}, \mathbf{c}^1)$ (here $\mathbf{x} = (x_1, \ldots, x_m)$, $\mathbf{c}^0 \in (D_{u_0})^k$, $(\mathbf{d}, \mathbf{c}^1) \in (D_{u_1})^{m+k}$, $u_0 \leq u_1$, $\mathbf{c}^0 \leq_k \mathbf{c}^1$).

If $D \notin \mathsf{L}(\mathbb{F}^{u_1})$ then $(A \vee D) \notin \mathsf{L}(\mathbb{F}^{u_1})$ (apply Lemma 4.6 - cf. the proof of Theorem 4). If $D \in \mathsf{L}(\mathbb{F}^{u_1})$, then Lemma 2.6 can be applied, e.g., if $A = J$, then $J \notin \mathsf{L}(\mathbb{F}^{u_1})$ implies $\mathbb{F}^{u_1} \nvdash J^{[1]}$ and thus $(J^{[1]} \vee D) \notin \mathsf{L}(\mathbb{F}^{u_0})$ (and similarly for Z and P_n).   ⊠

Analogously, by using Theorem 3 and the Remark after Lemma 2.6, one can prove the incompleteness (w.r.t the semantics of bundles) of the predicate logics $(\mathsf{H} + ((P_n)^{[2n]} \vee J))$ for $n > 1$. On the other hand in Figure 5 there is a bundle $\mathbb{F}$ such that $((P_n)^{[2n-1]} \vee J) \in \mathsf{L}(\mathbb{F})$ and $((P_n)^{[2n]} \vee J) \notin \mathsf{L}(\mathbb{F})$: $F = F_2$, $D_{u_j} = \{a_j\}$ (for $0 \leq j \leq 1$), $D_{u_2} = \{a_2\} \cup \{b_i, c_i : 1 \leq i \leq n\}$, $a_0 <_1 a_j$ (for $1 \leq j \leq 2$), $b_i \leq_1 c_i \leq_1 b_i$ (for $1 \leq i \leq n$).

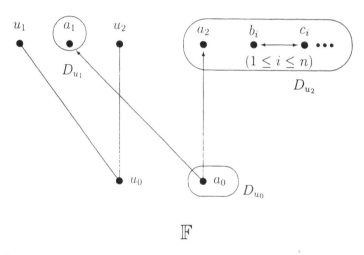

$$\mathbb{F}$$

Figure 5.

4.5 Finally, let us give another example of incompleteness in the semantics of bundles. The logic is not related to intersections, but the incompleteness proof is rather simple.

Proposition 4.7 *Let*

$$B_0' = \forall x \forall y\, (P'(x) \to P'(y) \vee Q(x) \vee Q(y))$$
$$B_1' = \forall x \forall y\, (P(x, y) \to P(x, x) \vee Q(x) \vee Q(y))$$
$$B_i = Q(z) \vee (Q(z) \to B_i')\ \ \text{for } i \leq 1.$$

Then

(1) $(H + B_0) \nvdash B_1$;

(2) for any bundle $\mathbb{F}$, $B_0 \in L(\mathbb{F}) \Rightarrow B_1 \in L(\mathbb{F})$.

Therefore, the predicate logic $(H + \forall z \, B_0)$ *is incomplete w.r.t. the semantics of bundles (this logic is similar to the incomplete modal logic from [7], Theorem 4.4.2).*

Proof. (1) Consider a category $\mathcal{C}$ with the frame representation F_1, $\mathcal{C}(u_1, u_1) = \{1_{u_1}, \mu^*\}$ (and one-element $\mathcal{C}(u_0, u_i)$ for $i \leq 1$). Take a $\mathcal{C}$-set $\mathbb{F}^* = (F, \overline{D}, E)$ such that $D_{u_0} = \{a_0\}$, $D_{u_1} = \{a_1, b_0, b_1\}$, $E_{\mu^*}(a_1) = a_1$, $E_{\mu^*}(b_i) = b_{1-i}$ and $E_{\mu}(a_0) = a_i$ for $\mu \in \mathcal{C}(u_0, u_i)$, $i \leq 1$. Thus in $(D_{u_1})^n$ we have

$$[(c_1, \ldots, c_n) \leq_n (d_1, \ldots, d_n)] \qquad \text{iff}$$

$$[\bigwedge_{i=1}^{k}(c_i = d_i)] \vee [\bigwedge_{i=1}^{k}((c_i = d_i = a_1) \vee (\{c_i, d_i\} = \{b_0, b_1\}))].$$

Let ξ be a valuation in $\mathbb{F}^*$ such that $\xi(Q) = \{a_1\}$, $\xi(P) = \{(b_0, b_1), (b_1, b_0)\}$, then $u_0 \nvDash B_1(a_0)$, since $u_0 \nvDash Q(a_0)$, $u_1 \vDash Q(a_1)$, $u_1 \vDash P(b_0, b_1)$, $u_1 \nvDash P(b_0, b_0)$, $u_1 \nvDash Q(b_0)$, $u_1 \nvDash Q(b_1)$.

On the other hand $\mathbb{F}^* \vDash B_0^{[m]}$. In fact consider any valuation in $\mathbb{F}^*$ and $(\mathbf{c}, c) \in (D_{u_1})^{m+1}$, $(\mathbf{a}_i, a_i) \in (D_{u_i})^{m+1}$ for $i \leq 1$. Then $u_1 \vDash (Q(\mathbf{c}, c) \vee \neg Q(\mathbf{c}, c))$ and $u_0 \vDash (B_0)^m(\mathbf{a}_0, a_0)$ (since $(\mathbf{a}_1, b_i) \leq_{m+1} (\mathbf{a}_1, b_{1-i})$ and therefore $u_1 \vDash P'(\mathbf{a}_1, b_i)$ implies $u_1 \vDash P'(\mathbf{a}_1, b_{1-i})$).

(2) Let $\mathbb{F}$ be a bundle, ξ be a valuation, $u_0 \in F$, $(\mathbf{a}, a) \in (D_{u_0})^{m+1}$ and suppose that $u_0 \nvDash (B_1)^m(\mathbf{a}, a)$. Then we have (for some $u_1 \geq u_0$ and $(\mathbf{b}, b, d_0, d_1) \in (D_{u_1})^{m+3}$): $(\mathbf{a}, a) \leq_{m+1} (\mathbf{b}, b)$, $u_0 \nvDash Q(\mathbf{a}, a)$, $u_1 \vDash Q(\mathbf{b}, b)$, $u_1 \vDash P(\mathbf{b}, d_0, d_1)$, $u_1 \nvDash P(\mathbf{b}, d_0, d_0)$ and $u_1 \nvDash Q(\mathbf{b}, d_i)$ for $i \leq 1$. Hence $\neg[(\mathbf{b}, d_0, d_0) \leq_{m+2} (\mathbf{b}, d_0, d_1)]$ and thus $\neg[(\mathbf{b}, d_1) \leq_{m+1} (\mathbf{b}, d_0)]$ (and $d_0 \neq d_1$). Taking a valuation $\xi'(Q) = \xi(Q)$, $\xi'(P') = \{(\mathbf{c}, c) : (\mathbf{b}, d_1) \leq_{m+1} (\mathbf{c}, c)\}$, we have $u_0 \nvDash Q(\mathbf{a}, a)$, $u_1 \vDash Q(\mathbf{b}, b)$, $u_1 \vDash P'(\mathbf{b}, d_1)$, $u_1 \nvDash P'(\mathbf{b}, d_0)$, $u_1 \nvDash Q(\mathbf{b}, d_i)$ for $i \leq 1$, and thus $u_0 \nvDash (B_0)^m(\mathbf{a}, a)$. $\boxtimes$

References

[1] S. Ghilardi, *Presheaf semantics and independence results for some non-classical first-order logics*, Arch. Math. Logic **29** (1989), 125–136.

[2] T. Hosoi, *On intermediate logics, I*, J. Fac. Sci. Univ. Tokyo, Sect. IA Math. **14** (1967), 293–312.

[3] V.A. Jankov, *Conjunctively indecomposable formulas in propositional calculi*, Math. USSR-Izv. **3** (1969), 17–35.

[4] S. Miura, *A remark on the intersection of two logics*, Nagoya Math. J. **26** (1966), 167–171.

[5] H. Ono, *A study of intermediate predicate logics*, Publ. Res. Inst. Math. Sci. **8** (1972–73), 619–649.

[6] V. Shehtman and D. Skvortsov, *Semantics of non-classical first order predicate logics*, Mathematical Logic (P. Petkov, ed.), Plenum Press, New York, 1990, (Proc. of Summer School and Conference in Mathematical Logic, Heyting '88), pp. 105–116.

[7] _____ , *Maximal kripke-type semantics for modal and superintuitionistic predicate logics*, Ann. Pure Appl. Logic **63** (1993), 69–101.

INSTITUTE OF SCIENTIFIC AND TECHNICAL INFORMATION, USIEVICHA UL. 20A, 125219, MOSCOW, RUSSIA

A Completeness Theorem for Formal Topologies

Silvio Valentini

Abstract

The main mathematical result of this work is a quite simple formulation and proof of a Rasiowa-Sikorski-like theorem for countable lattices. Then the paper suggests an interpretation of this mathematical result as a completeness theorem for the *formal topologies* introduced by G. Sambin in order to provide a constructive approach to topology which is expressible within Martin Löf's intuitionistic theory of types. This completeness theorem shows that, as long as one is interested in dealing only with the coverage relation between two open sets of a topology, within a constructive framework, a very simple mathematical structure is needed. It is necessary to stress that the completeness theorem holds because of the "weak" intuitionistic set and subset theory that we use when dealing with formal topologies.

1 Introduction

In this paper we will show a completeness theorem for formal topologies [9, 10], which we are briefly going to recall in the next section. It shows that, as long as one is interested in dealing only with the coverage relation between two open sets of a topology, in an intuitionistic framework, a very simple mathematical structure is needed.

Since formal topologies are intended to be a constructive approach to the classical, point-set topology, the first obliged step to define them is to introduce a theory of subsets which may be intuitionistically acceptable. To this purpose we will use a suitable extension of Martin-Löf's Intuitionistic Theory of Types (briefly ITT in the following) [7], which is described in full extent in [11]; here we will recall the definitions that we need at the beginning of the next section.

Then we will classically prove a Rasiowa-Sikorski-like theorem for the countable lattices which shows that it is possible to find a prime filter which respects a countable collection of suprema, in the same way the original theorem by Rasiowa and Sikorski does to prove the completeness theorem for first order

689

logic [8]. Such a theorem is the key to embed any countable distributive lattice into a point-set topology. It will be used in the last section to prove the wished completeness theorem for the formal topologies.

2   Formal topologies

In order to develop a constructive version of the classical topology we will need to deal with sets and subsets in an intuitionistic framework. While for the set theory we commit ourselves to Martin-Löf's Intuitionistic Theory of Types [7], in order to deal with subsets a specific notion for subsets must be adopted, together with some notation. Here we will illustrate only the main definitions that we are going to use, while for a deeper treatment the reader is referred to [11].

The distinction between sets and collections [1] is basic in ITT; a collection is a set only if one can effectively produce its elements. For instance N, i.e. natural numbers, is a set since its elements are (equivalent to) 0 or the successor of any element already known to be in N, whereas, in general, the collection of all the classical subsets of a set S cannot be a set insomuch as one cannot produce all of its elements. [2]

In general not only is the collection of all the subsets of a set not a set, but even *one* subset of a set can be only a collection but no set; as an example, consider the subset of N of the code numbers of the recursive functions which do not halt on 0: because of the unsolvability of the "halting problem", there is no way to effectively produce all the elements of this subset. On the other hand, being able to deal with subsets is necessary to develop almost any piece of mathematics. For this reason, we introduce the following notion of subset, which is suggested by the axiom of separation of ZF set theory. Let S be a set; then we put

$$U \subset S \equiv (x : S) \ U(x) \ prop$$

that is, we say that U is a *subset of* S whenever U is a propositional function on elements of S. In what follows also the alternative, usual, notation $\{x \in S : U(x) \ true\}$ will be used for the subset U of S, in order to make the exposition clearer. [3]

---

[1] In this paper we systematically use the word "collection" for what is called "category" in ITT [7].

[2] As a further example, consider that one can form the function space $N \rightarrow N$ in ITT but the type he obtains corresponds in no way to the collection of all the classical functions from N to N.

[3] The alternative possibility to identify the subset of the elements of S which satisfy the proposition $U(x) \ prop[x : S]$ with the ITT type $\Sigma(S, U)$ compels to define $a \varepsilon U$ by "*there is an element b such that $\langle a, b \rangle \in \Sigma(S, U)$*" and hence $a \varepsilon U$ cannot be a ITT proposition.

The subset-theory which results from this definition is a sort of local set theory, since all the relations and operations we introduce are always relativized to a set; to make this fact explicit, we will indicate the set as an index (even if in the following we will sometimes omit it when it is clear from the context).

The first definition is membership; this definition is an immediate consequence of the fact that U is a propositional function:

$$a \varepsilon_S U \ true \equiv a \in S \text{ and } U(a) \ true,$$
i.e. there is a proof of the proposition $U(a)$

The next step is the definition of inclusion between subsets of S, which is an obvious consequence of the previous definition of membership:

$$U \subseteq_S V \equiv (\forall x \in S) \ (U(x) \to V(x))$$

which in turn gives

$$U =_S V \equiv (\forall x \in S) \ (U(x) \leftrightarrow V(x)).$$

As usual, given a proposition over elements of S, i.e. $A(x) \ prop \ [x : S]$, universal quantification over a subset $U \subset S$, i.e. $(\forall x \varepsilon_S U) \ A(x)$, is nothing but an abbreviation for $(\forall x \in S) \ (U(x) \to A(x))$; similarly $(\exists x \varepsilon_S U) \ A(x)$ is just an abbreviation for $(\exists x \in S) \ (U(x) \ \& \ A(x))$.

Then subset operations can be introduced: in this paper we only need binary intersection and union and arbitrary union. Given two subsets U, V of S and a family $(V_i)_{i \in I}$ of subsets of S, i.e. a propositional function $V(i, x) \ prop \ [i : I, x : S]$, we put:

$$
\begin{aligned}
U \cap_S V &\equiv (x : S) \ U(x) \ \& \ V(x) \ prop \\
U \cup_S V &\equiv (x : S) \ U(x) \ \vee \ V(x) \ prop \\
\cup_{i \in I} V_i &\equiv (x : S) \ (\exists i \in I) V(i, x) \ prop
\end{aligned}
$$

It should be clear that, given any set S and a *fixed* countable language, as is the case for ITT, the number of "subsets" of such a set is at most countable since only a countable collection of propositional functions on its elements can be formed. While in general this fact gives rise to a weak set theory, in our approach it is an essential feature to prove the completeness theorem: using a little slogan we can say that "we win because of our weakness".

Let us now turn to formal topologies. Formal topologies were introduced by G. Sambin as a constructive approach to topology [9], in the tradition of the Johnstone's version of the *Grothendieck topologies* [6] and Fourman and Grayson's *Formal Spaces* [4], but using simpler technical devices and a constructive set theory based on Martin Löf's Intuitionistic Theory of Types.

Today, formal topologies are also a well developed device to present locales and to study their properties [1].

The main point in the *formal* approach to topology is the study, in a constructive framework, of the properties of a topological space $\langle X, \Omega(X) \rangle$, where $\Omega(X)$ is the family of the open subsets of the *collection* X, which can be expressed without any reference to the points, that is to the elements of the collection X.

Since a point-set topology can always be presented using one of its bases, the abstract structure that we will consider is a commutative monoid $\langle S, \bullet_S, 1_S \rangle$ where the *set* S corresponds to the set of the elements of the base of the point-set topology $\Omega(X)$, $\bullet_S$ corresponds to the operation of intersection between basic elements, and 1_S corresponds to the whole collection X.

In a point-set topology any open set is obtained as a union of elements of the base, but union does not make sense if we refuse any reference to points; hence we are naturally lead to think that an open set may directly correspond to a subset of the set S. Unfortunately this idea is not completely correct since there may be many different subsets of basic elements whose union is the same open set; hence, in order to better define what corresponds to an open set, we need to introduce also an equivalence relation $\cong_S$ between two subsets U and V of S such that $U \cong_S V$ holds if and only if, denoting by c^* the element of the base which corresponds to the formal basic open c, the opens $U^* \equiv \cup_{a \varepsilon U} a^*$ and $V^* \equiv \cup_{b \varepsilon V} b^*$ are equal. To this purpose we introduce an infinitary relation $\lhd_S$, which will be called *cover*, between a basic element a of S and a subset U of S whose intended meaning is that $a \lhd_S U$ when $a^* \subseteq U^*$. The conditions we will require of this relation are a straightforward rephrasing of the similar set-theoretic situation. It will be the aim of the completeness theorem to show that they are also sufficient.

Definition 2.1 (Formal topology) *Let S be a set, then a formal topology over S is a structure*

$$\mathcal{S} \equiv \langle S, \bullet_S, 1_S, \lhd_S \rangle$$

where $\bullet_S$ is an associative and commutative operation over S, 1_S is the unit of the operation $\bullet_S$ and $\lhd_S$ is an infinitary relation between an element of S and a subset of S such that, for any $a, b \in S$ and for any $U, V \subset S$ the following conditions hold

$$\text{(reflexivity)} \quad \dfrac{a \varepsilon_S U}{a \lhd_S U}$$

$$\text{(transitivity)} \quad \dfrac{a \lhd_S U \quad (\forall u \varepsilon_S U)\ u \lhd_S V}{a \lhd_S V}$$

$$\text{(•-Left)} \quad \dfrac{a \lhd_S U}{a \bullet b \lhd_S U}$$

$$\text{(•-Right)} \quad \dfrac{a \lhd_S U \quad a \lhd_S V}{a \lhd_S \{u \bullet v : u \varepsilon_S U, v \varepsilon_S V\}}$$

Let us introduce two abbreviations which will make the notation easier:

$$U \lhd_S V \quad \equiv \quad (\forall u \varepsilon_S U)\ u \lhd_S V$$
$$U \bullet V \quad \equiv \quad \{u \bullet v : u \varepsilon_S U, v \varepsilon_S V\}$$

These abbreviations allow a much more compact writing of the conditions above; besides the following derived conditions appear clearer by far.

$$\text{(global reflexivity)} \quad U \lhd_S U$$

$$\text{(global transitivity)} \quad \dfrac{U \lhd_S V \quad V \lhd_S W}{U \lhd_S W}$$

Moreover they suggest a way to define the relation $\cong_S$ that we were looking for. In fact by putting

$$U \cong_S V \equiv U \lhd_S V \text{ and } V \lhd_S U$$

we obtain an equivalence relation between subsets of S whose intended meaning is just that for any $u \varepsilon U$, $u^* \subseteq V^*$ and for any $v \varepsilon V$, $v^* \subseteq U^*$, which shows that $U^* = V^*$. We can rephrase somewhat more formally what we have achieved by identifying an open set with the equivalence class induced by the equivalence relation $\cong_S$, i.e. we will call $[U] \equiv \{V : U \cong_S V\}$ a *formal open* of the formal topology $\mathcal{S}$.

We can bring a *countable* distributive lattice $Open(\mathcal{S})$, actually a Heyting algebra, out of the collection of the formal opens of the formal topology $\mathcal{S}$ by putting:

$$\begin{aligned}
0_{Open(\mathcal{S})} &\equiv& [\emptyset] \\
1_{Open(\mathcal{S})} &\equiv& [S] \\
[U] \wedge_{Open(\mathcal{S})} [V] &\equiv& [U \bullet_S V] \\
[U] \vee_{Open(\mathcal{S})} [V] &\equiv& [U \cup V]
\end{aligned}$$

and we can extend the definition to the suprema over any indexed family of subsets by putting

$$\vee_{i \in I}[V_i] \equiv [\cup_{i \in I} V_i]$$

It is not difficult to check that $Open(\mathcal{S})$ is indeed a lattice which is distributive even with respect to suprema over any indexed family of subsets since an implication operation between formal opens can be defined by putting

$$[U] \rightarrow_{Open(\mathcal{S})} [V] \equiv [\{w \in S : \{w\} \bullet_S U \lhd_S V\}].$$

Moreover the lattice $Open(\mathcal{S})$ is countable since it has at most as many elements as the propositional functions on elements of S are.

It should be clear now the reason why we call our topologies *formal*: the open sets are in fact purely formal objects since they are not "made of points". So, proving the completeness theorem means showing that any formal open can be *filled* with points in order to obtain a point-set standard topology, i.e. to show that the formal topologies have *enough points* [4]. The idea, borrowed from the usual Stone representation theorems, is to look for the points among the prime filters of $Open(\mathcal{S})$. In fact let us recall how *formal points* can be introduced in formal topology (see [9]).

We start by looking for the conditions a point must satisfy which are expressible in the language of the formal topologies; to this aim, supposing $\mathcal{S}$ to be a formal topology and a one of its basic open subset, let us write $x \Vdash a$ to mean that the point x belongs to the basic open a. The first obvious condition is that any point belongs to the whole topological space and it can be stated by requiring that for any point x

(unit) $x \Vdash 1_S$

The second condition is that a point belongs to the intersection of two basic open sets if and only if it belongs to both, i.e. we require that for any point x and for any $a, b \in S$

(intersection) $x \Vdash a \bullet_S b$ iff $x \Vdash a$ and $x \Vdash b$

Finally we need to state that a point cannot be split by means of the basic open sets; to this aim we require that for any point x and for any $a \in S$ and $U \subset S$:

(completeness) $\dfrac{x \Vdash a \qquad a \lhd U}{(\exists u \varepsilon U)\ x \Vdash u}$

One can obtain the collection $Pt(\mathcal{S})$ of formal points of the formal topology $\mathcal{S}$ by considering all the subsets α of S which satisfy the following conditions for any $a, b \in S$ and for any $U \subset S$:

$$1_S \varepsilon \alpha, \qquad a \bullet_S b \varepsilon \alpha \text{ iff } a \varepsilon \alpha \text{ and } b \varepsilon \alpha, \qquad \frac{a \varepsilon \alpha \qquad a \lhd U}{(\exists u \varepsilon U)\ u \varepsilon \alpha}$$

---

[4]Some early attempts to construct points in a formal topology can be found in [12] but the lack of a suitable theory of subsets did not allow to obtain the completeness theorem.

since he can define the relation ⊩ by putting

$$\alpha \Vdash a \equiv a \varepsilon \alpha$$

Let us came back to the prime filters of the lattice $Sat(\mathcal{S})$. We can define a one–one correspondence ϕ between the collection of formal points $Pt(\mathcal{S})$ and the collection of completely prime filters of $Sat(\mathcal{S})$ by putting, for any $\alpha \in Pt(\mathcal{S})$:

$$\phi(\alpha) \equiv \{[U] : (\exists u \varepsilon U) \; \alpha \Vdash u\},$$

i.e. by associating to each (formal) point the completely prime filter of the open sets which contain it.

It should be clear now the reason why we are interested in the completely prime filters of the lattice $Sat(\mathcal{S})$ but before one begins any search for them he should be aware that using a standard theory of subsets there is, in general, no hope to find enough completely prime filters. For instance, let us consider the formal topology defined over a complete atomless boolean algebra $\mathcal{H}$ [5] by putting $a \lhd U \equiv a \leq \vee U$: in this case one can find no point at all in $Pt(\mathcal{H})$ since points are just the completely prime filters of $\mathcal{H}$ and it is well known that there is a one–one correspondence between atoms and completely prime filters of $\mathcal{H}$ [2]. Of course the problem in this case is that a prime filter must respect an uncountable quantity of subsets in order to be a point; the way out we propose here is to change the theory of subsets in such a way that the number of subsets of a given set is at most countable so that a Rasiowa-Sikorski-like theorem be sufficient in order to obtain completely prime filters.

3   A Rasiowa-Sikorski-like theorem for countable lattices

In this paragraph we will set up the technical device that we need in order to prove the completeness theorem. In particular we will show how any countable lattice with 0 and 1, $\mathcal{L} \equiv \langle L, 0, 1, \vee, \wedge \rangle$, can be embedded into a point-set topology in such a way that a countable quantity of suprema are respected, provided that $\wedge$ is distributive over such suprema. We must here stress on the fact that the proof we are going to show is classical and hence it is not possible to carry it on within ITT: for instance in case the lattice we are considering is countable we will show a theorem of existence of prime filters.

Let us begin by recalling some standard definitions. An order relation $\leq$ can be defined between two elements x and y of L by putting $x \leq y \equiv x = x \wedge y$, then the following definitions make sense.

---

[5]For instance one can consider the regular open algebra of the closed unit interval $[0, 1]$.

Definition 3.1 (Filter) *Let $\mathcal{L}$ be a lattice. Then a subset F of L is a* filter *if:*

$$1\varepsilon_{\mathcal{L}}F, \qquad \frac{x\varepsilon_{\mathcal{L}}F \quad x \leq y}{y\varepsilon_{\mathcal{L}}F}, \qquad \frac{x\varepsilon_{\mathcal{L}}F \quad y\varepsilon_{\mathcal{L}}F}{x \wedge y\varepsilon_{\mathcal{L}}F}$$

In this paper we will be interested only in *proper filters* of $\mathcal{L}$, i.e. filters which are proper subsets of L or, equivalently, which do not contain 0.

Definition 3.2 (Prime filter) *Let $\mathcal{L}$ be a lattice and F one of its proper filters. Then F is a* prime filter *of $\mathcal{L}$ if whenever $x \vee y\varepsilon_{\mathcal{L}}F$ then $x\varepsilon_{\mathcal{L}}F$ or $y\varepsilon_{\mathcal{L}}F$.*

On account of what we said in the previous section, we are interested in a particular collection of prime filters, i.e. those which respect a countable quantity of subsets of L.

Definition 3.3 *Let $\mathcal{L}$ be a lattice, F one of its filters and T a subset of L which has a supremum in $\mathcal{L}$. Then F* respects *T if whenever $\vee T\varepsilon_{\mathcal{L}}F$ there exists $b\varepsilon_{\mathcal{L}}T$ such that $b\varepsilon_{\mathcal{L}}F$.*

The main problem we have to solve now is to show that there exist prime filters which respect a countable quantity of subsets $T_1, \ldots, T_n, \ldots$. To this purpose we can prove the following theorem [6] which guarantees the existence in any distributive lattice of a proper filter (not necessarily a prime filter) which respects $T_1, \ldots, T_n, \ldots$.

Theorem 3.4 *Let $\mathcal{L}$ be a lattice, x, y two elements of L such that $x \not\leq y$ and $T_1, \ldots, T_n, \ldots$ a countable quantity of subsets of L which have a supremum in L such that $\wedge$ is distributive over them. Then there exists a filter F of $\mathcal{L}$ which contains x, does not contain y and respects all of the subsets $T_1, \ldots, T_n, \ldots$.*

Proof. We can classically construct such a filter in a countable number of steps, starting from the filter [7] $F_0 = \uparrow x \equiv \{z \in L : x \leq z\}$ and going on by extending it to a filter which respects all of the subsets $T_1, \ldots, T_n, \ldots$. Let us first construct a new countable list $W_1, \ldots, W_m, \ldots$ of subsets of S out of the list $T_1, \ldots, T_n, \ldots$ in such a way that any subset T_i appears a countable

---

[6] A similar result, even if bounded to the case of complete boolean algebras, can be found in [3], where it is called Tarski's lemma, but the proof we give here is different and more general of that of Tarski because the structures we are considering are weaker. Moreover J. Bell, after reading this paper, pointed out to me another related result for the problem of finding prime ideals in a countable Heyting algebra in [5].

[7] It is easy to check that F_0 is indeed a filter of $\mathcal{L}$.

number of times among $W_1, \ldots, W_m, \ldots$. Now put $c_0 = x$, and hence $F_0 = \uparrow c_0$, and suppose, by inductive hypothesis, that we have constructed an element c_n such that $c_n \not\leq y$ and we have defined $F_n = \uparrow c_n$, then put

$$c_{n+1} = \begin{cases} c_n & if \vee W_n \not\varepsilon_{\mathcal{L}} F_n \\ c_n \wedge b_n & if \vee W_n \varepsilon_{\mathcal{L}} F_n \end{cases}$$

where b_n is an element of W_n such that $c_n \wedge b_n \not\leq y$; in fact such an element exists because in the latter case $\vee W_n \varepsilon_{\mathcal{L}} F_n = \uparrow c_n$, and hence $c_n \leq \vee W_n$, then if for all $b \varepsilon_{\mathcal{L}} W_n, c_n \wedge b \leq y$ then, because of distributivity, $c_n = c_n \wedge \vee W_n = \vee_{b \varepsilon_{\mathcal{L}} W_n} c_n \wedge b \leq y$ which is contrary to the inductive hypothesis.

Now we can define the filter $F_{n+1} = \uparrow c_{n+1}$ and it is immediate to check that $F_n \subseteq F_{n+1}$ and that $y \not\varepsilon_{\mathcal{L}} F_{n+1}$ since $c_{n+1} \not\leq y$. Finally we put $F = \cup_{i \in \omega} F_i$. Then F is a filter since it is the union of a chain of filters; moreover $x \varepsilon_{\mathcal{L}} F$ since $x \varepsilon_{\mathcal{L}} F_0 \subseteq F$ while $y \not\varepsilon_{\mathcal{L}} F = \cup_{i \in \omega} F_i$ otherwise there would be an $i \in \omega$ such that $y \varepsilon_{\mathcal{L}} F_i$ which is contrary to the way we have constructed the filters; moreover F respects all of the subsets $T_1, \ldots, T_n, \ldots$ since if $\vee T_n \varepsilon_{\mathcal{L}} F = \cup_{i \in \omega} F_i$ then there is an $i \in \omega$ such that $\vee T_n \varepsilon_{\mathcal{L}} F_i$ and hence, since any T_n appears a countable number of times in the list $W_1, \ldots, W_m, \ldots$, for some $h \geq i$ it happens that $W_h = T_n$ and hence $\vee W_h = \vee T_n \varepsilon_{\mathcal{L}} F_i \subseteq F_h$ and so there exists $b_h \varepsilon_{\mathcal{L}} T_n$ such that $b_h \varepsilon_{\mathcal{L}} F_{h+1} \subseteq F$. [8]

It should be noted that the theorem ensures the existence of a proper filter which respects a countable quantity of suprema without any condition at all, since $1 \not\leq 0$ always holds and 1 is contained in any filter.

The previous theorem can be used to construct a prime filter in case we are dealing with a *countable* distributive lattice.

Corollary 3.5 *Let $\mathcal{L}$ be a countable distributive lattice, x, y two elements of L such that $x \not\leq y$ and $T_1, \ldots, T_n, \ldots$ a countable quantity of subsets of L which have a supremum in L such that $\wedge$ is distributive over them. Then there exists a prime filter which contains x, does not contain y and respects all of the subsets $T_1, \ldots, T_n, \ldots$.*

Proof. We have only to observe that in this case there is a countable quantity of binary suprema and hence we can use the previous theorem in order to obtain a filter which respects the countable quantity of binary suprema and the countable quantity of subsets $T_1, \ldots, T_n, \ldots$. It is then obvious that this filter is prime because it respects all the binary suprema.

---

[8] As the referee noted, the same proof works also for a slightly more general statement, i.e. the case $\mathcal{L}$ is a $\wedge$-semilattice instead of a lattice, but we think that the version given here better fit with the contents of the paper.

The existence of prime filters which respect a given countable quantity of suprema is the key point to construct a point-set topology in which a countable distributive lattice $\mathcal{L}$ can be embedded using a morphism which respects the countable quantity of suprema of the subsets $T_1, \ldots, T_n, \ldots$. In fact, let us consider the collection $Pf(\mathcal{L}) = \{P : P$ prime filter of $\mathcal{L}$ which respects $T_1, \ldots, T_n, \ldots\}$ and any topology τ on $Pf(\mathcal{L})$ such that the map defined by putting $ext(x) = \{P \in Pf(\mathcal{L}) : x \varepsilon_{\mathcal{L}} P\}$ associates an open set of the topology τ to any element of L. It is immediate to see that the map ext respects all the finitary operations:

$$ext(0) = \{P \in Pf(\mathcal{L}) : 0 \varepsilon_{\mathcal{L}} P\} = \emptyset$$
$$ext(1) = \{P \in Pf(\mathcal{L}) : 1 \varepsilon_{\mathcal{L}} P\} = Pf(\mathcal{L})$$

$$ext(x \vee y) = \{P \in Pf(\mathcal{L}) : x \vee y \varepsilon_{\mathcal{L}} P\}$$
$$= \{P \in Pf(\mathcal{L}) : x \varepsilon_{\mathcal{L}} P\} \cup \{P \in Pf(\mathcal{L}) : y \varepsilon_{\mathcal{L}} P\} = ext(x) \cup ext(y)$$

since $x \vee y$ is contained in a prime filter P if and only if $x \varepsilon_{\mathcal{L}} P$ or $y \varepsilon_{\mathcal{L}} P$, and

$$ext(x \wedge y) = \{P \in Pf(\mathcal{L}) : x \wedge y \varepsilon_{\mathcal{L}} P\}$$
$$= \{P \in Pf(\mathcal{L}) : x \varepsilon_{\mathcal{L}} P\} \cap \{P \in Pf(\mathcal{L}) : y \varepsilon_{\mathcal{L}} P\} = ext(x) \cap ext(y)$$

since $x \wedge y$ is contained in a filter P if and only if $x \varepsilon_{\mathcal{L}} P$ and $y \varepsilon_{\mathcal{L}} P$.

Moreover ext respects all the suprema of the subsets $T_1, \ldots, T_n, \ldots$ that we are considering. In fact

$$ext(\vee_{b \varepsilon_{\mathcal{L}} T_i} b) = \{P \in Pf(\mathcal{L}) : \vee_{b \varepsilon_{\mathcal{L}} T_i} b \varepsilon_{\mathcal{L}} P\}$$
$$= \cup_{b \varepsilon_{\mathcal{L}} T_i} \{P \in Pf(\mathcal{L}) : b \varepsilon_{\mathcal{L}} P\} = \cup_{b \in T_i} ext(b)$$

since $\vee_{b \varepsilon_{\mathcal{L}} T_i} b$ is contained in a prime filter P which respects all of the sets $T_1, \ldots, T_n, \ldots$ if and only if there is an element $b \varepsilon_{\mathcal{L}} T_i$ such that $b \varepsilon_{\mathcal{L}} P$.

Finally we note that also the order relation is respected since for any couple of elements $x, y \in L$, if $x \leq y$ then $ext(x) \subseteq ext(y)$.

To conclude the proof that ext is an embedding of $\mathcal{L}$ into the topology τ we have to show that it is injective, i.e. that if $x \neq y$ then $ext(x) \neq ext(y)$. But this is an immediate consequence of 3.5 since if $x \neq y$ then $x \not\leq y$ or $y \not\leq x$ must hold and hence we can construct a prime filter P which respects all of the considered suprema and "separates" x and y, i.e. P is an element which belongs to only one of the two sets $ext(x)$ and $ext(y)$.

The easiest way to define a topology on $Pf(\mathcal{L})$ such that $ext(x)$ is an open subset for each $x \in L$ is to consider the topology $\tau_{\mathcal{L}}$ whose base is any family of subsets $\{ext(x) : x \in \mathcal{B}\}$, where $\mathcal{B}$ is a *base* for the lattice $\mathcal{L}$ (see [1]), i.e. a subset of L, closed under $\wedge$, such that any element in L is the supremum of elements in $\mathcal{B}$. [9] In fact, if $\mathcal{L}$ is a countable distributive lattice, we can force the prime filters in $Pf(\mathcal{L})$ to respect all of the countable quantity of subsets of $\mathcal{B}$ which are needed in order to construct all the elements in L. Then, given $x \in L$ we know that $x = \vee\{b_1, \ldots, b_n, \ldots\}$ for a suitable choice of the elements $b_1, \ldots, b_n, \ldots \in \mathcal{B}$ and hence $ext(x) = ext(\vee\{b_1, \ldots, b_n, \ldots\}) = \cup_{b \in \{b_1, \ldots, b_n, \ldots\}} ext(b) \in \tau$. In the following we will call the topology $\tau_\mathcal{L}$ the point-set topology *associated* to $\mathcal{L}$.

4 The completeness theorem

Now we are going to apply the algebraic results of the previous section to show the completeness theorem for the formal topologies. Let us first give a formal definition of what a valuation of a formal topology into a point-set topology is.

Definition 4.1 (Valuation) *Let S be a formal topology and $\tau = \langle X, \Omega(X) \rangle$ be a point-set topology. Then a valuation V of the formal topology S into τ is a map of the elements of S into the open subsets of τ such that $V(1_S) = X$ and $V(a \bullet_S b) = V(a) \cap V(b)$ which is extended to a subset U of S by putting $V(U) = \cup_{u \in U} V(u)$.*

This definition allows to state in a mathematical way what, in the second section, had been called the intended meaning of the cover relation. In fact it is immediate to check that all the conditions we required on a cover relation are valid under any valuation V provided that $a \triangleleft_S U$ is interpreted as $V(a) \subseteq V(U)$. Hence the following validity theorem holds.

Theorem 4.2 (Validity) *Let S be a formal topology, τ a point-set topology and V a valuation of S into τ. Then if $U \triangleleft_S W$ then $V(U) \subseteq V(W)$.*

Our aim is to show that also the other implication holds, that is we want to prove the following theorem.

Theorem 4.3 (Completeness) *Let S be a formal topology, and suppose that for any point-set topology τ and for any valuation V of S into τ, $V(U) \subseteq V(W)$ holds. Then $U \triangleleft_S W$.*

---

[9]Note that any lattice $\mathcal{L}$ has at least one base, i.e. the collection L itself. The deep reason of this definition is that the base $\mathcal{B}$ for $\mathcal{L}$ may well be a ITT set even if L is only a collection.

In order to prove this theorem, we will show that there exists a suitable topology τ^* and a suitable valuation V^* of $\mathcal{S}$ into τ^* such that if $U \not\lhd_{\mathcal{S}} W$ then $V^*(U) \not\subseteq V^*(W)$, which is classically equivalent. The key point to find such a topology is to consider the topology $\tau_{Open(\mathcal{S})}$ associated to the countable distributive lattice $Open(\mathcal{S})$ (provided that we can construct it!).

The first step is then to find a base for the lattice $Open(\mathcal{S})$ but this is an easy task since one can see that the equivalence classes $\{[\{a\}] : a \in S\}$ form a base for the lattice $Open(\mathcal{S})$ since, for any subset U of S, $[U] = \vee_{u \varepsilon_S U}[\{u\}]$.

The second step is to observe that, as we work within an intuitionistic countable theory of sets and subsets, not only the distributive lattice $Open(\mathcal{S})$ but also the number of suprema we can construct on its elements is countable since they are at most as many as the sets of ITT which can be used as index-sets are and the sets of ITT are a countable collection since they are *exactly* [10] what one can form using the ITT formation rules [7]. Hence all the results of the previous section apply, i.e. we can construct the point-set topology $\tau_{Open(\mathcal{S})}$, associated with the lattice $Open(\mathcal{S})$, whose base are the subsets $a^* \equiv \{P \in Pf(Open(\mathcal{S})) : [\{a\}]\varepsilon_{Open(\mathcal{S})}P\}$, for any $a \in S$. Moreover we know that the lattice $Open(\mathcal{S})$ can be embedded into $\tau_{Open(\mathcal{S})}$ using the map ext defined on the base of $Open(\mathcal{S})$ by putting $ext([\{a\}]) = a^*$ and which respects all the countable quantity of existing suprema. So the topology $\tau_{Open(\mathcal{S})}$ is the point-set counterpart of the formal topology $\mathcal{S}$, i.e. it has the "same" base and the "same" open sets, the only novelty being that we have succeeded in *filling* the open sets of $\tau_{Open(\mathcal{S})}$ of points instead of dealing with purely formal objects.

These observations have finally led to the solution of our problem, i.e. we have found the valuation V^* we were looking for, since we can define it by simply putting $V^*(a) \equiv a^*$. We can immediately verify that V^* is indeed a valuation since $V^*(1_{\mathcal{S}}) = \{P \in Pf(Open(\mathcal{S})) : [\{1_{\mathcal{S}}\}]\varepsilon_{Open(\mathcal{S})}P\} = Pf(Open(\mathcal{S}))$ and $V^*(a \bullet_S b) = (a \bullet_S b)^* = ext([\{a \bullet_S b\}]) = ext([\{a\}]) \cap ext([\{b\}]) = a^* \cap b^* = V^*(a) \cap V^*(b)$. Moreover by using this definition we obtain that $V^*(U) = \cup_{u \varepsilon U} V^*(u) = \cup_{u \varepsilon U} ext([\{u\}]) = ext(\vee_{u \varepsilon U}[\{u\}]) = ext([\cup_{u \varepsilon U}\{u\}]) = ext([U])$ and hence if $U \not\lhd_{\mathcal{S}} W$ then $[U] \not\leq_{Open(\mathcal{S})} [W]$ and so $V^*(U) = ext([U]) \not\subseteq ext([W]) = V^*(W)$.

It is worth noting that the point-set topology $\tau_{Open(\mathcal{S})}$ and the valuation V^* do not depend on the particular subsets U and V we are considering but only on the formal topology $\mathcal{S}$, i.e. we have proved a result which is stronger than what we need: the point-set topology $\tau_{Open(\mathcal{S})}$ and the valuation V^* provide us with a *canonical model* of the formal topology $\mathcal{S}$.

---

[10]With respect to this point the status of ITT is unlike a standard set theory which is only a syntax to describe some intended *external* model which is in general not completely specifiable so that the fact that the language is countable does not guarantee that the number of sets in the model is countable.

5 Conclusions

Some comments on the proof of the completeness theorem may be useful. As it stands, the proof is carried on using a *classical* meta-mathematics and hence it is not possible to formalize it within the framework of ITT, or within any other constructive one. It may be considered as the proof a classically minded mathematician can produce when "playing" with formal topology. Moreover the result deeply depends on the particular theory of subsets we adopt here to deal with formal topology: in particular it holds for any countable theory of sets with a countable theory of subsets. Hence, from a strictly constructive point of view the provided proof can only exclude that it is impossible to obtain a completeness theorem. Anyhow from a constructive perspective such a theorem is still missing and this classical proof is the best information we have.

Acknowledgments. I'm indebted to G. Sambin and the referee which, besides many useful observations on the contents of the paper and the way the topics are presented, pointed out an error in the proof of 3.4 in a preliminary version of this work.

References

[1] G. Battilotti and G. Sambin, *Pretopologies and a uniform presentation of suplattices, quantales and frames*, to appear.

[2] J.L. Bell and M. Machover, *A course in Mathematical Logic*, North Holland Publishing Co., 1977.

[3] S. Feferman, *Review of Rasiowa and Sikorski*, J. Symbolic Logic **17** (1952), 72.

[4] M.P. Fourman and R.J. Grayson, *Formal spaces*, Proceedings of The L.E.J. Brouwer Centenary Symposium, North Holland, 1982, pp. 107–122.

[5] A. Horn, *Logic with truth values in a linearly ordered Heyting algebra*, J. Symbolic Logic **34** (1969), no. 3, 395–408.

[6] P.T. Johnstone, *Stone Spaces*, Cambridge Studies in Advanced Mathematics **3**, Cambridge University Press, 1982.

[7] P. Martin-Löf, *Intuitionistic type theory*, Bibliopolis, 1984, Notes by G. Sambin of a series of lectures given in Padua, June 1980.

[8] H. Rasiowa and R. Sikorski, *A proof of the completeness theorem of Gödel*, Fund. Math. **37**, 193–200.

[9] G. Sambin, *Intuitionistic formal spaces a first communication*, Mathematical logic and its applications (D. Skordev, ed.), Plenum, 1987, pp. 187–204.

[10] _____, *Intuitionistic formal spaces and their neighborhood*, Proceedings of Logic Colloquium '88 (C. Bonotto, R. Ferro, S. Valentini, and A. Zanardo, eds.), North-Holland, 1989, pp. 261–286.

[11] G. Sambin and S. Valentini, *Building up a toolbox for Martin-Löf intuitionistic type theory*, to appear.

[12] S. Valentini, *Points and Co-Points in Formal Topology*, Boll. Un. Mat. Ital. A (6) **7** (1993), 7–19.

DIPARTIMENTO DI MATEMATICA PURA ED APPLICATA, UNIVERSITÀ DI PADOVA, VIA G. BELZONI N.7,, I–35131 PADOVA, ITALY
e-mail address: VALENTINI@PDMAT1.MATH.UNIPD.IT